W9-CRL-700

OXFORD STUDIES IN NUCLEAR PHYSICS

GENERAL EDITOR

P. E. HODGSON

OXFORD STUDIES IN NUCLEAR PHYSICS

General Editor: P. E. Hodgson

1. J. McL. Emmerson: *Symmetry principles in particle physics* (1972)
2. J. M. Irvine: *Heavy nuclei, superheavy nuclei, and neutron stars* (1975)
3. I. S. Towner: *A shell-model description of light nuclei* (1977)
4. P. E. Hodgson: *Nuclear heavy-ion reactions* (1978)
5. R. D. Lawson: *Theory of the nuclear shell model* (1980)
6. W. E. Frahn: *Diffractive processes in nuclear physics* (1985)
7. S. S. M. Wong: *Nuclear statistical spectroscopy* (1986)
8. N. Ullah: *Matrix ensembles in the many-nucleon problem* (1987)
9. A. N. Antonov, P. E. Hodgson, and I. Zh. Petkov: *Nucleon momentum and density distributions in nuclei* (1988)
10. D. Bonatsos: *Interacting boson models of nuclear structure* (1988)
11. H. Ejiri and M. J. A. de Voigt: *Gamma-ray and electron spectroscopy in nuclear physics* (1989)

GAMMA-RAY AND ELECTRON SPECTROSCOPY IN NUCLEAR PHYSICS

H. EJIRI
Department of Physics
Osaka University, Japan

AND

M. J. A. de VOIGT
Technical University
Eindhoven, The Netherlands

CLARENDON PRESS · OXFORD
1989

Oxford University Press, Walton Street, Oxford OX2 6DP
Oxford New York Toronto
Delhi Bombay Calcutta Madras Karachi
Petaling Jaya Singapore Hong Kong Tokyo
Nairobi Dar es Salaam Cape Town
Melbourne Auckland
and associated companies in
Berlin Ibadan

Oxford is a trade mark of Oxford University Press

Published in the United States
by Oxford University Press, New York

British Library Cataloguing in Publication Data
Ejiri, H.
Gamma-ray and electron spectroscopy in
nuclear physics
1. Gamma ray spectroscopy 2. Electron spectroscopy
I. Title II. Voigt, M. J. A. de
537.5′352
ISBN 0–19–851723–8

Library of Congress Cataloging in Publication Data
Ejiri, H. (Hiroyasu), 1936–
Gamma-ray and electron spectroscopy in
nuclear physics
(Oxford studies in nuclear physics)
Bibliography: p. Includes index.
1. Nuclear structure. 2. Nuclear reactions.
3. Gamma ray spectrometry. 4. Electron spectroscopy.
I. Voigt, M. J. A. de II. Title. III. Series.
QC793.3.S8E35 1989 539.7 88–12532
ISBN 0–19–851723–8

Typeset by Macmillan India Ltd, Bangalore-25
Printed by St Edmundsbury Press,
Bury St Edmunds, Suffolk

PREFACE

This book is devoted to the role of gamma-ray and conversion-electron (γ–e) spectroscopy in developing our understanding of nuclear structure and nuclear reaction-mechanisms. The level of treatment is chosen so as to introduce graduate students into the field as well as to familiarize more experienced nuclear physicists with the new developments.

We decided to write this book because of the spectacular development in the last decade of new γ–e spectroscopic methods, and their application to various kinds of nuclear reactions. We also felt the need for a book which presents γ–e spectroscopy from the point of view of nuclear structure as well as of reaction mechanism. The importance of γ–e spectroscopy is due to the simplicity and familiarity of the electromagnetic interaction, which gives accurate values for many nuclear quantities and reveals special nuclear properties. We want to show how γ–e spectroscopy is applied to investigate static as well as dynamic nuclear properties over a wide range of excitation energies from the ground state to states of extreme temperatures and angular momentum, including some new degrees of freedom. Some parts of the book are based on lectures and seminars presented by the authors at their home universities and abroad.

We emphasize in this book the physics of nuclear structure and reaction mechanisms in relation to γ–e spectroscopic methods. In this sense the book differs from existing textbooks, which emphasize either the experimental techniques and methods or the nuclear theory. For special applications, technical details or theories we refer on several occasions to those textbooks. In some cases we also provide the relevant theoretical background. Extensive references are included to the original papers, mostly experimental.

We are indebted to Dr P. E. Hodgson for suggesting that we write this book and for his careful editing of the manuscript, and also to many of our colleagues for valuable discussions. We emphasize the importance of the work of nuclear physicists in many laboratories, who have contributed to the material used in this book. Our home laboratories and universities at Osaka and Groningen (KVI) as well as the Japanese Society for the Promotion of Science and the Dutch foundation for fundamental research of matter (FOM) provided excellent working conditions and support.

Finally we gratefully acknowledge, above all, the continuous stimulus and encouragement of Miyako Ejiri and An de Voigt.

July 1987

H. E.
M. J. A. de V.

CONTENTS

1

INTRODUCTION

1.1 Electromagnetic interactions in nuclei

The atomic nucleus is one of the most interesting quantum systems found in nature. It is a microscopic system with dimensions of the order of $2 \sim 10$ fm (10^{-15} m), which consists of a finite number ($1 \sim 250$) of nucleons with a density around 1.7×10^{-1} nucleon/fm^3. The most important interaction between the nucleons is the short-range strong (nuclear) interaction. These features of the nuclear system are in contrast to macroscopic matter, which consists of a large number ($N \approx 6 \times 10^{23}$) of atoms with a density around 10^{-16} atom/fm^3. The most important interaction in this case is the long-range electromagnetic interaction.

The nucleus in general is considered to be a finite system of interacting hadrons. Actually, most nuclear phenomena are described in a framework of the interacting nucleon system. Non-nucleonic degrees of freedom, however, manifest themselves explicitly or implicitly in some nuclear phenomena, and even play essential roles in particular cases. Meson (π, ρ, η, . . .) degrees of freedom are relevant to nuclear motions and to interactions associated with the characteristic meson quantum numbers (symmetry) and interaction ranges of the order of $\hbar/mc$, m being the meson mass. Strangeness degrees of freedom are involved in hypernuclei and in nuclear reactions associated with the K meson. Excited nucleons (Δ isobars, etc.) play important roles in nuclei with appropriate excitation energies and quantum numbers. Quark degrees of freedom may be effective in high momentum transfers and short-range correlations. The nuclear system may therefore be represented by a wave function $\psi(e)$ where e is the constituent hadron, as shown in Fig. 1.1.

The interactions involved in such hadrons are the strong, the electromagnetic (EM), and the weak interactions. Thus the Schrödinger equation of the nucleus is written as

$$\mathbf{H}\psi(e) = E\psi(e), \tag{1.1.1}$$

where the nuclear Hamiltonian

$$\mathbf{H} = \mathbf{H}_S + \mathbf{H}_{EM} + \mathbf{H}_W, \tag{1.1.2}$$

and $\mathbf{H}_S$, $\mathbf{H}_{EM}$ and $\mathbf{H}_W$ are the appropriate strong, EM and weak interactions, respectively. The most important interaction is the strong (nuclear) interaction, and it is responsible for a major part of the nuclear properties and motions involved in nuclear structure and nuclear reactions. The

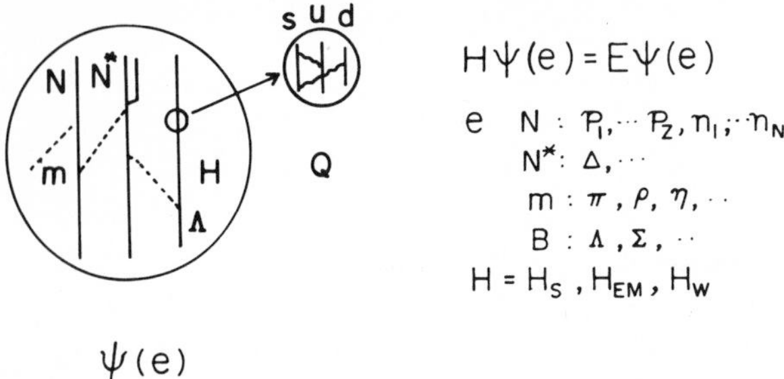

Fig. 1.1. Schematic diagram of a nuclear system containing a finite number of constituent hadrons, with e = N, N*, m and B, standing for nucleon, excited nucleon, meson, and baryon, respectively. $\mathbf{H}_S$, $\mathbf{H}_{EM}$, and $\mathbf{H}_W$ are the strong, EM and weak interactions, respectively. Q denotes s, u, and d quarks. As an example a Λ hypernucleus is shown. Most nuclei consist of nucleons.

EM interaction is weaker by three orders of magnitude than the nuclear interaction, and acts on electric charges, electric currents and on magnetic moments. The weak interaction is weaker again by another three orders of magnitude than the electromagnetic interaction and it is responsible for nuclear weak properties such as β decay.

The atomic nucleus, as a unique quantum system of interacting hadrons, may be regarded as an isolated microlaboratory for spectroscopic studies of fundamental interactions (strong, EM, and weak) and of the hadronic motions induced by those interactions. The nucleus indeed shows many different aspects of such interactions and hadronic motions. Some of them are associated with low energy nuclear properties of the bound hadron system, and others with medium and high energy hadron dynamics. The former is conventionally referred to as the nuclear structure and the latter as the nuclear reaction aspects of the system.

The static and dynamic properties of nuclei are studied experimentally by means of various probes. Nuclear phenomena are the observable response to the probes that act on the nucleus through some interaction $\mathbf{h}_\alpha$. In practice one observes the matrix elements of such interactions. Since the interactions used to probe nuclei are generally the strong, EM, and weak interactions, the observable quantities are

$$|\mathbf{M}_\alpha|^2 = |\langle \psi_f \| \mathbf{h}_\alpha \| \psi_i \rangle|^2, \tag{1.1.3}$$

where $\mathbf{h}_\alpha$ stands for the appropriate interaction characteristic of the nuclear probe with $\alpha = $ S, EM and W denoting the strong, EM, and weak interactions, respectively, and ψ_i and ψ_f are the initial and final wavefunctions of the system. These interactions used to probe nuclear properties have several advantages and disadvantages, depending on the aspect of the nucleus being studied.

The strong interaction provides the easiest way to study nuclei. Hadron beams such as nucleons, light and heavy ions, and mesons are used to probe nuclei via the strong interaction. Unfortunately its properties are not fully understood, so that effective nuclear interactions must be used in practice. These include many terms with different ranges such as the spin, isospin, spin–orbit and tensor terms. Meson-exchange interactions involve various kinds of mesons, whose properties are not well known. Quantum chromodynamic (QCD) theories have so far only been applied in quite a limited number of cases. Since the strong interaction strength is large, the hadron reaction is not necessarily a simple one-step process but may also have initial- and final-state interactions and multi-step contributions. The strong interaction using hadron probes has been widely used in nuclear reaction studies of nuclei.

The weak interaction $\mathbf{h}_W$ is so weak that its application as a nuclear probe is limited mainly to special cases where both particle emission via the strong interaction and gamma transitions via the EM interaction are forbidden by conservation laws and/or selection rules. Thus nuclear beta decays and hyperon weak decays are seen for unstable ground states and long-lived isomers. The particles involved in nuclear beta decay are leptons (neutrinos and electrons), and then the long wavelength approximation and the multipole expansion are valid as in case of the EM radiation. The weak interaction itself is interesting, as one of the fundamental interactions, and the nucleus provides a test ground for the weak interactions of elementary particles. Thus the weak interaction probe is very important for studying special nuclear responses to the weak field and the basic properties of weak nuclear processes.

The EM interaction $\mathbf{h}_{EM}$ is simple and rather weak compared with $\mathbf{h}_S$, but is much stronger than $\mathbf{h}_W$. Photons and electrons are used as the EM probe to study the EM response of nuclei. Electron scattering, and Coulomb excitation, photonuclear reactions (γ, X), and γ and conversion–electron decays as inverse processes of the (γ, X) are conventional tools and they interact with nuclei as shown in Fig. 1.2. The great advantage of the EM interaction is that such a simple and modest interaction plays unique and decisive roles in probing various aspects of nuclear structure, as emphasized by Bohr and Mottelson (1969). The e–γ spectroscopy utilizing the EM probe has thus been extensively used for studying nuclear structure and nuclear reactions. It is also used in many cases together with nuclear reactions and nuclear β decays.

The unique and useful features of the EM interaction used for probing nuclear properties are as follows.

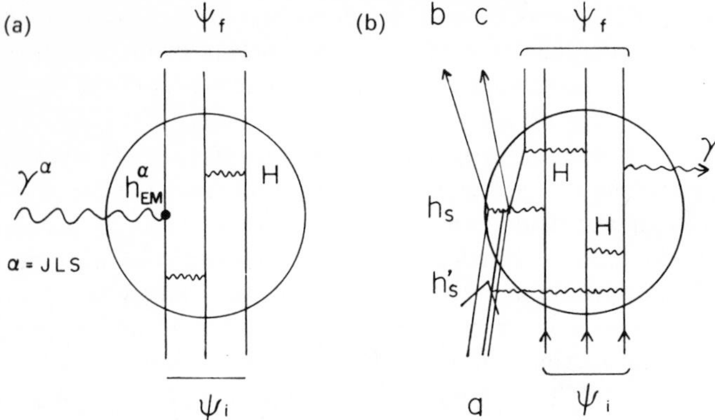

Fig. 1.2. Schematic diagram of nuclear systems and nuclear probes; ψ_i is the initial state and ψ_f is the final one. (a) EM probe with $\mathbf{h}^{\alpha}_{\mathrm{EM}}$ and γ^{α} being the α mode EM interaction and the γ ray, respectively. (b) hadron probe with strong interactions $\mathbf{h}_{\mathrm{S}}$, $\mathbf{h}'_{\mathrm{S}}, \ldots,$ and a γ decay following the hadron reaction.

1. The EM interaction $\mathbf{h}_{\mathrm{EM}}$ itself is well known. The matrix element is written as

$$M(\mathrm{EM}, \mathrm{if}) = \langle \psi_f | \mathbf{h}_{\mathrm{EM}} | \psi_i \rangle. \tag{1.1.4a}$$

Thus observation of $M(\mathrm{EM}, \mathrm{if})$ gives directly some information on ψ_i and ψ_f.

2. The EM interaction responds to the electric charges of the hadrons and their electric currents according to the expression (Bohr and Mottelson, 1969)

$$\mathbf{h}_{\mathrm{EM}} = \int \varphi(\mathbf{r}, t)\rho(\mathbf{r}, t)\,\mathrm{d}^3r - \frac{1}{c}\int \mathbf{j}(\mathbf{r}, t)\cdot A(\mathbf{r}, t)\,\mathrm{d}^3r, \tag{1.1.4b}$$

where (φ, A) is the four-vector potential and $(\rho, (1/c)\mathbf{j})$ is the four vector charge current. In the framework of a point-like nucleon system with the nucleon mass M, the nuclear density is

$$\rho(\mathbf{r}) = \sum_i e(\tfrac{1}{2} - t^i_z)\,\delta(\mathbf{r} - \mathbf{r}_i) \tag{1.1.4c}$$

and the charge current

$$\mathbf{j}(\mathbf{r}) = \sum_i e(\tfrac{1}{2} - t^i_Z)\tfrac{1}{2}\{\mathbf{v}_i\delta(\mathbf{r} - \mathbf{r}_i) + \delta(\mathbf{r} - \mathbf{r}_i)\mathbf{v}_i\}$$

$$+ \frac{e\hbar}{2M}\sum_i g^i_s\, \mathbf{\nabla} \times \mathbf{s}_i\,\delta(\mathbf{r} - \mathbf{r}_i), \tag{1.1.4d}$$

where t^i_Z is the isospin z component for the i-th nucleon, $\mathbf{v}_i$ the velocity vector and $\mathbf{s}_i$ the spin vector. The spin g factor g^i_{S} is written as $g^i_{\mathrm{S}} = \tfrac{1}{2}g_0 + t^i_Z g_1$ with the isoscalar factor $g_0 = g_p + g_n$ and isovector factor $g_1 = g_n - g_p$. Therefore $\mathbf{h}_{EM}$ probes directly the spins, isospins, charge densities, and velocities (currents) of nucleons, and accordingly the nuclear interactions responsible for these variables. In the case of hypernuclei $\mathbf{h}_{EM}$ responds to the hyperon magnetic moment and to the current. The finite size effect can be introduced by the electric and magnetic form factors. Meson currents are probed as well. Thus $\mathbf{h}_{EM}$ provides direct access to important variables of hadron motions in nuclei, and even to the renormalization of such variables and to the internal nucleon structure in nuclear fields.

3. The interaction strength of $\mathbf{h}_{EM}$ is quite weak compared with that of the strong (nuclear) interaction. Therefore, most of the EM responses are well described in terms of first order perturbation theory. This is in contrast with the nuclear (strong) interaction involved in nuclear reactions where higher order terms such as two-step and multistep processes contribute to some extent. Higher order EM processes are involved in some cases such as multiple Coulomb excitation, double γ decays, and so on, where the first order process is hindered by conservation laws and/or selection rules. They can be evaluated accurately as long as the EM interaction is well known.

4. The photon involved in the EM transition is mass-less, and the energies E_γ for most transitions are much smaller than the pion mass m_π. Thus the wavelength $\lambda\!\!\!^{-}_\gamma$ is much larger than the Compton wavelength $\lambda\!\!\!^{-}_\pi = \hbar/m_\pi c$ of the pion and also than the average radius $\langle r \rangle$ of the nucleon motion in the nucleus. It is given by

$$\lambda\!\!\!^{-}_\gamma = \hbar c/E_\gamma \gg \langle r \rangle \approx R, \qquad (1.1.5)$$

where R is the nuclear radius. For $E_\gamma = 1$ MeV $\lambda\!\!\!^{-}_\gamma$ is 200 fm, which is about two orders of magnitude larger than R for medium mass nuclei.

As a result, the long wavelength approximation, i.e. $r/\lambda\!\!\!^{-} = qr \ll 1$, is quite valid in most cases. For simplicity let us take a plane wave vector potential polarized in the x-direction. It is written by a sum of the circularly polarized components as (DeBenedetti, 1964),

$$A = \mathbf{x}e^{ikz} = \frac{1}{\sqrt{2}}(-\mathbf{e}_+ e^{ikz} + \mathbf{e}_- e^{ikz}). \qquad (1.1.6a)$$

Expanding the exponential part in terms of the spherical harmonics term $j_l(kr)\mathbf{Y}_l^0(\theta)$ and using the vector spherical harmonics expansion of $\mathbf{Y}_l^0(\theta)\mathbf{e}_\pm = \sum_L C(l\,1\,J';\,0\pm1)\,\mathbf{Y}_{J'l1}^{\pm1}$, one gets the multipole expansion,

$$\mathbf{e}_\pm e^{ikz} = \sum_{J'=1} i^{J'}\{2\pi(2J'+1)\}^{1/2}\{A(\mathrm{E}J') + A(\mathrm{M}J')\}, \qquad (1.1.6b)$$

where

$$A(\mathrm{E}J')=i^{-1}\left(\frac{J'+1}{2J'+1}\right)^{1/2} j_{J'-1}(kr)\mathbf{Y}^{\pm1}_{J',J'-1,1}$$

$$+i\left(\frac{J'}{2J'+1}\right)^{1/2} j_{J'+1}(kr)\mathbf{Y}^{\pm1}_{J',J'+1,1} \qquad (1.1.6\text{c})$$

and

$$A(\mathrm{M}J')=\mp j_{J'}(kr)\mathbf{X}^{\pm1}_{J'} \qquad (1.1.6\text{d})$$

stand for the electric and magnetic components, respectively. $A(\mathrm{E}J')$ has an orbital parity $(-)^{J'+1}$ and $A(\mathrm{M}J')$ has an orbital parity $(-)^{J'}$. The spherical Bessel function $j_{J'}(kr)$ is approximately given by $j_{J'}(kr)\approx a_{J'}(kr)^{J'}$, neglecting the higher order terms of $(kr)^{J'+2}, \ldots$. Consequently the multipole (J') expansion given by eqn (1.1.6b) is indeed useful since $j_{J'}(kr)\gg j_{J'+2}(kr)\gg j_{J'+4}(kr)$ in the long wavelength approximation. Although $A\{\mathrm{E(M)}J'\}$, $A\{\mathrm{E(M)}J'+2\}, \ldots$ contribute in principle to a transition between definite parity states, the lowest multipole component dominates the transition unless it is not forbidden by other selection rules (see Section 2.1). Noting that the photon spin-parity is 1^-, one gets for the lowest order term,

$$J=|I_\mathrm{i}-I_\mathrm{f}| \text{ for } I_\mathrm{i}\neq I_\mathrm{f}; \quad J=1 \text{ for } I_\mathrm{i}=I_\mathrm{f}>0 \qquad (1.1.7\text{a})$$

$$\pi=(-)^{J+K}=\pi_\mathrm{i}\pi_\mathrm{f}, \qquad (1.1.7\text{b})$$

where $K=0$ and 1 stand for electric and magnetic transitions, respectively. I_i, π_i are the spin and parity of the initial state and I_f, π_f those of the final states. The spin and parity of the γ-ray is given by $J(=\min.\ J')$ and $\pi=(-)^{J+K}$, respectively. Such simple multipole decomposition and selection rules make e–γ spectroscopy very useful. This feature is common for lepton and photon spectroscopies associated with light mass (e, β) and mass-less (γ, ν) probes.

1.2 Nuclear e–γ spectroscopy as a probe for nuclear structure and nuclear reactions

Nuclear e–γ spectroscopy is concerned with the EM response of nuclei. Using the unique and simple features of the EM interaction mentioned above, e–γ spectroscopy has been extensively applied to study both the static and the dynamic aspects of nuclei. The most important aspects of nuclear structure probed by nuclear e–γ spectroscopy are as follows.

1. Basic quantum numbers (eigenvalues) of nuclear states, such as energy (E), angular momentum (J), parity (π) etc., are obtained by measuring the corresponding variables E_γ, J_γ, π_γ, etc. using conservation rules (selection rules) for these quantities. High resolution semiconductor detectors make it possible to observe several cascade γ rays in the decay process and hence to

determine E_γ to an accuracy of $10 \sim 100$ eV, which is one part in $10^5 \sim 10^4$ for typical γ rays in the MeV region. This is small enough to detect slight shifts of energy levels due to small perturbations and also to resolve most of the energy levels. The multipolarity (J_γ) and parity (π_γ) are obtained from angular (directional and polarization) correlation measurements and/or conversion electron coefficients. Other quantum numbers (isospin, projection of J, etc.) may also be evaluated from those of the γ rays to the extent that conservation rules associated with those quantum numbers are valid.

2. The transition probabilities and the multipole mixing ratios, if the transition includes mixed multipoles, give the matrix element $\langle \mathbf{M}_{\mathrm{EM}}^{\mathrm{if}} \rangle$, i.e. the projection of ψ_{i} on ψ_{f} through $\mathbf{h}_{\mathrm{EM}}$. In a nuclear system of nucleons with point charges and magnetic moments the electromagnetic Hamiltonian $\mathbf{h}_{\mathrm{EM}}$ in eqn (1.1.4) is expanded in terms of tensor operators as

$$\mathbf{h}_{\mathrm{EM}} = \Sigma g(LSTJ)\boldsymbol{\tau}^T[\boldsymbol{\sigma}^S \times \mathbf{Y}_L]_J h_L(r) + \Sigma g(LlTJ)\boldsymbol{\tau}^T[\mathbf{l} \times \mathbf{Y}_L] h'_L(r), \quad (1.2.1)$$

where $\boldsymbol{\tau}$, $\boldsymbol{\sigma}$, and $\mathbf{l}$ are isospin, spin and orbital angular momentum operators, respectively. Then the matrix element is given as

$$\langle \mathbf{M}_{\mathrm{EM}}(\alpha) \rangle = \langle g(\alpha) h_L(r) \boldsymbol{\tau}^T [\boldsymbol{\sigma}^S \times \mathbf{Y}_L]_J \rangle + \langle g'(\alpha) h'_L(r) \boldsymbol{\tau}^T [\mathbf{l} \times \mathbf{Y}_L]_J \rangle, \quad (1.2.2)$$

where α represents the quantum numbers $LSTJ$ and $LlTJ$. Thus the transition matrix element $\langle \mathbf{M}_{\mathrm{EM}}(\alpha) \rangle$ gives the $\mathbf{M}_{\mathrm{EM}}(\alpha)$ component of ψ_{f} with respect to ψ_{i}, namely the overlap between ψ_{f} and the $\mathbf{M}_{\mathrm{EM}}(\alpha)$ mode excitation (or de-excitation) of ψ_{i}.

3. The long-range nature of the EM force makes e–γ spectroscopy quite sensitive to collective motions, which are frequently induced by the long-range components of the nuclear interactions. Thus one can study individual α-mode collective motions and accordingly the individual α-type nuclear interactions by investigating the corresponding α-mode γ transitions. In this sense e–γ spectroscopy acts as an antenna with a high quality tuner, which receives the signals from motions (vibrations) and nuclear interactions. Responding to an EM wave tuned to a particular frequency (E_γ) and in a particular mode (α), one can survey selectively the α-mode nuclear motion (vibration) and the α-type nuclear (strong) interaction, where α stands for the electric (isospin), magnetic (spin), dipole, quadrupole, and higher multipole modes.

4. Well-established properties and selection rules for the EM process are very useful for e–γ spectroscopy and provide a test of the fundamental symmetries.

5. Non-nucleonic degrees of freedoms, such as mesons, hyperons, and Δ isobars, have some EM moments (charge, magnetic moment). Thus e–γ

spectroscopy provides a sensitive test of the finite effects of such non-nucleonic degrees of freedom.

Since the EM interaction is weaker than the strong interaction, e–γ spectroscopy has limited applications either to the reaction residues left after the nuclear reactions associated with the strong interaction or to the weak γ branch during the fast reaction process. In spite of such minor roles of the γ process in nuclear reactions e–γ spectroscopy can contribute much to nuclear reaction studies because of its simple and clear features as a probe for nuclear phenomena. In what follows we list some unique and simple features of e–γ spectroscopy as used for studying nuclear reactions.

1. Nuclear reactions are in general written as

$$\mathrm{A}+\mathrm{a}\rightarrow\mathrm{B}^*+\mathrm{b};\quad \mathrm{B}^*\rightarrow\gamma_{\mathrm{B}},\gamma'_{\mathrm{B}},\cdots+\mathrm{B}. \tag{1.2.3}$$

The reaction residue B* is usually in an excited state and then it decays by emitting γ rays, $\gamma_{\mathrm{B}}, \gamma'_{\mathrm{B}}, \ldots$, which are indeed characteristic of B* specified by the proton (atomic) number Z_{B} and the mass number A_{B}. Note that every nucleus N is defined by A_{N} and Z_{N}, and has its own level scheme because A_{N} and Z_{N} (or the isospin $T_3=A/2-Z$, and the hypercharge in case of hyper-nuclei) are critical parameters for defining the level scheme. Thus measurements of the frequencies of the characteristic γ rays leads to identification of the reaction residue B and accordingly of the reaction channel. Knowing the incident channel and the reaction residue, one can get the mass number A_{b} and the charge Z_{b} for the reaction partner b using the conservation rules of mass and charge. This method can be used for cases of two and more reaction residues in excited states. Actually, in the case of heavy-ion reactions such as $\mathrm{A}+\mathrm{A}'\rightarrow\mathrm{B}+\mathrm{B}'+\mathrm{b}+\ldots$, the residual nuclei B and B′ are likely to be excited. These excited reaction residues are then easily identified by measuring the γ decays. Thus single and coincidence studies of γ rays (and/or conversion electrons) determine uniquely the reaction channel provided that the reaction residues are left in excited states. X-rays following conversion electrons in excited reaction residues are used to determine the atomic number Z_{B}.

2. Identification of the reaction channel (A_{B} and Z_{B}) can be extended to the other quantum numbers and quantities involved in the reaction by using appropriate conservation laws. The excitation energy $E_x(\mathrm{B})$ is obtained by measuring the total energy of the γ rays from B*, and then energy conservation gives the kinetic energy E_{b} for the reaction particle as

$$E_{\mathrm{b}}=(M_{\mathrm{A}}+M_{\mathrm{a}}-M_{\mathrm{B}}-M_{\mathrm{b}})c^2+E_{\mathrm{a}}-E_{\mathrm{B}}-E_x(B) \tag{1.2.4}$$

where M_i and E_i are the mass and kinetic energy of the particle i. If b is excited, the excitation energy $E_x(\mathrm{b})$ obtained by measuring γ-rays of b* should be added to the left-hand side of eqn (1.2.4).

3. The angular momentum $I(B^*)$ of B* is obtained by measuring the γ ray multiplicity and/or by measuring discrete γ rays following the decay of excited states with known spins. Then the angular momentum $I(b)$ carried away by b is obtained from the angular momentum conservation law. Where the target spin $I(A)$ is much smaller than the projectile angular momentum $I(a)$,

$$I(b) \approx I(a) - I(B^*). \tag{1.2.5}$$

The magnetic substate of B* is obtained from the angular distribution of the decay γ rays. The polarization of the angular momentum is obtained from the circular polarization of the γ rays. The isospin of B* is identified if one detects γ rays from a definite isospin state in B*. These quantities associated with the angular momenta, its z component, and the isospin are indeed very useful for studying the reaction mechanism connected with these quantum numbers.

4. The linear momentum of B* is evaluated from the Doppler effect of the γ rays, and then use of linear momentum conservation gives the linear momentum of b.

5. Even though the EM interaction is much weaker than the strong one, high efficiency e–γ spectroscopy can provide access to the weak γ branch in competition with strong particle decays in dynamic nuclear reaction processes. Study of the γ branch of a particular multipole mode then elucidates the particular mode of the motion at the high excitation region involved in the reaction.

6. Development of multidetector systems for both γ rays and particles, together with efficient multi-dimensional data acquisition systems, has made all kinds of particle correlation measurements feasible. Here γ rays with simple features cast light on rather complicated reaction mechanisms, and act in a way complementary to reaction particles so as to provide crucial information for understanding the reaction mechanism.

1.3 Recent developments of e–γ spectroscopy with nuclear reactions

Extensive application of e–γ spectroscopy employing nuclear reactions for studies of nuclear structures and reaction mechanisms requires high quality e–γ spectroscopic methods and devices such as high energy resolution e–γ detectors, high efficiency multi-detectors and high efficiency multi-data handling systems. Recent developments of these e–γ spectroscopic methods and devices have indeed made it possible for e–γ spectroscopy to explore the basic and important interactions and motions in both nuclear structure and nuclear reaction mechanisms.

The subjects covered by e–γ spectroscopy have been enlarged by extensive applications of various types of nuclear reactions. Statistical nuclear reactions have provided access to many nuclear phenomena for wide ranges of excitation energy, angular momentum and isospin z component $T_z = (N-Z)/2$ (nuclear species). They have also been used widely in in-beam γ-ray spectroscopy. Modern accelerators have provided high quality beams with improved characteristics such as high energy resolution, high intensity, small emittance, good stability and a variety of accelerated beams up to uranium. The advent of medium high energy (GeV) high intensity accelerators has made it quite feasible even to use new types of secondary projectiles such as π and K mesons, antiprotons, and radioactive nuclei. These developments have opened up for e–γ spectroscopy new scope in a multi-dimensional space. The scope spans extreme regions of angular momenta, excitation energy, isospin and its z component (far from stability) as well as many types of nuclear motion and new degrees of freedom.

The e–γ spectroscopic studies have revealed in the last decade various important aspects of nuclear interactions and nuclear motions. The fields studied by recent e–γ spectroscopy may be divided into three classes. The first is concerned with the basic properties of nuclei as a quantum system of a strongly-interacting nucleon ensemble. It includes new nuclear shapes, high-lying particle and hole motions, nuclear EM (spin, isospin, etc.) responses in the nuclear field, various collective (coherent) and individual nuclear motions, and the nuclear interactions associated with these nuclear properties. The second encompasses nucleon motions in extreme conditions such as high angular momentum (fast nuclear rotation, etc), high excitation energy and high temperature, and very fast reaction dynamics. Nuclei in these extreme conditions manifest unique and interesting properties. Interesting aspects of nuclear reaction mechanisms have been studied. The third deals with the fundamental symmetries of elementary particles in the quantum hadron system and the response of new degrees of freedom in the nuclear field.

We intend in this book to present recent developments of e–γ spectroscopy employing nuclear reactions to study nuclear structures and reaction mechanisms. Emphasis is put in the first place on the experimental facts found by means of e–γ spectroscopy and the methods used to obtain them, and secondly on the basic and important physical aspects of both nuclear structures and nuclear reactions studied by e–γ spectroscopy.

The general features of e–γ spectroscopy and the basic properties of the low-lying nuclear states are described in Chapter 2. It includes general aspects of the electromagnetic interaction and e–γ spectroscopy, nuclear shapes and nuclear potentials, particle and hole motions, and various modes of collective motion and the interplay between them. The third chapter describes recent developments in e–γ spectroscopic devices and methods. The multi-detector systems, medium-high energy γ detectors, conversion electron spectrometers, and methods for measuring EM diagonal and transition moments are described.

The use of e–γ spectroscopy following nuclear reactions necessarily involves an understanding of the reaction mechanism. Chapter 4 is concerned with interface between nuclear reaction mechanisms and e–γ spectroscopy. In that chapter we discuss the properties of the nuclear reactions used for e–γ spectroscopy, the nuclear reaction mechanisms studied by means of e–γ spectroscopy and the electromagnetic properties associated with dynamic nuclear reactions. The nuclear reactions described include statistical (equilibrium and preequilibrium) reactions, direct particle transfer and break-up reactions, inelastic scattering and radiative capture reactions.

Some specific subjects studied by recent e–γ spectroscopy are described in the second half of this book. As EM interactions are composed of spin, isospin and orbital angular momentum operators as given by eqn (1.2.1), the EM responses are particularly sensitive to spin–isospin effects in nuclei. Among these spin–isospin nuclear responses are spin–isospin polarizations, renormalizations of spin–isospin responses in nuclei, and spin isospin vibrations, and these have been studied by investigating EM multipole moments relevant to spin and isospin operators. They are described in Chapter 5.

Nuclei, when considered as ensembles of strongly-interacting nucleons, show interesting features in very high angular momentum states. Such high-spin nuclei are excited by heavy ion reactions and can be easily studied by modern e–γ spectroscopy. Chapter 6 reviews important nuclear features associated with high-spin states. These reactions provide unique opportunities to study a many-body quantum system under the influence of extremely fast rotation. Notable features of such reactions are associated with the interplay between the single-particle motions and the collective rotations, which are modulated by the fast rotation and even result in some cases in drastic changes of the nuclear shape. The e–γ spectroscopic study of high-spin states is presented both for discrete e–γ rays as well as for quasi-continuum γ rays.

In the final chapter (Chapter 7) brief descriptions are given of e–γ spectroscopy with exotic particles and of the fundamental interactions and symmetries studied by high precision e–γ spectroscopy. Nuclear properties related to the strangeness degrees of freedom are studied by hypernuclear γ spectroscopy, where K and π mesons are used to produce hypernuclei. As for fundamental symmetries and fundamental interactions, parity non-conservation in nuclear states and right-handed weak interactions have been investigated by means of high precision e–γ spectroscopy following nuclear reactions. The search for lepton number non-conservation and finite Majorana neutrino mass by means of ultra low-background double β–γ spectroscopy is briefly described at the end of the chapter.

Some fields of nuclear physics have been well developed by means of e–γ spectroscopy, as mentioned above. Aspects of nuclei under extreme conditions, where they show unique characteristics, may be further explored in the near future. Nuclear spectroscopy associated with new degrees of freedom such as hypernuclear γ spectroscopy has only recently begun. Such degrees of freedom offer new challenges for e–γ spectroscopy. The strangeness e–γ spectroscopy may be extended to more general quark flavour e–γ spectroscopy in the future. The recent development of QCD has certainly had a large impact on nuclear physics. It is interesting to search for finite QCD effects in nuclei by means of very refined spectroscopy. The bottomonium ($b\bar{b}$), toponium ($t\bar{t}$) and the quark structures certainly deserve to be explored by the e–γ spectroscopy of the next generation. As extensions of the standard electroweak theory, several grand unified theories, super-symmetry theories and other theories have recently been proposed. Evidence for such theories may be found in extremely rare events or in extremely small deviations from the standard model predictions. Nuclei, as a well-ordered quantum system of hadrons, are used to test such fundamental interactions by means of advanced techniques in e–γ spectroscopy, as currently applied to study double β decays, electron polarization, and so on.

There are many important aspects of e–γ spectroscopy that are not covered in this book. In particular most of the in-beam e–γ spectroscopy developed before the middle of the 1970's and the β–γ spectroscopy associated with radioactive isotopes are left out, since we prefer to emphasize the recent developments concerning nuclear reactions. Extensive treatments of conventional β–γ spectroscopic methods such as angular correlations, polarizations and EM moments etc. are not included. These subjects have been treated extensively in several books on nuclear spectroscopy (Ajzenberg-Selove, 1960; Siegbahn, 1965; Cerny, 1974; Morinaga and Yamazaki, 1976; Segrè, 1977). Neutron γ spectroscopy is also excluded. Our intention is to describe some crucial experimental data and the methods relevant to new aspects of nuclear physics rather than to give full theoretical descriptions. Detailed theoretical treatments and formulae are found in many books (DeBenedetti, 1964; deShalit and Talmi, 1963; Bohr and Mottelson, 1969, 1974; Marmier and Sheldon, 1970; Hodgson, 1971; deShalit and Feshbach, 1974; Hamilton, 1975; etc.).

Most of the important developments of e–γ spectroscopy are based on many careful and elaborate investigations supplementing each other in order to establish the facts of nuclear physics. We have chosen a consistent description of a certain number of important e–γ spectroscopic phenomena, rather than attempt to be complete. This implies that we have been very modest in referring to the available and relevant e–γ spectroscopic literature. The choice of experimental data used for the many illustrations in this book is largely based on the convenience with which we have access to those data, the original figures, and so on. We realize fully that other, even better, examples could be found in literature. We have also exploited fruitful presentations of new data at various workshops and conferences like the symposium on Nuclear Spectroscopy and Interactions, at Osaka, 1984 (Ejiri and Fukuda, 1984).

1.4 References

Ajzenberg-Selove, F. (1960). *Nuclear Spectroscopy.* Academic Press, New York.

Bohr, A. and Mottelson, B. R. (1969). *Nuclear Structure,* Vol. 1. W. A. Benjamin, New York.

Cerny, J. (1974). *Nuclear Spectroscopy and Reactions.* Academic Press, New York.

DeBenedetti, S. (1964). *Nuclear Interactions.* John Wiley, New York.

deShalit, A. and Talmi, A. (1963). *Nuclear Shell Theory.* Academic Press, New York.

deShalit, A. and Feshbach, H. (1974). *Theoretical Nuclear Physics.* John Wiley, New York.

Ejiri, H. and Fukuda, T. (1984). *Proc. Int. Symp. Nuclear Spectroscopy and Nuclear Interactions.* Osaka. World Scientific, Singapore.

Hamilton, W. D. (1975). *The Electromagnetic Interaction in Nuclear Spectroscopy.* North-Holland, Amsterdam.

Hodgson, P. E. (1971). *Nuclear Reactions and Nuclear Structure.* Clarendon Press, Oxford.

Marmier, P. and Sheldon, E. (1969) Vol 1. (1970) Vol. 2. *Physics of Nuclei and Particles.* Academic Press, New York.

Morinaga, H. and Yamazaki, T. (1976). *In-Beam Gamma-Ray Spectroscopy.* North-Holland, Amsterdam.

Sergrè, E. (1977). *Nuclei and Particles,* 2nd Edn. W. A. Benjamin, New York.

Siegbahn, K. (1965). *Alpha-, Beta- and Gamma-Ray Spectroscopy.* North-Holland, Amsterdam.

2

GAMMA-RAY AND ELECTRON SPECTROSCOPY AND NUCLEAR STRUCTURE

2.1 Electromagnetic transitions and e–γ spectroscopy

2.1.1 *Electromagnetic transitions and nuclear matrix elements*

Electromagnetic (EM) diagonal moments and EM transition moments lead directly to the nuclear matrix elements relevant to EM moments. Magnetic-dipole (M1) and electric-quadrupole (E2) moments give diagonal M1 and E2 matrix elements, respectively. Non-diagonal matrix elements are obtained from γ- (or conversion–electron) transition probabilities. As discussed in Chapter 1, these nuclear matrix elements, obtained by means of e–γ spectroscopic methods, give directly the required nuclear structure information. In this section we present briefly some specific concepts of EM moments such as transition probabilities, matrix elements, selection rules, and angular distributions that are needed in e–γ spectroscopic studies of nuclear structure. Details of the description of nuclear EM interactions such as derivations of multipole expansion formulae, EM transition probabilities, and EM multipole matrix elements are found in many standard text books (see, for example, Moszkowski, 1965; Bohr and Mottelson, 1969; de Shalit and Feshbach, 1974; and references given in Chapter 1).

2.1.1a *Transition probabilities and reduced matrix elements*

The EM transition moments for the multipolarity J are derived from the nuclear EM interaction as given by eqn (1.1.4) and the multipole expansion of the vector potential as given by eqn (1.1.6). They are written as (Bohr and Mottelson, 1969):

$$\mathbf{M}(\mathrm{E}, JM) = \frac{-i(2J+1)!!}{ck^{J+1}(J+1)} \int \mathbf{j}(\mathbf{r}) \cdot \nabla \times (\mathbf{r} \times \nabla)\{ j_J(kr) \mathbf{Y}_{JM}(\mathbf{r})\} \, \mathrm{d}\tau \quad (2.1.1\mathrm{a})$$

$$\mathbf{M}(\mathrm{M}, JM) = \frac{-(2J+1)!!}{ck^{J}(J+1)} \int \mathbf{j}(\mathbf{r}) \cdot (\mathbf{r} \times \nabla)\{ j_J(kr) \mathbf{Y}_{JM}(\mathbf{r})\} \, \mathrm{d}\tau, \quad (2.1.1\mathrm{b})$$

where $k = E_\gamma/\hbar c$ is the wave number, $\mathbf{M}(\mathrm{E}, JM)$ is the electric multipole transition operator with angular momentum J and parity $\pi^{\mathrm{E}} = (-)^J$, and $\mathbf{M}(\mathrm{M}, JM)$ is the magnetic one with angular momentum J and parity $\pi^{\mathrm{M}} = (-)^{J+1}$. As the intrinsic spin and parity of a photon (γ-ray) is 1^-, the electric and magnetic J-multipole γ-ray has orbital angular momenta $l_{\rm E}=J\pm 1$ and $l_{\rm M}=J$, respectively. Thus E1, M2, E3, . . . have odd parity and M1, E2, M3, . . . have even parity. In most cases of nuclear transitions $E_\gamma \ll m_\pi c^2$, where $m_\pi c^2 \approx 140$ MeV is the pion mass, and thus $kr = r/\lambda\!\!\!^{-}_\gamma \ll 1$. Then the spherical Bessel function $j_J(kr)$ is given by the first term of the $(kr)^n$ expansion as $j_J(kr) \approx (kr)^J/(2J+1)!!$. The multipole moments then reduce to the simple forms (Bohr and Mottelson, 1969)

$$\mathbf{M}(\mathrm{E}, JM) = \int \rho(\mathbf{r}) r^J \mathbf{Y}_{JM}(\mathbf{r})\, \mathrm{d}\tau \tag{2.1.2a}$$

$$\mathbf{M}(\mathrm{M}, JM) = \frac{-1}{c(J+1)} \int \mathbf{j}(\mathbf{r}) \cdot (\mathbf{r} \times \nabla) r^J \mathbf{Y}_{JM}(\mathbf{r})\, \mathrm{d}\tau. \tag{2.1.2b}$$

The γ transition probability $T\{\mathrm{E(M)}J, \mathrm{if}\}$ for the multipole transition $\mathbf{M}\{\mathrm{E(M)}J\}$ is then factorized into the product of the phase factor (external) term $G_J(k)$ and the nuclear structure (internal) term $B\{\mathrm{E(M)}J, \mathrm{if}\}$, and is written as

$$T\{\mathrm{E(M)}J, \mathrm{if}\} = \frac{\ln 2}{t_{1/2}\{\mathrm{E(M)}J, \mathrm{if}\}} = \frac{\Gamma_\gamma\{\mathrm{E(M)}J, \mathrm{if}\}}{\hbar}$$

$$= G_J(k) \cdot B\{\mathrm{E(M)}J, \mathrm{if}\} \tag{2.1.3a}$$

$$G_J(k) = \frac{8\pi(J+1)}{J\{(2J+1)!!\}^2} \cdot \frac{1}{\hbar} \cdot k^{2J+1} \tag{2.1.3b}$$

$$B\{\mathrm{E(M)}J, \mathrm{if}\} = \frac{1}{2I_{\rm i}+1} |\langle \psi_{\rm f} \| \mathbf{M}\{\mathrm{E(M)}, J\} \| \psi_{\rm i} \rangle|^2, \tag{2.1.3c}$$

where $t_{1/2}$ and Γ_γ are the half-life and γ-decay width, respectively, i and f refer to the initial and final states, and $I_{\rm i}$ is the spin of the initial state. $B\{\mathrm{E(M)}J, \mathrm{if}\}$ is called the reduced transition probability. It is conventionally expressed as $B\{\mathrm{E(M)}J\}$ (for example B(E1), B(M1), B(E2), and so on). The spin factor $(2I_{\rm i}+1)$ for the initial state may also be included in the external phase factor as $G_J(k)/(2I_{\rm i}+1)$. It is important to note that the nuclear structure term $B\{\mathrm{E(M)}J\}$ is decoupled from the phase factor term $G_J(k)$. $B\{\mathrm{E(M)}J\}$ is a simple projection of the $\psi_{\rm i}$ on $\psi_{\rm f}$ through the multipole operator $\mathbf{M}\{\mathrm{E(M)}J\}$, and $G_J(k)$ is a simple function of only the multipolarity J of the γ-ray and its wave number k. Thus experimental observation of the transition rate $T\{\mathrm{E(M)}J\}$ leads uniquely to the $B\{\mathrm{E(M)}J\}$ if the energy and multipolarity of the γ-ray are known.

Transition rates for bound particle states are obtained experimentally by direct measurements of half-lives $t_{1/2}$. Since strong decays are forbidden for such bound states the lifetimes are so long, ranging from 10^{-15} to $10^{3\sim 5}$ s, that they can be measured directly by various spectroscopic methods (see

Section 3.5). Lifetimes for unbound particle states are generally very short because of the dominant strong decay channel. In this case a direct half-life measurement is difficult and the transition rate is obtained by measuring instead the partial decay width Γ_γ involved in the EM process (see Section 4.5). Inverse processes such as γ-ray resonance absorption and Coulomb excitation yield the transition rates for both bound and unbound states.

Gamma-ray branching ratios give the ratio of the nuclear matrix elements as

$$\frac{\Gamma_\gamma\{\mathrm{E(M)}J,\mathrm{if}\}}{\Gamma_\gamma\{\mathrm{E(M)}J',\mathrm{if}'\}}=\frac{T\{\mathrm{E(M)}J,\mathrm{if}\}}{T\{\mathrm{E(M)}J',\mathrm{if}'\}}$$
$$=\frac{G_J(k)}{G_{J'}(k')}\cdot\frac{|\langle f\|\mathbf{M}\{\mathrm{E(M)}J\}\|\mathrm{i}\rangle|^2}{|\langle f'\|\mathbf{M}\{\mathrm{E(M)}J'\}\|\mathrm{i}\rangle|^2}. \tag{2.1.4}$$

In some cases absolute-value measurements of $T\{\mathrm{E(M)}J,\mathrm{if}\}$ are difficult, but branching ratios are easy to obtain.

The simple features mentioned above contrast with the rather complicated features associated with strong interactions. The nuclear reaction cross-section is not factorizable into a product of a nuclear structure term and a nucleus-independent term since particles entering into or leaving nuclei are still subject to strong nuclear fields. For example, the cross-sections of some simple particle transfer reactions are written as $\sigma=\sigma(\mathrm{DWBA})S_p$, where $\sigma(\mathrm{DWBA})$ is the cross-section for the particle transfer into a vacant orbit and S_p is the spectroscopic factor. The $\sigma(\mathrm{DWBA})$ is calculated by using distorted waves for the incoming and outgoing particles. Since it is sensitive to the nuclear forces (potentials) acting on them accurate evaluation of the absolute cross-sections is difficult. Moreover the strong interaction inevitably has higher order (multi-step) terms in addition to the first-order (one-step) direct term. Since the wave numbers of hadrons involved in nuclear reactions are of the order of 1 fm^{-1}, many angular momenta are involved, and consequently there are no simple selection rules for such strong hadron processes (reactions) except for some simple cases with well-defined nuclear structures.

2.1.1b *Selection rules*

On the basis of the simple structure of the EM interaction operator and of the multipole expansion, one can derive selection rules which are useful for e–γ spectroscopic studies. The γ transition between bound states with definite spins and parities I_i, π_i and I_f, π_f has, in principle, mixed multipolarities provided that $|I_\mathrm{i}-I_\mathrm{f}|\leqslant J\leqslant|I_\mathrm{i}+I_\mathrm{f}|$ and $\pi=\pi_\mathrm{i}\cdot\pi_\mathrm{f}$. Parity conservation restricts multipoles to $E(M)J$ with $J=J, J+2, \ldots$ for electric (magnetic) transitions of one type. Then the ratio of the transition rates for J and $J'=J+2$ is given by

$$\frac{T\{\mathrm{E(M)}J,\mathrm{if}\}}{T\{\mathrm{E(M)}J',\mathrm{if}\}}=C_1\cdot\frac{G_J(k)r^{2J+1}}{G_{J'}(k)r^{2J'+1}}=\frac{C_2}{(kr)^2}\gg 1, \tag{2.1.5}$$

where C_1 and C_2 are coefficients of the order of 1, and $(kr)^2 \ll 1$. Therefore the lowest multipole transitions dominate. Since there are two possible types (E and M) of transition, the signature numbers $K=0$ and $K=1$ are used to define the electric (E) and magnetic (M) transitions, respectively. The selection rules for γ decays are then written as

$$J = |I_i - I_f|, \qquad (-)^K = \pi_i \pi_f (-)^J \tag{2.1.6a}$$

$$J' = |I_i - I_f| + 1, \qquad K' = K + 1 \tag{2.1.6b}$$

$$J'' = |I_i - I_f| + 2, \qquad K'' = K \tag{2.1.6c}$$

$$|I_i + I_f| \geq J,\ J',\ J'' \geq 1, \tag{2.1.6d}$$

where J, J' and J'' stand for the lowest, second and third order multipoles, respectively, and K, K' and K'' are the corresponding signatures. The last condition eqn (2.1.6d) is imposed by the photon spin 1. Thus for $I_i = I_f \neq 0$ no γ decay with $J=0$ is allowed, the lowest multipole is $J'=1$ and the second one is J''. For $I_i = I_f = 0$ no single γ transition is possible. The third-order multipole term with J'' has the same signature K as the lowest one. Since $T\{\text{E(M)}J,\ \text{if}\} \gg T\{\text{E(M)}J+2,\ \text{if}\}$ for most bound-state transitions with $E_\gamma \lesssim 8$ MeV, the third order term with $J'' = J+2$ may be neglected unless the nuclear term $B\{\text{E(M)}J\}$ is reduced by some selection rules associated with the nuclear structure. On the other hand the second-order term has a different signature K' from the lowest one. Thus it contributes also to the transition, resulting in multipole mixing of (J, K) and $(J', K' = K+1)$. Since $T(\text{M}J)$ is in general smaller than $T(\text{E}J)$, one often has multipole mixing of $T(\text{M}J)$ and $T(\text{E}J+1)$ such as M1/E2, M2/E3 and so on.

The transition probability for the mixed multipole transition of J and J' is given by

$$T\{\text{E(M)}JJ',\ \text{if}\} = G_J(k)\ B\{\text{E(M)}J,\ \text{if}\}\ (1+\delta)^2, \tag{2.1.7a}$$

where

$$\delta = \{g_{J'}(k)/g_J(k)\}\ [\langle \text{f} \| \mathbf{M}\{\text{E(M)}J'\} \| \text{i} \rangle / \langle \text{f} \| \mathbf{M}\{\text{E(M)}J\} \| \text{i} \rangle] \tag{2.1.7b}$$

δ is the multipole mixing ratio and $\{g_J(k)\}^2 = G_J(k)$. The multipole assigment is discussed in Sections 2.1.2 and 2.1.3.

The selection rules related to the isospin are:

$$\Delta T_z = T_z(\text{i}) - T_z(\text{f}) = 0, \quad (N - Z = \text{constant}) \tag{2.1.8a}$$

$$\Delta T = 0,\ \pm 1, \tag{2.1.8b}$$

where $T_z(\text{i})$ and $T_z(\text{f})$ are the third components of the initial and final isospins, respectively. The first selection rule is imposed by charge conservation, and the baryon number $(N+Z)$ is also conserved. The second rule (2.1.8), based on pure isospin eigenstates of both initial and final states, does not hold strictly since the Coulomb interaction mixes the isospin a little.

The long wavelength approximation $\lambda_\gamma \gg r$ is not good in the medium high energy region $E_\gamma = 10 \sim 100$ MeV. In this energy region the states involved are mostly unbound with mixed spin and parity. Consequently one normally has mixed multipolarities like E1, E2, E3, . . . , M1, ... Nevertheless, it is important that e–γ spectroscopy, with knowledge of the phase factors $G_J(k)$, $G_{J'}(k)$ and so on, provides direct access to the nuclear multipole moment which reflects sensitively the nuclear motion with that multipole.

2.1.1c *Electromagnetic multipole matrix elements and nuclear models*
In the framework of the nuclear shell model the EM transition operators are written in terms of the single particle operator (Bohr and Mottelson, 1969)

$$\mathbf{M}(\mathrm{E}J) = \sum_i \mathbf{M}(\mathrm{E}J, i) = \sum_i e\{\tfrac{1}{2} - t_Z(i)\} r_i^J \mathbf{Y}_{JM}(\theta_i \phi_i) \tag{2.1.9a}$$

$$\mathbf{M}(\mathrm{M}J) = \sum_i \mathbf{M}(\mathrm{M}J, i) = \frac{e\hbar}{2Mc} \sum_i \left\{ g_s(i)\,\mathbf{s}_i + \frac{2g_l(i)}{J+1}\mathbf{l}_i \right\} \cdot \nabla_i \{ r_i^J \mathbf{Y}_{JM}(\theta_i \phi_i)\}, \tag{2.1.9b}$$

where M, g_s and g_l are the nucleon mass, the spin g factor and the orbital g factor, respectively, and **s** and **l** are the spin and orbital angular momentum operators. The isospin Z component is defined as $t_z(i) = 1/2$ for neutrons and $t_z(i) = -1/2$ for protons, and the summation runs over all nucleons of $i = 1, 2, \ldots, A$. The electric charge $e\{\frac{1}{2} - t_Z(i)\}$ in eqn (2.1.9a) must be corrected for the spurious motion of the centre of mass (deShalit and Feshbach, 1974). The corrected charge for the E1 transition is $e_P = Ne/A$ for protons and $e_n = -Ze/A$ for neutrons, and in the case of the E2 transition $e_P = e\{1 - (2/A) + (Z/A^2)\}$ and $e_n = eZ/A^2$. The corrections for higher multipole transitions are small. The reduced matrix element for a single-particle transition is given by

$$B_{\mathrm{SP}}\{\mathrm{E(M)}J, \mathrm{if}\} = \frac{1}{2I_\mathrm{i} + 1} |\langle \phi_\mathrm{f}(k) || \mathbf{M}\{\mathrm{E(M)}J, k\} || \phi_\mathrm{i}(k) \rangle|^2, \tag{2.1.10}$$

where $\phi_\mathrm{i}(k)$ and $\phi_\mathrm{f}(k)$ are the initial and final states, respectively, for the k-th single particle. The single-particle matrix element for a given J provides a standard unit to describe the magnitude of the matrix element $B\{\mathrm{E(M)}J, \mathrm{if}\}$, the simplest one being the Weisskopf unit (Bohr and Mottelson, 1969; deShalit and Feshbach, 1974; Moszkowski, 1965). The EM responses to charge currents and magnetic moments provide information on spin, and on the isospin and orbital motions of nucleons, and accordingly on the nuclear interaction relevant to such motions. The single-particle responses are renormalized in a nuclear field due to nuclear interactions. Renormalization of the $\mathbf{M}\{\mathrm{E(M)}J\}$ (α-mode) response is associated with modification of isospin (charge) and spin currents of the α-mode under the influence of the α-mode isospin spin nuclear field (see Chapter 5). Enhanced $B(\mathrm{E}J)$ over the single particle value of $B_{SP}(EJ)$ is attributed to the coherent (collective) motion of many nucleons in the electric $\mathbf{Y}_J$ mode, while enhanced $B(MJ)$ is due to the coherent motion in the magnetic (**s**, **l**) mode with $\mathbf{Y}_{J-1}$. Rotation of deformed nuclei necessarily induces a strong $B(EJ)$ with effectively a large charge of $e^{\text{eff}} \gg e$ (see Chapter 6). Coupling of a single particle with a collective motion is sensitively detected by observing deviation of the EM response from the pure (uncoupled) single particle value.

Single-particle motions are characterized by various quantum numbers such as the number of oscillator modes N, that of the radial node n, the orbital angular momentum l, the spin s, and the total angular momentum j. If single particles are in an axially symmetric deformed potential the projection of the angular momentum on the symmetry axis is a good quantum number. Collective motions are characterized by quantum numbers relevant to those motions. Selection rules associated with individual nuclear (nucleon) motions are derived in terms of the relevant quantum numbers. They are used for studying the extent to which such nuclear motions are associated with nuclear states characterized by a definite set of quantum numbers.

Single-particle motions and collective motions are discussed in view of their EM responses in Sections 2.2–2.5.

The strengths of γ-ray transitions in nuclei with $A = 21$–44 are compiled by Endt and van der Leun (1974), those with $A = 6$–44, 45–90, and 91–150 by Endt (1979a, 1979b, and 1981) and those with $A = 151$–190 by Andrejtscheff *et al.* (1975). The E2 and M1 transition probabilities are compiled by Stelson and Grodzins (1965), Venkova and Andrejtscheff (1981), Browne and Femenia (1971), and Krpič *et al.* (1976), and E2–M1 mixing ratios by Kranc (1975, 1976, 1977a, 1977b, and 1978). Half-lives for isomeric states ($T_{1/2} \geqslant 10^{-10}$ s) are listed by Kantele and Tannila (1968).

2.1.2 *Angular distributions of gamma rays*

The angular distributions of the γ rays play decisive roles in many studies of the properties of excited states and excitation (reaction) mechanisms. The γ ray multipolarity E(M)J, where J is the angular momentum (multipolarity), is restricted by selection rules eqn (2.1.6). In most cases the lowest one or two multipoles are allowed for a discrete transition $\psi_i(I_i\pi_i) \rightarrow \psi_f(I_f, \pi_f)$, where $(I_f\pi_i)$ and (I_f, π_f) are the spin and parities of the initial and final states, respectively. The angular distribution is written as (Frauenfelder *et al.*, 1965)

$$W(\theta_\gamma) = \sum_{\nu}^{n} \rho_\nu(I_i) f_\nu(I_i,\ I_f,\ \mathrm{E(M)}J,\ \mathrm{E(M)}J') P_\nu(\cos\theta_\gamma), \qquad (2.1.11)$$

where $P_\nu(\cos\theta_\gamma)$ are Legendre polynomials, θ_γ refers to the angle of the γ ray with respect to the axis of the quantization (spin orientation). E(M)J and E(M)J' are the multipolarities involved, and f_ν are functions of multipoles and spins associated with the γ transition. The spin-orientation factor is given by the statistical tensor term $\rho_\nu(I_i)$. The summation over ν extends to the maximum value given by

$$\nu_{max} \equiv n \leqslant \min (2I_i, (J+J')). \tag{2.1.12}$$

In the case of a pure multipole transition, $J=J'$ and $\nu \leqslant \min (2I_i, 2J)$. For directional distributions ν is restricted to even values, whereas for circular-polarization distributions odd ν powers are present.

The statistical tensor is written as

$$\rho_\nu(I) = (2I+1)^{1/2} \cdot \sum_m (-)^{J-m} \langle I\ m\ I\ -m|\nu 0\rangle a_m, \tag{2.1.13}$$

where a_m is the relative population of the magnetic substate m. It is normalized to $\Sigma a_m = 1$.

Since the angular distribution is given by a simple algebraic function of the spin-parities $(I_i\pi_i, I_f\pi_f)$, the multipolarities $\{E(M)J, E(M)J'\}$ and the spin orientation $\rho_\nu(I_i)$, measurement of the angular distribution gives directly some information on such variables relevant to the distribution as $I_i\pi_i$, $I_f\pi_f$, E(M)J, E(M)J', and $\rho_\nu(I_i)$. If the spin orientation is known, then $W(\theta_\gamma)$ is used to determine the quantum numbers, $I_i\pi_i$, $I_f\pi_f$, E(M)J, E(M)J', associated with the γ transition. On the other hand, $W(\theta_\gamma)$ for a known multipole transition between states with known $I\pi$ gives the substate distribution of the γ-emitting state ψ_i, which reflects the excitation (population) mechanism of ψ_i.

Spin orientation is achieved by various methods. A direct way is to polarize nuclear spins (I_i) by applying magnetic fields at low temperatures. Then the spin orientation (polarization) with respect to the magnetic field is obtained, and the statistical tensor $\rho_\nu(I_i)$ is evaluated from the magnetic interaction. Spins of excited states (I_i') populated by β and/or γ decays from the spin-oriented states with $\rho_\nu(I_i)$ are still oriented well. Then $\rho_\nu(I_i')$ for those states are obtained from the angular momentum of the preceding β and/or γ decays and $\rho_\nu(I_i)$. Angular distributions of γ-rays from these spin-oriented states are used as the β–γ and γ–γ angular correlations to determine spins and multipolarities relevant to the γ transition.

Nuclear reactions have been extensively used for exciting various kinds of states. Spins of such excited states are mostly oriented with respect to the quantum axis relevant to the reaction. Reaction types used for producing excited nuclei (states) B_i^* are expressed as

$$A(a,\ )B_a^* \tag{2.1.14a}$$

$$A(a, b)B_b^* \tag{2.1.14b}$$

$$A(a, b, c, \ldots, \gamma_1, \gamma_2, \ldots)B_c^*, \tag{2.1.14c}$$

where A is the target nucleus, a is the incident (projectile) particle (or photon), b, c, . . . are outgoing particles and γ_1, γ_2 . . . are outgoing γ rays. The spin of B_a^* is then oriented in a plane perpendicular to the beam axis and a_m in the statistical tensor—eqn (2.1.13)—is reduced to $a_m \approx \delta_{m0}$, by using the beam axis as the Z-axis. Spin orientations of residual nuclei (B_b^* and B_c^*) depend on whether intermediate particles (b, c, . . .) and γ rays ($\gamma_1, \gamma_2, \ldots$) are observed or not. If intermediate particles and γ rays are not observed, the spin orientation of residual nuclei is disturbed to some extent by such particle and γ ray emissions. When intermediate particles (or γ rays) are observed, the spin orientation and its axis are modified, depending on the angular momentum of the intermediate particle (γ ray) and its direction. Consequently the spin orientation of the residual state reflects the reaction mechanism, including all the particle and γ-ray emission processes at the preceding stage. Experimentally one can study the spin orientation by measuring either angular distributions of γ rays from B_i^* with respect to the incident beam direction or angular correlations of γ rays with respect to the direction of the outgoing particle b (or one of the intermediate particles). These distributions and correlations are called particle γ-ray angular correlations.

Angular distributions and angular correlations are discussed in detail in several nuclear spectroscopy articles (i.e. Frauenfelder *et al.*, 1965; Ferguson, 1965).

Some formulae for the angular distributions are given below for γ rays from nuclear levels with spin orientation (Frauenfelder *et al.*, 1965). The angular distribution of the γ rays without observation of the polarization is expressed by

$$W(\theta_\gamma) = \sum_\nu A_\nu(JJ', I_i I_f) P_\nu(\cos\theta_\gamma) \tag{2.1.15a}$$

$$A_\nu(JJ', I_i I_f) = \rho_\nu(I_i) B_\nu(JJ', I_i I_f). \tag{2.1.15b}$$

The coefficients B_ν are written using the F coefficients and the multipole mixing ratio δ as follows:

$$B_\nu(JJ', I_i I_f) = \frac{1}{1+\delta^2}\{F_\nu(JJ, I_i I_f) + \delta^2 F_\nu(J'J', I_i I_f) + 2\delta F_\nu(JJ', I_i I_f)\} \tag{2.1.16a}$$

$$F_\nu(JJ', I_i I_f) = (-)^{I_f - I_i - 1}\{(2J+1)(2J'+1)(2I_i+1)\}^{1/2} \times \langle J1J'-1|\nu 0\rangle W(I_i I_f JJ'; \nu I_f) \tag{2.1.16b}$$

$$\delta = \langle I_f \| \mathbf{M}(J') \| I_i\rangle / \langle I_f \| \mathbf{M}(J) \| I_i\rangle, \tag{2.1.16c}$$

where $W(abcd; ef)$ is the Racah coefficient, and $\mathbf{M}(J) = g_J(k)\mathbf{M}\{\mathrm{E(M)}J\}$—see eqn (2.1.7). In the case of a pure multipole transition B_ν is simply given by $F_\nu(JJ, I_i I_f)$. Note that $\rho_0 = F_0 = A_0 = 1$ for $\nu = 0$.

In the case where the initial state I_i for the γ transition is populated by preceding transitions $I_0 \to J^0 + I_1$, $I_1 \to J^1 + I_2, \ldots I_{i-1} \to J^{i-1} + I_i$, starting from a spin oriented state I_0, the statistical tensor $\rho_v(I_i)$ is written as

$$\rho_v(I_i) = \rho_v(I_0) \cdot U_v^0 \cdot U_v^1 \ldots U_v^{i-1}, \tag{2.1.17}$$

where $\rho_v(I_0)$ is the statistical tensor for the starting state with I_0 and U_v^k is an attenuation factor due to the change in the spin orientation due to the k-th preceding transition. The U coefficient for $I \to J + I'$ is given by

$$U_v = \{(2I+1)(2I'+1)\}^{1/2}(-)^{I-I'-J} W(III'I'; vJ). \tag{2.1.18}$$

Observation of the angular distribution coefficients A_v for a γ transition with known I_i, I_f and $J\,(J')$ leads to the statistical tensors $\rho_v(I_i)$, which are the product of the statistical tensors $\rho_v(I_0)$ and the attenuation factors $\Pi\, U_v^k$ if I_i is populated by the preceding γ cascades from I_0. Consequently, one gets information on the preceding γ cascades if $\rho_v(I_0)$ is known and vice versa. The concept of the attenuation coefficient can be extended to particle decays. In this case the statistical tensor of the residual state $\rho_v(I_0)$ is expressed in terms of the statistical tensor for the compound state C or the incident channel and the attenuation factor due to the change in the spin orientation caused by the particle decays. Here one sees interplays between the nuclear reaction aspects associated with incident and outgoing particles and the nuclear structure aspects associated with γ decays (see Section 4.2).

The polarization distributions of γ rays are also discussed by Frauenfelder *et al.* (1965).

So far γ transitions between definite spin-parity states have been discussed. Most of the bound states indeed have definite spins and parities because their level widths (γ-decay widths) are too small for them to overlap with each other. On the other hand, unbound (continuum) states involved in nuclear reactions are often mixed with each other except for isolated sharp resonance states. The γ decays from such states are thus given by the sum of amplitudes with multipoles J, J', J'', ... from many states with different spins and parities $J_i\pi_i$, $J_i'\pi_i'$, ... Transitions from mixed parity states necessarily include odd v powers in the angular distribution—eqn (2.1.11). Nevertheless, angular distributions are useful to get the major transition multipoles and giant resonances involved in the unbound high excitation region (see Section 4.5).

Angular distribution coefficients for γ transitions are given in various kinds of tables. The coefficients for angular distributions of γ rays from aligned nuclei are given by Yamazaki (1967), and those for γ rays from partially aligned nuclei by DerMateosian and Sunyar (1974). The coefficients for the angular correlations of γ rays from aligned nuclei and those for γ–γ directional correlations are tablated by Watson and Harris (1967) and Taylor *et al.* (1971). The angular distribution coefficients for γ–particle and particle–γ correlations are given in Carr and Baglin (1971) and Mante *et al.* (1974), and the polarization coefficients for them are given in Laszewski and Holt (1977) and Monahan *et al.* (1979).

2.1.3 *Internal conversion electrons*

The EM interaction of an excited nucleus with an atomic electron gives rise to emission of the electron to a continuum region. In the process, the nuclear excitation energy E_γ is converted internally to that of the ejected electron. The electron energy is given by

$$E_i^e = E_\gamma - B_i > 0, \qquad (2.1.19)$$

where the subscript i refers to the electron shell and B_i is the electron binding energy of that shell. Then the EM transition probability is given by

$$\begin{aligned} \Gamma &= \Gamma_\gamma + \Sigma \Gamma_i \\ &= \Gamma_\gamma (1 + \Sigma \alpha_i), \end{aligned} \qquad (2.1.20)$$

where Γ_i is the i-shell electron conversion probability and α_i is the i-shell internal conversion coefficient. Actually the EM interactions for inner-shell electrons with $i = K, L, M, \ldots$ are stronger and accordingly these are more likely to be converted than the outer-shell ones, provided that E_γ substantially exceeds the binding energies.

Internal conversion electrons provide useful data, which are complementary to the γ-ray data in many cases and even provide crucial additional information in some cases. E0 transitions with $J = 0$, $\pi = +1$ are allowed for conversion electrons—note that E0 γ decay is forbidden because of the photon spin 1: see eqn (2.1.6d). Thus $0^+ \to 0^+$ transitions are possible for E0 internal conversion electrons and transitions between the same spin-parity states of $J_\mathrm{i} = J_\mathrm{f}$, $\pi_\mathrm{i} = \pi_\mathrm{f}$ include E0 conversion electrons as well as M1 and E2 ones. In such cases one gets E0 nuclear matrix elements for the conversion electron probabilities. The conversion coefficients α_i are quite sensitive to the multipolarity, E(M)J, but they are rather independent of the nuclear transition matrix elements, as will be shown later. Thus multipolarities can be uniquely assigned by observing the absolute values of α_i and their ratios α_i/α_j such as K/L, $L_\mathrm{I}/L_\mathrm{II}/L_\mathrm{III}$, and so on. The electron orbits may partly penetrate into the nucleus, and this penetration effect provides important information on nuclear dynamics (Church and Weneser, 1956).

From the experimental side, conversion electrons have several unique features. Low-energy and/or high-multipole transitions in medium and heavy nuclei are largely converted to electrons, and thus are detected efficiently by means of the conversion electrons. High resolution measurements of electrons are made by means of electron spectrometers and/or semiconductor detectors, which are free of the Compton tails appearing in most of the γ-ray detectors. The conversion coefficient α_i is easily obtained from the ratio of the conversion electron yield to the γ-ray yield and the ratio α_i/α_j from the yield ratio of the i-shell electron yield to the j-shell one. Conversion electrons leave electron holes which are filled by emission of X-rays or Auger electrons. In this way X-rays, characteristic of the atomic number Z, can be used to assign the value of Z to the residual nucleus. (See Section 3.4 for details of the conversion electron measurements, and Section 4.3 for the X-rays following conversion electrons.)

The theoretical aspects of measurements of conversion electrons have been discussed in many articles on γ-ray spectroscopy (Rose, 1955; 1965; Marmier and Sheldon, 1969). Here we give some of the basic expressions for conversion electrons according to Rose (1965), and discuss important features of them. The electron transition probability is written as

$$T_e = \frac{2\pi e^2}{\hbar c}\left|\iint \mathrm{d}\tau_\mathrm{n}\mathrm{d}\tau_\mathrm{e}(\mathbf{j}_\mathrm{n}\cdot\mathbf{j}_\mathrm{e} - \rho_\mathrm{n}\rho_\mathrm{e})\frac{\exp(ik|\mathbf{r}_\mathrm{n}-\mathbf{r}_\mathrm{e}|)}{|\mathbf{r}_\mathrm{n}-\mathbf{r}_\mathrm{e}|}\right|^2, \tag{2.1.21}$$

where the subscripts n and e refer to the nucleon and the electron, respectively, and $k = E_\gamma/\hbar c$. $\mathbf{j}_\mathrm{e}$ and ρ_e are the components of the electron transition 4-current, and $\mathbf{j}_\mathrm{n}$ and ρ_n are the dynamic nucleon 4-vector components. The integral $\int \mathrm{d}\tau$ stands for the volume integral over all space. Using the expansion,

$$\frac{\exp(ik|\mathbf{r}_\mathrm{n}-\mathbf{r}_\mathrm{e}|)}{|\mathbf{r}_\mathrm{n}-\mathbf{r}_\mathrm{e}|} = 4\pi i \Sigma j_L(kr_<)h_L(kr_>)\mathbf{Y}^*_{LM}(\mathbf{r}_\mathrm{n})\mathbf{Y}_{LM}(\mathbf{r}_\mathrm{e}), \tag{2.1.22}$$

where j_L and h_L are the Bessel and Hankel functions, respectively. The matrix element for the $\mathbf{j}_\mathrm{n}\cdot\mathbf{j}_\mathrm{e}$ term in eqn (2.1.21) becomes

$$H'_\mathrm{e} = 4\pi ik\left\{\int_0^\infty \mathrm{d}\tau_\mathrm{n}\mathbf{j}_\mathrm{n}\cdot\mathbf{A}^*_{LM}\int_0^\infty \mathrm{d}\tau_\mathrm{e}\mathbf{j}_\mathrm{e}\cdot\mathbf{B}_{LM} + \int_0^\infty \mathrm{d}\tau_\mathrm{n}\mathbf{j}_\mathrm{n}\cdot\mathbf{A}^*_{LM}\int_0^{r_\mathrm{n}} \mathrm{d}\tau_\mathrm{e}\left(-\mathbf{j}_\mathrm{e}\cdot\mathbf{B}_{LM} + \frac{h_L(kr_\mathrm{n})}{j_L(kr_\mathrm{n})}\mathbf{j}_\mathrm{e}\cdot\mathbf{A}_{LM}\right)\right\}, \tag{2.1.23}$$

where $\mathbf{A}_{LM}$ is the standing wave Maxwell vector potential and $\mathbf{B}_{LM}$ is that for the outgoing wave. It will be noted that $\mathbf{B}_{LM}$ contains h_L in place of j_L in $\mathbf{A}_{LM}$.

The first term of eqn (2.1.23) dominates in the conversion processes. The second term is due to overlap of the electron orbit with the nucleus and vanishes if the overlap is negligible in the limit of a point nucleus. The first term is expressed as a simple product of the nuclear term and the electron term. The nuclear term is just the γ radiation matrix element

$$\mathrm{H}_\gamma = \int_0^\infty \mathrm{d}\tau_\mathrm{n} \mathbf{j}_\mathrm{n} \cdot \mathbf{A}^*_{LM}. \tag{2.1.24}$$

Consequently the conversion electron probability for the first term is given by

$$T_\mathrm{e} = \alpha_i^0 T_\gamma, \tag{2.1.25a}$$

$$\alpha_i^0 = c_i \left| \int_0^\infty \mathrm{d}\tau_\mathrm{e} \mathbf{j}_\mathrm{e} \cdot \mathbf{B}_{LM} \right|^2, \tag{2.1.25b}$$

where α_i^0 is the conversion coefficient. The $\rho_\mathrm{n}\rho_\mathrm{e}$ term in eqn (2.1.21) contributes to electric transitions. Inclusion of the $\rho_\mathrm{n}\rho_\mathrm{e}$ term in eqn (2.1.23) does not change the basic feature of the factorization of T_e as $T_\gamma \cdot \alpha_i$. It is important to note that the nuclear part T_γ is completely factorized out and accordingly the conversion coefficient is independent of the nuclear matrix element. The coefficient α_i^0 depends on the γ and electron phase factors and electron wave functions. Thus it is expressed as a function of E_γ, J, k, Z and $\psi_\mathrm{i}(\mathbf{r}_\mathrm{e})$, and is used to determine the multipolarity E(M)J. Numerical values of conversion coefficients are given in several tables (Sliv and Band, 1965; Hager and Seltzer, 1968, 1969; Dragoun *et al.*, 1969, 1971; and Trusov, 1972).

The second term in eqn (2.1.23) gives the dynamic effect of the finite nuclear size. It arises from the electron penetration in the nuclear region $r_\mathrm{e} < r_\mathrm{n}$. The related matrix element is given by

$$M_\mathrm{e}^P = \int_0^\infty \mathrm{d}\tau_\mathrm{n} \mathbf{j}_\mathrm{n} \cdot \mathbf{A}^*_{LM} f(r_\mathrm{n}), \tag{2.1.26}$$

where $f(r_\mathrm{n})$ is expressed as the integral in the nuclear region $r < r_\mathrm{n}$. Then the conversion coefficient is

$$\alpha_i = \alpha_i^0 |1 + \lambda|^2 \tag{2.1.27a}$$

$$\lambda = \frac{M_\mathrm{e}^P}{\int \mathbf{j}_\mathrm{n} \cdot \mathbf{A}^*_{LM} \mathrm{d}\tau_\mathrm{n} \int \mathbf{j}_\mathrm{e} \cdot \mathbf{B}_{LM} \mathrm{d}\tau_\mathrm{e}}, \tag{2.1.27b}$$

where λ is the penetration factor (Church and Weneser, 1956). One can get the penetration factor λ by comparing the observed coefficient α_i with the calculated value α_i^0. The penetration factor is rewritten as

$$\lambda = \frac{M_\mathrm{e}^P}{C M_\gamma \cdot \sqrt{\alpha_i^0}}, \tag{2.1.28}$$

where M_γ is the γ transition matrix element and C is a coefficient. Thus, by using M_γ obtained from the γ decay width (probability), one can get the new nuclear matrix element M_e^P, which cannot be derived from γ decay experiments. When M_γ is very small λ becomes quite important as shown in eqn (2.1.28). Because of the different structure of M_e^P and M_γ, there are cases in which M_γ is greatly reduced by selection rules, but M_e^P is not. An example of this is the l-forbidden transition discussed in Chapter 5.

Angular distributions involving conversion electrons are modified compared with those for γ rays by the particle parameter b_v. The electron distribution for the transition $I_i \rightarrow I_f + J$ is

$$W(\theta_e) = \Sigma b_v \rho_v(I_i) F_v(JJ, I_i I_f)\; P_v(\cos\theta_e)$$
$$= \Sigma b_v A_v(JJ, I_i I_f)\; P_v(\cos\theta_e), \qquad (2.1.29)$$

where ρ_v, F_v and A_v are as defined in Section 2.1.2 for γ rays, and the particle parameters b_v are defined by Biedenharn and Rose (1953). Numerical values of the particle parameters are given in tables (Band *et al.*, 1965).

2.2 Single-particle excitations in nuclei

2.2.1 *Single-particle potentials*

The motion of the individual nucleons in nuclei is determined in principle by their interactions with the other nucleons, but the nucleon–nucleon interaction in nuclei is not well known and many-body problems are hard to solve. In the independent-particle model it is assumed that the sum over all the nucleon–nucleon interactions can be replaced by an effective one-body potential. Several forms for this potential have been proposed.

The simplest potential is the harmonic oscillator (h.o.) potential. The corresponding Hamiltonian is

$$\mathbf{H} = \frac{1}{2m}\mathbf{p}^2 + \frac{1}{2}m\omega^2\mathbf{r}^2, \qquad (2.2.1)$$

where $\mathbf{p}$ is the momentum, m the mass of the nucleon, $\mathbf{r}$ the distance of the nucleon from the centre of the nucleus and ω the angular (oscillator) frequency. The advantage of this potential is that the eigenvalue problem can be solved analytically. The energy eigenvalues are expressed in terms of the oscillator quantum number N as

$$E = (N + \tfrac{3}{2})\hbar\omega, \qquad N = 0, 1, 2, \ldots \qquad (2.2.2)$$

The energy of the oscillator quantum, $\hbar\omega$, is approximately given by $41\text{A}^{-1/3}$ MeV. The maximum number of nucleons that can occupy the shell N is given by the degree of degeneracy $2(N+1)(N+2)$. A subdivision of N is made into the radial quantum number n (number of nodes in the radial wave function) and the orbital (angular momentum) quantum number l.

$$N = 2(n-1) + l, \quad n = 1, 2, 3, \ldots, \quad l = N, N-2, \ldots 0 \text{ or } 1. \qquad (2.2.3)$$

The degeneracy with respect to l (or n) is removed when the potential deviates from that of the h.o. An example of such a potential is the Woods–Saxon (W.S.) potential

$$U(r) = U_0/\{l + \exp(r - R_0)/a\} \qquad (2.2.4)$$

where the depth of the potential is $U_0 \simeq -50$ MeV, the nuclear radius $R_0 = r_0 A^{1/3}$ with $r_0 \simeq 1.2$ fm and the surface diffuseness parameter $a \simeq 0.7$ fm. The energy levels of these two potentials are shown in Figs 2.1(a) and 2.1(b). As shown in Fig. 2.1(b) the energy levels with higher-l values are lowered in energy with respect those with the low-l values for the same N, and this is attributed to the flat bottom of the W.S. potential.

The strong spin–orbit coupling terms, which is unique to the nuclear potential, is included in the single-particle Hamiltonian. It is written as

$$U_{\text{s.o.}}(f) = U(r)\mathbf{l}\cdot\mathbf{s}. \qquad (2.2.5)$$

The single-particle energy is then shifted by $-\langle U(r)\rangle(l+1)/2$ for spin $j = l - 1/2$ levels and by $\langle U(r)\rangle l/2$ for $j = l + 1/2$ levels, with $\langle U(r)\rangle = -20\text{A}^{-2/3}$ MeV (Brussaard and Glaudemans, 1977). The level scheme with the spin–orbit term included is shown in Fig. 2.1.

The independent-particle model with such one-body potentials is rather simplified, and in fact residual two-body interactions have to be taken into account. Modern shell-model calculations include many valence nucleons distributed over several shells and subshells with several kinds of residual interactions. Refined spectroscopic techniques are needed to elucidate important features of the nuclear shell potentials and of the nuclear interactions so as to provide sensitive tests of the calculations. We will present here some details of a recent investigation of the ^{25}Mg$(p,\gamma)^{26}$Al reaction by Endt *et al.* (1986) as an example of the sort of detailed spectroscopic information needed to test the nuclear shell potential.

The γ decay has been measured of 75 resonances in the $E_p = 0.31$–1.84 MeV region by means of a 100 cm^3 Ge detector with a Compton suppression shield (CSS). High sensitivity for weak, especially low-energy, γ rays was achieved by the high-resolution Ge detector with a CSS, together with the large intensity ($\sim$10 Coulomb) of proton beam used to measure one spectrum. The CSS reduces the Compton background of the spectra and the intensities of single- and double-escape peaks by at least a factor of 12. Figure 2.2 shows, as an example, seven previously unobserved weak branches of the decay of the $E_x = 4548$ keV ^{26}Al bound state, with intensities of only 0.2–1.3 per cent of the total. The total numbers of observed decay lines from bound states and from resonances amount to 2770, including ultra-weak lines with intensities even

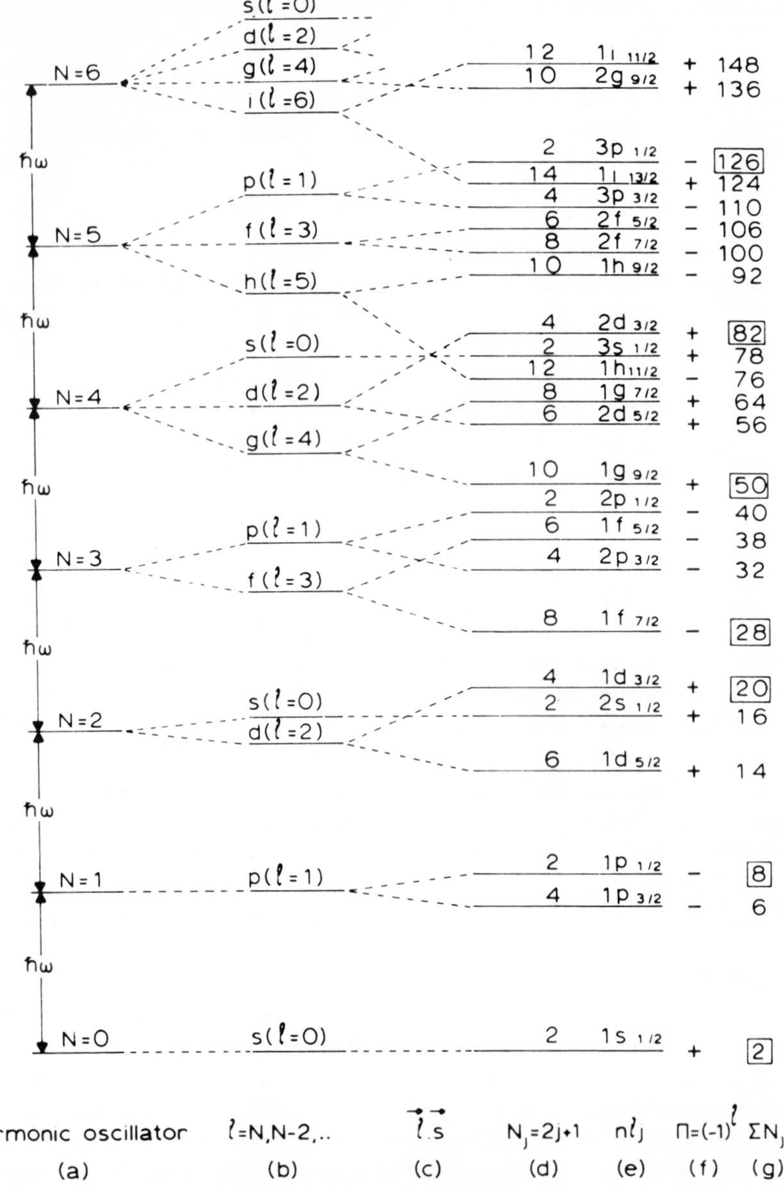

Fig. 2.1. (a) The single-particle energies of a harmonic-oscillator potential as a function of the oscillator quantum number N. (b) A schematic representation of the single-particle energies of a Woods–Saxon potential. (c) A schematic illustration of the level splitting due to the spin–orbit coupling term. (d) The number $N_j = 2j+1$ of identical particles that can occupy each state. (e) The spectroscopic notation of the

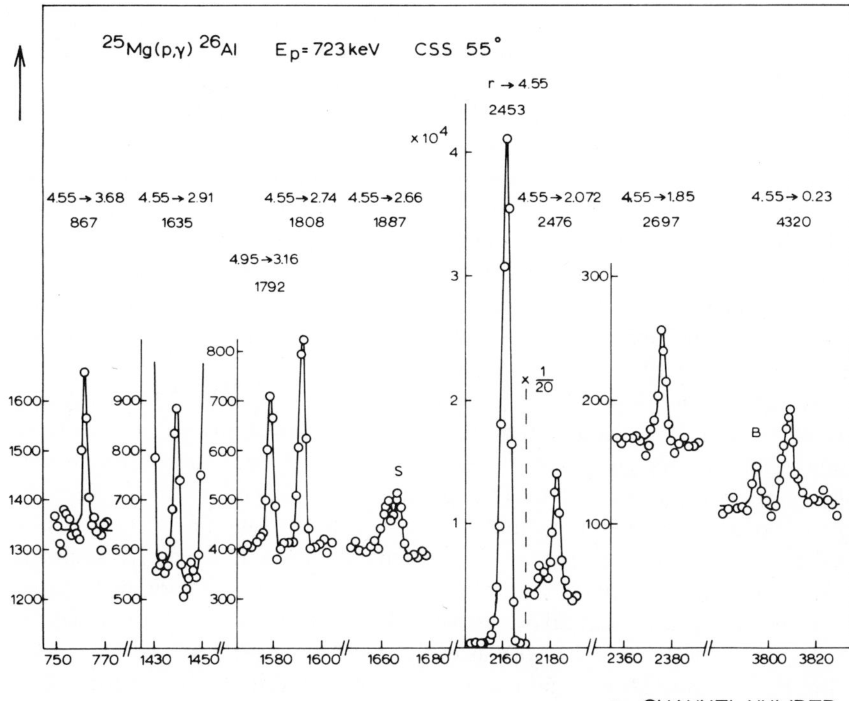

Fig. 2.2. Selected parts of the γ-ray spectrum measured at the $E_p = 723$ keV ^{25}Mg(p,γ)^{26}Al resonance with a Compton suppression spectrometer (CSS). It shows the primary transition to the $E_x = 4548$ keV level ($E_\gamma = 2453$ keV) and seven previously unobserved decay branches of the latter level; a single-escape peak is marked s, a background peak B. (Endt *et al.*, 1986, 1988.)

Fig. 2.1 (*Contd.*)

single-particle quantum numbers *n*, *l* and *j*. (f) The parity of each state. (g) The magic numbers are seen to appear at the energy gaps as the subtotals of the number of particles. The level pattern given above represents qualitative features only. This holds especially for states with $N \geqslant 4$, where the single-particle level-order differs for protons and neutrons and depends also on the number of nucleons occupying lower states. (Brussaard and Glaudemans, 1977.)

Important

below 0.01 per cent of the total decay. The construction of the decay scheme of ^{26}Al is based on an extensive data set of γ rays with extremely accurate energies. The uncertainties in the resulting excitation energies range between 3 and 260 eV. Lifetimes (or lifetime limits) have been deduced from the energy differences, due to the Doppler shifts, of the γ-rays observed at $\theta = 55°$ and 90°. Assignments of spins (J), parities (π) and isospins (T) to the levels are made mostly by the accurate γ-ray branching ratios and the lifetimes, on the basis of the simple selection rules associated with J, π and T in the γ decays. Other investigations have also contributed to the assignments, in particular proton stripping and neutron pick-up reactions (see Endt and van der Leun, 1974), and the ^{25}Mg(p, p_0) reaction (Adams *et al.*, 1984) which all provide l-values and thus parities, the former two reactions mainly for the bound states and the latter for a number of resonances. The ^{28}Si(d, α)^{26}Al reaction with tensor polarized deuterons (Boerma *et al.*, 1986) has also provided the parities (natural or unnatural) of bound states.

A comparison of the 44, $T = 1$, levels in ^{26}Al with the ^{26}Mg level scheme is shown in Fig. 2.3. It is seen that almost all those levels in one nucleus have counterparts in the other one and that the level ordering in these mirror nuclei is approximately the same. The present extensive information on ^{26}Al provides an ideal testing ground for large-scale shell-model calculations. A calculation has been carried out in an extended basis with ten active nucleons (5 protons and 5 neutrons), outside an assumed inert ^{16}O core, distributed over the $1d_{5/2}$, $2s_{1/2}$ and $2d_{3/2}$ subshells (Wildenthal, 1984). All the positive-parity states below $E_x = 7.1$ MeV as well as most of the positive-parity levels in the $E_x = 7.1$–8.1 MeV region are nicely reproduced by the theory. The calculated γ-ray strengths agree with experiment for many hundreds of transitions, so that they become a valuable aid in establishing the correspondence between theoretical and experimental levels.

Single-particle levels in deformed nuclei are described by the Nilsson model with the deformed h.o. potential. The total one-particle Nilsson Hamiltonian for a spheroidal potential with axial symmetry has a central part with $\omega_x = \omega_y \neq \omega_z$, the spin–orbit term with coefficient $U(r) = v_{ls}\hbar\omega_0$ and the term $v_{ll}\hbar\omega_0\mathbf{l}^2$—see eqn (6.3.34a).

$$\mathbf{H}_{\rm sp} = \frac{1}{2m}\mathbf{p}^2 + \frac{1}{2}m(\omega_x^2 x^2 + \omega_y^2 y^2 + \omega_z^2 z^2) + v_{ls}\hbar\omega_0\,\mathbf{l}\cdot\mathbf{s} + v_{ll}\hbar\omega_0(\mathbf{l}^2 - \langle \mathbf{l}^2 \rangle). \tag{2.2.6}$$

(See Section 6.3.3a for the meaning of the various quantities). The last term is added artificially to make the potential more flat in the centre ($r = 0$) and more attractive at the surface; it is not quite adequate, particularly for high-spin states. In such cases one may use an optimized Woods–Saxon potential (Dudek *et al.*, 1979). A comparison between the Nilsson and Woods–Saxon potentials is discussed by de Voigt *et al.* (1983).

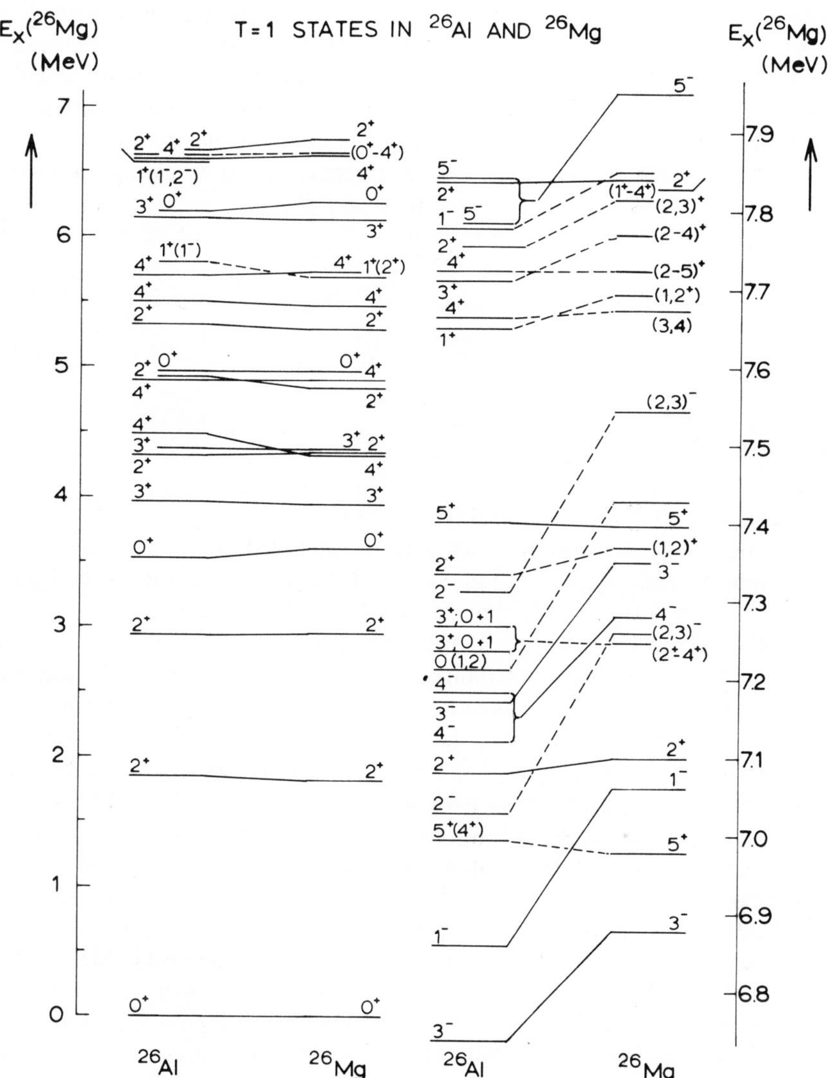

Fig. 2.3. Comparison of $T=1$ states in ^{26}Al and ^{26}Mg, with the ^{26}Al first-excited state at $E_x = 228$ keV, lined up with the ^{26}Mg ground state. Note the large difference in energy scales for the two halves of the figure, because the right-hand side is a continuation of the left-hand side with a much more expanded scale. (Endt *et al.*, 1986.)

The deformation is introduced by two different h.o. frequency components of ω_0; one for the oscillation in the direction of the symmetry axis and one for the oscillation perpendicular to that axis. The two frequencies are given in eqns (6.3.34b, c) as functions of the deformation parameter ε. In a deformed potential, l and j are not good quantum numbers. In the limit of large deformation the states are characterized by the quantum numbers along the symmetry axis $[NN_z\Lambda]\Omega$, where N_z, Ω, and Λ stand for the projections of N, j, and l on the symmetry axis, respectively. Here N_z, Ω and Λ have the following values

$$N_z = 0,\ 1,\ 2,\ \ldots\ N \tag{2.2.7}$$

$$|\Lambda| = N - N_z,\ N - N_z - 2,\ N - N_z - 4,\ \ldots\ 0 \text{ or } 1 \tag{2.2.8a}$$

$$\Omega = \Lambda \pm 1/2. \tag{2.2.8b}$$

The parity is

$$\pi = (-1)^N. \tag{2.2.9}$$

A Nilsson level diagram is displayed in Fig. 2.4.

The effect of deformation can also be included in the Woods–Saxon potential by replacing $(r - R_0)$ in eqn (2.2.4) by a vector $\mathbf{r}$, with $|\mathbf{r}|$ being the distance from the centre of the nucleus to the nuclear surface. For non-axially symmetrical nuclei the nuclear shape is given as a function of the deformation parameter β and of the non-axiality parameter γ (see eqn 6.3.30). Besides the quadrupole deformation β_2 one may also include higher-order deformations, such as the hexa-decapole β_4 (see e.g. Szymański, 1983). An extensive description of the behaviour of deformed nuclei, such as rotation, vibration, Coriolis coupling and single-particle aspects is given by Bohr and Mottelson (1969, 1975). Some of those features are discussed in Section 6.3.

A nice example of an extensive e–γ spectroscopic study, testing the Nilsson configurations for ^{161}Dy, is given by Schmidt *et al.* (1986). They employed a thermal–neutron induced reaction and measured both the γ rays and the conversion electrons. With their curved-crystal spectrometers GAMS, used at the high-flux reactor of the ILL, Grenoble, an extremely good energy resolution of 30 eV has been obtained. The level scheme includes 180 γ transitions, and 13 rotational bands have been identified. As shown in Fig. 2.5, those Nilsson configuration assignments are based on the e–γ data as well as on the systematics in neighbouring nuclei and on (d, p) and (d, t) reaction cross-sections.

Above 500 keV in excitation energy the bands are frequently mixed due to the Coriolis coupling. It is then appropriate to introduce a (Coriolis) decoupling parameter, a, in addition to the effective rotational parameter A. The excitation energies of the rotational bands can then be written in terms of

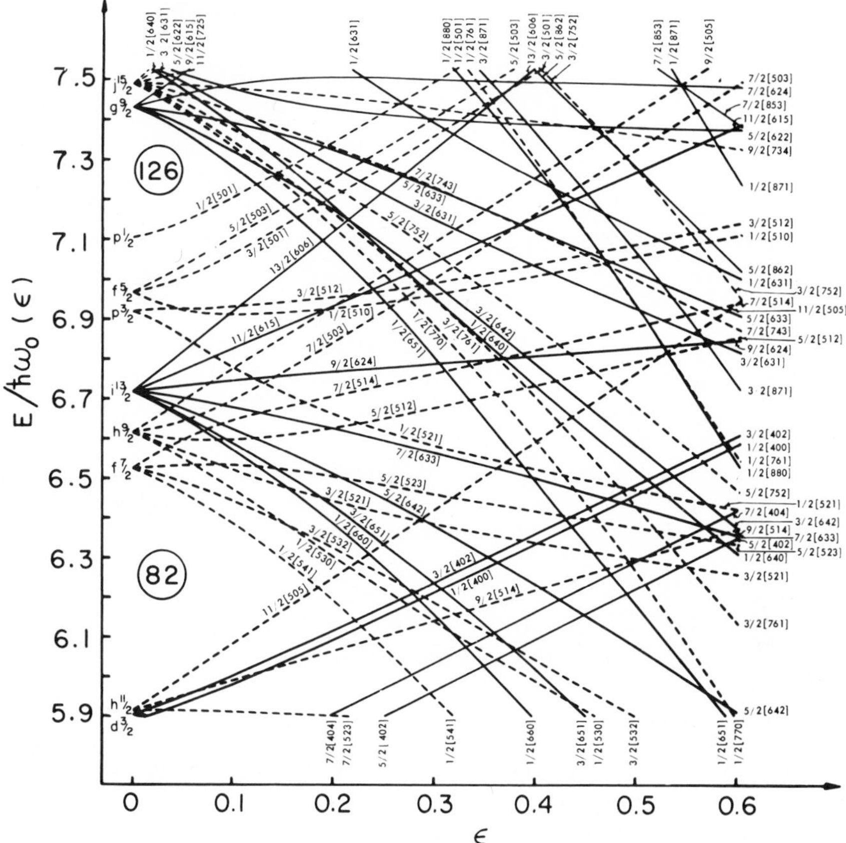

Fig. 2.4. The Nilsson diagram for neutrons in nuclei with $82 \leqslant N \leqslant 126$. The energy is given in units of the harmonic oscillator energy and ε is the deformation parameter. The levels are identified by the quantum numbers $\Omega[NN_Z\Lambda]$ as explained in the text. (Lederer and Shirley, 1978.)

the parameters A and a as

$$E = E_0 + A\{I(I+1) + (-1)^{I+1/2}a(I+\tfrac{1}{2})\}, \tag{2.2.10}$$

where E_0 represents the energy of the band-head with a particular (mixed) Nilsson configuration. The parameters deduced from two or three adjacent levels in each band are summarized in Table 2.2.1.

The systematics of single-particle excitations deduced for the odd Dy isotopes are presented in Fig. 2.6. Such systematics, derived from the data of many different experiments, provide the identification of the Nilsson levels. The experiments thus play an essential role in e–γ spectroscopy, carried out in

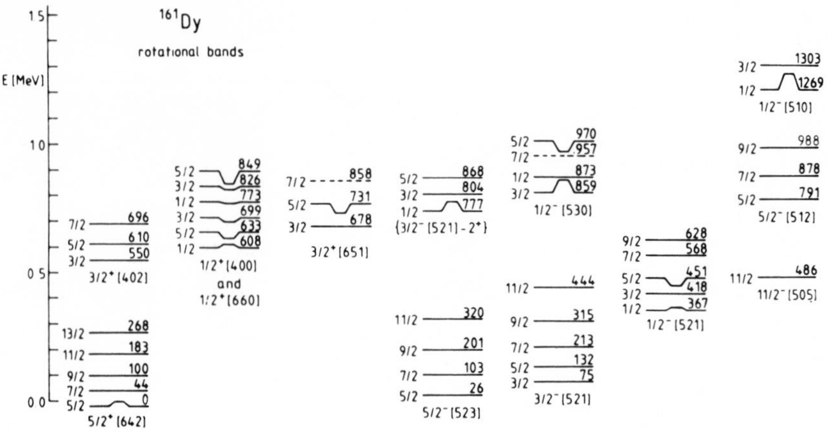

Fig. 2.5. Rotational bands in ^{161}Dy with level energies in keV. The dominant Nilsson (neutron) configuration is given for each band. See Fig. 2.4 for the Nilsson scheme. (Schmidt *et al.*, 1986.)

order to elucidate nuclear structure. The properties of high-spin states in deformed nuclei are described in detail in Chapter 6.

2.2.2 *Shell closures*

The study of nuclei around doubly-closed shells is interesting because of the possibility of observing simple few-particle or few-hole excitations in a clean way. In the doubly-magic nuclei the first excited state has a high excitation energy and is not necessarily the 2^+ state, as in most other even nuclei. The first excited state, for instance, in ^{208}Pb is 3^- at 2.61 MeV, and 0^+ in ^{16}O and in ^{40}Ca.

A special doubly closed-shell nucleus is $^{132}_{50}\mathrm{Sn}_{82}$, which is situated very far from the valley of stability. It is hard to produce this nucleus in a common nuclear reaction with any available target–projectile combination. However, it has been obtained from the β decay of ^{132}In, following fission of uranium, induced by 600 MeV protons at Isolde, CERN (Björnstad *et al.*, 1986). The level scheme is derived from extensive $\gamma\gamma$ and $\beta\gamma$ coincidence experiments and conversion electron measurements, and is shown in Fig. 2.7. The first excited 2^+ state occurs at a high excitation energy of $E_x = 4.04$ MeV and the second excited state with 3^- at $E_x = 4.35$ MeV.

The ^{132}Sn nucleus is very interesting because of the unusually high neutron–proton ratio. Shell-model calculations indicate a $\pi g_{9/2}^{-1}$ hole configuration for the ground state of the adjacent odd nucleus ^{131}In and a $\nu f_{7/2}$ particle configuration for the ground state of ^{133}Sn (Blomqvist, 1981). The

Table 2.2.1 Effective rotational parameters A and decoupling parameters a of the rotational bands

Band	Band head energy	*A*-values and *a*-values in parentheses					
		$\frac{1}{2}-\frac{3}{2}(-\frac{5}{2})$	$\frac{3}{2}-\frac{5}{2}(-\frac{7}{2})$	$\frac{5}{2}-\frac{7}{2}(-\frac{9}{2})$	$\frac{7}{2}-\frac{9}{2}$	$\frac{9}{2}-\frac{11}{2}$	$\frac{11}{2}-\frac{13}{2}$
$\frac{5}{2}^+[642]$	0.0			6.26	6.29	7.54	6.50
$\frac{5}{2}^-[523]$	25.65			11.06	10.89	10.80	
$\frac{3}{2}^-[521]$	74.56		11.44	11.60	11.33	11.70	
$\frac{1}{2}^-[521]$	366.97	11.87(0.439)	11.65(0.428)	11.67(0.426)			
$\frac{3}{2}^+[402]$	550.25		11.92	12.32			
$\frac{3}{2}^+[651]$	678.32		10.52	(18.08)			
$\{\frac{3}{2}^-[521]-2^+\}$	777.13	10.89(−0.166)					
$\frac{5}{2}^-[512]$	790.65			12.55	12.22		
$\frac{1}{2}^-[530]$	873.09	8.69(−1.548)	10.18(−1.175)?				

A is calculated from two adjacent levels; for $K=\frac{1}{2}$ bands, three adjacent levels are used for the calculation of A and a. All energies are given in keV.

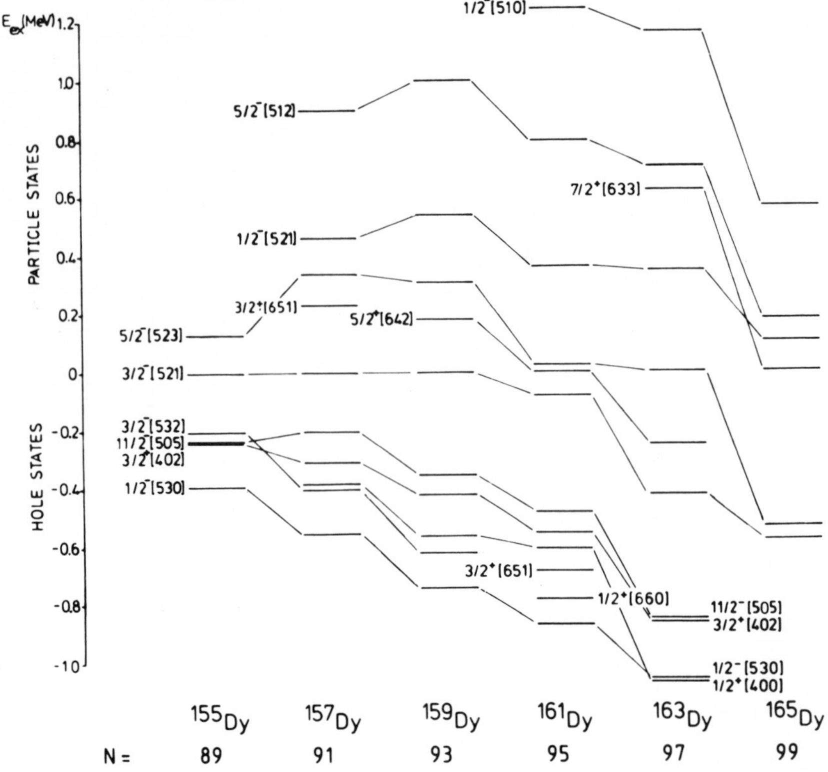

Fig. 2.6. Systematics of the quasi-particle band head energies of the Dy isotopes. The dominant Nilsson configuration is given according to the notation in Fig. 2.4. (Schmidt *et al.*, 1986.)

ground-state configuration of ^{132}In is then assumed to be $\nu f_{7/2}\pi g_{9/2}^{-1}$. The calculated β-decay value $Q_\beta(\nu f_{7/2}\pi g_{9/2}^{-1}\rightarrow 0)=13.68\pm 0.17$ MeV agrees very well with the measured value of $Q_\beta=13.6\pm 0.4$ MeV. The negative–parity excited states, shown in Fig. 2.7, are well explained by the $\nu f_{7/2}\nu g_{7/2}^{-1}$ or the $\nu f_{7/2}\nu d_{3/2}^{-1}$ particle–hole configuration. The positive–parity states are assigned to the $\nu f_{7/2}\nu h_{11/2}^{-1}$ configuration on the basis of B(E2) values, as deduced from the measured lifetimes (Björnstad *et al.*, 1986).

The ^{146}Gd nucleus has the 3^- first excited state at $E_x(3^-)=1.58$ MeV, similar to ^{208}Pb and accordingly has also a doubly magic character (Kleinheinz *et al.*, 1978). The E3 transition probability for the ground-state decay of the 3^- state has been measured as $B(\mathrm{E3})/B(\mathrm{E3})_{\mathrm{sp}}=37\pm 4$ Wu. This is the same strength as for the 3^- decay in ^{208}Pb. Further evidence for the $Z=64$ shell closure in ^{146}Gd has been obtained from the location of the first excited

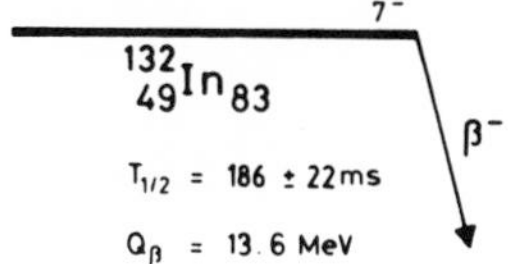

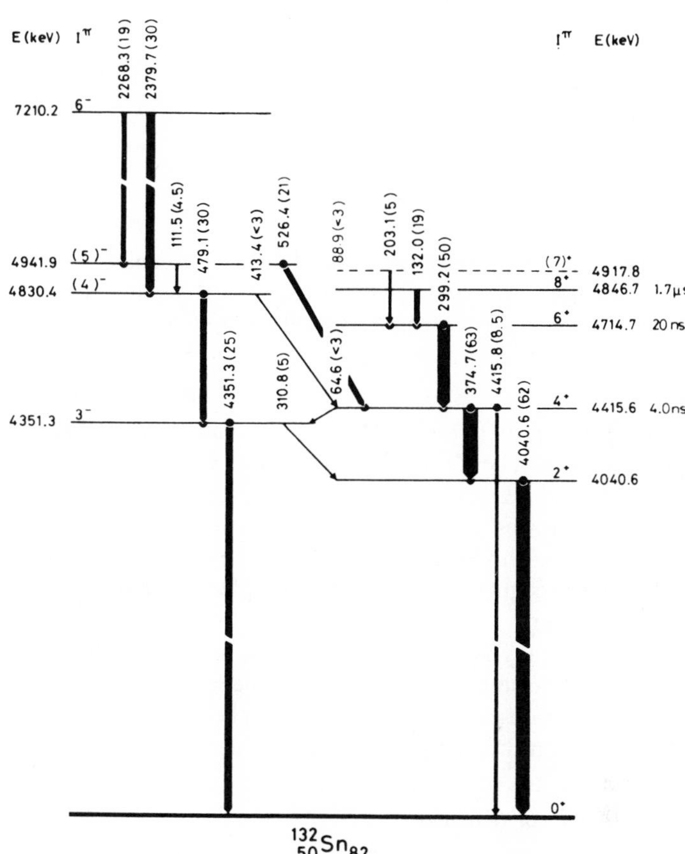

Fig. 2.7. The level scheme of the doubly magic nucleus ^{132}Sn as deduced from the β decay of ^{132}In, produced in fission. The transition energies are given in keV with the corresponding intensities within brackets. (Bjørnstad *et al.*, 1986.)

2^+ state at 1.97 MeV, more than 300 keV higher than in other $N=82$ nuclei. The energies of the single-particle levels can be derived from the nucleon separation energies. The proton and neutron single-particle energies are given in Fig. 2.8. Besides the significant gap of $\delta_n = 3.68$ MeV at $N=82$, a gap of $\delta_p = 3.38$ MeV also occurs at $Z=64$. This gap in effect is smaller because of the

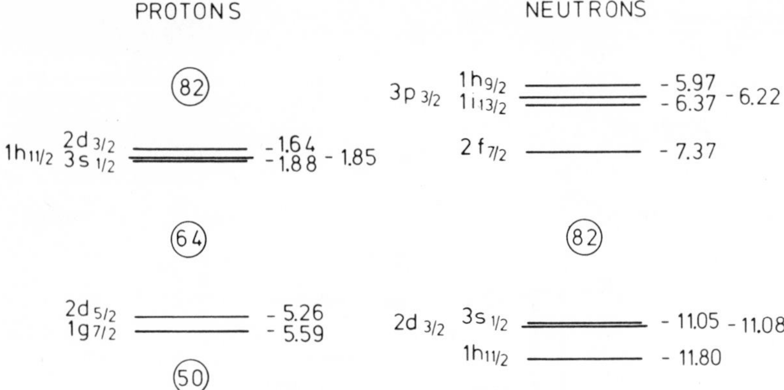

Fig. 2.8. Empirical single-particle level energies around ^{146}Gd as deduced from observed single-particle states in the odd nuclei and thus from the nucleon separation energies. (Kleinheinz *et al.*, 1979.)

pairing interaction ($\Delta_p = 1.2$ MeV), which causes scattering of nucleon pairs across the gap. The $Z=64$ gap then becomes $\delta_p = 2.4$ MeV.

A simplified level scheme of the low-lying states in ^{146}Gd is presented in Fig. 2.9. The lowest one-particle one-hole excitation occurs at 2.66 MeV, and is the 5^- state. This energy is just above the value of the gap $\delta_p = 2.4$ MeV at $Z=64$. The assigned particle–hole multiplets are also indicated in Fig. 2.9.

Nuclei around ^{146}Gd are well studied by means of in-beam e–γ spectroscopy with heavy-ion induced reactions, in contrast to the region around ^{208}Pb where no stable targets are available to perform such experiments. The region of the rare-earth nuclei thus offers the possibility of obtaining detailed information on the high-spin excitations across doubly-closed shells.

2.2.3 *Shape coexistence*

So far we have discussed separately the spherical shell-model states and the deformed ones. The deformation is mainly due to the proton–neutron quadrupole interaction (see Section 2.5). The number of valence nucleons is largest half-way between the closed shells and consequently nuclei located there have the largest deformations. Near the closed shells only a few valance nucleons are active and thus spherical shapes prevail. However, it is also possible to have both a deformed and a spherical shape in the same nucleus at a certain excitation energy. This phenomenon, referred to as shape coexistence, means the existence of two stable shapes at some excitation region, rather than the shape change as the excitation energy changes.

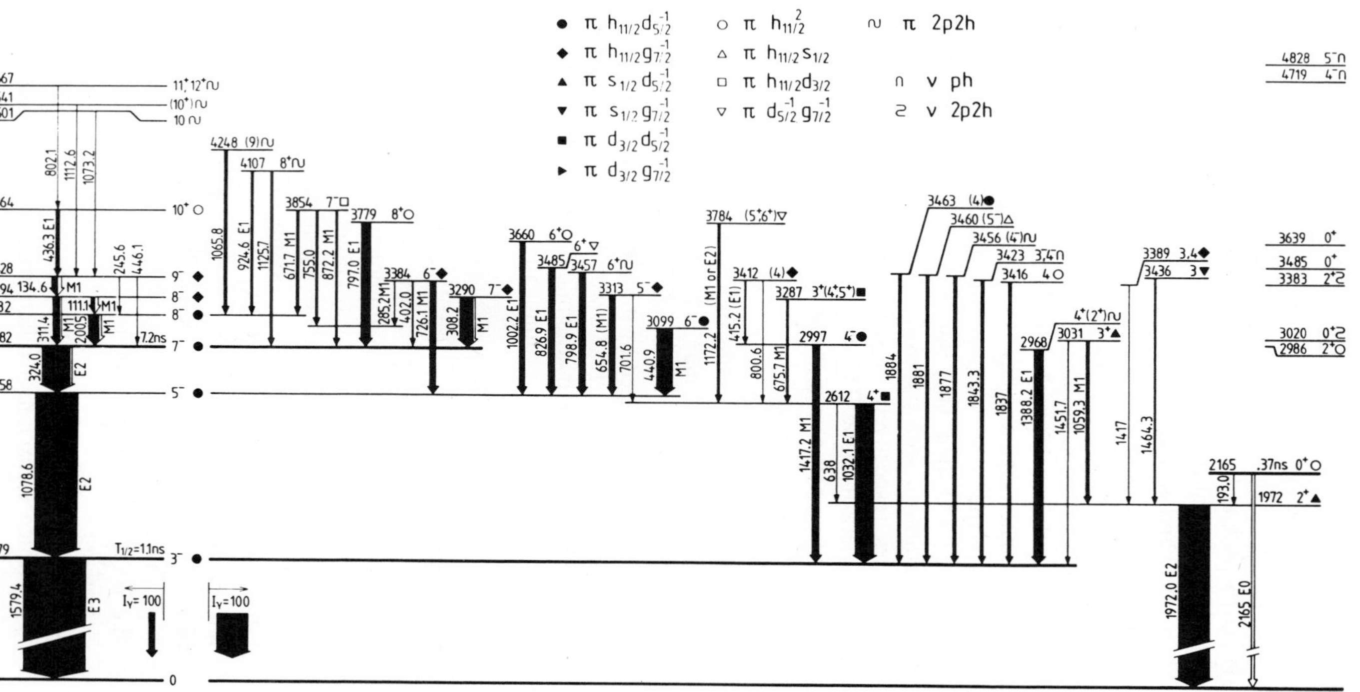

Fig. 2.9. Partial level scheme of ^{146}Gd (below 5 MeV). The configuration assignments were attained from phenomenological diagonal particle-hole matrix elements. (Kleinheinz *et al.*, 1979; Yates *et al.*, 1986.)

In the case of coexistence of spherical and deformed shapes, single-particle states based on the spherical shape and the rotational band-heads due to the deformed shape appear at approximately the same excitation energy. Here the deformed shape is mostly prolate, but in some cases it is oblate or triaxial. The phenomenon of shape coexistence is interesting in view of the interplay between the single-particle and collective degrees of freedom.

One particular possibility for shape coexistence is associated with the occurrence of intruder states. An intruder state is a state with an intrinsic structure that differs significantly from that of the other states in the same excitation region. In the odd nuclei intruder single-particle states come in most cases from the next higher shell and have thus a parity opposite to the other states. In even nuclei intruder states are associated with the excitation of one or two particles into the next shell. They thus have one-particle one-hole (1p–1h) or 2p–2h configurations, with opposite or the same parities with respect to the normal states, respectively. In semi-magic nuclei the normal states are based on the spherical shape. The excitations involve mainly the valence nucleons of one type; i.e. neutrons in the non-closed shell in the case of proton-magic nuclei. If a few particles of the other type (protons) are promoted across the closed shell, the newly created few (proton) particles and holes interact with the many valence nucleons of the other type (neutrons). Then the proton–neutron interaction may induce deformations. The particle–hole states of the above nature induce deformation in an excitation region which is predominantly characterized by the spherical symmetry. The excitation energy of the intruder state decreases with increasing number of valence nucleons of the other type towards the middle of the shell. Shape-coexistence is thus expected to occur most prominently in semi-magic nuclei with the non-closed shell half-filled.

An early γ-spectroscopic investigation of semi-magic Sn nuclei (with $Z=50$) established shape coexistence in the even-mass $^{112-118}$Sn isotopes (Bron *et al.*, 1979; see also Fielding *et al.*, 1977). The neighbouring odd-mass In ($Z=49$) and Sb (Z=51) nuclei have rotational bands built upon the 1p–2h ($I^\pi=1/2^+$) and 2p–1h ($I^\pi=9/2^+$) intruder states, respectively. Potential-energy calculations show that the excitation energies of these p–h states, which are associated with the [431]$1/2^+$ and [404]$9/2^+$ Nilsson orbitals, decrease considerably in a deformed potential (van Isacker *et al.*, 1977). Similar calculations for the even Sn isotopes show that excitation energies of the 2p–2h states, based on the same Nilsson orbitals, become minimal at a deformation of $\varepsilon\approx0.11$. Therefore one may expect to observe the intruder states, as well as the rotational bands built upon them, in the low-lying spectrum.

The extensive search for such deformed bands in the Sn isotopes employed measurements of γ–γ coincidences, excitation functions, angular distributions, linear polarizations, and conversion electrons, using the Cd(α, 2nγe)Sn reactions with 17–33 MeV alpha beams. Partial level schemes are shown in Fig. 2.10 for four tin isotopes (Bron *et al.*, 1979). All these nuclei have a rotational band, with the 0^+ band-heads located at an energy of about 2 MeV. The 0^+ state in ^{116}Sn has been firmly established. The rotational character of the bands in the Sn isotopes has been established on the basis of the level spacings and the enhanced intraband decay as compared with the interband transitions. The B(E2) values are indicated in brackets in Fig. 2.10.

Many studies have been carried out on semi-magic nuclei, and particular attention has been devoted to the nuclei around the closed proton shells, $Z=50$ (Sn, Ag, In, Sb, I) and $Z=82$ (Pb, Pt, Au, Tl, Bi). The data for odd nuclei are reviewed and theoretically interpreted in terms of intruder states by Heyde *et al.* (1983).

As an illustration of the power of electron–gamma spectroscopy we discuss the results for the neutron-deficient even–even Pb isotopes. The excited states in the neutron-deficient Pb isotopes with closed $Z=82$ shell are to a large extent determined by the valence neutrons. The $2^+, 4^+, \ldots 12^+$ excited states in the even-mass isotopes are explained by the coupling of two $i_{13/2}$ neutrons. Those states are strongly excited in most nuclear reactions, including the heavy-ion induced fusion reactions.

Proton 2p–2h excitations across the energy gap at $Z=82$ would require as much as ~ 7 MeV in a spherical potential. The π(2p–2h)0^+ intruder state in a deformed potential, however, can be considerably lower in excitation energy, due to the proton–neutron quadrupole interaction and to the energy gained by the breaking the pairing correlation. It is thus expected that the intruder state will come down in energy when the number of valence neutrons increases from ^{208}Pb going to the neutron deficient isotopes ($N=104$, ^{196}Pb, at midshell).

Investigation of these π(2p–2h) states in the light Pb isotopes is difficult because of a lack of stable targets to perform few-nucleon transfer reactions to reach such states. Alpha- or heavy ion-induced fusion reactions are not suitable since they populate predominantly the yrast levels with neutron quasi-particle configurations. Moreover the large angular momentum associated with heavy ion-induced reactions, creates a favourable situation for fission of the compound nuclei, and produces disturbing backgounds.

A successful population of those π(2p–2h) states in the light Pb isotopes has been achieved by the β^+/EC decay of Bi isotopes after "isotope separation on line (ISOL)", following 110–180 MeV ^{16}O induced fusion reactions (Van Duppen *et al.*, 1984). Here the Bi activities, being mass-separated, were transported by a tape to perform off-line measurements of e, γ, and X-ray spectra including γ–γ and γ–X coincidences. A pronounced peak is seen in the electron spectrum for ^{196}Pb, but not in the γ spectrum (see Fig. 2.11). This

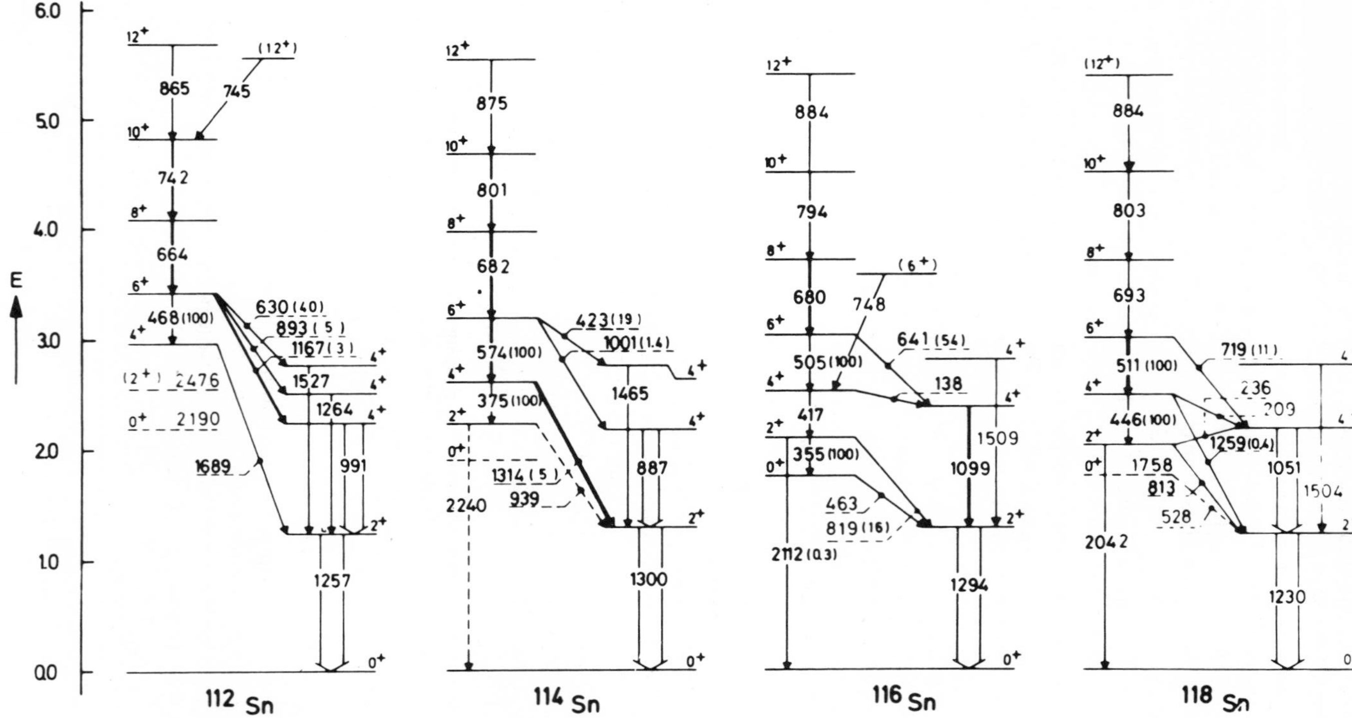

Fig. 2.10. Partial level schemes showing the positive parity bands in ^{112,114,116,118}Sn. The numbers in parentheses are relative B(E2) values. The rotational bands, starting around 2 MeV, are clearly seen in all cases. (Bron *et al.*, 1979.)

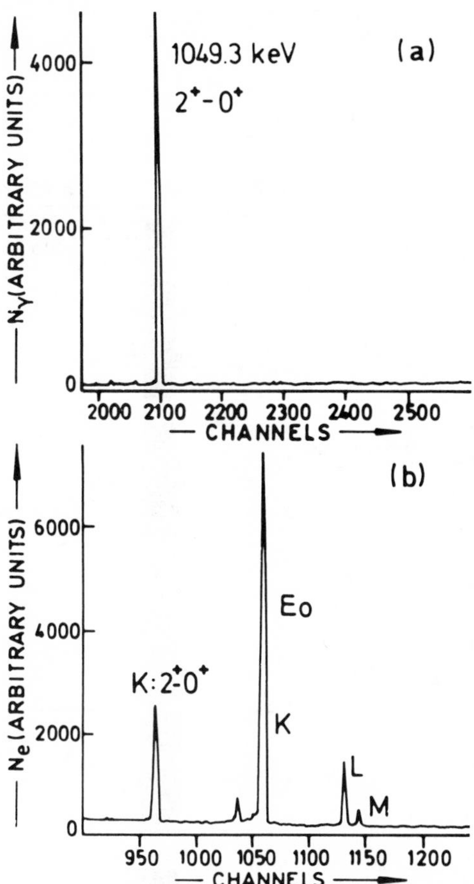

Fig. 2.11. A comparison of parts of the gamma (a) and electron (b) spectra of ^{196}Pb, observed in the β decay of ^{196}Bi. (van Duppen *et al.*, 1984.)

peak is attributed to the E0 transition from the1143 keV 0_2^+ state to the ground state in ^{196}Pb. The 0_2^+ state lies only 94 keV above the 2_1^+ state in ^{196}Pb, as can be seen in Fig. 2.12. In the same investigation such anomalously low-lying 0^+ states were discovered in ^{192,194,198}Pb as well. The 2_2^+ excited state and a candidate for the 4_2^+ state have been found in ^{196}Pb, as shown in Fig. 2.12. This suggests a rotational band built on the 0_2^+ intruder state, which would then establish the deformed shape. This method is limited to only a few cases where low-spin isomeric states have sufficiently long lifetimes and are populated strongly enough for the ISOL measurements.

To investigate excited states at somewhat higher spin one has to use the in-beam spectroscopy with fusion reactions. The ^{198}Hg(α, 6n)^{196}Pb reaction was

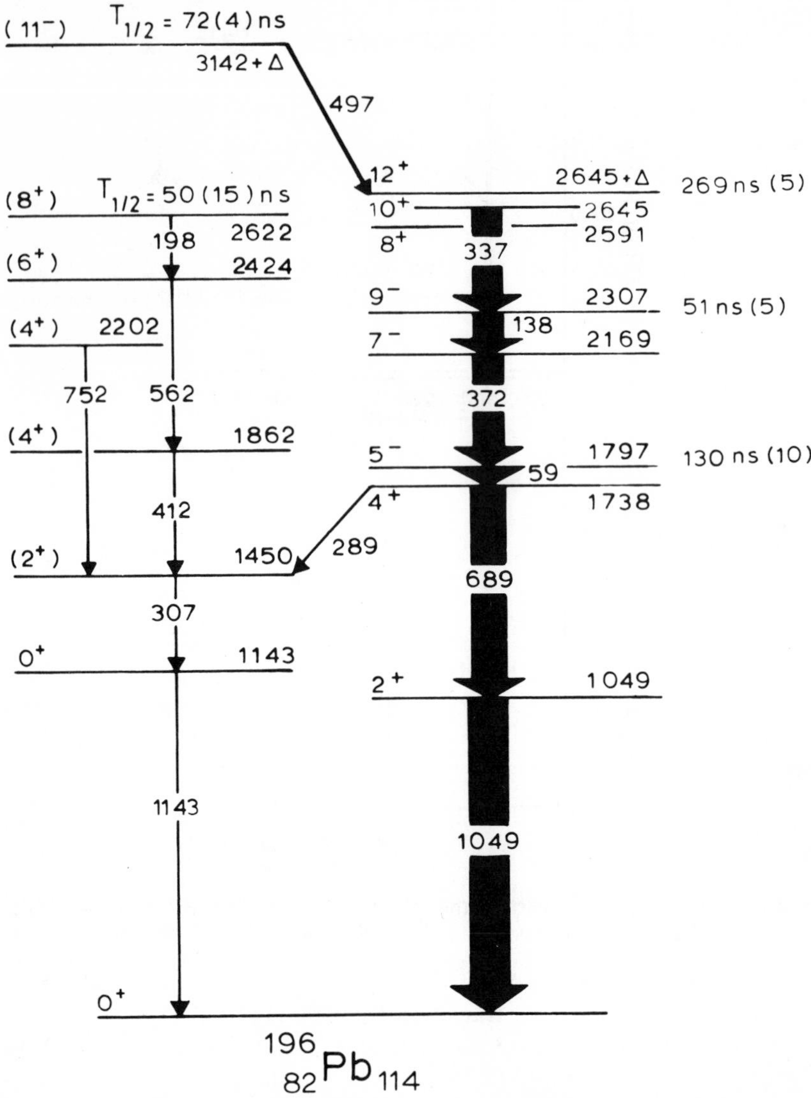

Fig. 2.12. Partial level scheme of ^{196}Pb, which shows a rotational band besides the common two quasi-particle structure. (Pennings *et al.*, 1987.)

used to populate levels in ^{196}Pb for further investigation (Penninga *et al.*, 1987). The method consisted of γ–e coincidences with prompt–prompt, prompt–delayed and delayed–delayed time requirements to identify γ-ray cascades in coincidence with the $0_2^+ \rightarrow 0_1^+$ E0 conversion electrons. As shown in Fig. 2.12, the sensitivity of the method revealed a collective band built on the deformed 0_2^+ state up to (6^+), where the band terminates with a nearby lying (8^+) isomeric state.

In the work of Van Duppen *et al.* (1984) it has been pointed out that for an oblate deformation the $\pi[514]9/2^-$ and $\pi[606]13/2^+$ Nilsson orbitals are strongly down-sloping in energy as the deformation ε becomes more negative above $Z=82$, whereas the $\pi[440]1/2^+$ is up-sloping. Thus exciting two protons from the latter orbital into one of the two former orbitals creates a proton intruder state, which may even compete in energy with the regular neutron excitations. Indeed, the observed 0_2^+ states in $^{192,194}Pb$ become even the first excited states.

In this section it has been demonstrated with a few selected examples that γ–e spectroscopy contributes in an essential way to our understanding of the nuclear potentials and shapes, and that the proton–neutron interaction plays an essential role.

2.3 Single particle motion and electromagnetic transitions

2.3.1 *Gamma transition schemes and shell structure*

Electromagnetic (EM) transitions with simple selection rules provide very powerful tools for investigating single particle motions in nuclei. Nuclear e–γ spectroscopy has indeed played an essential role in the study of nuclear shell structure in the same way as atomic spectroscopy has done for atomic shell structure. Gamma transition probabilities (γ-decay widths), angular correlations and selection rules are very useful for obtaining the quantum numbers associated with shell-model states and single particle strengths (see Section 2.1).

Nuclear shells are classified as $1\hbar\omega$, $2\hbar\omega$, . . . , $N\hbar\omega$ shells in the framework of a harmonic oscillator (h.o.) model. The $N\hbar\omega$ shell has orbital motions defined by the quantum numbers (n, l), where the orbital angular momentum $l=N, N-2, \ldots, 1$ or 0, and the corresponding number of radial nodes $n=1, 2, \ldots, (N-l+2)/2$ (see Section 2.2). Introducing the spin–orbit (**ls**) interaction the shell model orbit is defined by the quantum numbers (n, l, j) with the total angular momentum $j=l\pm\frac{1}{2}$. A part of the nuclear model shell orbits is illustrated in Fig. 2.13. All the states in the $N\hbar\omega$ shell have the same parity π_N, while states in the $(N+1)\hbar\omega$ shell have the opposite parity $\pi_{N+1}=-\pi_N$. In medium heavy nuclei the lowest state of $(n, l, j)=(1, l'+1, l'+1+\frac{1}{2})$ with $l'=N$ in the $(N+1)\hbar\omega$ shell comes down into the $N\hbar\omega$ shell as an intruder state, forming the new M-th major shell as shown in Fig. 2.13. Systematic features of the nuclear shells, combined with the simple selection rules for γ transitions, lead to a general γ-decay scheme as illustrated in Fig. 2.13. Single-particle transitions are to some extent modified by admixture of other states. In

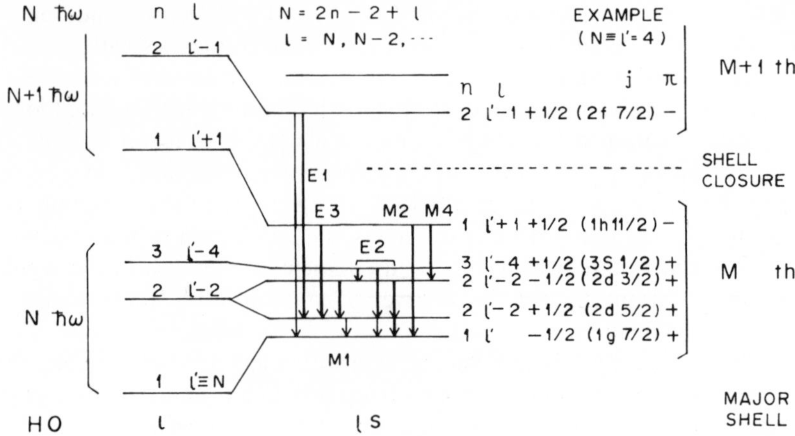

Fig. 2.13. Gamma-decay scheme of the Mth major shell consisting of the $N\hbar\omega$ h.o. shell states and the intruder state with $(n, l, j) = (1, l' + 1, l' + 1 + \frac{1}{2})$ from the $(N + 1)\hbar\omega$ h.o. shell, where $l' = N$ is the maximum orbital angular momentum of the $N\hbar\omega$ shell states: h.o., l, and ls stand for the harmonic oscillator, splitting by the orbital angular momentum and splitting by the ls coupling interaction, respectively.

particular, the collective states greatly affect the single-particle transition rates. Core polarization effects on EM transition rates are found to be very important (Arima and Horie, 1954; Bohr and Mottelson, 1969). The general experimental aspects of EM transitions are given in many books on nuclear spectroscopy (see references in Chapter 1). We will describe briefly some features of EM transitions in the framework of the nuclear shell model.

2.3.1a *E1 transitions*

Electric dipole (E1) transitions with $\Delta l = l_f - l_i = \pm 1$ and $\Delta\pi = 1$ occur between the $(N + 1)\hbar\omega$ and $N\hbar\omega$ shells. Transitions $(n_i, l_i, j_i) \rightarrow (n_f, l_f, j_f)$ with $\Delta n = n_f - n_i = 0$, and $\Delta j = j_f - j_i = \pm 1$ are favoured because of the large radial overlap-integral without spin–flip. Transitions with $\Delta j = 0$ are retarded due to spin–flip, and those with $\Delta n = \pm 1$ due to the change in the number of radial nodes. The $(N + 1)\hbar\omega$ states are located at a high excitation energy region in the $N\hbar\omega$ shell nucleus. Some of them are unbound, and they appear as resonance states in nuclear reactions. Examples of E1 transitions in a light nucleus are shown in Fig. 2.14.

The energies of favoured E1 decays in medium heavy nuclei are around $\hbar\omega = 8 \sim 12$ MeV. It is hard to find examples of discrete E1 transitions in medium heavy nuclei since the level density in the one-$\hbar\omega$ excitation region is quite high. The exceptions are isobaric analogue states with the upper isospin of $T + 1$ with respect to the ground-state isospin T (see Fig. 2.14). Indeed,

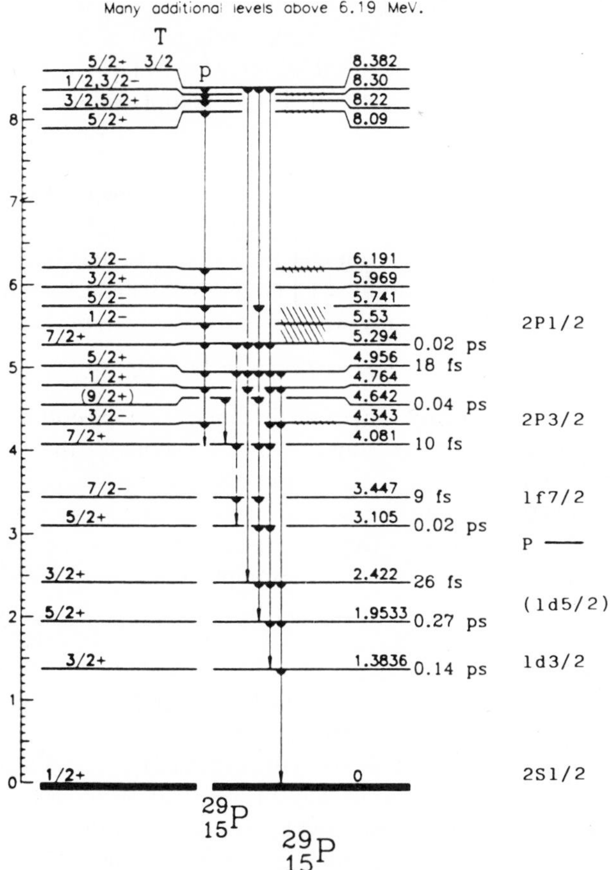

Fig. 2.14. Level scheme of ^{29}P with the ground state isospin $T=1/2$. (Lederer and Shirley, 1978.) Major single particle orbits and the proton binding energy (P) are shown at the right hand side. The $2s_{1/2}$, $1d_{3/2}$ and $1d_{5/2}$ single proton states are bound states, while the $1f_{7/2}$ and $2p_{3/2}$ states are resonance (unbound) states. The $2p_{3/2}\rightarrow 2s_{1/2}$ and $1f_{7/2}\rightarrow 1d_{5/2}$ are single particle E1 transitions. In fact, these states have some mixed configurations. (Ejiri, 1964.)

discrete E1 transitions from isolated isobaric analogue resonances have been studied, as discussed in Chapter 5.

The intruder state of the configuration $(n, l, j)=(1, l'+1, l'+1+\frac{1}{2})$ with $l'=\mathrm{N}$, coming down from the $(\mathrm{N}+1)\hbar\omega$ h.o. shell into the M-th major shell (mostly $\mathrm{N}\hbar\omega$ h.o. shell), hardly ever decays by E1 transitions to any states in the M-th shell because $\Delta j \geqslant 2$ (note that the state with $(1, l', l'+\frac{1}{2})$ in medium heavy nuclei lies in the lower $(M-1)$th major shell). Moreover, the E1 strength is mainly concentrated in the E1 giant resonance (GR) in the higher excitation region (see Chapter 5). Consequently most of the E1 transitions between states of the same major shell are retarded by factors of 10^{-3} to 10^{-6} compared with the typical single-particle (Weisskopf) value.

2.3.1b *E2 transitions*

Electric quadrupole (E2) transitions $(n_i, l_i, j_i) \rightarrow (n_f, l_f, j_f)$ with $|\Delta n| \leqslant 2$, $|\Delta l| \leqslant 2$, $|\Delta j| \leqslant 2$ are frequently seen between states in one major shell, as shown in Fig. 2.13. Since low-lying states in medium heavy nuclei have the same isospin, one find many enhanced E2 transitions with large isoscalar effective charges. The effective charge is conventionally written as

$$e^{\text{eff}}(\text{E}L, \tau) = e(\text{E}L, \tau) + e^{\text{pol}}(\text{E}L, \tau), \tag{2.3.1}$$

where e and e^{pol} stand for the single-particle charge (see Section 2.1) and the polarization charge, respectively, and $\tau = 0$ and 1 stand for the isoscalar and isovector transitions, respectively. The large polarization charge e^{pol} (E2, 0) for the isoscalar mode is due to the large E2 core polarization induced by the attractive E2-type interaction. Detailed discussions of the E2 polarizations are given in several textbooks (see, for example, Bohr and Mottelson, 1969, 1975).

2.3.1c *E3 transitions*

The intruder state of $(n, l, j) = (1, l'+1, l'+1+\frac{1}{2})$ with $l' = N$ can decay by the electric octupole (E3) transition to the state with $(2, l'-2, l'-2+\frac{1}{2})$, as shown in Fig. 2.13. The E3 transition is normally enhanced to some extent by the isoscalar-type E3 core polarization. The intruder state also decays by the M2 transition to another state as discussed later. Because of the reduction of the normal M2 transition and the enhancement of the E3 one, the E3 transition can compete with the lower-multipole M2 transition. Figure 2.15 shows

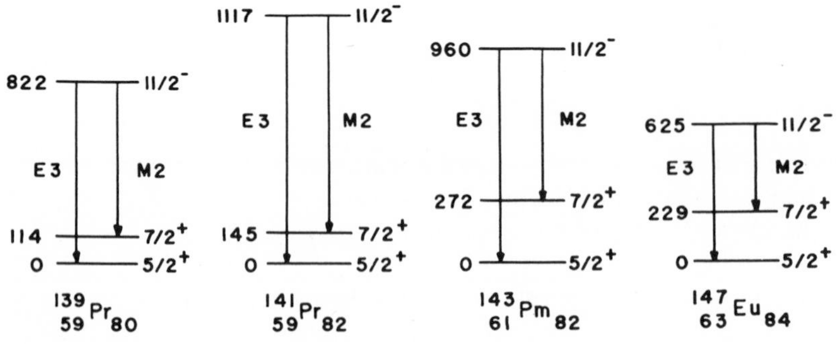

Fig. 2.15. The M2 and E3 transition schemes in nuclei with $N = 82$. (Ejiri *et al.*, 1973.)

examples of E3 transitions from the $1h_{11/2}$ intruder states in nuclei with $N \approx 82$. The effective charge $e^{\text{eff}}(\text{E3})$ is obtained as

$$e^{\text{eff}}(\text{E3})/e(\text{E3}) = \langle M(\text{E3}) \rangle_{\text{exp}} / \langle M(\text{E3}) \rangle_{\text{sqp}}, \tag{2.3.2}$$

where $e(\text{E3})$ is the single-particle charge for E3 transitions, and $\langle M(\text{E3}) \rangle_{\text{exp}}$ and $\langle M(\text{E3}) \rangle_{\text{sqp}}$ are the observed E3 matrix element and the evaluated one for the single quasi-particle transition, respectively, see eqn (2.3.6).

The single particle E3 transitions between $1h_{11/2}$ and $2d_{5/2}$ states are indeed enhanced by factors $e^{\text{eff}}(\text{E3})/e(\text{E3}) = 2$ to 4 as shown in Fig. 2.16. The enhancement is due to the constructive effect of the attractive $\mathbf{Y}_L$ interaction on the isoscalar electric transition. Actually an enhancement factor of around 3 is obtained by coupling (admixture) of the octupole (E3) vibration with the single-particle state (Shibata *et al.*, 1975).

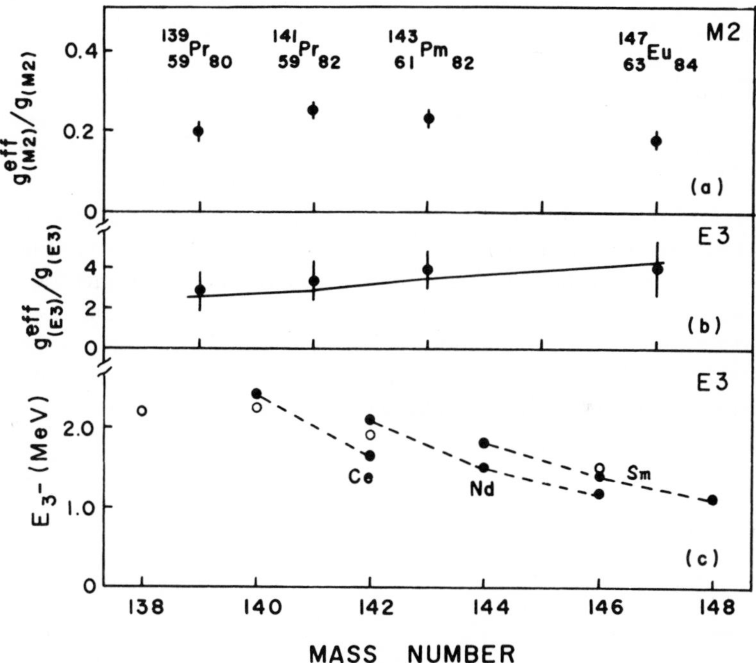

Fig. 2.16. (a): Effective coupling constants for the M2 transitions. (b): Effective coupling constants for the E3 transitions. The solid line is the calculation. (c): Energies of the E3 phonons in even nuclei. The solid circles are experimental values. The calculated values for ^{138}Ce, ^{140}Ce, ^{142}Nd, and ^{146}Sm, which are the even cores of the odd-Z nuclei ^{139}Pr, ^{141}Pr, ^{143}Pm, and ^{147}Eu, respectively, are shown by open circles. (Ejiri *et al.*, 1973.)

2.3.1d *M1 transitions*

Magnetic dipole (M1) transitions between *ls*-pair states with $(n, l, j=l-\frac{1}{2})$ and $(n, l, j=l+\frac{1}{2})$ are favoured (spin–flip) M1 transitions. The major operator for the spin–flip process is

$$g\boldsymbol{\sigma}=\tfrac{1}{2}g_0\boldsymbol{\sigma}+\tfrac{1}{2}g_1\tau_3\boldsymbol{\sigma}, \tag{2.3.3}$$

where g_1 and g_0 are the isovector and isoscalar spin g-factors, respectively. Since $g_1=g_n-g_p$ is much larger than $g_0=g_n+g_p$, the second term of eqn (2.3.3) is dominant. The strong repulsive interaction with the spin–isospin ($\sigma\tau$) mode, however, gives rise to the $\sigma\tau$ type M1 giant resonance (M1GR) in *ls* non-closed-shell nuclei, and thereby a large fraction of the $\sigma\tau$ type M1 strength is absorbed into the M1GR.

M1 transitions between states with $(2, l'-2, l'-2+\frac{1}{2})$ and $(1, l', l'-\frac{1}{2})$ are l-forbidden because $\Delta l=2\geqslant 1$. Such l-forbidden M1 transitions are analysed by introducing an effective M1 operator $g_p\,(\boldsymbol{\sigma}\times\mathbf{Y}_2)_1$, with the $l=2$ spherical harmonics, which connects directly those two states. M1 transitions are discussed in Chapter 5.

2.3.1e *M2 and M4 transitions*

Magnetic quadrupole (M2) and magnetic hexadecapole (M4) transitions with $\Delta\pi=$ yes from the intruder state of $(n, l, j)=(1, l'+1, l'+1+\frac{1}{2})$ with $l'=N$ to the low-lying states with $(1, l', l'-\frac{1}{2})$ and $(2, l'-2, l'-2-\frac{1}{2})$, respectively, are shown in Fig. 2.13. These spin–flip M2 and M4 transitions are favourable (spin–flip) magnetic transitions. Some M2 transitions of this type are shown in Fig. 2.15 and some M4 transitions in Fig. 2.17. M2 and M4 transition rates are much smaller than E1, E2, and M1 transition rates. Thus most of the states, which can decay only by the M2 or M4 transition, are isomeric states. If they are excited by nuclear reactions the isomeric transitions are observed in delayed spectra. It is interesting to note that systematic trends of such low-lying states with $(1, l', l'-\frac{1}{2})$ and $(2, l'-2, l'-2-\frac{1}{2})$ are surveyed through M2 and M4

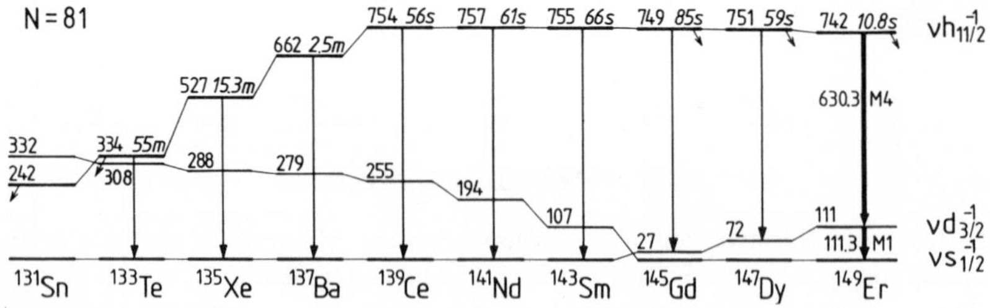

Fig. 2.17. The three lowest single neutron–hole states, $h^{-1}_{11/2}\to d^{-1}_{3/2}$ M4 transitions and half-lives of the $h^{-1}_{11/2}$ states in $N=81$ isotones. The excitation energies of levels are given in units of keV. (Zuber *et al.*, 1986.)

transitions from the intruder state, respectively, as shown in Figs 2.13, 2.15 and 2.17.

The major transition operators for the spin–flip M2 and M4 transitions are the isovector spin operators that are written as $g_1\boldsymbol{\tau}[\boldsymbol{\sigma}\times\mathbf{Y}_1]_2$ and $g_1\boldsymbol{\tau}[\boldsymbol{\sigma}\times\mathbf{Y}_3]_4$, respectively. Thus M2GR and M4GR, which are caused by $\boldsymbol{\tau\sigma}\mathbf{Y}_L$-type repulsive interactions, absorb large fractions of the single particle M2 and M4 strengths of low-lying states, as in case of M1 transitions. Effective MJ coupling constants $g^{\mathrm{eff}}(\mathrm{M}J)$ are defined, in the same way as the $e^{\mathrm{eff}}(\mathrm{EJ})$ in eqn (2.3.2), by

$$g^{\mathrm{eff}}(\mathrm{M}J)/g(\mathrm{M}J)=\langle M(\mathrm{M}J)\rangle_{\mathrm{exp}}/\langle M(\mathrm{M}J)\rangle_{\mathrm{sqp}}. \tag{2.3.4}$$

Actually $g^{\mathrm{eff}}(\mathrm{M}J)/g(\mathrm{M}J)$ is indeed about 0.2 to 0.4 for most of the M2 and M4 transitions, in contrast to $e^{\mathrm{eff}}(\mathrm{E}J)/e(\mathrm{E}J)=2$ to 3 for E2 and E3 transitions (Ejiri *et al.*, 1973). The reduction of the magnetic transition strength reflects the destructive effect of the repulsive $\boldsymbol{\tau\sigma}\mathbf{Y}_L$ interaction. Detailed discussions of MJ transitions are given in Chapter 5.

2.3.2 *Particle and hole transitions*

In real nuclei the shell orbits are filled partly by neutrons and partly by protons. In a nucleus with one nucleon outside the closed $(M-1)$th shell, one-particle transitions between shell orbits in the Mth shell are expected, as discussed in Section 2.3.1. Similarly, a nucleus with one hole in the M-th closed shell shows one-hole transitions between the M-th shell orbits. Examples of such particle and hole transitions are shown in Fig. 2.18. Here low-lying states in ^{147}Gd and ^{145}Gd constitute one particle states and one-hole states, respectively, with respect to the doubly closed ^{146}Gd core.

Shell model orbits in a non-closed shell are partially filled by particles. The Fermi surface is no longer sharp because the pairing interaction excites nucleon pairs above the Fermi surface. Then the shell orbit i is partially occupied with an occupation probability V_{i}^2 and is partially vacant with a vacancy probability $U_{\mathrm{i}}^2=1-V_{\mathrm{i}}^2$. The quasi-particle operator is then expressed as

$$\alpha_{\mathrm{i}}^{+}=U_{\mathrm{i}}b_{\mathrm{i}}^{+}-V_{\mathrm{i}}b_{\bar{\mathrm{i}}}^{-}, \tag{2.3.5}$$

where b_{i}^{+} (b_{i}) and $\alpha^{+}(\alpha)$ are the creation (annihilation) operators for the single particle i and the quasi-particle i, respectively. The subscripts i and $\bar{i}$ refer to the $\psi(jm)$ and $(-)^{j-m}\psi(j-m)$ states, respectively. Then the γ transition between quasi-particle states involves the particle transition and the hole (time-reversed particle) transition as (Kisslinger and Sorensen, 1963)

$$\langle\alpha(j_{\mathrm{f}}m_{\mathrm{f}})|\mathbf{M}\{\mathrm{E(M)}J\}|\alpha^{+}(j_{\mathrm{i}}m_{\mathrm{i}})\rangle=(U_{\mathrm{i}}U_{\mathrm{f}}-(-)^{T}V_{\mathrm{i}}V_{\mathrm{f}})\times\langle j_{\mathrm{f}}m_{\mathrm{f}}|\mathbf{M}\{\mathrm{E(M)}J\}|j_{\mathrm{i}}m_{\mathrm{i}}\rangle, \tag{2.3.6}$$

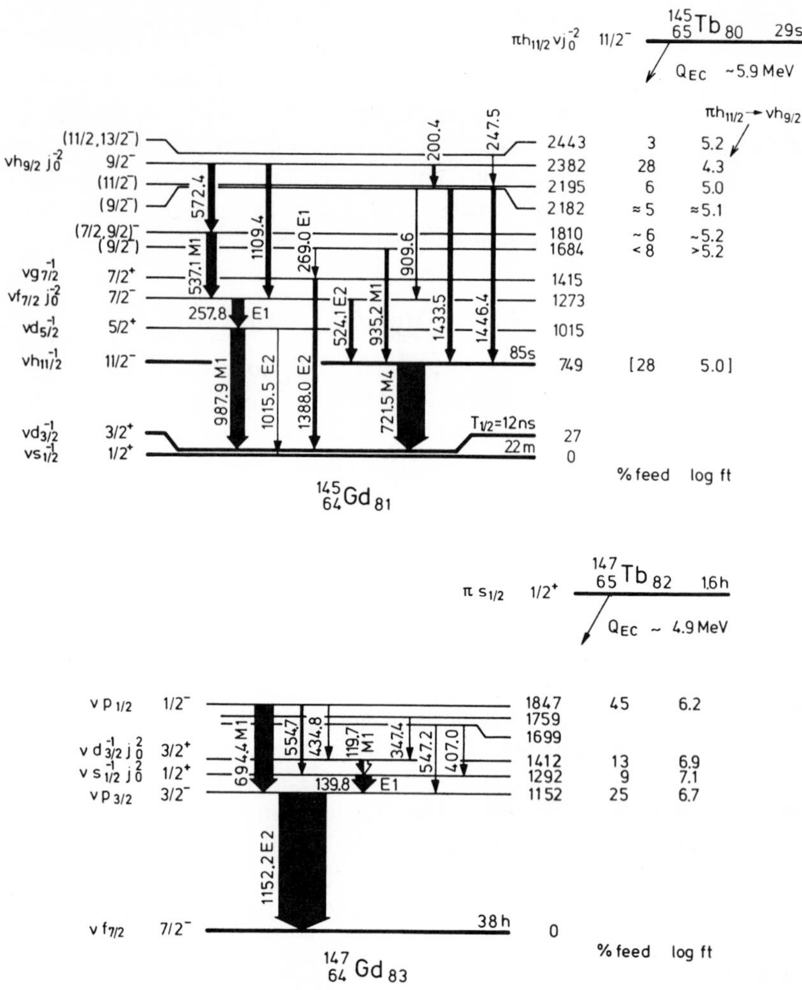

Fig. 2.18. Gamma transition schemes. Top: ^{145}Gd with one hole in the doubly closed shell nucleus ^{146}Gd. Bottom: ^{147}Gd (one particle + ^{146}Gd). (Styczen *et al.*, 1982.)

where $(-)^T$ denotes the sign change under time reversal, $T=0$ for electric transitions and $T=1$ for magnetic transitions. Thus the pairing factor for electric transitions is given by the destructive sum $U_i U_f - V_i V_f$, and in the case of magnetic transitions by the constructive sum $U_i V_f + V_i V_f$. Electric transitions are sensitive to the values of U_k and V_k and are greatly retarded for transitions between half-filled orbits as shown in Fig. 2.19, while magnetic transitions remain rather stable (see Chapter 5). The reduction of E2 and E3

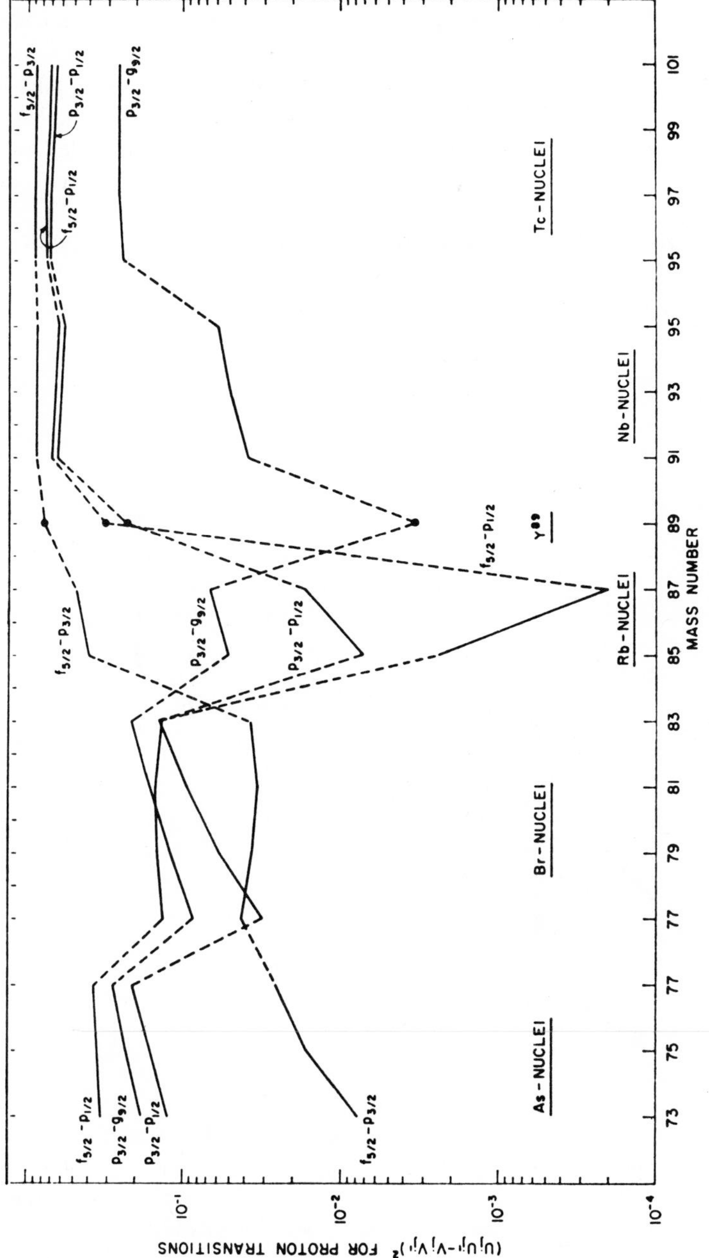

Fig. 2.19. $(U_iU_f - V_iV_f)^2$ factors for E2 proton transitions $f_{5/2} \to p_{3/2}$, $f_{5/2} \to p_{1/2}$, and $p_{3/2} \to p_{1/2}$, and for E3 ones of $p_{3/2} \to g_{9/2}$ in nuclei around the region of the neutron number $N = 50$. (Ikegami and Udagawa, 1964.)

transitions by the pairing factor have been studied by Ikegami and Udagawa (1964).

So far single-particle γ transitions between different shell model orbits have been discussed. In the case of two particle states, γ transitions may occur between the same type of two particle states of $(a_i \times a_k)_J$ with different couplings of $J = J_{\mathrm{i}}$ and J_{f}. Here a_i is either a particle or a hole in the shell i and a_k is the same in the shell k. States with $[a_i \times a_k]_J$ constitute multiplet states with $J = |j_i - j_k|, |j_i - j_k| + 1, \ldots, |j_i + j_k|$ (the possible values for J for two like particles in the same $(i = k)$ orbit are restricted by the Pauli principle). Then E2 and M1 transitions are seen between these multiplet states. Figure 2.20 shows

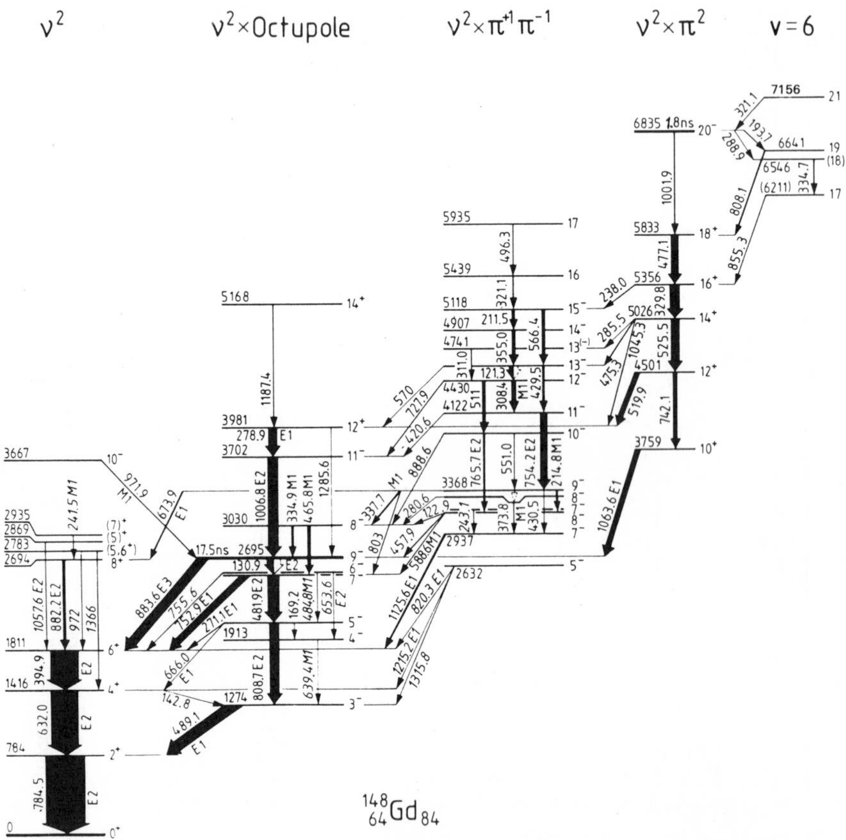

Fig. 2.20. Energy levels of ^{148}Gd populated in the (α, 4n) reaction. The proposed shell model configurations in terms of 2 neutrons (ν^2) states, also coupled to octupole, proton particle-hole ($\pi^{+1}\pi^{-1}$), and 2 proton (π^2) states, as well as seniority $\nu = 6$ states as indicated at the top. The in-beam γ-ray intensities as measured at 51 MeV bombarding energy are also shown. (Lunardi *et al.*, 1980; Piiparinen *et al.*, 1986.)

γ transitions between two-neutron states in ^{148}Gd, and between states with proton particle–hole configurations.

The matrix element of the γ transition associated with one (i) particle in the two-particle state with $(a_i \times a_k)$ is expressed in terms of the one-particle matrix element (deShalit and Talmi, 1963),

$$\langle (a_i' \times a_k')J_f \| \mathbf{M}_i\{\mathrm{E(M)}J\} \| (a_i \times a_k)J_i \rangle$$

$$= (-)^{j_i'+j_k'+J_i+J}(2J_i+1)^{1/2}(2J_f+1)^{1/2}\begin{bmatrix} j_i' J_f j_k' \\ J_i j_i J \end{bmatrix}$$

$$\times \langle a_i' \| \mathbf{M}_i\{\mathrm{E(M)}J\} \| a_i \rangle \delta(a_k', a_k), \tag{2.3.7}$$

where j_c refers to the angular momentum of the appropriate single particle a_c, and $\mathbf{M}_i$ is the γ transition operator acting on the single particle a_i.

2.4 Collective motions and e–γ spectroscopy

2.4.1 *Collective vibrations and gamma transitions*

Nuclear interactions are not fully represented by the one-body shell model Hamiltonian $\mathbf{H}_0$, and there remain some residual two-body interactions $\mathbf{H}_I$. The single-particle wave functions $|i\rangle_0$ are more or less affected by $\mathbf{H}_I$, and accordingly γ-transition probabilities $\mathbf{T}\{\mathrm{E(M)}J_\gamma, \mathrm{if}\}$ are modified to some extent compared with the single particle ones. Moreover, the residual interactions give rise to collective states $|C\rangle$, which have EM properties quite different from those of the shell-model states. For simplicity we consider first a schematic model with a separable form of the α-type residual interaction $\mathbf{H}_I^\alpha$, where α stands for the type of the interaction (i.e. quadrupole interaction, spin–isospin interaction, and so on). The Hamiltonian is written as

$$\mathbf{H} = \mathbf{H}_0 + \mathbf{H}_I^\alpha, \tag{2.4.1a}$$

$$\mathbf{H}_0 |i\rangle_0 = E_i^0 |i\rangle_0 \quad i = 1, 2, \ldots, n, \tag{2.4.1b}$$

$${}_0\langle i | \mathbf{H}_I^\alpha | i' \rangle_0 = \chi_\alpha G_{ii'}^\alpha = \chi_\alpha G_i^\alpha G_{i'}^\alpha. \tag{2.4.1c}$$

Here χ_α is the interaction strength, and $G_{ii'}^\alpha$ is the interaction matrix element which is expressed in a separable form as a product of G_i^α and $G_{i'}^\alpha$. The separable form of eqn (2.4.1c) for the α-type interaction is introduced in order to take into account explicitly the type of interaction that causes the α-mode collective (vibrational) motion. The repulsive interaction, with $\chi_\alpha > 0$, pushes up one collective state $|C^\alpha\rangle$ above the other states and the attractive one, with $\chi_\alpha < 0$, pushes it down below the others. It is expressed as

$$|C^\alpha\rangle = \frac{1}{N} \sum_i a_i |i\rangle_0, \tag{2.4.2a}$$

where

$$a_i = G_i^\alpha/(E_c^\alpha - E_i^0), \quad N = (\Sigma a_i^2)^{1/2}. \tag{2.4.2b}$$

The energy is given by the dispersion equation,

$$\sum_i \frac{\chi_\alpha (G_i^\alpha)^2}{E_c^\alpha - E_i^0} = 1. \tag{2.4.2c}$$

If the unperturbed states are almost degenerate in excitation energy around $E_i^0 \approx E_0$, $|C^\alpha\rangle$ is reduced to a simple form,

$$|C^\alpha\rangle \approx \frac{1}{N'}\sum_i G_i^\alpha |i\rangle_0 \tag{2.4.3a}$$

$$E_c^\alpha \approx E_0 + \chi_\alpha n (G^\alpha)^2, \tag{2.4.3b}$$

where $(G^\alpha)^2 \equiv \Sigma(G_i^\alpha)^2/n$. As shown in eqns (2.4.2) and (2.4.3), $|C^\alpha\rangle$ is the coherent sum of the unperturbed states with coefficients G_i^α. In the case where $|i\rangle_0$ is the α-mode i-particle i'-hole state, $|C^\alpha\rangle$ is rewritten as

$$|C^\alpha\rangle = \frac{1}{N}\sum_i \frac{G_i^\alpha}{E_c^\alpha - E_i^0} B^+(\alpha, ii')|0\rangle \tag{2.4.4}$$

where $B^+(\alpha, ii')$ is the creation operator for the α-mode i-particle i'-hole excitation and $|0\rangle$ is the ground state.

The α-mode collective state $|C^\alpha\rangle$decays strongly to $|0\rangle$ by the similar α-mode γ transition. If the γ-transition operator $\mathbf{M}(\alpha)$ is written in the same form as the interaction, one can write

$$\langle 0|\mathbf{M}(\alpha)|i\rangle_0 = C G_i^\alpha, \tag{2.4.5a}$$

$$\mathbf{M}(\alpha) = \sum_i C G_i^\alpha \mathbf{B}(\alpha, ii'). \tag{2.4.5b}$$

Then, the γ-transition strength for $|C^\alpha\rangle$ is

$$\langle 0|\mathbf{M}(\alpha)|C^\alpha\rangle = \frac{1}{N}\sum_i \frac{C(G_i^\alpha)^2}{E_c^\alpha - E_i^0}. \tag{2.4.6}$$

For the simple case given by eqn (2.4.3), the matrix element becomes $\langle 0|\mathbf{M}(\alpha)|C^\alpha\rangle \approx C\sqrt{\{\sum(G_i^\alpha)^2\}}$ and its square amounts nearly to the sum rule limit $C^2\sum(G_i^\alpha)^2$. Since the collective state is either the highest or the lowest state in excitation energy, depending on whether $\chi_\alpha > 0$ or $\chi_\alpha < 0$, the denominators $E_c^\alpha - E_i^0$ in eqn (2.4.6) have the same sign, positive in the case of $\chi_\alpha > 0$ and negative in the case of $\chi_\alpha < 0$. Consequently, the matrix element $\langle 0|\mathbf{M}(\alpha)|C^\alpha\rangle$ is given by the coherent sum of $N^{-1}\cdot C(G_i^\alpha)^2/(E_c^\alpha - E_i^0)$, all with the same sign. Note that the product of the coefficient G_i^α for $|i\rangle_0$ in the collective state $|C\rangle$ (eqn 2.4.2) and the γ-transition matrix element

$\langle 0|\mathbf{M}(\alpha)|i\rangle_0 = CG_i^\alpha$ leads to $C(G_i^\alpha)^2$ with the same sign as C. In other words the α-mode collectivity (enhancement of the α-mode transition) is due to the similarity of the matrix elements of $\langle 0|\mathbf{H}_I^\alpha|i\rangle_0 = G_i^\alpha$ for the nuclear interaction and those of $\langle 0|\mathbf{M}(\alpha)|i\rangle_0 = CG_i^\alpha$ for the EM interaction γ transition. For a different mode (α') of γ decays there is an α'-mode collective state which absorbs a large fraction of the α'-mode γ-decay strength, provided that the α'-type residual interaction is strong enough to produce one collective state far away from the others.

The collectivity depends on the number of states involved and on the interaction strength χ_α. If the interaction strength is much larger than the energy spread of the unperturbed states, then the collective state is established far from the unperturbed states. In this case the transition amplitude becomes large because of the coherent contributions, and the probability is enhanced by nearly a factor n over the single-particle value of $C^2(G_i^\alpha)^2$.

A central type residual interaction contains in general a separable interaction written as a tensor product,

$$\mathbf{H}_I^\alpha = \chi_a \mathbf{G}_\alpha \mathbf{G}_\alpha^+ \tag{2.4.7a}$$

$$\mathbf{G}_\alpha = \sum_i f(r_i)\boldsymbol{\tau}^T\{\boldsymbol{\sigma}^S \times \mathbf{Y}_L(\theta_i)\}_J \mathbf{B}(\alpha, ii'), \tag{2.4.7b}$$

where T, S, L, and J are the isospin, spin, orbital angular momentum and total angular momentum, respectively, and α stands for these quantum numbers. The operator of the EM transition in the α-mode has a form similar to eqn (2.4.7b) for the α-type separable interaction. It is expressed as

$$\mathbf{M}(\alpha) = \sum_i g_\alpha j(r_i)\boldsymbol{\tau}^T\{\boldsymbol{\sigma}^S \times \mathbf{Y}_L(\theta_i)\}_J \mathbf{B}(\alpha, ii'). \tag{2.4.8}$$

The nuclear operators $\mathbf{M}(\alpha)$ associated with the collective states $|C^\alpha\rangle$ are not necessarily restricted to γ transitions, but are extended to other kinds of nuclear operators such as β decays, inelastic scatterings, charge exchange reactions, nucleon-pair transfer reactions, and so on. Using the isospin z-component τ_3, the isospin raising operator τ^+ and the isospin lowering operator τ^-, the operators in eqn (2.4.8) with τ_3, τ^+ and τ^- are transition operators for γ, β^+, and β^- decays, respectively. Therefore collective states associated with β^- and β^+ decays are expected in the same way as the collective state (giant resonance) associated with the γ decay with $T = 1$ (isovector γ decay).

The collective states and separable interactions discussed so far are rather too simple and schematic to describe the basic concepts of collective motion, and more realistic models with refined interactions are needed to reproduce well the observed data. Quasi-particles have to be used in place of particles in order to incorporate the pairing interaction, and backward (ground state) correlations have to be taken into account in addition to forward correlations. General descriptions of collective states and γ transitions are given in many textbooks (see Bohr and Mottelson, 1969, 1975, and references in Chapter 1).

2.4.2 *Enhancement of gamma transitions from collective states*

The collective state $|C^{\alpha}\rangle$ excited by the interaction $\mathbf{H}_{\alpha}=\chi_{\alpha}\mathbf{G}_{\alpha}\mathbf{G}_{\alpha}$ is characterized by the enhanced α-mode γ transition with the transition operator $\mathbf{M}(\alpha)$. There are various types of collective states in a wide excitation energy region. Collective states located in the low excitation region are called vibrational (phonon) states, and those in the high excitation region appear as giant resonances (GR), see van der Woude (1987).

2.4.2a *Isoscalar electric vibrations*

Electric multipole (EJ_{γ}) vibrations with $T=0$ are induced by the isoscalar EJ_{γ} interactions. Quadrupole vibrations ($Q^{+}|0\rangle$) with $\alpha=(TSLJ)=(0022)$, which are induced by the quadrupole–quadrupole ($\mathbf{Q}\cdot\mathbf{Q}$) interaction, have been extensively studied in many nuclei. The quadrupole operator used conventionally is similar to the $E2$ γ operator. It is written as

$$\mathbf{Q}=\Sigma\langle j'm'|r^{2}\mathbf{Y}_{2}|jm\rangle a^{+}_{j'm'}a_{jm}. \tag{2.4.9}$$

The 2^{+} vibrational state ($|C^{Q}\rangle$) is pushed down in excitation energy from other 2^{+} states by the attractive $\mathbf{Q}\cdot\mathbf{Q}$ interaction. It appears as the first excited state in most even nuclei except for doubly closed nuclei. The excitation energy is given by the quadrupole phonon energy $E_{v}(2^{+})=\hbar\omega_{2}$. The two-phonon multiplets of $|C^{QQ}\rangle=[\mathbf{Q}'\times\mathbf{Q}]_{J}|0\rangle$ with $J=0$, 2, 4 may appear around $E_{v}(2^{+},2)=2\hbar\omega_{2}$. Many-phonon multiplets and other types of vibrations may also be expected (see Section 2.5). Gamma-transition rates for the 2^{+} vibrational states have been studied by measuring Coulomb excitations, resonance fluorescence, delayed coincidence spectra, Doppler effects, etc. As shown in Fig. 2.21, reduced transition probabilities are greatly enhanced for nuclei between closed shells, indicating the strong collectivity (enhancement) in these regions. Some of the nuclei between closed shells are well deformed, and their first 2^{+} states are treated as rotational states. In such cases the $B(E2)$ value is related to the deformation parameter β_{2} by

$$\beta_{2}^{2}=\{(4\pi/3)ZeR_{0}^{2}\}^{2}\{B(\mathrm{E2},0^{+}\rightarrow 2^{+})\}. \tag{2.4.10}$$

Note that here the B(E2) for the transition $0^{+}\rightarrow 2^{+}$ is used—see eqn (2.1.3 c). Observed values for β_{2} in units of the single particle value of $\beta_{2-\mathrm{sp}}$ are plotted in Fig. 2.22 as a function of a product of the valence proton (or proton–hole) number (N_{P}) and the valence neutron (or neutron–hole) number (N_{n}). They fall near a single smooth line, increasing as $N_{P}N_{n}$ increases (Hamamoto, 1965; Raman *et al.*, 1987; Casten, 1985).

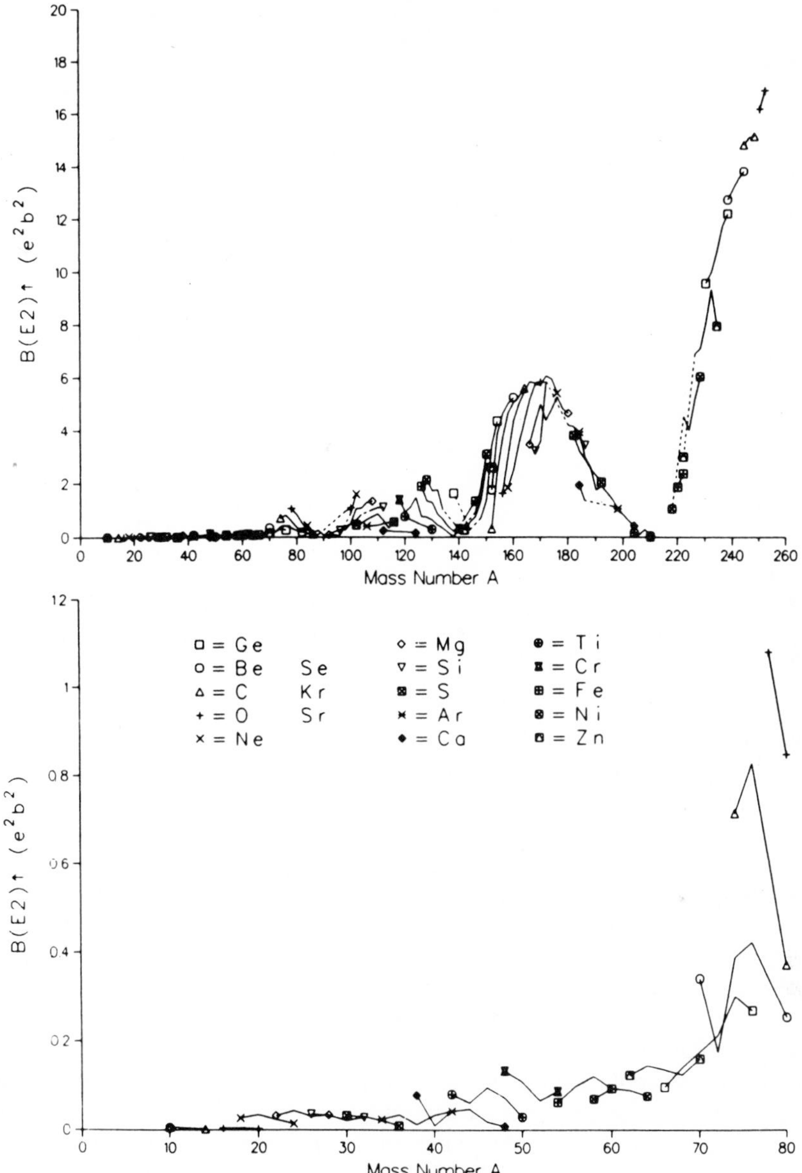

Fig. 2.21. The reduced transition probability, $B(\mathrm{E2})\uparrow$. (Raman *et al.*, 1987.)

Octupole (E3) vibrational states are located around 5 to 1.2 MeV for nuclei with mass numbers A = 50–200. The observed transition rates for the octupole vibrational states are typically larger than the single particle value by factors $(\beta_3/\beta_{3-\mathrm{sp}})^2 = 10$ to 40 (Bohr and Mottelson, 1975). Both the $E_v(3^-)$

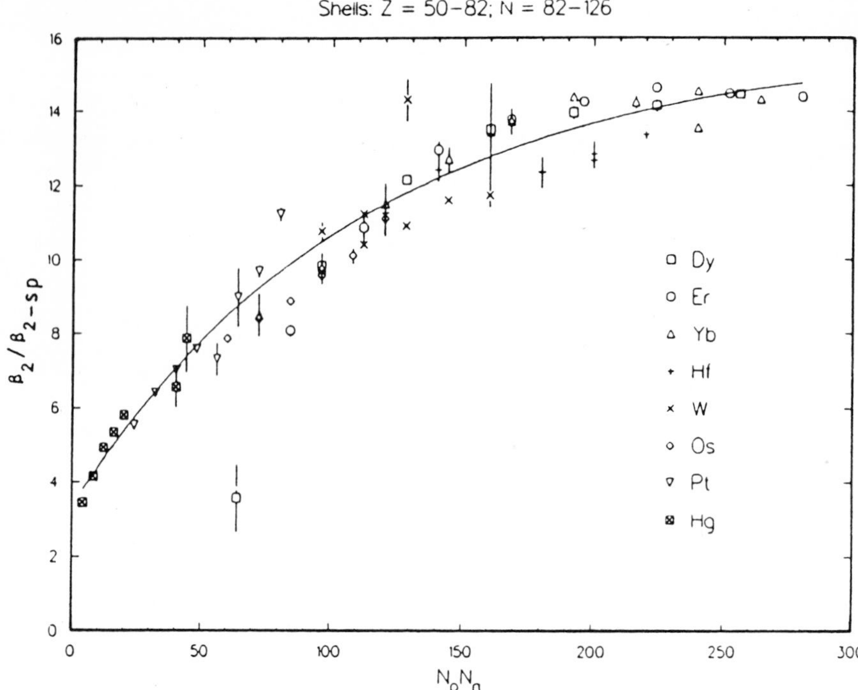

Fig. 2.22. The quadrupole deformation parameter, β_2, divided by $\beta_{2-\mathrm{sp}}$ ($\equiv 1.59/Z$) as a function of the quantity $N_\mathrm{p}N_\mathrm{n}$ for even-Z isotopes in the $64<Z<82$ region. The curve is drawn to guide the eye. (Raman *et al.*, 1987.)

and B(E3) do not show such strong shell effects as those of the 2^+ states since E3 transitions involve parity-change particle–hole excitations. Hexadecapole (E4) vibrational states are known in some nuclei, and they have been studied by Coulomb excitation and inelastic scattering.

2.4.2b *Magnetic dipole states*

Magnetic multipole (MJ_γ) states with $T=0$, 1 and $S=1$ are interesting in view of spin–isospin interactions and spin–isospin vibrations in nuclei. There are two kinds of M1 states in *ls* non-closed nuclei. One is the $\{\pi(j=l-\frac{1}{2}), \pi(j=l+\frac{1}{2})^{-1}\}_1$, and the other is the $\{\nu(j=l'-\frac{1}{2}), \nu(j=l'+\frac{1}{2})^{-1}\}_1$, where π and ν stand for proton and neutron, respectively. Since they are close in excitation energy, they mix with each other, forming one isovector $T=1$ state and one isoscalar $T=0$ state. The 5.846 MeV M1 state in ^{208}Pb has been studied by nuclear resonance fluorescence scattering of linearly polarized photons (Wienhard *et al.*, 1982). The azimuthal asymmetry of the linearly polarized photons scattered by ^{208}Pb shows one positive–parity state at 5.846 MeV and other negative–parity states below 7 MeV. The γ-ray spectra and the observed asymmetries are shown in Fig. 2.23. The 5.846 MeV double-escape peak is much reduced at $\phi = 90°$, indicating a positive parity transition. The assignment of $J_\gamma = 1$ is based on the angular distribution of the unpolarized photons. It has an M1 strength of $B(\mathrm{M1}, 0^+ \to 1^+) = (1.6 \pm 0.5)\mu_N^2$. The 5.846 MeV state is shown to be the isoscalar M1 state from the angular distribution of the inelastically scattered protons (Hayakawa *et al.*, 1982).

The collective M1 state due to the orbital (convection) part of the M1 operator has recently been found in (e, e') reactions. A strong M1 state in ^{156}Gd is located at 3.075 MeV with $B(\mathrm{M1}, 0^+ \to 1^+) = 1.3 \pm 0.2\ \mu_N^2$ (Bohle *et al.*, 1984), and in other nuclei in the mass region of A = 154 ~ 174 with B(M1, $0^+ \to 1^+$) = 0.8 to 1.5 μ_N^2 as shown in Fig. 2.24 (Bohle *et al.*, 1984). Absence of

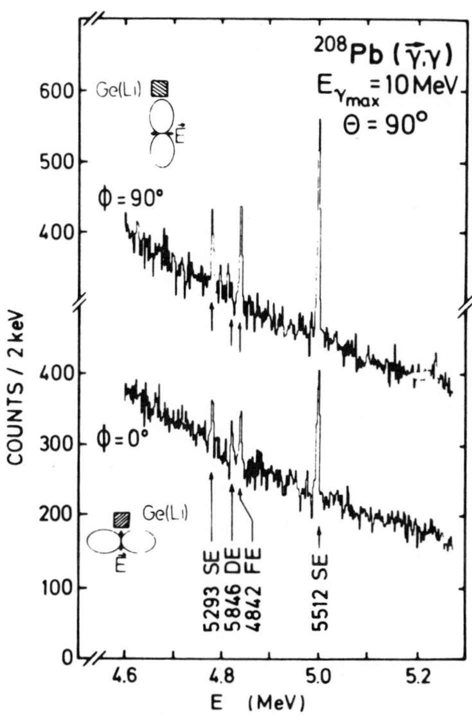

Fig. 2.23. Part of ^{208}Pb(γ_{pol}, γ) spectra in the 5-MeV region. In the upper part the electric vector **E** of the incoming photons was perpendicular and in the lower part parallel to the scattering plane as shown in the inset. FE, SE, and DE stand for the full energy, single-escape, and double-escape peaks, respectively. (Wienhard *et al.*, 1982.)

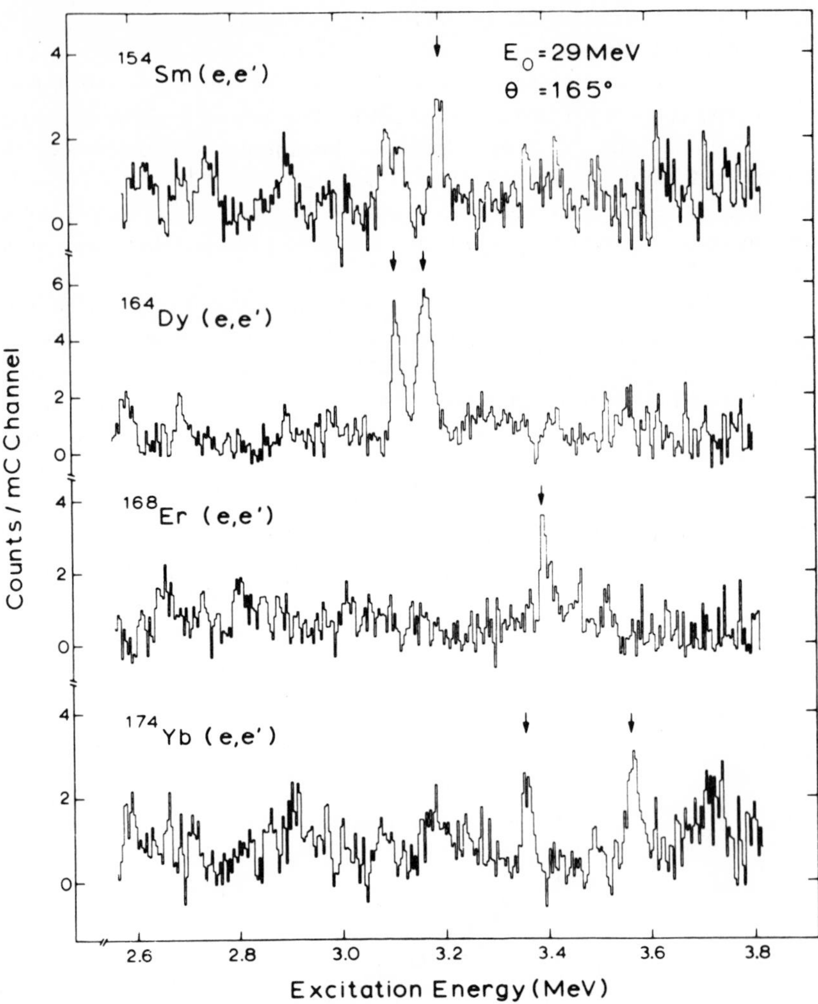

Fig. 2.24. Inelastic electron-scattering spectra on the four deformed nuclei ^{154}Sm, ^{164}Dy, ^{168}Er and ^{174}Yb all taken at kinematical conditions favouring magnetic dipole excitation. Peaks denoted by arrows are interpreted as $J^{\pi}=1^{+}$ states. (Bohle *et al.*, 1984.)

the 1^+ state in the inelastic proton spectrum indicates that the 1^+ state is not a spin mode excitation but is rather an orbital mode 1^+ state, excited only weakly by the **ls** interaction (Djalali *et al.*, 1985; Wesselborg *et al.*, 1986).

The two types of M1 states discussed above are good examples of experimental studies with the EM probe (e, e', or γ, γ') and the hadron probe (p, p') complementing each other.

2.4.2c *EM multipole giant resonances*

EJ_γ-type giant resonances (GR's) have been extensively studied by inelastic scattering of hadrons and electrons (van der Woude, 1987). Isovector and/or spin mode Y_L-type GR's are considered as isospin and/or spin vibrations coupled with the Y_L-type vibration in the space coordinates. Vibrational GR modes are schematically illustrated in Fig. 2.25. Dipole ($L=1$) resonances with $\Delta\pi=$yes are based on particle–hole (p–h) excitations over one $\hbar\omega$ shell, while monopole ($L=0$) and quadrupole ($L=2$) ones with $\Delta\pi=$no are based on p–h excitations over two $\hbar\omega$ shells. Isovector and/or spin mode GR's are pushed up from those unperturbed *p–h* states because of the repulsive ($\chi>0$) interactions. These GR's lie in the unbound excitation region of $E_x=10$ to 40 MeV and their widths are as large as several MeV due to the contributions of the escape and spreading widths. Some GR's overlap with each other, and so e–γ spectroscopy with these GR's is confronted with small γ-branching ratios in competition with particle decays and with mixed multipolarities in the case of overlapping GR's. Simple features of EM interactions, however, make e–γ spectroscopy quite powerful for studying individual modes of the EM strengths. Gamma decays of the electric quadrupole GR are described in Section 4.4, and details of GR associated with isospin and/or spin are discussed in Chapter 5.

2.4.3 *Coupling of single particles with collective motions*

The coupling of a single-particle state with a collective state (vibration) produces multiplet states expressed as

$$|jJ_{c'}\, JM\rangle = \sum_{mM_c} (jmJ_cM_c|JM)\, a^+_{jm}A^+_{J_cM_c}$$
$$\equiv [a^+_j \times A^+_{J_c}]_{JM}, \qquad (2.4.11)$$

where $a^+_{jm}|0\rangle=|jm\rangle$ is the single particle state and $A^+_{J_cM_c}|0\rangle=|C^\alpha, J_cM_c\rangle$ is the α-mode collective state. If the particle–vibration (phonon) coupling interaction is weak, the multiplet states with $J=|J_c-j|, |J_c-j|+1, \ldots, |j+J_c|$ are almost degenerate. In fact, they are split to some extent by the coupling interaction as well as by the coupling with other excited states. The pairing interaction can be taken into account by introducing the quasi-particle operator α^+_{jm} in place of the particle operator a^+_{jm}.

Multiplet states composed of the $1g_{9/2}$ proton with the quadrupole (E2 vibrational) phonon are shown in Fig. 2.26. The multiplet states in ${}^{93}_{41}\mathrm{Nb}_{52}$ are located around the 2^+ state in the core of ${}^{92}_{40}\mathrm{Zr}$ (Stelson *et al.*, 1971). On the other hand, multiplet states in ${}^{95}_{43}\mathrm{Tc}$ are found to be widely spread in excitation energy from a study of the ${}^{93}\mathrm{Nb}(\alpha, 2n\gamma){}^{95}\mathrm{Tc}$ reaction (Shibata *et al.*, 1975b). ${}^{93}_{41}\mathrm{Nb}$ has only one proton in the $1g_{9/2}$ shell outside the subclosed core with

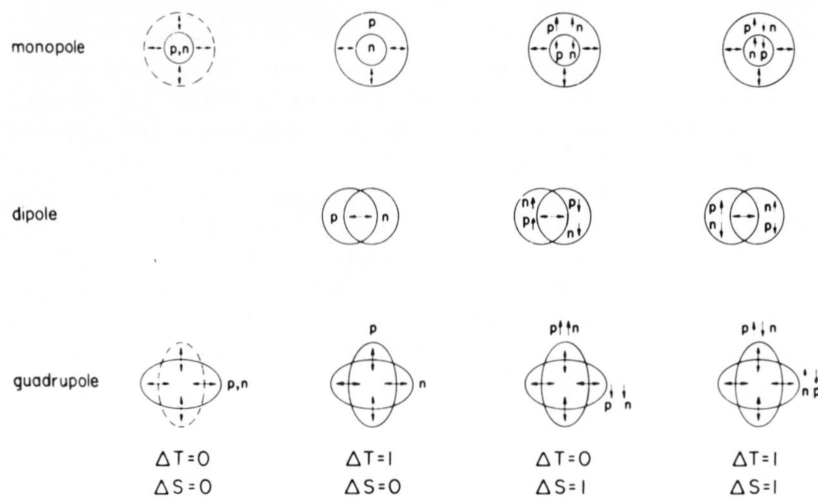

Fig. 2.25. Qualitative picture of giant resonance modes of the nucleus. (van der Woude, 1987.)

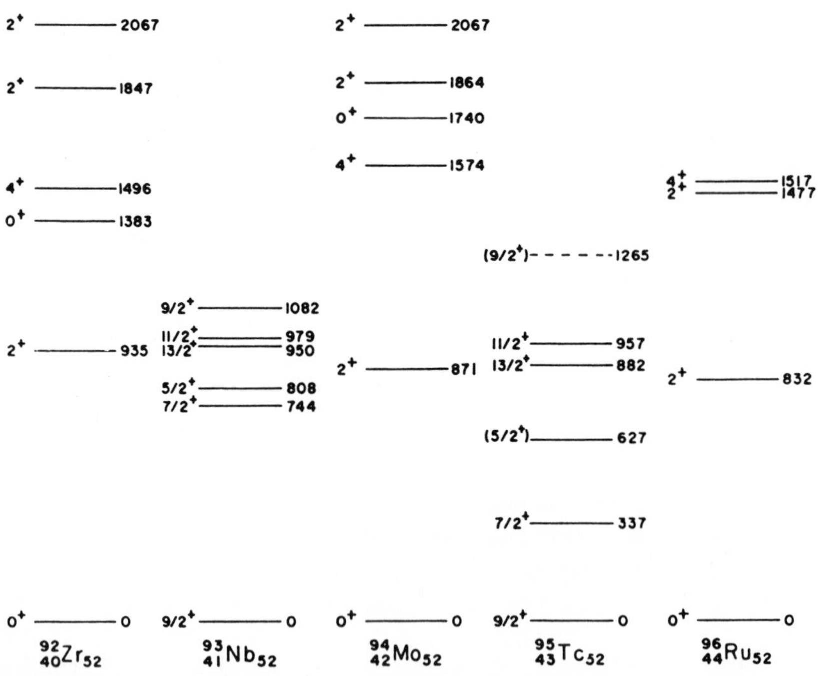

Fig. 2.26. The particle-core multiplets in ^{93}Nb and ^{95}Tc resulting from the coupling of a 1 $g_{9/2}$ proton to the first 2^+ state of the core. (Shibata *et al.*, 1975.)

$Z=40$, while $^{95}_{43}$Tc has three protons in the $1g_{9/2}$ shell. The large spreading of the multiplet states and the low excitation energy of the $7/2^+$ state in $^{95}_{43}$Tc are accounted for in terms of the three quasi-particle correlation (Kuriyama *et al.*, 1974).

Multiplet states consisting of one particle coupled with a octupole (E3) phonon are clearly seen in nuclei with one nucleon outside the doubly closed-shell core, where the 3^- phonon state appears as the first excited state. The multiplet states in ^{147}Gd, which is one proton outside the doubly closed ^{146}Gd, have been studied by the (^{3}He, $3n\gamma$) reaction (Piiparinen *et al.*, 1982). The septuplet states with the structure $\{(2f_{7/2})_n \times 3^-\}_J$ and energies around $E(3^-)=1.579$ MeV of the 3^- phonon in ^{146}Gd and the $2f_{7/2}$ neutron state coupled to the two 3^- phonons are shown together with one quasi-neutron states and three quasi-neutron states in Fig. 2.27. $^{209}_{83}$Bi is composed of one $1h_{9/2}$ proton outside the doubly closed ^{208}Pb. The multiplet states with the $\{(1h_{9/2})_n \times 3^-\}_J$ in $^{209}_{83}$Bi have been studied by means of (α, $\alpha'\gamma$), (^{16}O, ^{16}O$'\gamma$), and (^{7}Li, $\alpha 2n\gamma$) reactions (Häusser *et al.*, 1972). Theoretical analysis of the particle–vibration (3^-) coupling has been made for the multiplet states in ^{209}Bi and ^{209}Pb (Hamamoto, 1969).

2.4.4 *Rotation vibration coupling in deformed nuclei*

In deformed nuclei the rotational states $|I^g\rangle$ based on the ground state and those $|I^e\rangle$ based on a vibrational state coexist. These two kinds of states couple with each other through the rotation–vibration interaction. The beta-vibrational state with $I^\pi=0^+$ in the well-deformed nucleus ^{174}Hf is located in a fairly low excitation region, being isolated from other 0^+ levels in excitation energy. Thus ^{174}Hf is a good nucleus for studying rotation–vibration coupling.

The EM properties of rotational states $|I^\beta\rangle$ built on the β-vibrational state in ^{174}Hf have been extensively studied by e–γ spectroscopy, supplemented by the (α, α') reaction (Ejiri and Hagemann, 1971). The γ-ray spectrum from the ^{175}Lu(p, $2n\gamma$) reaction is shown in Fig. 2.28. In addition to the major transition flow along the ground-state rotational levels $|I^g\rangle$, weak γ rays from the states $|I^\beta\rangle$ are also seen. The γ-decay scheme for $|I^g\rangle$ and $|I^\beta\rangle$ in ^{174}Hf is shown in Fig. 2.29.

The rotation–vibration interaction $\mathbf{H}_{\beta g}$ leads to the coupled states,

$$|I^g\rangle=(|I^g\rangle_0+\alpha_I|I^\beta\rangle_0)/(1+\alpha_I^2)^{1/2} \tag{2.4.12a}$$

$$|I^\beta\rangle=(|I^\beta\rangle_0-\alpha_I|I^g\rangle_0)/(1+\alpha_I^2)^{1/2}, \tag{2.4.12b}$$

where $|I^x\rangle_0$ is the unperturbed state and α_I is the mixing amplitude. The mixing amplitude is given as $\alpha_I=-\langle\mathbf{H}_{\beta g}\rangle/E_\beta^0\approx-\langle\mathbf{H}_{\beta g}\rangle/E_\beta$, where E_β^0 and E_β are excitation energies for $|J^\beta\rangle_0$ and $|J^\beta\rangle$, respectively. The interaction matrix

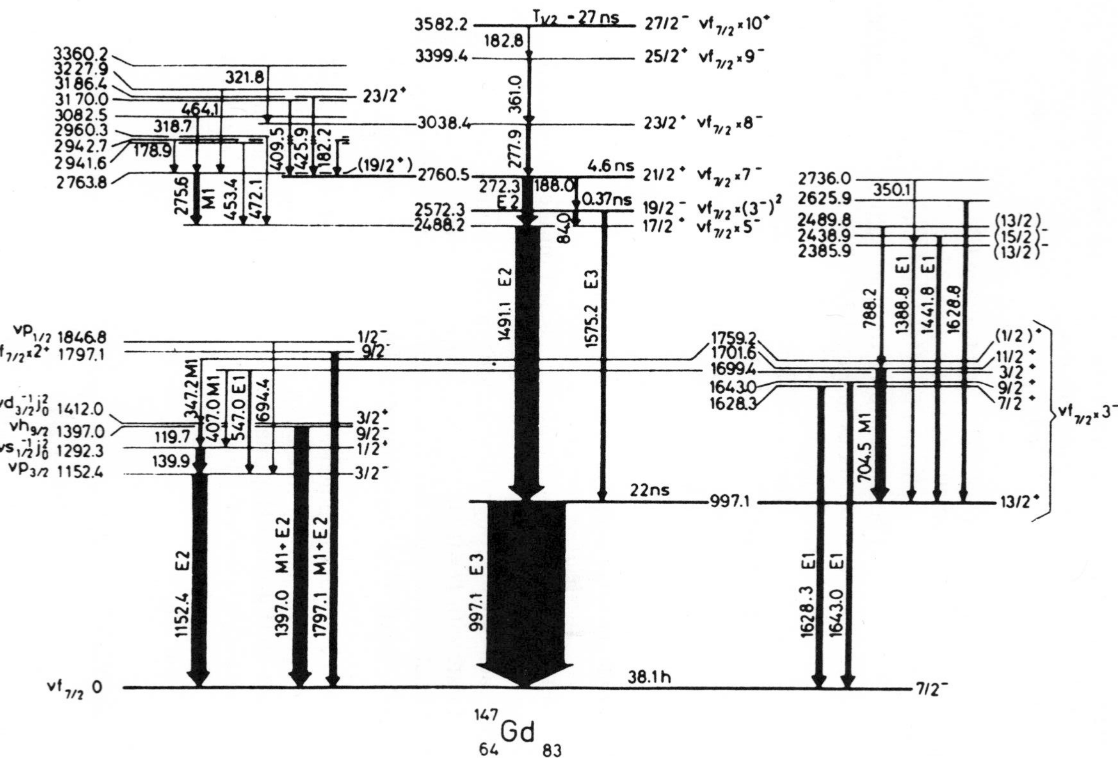

Fig. 2.27. Level scheme of ^{147}Gd. The transition intensities are as observed by the (^{3}He, 3n) reaction at 22 MeV. (Piiparinen *et al.*, 1982.)

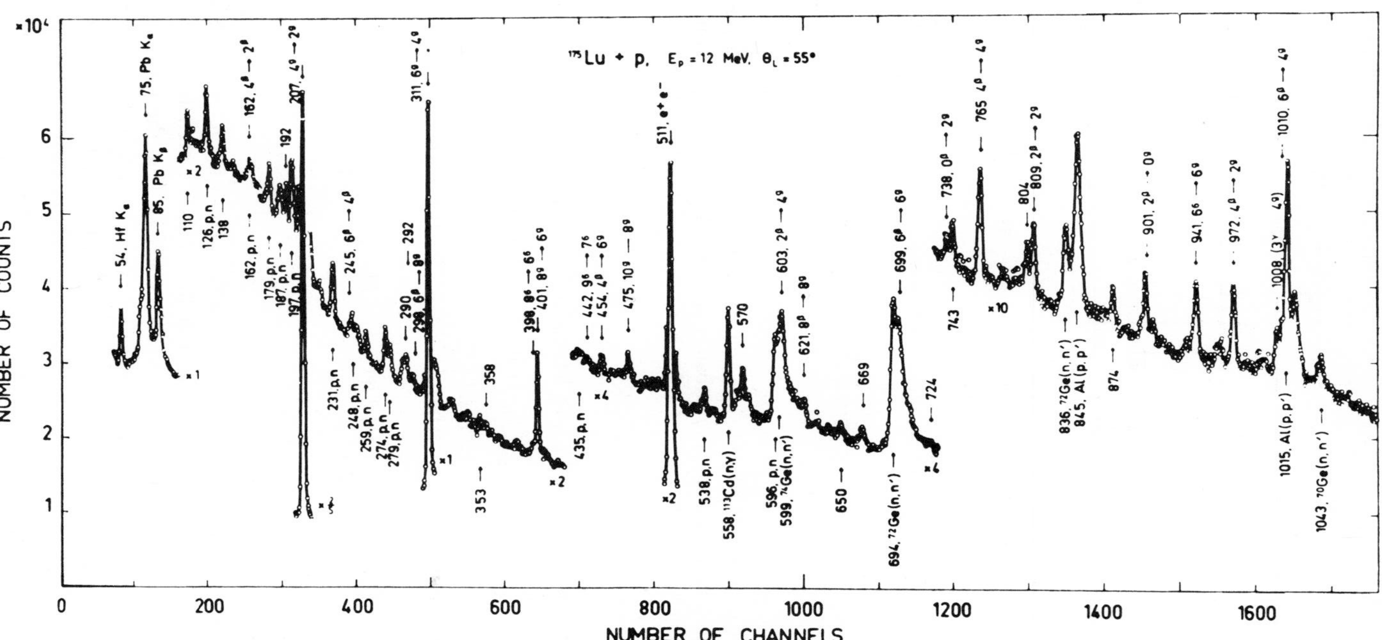

Fig. 2.28. Gamma-ray spectrum at $\theta_L = 55°$ from the ^{175}Lu(p, 2nγ)^{174}Hf reaction. I^g, I^β and I^6 are states with spin I in the ground, beta and the 1.55 MeV $K = 6$ bands, respectively (Ejiri and Hagemann, 1971.)

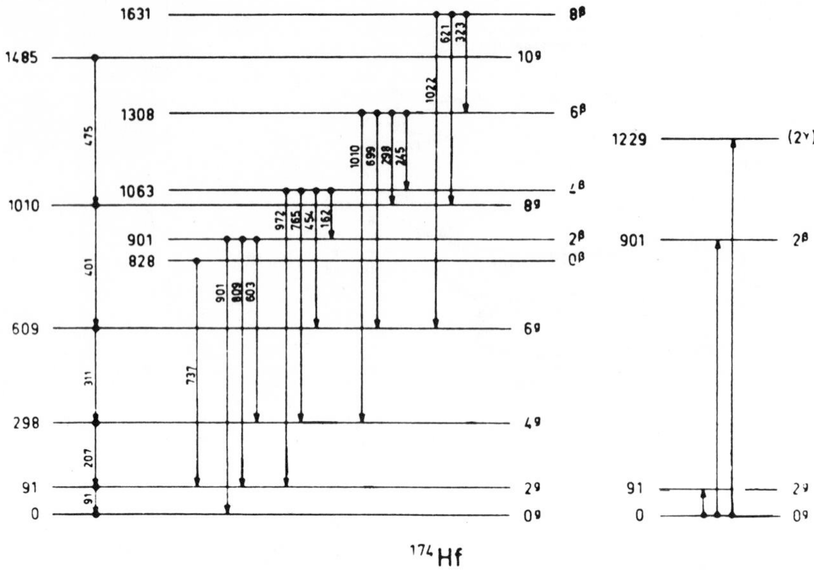

Fig. 2.29. Excitation (right-hand side) and de-excitation (left-hand) scheme of the levels in the ground, beta, and gamma bands of ^{174}Hf. The excitation and transition energies are in units of keV. (Ejiri and Hagemann, 1971.)

element is expanded as (Bohr and Mottelson, 1975)

$$\langle \mathbf{H}_{\beta g} \rangle = {}_0\langle I^g | \mathbf{H}_{\beta g} | I^\beta \rangle_0 = hI(I+1) + h'I^2(I+1)^2 + \ldots . \quad (2.4.13)$$

Taking the principal term in eqn (2.4.13), we obtain $\alpha_I = -hI(I+1)/E_\beta$. The E2 transition moment for the interband $|I'^\beta\rangle \to |I^g\rangle$ transition is then

$$Q_{\beta g}(I'I) = Q^0_{\beta g}\left[1 + \{I(I+1) - I'(I'+1)\}\frac{-h}{E_\beta}\cdot\frac{Q^0_{\beta\beta}}{Q^0_{\beta g}} + I'(I'+1)\frac{-h}{E_\beta}\cdot\frac{Q^0_{\beta\beta} - Q^0_{gg}}{Q^0_{\beta g}}\right] \quad (2.4.14)$$

where $Q_{\lambda\nu}(I'I)$ is defined by $B_{\lambda\nu}(\text{E2}, I'K' \to IK) = |\langle I'K'2\ \ K-K'|IK\rangle|^2 |Q_{\lambda\nu}(I'I)|^2$. Since the intraband transition moment is much larger than the interband one, the small mixing amplitude $(-h/E_\beta \ll 1)$ in the wave functions is enhanced in the E2 transition moment by the factor $Q^0_{\beta\beta}/Q^0_{\beta g} \gg 1$ so that the second term in eqn (2.4.14) becomes important. The third term is small because of $Q^0_{\beta\beta} - Q^0_{gg} \ll Q^0_{\beta\beta}$ for ^{174}Hf with a well-defined deformation. The observed values for $Q_{\beta\beta}$ (I', I) increases linearly with $I(I+1) - I'(I'+1)$, as shown in Fig. 2.30. The interaction strength is obtained as $h = -2.55$ keV.

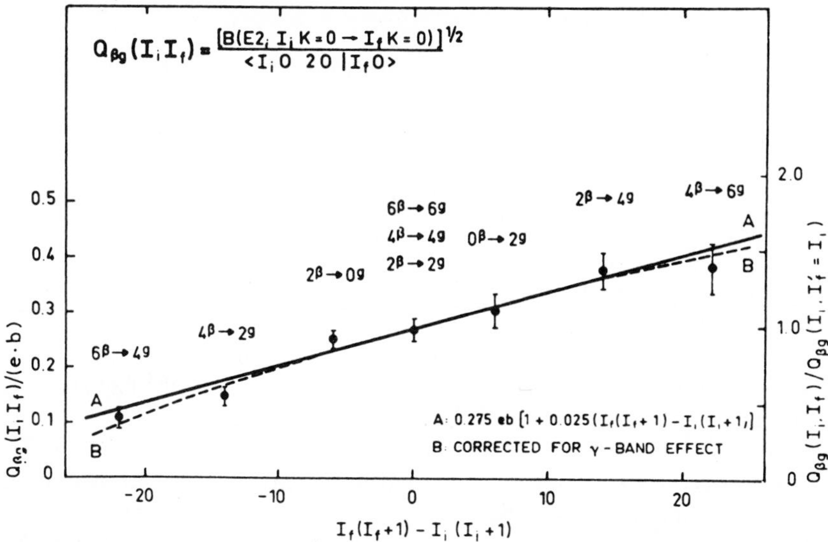

Fig. 2.30. The ratios of the interband E2 transition matrix elements $Q_{\beta g}(I'I)/Q_{\beta g}(I'I')$ (right-hand scale) plotted against the quantity $\{I(I+1)-I'(I'+1)\}$. The mean experimental values from the ^{175}Lu(p, 2n)^{174}Hf and the ^{172}Yb(α, 2n)^{174}Hf reactions are plotted. The solid line (A) is the best fit to the data. The dotted line is obtained by including small effects due to the coupling with the γ-band. By using the measured value for $Q_{\beta g}(20)$ the absolute values of $Q_{\beta g}(I'I)$are obtained as shown by the left-hand scale. Here we have assumed $Q_{\beta g}(22)=Q_{\beta g}(44)=Q_{\beta g}(66)$. (Ejiri and Hagemann, 1971.)

The simple dependence of the interband transition moment on $I(I+1)-I'(I'+1)$ was first discussed by Bohr and Mottelson (1975) with examples of the rotational states built on the γ-vibrational state in ^{166}Er as well as those built on the β-state in ^{174}Hf as mentioned above.

Deformed nuclei have rotational levels based on various kinds of vibrational states such as β, γ and octupole vibrations and those based on quasi-particle states in addition to those based on the ground state. Interactions between different kinds of excitation modes are studied by investigating interband and intraband γ transitions. Rotational levels based on several kinds of excitation modes in ^{160}Dy have been studied by means of the (α, 2n) reaction on ^{158}Gd (Riezebos *et al.*, 1987). Seven rotational bands are identified as shown in Fig. 2.31. The coupling between them is shown by the moments of inertia plotted as a function of spin in Fig. 2.32. The odd–even staggering of the moments of inertia for the γ-band ($K=2^+$) indicates the $K=0$ (s-band and β-band) mixing in the even spin levels, and that for the octupole–vibrational bands with $K=1^-$ and 2^- suggests interaction with higher-lying $K^\pi=0^-$ and 1^- (octupole) bands.

2.4.5 *Shape isomers and superdeformed shapes*

The single-particle shell effects in the nuclei with a finite number of nucleons may sometimes cause a second minimum in the potential, corresponding to a strongly deformed shape. The discovery of spontaneous fission isomerism in heavy nuclei revealed a double-humped fission barrier (Polikanov *et al.*, 1962). Here the single-particle effects modulate the deformed potential so as to produce the second minimum at a 'super' deformation around $\varepsilon \simeq 0.6$. The moment of inertia has been obtained by measuring E2-cascade γ transitions from rotational levels built on the second minimum isomer (Konecny *et al.*, 1972; Specht *et al.*, 1972). The energy spectrum of the conversion electrons in delayed coincidence with fission fragments from the 4 ns ^{240}Pu fission isomer is shown in Fig. 2.33. The measured moment of inertia is $\mathscr{I} = (\hbar^2/6.66)\ \text{keV}^{-1}$, which is twice as large as the normal rigid sphere one of $\mathscr{I} = (\hbar^2/14.3)\ \text{keV}^{-1}$ and corresponds to the deformation $\varepsilon = 0.6$. This is called superdeformation.

The two deformed shapes are connected via γ decays if the wave functions in the second minimum potential can penetrate through the inner barrier. Gamma decays from the 200 ns ^{238}U shape isomer were observed by

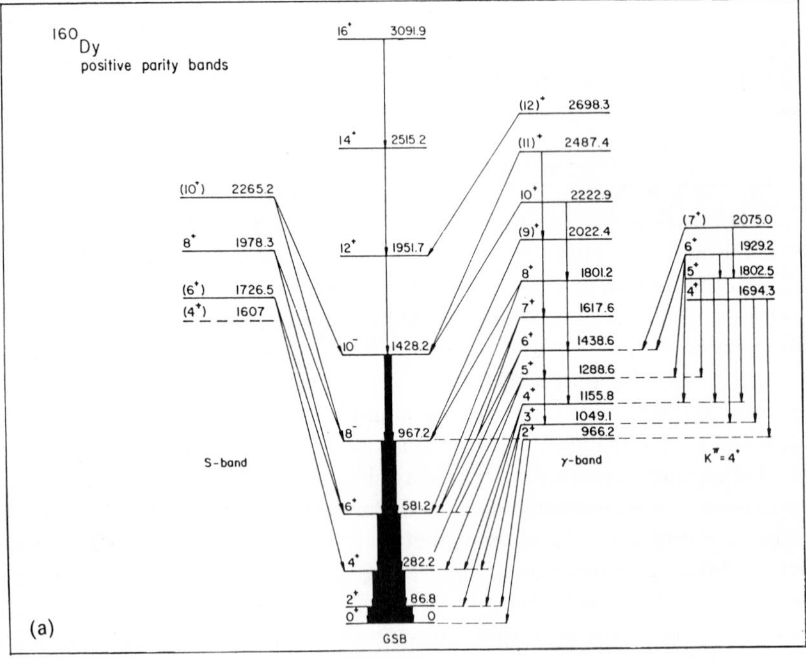

Fig. 2.31(a).

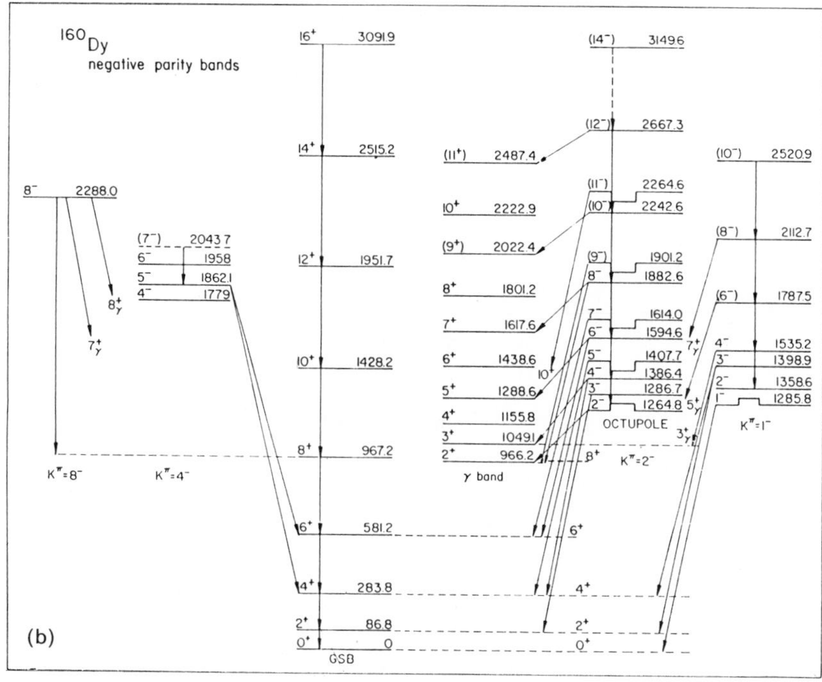

Fig. 2.31. The level scheme of ^{160}Dy. a; for positive parity bands, and b; for negative parity bands. (Riezebos *et al.*, 1987).

measuring delayed γ rays from the isomeric state populated by the 18 MeV (d, pn) reaction (Russo *et al.*, 1975). A strong γ transition from the 2.559 MeV 0^+ shape isomer to the 0.045 MeV 2^+ rotational level based on the ground state was observed, as shown in Fig. 2.34(a). The total γ-decay branch is about five times as large as the fission branch and the γ-branch half-life is 200 ns. The γ-decay probability between the two shapes gives the interaction between them. The potential–energy is plotted as a function of the deformation in Fig. 2.34(b).

Angular momentum also affects the nuclear shape. With increasing angular momentum the nuclear shape undergoes changes in the degree of deformation (moment of inertia), its sign (oblate, prolate) and the symmetry axis (spheroidal, triaxial). Recently a superdeformation with the long axis almost twice as large as the short axis has been found by measuring high-spin rotational γ transitions following heavy-ion *xn* reactions (Nyakó *et al.*, 1984; Twin *et al.*, 1986). The superdeformed shape at high spin is discussed extensively in Section 6.5.

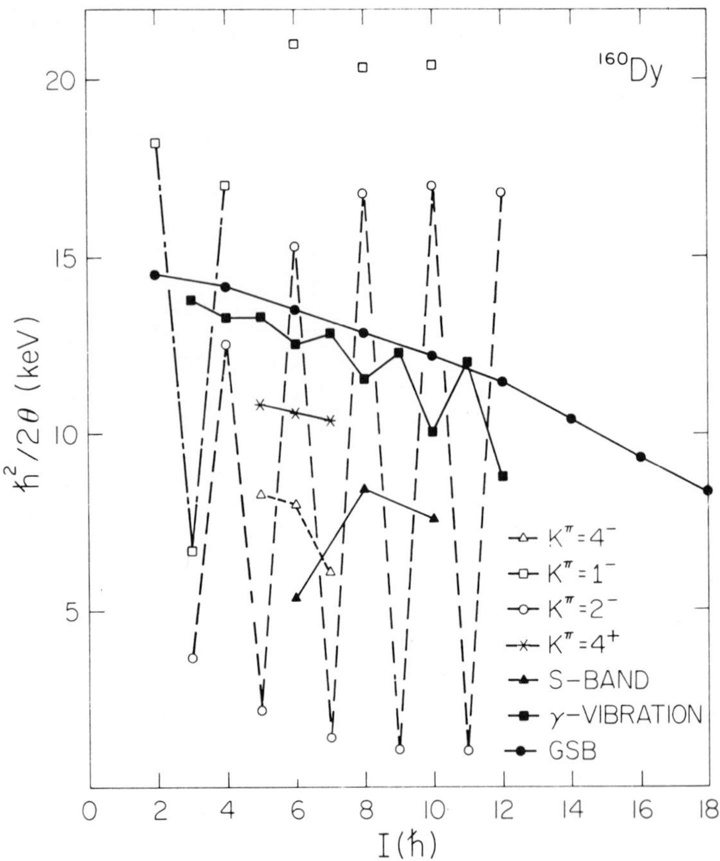

Fig. 2.32. The moment-of-inertia parameter plotted as a function of I for various bands in ^{160}Dy. (Riezebos *et al.*, 1987.)

2.4.6 *Systematics of E2 vibrational states and rotational states*

E2 vibrational states in nearly spherical nuclei and rotational states in quadrupole-type deformed nuclei are characterized by enhanced E2 γ transitions. Pure vibrational states have energies $E(I, N) = N\hbar\omega_2$ for N phonon states, and angular momenta $I=2$ for the one phonon ($N=1$) state, $I=0$, 2 and 4 for the two phonon ($N=2$) states and so on. A set of vibrational states with the maximum angular momentum $I=2N$ for a given number (N) of phonons forms a ground-state band (GSB). It is defined as the set of states with minimum energy for each angular momentum. Thus the energies of the

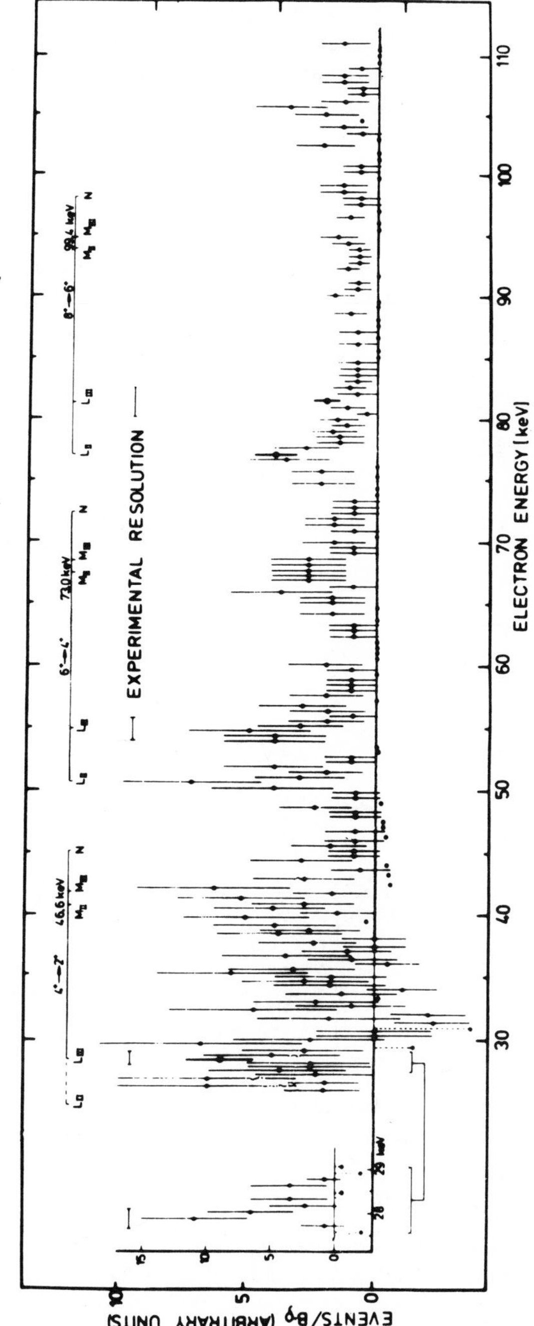

Fig. 2.33. The energy spectrum of the conversion electrons preceding isomeric fission with chance coincidence events substracted. The line at 28 keV has, in addition, been measured with improved resolution as shown in the inset. (Specht *et al.*, 1972.)

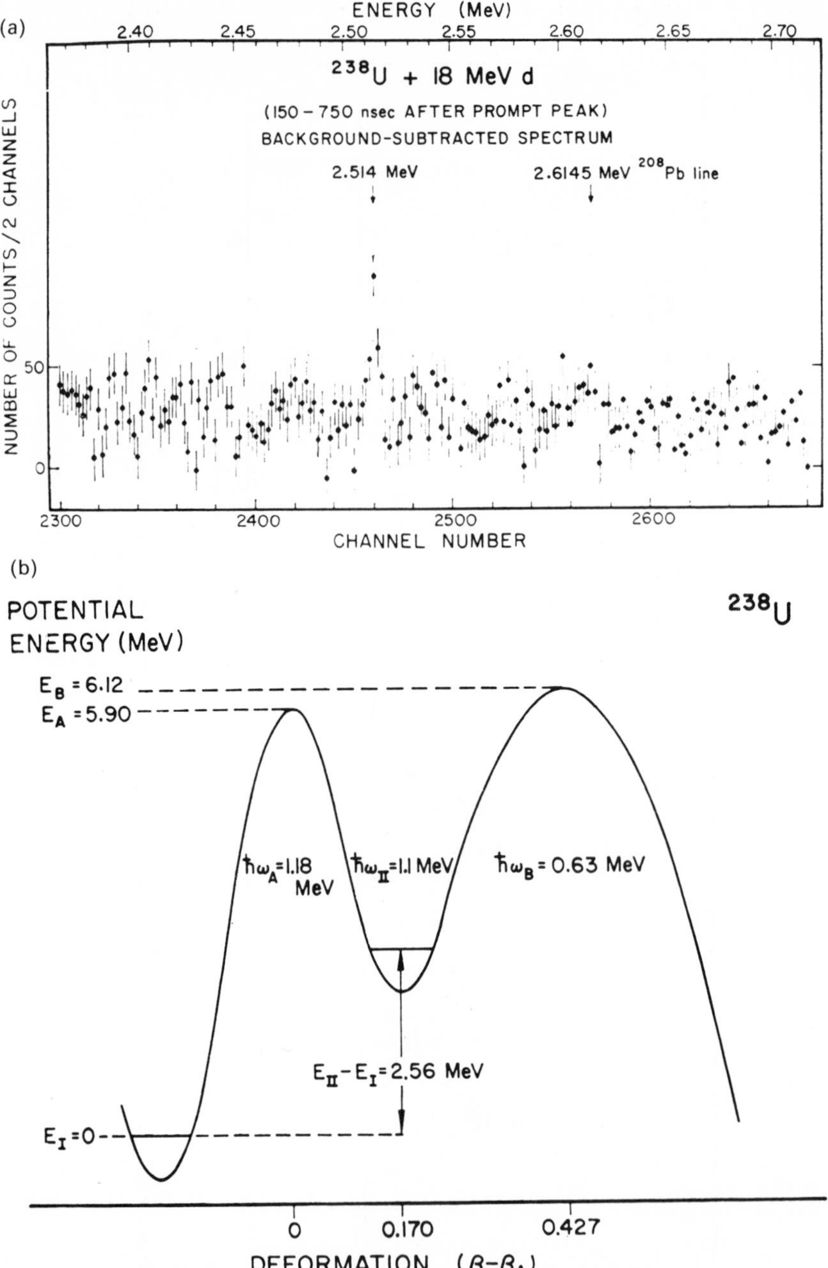

Fig. 2.34. (a) Background-subtracted γ-energy spectrum near 2.5 MeV for the 18 MeV deuteron bombardment of uranium. (Russo *et al.*, 1975.) (b) The ^{238}U fission barrier constructed from two inverted parabolas smoothly joined by a parabola. E_{II}, $\hbar\omega_A$ and $\hbar\omega_B$ are the values deduced from that experiment. (Russo *et al.*, 1975.)

GSB for vibrational states are given by

$$E(I) = E(I = 2N, N) = aI \tag{2.4.15a}$$

with
$$I = 0, 2, 4, 6, \ldots, \quad a = \frac{\hbar\omega}{2}. \tag{2.4.15b}$$

The rotational states based on the deformed ground-state constitute the GSB of the deformed nucleus, and their energies are

$$E_R(I) = kI(I+1) \tag{2.4.16a}$$

with
$$I = 0, 2, 4, 6, \ldots, \quad k = \hbar^2/2\mathscr{I}, \tag{2.4.16b}$$

where $\mathscr{I}$ is the moment of inertia.

Levels in the GSB have been established up to fairly high-spin with $I = 10$ to 60 (see Chapter 6) by means of e–γ spectroscopy using nuclear reactions such as heavy-ion *xn* reactions (see Section 4.2) and Coulomb excitations (see Section 6.2). They are compiled in tables, together with other bands based on excited states (Sakai, 1984).

The observed energies are somewhat modified compared with the simple formulae of eqns (2.4.15) and (2.4.16). GSB's in transitional nuclei, intermediate between the vibrational motion and the rotational motion, are shown in Fig. 2.35. The energy levels change gradually from the vibrational limit (^{150}Gd) to the rotational limit (^{160}Gd). The gradual change from the pure vibrational phase to the pure rotational phase is characteristic of nuclei composed of a finite number of nucleons. This feature is in contrast to the sudden phase transition between two fairly pure phases encountered in macroscopic systems which are composed of almost an infinite number ($\sim 10^{24}$) of atoms.

The systematic behaviour of GSB's over a wide range of mass number have been studied by many investigators, and some of them are reviewed by Sood (1968). One of the most natural ways of doing this is to express the excitation energy by a power series in $I(I+1)$, starting from the pure rotational limit of eqn (2.4.16), as follows;

$$E(I) = hI(I+1) - BI^2(I+1)^2 + \ldots. \tag{2.4.17}$$

Harris (1965) has proposed a power-series expansion on the basis of the cranking model. This is written in terms of the angular velocity ω as

$$E(I) = \tfrac{1}{2}\omega^2(\mathscr{I}_0 + 3C\omega^2 + 5D\omega^4 + 7F\omega^6 + \ldots) \tag{2.4.18a}$$

$$I(I+1) = \omega^2(\mathscr{I}_0 + 2C\omega^2 + 3D\omega^4 + 4F\omega^6 + \ldots)^2. \tag{2.4.18b}$$

The expansion converges quite well, and even the two-parameter approximation shows good fits to the observed energy levels in rotational nuclei.

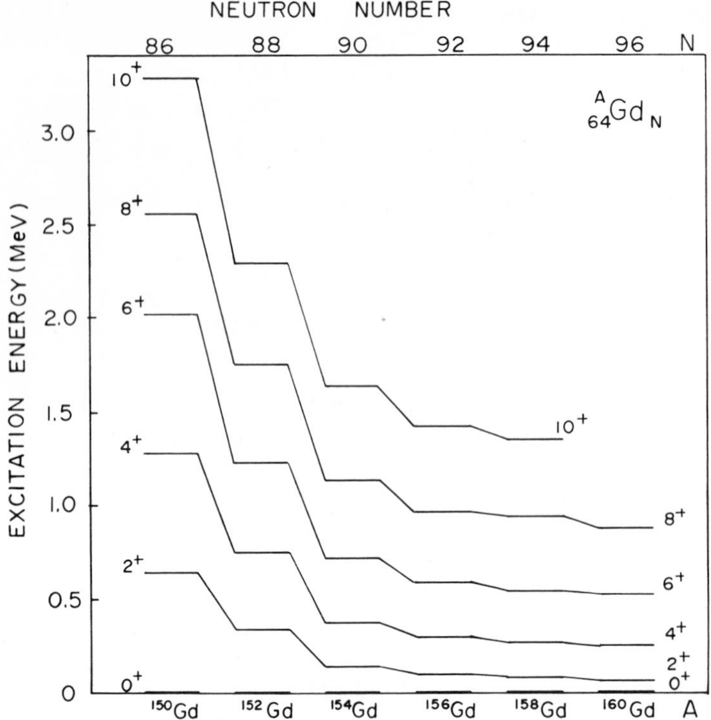

Fig. 2.35. The energy levels of ground state bands (GSB) in Gd isotopes.

Equation (2.4.15) can be improved for actual vibrational nuclei in a similar way by adding an anharmonic term.

Ejiri (1966) has derived a new formula to express in a uniform way the GSB energy levels in the entire region of vibrational, transitional and rotational nuclei. This formula is a simple sum of the terms for the two ideal cases,

$$E(I) = aI + kI(I+1). \tag{2.4.19}$$

Analyses of GSB energy levels with this expression shows a gradual change from the vibrational limit with a large $a(\gg k)$ to the rotational limit with a small $a(\ll k)$ as a function of a. In order to compare the predictions of eqn (2.4.19) with observed energy levels, the values $E(I)/I = (a+k) + kI$ are plotted for transitional nuclei in Fig. 2.36. Most of the data points indeed lie on straight lines. It is interesting to note that the fitted lines are almost parallel to each other, indicating a constant k for such transitional nuclei. In this sense eqn (2.4.19) is a one-parameter expression with one variable a. As shown in Fig. 2.37 the parameter a changes substantially as the neutron number

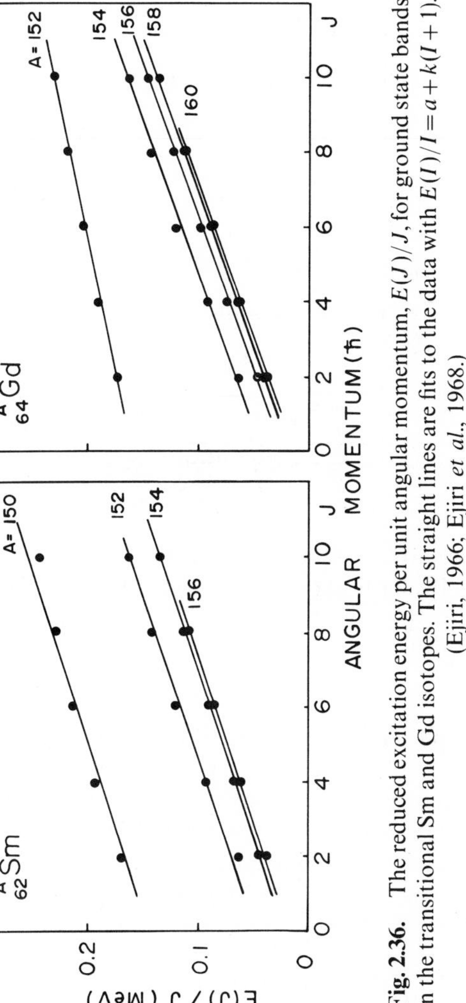

Fig. 2.36. The reduced excitation energy per unit angular momentum, $E(J)/J$, for ground state bands in the transitional Sm and Gd isotopes. The straight lines are fits to the data with $E(I)/I = a + k(I+1)$. (Ejiri, 1966; Ejiri *et al.*, 1968.)

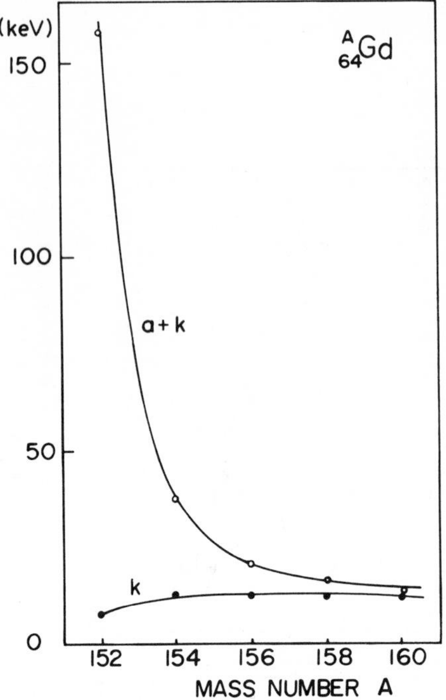

Fig. 2.37. The parameters $a+k$ and k used in the formula $E(I)/I = aI + kI(I+1)$, see eqn (2.4.19), for the ground state bands (GSB) in Gd isotopes.

increases, while k stays rather constant. Thus a is a measure of the collectivity, which depends on the number of valence nucleons in one major shell. The formula is able to explain a vast amount of data on GSB energy levels in medium and heavy nuclei (Ejiri, 1966, 1967; Ejiri *et al.*, 1968).

The energy levels of GSB's in deformed nuclei have been analysed in terms of a variable-moment-of-inertia (VMI) model (Mariscotti *et al.*, 1969). The two-parameter form of the VMI model is

$$E(I) = \frac{1}{2\mathscr{I}(I)} I(I+1) + \tfrac{1}{2} C_0 \{\mathscr{I}(I) - \mathscr{I}(0)\}^2, \tag{2.4.20}$$

$$\frac{\partial E(I)}{\partial \mathscr{I}(I)} = C_0 \{\mathscr{I}(I) - \mathscr{I}(0)\} - \frac{I(I+1)}{2\{\mathscr{I}(I)\}^2} = 0 \tag{2.4.21}$$

An analysis has been made of GSB's on the basis of the quadrupole boson operator B^{+}_{2m} (Das *et al.*, 1970). The Hamiltonian consists of a harmonic term and anharmonic terms,

$$\mathbf{H}_I = b(B^+_{2m}B_{2m'})_{00} + \sum_I C_I\{(B^+_{2m}B^+_{2m'})_{IM}(B_{2k}B_{2k'})_{IK}\}_{00}. \qquad (2.4.22)$$

The excitation energy of the GSB (states with the stretched angular momentum of $I=2N$ for N-phonons) is

$$E(I) = E(I=2N, N) = a'I + b'I(I-2), \qquad (2.4.23)$$

where $a' = b/(2\sqrt{5})$ and $b' = C_4/12$. It is remarkable that eqn (2.4.23) agrees with the empirical formula eqn (2.4.19) with the rearrangement of parameters $a = a' - 3b'$ and $k = b'$. It is found that the boson expansion model with the boson–boson interaction, eqn (2.4.23), and the next higher-order term reproduces excitation energies of GSB in even–even nuclei. GSB's and other bands have been analysed extensively in terms of interacting boson models (IBM) as described in the following section. It is pointed out that the prediction of the IBM for GSB's is summarized by the two-parameter formula of eqn (2.4.23)—equivalent to eqn (2.4.19)—with $|b'| \ll a'$ for the SU(5) (vibrational) limit and $b'/a' = 1/3$ for the SU(3) (rotational) limit (Klein, 1980). A generalized VMI formula, which reduces to eqn (2.4.23) for low I, has been proposed to give over-all fits to GSB's in even–even nuclei (Bonatsos and Klein, 1984; Klein, 1980).

2.5 The interacting boson model

2.5.1 *General features of the IBM*

The interacting boson model (IBM) provides a unified description of low-lying collective states in terms of interacting bosons. Nuclear collective motions such as quadrupole vibrations and rotations are associated with strong long-range quadrupole interactions. Since the nucleus is a system of a finite number of fermions, these collective motions are to a large extent subject to shell effects and depend also on the number of nucleons involved in such motions. The vibrational features change gradually to rotational ones, and vice versa, when the neutron and proton numbers change (see Section 2.4.6). A single-particle motion emerges as well as collective motions, and they interfere with each other. Characteristic aspects and interplays of these single-particle motions, collective vibrations and rotations have been studied extensively by Bohr and Mottelson (1969, 1975). The vibrations are originally described geometrically as surface vibrations (Bohr and Mottelson, 1953). In the geometrical approach the nuclear surface is parametrized by five quadrupole-shape variables and the average radius.

The IBM describes the vibrational and rotational limits as well as intermediate cases such as asymmetric rotations and transitional nuclei

(Iachello, 1981). In the IBM the symmetries of the collective motion are emphasized rather than the geometries, making use of algebraic formulations in terms of boson operators. From a phenomenological point of view some limiting cases of the IBM are similar and sometimes even identical to the geometrical description. Following early suggestions of Belyaev and Zelevinski (1962) and of Marumori *et al.* (1964) many authors have discussed boson representations of collective motion. More recent overviews of those works can be found in Das *et al.* (1970), Kishimoto and Tamura (1972), Sørensen (1973), Jansen *et al.* (1974), Marshalek (1974), and Holzwarth and Lie (1975).

The IBM has been developed to provide a unified description of low-lying collective states in terms of interacting bosons. It accommodates vibrations, rotations and intermediate motions. The IBM, as developed by Arima and Iachello (1975), describes the simple limiting situations by analytical forms. It provides a simple and transparent classification of nuclear excitations, which is very useful for analysing experimental data. The IBM, as in microscopic collective models, deals with the finite number of active nucleons, while the geometrical description treats rather idealized cases of infinite dimensional collective motions. Moreover the algebraic description of the IBM allows a direct connection with the shell model, thus bridging the gap between macroscopic (collective) and microscopic (single-particle) descriptions of nuclei.

The Hamiltonian in the IBM is written in the second quantized formalism in terms of the six boson creation and annihilation operators

$$b_{l\mu}^{+}, b_{l\mu} \quad l=0,2 \quad \mu=l, l-1, \ldots -l. \tag{2.5.1}$$

The $l=0$ stands for the s and $l=2$ for the d-boson. The corresponding magnetic substates are $\mu=-2,-1,0,1,2$, respectively. The operators $b_{l\mu}^{+}$ and $\tilde{b}_{l\mu}$ transform as spherical tensors under rotation, but $b_{l\mu}$ does not. The operator $\tilde{b}_{l\mu}$ is defined as

$$\tilde{b}_{l\mu}=(-1)^{l+\mu} b_{l-\mu}, \tag{2.5.2}$$

where $b_{l-\mu}$ represents the boson annihilation operator as given in eqn (2.5.1). The Hamiltonian can now be written in terms of tensor products $G_{\kappa}^{(k)}(ll')$ $=[b_{l}^{+} * \tilde{b}_{l'}]_{\kappa}^{(k)}$ of the operators defined in eqns (2.5.1) and (2.5.2) as

$$\begin{aligned} \mathbf{H}= & \Sigma_{l} \varepsilon_{l}(b_{l}^{+} \cdot \tilde{b}_{l})_{o}^{(o)}+\Sigma_{k, l_{1} l_{2} l_{3} l_{4}} u_{l_{1} l_{2} l_{3} l_{4}}^{(k)} \\ & \cdot \mathbf{G}_{\kappa}^{(k)}(l_{1}, l_{2}) \cdot \mathbf{G}_{\kappa}^{(k)}(l_{3}, l_{4}), \end{aligned} \tag{2.5.3}$$

where the labels l_1, l_2, l_3, l_4 stand for the s and d bosons and $k=0,1,2,3,4$. Only one- and two-body terms contribute to the Hamiltonian, with the first terms representing the s- and d-boson energies and the other terms the boson–boson interaction. The full Hamiltonian is built out of all possible 36 operators $G_\kappa^{(k)}(ll')$, with $l = s^+, d^+, l' = \tilde{s}, \tilde{d}, k = 0, 1, 2, 3, 4, \kappa = k, k-1, \ldots -k$. Thus the full Hamiltonian has the group structure U(6) of unitary transformations in six dimensions. The $36 = 6^2$ operators are called the generators of U(6). The most general one- and two-body Hamiltonian **H** within the boson space can thus be written in terms of generators of the six-dimensional special unitary group SU(6). The eigenvalues of **H**, when expressed in terms of SU(6), are given analytically. This is a limiting case with dynamical symmetry, where transition probabilities and other properties are also given analytically. These limiting situations are especially useful in the analysis of the vast body of experimental data (see e.g. Sakai, 1975) contributing to the understanding of the gross features of important nuclear phenomena. In a simple form one may write the Hamiltonian as

$$\mathbf{H} = \Sigma_i \varepsilon_i + \Sigma_{i<j} V_{ij}. \tag{2.5.4}$$

The first term represents the s- and d-boson energies and the second term the boson–boson interactions.

The Hamiltonian of eqns (2.5.3) and (2.5.4) is specified by nine parameters, two of them appearing in the one-body terms (for s- and d-boson energies) and seven in the two-body terms. Noting that the number of bosons is conserved ($N = n_s + n_d$) a rearrangement is possible such that two terms contribute only to the binding energies and six parameters determine the excitation energies. The most general Hamiltonian used to describe the excitation energies is then written as

$$\mathbf{H} = \varepsilon n_d + a_0(\mathbf{P}^+ \cdot \mathbf{P}) + a_1(\mathbf{L} \cdot \mathbf{L}) + a_2(\mathbf{Q} \cdot \mathbf{Q}) + a_3(\mathbf{T}_3 \cdot \mathbf{T}_3) + a_4(\mathbf{T}_4 \cdot \mathbf{T}_4), \tag{2.5.5}$$

where the operators for d-boson number n_d, pairing **P**, angular momentum **L**, quadrupole moment **Q**, and $\mathbf{T}_3$, $\mathbf{T}_4$ are given by

$$\begin{aligned}
n_d &= (d^+ \cdot \tilde{d}) \\
\mathbf{P} &= 1/2\{(\tilde{d} \cdot \tilde{d}) - (\tilde{s} \cdot \tilde{s})\} \\
\mathbf{L} &= 10^{1/2}[d^+ \tilde{d}]^{(1)} \\
\mathbf{Q} &= [d^+ \tilde{s} + s^+ \tilde{d}]^{(2)} - (1/2)(7)^{1/2}[d^+ \tilde{d}]^{(2)} \\
\mathbf{T}_3 &= [d^+ \tilde{d}]^{(3)} \\
\mathbf{T}_4 &= [d^+ \tilde{d}]^{(4)}.
\end{aligned} \tag{2.5.6}$$

This form of the Hamiltonian in terms of six free parameters is very convenient because it has been found empirically that often only two terms (the n_d and $\mathbf{Q} \cdot \mathbf{Q}$ terms) are sufficient to give the basic structure of the excitation. After finding the Hamiltonian the energy levels can be obtained by diagonalizing **H** in an appropriate basis. Such a basis is most conveniently constructed by using the powerful techniques of group theory.

There are three possible chains of sub-algebras, each with a subset of generators which is closed under commutation (Iachello 1980, 1981). The first subset is formed by the $5^2 = 25$ generators $G_\kappa^{(k)}$ (dd), with $k = 0, 1, 2, 3, 4$, which close under the algebra U(5). The group of special unitary transformations in five dimensions SU(5) represents the vibrational limit, as discussed in subsection 2.5.2. The second subset is formed by the $3^2 = 9$ generators $G_0^{(0)}$ (ss), $G_\kappa^{(1)}$ (dd), $G_\kappa^{(2)}$(ds), $G_\kappa^{(2)}$ (sd), and $G_\kappa^{(2)}$ (dd) which close under commutation and form the algebra of U(3). The group of special unitary transformations in three dimensions SU(3) represents the limit of axially symmetrical rotations, discussed in Section 2.5.3. Note that the obvious subalgebras are in this case the group of ordinary rotations in three dimensions O(3) formed by the three generators $G_\kappa^{(1)}$(dd) and the group of rotations around the z-axis O(2) formed by the generator $G_0^{(1)}$(dd). The third subset is obtained from the $6 \times 5/2 = 15$ generators $G_\kappa^{(1)}$(dd), $G_\kappa^{(3)}$(dd), $G_\kappa^{(2)}$(ds), and $G_\kappa^{(2)}$(sd), which form the orthogonal group in six dimensions O(6). This group represents the γ-unstable limit, as discussed in Section 2.5.4.

In the IBM it is assumed that the nuclear structure is dominated by the valence particles outside the closed shells 2, 8, 20, 28, 50, 82, 126. Bosons in even–even nuclei are pairs of identical particles with angular momentum $\mathbf{L} = 0$ and 2, s and d bosons, respectively. In the IBA-1 model no distinction is made between protons (π) and neutrons (ν). The total number of bosons is then $N = N_\pi + N_\nu$. In the IBA-2 model the protons and neutrons are treated separately. In the following four subsections we will discuss the limiting and intermediate cases within the first approximation. The discussion is extended in Section 2.5.6 to odd–even nuclei in terms of the interacting boson–fermion model (IBFM). An even closer connection to a microscopic description is offered by the neutron–proton interacting boson–fermion model, IBFA-2, as discussed in Section 2.5.7. Finally, a boson-number systematics is deduced from experimental data in Section 2.5.8.

2.5.2 *The vibrational limit, SU(5)*

In the vibrational limit of SU(5) the excited states are characterized by the number of d bosons ($L = 2$) and their interactions; the quantum number n_s is not needed. The d-boson energy term dominates most of the low-lying states in vibrational nuclei. A boson interaction term is important to account for the large splitting of the 0^+, 2^+, 4^+ states, belonging to the two-phonon triplet ($n_d = 2$) in deformed nuclei. Therefore Arima and Iachello (1976) proposed to simplify the general Hamiltonian of eqn (2.5.5) to

$$\mathbf{H} = \varepsilon \Sigma_m d_m^+ d_m + \Sigma_{L=0,2,4}(2L+1)^{1/2} c_L/2 \times \{[\mathrm{d}^+ \mathrm{d}^+]^{(L)} [\tilde{\mathrm{d}} \tilde{\mathrm{d}}]^{(L)}\}^{(0)}. \tag{2.5.7}$$

The first term corresponds to the Bohr Hamiltonian and the anharmonicity is introduced by the boson–boson interaction terms.

Five quantum numbers are needed to classify states of n_d bosons, which form the basis for the group SU(5). The trivial quantum numbers are n_d, the angular momentum L and its third component M. The angular momentum is built up by the bosons not coupled to zero. We have thus to count the number of boson pairs n_β and boson triplets n_Δ, which are coupled to zero angular momentum. The relation between those numbers is given by

$$n_d = 2n_\beta + 3n_\Delta + \lambda. \tag{2.5.8}$$

The angular momentum is then given by λ as

$$L = \lambda, \lambda + 1, \ldots 2\lambda, \tag{2.5.9}$$

with the exception of $2\lambda - 1$, due to angular–momentum coupling of similar bosons. Note that the seniority v, the number of unpaired bosons, is written as

$$v = n_d - 2n_\beta. \tag{2.5.10}$$

The eigenvalue of the interacting d-boson Hamiltonian in eqn (2.5.7) is obtained as

$$E(n_d, v, n_\Delta, \mathbf{L}, M) = \varepsilon n_d + \alpha n_d(n_d - 1)/2 + \beta(n_d - v)(n_d + v + 3) + \gamma\{L(L+1) - 6n_d\}. \tag{2.5.11}$$

We notice that the excitation energies are determined by only four parameters, ε, α, β, and γ. The excited states are all positive–parity states grouped into various bands, characterized by the quantum numbers n_d, v, and n_Δ. The GSB(Y), X and Z-bands, etc. are given by (n_d, $v = n_d$, $n_\Delta = 0$, $L_{max} = 2n_d$, $2n_d - 2$, $2n_d - 3, \ldots$), the β band by (n_d, $v = n_d - 2$, $n_\Delta = 0$, $L = 2n_d - 4$), and the Δ band by (n_d, $v = n_d$, $n_\Delta = 1$, $L = 2n_d - 6$) as illustrated in Fig. 2.38.

An interesting feature of energy levels for all bands is that the γ-ray energies of intraband E2 transitions are simply given by $E_\gamma(L+2 \rightarrow L) = C + (\alpha + 8\gamma)L/2$, with the proportionality constant $\alpha + 8\gamma$ arising from the boson–boson interaction. A comparison of the theoretical and experimental level schemes for ^{110}Cd is presented in Fig. 2.39.

One can also couple other degrees of freedom, like the octupole vibration and quasi-particle excitation, to the d-boson excitation. The octupole degree of freedom is represented by the f boson with $L = 3$ which can occupy the $L = 0$ and 3 levels. Usually only one f boson is added to the d bosons. The Hamiltonian of eqn (2.5.7) is then extended with the f-boson Hamiltonian and the f–d boson interaction terms. A simple expression for the eigenvalues is obtained in the case of the totally aligned spin values $L = 2n_d + 3$

$$E(n_d, L_d = 2n_d; n_f, L = 2n_d + 3) = E(n_d, L_d = 2n_d) + \varepsilon_f + n_d x_5 \tag{2.5.12}$$

and for the $L = 2n_d + 2$ sequence an extra term $\{(2n_d + 3)/5\}\Delta_4$ is added to eqn

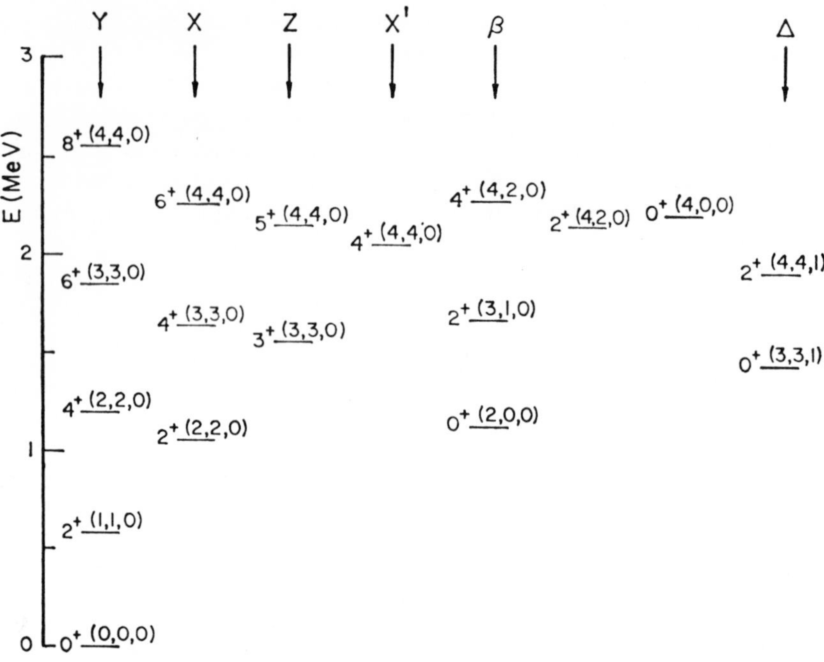

Fig. 2.38. A typical spectrum in the d-boson limit. The numbers in parenthesis are the SU(5) quantum numbers (n_d, v, n_Δ). The angular momentum quantum number is explicitly written to the left of each level. The parameters used are $\varepsilon = 579$ keV, $c_4 = 39.4$ keV, $c_2 = -95.3$ keV, $c_0 = -27.4$ keV. (Arima and Iachello, 1976.)

(2.5.12). Thus the fully aligned octupole band is described by two parameters in addition to the d-boson band, with one extra parameter for each additional band. Thus the γ-ray energies of the intraband $E2$ transitions for the octupole-coupled states are the same as those for the corresponding transitions in the GSB with the same n_d. An early experimental identification of the octupole band in ^{150}Sm confirms the validity of the above concepts, as illustrated in Fig. 2.40.

Two quasi-particle (QP) states may occur in the spectrum at an excitation energy of twice the pairing gap $2\Delta \approx 2$ MeV and at higher energies. When these states lie below the collective states they may become yrast states and thus may be populated in heavy-ion induced reactions. This situation occurs when the two particles with spin j_1 and j_2 couple to a relatively high spin l. In a drastic approximation the two quasi-particle mode is treated as a boson mode. The QP Hamiltonian for one quasi-particle pair and the QP d-boson interaction terms are added to the d-boson Hamiltonian. The eigenvalues for the fully aligned $L = 2n_d + l$ state is written in a similar way as for the octupole

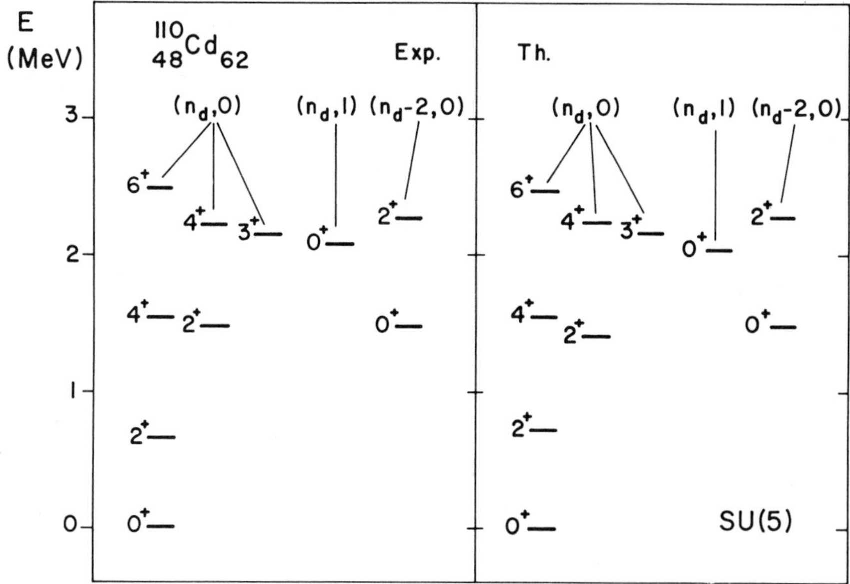

Fig. 2.39. An example of a spectrum with SU(5) symmetry: ${}^{110}_{48}\mathrm{Cd}_{62}$, $N_\pi = 1$, $N_\nu = 6$, $N = 7$. (Iachello, 1980.)

bands given above,

$$E(n_d, L_d = 2n_d;\ (j_1 j_2)l,\ L = 2n_d + l) = E(n_d, L_d = 2n_d) + \varepsilon_l + n_d y_{l+2}. \tag{2.5.13}$$

An extra term $\{(2n_d + l)/(l+2)\}\Delta_{l-1}$ is added for the $L = 2n_d + l - 1$ sequence. Again the γ-ray energies show exactly the same linear dependence on n_d as discussed above for the other bands. A comparison of experimental and theoretical results for ^{156}Er is presented in Fig. 2.41.

The transitions between the QP and ground-state bands (GSB) are highly forbidden because a large change of the internal structure is involved, such as the type of bosons and the number of bosons. Thus the QP intraband transitions are favoured until the QP band head is reached and a retarded QP-GSB transition occurs. The γ-ray energies of the GSB and the QP bands are given by

$$E_\gamma(\mathrm{GSB}) = \varepsilon_d + [I/2](\alpha + 8\gamma) \tag{2.5.14}$$

$$E_\gamma(\mathrm{QP}) = \varepsilon_d + [(I - l)/2](\alpha + 8\gamma) + y_{l+2}, \tag{2.5.15}$$

where I is the angular momentum.

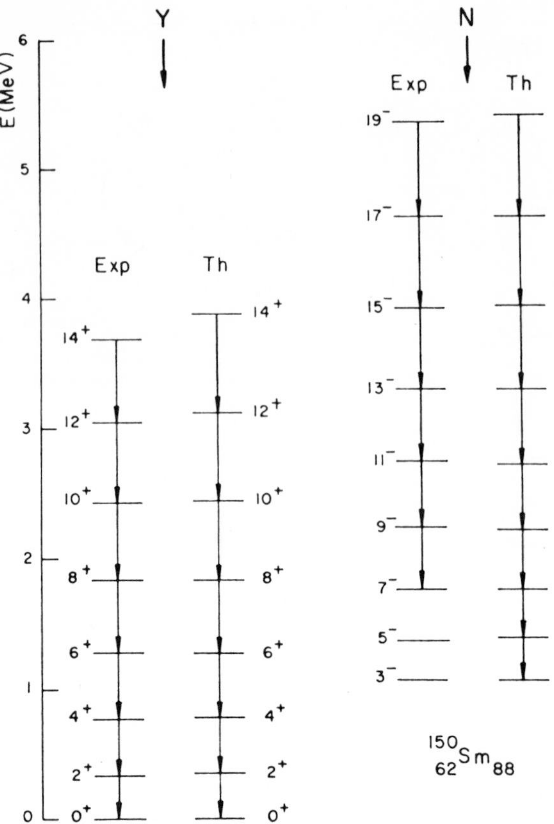

Fig. 2.40. A comparison between the experimental (de Voigt *et al.*, 1975) and theoretical spectrum in ^{150}Sm. The parameters in the theoretical spectrum are $\varepsilon_d = 356$ keV, $c_4 = 66$ keV, $\varepsilon_f = 1077$ keV, $x_5 = -47$ keV. (Arima and Iachello, 1976.)

The results for ^{156}Er are compared with the experimental data in Fig. 2.42. It is seen in this figure that the theoretical curve follows the 'backbending' curve of the experimental data. Here the backbending of the IBM curve is simply due to the lower energies of the QP-states with respect to the corresponding states in the GSB, which have larger n_d values. A more extensive treatment of the backbending phenomena in terms of the IBFA-2 (see Section 2.5.7a) can be found in the work of Arima (1983).

Also, the electromagnetic transition probabilities can be calculated very easily in this limit. Assuming that the transition operators are at most one-body operators we can write the E2 transition operator, in the second

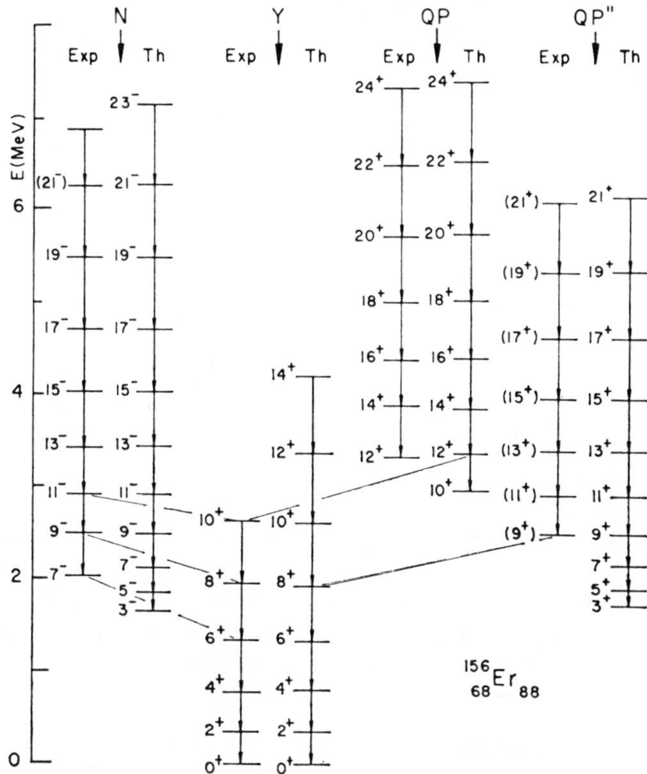

Fig. 2.41. The probable band structure of ^{156}Er. In the absence of detailed information on the decay properties of the QP″ band, the assignment QP″ $=3^+ \times d^n$ must be regarded as tentative. The parameters in the theoretical spectrum are $\varepsilon = 365$ keV, $c_4 = 78$ keV, $\varepsilon_f = 1647$ keV, $x_5 = -162$ keV, $\varepsilon_{10+} = 2965$ keV, $y_{12} = 32$ keV, $\varepsilon_{3+} = 1728$ keV, $y_5 = -189$ keV. (Arima and Iachello, 1976; experimental data from Stephens, 1985.)

quantized form, as

$$T_m(\text{E2}) = \alpha_2[\text{d}^+\tilde{\text{s}} + \text{s}^+\tilde{\text{d}}]_m^{(2)} + \beta_2[\text{d}^+\tilde{\text{d}}]_m^{(2)}. \tag{2.5.16}$$

Operators for radiation of other multipolarities have similar forms (Iachello, 1980). The quadrupole operator acting on the d-boson states has the selection rule $\Delta n_{\text{d}} = 0, \pm 1$. Since the states in this limit are characterized by a fixed number of d bosons, n_{d}, only the first term contributes to the B(E2) value, which takes the simple form for the GSB, with $L' = 2n_{\text{d}}' = 2(n_{\text{d}} + 1) \rightarrow L = 2n_{\text{d}}$,

$$B(\text{E2}) = \alpha_2^2(L+2)(2N-L)/4. \tag{2.5.17}$$

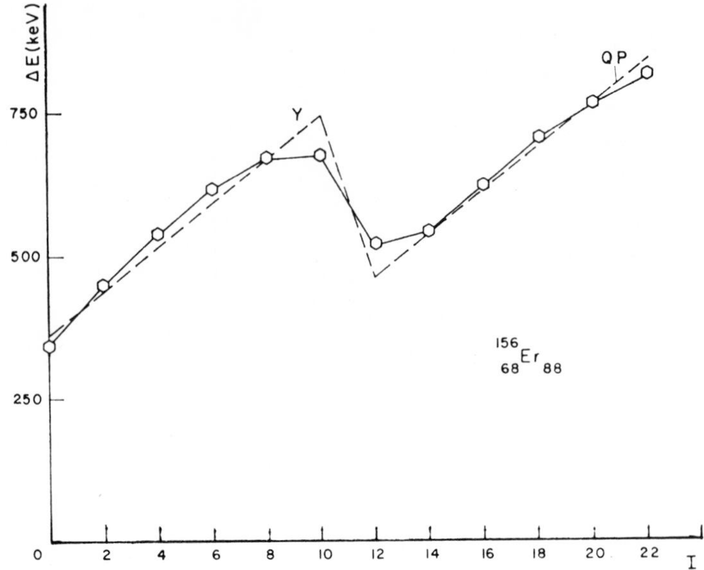

Fig. 2.42. Plot of ΔE versus I for the yrast bands in ^{156}Er, where $\Delta E = E(I+2) - E(I)$. The experimental data (full line) are the points (from Stephens, 1985). The broken line is the theoretical prediction. (Arima and Iachello, 1976.)

Note that the total number of bosons N is given by the d bosons because n_s is not relevant in this limit (N is assumed to be large).

The quadrupole moment is instead given by the second term in eqn (2.5.16). It is given for states of the GSB with spin L as

$$Q_L = \beta_2(16\pi/5)^{1/2}(1/14)^{1/2}L. \tag{2.5.18}$$

It is seen from eqn (2.5.17) that the B(E2) values increase linearly with L. In practice, the symmetry breaking causes deviations from this linear dependence, particularly at high spin, and in most cases the B(E2) values decrease with L for high-spin states.

2.5.3 *The axially symmetric rotor; SU(3) limit*

In this limit (Arima and Iachello, 1978) the single boson energy ε in the general Hamiltonian of eqn (2.5.5) is relatively small with respect to the other terms and another form can be used which is identical to the one introduced by Elliott (1958) for fermions or to the geometrical description of Bohr and Mottelson (1975). Taking the β- and γ-bands as degenerate.

$$\mathbf{H} = -\kappa\mathbf{Q}\cdot\mathbf{Q} - \kappa'\mathbf{L}\cdot\mathbf{L} = -\kappa\Sigma_{ij}\mathbf{Q}_i\mathbf{Q}_j - \kappa'\Sigma_{ij}\mathbf{L}_i\mathbf{L}_j. \tag{2.5.19}$$

Here the three angular-momenta and the five quadrupole generators of SU(3) are given by eqn (2.5.6). The parameters κ and κ' are the quadrupole–quadrupole and the angular momentum interaction strengths, respectively.

It is convenient to introduce the quantum numbers λ and μ, which are obtained from a decomposition of N, according to Elliott (1958) as

$$\begin{aligned}[N] = &(2N, 0), (2N-4, 2), \ldots, (0, N) \text{ for } N=\text{even or } (2, N-1) \text{ for } N=\text{odd}\\ &(2N-6, 0), (2N-10, 2), \ldots, (0, N-3) \text{ for } N-3=\text{even}\\ &\qquad\qquad \text{or } (0, N-4) \text{ for } N-3=\text{odd}.\\ &\ldots\end{aligned} \tag{2.5.20}$$

The angular momenta contained in each (λ, μ) in the Elliott basis are given by

$$L = K, K+1, K+2, \ldots K+\max(\lambda, \mu) \tag{2.5.21}$$

with

$$K=\text{integer}=\min(\lambda, \mu), \min(\lambda, \mu)-2, \ldots 1 \text{ or } 0 \tag{2.5.22}$$

with the exception for $K=0$ for which we have

$$L=\max(\lambda, \mu), \max(\lambda, \mu)-2, \ldots 1 \text{ or } 0. \tag{2.5.23}$$

The eigenvalue of the Hamiltonian in eqn (2.5.19) is given by

$$E\{[N](\lambda, \mu), K, L, M\} = (3\kappa/4-\kappa')L(L+1)-\kappa[\lambda^2+\mu^2+\lambda\mu+3(\lambda+\mu)]. \tag{2.5.24}$$

A spectrum generated by this equation for $N=8$ in the decomposition into (λ, μ) according to eqn (2.5.20) is given in Fig. 2.43. The various bands observed, for instance, in ^{156}Gd can be explained rather well by this approximation, as illustrated in Fig. 2.44.

The E2 transition operator is given by eqn (2.5.16), but the constant in the second term has to be replaced by $\beta_2+7^{1/2}\alpha_2/2$. It turns out that the first term in this limit is much larger than the second one. The $Q^{(2)}$ generator of SU(3) cannot connect different (λ, μ) representations, which gives the selection rule

$$\Delta\lambda = \Delta\mu = 0. \tag{2.5.25}$$

The B(E2) values along the GSB, defined by $\lambda = 2N$, $\mu = 0$, $L' = L+2 \rightarrow L$ are given by

$$B(\text{E2}) = \alpha_2^2 3(L+2)(L+1)(2N-L)(2N+L+3)/\{4(2L+3)(2L+5)\}, \tag{2.5.26}$$

and the quadrupole moments by

$$Q_L = -\alpha_2(16\pi/40)^{1/2}L(4N+3)/(2L+3). \tag{2.5.27}$$

It thus appears that the B(E2) values are quadratic in the boson number N in

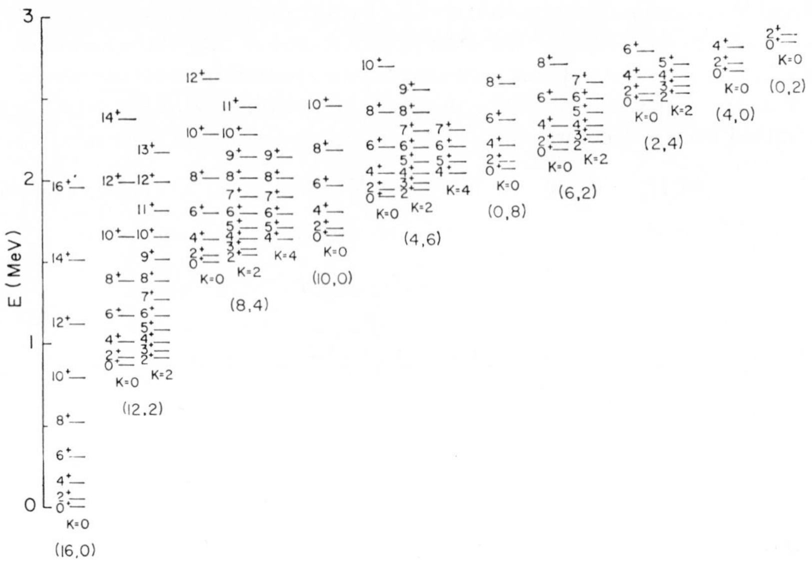

Fig. 2.43. The spectrum of eqn (2.5.24) for $N=8$. The decomposition of the totally symmetric representation ($N=8$) of SU(6) into representation (λ, μ) of SU(3) is given by eqn (2.5.20). The bands are labelled by the quantum numbers K and (λ, μ). (Arima and Iachello, 1978.)

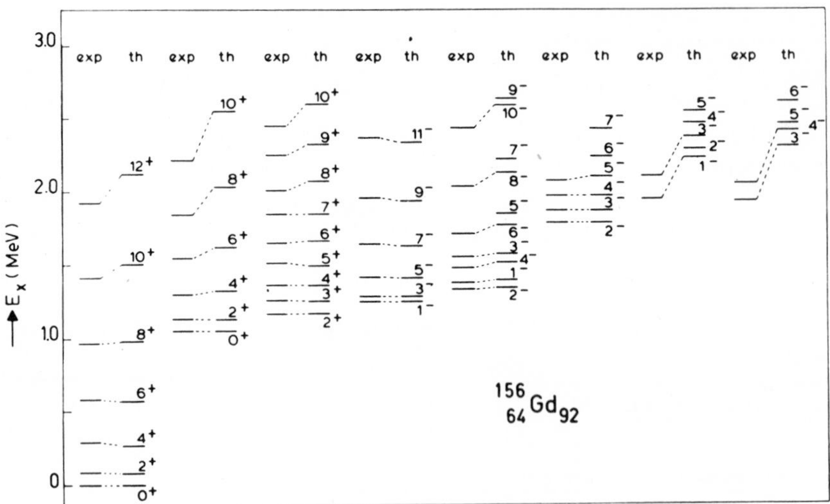

Fig. 2.44. A comparison of energy levels in ^{156}Gd, calculated within the SU(3) limit, with the experimental data. (Konijn *et al.*, 1981.)

this SU(3) limit and linear in the SU(5) limit as discussed in Section 2.5.2. This is illustrated in Fig. 2.45. B(E2) reaches much larger values in the rotational limit than in the vibrational one, because of the large number of active bosons.

The correspondence with the geometrical model of Bohr and Mottelson (1975) is expressed by

$$B(\mathrm{E2})/B(\mathrm{E2};\,\mathrm{BM}) = (2N-L)(2N+L+3)/(2N+3/2)^2 \text{ for large } N \tag{2.5.28}$$

For the simple case of $K=0$ the geometrical quadrupole moment is obtained from eqn (6.3.71) as

$$Q_L(\mathrm{BM}) = -eQ_0L/(2L+3). \tag{2.5.29}$$

From a comparison of this expression with eqn (2.5.27) we find that the intrinsic quadrupole moment eQ_0 in the IBM is given by

$$eQ_0 = \alpha_2(16\pi/40)^{1/2}(4N+3). \tag{2.5.30}$$

The IBM description yields B(E2) values which are sensitive to L, in contrast to the geometrical model, as illustrated in Fig. 2.46.

2.5.4 *The γ unstable limit*; O(6) *limit*

This limit is useful to describe nuclei at the end of the major closed shells. It is similar to the geometrical γ-unstable model of Wilets and Jean (1956)

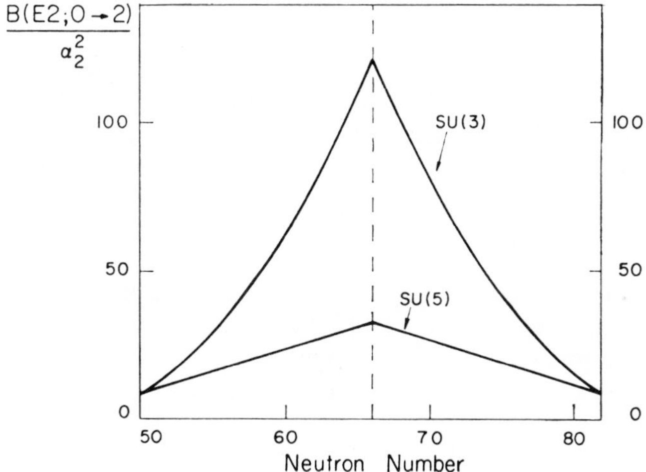

Fig. 2.45. Schematic behaviour of the B(E2) values as a function of the number of bosons N for the vibrational (SU(5)) and the rotational (SU(3)) limits according to eqns (2.5.17) and (2.5.26), respectively. (Arima and Iachello, 1978.)

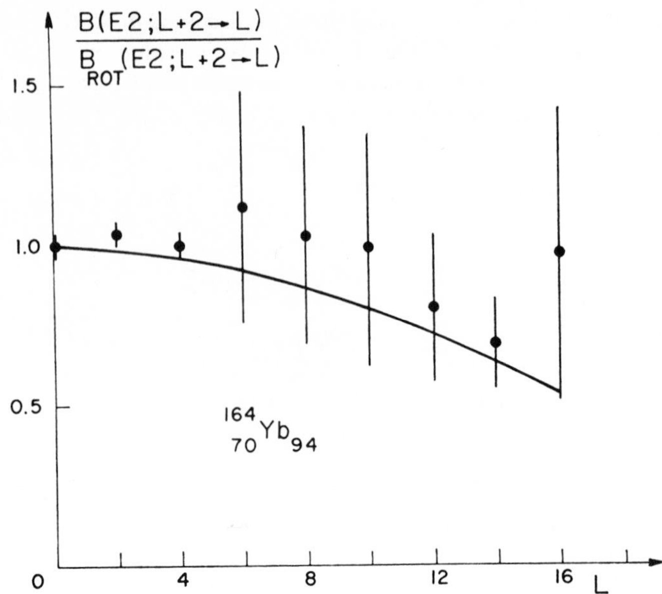

Fig. 2.46. The experimental ratio $B(\mathrm{E}2;\ L+2\rightarrow L)/B_{\mathrm{ROT}}(\mathrm{E}2;\ L+2\rightarrow L)$ in ${}^{164}_{70}\mathrm{Yb}_{94}$ $(N=N_\pi+N_\nu=12)$ as compared with the SU(3) prediction. The values $B_{\mathrm{ROT}}(\mathrm{E}2;\ L+2\rightarrow L)=\{(2N)(2N+3)/(2N+3/2)^2\}B_{\mathrm{BM}}(\mathrm{E}2;\ L+2\rightarrow L)$ are used here rather than B_{BM} in order that the ratio be normalized to 1 at $L=0$. (Arima and Iachello, 1978.)

and under certain conditions also to the triaxial rotor with $\gamma=30°$. The Hamiltonian can be written in terms of the 15 generators of the group O(6), which are conveniently expressed in the pairing operator P_6 being the quadratic Casimir operators of O(6) and C_5 the one of the group O(5) and C_3 of O(3) (Arima and Iachello 1979).

$$\mathbf{H}=\mathrm{AP}_6+\mathrm{BC}_5+\mathrm{CC}_3. \qquad (2.5.31)$$

In this limit, five quantum numbers, N, σ, τ, L, M, are needed to label the states. The quantum numbers σ and τ characterize the representations of O(6) and O(5), respectively. They are obtained as

$$\sigma=N,\ N-2,\ \ldots,\ 0 \text{ or } 1 \text{ (for } N \text{ even or odd)} \qquad (2.5.32)$$

$$\tau=\sigma,\ \sigma-1,\ \ldots\ 0. \qquad (2.5.33)$$

When we introduce the number of boson triplets, n_Δ, coupled to zero angular momentum like in Section 2.5.2, then the angular momentum L is determined by λ as

$$\tau = 3n_{\Delta} + \lambda \tag{2.5.34}$$

$$L = 2\lambda, 2\lambda - 2, \ldots \lambda + 1, \lambda. \tag{2.5.35}$$

The eigenvalue of **H** can now be given in a closed form in terms of those quantum numbers as

$$E\{[N]\sigma, \tau, n_{\Delta}, L, M\} = A(N-\sigma)(N+\sigma+4)/4 + B\tau(\tau+3)/6 + CL(L+1). \tag{2.5.36}$$

The spectrum of this expression shows repeating patterns of 0^+, 2^+, 4^+, 2^+, ... states, as shown in Fig. 2.47. A comparison of the calculated level scheme of ^{196}Pt with experimental data is presented in Fig. 2.48. The energy spacings between states with the same σ are given by the second term in eqn (2.5.36) when $C=0$, which is identical to the model of Wilets and Jean (1956).

The transition probabilities can be calculated using the operator of eqn (2.5.16). In this limit one may put $\beta_2 = 0$ because the second term is not a generator of O(6). The operator then becomes a generator of O(6) and thus cannot connect different representations of O(6). This leads to the following selection rules

$$\Delta\sigma = 0, \quad \Delta\tau = \pm 1. \tag{2.5.37}$$

The B(E2) values along the GSB with $L' = 2\tau + 2 \rightarrow L = 2\tau$ are given by

$$B(\mathrm{E2}) = \alpha_2^2 (L+2)(2N-L)(2N+L+8)/\{8(L+5)\}. \tag{2.5.38}$$

The quadrupole moments for all states in the representation of O(6) are zero because of the $\Delta\tau = \pm 1$ rule (eqn 2.5.37). In reality, quadrupole moments have

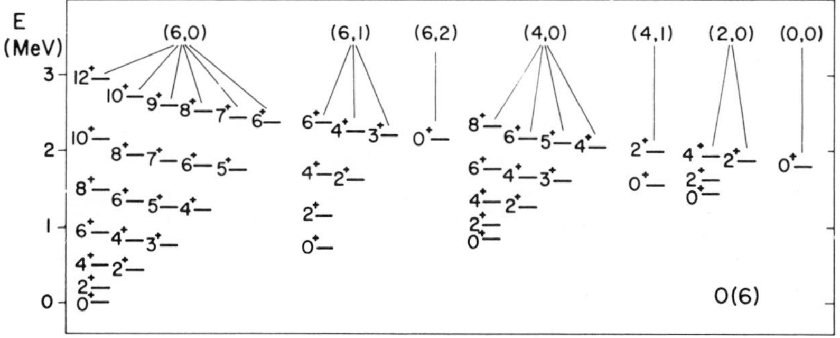

Fig. 2.47. A typical spectrum with O(6) symmetry and $N=6$. The energy levels are given by eqn (2.5.36) with $A = 100$ keV, $B = 240$ keV, $C = 5$ keV. In parenthesis are the values of σ and n_{Δ}. (Arima and Iachello, 1979.)

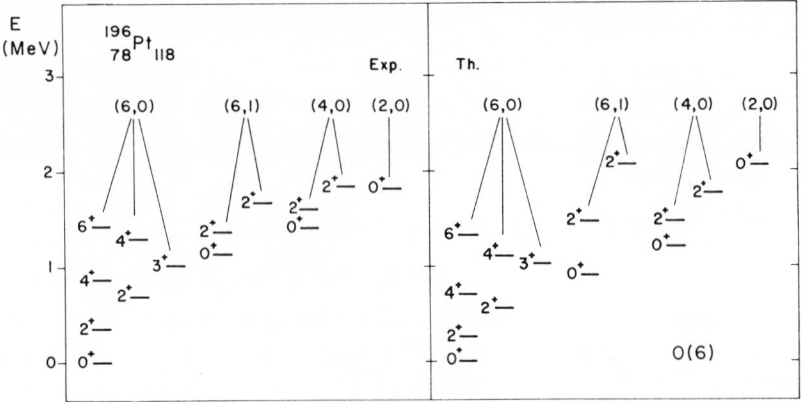

Fig. 2.48. An example of a spectrum with O(6) symmetry: $^{196}_{78}\mathrm{Pt}_{118}$, $N_\pi = 2$, $N_\nu = 4$, $N = 6$. (Iachello, 1980.)

finite values, which can be calculated by taking $\beta_2 \neq 0$ in eqn (2.5.16), which breaks the O(6) symmetry.

2.5.5 *Transitional nuclei*

There are many nuclei which have properties that cannot be described simply within one of the three limiting cases, as discussed above, but are rather intermediate between them. We will discuss here two main cases, namely the nuclei in between the SU(5) and SU(3) limits (Scholten *et al.*, 1978) and those in between SU(3) and O(6).

A combination of the SU(5) and SU(3) Hamiltonians, according to eqns (2.5.7) and 2.5.19), respectively, gives

$$\mathbf{H} = \varepsilon n_{\mathrm{d}} - \kappa \mathbf{Q} \cdot \mathbf{Q} - \kappa' \mathbf{L} \cdot \mathbf{L}. \tag{2.5.39}$$

The vibrational limit is relevant in case of $\varepsilon \gg \kappa$, κ', while the rotational one dominates in case of $\varepsilon \ll \kappa$, κ'. In reality ε changes as the neutron and proton numbers change and accordingly the properties of the low-lying states change from a vibrational to a rotational character. The energy levels in the Sm isotopes with $N = 84$–94, indeed, show the vibrational feature of $E(4^+) \simeq 2E(2^+)$ at the lower N (larger ε) and the rotational behaviour of $E(4^+) \simeq (10/3)E(2^+)$ in the heavier mass region with small ε (see Fig. 2.49). The second 0^+ state is a part of the two-phonon multiplet at low N; its energy thus decreases linearly with N, but increases again when the character completely changes to the SU(3) limit, when it becomes the band-head of the β-band ($\lambda = 2N - 4$, $\mu = 2$, $K = 0$).

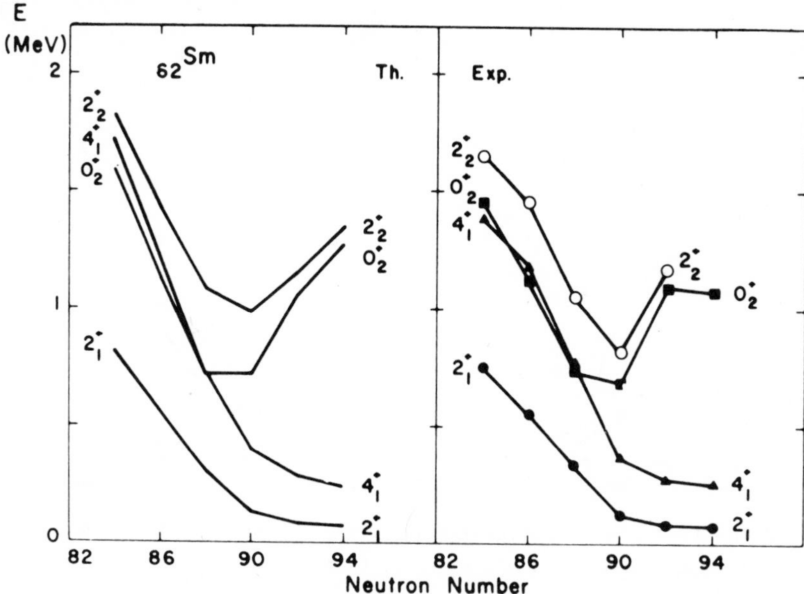

Fig. 2.49. Comparison between the calculated (full line, left) and experimental (points, right) energies of the 2^+_1, 4^+_1, and 0^+_2 states in the Sm isotopes. (Iachello, 1980.)

The changes in the transition probabilities can be treated in a similar way as the energies. The transition rates also clearly illustrate the transition from SU(5) at low N to SU(3) at large N, as shown in Fig. 2.50. Note that $N=82$ is at the closed shell and $N=96$ is already 14 neutrons away from that shell.

The transitional nuclei intermediate between SU(3) and O(6) can be described by a combination of the two Hamiltonians of eqns (2.5.19) and (2.5.31) in the form

$$\mathbf{H}=AP_6+BC_5+CC_3-\kappa\mathbf{Q}\cdot\mathbf{Q}. \tag{2.5.40}$$

For large κ the SU(3) limit is recovered and for small κ the O(6) limit. Examples of such a transition can be found in the $^{186-192}$Os isotopes.

2.5.6 *The interacting boson–fermion model (IBFM); Spin(6) limit*

2.5.6a *The spinor symmetry*

Spinor symmetry arises when boson and fermion group structures are isomorphic; i.e. the elements of the different groups can be mapped onto each other. The three dynamical symmetries, discussed in the previous sections, are associated with the boson groups SU(5), SU(3) and SO(6). Here we shall

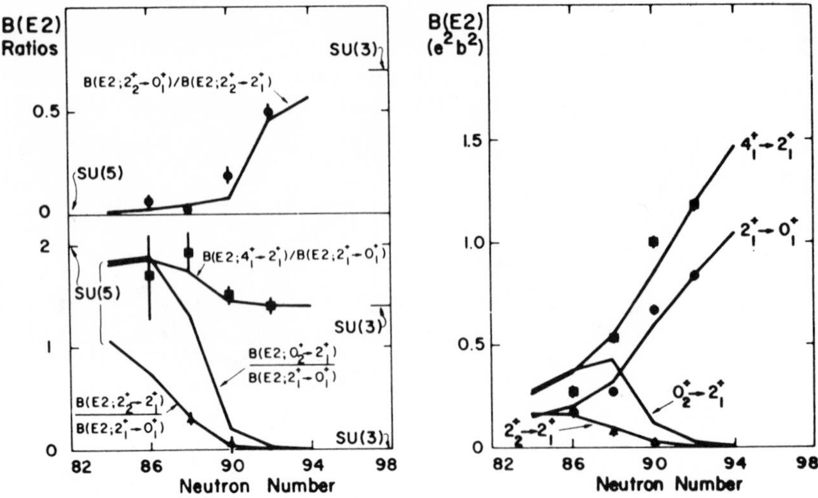

Fig. 2.50. Typical features of the transitional nuclei in between SU(5) at low N and SU(3) at large N: electromagnetic transition rates for Sm isotopes. (Iachello, 1980.)

consider as an example only the spinor symmetry associated with the group O(6), called the Spin(6) limit (Iachello and Kuyucak, 1981).

For the description of odd–even nuclei one has to couple the fermion degrees of freedom to the boson excitations. In addition to the boson operators $(b^+_{\lambda\mu}, \tilde{b}_{lm})$ for the s and d bosons, m fermion operators a_i^+ with $i=1, \ldots m$ are introduced. The number m is given by the sum over all single-particle levels as

$$m=\Sigma_i(2j_i+1). \tag{2.5.41}$$

Here all single-particle levels are considered within a major shell. The maximum number for the $Z=50$–82 shell, for instance, is $m=32$ and for the $d_{3/2}$ level only $m=4$.

The Hamiltonian consists of the boson Hamiltonian $\mathbf{H}^{\mathrm{B}}$ (see eqn (2.5.31) for this limit), the fermion Hamiltonian $\mathbf{H}^{\mathrm{F}}$ and the boson fermion interaction V^{BF}:

$$\mathbf{H}=\mathbf{H}^{\mathrm{B}}+\mathbf{H}^{\mathrm{F}}+\mathrm{V}^{\mathrm{BF}}. \tag{2.5.42}$$

The general algebraic structure of eqn (2.5.42) is that of the group $\mathrm{U}^{\mathrm{B}}(6)*\mathrm{U}^{\mathrm{F}}(m)$. Here $\mathbf{H}^{\mathrm{F}}$ contains only one-body terms with proton and neutron creation and annihilation operators, factorized by the single-particle energies. In general, the eigenvalue problem must be solved numerically. However, in the case where dynamical symmetries exist in the boson–fermion system an analytical solution is possible. In the case of $m=4$ with only a fermion in the $j=3/2$ orbital the group $SU^F(4)$ is isomorphic to $SO^B(6)$, resulting in the spinor symmetry of the spin(6) group.

In order to generate energy spectra of the Spin(6) symmetry we have to find the proper quantum numbers that specify the states. The totally symmetric irreducible representation of $U^B(6)$ is labelled by the number of bosons N and the totally antisymmetric irreducible representation of $U^F(4)$ by the number of fermions M. For SO(6) the label Σ is introduced as

$$\Sigma = N,\ N-2, \ldots 1 \text{ or } 0 \text{ (for } N \text{ odd or even).} \tag{2.5.43}$$

The irreducible representation of Spin(6) has three labels $(\sigma_1, \sigma_2, \sigma_3)$. For the simple cases of $M=0$ or 1 we obtain the following values: For the even–even nuclei, $M=0$

$$\sigma_3=\sigma_2=0, \quad \sigma_1=\Sigma=N,\ N-2, \ldots 1 \text{ or } 0 \text{ (for } N \text{ odd or even).} \tag{2.5.44}$$

For the odd–even nuclei, $M=1$ the numbers are

$$|\sigma_3|=1/2, \quad \sigma_2=1/2, \quad \sigma_1=N+1/2,\ N-1/2,\ N-3/2, \ldots 1/2. \tag{2.5.45}$$

Two additional labels (τ_1, τ_2) are given as

$$\tau_2=0, \qquad \tau_1=\sigma_1,\ \sigma_1-1, \ldots, 0 \qquad (\text{for } M=0), \tag{2.5.46}$$

$$\tau_2=1/2, \qquad \tau_1=\sigma_1,\ \sigma_1-1, \ldots, 1/2 \qquad (\text{for } M=0). \tag{2.5.47}$$

The spin J of a state is then obtained from the quantum numbers (τ_1, τ_2) and from a label n_Δ

$$n_\Delta=0,\ 1,\ 2, \ldots \qquad \text{for } M=0, \tag{2.5.48}$$

$$n_\Delta=0,\ 1/2,\ 1,\ 3/2, \ldots \qquad \text{for } M=1. \tag{2.5.49}$$

The spins of the states for even–even nuclei ($M=0$) are given by

$$J=(2\tau_1-3n_\Delta),\ (2\tau_1-3n_\Delta-2),\ (2\tau_1-3n_\Delta-3), \ldots, (\tau_1-3n_\Delta+1),\ (\tau_1-3n_\Delta), \tag{2.5.50}$$

and for odd–even nuclei with $M=1$ by

$$J=\{2(\tau_1-\tau_2)-6n_\Delta+3/2\}, \{2(\tau_1-\tau_2)-6n_\Delta+1/2\}, \ldots$$
$$[(\tau_1-\tau_2)-3n_\Delta-1/4\{1-(-)^{2n_\Delta}\}+3/2]. \tag{2.5.51}$$

The excitation energies can now be expressed in closed form in terms of those quantum numbers as

$$E\{N,\ M,\ \Sigma, (\sigma_1,\sigma_2,\sigma_3), (\tau_1,\tau_2),\ n_\Delta,\ J,\ M_J\} = (A_1/4)\Sigma(\Sigma+4) - (A_2/4)\{\sigma_1(\sigma_1+4)$$
$$+\sigma_2(\sigma_2+2)+\sigma_3^2\} + (B/6)\{\tau_1(\tau_1+3)$$
$$+\tau_2(\tau_2+1)\} + CJ(J+1). \tag{2.5.52}$$

The spectrum of ^{191}Ir provides a nice example of spinor symmetry as shown in Fig. 2.51. In the comparison between experiment and theory only those states are shown which belong to the lowest Spin(6) representation. Therefore only two parameters, B and C, are needed to describe these states because the other two parameters A_1 and A_2 determine the splitting between the different $(\sigma_1, \sigma_2, \sigma_3)$ multiplets of Spin(6). The specific relations between the three terms of the Hamiltonian in eqn (2.5.42) needed for spinor symmetry have been discussed by Iachello and Kuyucak (1981).

2.5.6b *Supersymmetry*

Supersymmetry occurs when nuclear states in both even–even and odd–even nuclei can be described in a single framework. This is a rather special situation since most symmetries commonly found in nature apply either to a system of bosons or to a system of fermions. It has been shown by Balantekin *et al.* (1981) that in some nuclei, particularly in the Os–Pt region, bosonic and fermionic states can be placed in one and the same multiplet. The symmetry operations within such a supermultiplet transform bosons into fermions and vice versa. A supermultiplet consists of states belonging to a group of nuclei with a fixed number of bosons plus fermions $N+M$.

The parameters B and C in eqn (2.5.52), obtained in a fit to the spectrum of the odd–even nucleus ^{191}Ir turn out to be very close to the values obtained from fits to the adjacent even–even nuclei. This suggests that those nuclei in that mass region belong to the same multiplet of a group larger than

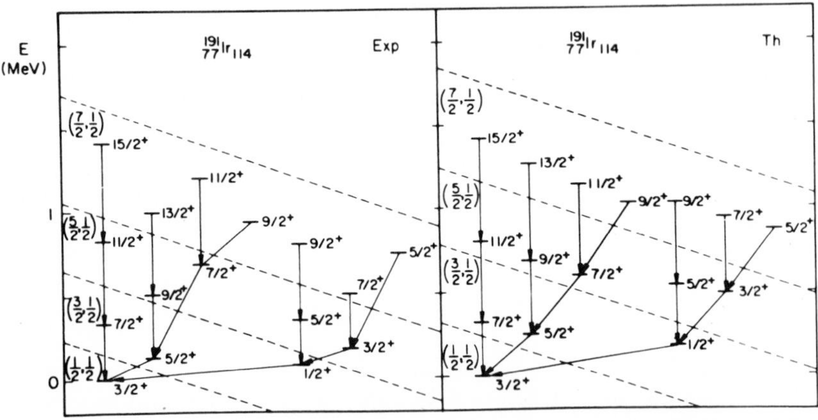

Fig. 2.51. An example of a spectrum with Spin(6) symmetry: $^{191}_{77}\mathrm{Ir}_{114}$, $(N=8, M=1)$. The energy levels in the theoretical spectrum are calculated using eqn (2.5.52) with $\frac{1}{6}B=40$ keV, $C=14$ keV. All states belong to the representation $(\sigma_1, \sigma_2, \sigma_3)=(\frac{17}{2}, \frac{1}{2}, \frac{1}{2})$ of Spin (6). The labels on the left of the spectrum are (τ_1, τ_2). (Iachello and Kuyucak, 1981.)

$U^B(6) * U^F(4)$, discussed above in Section 2.5.6a, and called a supergroup. The low-lying collective states in even–even nuclei ($M=0$) in the Os–Ir region are known to exhibit $SO^B(6)$ symmetry. The odd proton in the adjacent odd–even nuclei occupies the single-particle $2d_{3/2}$ level, thus forming the $U^F(4)$ group. As discussed above, the boson and fermion groups are then combined into a spinor Spin(6) group. The appropriate supergroup is denoted by O(6/4). Families of states can thus be built in the above-mentioned isotopes with supersymmetry properties. An example of such a family is given in Fig. 2.52.

The excitation energies in the supersymmetry scheme is given by eqn (2.5.52). For the ground-state configurations of even–even nuclei we have; $M=0$, $\sigma_2=\sigma_3=\tau_2=0$ and $\Sigma=\sigma_1=\sigma$ (with $\sigma=N, N-2, \ldots$), $\tau_1=\tau$. In this case eqn (2.5.52) can be simplified to

$$E(\text{even–even}) = -(1/4)(A_1+A_2)\sigma(\sigma+4) + (1/6)B\tau(\tau+3) + CJ(J+1). \tag{2.5.53}$$

The spectra of the pair of nuclei ^{190}Os ($Z=76$) and ^{191}Pt ($Z=77$), as deduced from experiment, are compared with the results of the supersymmetry scheme of eqns (2.5.52) and (2.5.53) in Fig. 2.53. All the states shown in the figure belong to the maximum allowed representation of Spin(6), namely with $\sigma=N$ in the even–even nucleus and $\sigma=N+1/2$ in the odd–even nucleus. The only

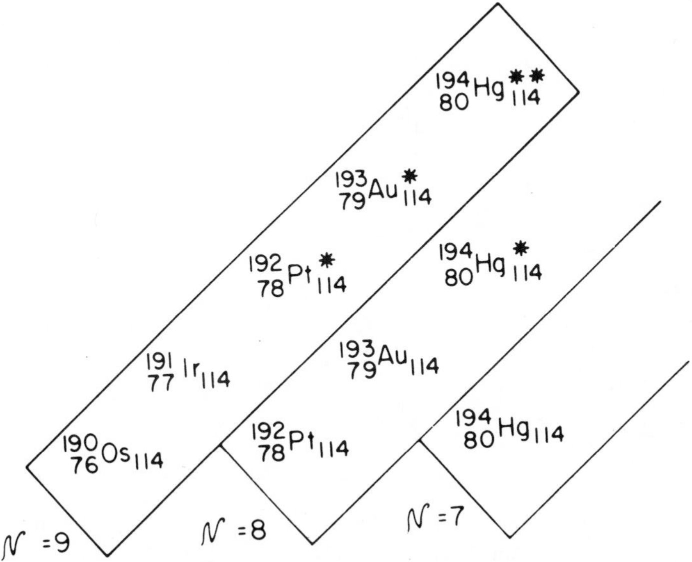

Fig. 2.52. Possible supersymmetric multiplets in the Os–Pt region. N denotes the total number of bosons plus fermions. Excited configurations are labelled by one or two stars. (Balantekin *et al.*, 1981.)

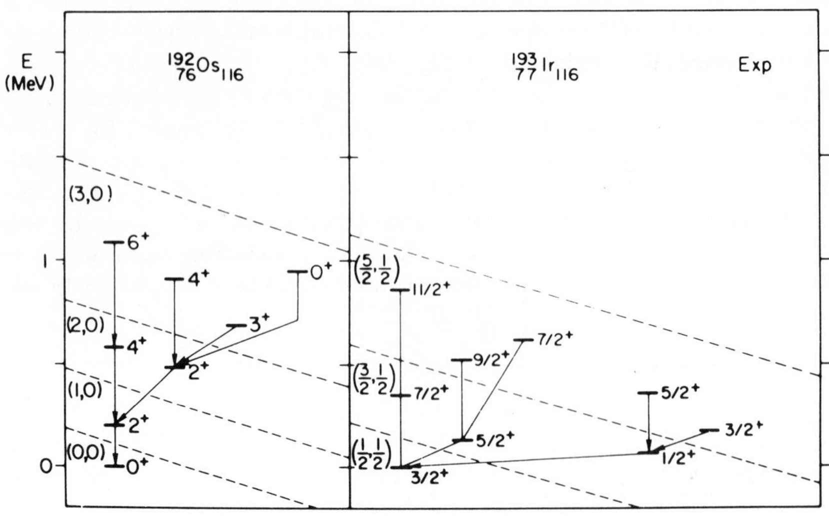

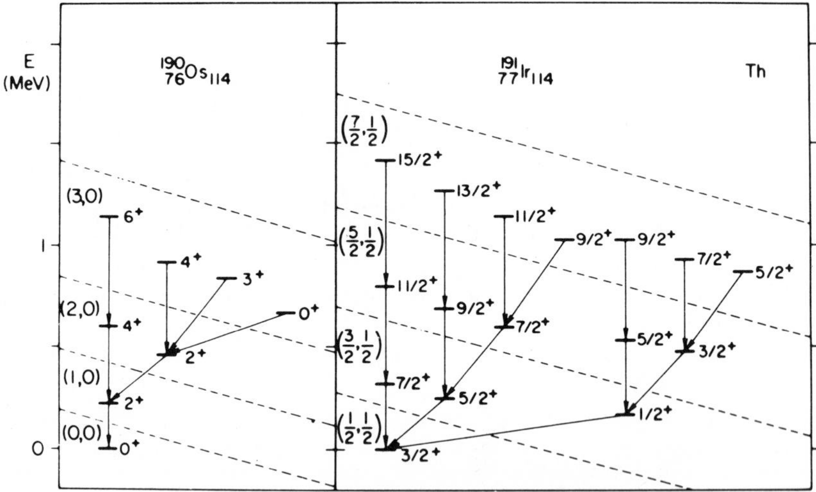

Fig. 2.53. Experimental (above) and theoretical spectra (below) of the pair of nuclei $^{190}_{76}Os_{114}-^{191}_{77}Ir_{114}$ obtained using eqns (2.5.52) and (2.5.53) with $\frac{1}{6}B=40$ keV, $C=10$ keV for both nuclei. The numbers in parentheses denote the Spin(5) labels (τ_1, τ_2). (Balantekin *et al.*, 1981.)

two parameters B and C, needed to describe the spectra of the even–even and odd–even nuclei simultaneously, appear to be rather constant for systems with $N+M=8$ and 9. This may indicate an even richer symmetry in this mass region than that of U(6/4).

2.5.7 *The interacting proton–neutron boson (-fermion) approximation*

2.5.7a *The IBA-2 and IBFA-2*

It is well-known that the proton–neutron quadrupole interaction is most important for low-lying collective states in even–even nuclei. In odd–even nuclei the odd proton interacts with the neutrons in the even–even core and vice versa. In the IBA-2 version of the IBM the proton–neutron interaction is included in the boson Hamiltonian (Otsuka *et al.*, 1978; Iachello *et al.*, 1979). With the IBFA-2 the neutron–proton effects in odd–even nuclei can be studied by comparing the nuclei with an odd proton and an odd neutron with the same even–even core (Bijker, 1984). In this way the microscopic theories of collective states can be tested because the parameters appearing in the interacting proton–neutron Hamiltonian are related to the nucleon–nucleon interaction.

The proton–neutron boson Hamiltonian $\mathbf{H}^{\mathrm{B}}$ is written as

$$\mathbf{H}^{\mathrm{B}}=\varepsilon_{\nu}n_{\mathrm{d},\nu}+\varepsilon_{\pi}n_{\mathrm{d},\pi}+\kappa\mathbf{Q}_{\nu}^{(2)}\cdot\kappa'\mathbf{Q}_{\pi}^{(2)}+V_{\nu\nu}+V_{\pi\pi}+M_{\pi\nu}. \qquad (2.5.54)$$

The various terms are specified for protons (π) and neutrons (ν) separately, and the last three terms are added to account for the remaining π–π, ν–ν, and π–ν fermion interactions. The Hamiltonian in eqn (2.5.54) represents the IBA-2 version of IBM. The last term $M_{\pi\nu}$ represents the Majorana force, which may be the most important term in some cases. The nucleon–nucleon interaction (Wigner type attractive two-body interaction) gives rise to a strong pairing component in the identical particle channel which determines ε, taken for simplicity as $\varepsilon=\varepsilon_{\pi}=\varepsilon_{\nu}$. The strong quadrupole component in the non-identical particle channel determines κ and the proton and neutron parameters in $Q^{(2)}$. The calculated energy spectra of low-lying states in the even–even Xe isotopes are compared with experimental data in Fig. 2.54.

The description of odd–even nuclei can be given in terms of the IBFA-2 by adding to $\mathbf{H}^{\mathrm{B}}$ the fermion Hamiltonian $\mathbf{H}^{\mathrm{F}}$ and the boson–fermion interaction term V^{BF} as in eqn (2.5.42). The energy spectra of odd–even nuclei are largely determined by the (three) parameters in V^{BF}. The results of a calculation for the negative parity states of the odd–A Xe isotopes are compared with experiment in Fig. 2.55 (Bijker, 1984).

2.5.7b *F-spin*

The concept of F-spin, introduced by Arima *et al.* (1977), accounts explicitly for the different constituents of the nucleus, the protons and the neutrons. The

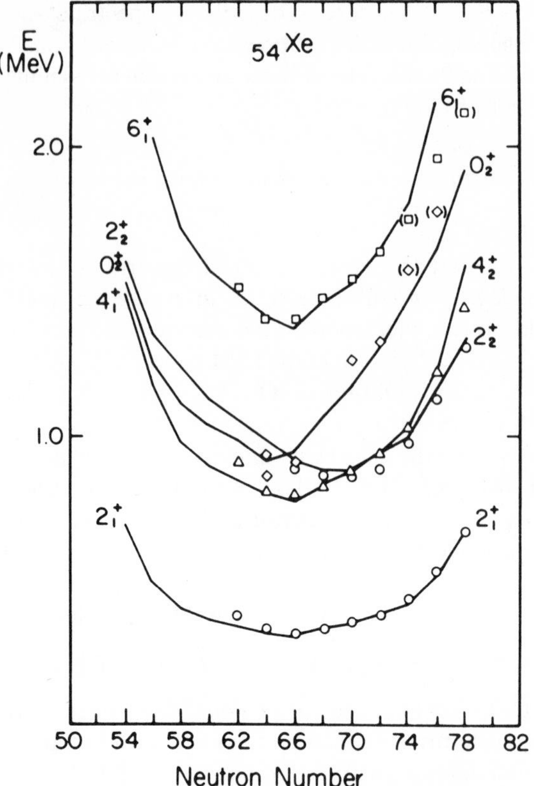

Fig. 2.54. Calculated energy spectra in the even–even $_{54}$Xe isotopes. The circles, squares, triangles are the experimental values. (Iachello *et al.*, 1979.)

F-spin plays the same role in boson systems as the isospin does in fermion systems. It is defined in terms of the number of protons N_π and the number of neutrons N_ν outside an assumed closed shell.

$$F_{\max} = (N_\pi + N_\nu)/2, \tag{2.5.55}$$

$$F_0 = (N_\pi - N_\nu)/2. \tag{2.5.56}$$

$F_{\max}$ is the total number of active bosons in even nuclei. The component F_0 of F is thus the minimum possible value of F. F-spin multiplets are defined as a series of nuclei which have the same total boson number (the same $F_{\max}$), but which differ in the number of proton and neutron bosons (different F_0). In this sense F-spin is analogous to isospin, but not identical to it because the F-spin multiplet does not contain $2F+1$ nuclei. This limitation is due to the fact that the particle-type bosons change to hole-type bosons at midshell, thus

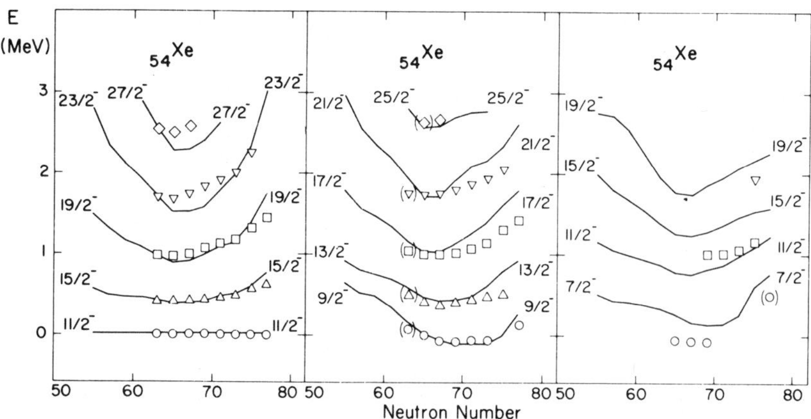

Fig. 2.55. Comparison between experimental and calculated negative parity spectra in the odd-A Xe isotopes. All states are plotted relative to the $11/2^-$ state. (Bijker, 1984.)

changing the counting of the number of bosons. A search for F-spin multiplets would thus reveal to which extent the nuclear structure is determined by the neutron and proton boson numbers. This can be the case for only a limited number of chains of nuclei. Moreover, it seems necessary to define precisely the number of active or valence bosons outside a closed shell. Conversely, the discovery of a chain of nuclei with similar spectra might reveal the relevant closed shells.

A nice example of F-spin multiplets in the rare-earth nuclei is shown in Fig. 2.56. Those multiplets have been analysed by von Brentano *et al.* (1985) in terms of IBA-2 and a projected IBM approach. It was found that the remarkable constancy of the energies in the ground bands is due to the rather constant product $N_p N_n$. As a consequence the parameters in the IBA-2 Hamiltonian must remain constant within the multiplet. The role of the product $N_\pi N_\nu$ is further discussed in the next section.

2.5.8 *Boson-number systematic from experimental data*

The structure of nuclei is largely determined by the number of nucleons, in particular by the number of protons with respect to the number of neutrons. It has been believed for a long time that the rigidity to deformation of singly magic nuclei is due to the proton–neutron interaction (Talmi, 1962). In case of valence nucleons of only one kind the proton–neutron (π–ν) interaction is very weak, and accordingly deformations disappear. A measure of the π–ν interaction strength is the number of valence fermions N_π, N_ν or the number

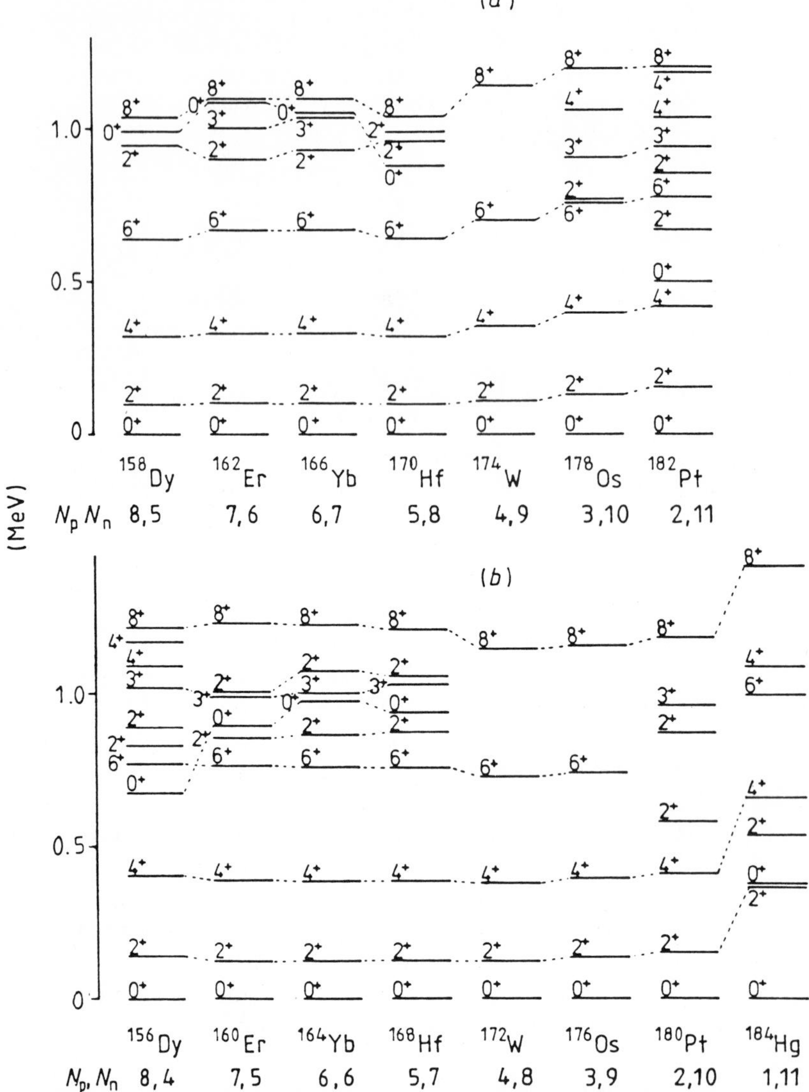

Fig. 2.56. F-spin multiplets with (a) $F=13/2$ and (b) $F=6$. N_p and N_n are the proton and neutron boson numbers. The levels are shown of the ground band and of the quasi-beta-band up to an excitation energy around 1 MeV. (von Brentano *et al.*, 1985.)

of valence bosons N_p, N_n. This is clearly demonstrated by examining the properties of low-lying states as a function of the product of the proton and neutron numbers $N_pN_n=4N_\pi N_\nu$, where $N_p(N_n)$ and $N_\pi(N_\nu)$ refer to the numbers of valence bosons and fermions, respectively (Casten, 1985).

We can determine experimentally a number of quantities which characterize the structure of nuclear states. For instance, the ratio of the excitation energies of the 4^+ and the 2^+ states is in general <2 for few-particle (shell-model) states, ~ 2 for SU(5) states (vibrator), ~ 2.5 for O(6) states (γ unstable rotor) and ~ 3.3 for SU(3) states (axially symmetric rotor). Also the excitation energy of the 2^+ state is indicative; i.e. ~ 1 MeV for closed shell nuclei and ~ 100 keV for deformed nuclei. The ratios $E(4^+)/E(2^+)$ are plotted for some nuclei in the mass region $A = 100$ in Fig. 2.57 as functions of Z or N and of the products of the boson and fermion numbers. The left-hand sides of the figure shows that particularly the Sr ($Z = 38$) and Zr ($Z = 40$) isotopes exhibit a phase transition from SU(5) for low N to SU(3) for the heavier isotopes. The curves representing the functions of Z or N are obviously very different for each isotope. In contrast with this complicated behaviour is the single curve representing all those data as a function of the product $N_\pi N_\nu$. Such a curve can therefore be called a universal curve (Casten, 1985).

The phase transitions from spherical to deformed shapes for proton-closed ($Z = 50$) nuclei in the mass region $A = 100$ is due to the proton–neutron interaction between the "spin–orbit partners" (SOP); the $g_{7/2}$ neutrons ($l = 4$, spin down) and the $g_{9/2}$ protons ($l = 4$, spin up). The overlap between the two orbitals is then very large and thus the interaction very strong. Such a situation may be created when for instance the neutrons fill the $g_{7/2}$ shell, which lowers the energy of the $\pi g_{9/2}$ level and causes the disappearance of the singly-magic character of the $Z = 38$ shell. Protons can then easily be excited

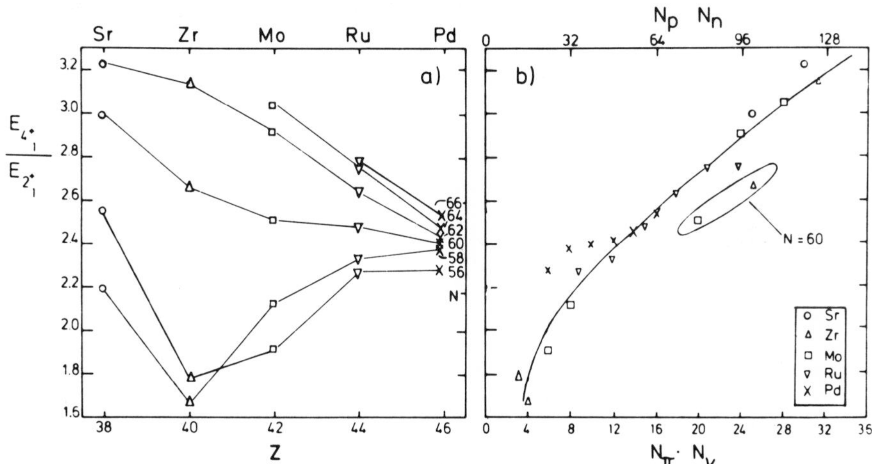

Fig. 2.57. The $E_{4_1^+}/E_{2_1^+}$ systematics for the $A = 100$ region. In part (b) the valence nucleon product is given both in terms of a fermion product $N_p N_n$ and the equivalent boson product $N_\pi N_\nu$. The data points corresponding to $N_p N_n = 0$ are omitted. (Casten, 1985.)

to the $g_{9/2}$ level and constitute SOP with the $g_{7/2}$ neutrons. This seems also the mechanism that can explain the occurrence of intruder states—see e.g. Heyde *et al.*, 1983.

In the construction of the universal curve it is essential to locate the proper closed shell, which determines the number of valence particles. For instance Zr has two $g_{9/2}$ protons outside the closed $Z = 38$ shell in the case of low neutron numbers, $N < 60$. However, when the number of valence neutrons increase, $N \geqslant 60$, the singly magic shell $Z = 38$ disappears and one has to count the protons with respect to the $Z = 28$ shell. In that case the number of protons has passed the middle of the $Z = 28$–50 shell and one has to take 10 proton holes with respect to the $Z = 50$ shell. The data points in the Fig. 2.57 enclosed by an ellipse are calculated with respect to the ordinary ($Z = 38$) shell, but they shift to the universal curve when the modification of the shell structure is taken into account. These $N_p N_n$ plots thus provide a remarkable simplification of the complex nuclear transitional regions. Similar curves have been constructed in other mass regions; i.e around $A = 130$ where a smooth transition U(5) → O(6) → SU(3) occurs, in the transitional region $A = 150$ with the disappearance of the $Z = 64$ gap, and around $A = 190$ with the SU(3) → O(6) transition modifying the neutron gap to $N = 114$ (Casten, 1985).

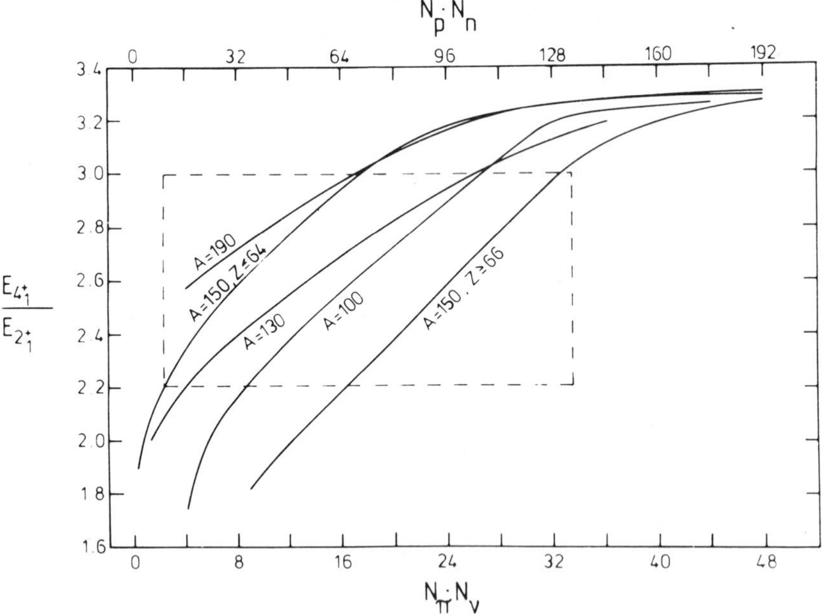

Fig. 2.58. Plot of five smooth curves through the $E_{4_1^+}/E_{2_1^+}$ data representing data in the mass regions indicated, on the same scale. The dashed box encloses the crucial region in $E_{4_1^+}/E_{2_1^+}$ between 2.2 and 3.0. (Casten, 1985.)

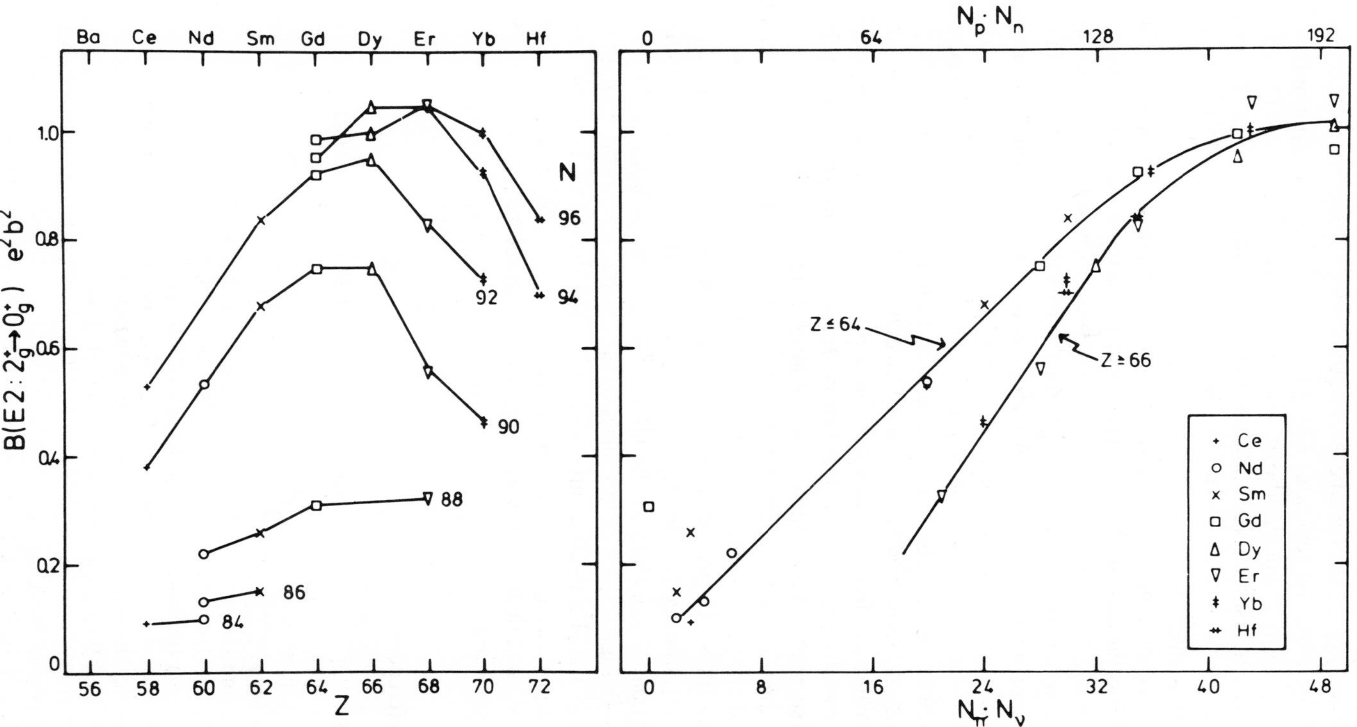

Fig. 2.59. $B(\mathrm{E2}\colon 2_g^+ \to 0_g^+)$ systematics for the $A = 150$ regions. (Casten, 1985.)

A summary of the universal curves in the various mass regions is presented in Fig. 2.58. Note the region of $E(4^+)/E(2^+)=2.2$–3.0, corresponding to the phase transitions discussed above. The smooth representation of many data on a few universal curves indicate that the proton–neutron interaction provides a simple mechanism for the driving force towards deformation if it is caused by a single process.

A very sensitive measure for deformation is the transition probability B(E2). The B(E2) values for the $2^+ \rightarrow 0^+$ GSB transitions in the rare-earth region are plotted in Fig. 2.59 versus Z and versus the product number $N_\pi N_\nu$. In the left figure we note the transition from the spherical to the deformed domain when the neutron number increases. All those data points fall on two universal curves, one for $Z \leqslant 64$ and one for $Z \geqslant 66$, which illustrates the significance of the $Z = 64$ gap and its disappearance for the heavier rare-earth isotopes. A similar universal curve is shown in Fig. 2.22 (Raman and Nestor, 1986), which represents the observed deformation parameters β_2 in units of the single-particle value as a function of $N_p N_n$, following an idea of Hammamoto (1965). It is indeed striking that the data (for both energies and transition probabilities) of so many nuclei show such a smooth behaviour.

We hope that we have illustrated in this section the importance of gamma and electron spectroscopy for the investigation of the nuclear structure of the low-lying states. To provide a rigorous test-ground for nuclear models, such as the IBM, one needs more details of the lateral bands such as the interband and intraband transition probabilities. The advent of large multi-detector arrays with high resolution promises to satisfy this need in the coming years. To spread the effort sufficiently, so as to cover the whole periodic table, it seems important that many (including small) groups collaborate on an international scale at the several large nuclear-structure facilities available in various countries.

2.6 References

Adams, G., Bilpuch, E. G., Mitchell, G. E., Nelson, R. O. and Westerfeldt, C. R. (1984). *J. Phys.* **G10,** 1747.

Andrejtscheff, W., Schilling, K. D. and Manfrass, P. (1975). *Atomic Data and Nucl. Data Tables* **16,** 515.

Arima, A. and Horie, H. (1954). *Progr. theor. Phys., Kyoto* **11,** 509.

Arima, A. and Iachello, F. (1975). *Phys. Rev. Lett.* **35,** 1069.

Arima, A. and Iachello, F. (1976). *Ann. Phys.* **99,** 253.

Arima, A., Otsuka, T., Iachello, F. and Talmi, I. (1977). *Phys. Lett.* **66B,** 205.

Arima, A. and Iachello, F. (1978) *Ann. Phys.* **111,** 201.

Arima, A. and Iachello, F. (1979). *Ann. Phys.* **123,** 468.

Arima, A. (1983). *Progress in particle and nuclear physics* (ed. D. Wilkinson), Vol. 9, p. 51. Pergamon Press, Oxford.

Balantekin, A. B., Bars, I. and Iachello, F. (1981). *Nucl. Phys.* **A370,** 284.

Band, I. M., Listengarten, M. A., Sliv, L. A. and Thun, J. E. (1965). *Alpha-, Beta-, and Gamma-ray Spectroscopy* (ed. K. Siegbahn), p. 1683. North-Holland, Amsterdam.

Belyaev, S. T. and Zelevinski, Z. G. (1962). *Nucl. Phys.* **39,** 582.

Biedenharn, L. C. and Rose, M. E. (1953). *Rev. mod. Phys.* **25,** 729.

Bijker, R. (1984). Thesis, University of Groningen, The Netherlands.

Björnstad *et al.*, (1986). *Nucl. Phys.* **A453,** 463.

Blomqvist, J. (1981). *Proc. Int. Conf. on Nuclei far from Stability.* Helsingör, CERN 81-09.

Boerma, D. O., Arends, A. R., Endt, P. M., Grüebler, W., König, V., Schmelzbach, P. A. and Risler, R. (1986). *Nucl. Phys.* **A449,** 187.

Bohle, D., Richter, A., Steffen, W., Diperink, A. E. L., LoIudice, N., Palumbo, F. and Scholten, O. (1984). *Phys. Lett.* **137B,** 27.

Bohle, D., Küchler, G., Richter, A. and Steffen, W. (1984). *Phys. Lett.* **148B,** 260.

Bohr, A. and Mottelson, B. (1953). *K. Dan. Vidensk. Selsk. Mat. Fys. Medd.* **27,** No. 16, 1.

Bohr, A. and Mottelson, B. (1969, 1975). *Nuclear Strucutre*, Vols 1 and 2, Benjamin Press, New York.

Bonatsos, D. and Klein, A. (1984). *Atomic Data and Nucl. Data Tables* **30,** 27.

Brentano von, P., Gelberg, A., Harter, H. and Sala, P. (1985). *J. Phys. G.* **11,** L85.

Broda, R., Kleinheinz, P., Lunardi, S., Styczen, J. and Blomqvist, J. (1983). KFA Annual Report 1982, Jülich. p. 50.

Bron, J., Hesselink, W. H. A., van Poelgeest, A., Zalmstra, J. J. A., Uitzinger, M. J. and Verheul, H. (1979). *Nucl. Phys.* **A318,** 335.

Browne, E. and Femenia, F. R. (1971). *Atomic Data and Nucl. Data Tables* **10,** 81.

Brussaard, P. and Glaudemans, P. W. M. (1977). *Shell-model Applications in Nuclear Spectroscopy.* North-Holland, Amsterdam.

Carr, R. W. and Baglin, J. E. E. (1971). *Nucl. Data Tables,* **10,** 143.

Casten, R. F. (1985). *Nucl. Phys.* **A443,** 1.

Church, E. L. and Weneser, J. (1956). *Phys. Rev.* **104,** 1382.

Das, T. K., Dreizler, R. M. and Klein, A. (1970). *Phys. Rev.* **C2,** 632.

DerMateosian, E. and Sunyar, A. W. (1974). *Atomic Data and Nucl. Data Tables* **13,** 391.

DeShalit, A. and Talmi, A. (1963). *Nuclear Shell Theory.* Academic Press, New York.

DeShalit, A. and Feshbach, H. (1974). *Theoretical Nuclear Physics*, Ch. VIII. John Wiley, New York.

Djalali, C., Marty, N., Morlet, M., Willis, A., Jourdain, J. C., Bohle, D., Hartmann, V., Küchler, G., Richter, A., Caskey, G., Crawley, G. M. and Galonsky, A. (1985). *Phys. Lett.* **164B,** 269.

Dragoun, O., Pauli, H. C. and Schmutzler, F. (1969). *Nucl. Data Tables* **A6,** 235.

Dragoun, O., Plajner, Z. and Schmutzler, F. (1971). *Nucl. Data Tables* **A9,** 119.

Dudek, J., Majhofer, A., Skalski, J., Werner, T., Cwiok, S. and Nazarewicz, W. (1979). *J. Phys.* **G5,** 1359.

Duppen Van, P., Coenen, E., Deneffe, K., Huyse, M., Heyde, K. and van Isacker, P. (1984). *Phys. Rev. Lett.* **52,** 1974.

Ejiri, H. (1964). *Nucl. Phys.* **52,** 561, 578.

Ejiri, H. (1966, 1967). Institute for Nuclear Studies Journal (Tokyo) INSJ-101, and INSJ-104.

Ejiri, H., Ishihara, M., Sakai, M., Katori, K. and Inamura, T. (1968). *J. Phys. Soc. Japan* **24,** 1189.

Ejiri, H., Ikeda, K. and Fujita, J. I. (1968). *Phys. Rev.* **176,** 1277.
Ejiri, H. and Hagemann, G. B. (1971). *Nucl. Phys.* **A161,** 449.
Ejiri, H., Shibata, T. and Fujiwara, M. (1973). *Phys. Rev.* **C8,** 1892.
Endt, P. M. and Van der Leun, C. (1974). *Atomic Data and Nucl. Data Tables* **13,** 67.
Endt, P. M. (1979a). *Atomic Data and Nucl. Data Tables* **23,** 3.
Endt, P. M. (1979b). *Atomic Data and Nucl. Data Tables* **23,** 547.
Endt, P. M. (1981). *Atomic Data and Nucl. Data Tables* **26,** 47.
Endt, P. M., de Wit, P. and Alderliesten, C. (1986, 1988). *Nucl. Phys.* **A459,** 61; **A476,** 333.
Elliott, J. P. (1958). *Proc. R. Soc. Ser.* **A245,** 128, 562.
Ferguson, A. J. (1965). *Angular Correlation Methods in γ-ray Spectroscopy*. North Holland, Amsterdam.
Fielding, H. W., Anderson, R. E., Zafiratos, C. D., Lind, D. A., Cecil, F. E., Weiman, H. H. and Alford, W. P. (1977). *Nucl. Phys.* **A281,** 389.
Frauenfelder, H., Steffen, R. M., De Groot, S. R., Tolhoek, H. A. and Huiskamp, W. J. (1965). *Alpha-, Beta- and Gamma-ray Spectroscopy* (ed. K. Siegbahn). North-Holland, Amsterdam.
Hager, R. S. and Seltzer, E. C. (1968). *Nucl. Data Tables* **A4,** 1; **A4,** 397.
Hager, R. S. and Seltzer, E. C. (1969). *Nucl. Data Tables* **A6,** 1.
Hamamoto, I. (1965). *Nucl. Phys.* **73,** 225.
Hamamoto, I. (1969). *Nucl. Phys.* **A126,** 545.
Harris, S. M. (1965). *Phys. Rev.* **138,** B509.
Häusser, O., Khanna, F. C. and Ward, D. (1972). *Nucl. Phys.* **A194,** 113.
Hayakawa, S. I., Fujiwara, M., Imanishi, S., Fujita, Y., Katayama, I., Morinobu, S., Yamazaki, T., Itahashi, T. and Ikegami, H. (1982). *Phys. Rev. Lett.* **49,** 1624.
Heyde, K., van Isacker, P., Waroquier, M., Wood, J. L. and Meyer, R. A. (1983). *Phys. Reports* **102,** 291.
Holzwarth, G. and Lie, S. G. (1975). *Phys. Rev.* **C12,** 1035.
Iachello, F., Puddu, O., Scholten, O., Arima, A. and Otsuka, T. (1979). *Phys. Lett.* **89B,** 1.
Iachello, F. (1980). Group Theory and Nuclear Spectroscopy, in: Lecture Notes in Physics, *Nuclear Spectroscopy*, Vol. 119. Springer, Berlin.
Iachello, F. (1981). In *Nuclear Structure* (ed. K. Abrahams, K. Allaart and A. E. L. Dieperink), p. 53. Plenum.
Iachello, F. and Kuyucak, S. (1981). *Ann. Phys.* **136,** 19.
Ikegami, H. and Udagawa, T. (1964). *Phys. Rev.* **133,** B1388.
Isacker, P. van, Waroquier, M., Vinx, H. and Heyde, K. (1977). *Nucl. Phys.* **A292,** 125.
Jansen, D., Jolos, R. V. and Dönau, F. (1974). *Nucl. Phys.* **A224,** 93.
Kantele, J. and Tannila, O. (1968). *Nucl. Data Tables* **A4,** 359.
Kishimoto, T. and Tamura, T. (1972). *Nucl. Phys.* **A192,** 246.
Kisslinger, L. S. and Sorensen, R. A. (1963). *Rev. mod. Phys.* **35,** 853.
Klein, A. (1980). *Phys. Lett.* **93B,** 1; *Nucl. Phys.* **A347,** 3.
Kleinheinz, P., Ogawa, M., Broda, R., Daly, P. J., Haenni, D., Beuscher, H. and Kleinrahm, A. (1978). *Z. Phys.* **A286,** 27.
Kleinheinz, P., Broda, R., Daly, P. J., Lunardi, S., Ogawa, S. and Blomqvist, J. (1979). *Z. Phys.* **A290,** 279.
Konecny, E., Specht, H. J., Weber, J., Weigmann, H., Ferguson, R. L., Osterman, P., Waldschmidt, W. and Siegert, G. (1972). *Nucl. Phys.* **A187,** 426.

Konijn, J., de Boer, F. W. N., van Poelgeest, A., Hesselink, W. H. A., de Voigt, M. J. A. and Verheul, H. (1981). *Nucl. Phys.* **A352,** 191.

Krane, K. S. (1975; 1976; 1977a; 1977b; and 1978). *Atomic Data and Nuclear Data Tables* **16,** 383; **18,** 137; **19,** 363; **20,** 211; and **22,** 269, respectively.

Krpić, D. K., Savić, I. M. and Aničin, I. V. (1976). *Atomic Data and Nucl. Data Tables* **18,** 509.

Kuriyama, A., Marumori, T. and Matsuyanagi, K. (1974). *Progr. theor. Phys.* **51,** 779.

Laszewski, R. M. and Holt, R. J. (1977). *Atomic Data and Nucl. Data Tables* **19,** 305.

Lederer, C. M. and Shirley, V. S. (1978). *Tables of Isotopes*, p. 58. John Wiley, New York.

Lunardi, S., Ogawa, M., Meier, M. R. and Kleinheinz, P. (1980). KFA Annual Report 1979, Jülich, p. 51.

Mante, R. E., D'Amato, D. P. and Blatt, S. L. (1974). *Atomic Data and Nucl. Tables* **13,** 499.

Mariscotti, M. A. J., Scharff-Goldhaber, G. and Buck, B. (1969). *Phys. Rev.* **178,** 1864.

Marmier, P. and Sheldon, E. (1969). *Physics of Nuclei and Particles.* Academic Press, New York.

Marshalek, E. R. (1974). *Nucl. Phys.* **A224,** 245.

Marumori, T., Yamamura, M. and Tokunaga, A. (1964). *Progr. theor. Phys.* **31,** 1009.

Monahan, J. E., Holt, R. J. and Laszewski, R. M. (1979). *Atomic Data and Nucl. Tables* **23,** 97.

Moszkowski, S. A. (1965). *Alpha-, Beta- and Gamma-ray Spectroscopy* (ed. K. Siegbahn), p.863. North-Holland, Amsterdam.

Nyakó, B. M., Cresswell, J. R., Forsyth, P. D., Howe, D., Nolan, P. J., Riley, M. A., Sharpey-Schafer, J. F., Simpson, J. and Ward, N. J. (1984). *Phys. Rev. Lett.* **52,** 507.

Otsuka, T., Arima, A., Iachello, F. and Talmi, I. (1978). *Phys. Lett.* **76B,** 139.

Penninga, J., Hesselink, W. H. A., Balanda, A., Stolk, A., Verheul, H., van Klinken, J., Riezebos, H. J. and de Voigt, M. J. A. (1987). *Nucl. Phys.* **A471,** 535. See also Penninga, J. (1986). Thesis, Free University, Amsterdam.

Piiparinen, M., Komppa, T., Komu, R. and Pakkanen, A. (1982). *Z. Phys. A, Atoms and Nuclei* **309,** 87.

Piiparinen, M., Ogawa, M., de Angelis, G., Kleinheinz, P. and Blomqvist, J. (1986). *Proc. Int. Conf. Nucl. Phys.*, Hasrogate, UK, 25–30 August 1986, B141.

Polikanov, S. M., Druin, V. A., Karnaukhov, V. A., Mikheev, V. L., Pleve, A. A., Skobelev, N. K., Subbotin, V. G., Ter-Akopyan, G. M. and Fomichev, V. A. (1962). *JETP* (*Soviet Phys.*), **15,** 1016.

Raman, S. and Nestor Jr., C. W. (1986). *Proc. Int. Symp. on Weak and Electromagnetic Interactions in Nuclei*, Heidelberg.

Raman, S., Malarkey, C. H., Milner, W. T., Nestor, C. W. Jr. and Stelson, P. H. (1987). *Atomic Data and Nucl. Data Tables*, **36,** 1.

Riezebos, H. J., de Voigt, M. J. A., Fields, C. A., Cheng, X. W., Peterson, R. J., Hagemann, G. B. and Stock, A. (1987). *Nucl. Phys.* **A465,** 1.

Rose, M. E. (1955). *Multipole Fields.* John Wiley, New York.

Rose, M. R. (1965). *Alpha-, Beta- and Gamma-ray Spectroscopy* (ed. K. Siegbahn), p. 887. North-Holland, Amsterdam.

Russo, P. A., Pedersen, J. and Vandenbosch, R. (1975). *Nucl. Phys.* **A240,** 13.

Sakai, M. (1975). *Nucl. Data Tables*, **15,** 513.

Sakai, M. (1984). *Atomic Data and Nucl. Data Tables* **31,** 399.

Schmidt, H. H. *et al.* (1986). *Nucl. Phys.* **A454,** 267.
Scholten, O., Iachello, F. and Arima, A. (1978). *Ann. Phys.* **115,** 325.
Shibata, T., Ejiri, H. and Sano, M. (1975a). *Nucl. Phys.* **A254,** 7.
Shibata, T., Itahashi, T. and Wakatsuki, T. (1975b). *Nucl. Phys.* **A237,** 382.
Sliv, L. A. and Band, I. M. (1965). *Alpha-, Beta- and Gamma-ray Spectroscopy* (ed. K. Siegbahn), p. 1639. North-Holland, Amsterdam.
Sood, P. C. (1968). *Nucl. Data Tables* **A4,** 281.
Sørensen, B. (1973). *Nucl. Phys.* **A217,** 505.
Specht, H. J., Weber, J., Konecny, E. and Heunemann, D. (1972). *Phys. Lett.* **41B,** 43.
Stelson, P. H. and Grodzins, L. (1965). *Nucl. Data Tables* **A1,** 21.
Stelson, P. H. Robinson, R. L., Milner, W. T., McGowan, F. K. and Ludington, M. A. (1971). *Bull. Am. phys. Soc.* **16,** 619.
Stephens, F. S. (1985). *Niels Bohr Centennial Conf. on Nuclear Structure*, p. 363. Copenhagen.
Styczen, J., Kleinheinz, P., Piiparinen, M. and Blomqvist, J. (1982). KFA Annual Report 1981, Jülich, p. 56.
Talmi, I. (1962). *Rev. mod. Phys.* **34,** 704.
Taylor, H. W., Singh, B., Prato, F. S. and McPherson, R. (1971). *Nucl. Data Tables* **A9,** 1.
Trusov, V. F. (1972). *Nucl. Data Tables,* **10,** 477.
Twin, P. J., Nyakó, B. M., Nelson, A. H., Simpson, J., Bentley, M. A., Cranmer-Gordon, H. W., Forsyth, P. D., Howe, D., Mokhtar, A. R., Morrison, J. D., Sharpey-Schafer, J. F. and Sletten, G. (1986). *Phys. Rev. Lett.* **57,** 811.
Venkova, Ts. and Andrejtscheff, W. (1981). *Atomic Data and Nucl. Data Tables* **26,** 93.
Voigt de, M. J. A., Sujkowski, Z., Chmelewska, D., Jansen, J. F. W., van Klinken, J. and Feenstra, S. J. (1975). *Phys. Lett.* **59B,** 137.
Voigt de M. J. A., Dudek, J. and Szymanski, Z. (1983). *Rev. mod. Phys.* **55,** 949.
Watson, D. D. and Harris, G. I. (1967). *Nucl. Data Tables* **A3,** 25.
Wesselborg, C., Schiffer, K., Zell, K. O., Von Brentano, P., Bohle, D., Richter, A., Berg, G. P. A., Brinkmöler, B., Römer, J. G. M., Osterfeld, F. and Yabe, M. (1986). *Z. Physik.* **A323,** 485.
Wienhard, K., Ackermann, K., Bangert, K., Berg, U. E. P., Bläsing, C., Naatz, W., Ruckelshausen, A., Rück, D., Schneider, R. K. M. and Stock, R. (1982). *Phys. Rev. Lett.* **49,** 18.
Wildenthal, B. H. (1984). *Progr. part. nucl. Phys.* **11,** 5.
Wilets, L. and Jean, M. (1956). *Phys. Rev.* **102,** 788.
Woude, A. van der (1987). *Progr. part. nucl. Phys.* **18,** 217.
Yamazaki, T. (1967). *Nucl. Data Tables* **A3,** 1.
Yates, S. W., Julin, R., Kleinheinz, P, Rubio, B., Mann, L. G., Henry, E. A., Stöffl, W., Deckman, D. J. and Blomqvist, J. (1986). *Z. Phys.* **A327,** 417.
Zuber, K., Kleinheinz, P., Schardt, D., Larsson, P. O., Kirchner, R., Klepper, O., Koslowsky, V. T., Roeckl, E., Rykaczewski, K. and Blomqvist, J. (1986). KFA Annual Report 1985, Jülich, p. 122.

3

INSTRUMENTS AND EXPERIMENTAL METHODS

3.1 Development of experimental methods for in-beam e–γ spectroscopy

The newly-developed experimental methods and devices for in-beam e–γ spectroscopy have played an important role in the progress of nuclear physics. In-beam e–γ spectroscopic experiments consist basically of the excitation of nuclear states by means of accelerated beams and the detection of the de-excitation e–γ rays.

Recent developments of accelerators have provided in-beam e–γ spectroscopy with a variety of high-quality beams. High intensity beams are essential for detecting weak γ rays and conversion electrons, and beams with good emittance are important for removing all kinds of background. A variety of accelerated beams such as light ions, heavy ions, polarized beams, electrons and even unstable nuclei (nuclear fragments) has greatly enlarged the range of nuclear states studied, and this now includes wide regions of linear and angular momenta, excitation energies, isospins and their z components. We do not discuss accelerators and beam qualities in Chapter 3, but concentrate mainly on detectors and methods for in-beam e–γ spectroscopy.

Electron–γ spectroscopic methods for in-beam studies have recently developed greatly in both quantity and quality. The rapid increase of the number of observed nuclear properties in the last two decades is mainly correlated with the development of new detection techniques, as well as the progress in accelerator technology. This is illustrated in Fig. 3.1, where the number of observed nuclear levels, spins and parities are plotted as a function of time. A marked increase is noted around 1965 when the high-resolution Ge(Li) detector was introduced. For e–γ spectroscopy important components are e–γ detectors, spectroscopic methods and data-acquisition systems. In this chapter we discuss in detail some important developments of e–γ detectors and spectroscopic methods, but leave out those of data acquisition. In recent years the developments of computers for on-line and off-line applications have made it possible to control many detectors, to monitor various components of experiments, and to perform complicated on-line and off-line data analyses. Thus most of the advanced devices for the detectors and spectroscopic analyses are effectively operated for spectroscopic studies by means of such on-line and off-line computers.

Detector assemblies composed of many detectors have made it possible to measure efficiently the multiplicity of the γ rays, their angular distributions, and coincidences (correlations) of cascade γ rays. Combinations of detectors

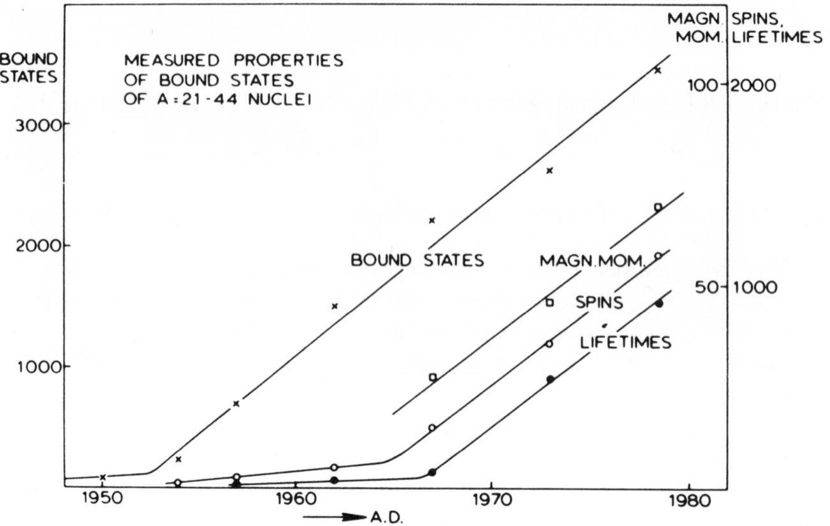

Fig. 3.1. The measured properties of bound states in $A = 21$–44 nuclei as a function of time starting in 1950. (Endt and van der Leun, 1978.)

are used to suppress Compton tails in the γ-ray energy spectrum and to select low-energy or medium-energy γ rays. Large detectors covering nearly all the 4π solid angle are used to measure the total energy of cascade γ rays. These features of γ-ray detectors provide very powerful tools to study low-energy and high-multiplicity γ rays from bound states. Enlargement of the single-detector volume is important for detecting medium-high energy γ rays in the 10 to 100 MeV region. These energetic γ rays, which are associated with highly-excited unbound states, are used for studying the electromagnetic properties of highly excited states, nuclear reaction dynamics, and high-momentum transfer mechanisms. High energy γ rays above 100 MeV are detected efficiently by an ensemble of detector modules. The total γ-ray energy is obtained from the sum of all signals from the modules and the direction of the γ ray from their spatial (angular) distribution.

Developments of conversion electron spectrometers have been made in two ways: one type is the multi-gap spectrometer, the other is the electron transporter. In the former case the electron energy is analysed by the magnetic spectrometer. Spectrometers with multi-gaps are used to measure the angular distributions simultaneously. The electron transporter is used to transport conversion electrons over a large momentum range from the target area to the detector area to remove γ rays and other background. The electron energy is analysed by the detector, usually silicon semiconductor detectors. Various kinds of efficient electron spectrometers have been developed.

Nuclear reactions have many features that may be studied by in-beam e–γ spectroscopy methods. Such features, analysed with appropriate e–γ detectors, have been used extensively to study nuclear structures and reactions. The large linear momenta given to the recoil nuclei by nuclear reactions are useful for lifetime measurements made by using the Doppler effect on the γ-ray energy. Multiple Coulomb excitation is used to determine electromagnetic transition moments up to high-spin states. The alignment of the angular momenta of residual nuclei is used for angular correlation studies. These give information on spins and multipolarities, and may be extended to electromagnetic moments by applying electromagnetic fields. Transient fields are quite unique phenomena associated with residual nuclei moving with high recoil velocities through matter. The recoiling nuclei emitted from inclined (tilted) foils may have spin polarization.

Detector assemblies and their applications are discussed in Section 3.2. Large γ-ray detectors and large-scale detector ensembles for medium- and high-energy γ rays are described in Section 3.3. Various kinds of electron spectrometers and electron transporters are presented in Section 3.4. In the last Section (3.5) of this chapter we discuss some of the e–γ spectroscopic methods characteristic of in-beam studies, such as the measurement of electromagnetic transition moments by using Doppler effects, multiple Coulomb excitations, and measurements of electromagnetic moments by using the spin alignment and the transient fields.

3.2 Detector assemblies

In modern multi-detector assemblies one aims to combine several qualities; high efficiency, good energy resolution and a reduction of the γ and neutron backgrounds. The best energy resolution is achieved with Ge(Li) or (hyper) pure Ge detectors. The peak efficiency, however, is normally not so large, for example up to thirty per cent for a 7.6 × 7.6 cm NaI(Tl) crystal at a distance of twenty-five centimetres. The use of N detectors in the multi-detector assembly increases the total efficiency for single γ-ray measurements by a factor of N and the efficiency for two-parameter coincidences by a factor of $N(N-1)$.

Gamma-ray spectra consist in general of discrete lines as well as of a continuum background. The γ-ray background in the spectra is caused by scattered γ rays and neutrons from beam lines and target and also by mainly two different kinds of processes; Compton scattering of γ rays and statistical decay of compound nuclei causing a quasi-continuum background. The first type of background does not contain interesting physical information. Compton-Suppression-Spectrometers (CSS) are designed to reduce the background due to Compton scattered γ rays and they will be discussed in detail in Section 3.2.1. The quasi-continuum background, however, is related to the physics of the reaction and cannot be suppressed by the CSS. If desired, the contribution of that component can be diminished by adequate selection techniques on the total angular momentum (see Section 3.2.2) and on the total energy (Section 3.2.3). The special devices which consist of many scintillator crystals, called "crystall balls", will be discussed in Section 3.2.4. The multi-detector CSS combined with a multiplicity filter and/or a sum spectrometer will be discussed in Section 3.2.5.

3.2.1 *Compton suppression spectrometers (CSS)*

NaI(Tl) scintillators have been extensively used as CSS for a long time (Konijn *et al.*, 1973). The basic idea of the CSS is to detect all the scattered photons from the Ge detector by means of a large crystal surrounding the Ge detector and subsequently providing a veto signal in all those cases. The CSS with the NaI(Tl) crystals have the very good suppression of a factor of 10 to 12, but their dimensions are too large for application in multi-detector assemblies. This problem has been solved by the introduction of a high-density scintillation material, bismuth gérmanate, $Bi_4Ge_3O_{12}$, commonly abbreviated as BGO. Its density is about a factor of 2 higher than that of NaI(Tl) and therefore it has a stronger absorptive power for γ rays. Some specific properties of the three scintillation materials, NaI(Tl), BGO, and BaF_2, are compared in Table 3.2.1.

A 1 MeV γ ray, for example, will be 95 per cent absorbed by a 6 cm thick BGO, a 10.5 cm BaF_2 and a 14 cm NaI(Tl). The most interesting property of BaF_2 is the fast timing component, which will find an important application in discriminating between γ rays and neutrons as is done in the crystal balls (see Section 3.2.4). The light output for a certain γ ray is in NaI(Tl) much larger than in BGO or BaF_2 because the scintillation conversion efficiency of BGO is only 8 per cent of that of NaI, and for BaF_2 it is only 20 per cent.

The importance of Compton suppression, particularly in many-fold coincidence experiments, can be illustrated in the following way: Let p_0 be

Table 3.2.1 Properties of different scintillation materials

Material	Wave length at maximum emission (nm)	Decay constant (ns)	Density (kg/dm^3)	Hygroscopic	Resolution (1 MeV γ rays) Energy (%)	Time (ns)
NaI	410	230	3.67	yes	$7\sim8$	~3
BGO	480	300	7.13	no	~15	~4
BaF_2	325	620	4.88	no	~10	~0.6
	220	0.6				

the peak to total ratio (ratio of the number of counts in the photopeak to the total number) in the case without suppression for a certain γ ray. The peak to total ratio in the case of a suppression with an average factor r over the whole Compton region can be written

$$p = p_0/\{p_0 + (1 - p_0)/r\}. \tag{3.2.1}$$

We may define a quality improvement factor Q_n for a n-fold coincidence ($n = 1$ means singles) as

$$Q_n = (p/p_0)^n = \{1/(p_0 + (1 - p_0)/r\}^n. \tag{3.2.2}$$

This factor is plotted as a function of r in Fig. 3.2. The many-fold coincidence spectra are seen to profit most from good suppression.

To design a Compton suppression shield one has to calculate the scattering of the γ rays from the central Ge detector in all directions. The energy of a scattered photon at an angle θ with initial energy E_0, is given by

$$E = E_0/\{1 + E_0(1 - \cos\theta)/mc^2\}. \tag{3.2.3}$$

From this formula it is seen that in forward direction the scattered photon carries the largest energy and in backward direction the lowest energy. Therefore anti-Compton material placed in the forward direction improves mainly the low-energy part of the spectrum taken with the Ge detector, while material in the backward direction improves the high-energy part of the spectrum, i.e. diminishes the sharp Compton edges.

In practice, one has to make a Monte Carlo calculation in order to take into account the multiple scattering and absorption in the dead layers of the crystal and in other materials. Because these calculations are not so simple we will present here some details of the results for a few commonly used CSS. First, we present in Fig. 3.3 the most simple case of the absorption of 1.5 MeV and 3 MeV γ rays in BGO after Compton scattering in all directions. It is seen that in the forward direction much material is needed and in backward direction only a few centimetres provide adequate Compton suppression. Calculations also show that the suppression is very sensitive for inactive materials which absorb γ rays but do not provide veto signals. So one has to reduce as much as possible the dead layers such as the heat shields and aluminum caps of the Ge detector, and the aluminum covers and MgO reflectors of the CSS. The most favourable material for the detector itself seems to be (hyper) pure N-type germanium because it has no dead layer on the outside like the Ge(Li) and P-type detectors.

The CSS can be constructed in two ways, symmetrically or asymmetrically. A symmetric configuration is shown in Fig. 3.4(a) and the results of a Monte Carlo calculation in Fig. 3.4(b) (Lieder *et al.*, 1984). The asymmetric type with the Monte Carlo results are shown in Fig. 3.5. The main difference between the two types of CSS is the material in front and behind the Ge detector in the

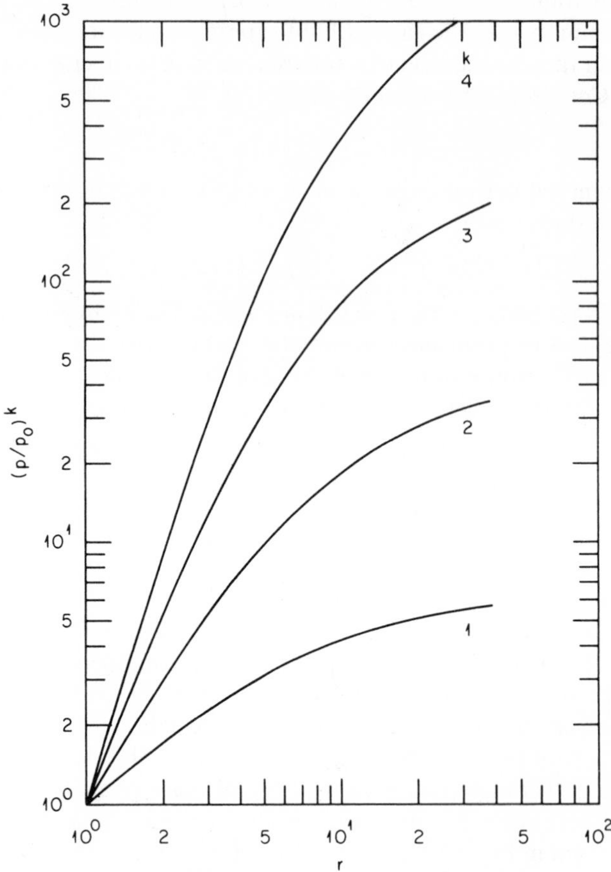

Fig. 3.2. The improvement factor $Q = (p/p_0)^k$ of the peak to total ratio as a function of the suppression factor r for different values of the coincidence fold $k (= n$ in the text). (Johnson *et al.*, 1984.)

asymmetric one, which is absent in the symmetric case. This is clearly reflected in the difference between the two spectra. The effect on the suppression of the scintillation material in the asymmetric type on the side of the central Ge detector—indicated by A in Fig. 3.5(a)—is shown in Fig. 3.6(a). It can be seen here that a thicker layer of BGO on the side (i.e. a larger diameter) affects positively the suppression over the whole energy range of the spectrum and is thus very effective in most of the applications. The effect of material behind the detector (B) is shown in Fig. 3.6(b). This material works mainly effectively in the low-energy part of the spectrum, up to 500 keV for a ^{60}Co source. Thus,

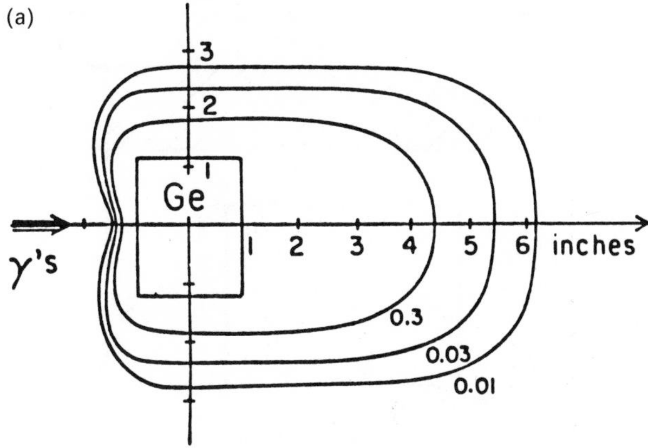

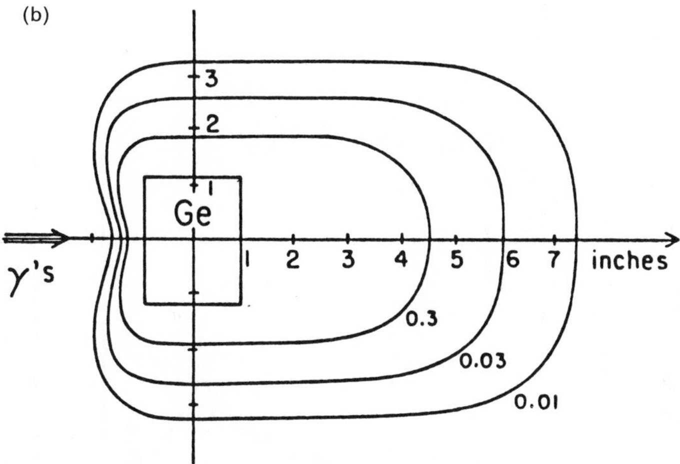

Fig. 3.3. The contour lines for BGO shields corresponding to the indicated fractions of Compton-scattered γ's that escape undetected. (a) for 1.5 MeV γ's; (b) for 3 MeV γ's. (Saladin *et al.*, 1980.)

much scintillation material behind the Ge detector is necessary when one is interested in low-energy transitions in the presence of high-energy ones.

From the above discussion it seems clear that the asymmetric type of CSS has a superior suppression compared with the symmetric one. On the other

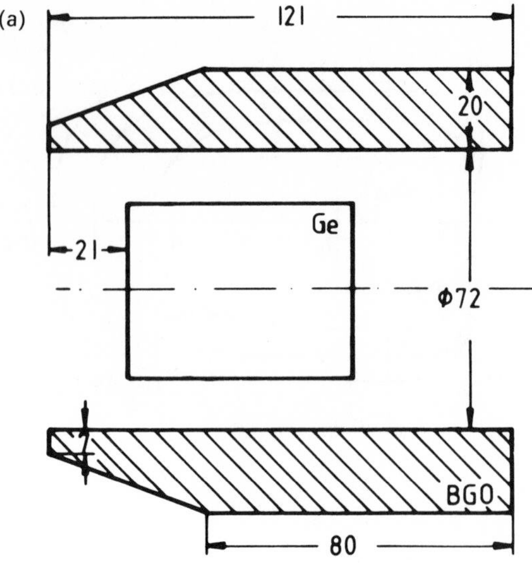

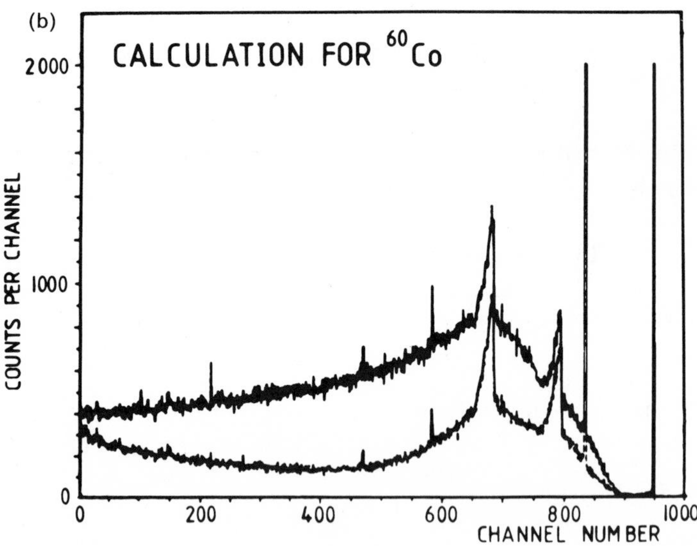

Fig. 3.4. (a) Layout of the symmetric anti-Compton spectrometer considered in the calculations. The ACS consists of a cylinder with a conical front end. The dimensions are given in millimetres. (b) Monte Carlo calculations of unsuppressed and suppressed ^{60}Co spectra of the symmetric anti-Compton spectrometer. The ^{60}Co lines of 1173 and 1333 keV have 5×10^4 counts in the respective peak channels. (Lieder *et al.*, 1984.)

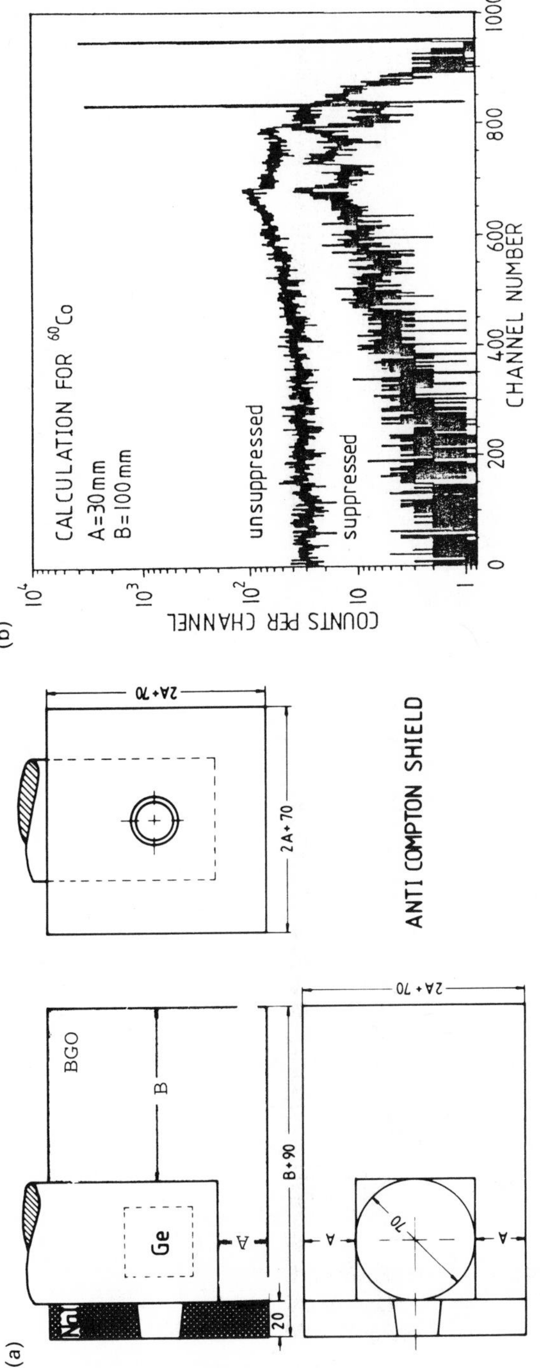

Fig. 3.5. (a) Design of an anti-Compton spectrometer of the asymmetric type. The dimensions are given in millimetres. (b) Monte Carlo calculations of unsuppressed and suppressed ^{60}Co spectra for the asymmetric anti-Compton spectrometer. (Lieder *et al.*, 1984.)

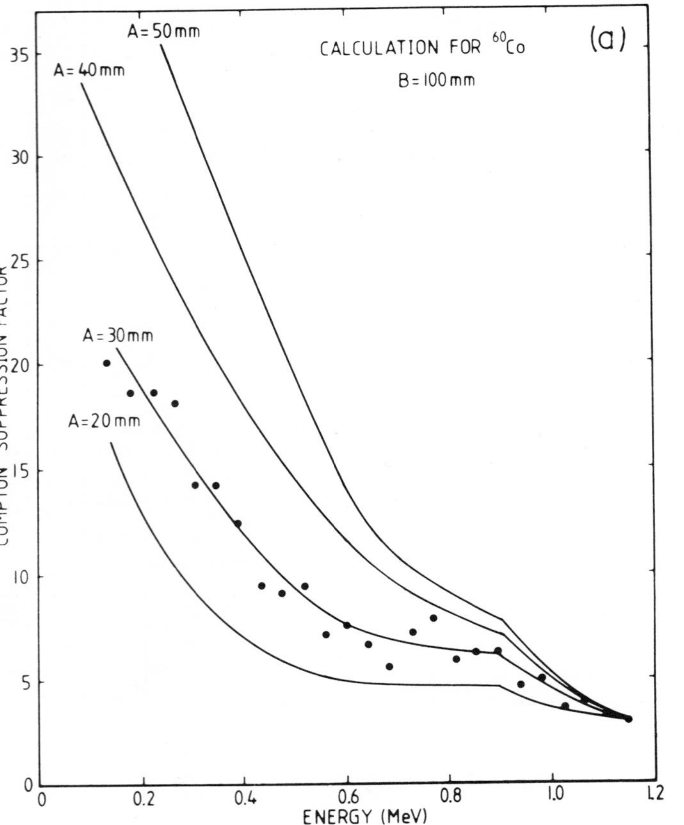

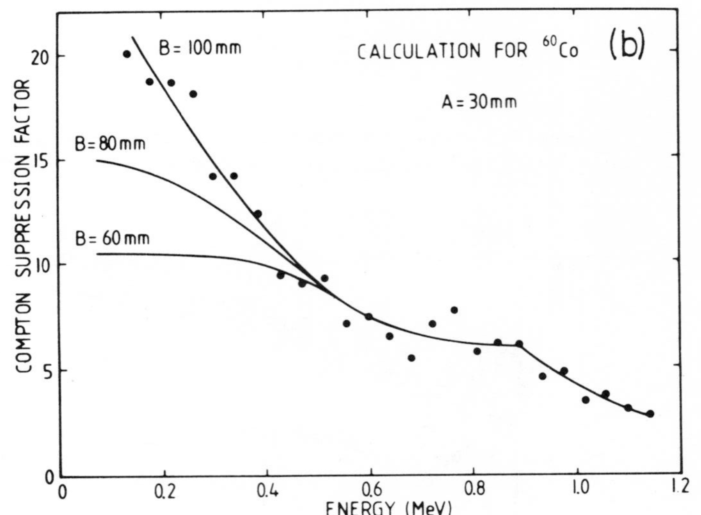

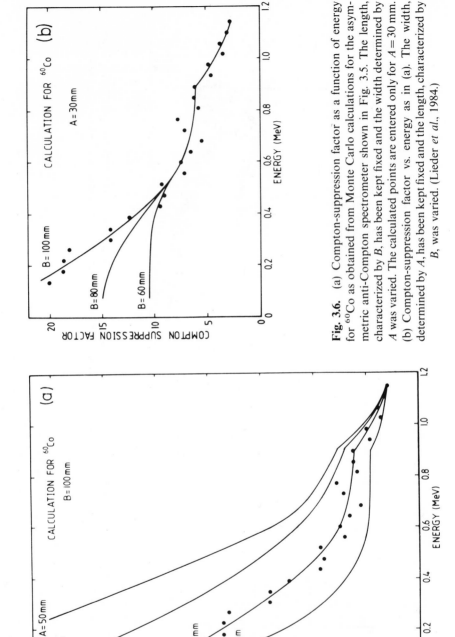

Fig. 3.6. (a) Compton-suppression factor as a function of energy for ^{60}Co as obtained from Monte Carlo calculations for the asymmetric anti-Compton spectrometer shown in Fig. 3.5. The length, characterized by B, has been kept fixed and the width determined by A was varied. The calculated points are entered only for $A = 30$ mm. (b) Compton-suppression factor vs. energy as in (a). The width, determined by A, has been kept fixed and the length, characterized by B, was varied. (Lieder *et al.*, 1984.)

hand the symmetric type is much more suitable for many-detector assemblies. For that reason the symmetric CSS is much more popular than the asymmetric type, and much attention has been devoted to the improvement of its suppression. A design of the Liverpool group (Nolan *et al.*, 1985) and built by the Harshaw Company in The Netherlands, is shown in Fig. 3.7. Such units are incorporated in the TESSA III spectrometer at Daresbury, as discussed in Section 3.2.5. In these units important additions to the cylindrical BGO crystal are a NaI(Tl) piece in front of the Ge detector and a BGO catcher behind it. To make such a configuration, a special cryostat had to be developed for the Ge detector which leaves enough space free for the catcher. The material at the front can be the low-density NaI(Tl) which produces more light than BGO. This appeared to be important because the photo-tubes are mounted behind the spectrometer. The performance is very good with a suppression factor that reaches 10 around 700 keV and is still about 5 at low energy in the case of γ rays from a ^{60}Co source, as illustrated in Fig. 3.8.

A different philosophy has been followed by Emling *et al.* (1986) who developed the first triple (*P*-type) Ge telescope in one cryostat, produced by DSG, Darmstadt. The triple and double (N-type) Ge telescopes, each in one cryostat, are used by an Amsterdam–Groningen collaboration in a symmetrical BGO CSS (Penninga, 1986). The CSS, shown in Fig. 3.9, is very similar to the Liverpool one and is also produced by the Harshaw company in The Netherlands. The Ge telescopes, produced by Canberra Belgium, consist of two (hyper) pure Ge detectors ($\sim$20 per cent relative efficiency each). The front crystal (*N*-type) acts as a normal detector while the second one (*P*-type to cut the costs) can be used as a veto detector, like the catcher in the Liverpool design. One spectrometer contains a triple telescope as shown in Fig. 3.10. The low-energy planar detector in front of the two Ge crystals suppresses the sharp Compton edges at the high-energy side. Apart from the suppression those telescopes offer the advantage of having much high-quality detection material near the target which can be used in various coincidence combinations as well as in the veto mode. For instance, the first detector may record data with the second detector in coincidence or anticoincidence. Much high-resolution material near the target would stimulate the recording of data rather than its rejection. The measured spectra of a ^{60}Co source with the triple telescope is shown in Fig. 3.11. Here the spectrum of the central crystal is shown without any suppression and with a veto signal from the front-planar, the back *P*-type detector and from the BGO(NaI) shield. It should be noticed that a fair comparison with a single bare Ge detector is difficult because the multiple Ge detectors cause more mutual Compton scattering which is also suppressed. The suppression factor for the triple telescope and for the Liverpool detector are compared in Fig. 3.8. Both detector systems reach suppression factors of 9 to 10 at the most.

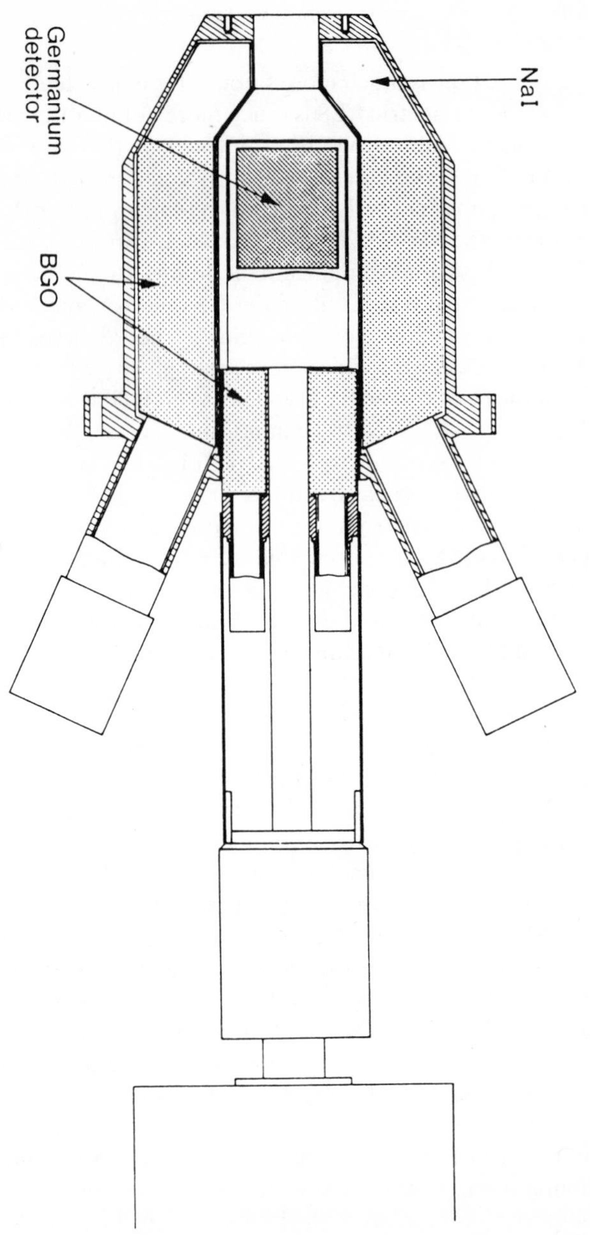

Fig. 3.7. The Liverpool design of a BGO cylinder, NaI front-end and the BGO catcher as a CSS for a Ge detector placed in the centre, manufactured by Harshaw Chemie, The Netherlands. In front of the NaI(Tl) crystal a 4-cm-thick heavy-metal collimator shields the scintillators from direct γ-radiation from the source in front of the detector. Eight photomultipliers are mounted on the back of the BGO cylinder. (Nolan *et al.*, 1985.)

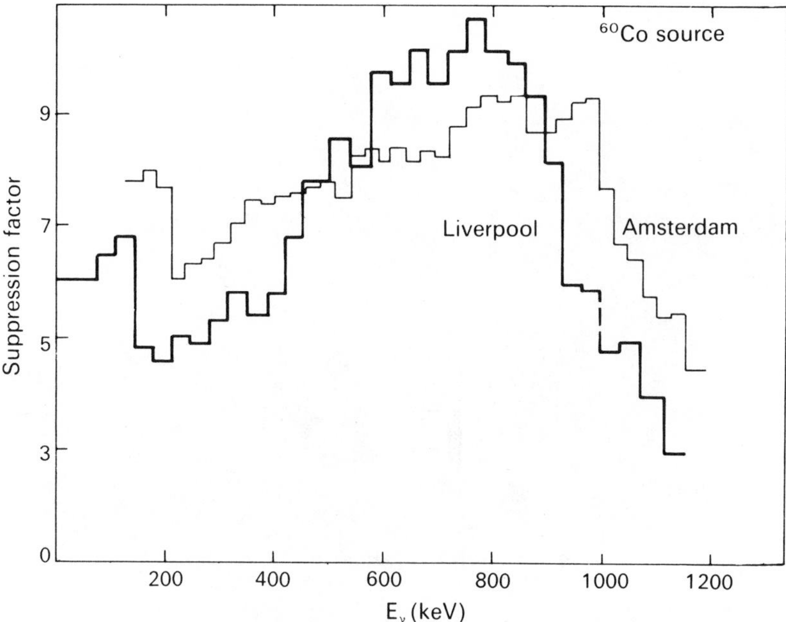

Fig. 3.8. Compton suppression factors for a ^{60}Co source measured with the symmetric CSS of the Amsterdam and Liverpool designs as shown in Figs 3.10 and 3.7, respectively.

3.2.2 *Multiplicity filters*

Multiplicity filters were introduced for the first time by Hagemann *et al.* (1975) to study specific properties of quasi-continuum γ rays emitted from high-spin states (see Section 6.5). The main purpose of those instruments is to determine the number of γ rays, the multiplicity of a γ-ray cascade or alternatively to select, as by a filter, cascades within a certain range of multiplicities.

The experimental arrangement consist of a number of similar detectors N, each having an average probability $P = \Omega$ of detecting a γ ray. This probability or total efficiency is given by $\Omega = \varepsilon\omega/4\pi$ where ε is the average detection efficiency in a certain energy range and ω is the solid angle (in sterradians). The addition of more detectors has two effects; firstly the efficiency will be larger for a certain k-fold coincidence (k runs from 0 to the number of detectors N) and secondly higher k-fold coincidences can be measured. This probability also depends on the number of γ rays in the cascade, which is the basis for the determination of the multiplicity M. Therefore we have to derive a formalism that relates the measured coincidence probabilities to M. We will first summarize the properties of one detector and then derive a simple

Fig. 3.9. BGO CSS of the symmetric type used by the Amsterdam–Groningen collaboration, manufactured by Harshaw Chemie, The Netherlands.

formula for a two-detector arrangement before considering an assembly with an arbitrary number of detectors.

The probability of not detecting a γ ray by a particular detector, with a total efficiency Ω, is given by $\bar{P} = 1 - P = 1 - \Omega$. The probability that γ rays with multiplicity M are not detected by that detector is then given by $\bar{P} = (1 - \Omega)^M$. Thus the probability for producing a signal in that detector by at least one

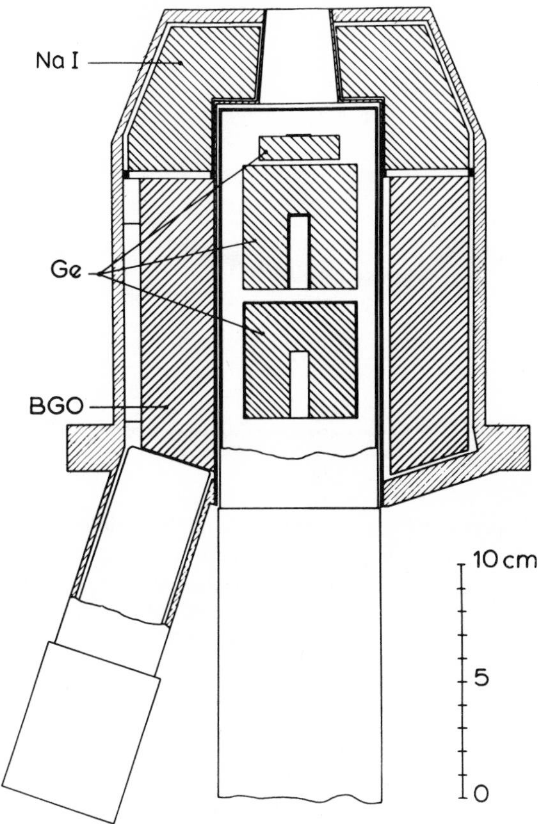

Fig. 3.10. A schematical representation of the Dutch triple Ge-telescope, manufactured by Canberra, Belgium, with BGO(NaI) CSS of which a picture is shown in Fig. 3.9.

γ ray out of a total of M is given by

$$P_1^M = 1-(1-\Omega)^M. \tag{3.2.4}$$

In the case of two detectors consider a high resolution Ge detector, which detects a discrete γ ray of the cascade, and a NaI(Tl) detector with total efficiency Ω for the detection of at least one of the remaining $M-1$ γ rays in the cascade. The detection probability for at least one γ ray in the second detector is P_1^{M-1} as given by eqn (3.2.4). The coincidence count-rate N^c between the two detectors can be expressed in terms of the probability P_1^{M-1} and the singles count-rate N^s in the Ge detector as

$$N^c = N^s P_1^{M-1}. \tag{3.2.5}$$

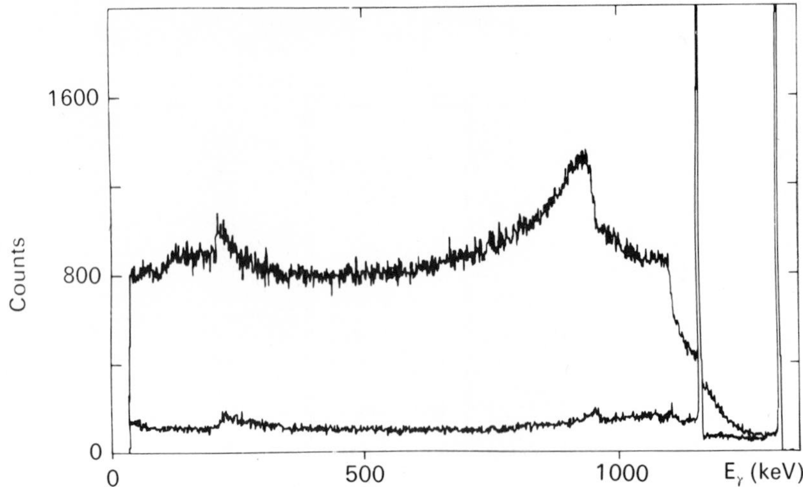

Fig. 3.11. Singles spectra of ^{60}Co source taken with the middle N-type Ge detector of the Dutch telescope (see Fig. 3.10). The upper curve is the unsuppressed spectrum and the lower one with suppression of Compton events scattered to the BGO–NaI shield and to the planar front detector and the P-type back detector. The suppression factor is shown in Fig. 3.8.

Both N^c and N^s are counts in the same photopeak occurring in the two spectra measured with the same Ge detector. The NaI detector is operated in the full energy range with an average efficiency Ω above a certain energy threshold.

If $\Omega \ll 1$ then we can expand $(1-\Omega)^{M-1}$ in powers of Ω

$$(1-\Omega)^{M-1} = 1-(M-1)\Omega+(M-1)(M-2)\Omega^2/2+ \ldots \approx 1-(M-1)\Omega. \tag{3.2.6}$$

Combining eqns (3.2.4), (3.2.5) and (3.2.6) yields

$$N^c = N^s(M-1)\Omega. \tag{3.2.7}$$

With this simple formula the average multiplicity can be determined from a measurement with two detectors (see also Section 6.5.4).

Let us now derive the probability P_{Nk}^M that k out of N similar detectors detect γ rays from a cascade with multiplicity M (see also Ockels, 1978 and van der Werf, 1978, and references quoted therein). This probability can be written as a product of two probabilities; firstly that N detectors detect m γ rays and secondly that these m γ rays are distributed over k detectors. The first probability is given by the binomial distribution, assuming the same average efficiency Ω for all detectors and neglecting effects of angular distributions

$$P(N \text{ dets. fired by } m\ \gamma\text{'s}) = \binom{M}{m}(N\Omega)^m(1-N\Omega)^{M-m}. \qquad (3.2.8)$$

Here $\binom{M}{m}$ is the binomial coefficient and represents the number of possible m γ rays selected out of a total of M. The second probability can be obtained from Stirling's number of the second kind S_m^k, which gives the number of ways that m γ rays are distributed over k detectors ($m \geqslant k$)

$$S_m^k = (1/k!) \sum_{n=0}^{k} (-1)^{k-n}\binom{k}{n} n^m. \qquad (3.2.9)$$

Thus the probability of hitting k detectors out of N by m γ rays is given by

$$P(m\ \gamma\text{'s hit } k \text{ dets}) = \binom{N}{k}(S_m^k k!/N^m). \qquad (3.2.10)$$

The division by N^m is done to account for the number of ways that m γ rays can be distributed over N equivalent detectors. The probability P_{Nk}^M can now be given by multiplying the two probabilities of eqns (3.2.8) and (3.2.10), using the explicit form of eqn (3.2.9) for the Stirling numbers and summing over all possible m γ rays.

$$P_{Nk}^M = \sum_{n=0}^{k} (-1)^{k-n}\binom{k}{n}\binom{N}{k}\left\{\sum_{m=o}^{M}\binom{M}{m}(n\Omega)^m(1-N\Omega)^{M-m}\right\}. \qquad (3.2.11)$$

Here the sum over m can be started at 0 because Stirling's numbers are zero for $m < k$. Since the term in curly brackets is the binomial expansion of $\{1-(N-n)\Omega\}^M$, eqn (3.2.11) can be written as

$$P_{Nk}^M = \binom{N}{k}\sum_{n=0}^{k}(-1)^{k-n}\binom{k}{n}\{1-(N-n)\Omega\}^M. \qquad (3.2.12)$$

Note that this probability includes the multiple hits of one detector, which contributes to the statistical uncertainty. This problem can only be solved when $N \gg M$ as is the case with the crystal balls (see Section 3.2.4).

This k-fold coincidence probability is plotted in Fig. 3.12 for three combinations of N and Ω. We see from those curves that the 0-fold probability (i.e. only a Ge-detector fires without any signal from N detectors of the multiplicity filter) decreases exponentially with multiplicity M. The higher-order probabilities increase with M, in particular very strongly for low M values. The increase from eight to sixteen detectors with the same Ω affects most strongly the many-fold probabilities. This is caused by the fact that sixteen detectors form many more combinations than do eight, for instance for a 5-fold coincidence by a factor of $\binom{16}{5}/\binom{8}{5} = 78$ more. Comparing the two

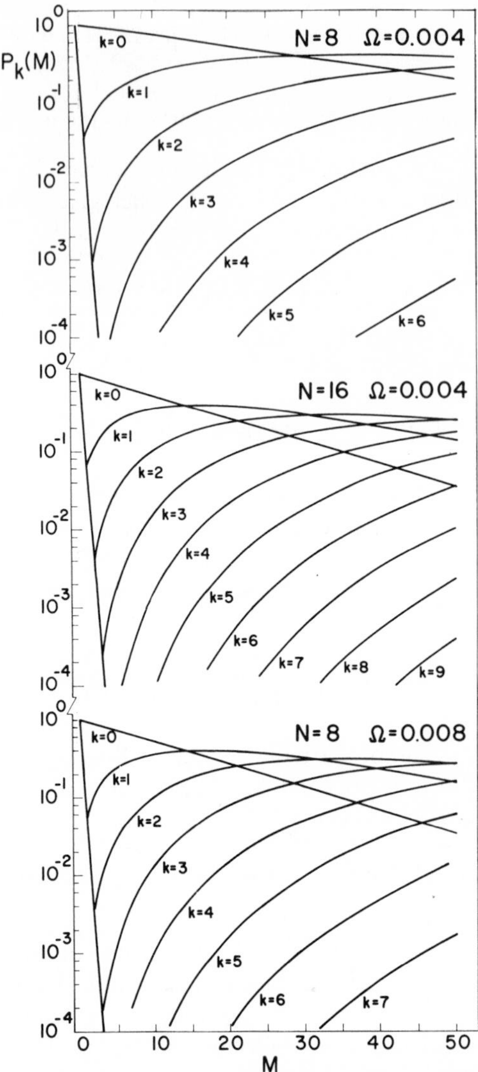

Fig. 3.12. Some k-fold coincidence probabilities $P_k(M)$ as function of the number of γ-rays emitted, M, calculated for three different detector arrangements: the number of detectors $N = 8$ each with an efficiency $\Omega = 0.004$ in the top figure; $N = 16$ detectors with an efficiency $\Omega = 0.004$ in the middle figure and $N = 8$ detectors with $\Omega = 0.008$ in the bottom figure. The last two figures are presented to show the effect of the different number of detectors in case of the same total efficiency. (Ockels, 1978.)

lower graphs in Fig. 3.12 shows that decreasing the number of detectors while keeping the product $N\Omega$ constant, does not very much affect the low-fold probabilities, but more so the many-fold ones. We will show below that in a non-linear expansion method the many-fold coincidence probabilities are less important than in a linear expansion method. This is caused by the fact that in the non-linear method multiplicity information is more concentrated in lower-fold coincidence probabilities than in the linear method, as will be discussed below.

Next we want to discuss the problem of deriving the multiplicity from the observed coincidence probabilities. First, we have to realize that the multiplicity M is not a fixed number in most reactions, but has a certain distribution $\rho(M)$ over a limited range of M-values. It is therefore more appropriate to discuss the mean value $\langle M \rangle$ and the central moments μ_j defined by

$$\langle M \rangle = \sum_M M\rho(M), \tag{3.2.13}$$

$$\mu_j = \langle (M - \langle M \rangle)^j \rangle = \sum_M (M - \langle M \rangle)^j \rho(M), \tag{3.2.14}$$

with

$$\sum_M \rho(M) = 1. \tag{3.2.15}$$

Note that $\mu_0 = 1$ and $\mu_1 = 0$. The experimental value for the k-fold coincidence probability P_k is derived from the number of counts Y_k in the k-fold spectrum, normalized to the sum over all spectra

$$P_k(\exp) = Y_k/N, \tag{3.2.16}$$

with

$$N = \sum_{k=0}^{N} Y_k. \tag{3.2.17}$$

The expectation value $\langle P_k \rangle$ for a sharp multiplicity, $\rho(M) = \delta_M$, is $\langle P_k \rangle = P^M_{Nk}$. If the experiment is designed correctly then the measured probability is $P_k(\exp) = \langle P_k \rangle$. For a certain multiplicity distribution $\rho(M)$ the relation between the observed coincidence probabilities and the P^M_{Nk} of eqn (3.2.12) can be written as

$$P_k(\exp) = \langle P_k \rangle = \sum_M \rho(M) P^M_{Nk}. \tag{3.2.18}$$

The fundamental problem in evaluating eqn (3.2.18) is that the central moments of $\rho(M)$ are non-identifiable. Without any external constraint we have to sum over all possible M-values up to infinity in eqns (3.2.13, 14, 15, and 18). From a strictly mathematical point of view this problem cannot be solved (see Bron *et al.*, 1985 for some methodological aspects of estimating the first few multiplicity moments). In practical applications, however, one may assume a simple distribution $\rho(M)$, e.g. a Gaussian one extending from $M = 0$ to a maximum multiplicity imposed by the maximum angular momentum transfer in the reaction. This implies that the above summations have to be carried out only up to this maximum multiplicity. The statistical accuracy of many-fold coincidences is often very poor and one simply sets them equal to zero (e.g. 7-fold and higher in the case that $M \leqslant 30$ and $N \leqslant 16$). One is then in most cases satisfied to estimate besides $\langle M \rangle$ only the lowest two central moments related to the variance σ_M and skewness s_M of $\rho(M)$,

$$\sigma_M = (\mu_2)^{1/2} \tag{3.2.19}$$

$$s_M = \mu_3/(\mu_2)^{3/2}. \tag{3.2.20}$$

Note that $s_M = 0$ for an exact Gaussian distribution.

First, we present a commonly used method to expand P^M_{Nk} in eqn (3.2.12) linearly in powers of Ω (Hagemann *et al.*, 1975). In the derivation of eqn (3.2.12) we used eqns (3.2.8–10), which can be combined in a slightly different way as

$$P^M_{Nk} = \binom{N}{k} k! \sum_{j=k}^{M} \binom{M}{j} \Omega^j \sum_{m=k}^{j} \binom{j}{m} (-N)^{j-m} S^k_m. \tag{3.2.21}$$

In this summation use is made of the property of the Stirling numbers that $S^k_m = 0$ for $0 \leqslant m < k$. We also notice that terms with orders of Ω lower than Ω^k cancel. Thus the linear expansion has the form

$$P^M_{Nk} = \sum_{j=k}^{M} a_{kj} \binom{M}{j} \Omega^j, \tag{3.2.22}$$

with

$$a_{kj} = k! \binom{N}{k} \sum_{m=k}^{j} \binom{j}{m} (-N)^{j-m} S^k_m. \tag{3.2.23}$$

Similarly, the observed coincidence probabilities or the expectation values $\langle P_k \rangle = P_k(\exp)$ can be expanded as

$$\langle P_k \rangle = \sum_{j=k}^{M} a_{kj} \left\langle \binom{M}{j} \right\rangle \Omega^j, \tag{3.2.24}$$

with

$$\left\langle \binom{M}{j} \right\rangle = \textstyle\sum_M \rho(M) \binom{M}{j}. \tag{3.2.25}$$

The latter two formula give an expression for $P_k(\exp)$ which is the same as given by eqn (3.2.18). Equation (3.2.24) shows explicitly the expansion of the k-fold probabilities in terms of factorial moments

$$\left\langle \binom{M}{j} j! \right\rangle = \langle M(M-1) \ldots (M-j+1) \rangle. \tag{3.2.26}$$

The expectation values of $\binom{M}{j}$ in eqn (3.2.25) are directly related to the first shape parameters (central moments) of $\rho(M)$:

$$\langle M\rangle=\left\langle\binom{M}{1}\right\rangle \tag{3.2.27}$$

$$\mu_2=2\left\langle\binom{M}{2}\right\rangle+\langle M\rangle-\langle M\rangle^2 \tag{3.2.28}$$

$$\mu_3=6\left\langle\binom{M}{3}\right\rangle-3(\langle M\rangle-1)\mu_2-2\langle M\rangle+3\langle M\rangle^2-\langle M\rangle^3. \tag{3.2.29}$$

Note that in eqn (3.2.24) the dependence of the k-fold probabilities on $\rho(M)$ is separated from that on Ω. Because a_{kj} is a triangular matrix we can write the inverse of eqn (3.2.24) as

$$\left\langle\binom{M}{j}\right\rangle\Omega^j=\sum_{k=j}^{M} b_{jk}\langle P_k\rangle, \tag{3.2.30}$$

with

$$b_{jk}=-1/a_{jj}\sum_{n=j+1}^{k} a_{jn}b_{nk},\quad k>j, \tag{3.2.31}$$

and

$$b_{jj}=1/a_{jj}. \tag{3.2.32}$$

Using the first few Stirling numbers as given by the explicit form of eqn (3.2.9) we can work out the expansion of the factorial moments in terms of the k-fold probabilities of eqn (3.2.30) as

$$\begin{aligned}\left\langle\binom{M}{j}\right\rangle\Omega^j j! &= \langle Q_j\rangle+(N-j/2)\langle Q_{j+1}\rangle+\{N(N-j-1)/2\\ &\quad+(3j^2+5j)/24\}\langle Q_{j+2}\rangle\\ &\quad+[\{(2N+j)(N-j-1)/12+(3j^2+5j)/24\}(N-j-2)\\ &\quad+(j^3+5j^2+6j)/48]\langle Q_{j+3}\rangle+\ldots\end{aligned} \tag{3.2.33}$$

where $Q_k=\binom{N}{k}^{-1}P_k$, with $\langle P_k\rangle$ being the experimental k-fold coincidence probabilities.

The series of eqn (3.2.33) converges slowly in k. This is a serious problem because in most cases the high k-fold probabilities are not known or have very large experimental uncertainties. This problem is discussed by Kohl *et al.*, (1978). Another problem is the determination of the errors in the deduced factorial moments and via eqns (3.2.27, 28, 29) in the central moments. The $\langle P_k\rangle$ values for different k depend on each other, and only the experimental count rates Y_k and N used in eqn (3.2.16) are independent of each other.

When higher k-fold probabilities are not measured and thus putting $P_{k_{\max}+1} - P_N = 0$, for $k_{\max} \leqslant N$, one may observe a lack of convergence already in the second and third factorial moments. The deduced higher moments are then progressively too small, because the elements b_{jk} in eqn (3.2.30) are all positive. A practical solution for this problem has been proposed by van der Werf (1978) by imposing lower boundaries on the values that the higher moments can take. In this "bootstrap method" the moments around the origin, $\langle M^j \rangle$, are forced to satisfy the following inequality (van der Werf, 1978)

$$\langle M^{j+2} \rangle / \langle M^{j+1} \rangle \geqslant \langle M^{j+1} \rangle / \langle M^j \rangle. \tag{3.2.34}$$

In practice, it is sufficient to check this inequality up to the power $j = 20$. A probability distribution, defined for $M \geqslant 0$ only, that does not obey eqn (3.2.34) necessarily contains negative probabilities and thus cannot correspond to a physical situation. The calculation procedure is then as follows: The factorial moments defined by eqn (3.2.26) and calculated with eqn (3.2.30) or (3.2.33) are converted to moments around the origin, $\langle M^j \rangle$. The inequality eqn (3.2.34) is checked and one gives $\langle M^j \rangle$ the value $\langle M^j \rangle \rightarrow \max\{\langle M^j \rangle, \langle M^{j-1} \rangle^2 / \langle M^{j-2} \rangle\}$. Then, a value for $P_{k_{\max}+1} - P_N$ is calculated from eqn (3.2.24) using the full matrix a_{kj} and maximized moments. This procedure is repeated until convergence is reached for the estimates of the higher k-fold probabilities; this occurs rather rapidly.

A quite different approach has been proposed by Ockels (1978), which is non-linear in $\langle P_k \rangle$ and rapidly convergent. The probability that none of the $(N-n)$ detectors detect a γ ray out of M is expressed by

$$\langle G_n \rangle = \langle \{1 - (N-n)\Omega\}^M \rangle. \tag{3.2.35}$$

The left-hand value can be deduced from experiment as

$$\langle G_n \rangle_{\exp} = \sum_{k=0}^{n} \binom{n}{k} \binom{N}{k}^{-1} \langle P_k \rangle_{\exp}. \tag{3.2.36}$$

It is convenient to derive a logarithmic expression for G_n

$$\ln \langle G_n \rangle = \ln \langle x^M \rangle, \tag{3.2.37}$$

with

$$x = \{1 - (N-n)\Omega\}.$$

For a certain distribution $\rho(M)$ we can define the expectation value $\langle x^M \rangle$ as follows

$$\langle x^M \rangle = x^{\langle M \rangle} \langle \exp\{(M - \langle M \rangle) \ln x\} \rangle. \tag{3.2.38}$$

The expectation value of the exponential factor can be written explicitly as a sum over all M values, similar to eqn (3.2.13). Using the expansion of the exponential and introducing μ_j by the substitution of eqn (3.2.14) we obtain

$$\langle x^M \rangle = x^{\langle M \rangle} \sum_{j=0} (\ln x)^j \mu_j/j!. \qquad (3.2.39)$$

This leads directly to the logarithmic form of G_n

$$\ln\langle G_n \rangle = y_n \langle M \rangle + \ln \sum_{j=0} y_n^j \mu_j/j!, \qquad (3.2.40)$$

where $y_n = \ln\{1-(N-n)\Omega\}$.

The final expansion can now be written as

$$\begin{aligned}\ln\langle G_n \rangle_{\text{exp}} &= y_n\langle M \rangle + y_n^2\mu_2/2! + y_n^3\mu_3/3! + y_n^4(\mu_4 - 3\mu_2^2)/4! \\ &+ y_n^5(\mu_5 - 10\mu_3\mu_2)/5! \\ &+ y_n^6(\mu_6 - 15\mu_4\mu_2 - 10\mu_3^2 + 30\mu_2^3)/6! + \ldots \qquad (3.2.41)\end{aligned}$$

This series converges much faster than the one given by eqn (3.2.33). For a Gaussian-shape distribution $\rho(M)$ all terms are zero, except the first two terms. The higher-order central moments ($j \geqslant 3$) in eqn (3.2.41) are only of influence if they differ from those of a Gaussian shape. Therefore it is fairly easy to extract the mean value and the width of $\rho(M)$, while this method yields also rapidly converging values for the higher moments.

The results obtained with the linear and the non-linear methods are compared in Fig. 3.13. Here a hypothetical beta binomial distribution was chosen for $\rho(M)$, characterized by the shape parameters $\langle M \rangle = 15$, $\sigma = 5$ and the skewness $s = -0.5$. A multiplicity filter with $N = 16$ detectors, each with an efficiency of $\Omega = 0.005$ was assumed. The shape parameters have been derived as a function of the highest fold (k_{max}) coincidence probability. It is clearly demonstrated that even the skewness is reproduced correctly in the non-linear method already with $k_{\text{max}} = 3$, whereas in the linear method convergence is not reached until k_{max} exceeds 5–6. It should be noted that with the bootstrap method, discussed above, similar results are obtained as with the non-linear method. Another comparison with two sets of experimental data is made in Fig. 3.14. Coincidence probabilities up to $k_{\text{max}} = 6$ have been measured for alpha-particle and ^{12}C-induced reactions with the 16-detector multiplicity filter as shown in Fig. 3.15 (Ockels, 1978). Again, the non-linear method enables us to deduce the mean value $\langle M \rangle$ from the 0 and 1-fold probabilities, σ from the 0, 1, and 2-fold values and s from the 0, 1, 2, and 3-fold probabilities.

The higher moments can in principle be extracted from a multiplicity filter experiment with an adequate number of detectors, as discussed above. However, the statistical errors become progressively larger for higher moments. It is already difficult to derive the third moment with an accuracy better than fifty per cent, even with a 16-detector multiplicity filter.

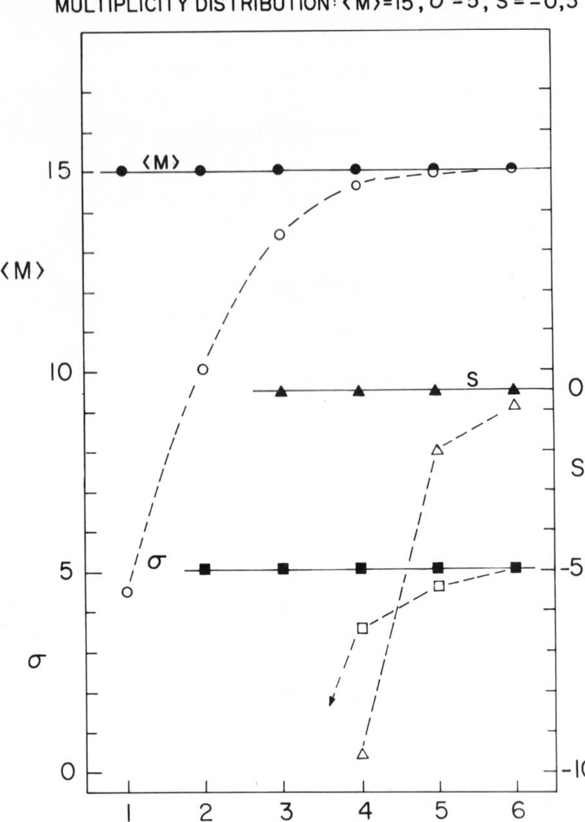

Fig. 3.13. Calculated shape parameters $\langle M \rangle$, $\sigma = \mu_2^{1/2}$, $s = \mu_3/\mu_2^{3/2}$ deduced from the k-fold coincidence probabilities which were calculated using a theoretical multiplicity distribution with $\langle M \rangle = 15$, $\sigma = 5$ and $s = -0.5$. The non-linear method (solid lines) is compared with the linear method (dashed lines). The k-fold coincidence probabilities were set equal to zero for $k > k_{max}$, and the same normalization was applied in both methods. (Ockels, 1978.)

The multiplicity moments derived from measured coincidence probabilities, according to the formalism discussed above, have still to be corrected for various disturbing effects. These effects concern the angular distributions of γ rays and also effects due to neutrons, charged particles, conversion electrons, X-rays and delayed γ rays. When isomers are present in the decay of a compound nucleus and no time selection is applied, then one should take into account explicitly the intensity of delayed γ rays with their specific lifetimes. In

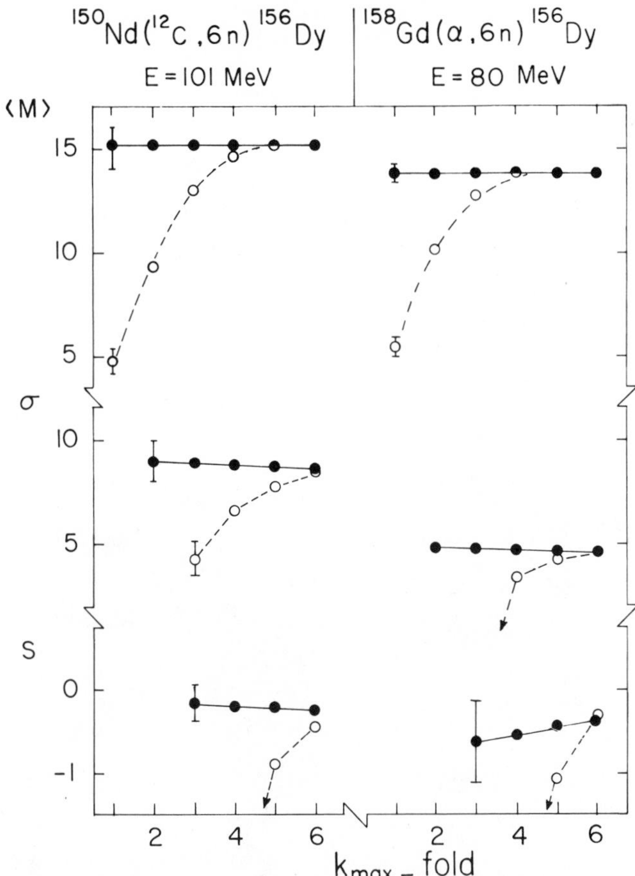

Fig. 3.14. Calculated shape parameters $\langle M \rangle$, σ, s as in Fig. 3.13. The k-fold coincidence probabilities were obtained for the $4^+ \to 2^+$ γ-ray transition in ^{156}Dy observed in ^{150}Nd(^{12}C, 6n)^{156}Dy and the ^{158}Gd(α, 6n)^{156}Dy reactions. The error bars roughly indicate the statistical erros involved. (Ockels, 1978.)

this case the decay scheme as well as the lifetimes involved have to be known for a proper analysis. The multiplicity of a delayed γ-ray cascade below an isomer can be measured by recording the time spectra of both the Ge detector and the multiplicity filter with respect to a reference (RF) signal. The coincidence probabilities can then be deduced for proper time gates, i.e. delayed Ge detector and delayed filter (Hageman, 1981). The problem of X-rays is easily dealt with by proper absorption material. Only in the case of high-energy X-rays does one have to apply a correction. The same statement can be made for protons, which are usually absorbed up to $\sim$10 MeV. A

Fig. 3.15. The Groningen multiplicity filter consisting of 16 NaI(Tl) detectors of 7.6×5.1 cm in size. They are positioned in lead cones, to prevent cross talk, at a distance of 15 cm from the target. The target is visible in the centre of a glass scattering chamber. (Ockels, 1978.)

correction for conversion electrons can be applied when the decay scheme is known. In the case of quasi-continuum γ rays one has to measure both the spectral distribution and the multipolarity in order to estimate the correction for conversion electrons.

An important feature of the γ rays emitted from aligned nuclear states is their asymmetric angular distribution. In the formalism derived above one may generalize the expressions by including the possibilities of different efficiencies for each of the detectors. In this way one can also include a correction for certain angular correlations (van der Werf, 1978). Another possibility is to design the multiplicity filter such that the most commonly occurring angular correlations are averaged out. This is done in the case of cascades of stretched E2 transitions for the Groningen multiplicity filter shown in Fig. 3.15 (Ockels, 1978). Corrections have then to be calculated for other types of transitions.

Slow neutrons can be discriminated by means of the time of flight, if the distance of the detectors to the target is large enough. For the detected fast neutrons one has to correct on the basis of the estimated neutron multiplicity and of the neutron efficiency of the NaI detectors (see Section 6.5). The effect of neutrons depends on the efficiency of the NaI detectors for neutrons, Ω_n, relative to that for γ rays, Ω. If the neutron multiplicity in the reactions used is given by x then the effect may be corrected for as follows:

$$\langle M \rangle_{\rm corr} = \langle M \rangle - x\Omega_n/\Omega. \tag{3.2.42}$$

The correction concerns only the relatively fast neutrons, which reach the detector within the time interval of the γ–γ coincidence. For a commonly used 7.6×5.1 cm NaI crystal the efficiency ratio is $\Omega_n/\Omega = 0.1$.

3.2.3 *Sum spectrometers*

Sum spectrometers are used for measuring the total γ energy released in nuclear reactions. They are most commonly applied in heavy-ion induced complete fusion reactions for studying the deexcitation of high-spin states (see Sections 4.2 and 6.5). The γ cascades consist in most cases of many γ rays, with multiplicities of the order of 10–30. The γ-ray energies range from ~ 100 keV to several MeV with average values around $\langle E_\gamma \rangle = 1$ MeV. The sum spectrometer has to detect most of the emitted γ rays with a large efficiency. Therefore, the spectrometers have to be very large and surround the target in approximately a 4π geometry. When the multiplicity is very large, one can also obtain a good indication of the total energy with spectrometers which do not have a full 4π solid angle. In that case, however, the spread in energy increases with decreasing Ω, as discussed below. The energy resolution has to be reasonably good in order to select effectively a certain interval in the total γ energy. A large size NaI(Tl) crystal is most commonly used as a sum spectrometer because of the large solid angle (0.7–0.9 of 4π) with a reasonable energy resolution of $\Delta E/E \leqslant 10\%$ at 661 keV.

The intensity transmission through the crystal can be calculated with the simple formula

$$N_\gamma = N_{\gamma 0} \exp(-\lambda x), \tag{3.2.43}$$

where λ is the linear absorption coefficient or $1/\lambda$ is the mean free path before an interaction (collision) occurs. Such a collision removes the γ ray from the original beam and decreases the intensity as expressed by eqn (3.2.43). The typical values for NaI(Tl) are $\lambda = 0.36\ \mathrm{cm}^{-1}$ for 0.5 MeV γ rays, 0.22 for 1 MeV, and 0.13 for 5 MeV. This means that 0.5 MeV γ rays are absorbed by 99.7 per cent in 16 cm NaI(Tl) and by 94 per cent in 8 cm, while for 1 and 5 MeV γ rays these numbers become 97 per cent, 83 per cent and 88 per cent, 65 per cent, respectively. Note that absorption here means at least one interaction, i.e. a Compton scattering or more scatterings or photoabsorption. This implies that for a good spectrometer one needs a crystal thickness of at least 8 cm, but preferably 16 cm when γ rays with an energy of a few MeV are also expected.

A crucial feature in the design of a sum spectrometer is the detection efficiency Ω. One can write Ω as

$$\Omega = \varepsilon\omega/4\pi, \tag{3.2.44}$$

where ω is the solid angle in steradians and ε is the fraction of the total γ energy released in the detector. The fraction ε depends on the energies $E_{\gamma i}$ of the individual γ rays

$$\varepsilon = (\sum E_{\gamma i}\varepsilon_i)/\sum E_{\gamma i}, \tag{3.2.45}$$

where $\varepsilon_i = fN/N_0$, with f denoting the probabilities for full absorption; i.e. $f = 1$ for photoeffect and $f < 1$ for the Compton and other scatterings.

The average sum energy $\langle E \rangle$ deposited in the crystal, which is a fraction of the total γ energy, depends on the multiplicity M of the γ-ray cascades, on the average γ-ray energy $\langle E_\gamma \rangle$ of the individual γ rays and on the efficiency Ω.

$$\langle E \rangle = \Omega M \langle E_\gamma \rangle. \tag{3.2.46}$$

The spread in this measured sum energy $\langle E \rangle$ also depends on the multiplicity of the cascades and on the efficiency Ω as expressed by the standard deviation of the energy

$$\sigma = \langle E_\gamma \rangle \{M\Omega(1-\Omega)\}^{1/2}. \tag{3.2.47}$$

This expression can be understood by noticing that $M\Omega$ is the number of observed γ rays and $(1-\Omega)$ is a scaling factor. For $\Omega = 1$ all γ rays are detected and no fluctuation in the number is measured. For very small Ω the fluctuation becomes largest and approaches the common statistical uncertainty $(M\Omega)^{1/2}$. In expression (3.2.47) it is assumed that all cascades have the same $\langle E_\gamma \rangle$. In practice, however, a spread in this average γ-ray energy also contributes to σ.

The relation between the standard deviation and the full width at half maximum (fwhm) of the total energy is, for large M, $\mathrm{fwhm} = 2(2\ln 2)^{1/2}\ \sigma = 2.35\sigma$. Thus the relative energy spread is

$$\mathrm{fwhm}/\langle E \rangle = 2.35\ \{(1-\Omega)/M\Omega\}^{1/2}. \tag{3.2.48}$$

As an example, take a cascade with multiplicity $M = 30$ and an average γ-ray energy of 1 MeV. For an efficiency of $\Omega = 0.9$ we find from eqns (3.2.46) and (3.2.48) an average energy of $\langle E \rangle = 27$ MeV and fwhm = 14.3 per cent, while these figures become 24 MeV and 21.5 per cent for $\Omega = 0.8$. The corresponding spreads in energy are 3.9 and 5.1 MeV, respectively. The overall spread has still to be enlarged by the intrinsic energy resolution. This resolution of $\leqslant 10$ per cent, however, is smaller than the spreads discussed above. In practice, however, the spreads are larger than the calculated values, obtained from the above expressions, because of the spread in multiplicities and γ-ray energies. This is mainly due to the presence of various reaction channels, each with a different total γ-ray energy, thus giving rise to a spread in $M\langle E_\gamma \rangle$ as well. On the other hand, this feature also offers the possibility of selecting a particular exit channel of interest, as will be demonstrated below.

The mechanical construction can be carried out in various ways, but one has always to provide as much crystal as possible around the target to ensure a sufficiently large detection efficiency. Two separated large crystals below and above the target are commonly used, which leave enough space for the beam pipe and additional detectors. The spectrometer in use by the Amsterdam–Groningen collaboration is shown schematically in Fig. 3.16 with a picture in Fig. 3.17. The design is based on six separated segments, as in the spectrometer at GSI Darmstadt (Simon, 1984). The segments are assembled so that they form a cylinder 40 cm in diameter and 40 cm in length. An axial hole 8 cm in diameter leaves enough space for the beam pipe and for one or two Ge detectors, one with a hole allowing the beam to pass through to the target. The separation of the segments provides a good light collection in each of the photo tubes and enables a flexible arrangement in different experiments. The spectrometers are made by the Harshaw company in The Netherlands. The configuration shown in Figs 3.16 and 3.17 consists of three NaI(Tl) segments above the beam pipe separated from three segments below it. In this way about six Ge detectors can be placed in the horizontal plane close to the target. The loss of detection efficiency may be compensated partly by a few small (BGO) crystals placed near the target. Those detectors serve at the same time as a simple multiplicity filter. In this respect one should realize that in a sum spectrometer, detection efficiency is emphasized rather than use as a multiplicity filter. The number of segments is usually much smaller than the multiplicity and the cross-talk very large. This makes the sum spectrometer a rather poor multiplicity filter.

Examples of sum spectra are presented in Fig. 3.18 for ^{12}C and ^{40}Ar-induced reactions, both leading to ^{152}Dy as the main evaporation product. The sum energy is ~ 11 MeV in the first case and ~ 17 MeV in the latter one. This difference is most likely due to the larger excitation energy above the yrast line of the compound nucleus ^{156}Dy, formed with the ^{40}Ar beam, $E^*_{\text{cn}} = 68.2$ MeV, as compared to 55.8 MeV for the ^{12}C-induced reaction.

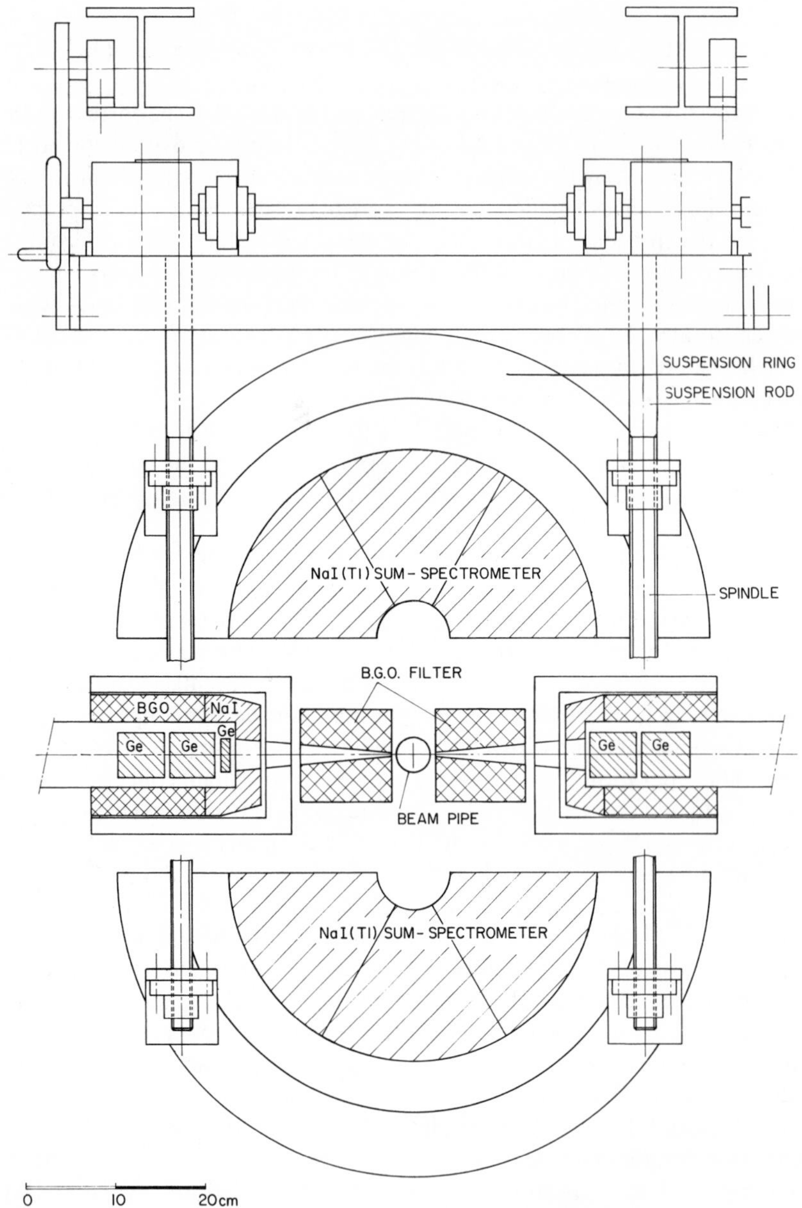

Fig. 3.16. Schematical representation of the sum spectrometer, separated in the horizontal plane to allow space for the Ge–BGO assemblies. In the centre, near the beam pipe, a few BGO crystals are designed to enhance the total efficiency. The assembly is in use by the Amsterdam–Groningen γ-ray spectroscopy groups.

Fig. 3.17. Picture of the sum spectrometer with other detectors in the horizontal plane as shown schematically in Fig. 3.16. In the centre (at 90°) we see a BGO cylinder with 8 phototubes around a Ge detector. The sum spectrometer is manufactured by Harshaw Chemie, The Netherlands.

Because of the dependence of the sum energy on the multiplicity we may select the high-spin region with a high sum–energy gate on the total energy released in the sum crystal (see Sections 4.3 and 6.5.7).

An example of exit-channel selection is given in Fig. 3.19 for ^{12}C-induced reactions. A reaction channel with a lower neutron multiplicity is associated with a higher total γ-ray energy. Thus a high sum–energy gate usually selects evaporation residues with low neutron multiplicity. Another valuable use of the multi-segment sum spectrometer is to discriminate low-multiplicity channels such as radioactive decays, or transfer reactions and Coulomb excitations. This can be done by requiring a minimum number of segments to fire and/or to impose a certain energy threshold. Also, X-ray emission can be discriminated in this way. The sum spectrometer can also serve as a time trigger with a high efficiency, because of the fast response of the NaI(Tl) crystal

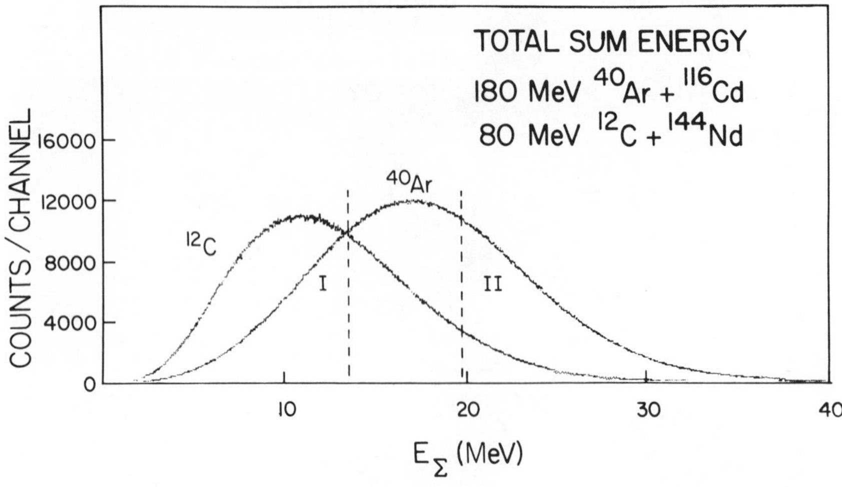

Fig. 3.18. Sum-energy γ-ray spectra measured with the assembly shown in Fig. 3.17, for ^{12}C and ^{40}Ar-induced reactions both leading to ^{152}Dy as the main reaction residue.

with respect to the Ge detector. Note that the time resolution is determined by the time response of the crystal and not by the passing time of the photons.

3.2.4 *Crystal balls*

In the previous sections we discussed multiplicity filters and sum spectrometers, which emphasize the measurements of the multiplicity moments and total γ-ray energy, respectively. It is shown that the statistical uncertainties in those quantities are rather large. In fact, no specific multiplicity of a single γ-ray cascade can be measured; but only the average value and the first one or two central moments of a multiplicity distribution, which is determined as an average over many cascades. This is due to the fact that for each reaction event only some of the γ rays are detected in a few crystals and that there is a certain probability for multiple hits in those crystals. Part of this problem can be solved by a crystal ball, which covers in principle 4π of the solid angle with many detectors.

The special property of a crystal ball is that one can measure in an event-by-event mode several quantities of the individual γ-ray cascades simultaneously. These quantities are the γ-ray energies, multiplicities, multipolarities, angular correlations, energy–energy correlations, total energies and times of emission in the nanoseconds range. Because $N \gg M$ one measures quantities of

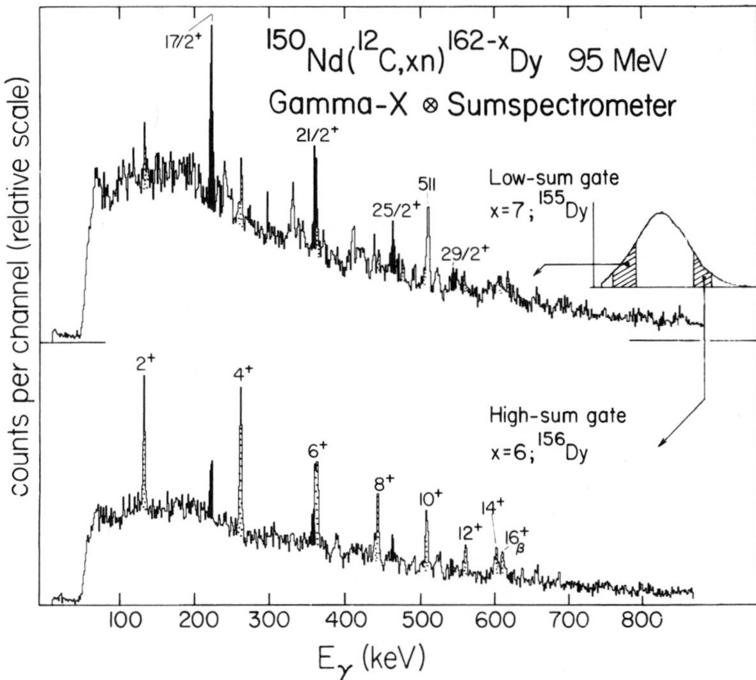

Fig. 3.19. An example of exit-channel selection with the sum spectrometer of Fig. 3.17 for a ^{12}C-induced reaction. In this case the gap between the two halves was closed and a Ge detector was put at 0°.

individual cascades rather than distributions of them. Consequently it is possible to set gates on multiplicities or to measure the multiplicity spectra directly. It is even possible to obtain linear polarizations of γ rays by measuring the correlations of Compton scattering. The possibility to set gates on both the multiplicity and total energy enables one to select the entry points (regions) in the energy–angular momentum plane much more precisely than is possible with the other devices (see Sections 4.3 and 6.5). For measurements of large multiplicities with a high efficiency we need many large-size detectors. The NaI(Tl) crystals satisfy this criterion for a reasonable price, but one sacrifices energy resolution in the detection of the individual γ rays. This has been experienced as such a serious set-back for γ-ray spectroscopy that newer spectrometers have been developed that include many Ge detectors, as will be discussed in the next subsection.

As mentioned above it is important to avoid multiple γ rays triggering one detector in the same event. This requires that the number of detectors N is much larger than the maximum multiplicity M_{max}. In heavy-ion induced fusion reactions M_{max} may reach values of 30 to 40 and in deep inelastic processes, for instance, with two highly excited nuclear fragments, M_{max} may reach values over 50. Therefore most crystal balls have $N > 50$ and the one at Heidelberg–GSI has as many as 162 NaI(Tl) detectors (Habs *et al.*, 1979). Some properties of the three major crystal balls presently in use are summarized in Table 3.2.2. They all consist of specially shaped scintillation crystals of NaI(Tl), BaF_2, or BGO (the latter material is used in spectrometers as discussed in the next subsection).

An overview of the crystal ball at Heidelberg–Darmstadt is shown in Fig. 3.20. A detail of tapered NaI(Tl) crystals with a pentagon surrounded by five hexagons is shown in Fig. 3.21(a). One of the problems in such a closely packed assembly is the scattering from one to another detector (cross-talk). Figure 3.21(b) shows a spectrum of the 1.172 MeV γ ray measured with the central hexagon with and without rejection of the Compton scattered events into the adjacent crystals.

The formula that describes the k-fold coincidence probability P_{Nk}^{M} as a function of multiplicity M is the same as that given by eqn. (3.2.12). This formula has now to be generalized to include Compton scattering. Let f be the fraction of γ rays that produces a secondary γ ray via Compton scattering and enters an adjacent detector. For $f = 0$ no additional γ rays are produced

Table 3.2.2 Some properties of three crystal balls

Property	Oak-Ridge Spin Spectrometer	Heidelberg–Darmstadt Crystal ball	France Chateau de Cristal
Material	NaI(Tl)	NaI(Tl)	BaF_2
Shapes	pentagons hexagons	pentagons hexagons	hexagons
Number	72 fixed	162 fixed	74 extensible
Radius (cm) (inner–outer)	17.8–35.6	22.8–38.1	21.6–31.6 (cm)
Mean solid angle/4π	1.38%	0.62%	1.35%
Resolution			
energy[†]	8%	7%	10%
time[†]	2.8 ns	3.2 ns	<1 ns

[†] For 661 keV.

Fig. 3.20. The Heidelberg–Darmstadt crystal ball which consists of 162 NaI(Tl) crystals in a 4π geometry (see Fig. 3.21 for details). (Annual Report, 1981, Max Planck Institut, Heidelberg.)

and for $f=1$ each initial γ ray yields two γ rays of which one enters an adjacent detector. The probability that i additional γ quanta are produced out of M original ones via Compton scattering into the adjacent crystals is given by the binomial element g_i on the basis of the same reasoning as for eqn. (3.2.8).

$$g_i = \binom{M}{i} f^i (1-f)^{M-i}. \tag{3.2.49}$$

The probability P_{Nk}^{M} including Compton scattering can be written as a product of the probabilities that $m+i$ γ rays trigger k detectors, given by eqn (3.2.12), and that i of them are Compton scattered, according to eqn (3.2.49), summed over all possible additional quanta i from 0 to M.

$$P_{Nk}^{M} = \sum_{i=0}^{M} P_{Nk}^{M+i} g_i. \tag{3.2.50}$$

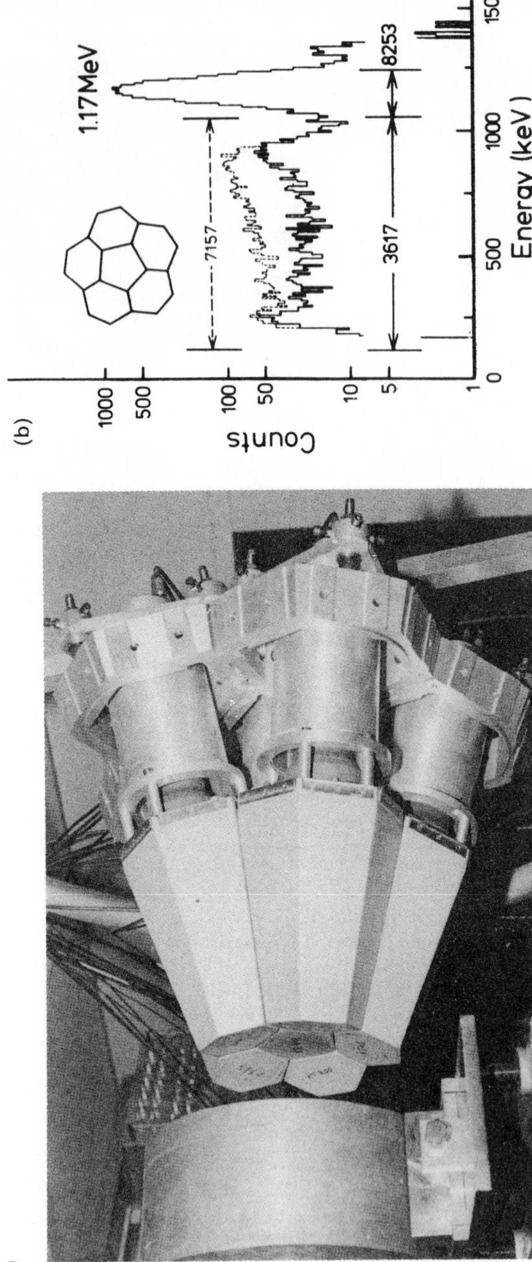

Fig. 3.21. Detail of the crystal ball at Heidelberg, showing a central pentagon crystal surrounded by five hexagons, all 20 cm long (a). The response function for the central pentagon for the 1.17 MeV γ-ray from a ^{60}Co source without Compton suppression (upper curve) and with suppression (lower curve) by the outer hexagons (b). (Albrecht *et al.*, 1980.)

Combining eqn (3.2.12) and (3.2.50) gives

$$P_{Nk}^{M}=\binom{N}{k}\sum_{n=0}^{k}(-1)^{k-n}\binom{k}{n}\left[\sum_{i=0}^{M}\{1-(N-n)\Omega\}^{M+i}\binom{M}{i}f^{i}(1-f)^{M-i}\right]. \tag{3.2.51}$$

Note that the term in the square brackets is $(1-x)^M$ times the binomial expansion of $\{(1-f)+f(1-x)\}^M$, with $x=(N-n)\Omega$. The final result can then be written as

$$P_{Nk}^{M}=\binom{N}{k}\sum_{n=0}^{k}(-1)^{k-n}\binom{k}{n}\Big[\{1-(N-n)\Omega\}\{1-f(N-n)\Omega\}\Big]^{M}. \tag{3.2.52}$$

This probability as a function of the number of hits is given for various multiplicities in Fig. 3.22 in the case of 50 detectors, 100 per cent efficiency and 20 per cent Compton scattering between the crystals. We notice that for low

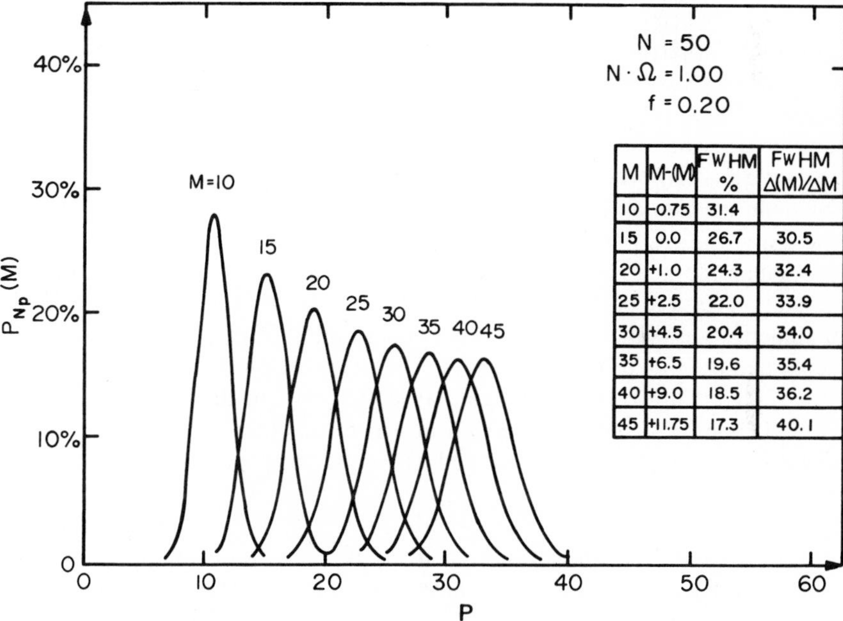

M	M-⟨M⟩	FWHM %	FWHM Δ(M)/ΔM
10	-0.75	31.4	
15	0.0	26.7	30.5
20	+1.0	24.3	32.4
25	+2.5	22.0	33.9
30	+4.5	20.4	34.0
35	+6.5	19.6	35.4
40	+9.0	18.5	36.2
45	+11.75	17.3	40.1

Fig. 3.22. The coincidence probabilities are calculated as a function of the number of hits $k=p$ in $N=50$ detectors, for various multiplicities. The insert shows at each multiplicity value used the deviation of the mean value $\langle M\rangle$ from the taken value, the fwhm in per cent and the resolution as the fwhm divided by the non-linearity of the multiplicity scale; see text. (Habs *et al.*, 1979.)

multiplicity the number of hits k is larger than the multiplicity M and smaller for large M, due to the multiple scattering and the finite efficiency. It is also apparent that for each event the k-fold probability yields a certain multiplicity within a certain resolution. The relative fwhm (in %) of the bell-shaped multiplicity response function decreases with increasing M, but unfortunately the resolution also decreases. Therefore, as a figure of merit for the assembly, one divides the fwhm by the non-linearity of the multiplicity scale $\Delta\langle M\rangle/\Delta M$ (the ratio of the measured and real multiplicity differences). This figure of merit is then taken as the multiplicity resolution. The resolution improves strongly with increasing N and also with efficiency Ω, but for $N > 200$ the pay-off becomes less significant due to other limiting factors. The negative influence of Compton scattering decreases with increasing N, but remains rather constant after f has reached a value of ~ 0.15. For the Heidelberg crystal ball with $N = 162$, $N\Omega = 90$ per cent and with $f = 25$ per cent a multiplicity resolution is obtained of about 20 per cent. The same resolution is found for the total energy, using eqn (3.2.48). These figures imply that with such a spectrometer, for instance, entry points in the energy–angular momentum space can be selected in an event-by-event mode with a resolution of 20 per cent in both dimensions.

The spin spectrometer at Oak Ridge has a fixed number of 72 detectors and other properties similar to the one at Heidelberg–Darmstadt, as can be seen in Table 3.2.2. A picture of the spin spectrometer is given in Fig. 3.23. Both crystal balls have a comparable time resolution of 2–3 ns. A good time-resolution is important for the discrimination between γ rays and neutrons. Neutron capture and inelastic scattering in the NaI(Tl) crystals distort the measured γ-ray multiplicity, total energy and angular distributions. In particular the neutron capture deposits a lot of energy (8–10 MeV) in the crystals. The time of flight of 1 MeV neutrons, for instance, is 72 ns/m. With an inner radius of the crystal balls of approximately 20 cm, those neutrons are separated by ~ 14 ns from the γ rays. Thus with the time resolution of ~ 3 ns fast neutrons up to ~ 20 MeV can, at least partly, be discriminated from γ rays. This means that all evaporation neutrons, which have energies up to only ~ 4 MeV can be discriminated as well as most of the pre-equilibrium neutrons (see Section 4.2).

The importance of the time resolution is emphasized in the design of the Crystal Castle in France. There, a different scintillation material is used, namely BaF_2, arranged in rings of hexagons around the target, as is illustrated in Fig. 3.24. The number of crystals per ring are 1, 6, 12, 18, 24, etc. One can place two of each ring around the target chamber, one ring above and one below the target. Thus the total number of detectors becomes 2, 14, 38, 74, 122, etc. The first design aims at 74 detectors and can be extended to 122. As is shown in Table 3.2.1, BaF_2 produces a $\lambda = 220$ nm light component with a very fast decay constant. A time resolution of 112 ps (fwhm) has been achieved for a ^{60}Co source with a small BaF_2 crystal as a stop detector and a small

Fig. 3.23. A view of the spin spectrometer at Oak Ridge which can be separated into two halves. (Annual Report ORNL, 1981.)

plastic scintillator as a start detector. The time spectra of a Californium source at 12 cm distance from the BaF_2 and NaI detectors are compared in Fig. 3.25. The separation between the γ rays and the (low energy) neutrons can only be achieved in this case with BaF_2 and not with NaI.

With the crystal balls discussed in this subsection it is possible to replace one or more crystals by a high-resolution Ge detector and to use the surrounding crystals as Compton suppression shields. This, however, implies that some of the unique features which have been obtained at high cost are sacrificed. Therefore the crystal balls are not used extensively for high-resolution γ-ray spectroscopy, but rather for reaction–mechanism studies. In these studies one can easily place particle detectors inside the balls, most favourably low-Z and low-mass detectors such as parallel-plate avalanche detectors to avoid absorption of γ rays. Gamma-ray spectroscopy is nowadays carried out with a combination of detectors as will be discussed below.

3.2.5 *High-resolution multi-detector assemblies*

It has become clear in the last few years that a full elucidation of nuclear properties, for instance at high spin, can only be obtained with a high resolution of the γ-ray spectrum. This has motivated several groups to

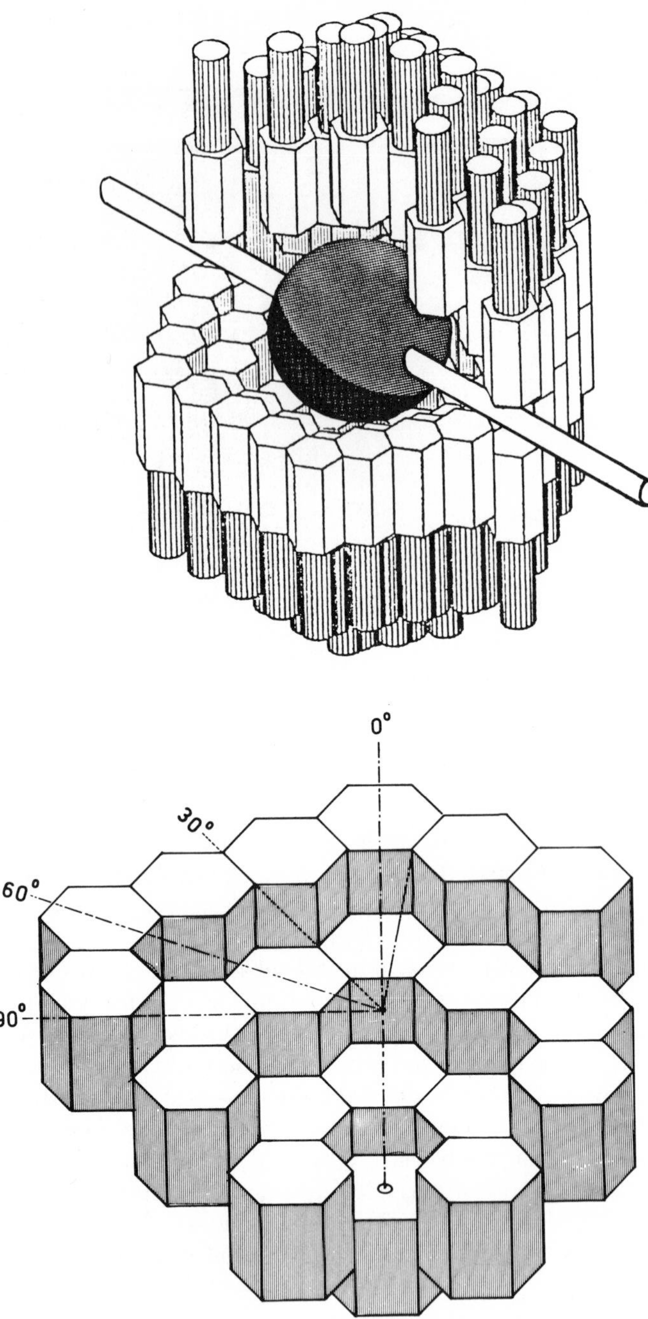

Fig. 3.24. Sketch of the 'Château de Cristal' array showing (a) partial view of the BaF_2 scintillators and phototubes with the beam entrance and reaction chamber; (b) part of the arrangement of the BaF_2 crystals into six concentric rings plus two detectors. (Vivien, 1983.)

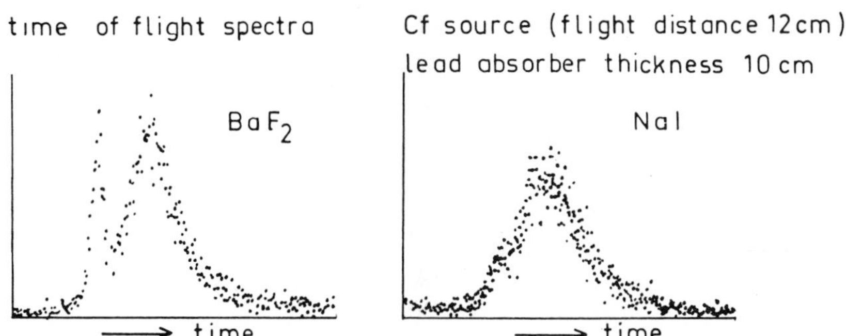

Fig. 3.25. The time spectra of a Cf source measured with a small BaF_2 and a NaI(Tl) crystal versus a small plastic scintillator (relative time scale). The γ-peak is only separated from the broad neutron distribution with the BaF_2 crystal. (Vivien, 1983.)

develop multi-detector systems with a higher resolving power than is achieved by scintillation material like NaI(Tl). The state of the art Ge detectors with a high resolution of about 1.8 keV at fwhm for 1.3 MeV γ rays are the most favourable. The high resolution, which is about a factor of 30 better than for NaI(Tl) detectors, can only be achieved by semi-conductor detectors which are relatively small in size. The best efficiency for commonly used N-type Ge detectors amounts to 25–30 per cent of that of a 7.6 × 7.6 cm NaI detector. Therefore one needs many of those costly spectrometers.

Since the high-resolution detectors cover only a small fraction of the solid angle, one still needs a highly efficient crystal ball for selection of the reaction products, total energy and multiplicity. This ball has to be very compact in order to position the Ge detectors close enough to the target. The BGO scintallation material is dense enough to construct compact assemblies with a moderately large number of crystals. A BGO ball of about forty-five detectors and an outer diameter of about twenty centimetres reaches efficiencies slightly smaller than those of the big NaI crystal balls, namely 80–90 per cent for 1 MeV γ rays. In this case the resolution for the total energy is about 20 per cent, but for the multiplicity about fifty per cent, due to the relatively small number of detectors, which is considerably worse than for the crystal balls. Thus the actual trade-off is between γ-ray energy and multiplicity resolution when comparing those assemblies and crystal balls.

An arrangement of this type already in use for several years is the TESSA II spectrometer at Daresbury, shown schematically in Fig. 6.17a (Twin *et al.*, 1983). It consists of 6 Ge detectors surrounded by NaI(Tl) anti-Compton shields and a BGO ball of 62 hexagonal pieces inside. Because of the large NaI shields the distance of the centres of the Ge detectors to the target is 27 cm.

Although this large distance reduces the efficiency of a single Ge detector, the small solid angle of 0.16 per cent, for one Ge detector after collimation, has the advantage of a limited Doppler broadening (see Section 3.5). The BGO ball consists of 62 crystals of hexagonal shape (38 mm face to face, 50 or 75 mm in length) arranged in a honeycomb. It is intended that the NaI anti-Compton shields will be replaced by BGO shields of the Liverpool design (see Section 3.2.1). In this way one can greatly increase the number of Compton-suppressed Ge detectors, with obvious advantages discussed below.

The largest number of Ge detectors has been achieved so far at Berkeley. The High-Energy-Resolution-Assembly (HERA) comprises three rings each with seven Ge detectors. A BGO cylinder which consists of forty-four elements is placed in the centre. A schematic picture of part of the arrangement is shown in Fig. 3.26, while a picture of the assembly without the central BGO ball is shown in Fig. 3.27. The Ge detectors view the target through gaps

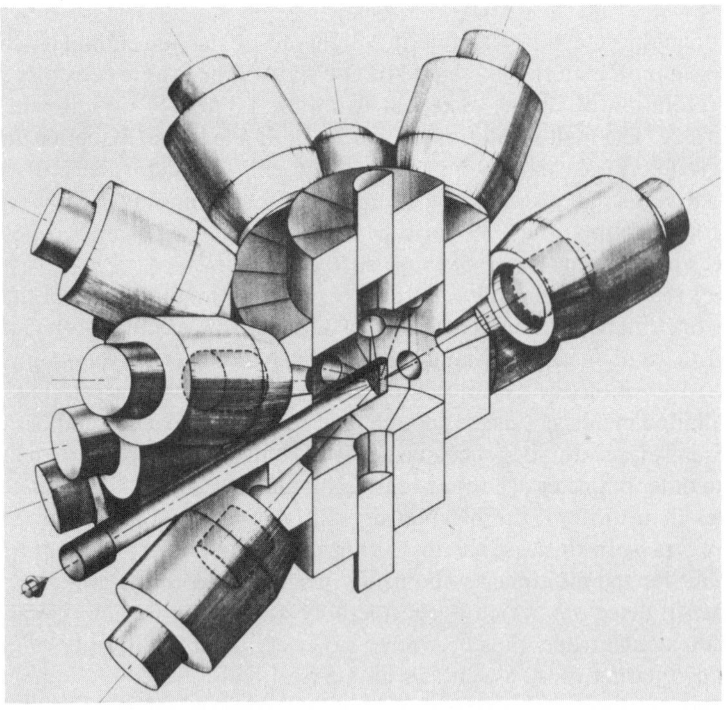

Fig. 3.26. A perspective view of the HERA spectrometer at Berkeley, which consists of 21 Ge detectors with BGO CSS and a cylindric assembly of 44 BGO crystals in the centre. (Stephens, 1982.)

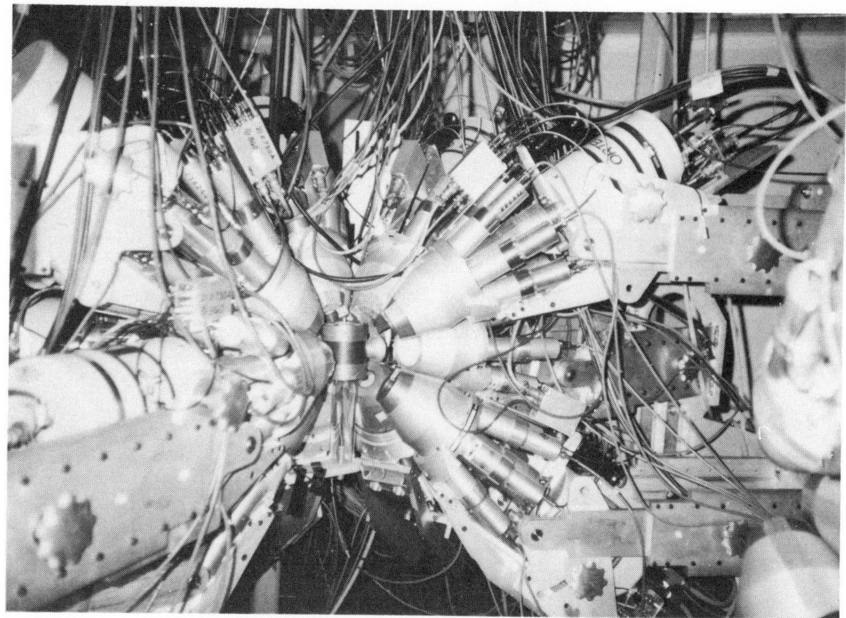

Fig. 3.27. A picture of HERA (see Fig. 3.26) which is separated into two halves. In the centre the small scattering chamber is visible. No BGO assembly at the inner space is placed in this picture.

in the BGO elements. In this way the BGO elements serve as active collimators for the anti-Compton spectrometers of which the BGO cylinders have to be screened from direct radiation of the target. The front surfaces of the Ge detectors are only fourteen centimetres from the target. The BGO ball reaches a solid angle of ~80 per cent, the loss of ~20 per cent is mainly due to the gaps.

The BGO anti-Compton cylinders are 12 cm long and 10.5 cm in diameter. With a Ge cryostat cap of 6 cm in diameter a BGO wall thickness of just over 2 cm has been achieved. This does not yield the most favourable suppression factor (see Section 3.2.1), but one still achieves a peak-to-total ratio of 0.5. This has to be compared with a ratio of 0.15 for a bare Ge detector. This improvement is particularly impressive for the many fold coincidences, as can be seen from Table 3.2.3. It is believed that new physics has to be deduced from triple and higher-fold coincidences. In the important measurements of γ–γ energy correlation (see Chapter 6) the number of full-energy events improves by a factor of 10 from only 2.2 per cent to 25 per cent with CSS.

In the lower part of Table 3.2.3 the event rates are given for an assumed reaction rate of 10^5 per second and a multiplicity of twenty. It can be seen that

Table 3.2.3 Performance of bare and Compton-suppressed Ge detectors†

	Singles	2-fold	3-fold	4-fold
Bare	0.15	0.022	0.0034	0.0005
Peak/total				
Compton supp.	0.5	0.25	0.13	0.06
Improvement	3	10	30	100
event rates for 10^5 triggers/s and $M_\gamma = 20$				
5 detectors at 17 cm		230	3.3	0.02
5 detectors at 12 cm		870	25	0.35
21 detectors at 12 cm		11000	2200	280

† Ge detector 20 per cent efficient at 1.33 MeV; c.f. $7.6\phi \times 7.6$ cm for NaI detector

only with twenty-one detectors can a reasonable triples count rate of 2200 per second be achieved and only 1.3 days are needed to collect 250 million triple events.

Several ambitious projects have been developed by a number of European countries; such as the Essa-30 spectrometer at Daresbury, the Nordball at Copenhagen and more recently the Superball at the USA and Euroball at Europe, both with over one hundred Ge detectors. The design of the Nordball allows one to assemble many different types of detectors in order to carry out a wide variety of experiments (Herskind, 1985). The modular Nordball frame has a truncated icosahedron geometry with a diameter of 66 cm, and can be separated into two parts as shown in Fig. 3.28(a). The holes for hexagon-shaped detectors are intended for twenty large units of a BGO–Ge CSS with an outer diameter of 20 cm ϕ, while twelve smaller pentagon-shaped detectors with a diameter of 17 cm ϕ can be placed in the smaller holes. A cross-section through the centre of the ball is shown in Fig. 3.28(b).

A new aspect of the system is the inner BaF_2 ball, acting as a calorimeter for the measurement of the γ-ray sum energy and multiplicity. The sixty identical concentric elements of BaF_2 crystals form a shell around the target with a thickness of 8 cm and with an inner diameter of 12 cm, as is shown in Fig. 3.28(c). The measured time resolution of $\Delta T < 300$ picoseconds makes it possible to discriminate against neutrons by means of the time-of-flight technique. This timing property is much better than that of the commonly used BGO material, as discussed in Section 3.2.4. Also the tolerable count rate in BaF_2 is an order of magnitude higher than in BGO because the pile-up rejection time can be diminished to a few nanoseconds. This opens up the possibility of studying extremely weak reaction channels in the presence of a high background count rate, as is the case with deep–inelastic scattering. Such type of reactions may give access to high-spin states in nuclei near the stability line, which can only be studied with multiple Coulomb excitation today or are not produced at all. Moreover, one can excite states with higher spin in the high angular–momentum transfer reactions than is possible in multiple Coulomb excitations.

The CSS are also improved in the design by utilizing Monte Carlo calculations. The BGO cylinders have a conical outer shape to reduce internal trapping of scattered light. The NaI nose in the Liverpool and Dutch designs (see Figs 3.7 and 3.9, respectively) is replaced by BGO as to improve the rejection in the Compton-edge region. Segmented Ge detectors are used because of the advantages pointed out in Section 3.2.1. In this case, however, a double telescope consists of a 1-cm-thick detector in front of a large N-type

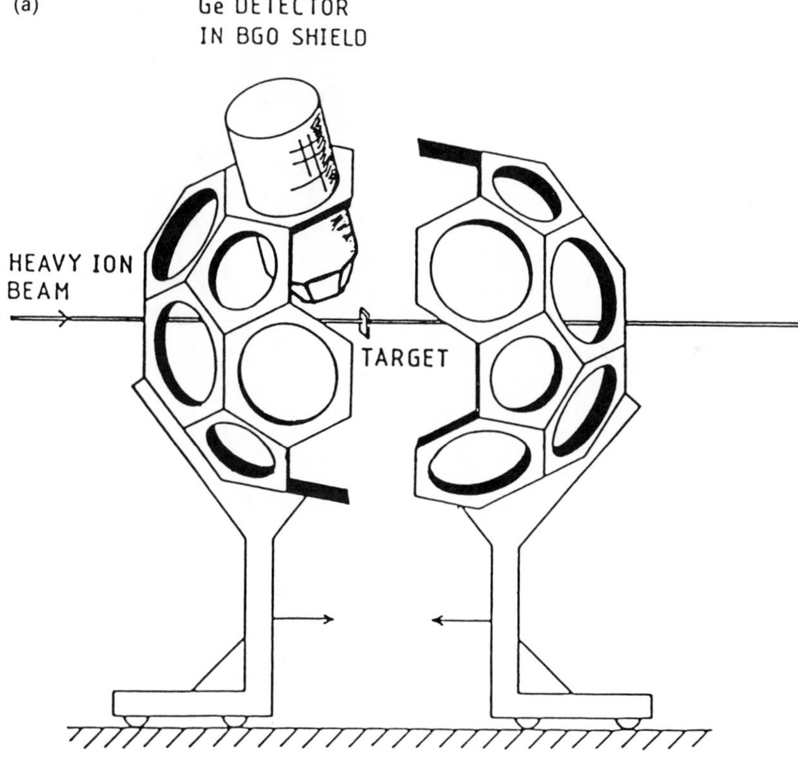

Fig. 3.28(a).

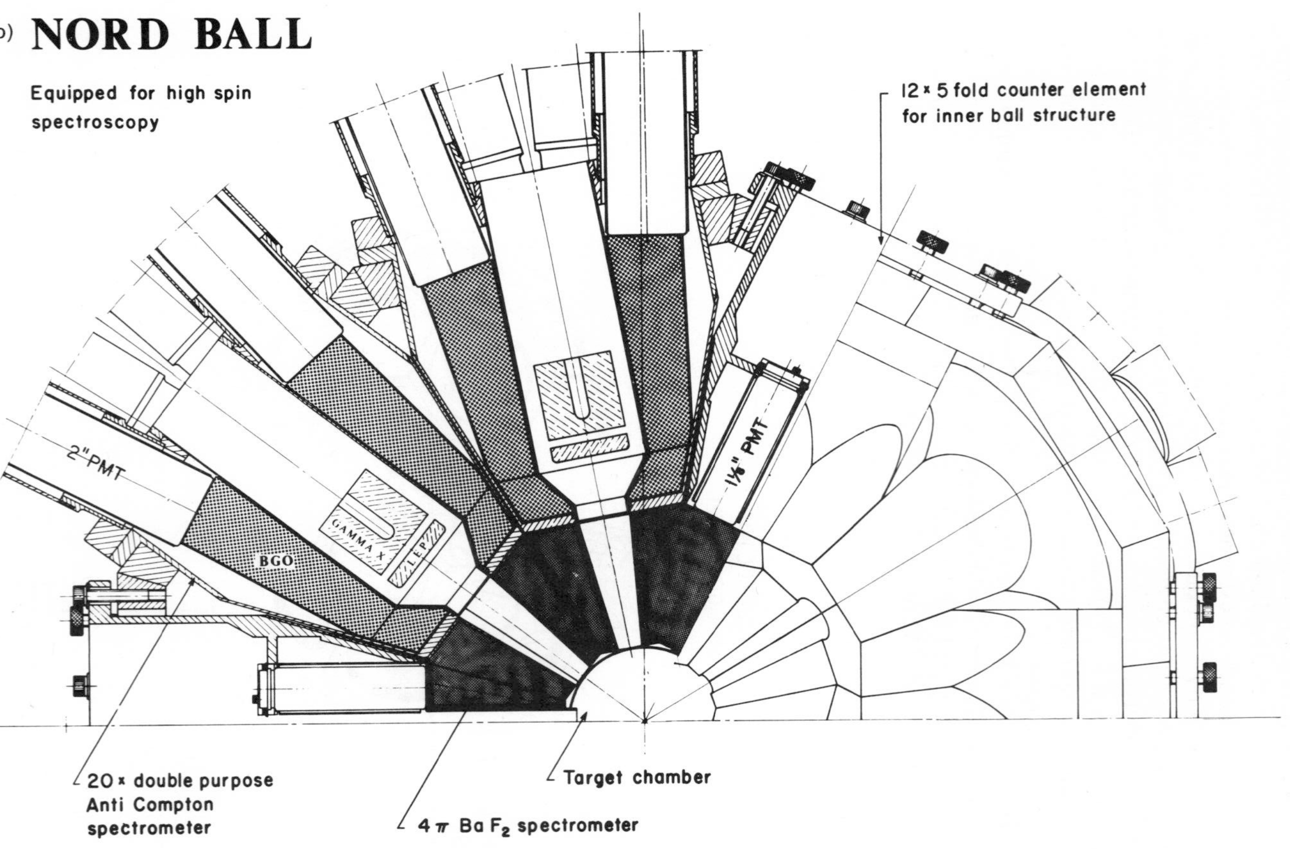
(b) NORD BALL
Equipped for high spin spectroscopy
12 × 5 fold counter element for inner ball structure
2" PMT
BGO
GAMMA X
LEP
1⅝" PMT
20 × double purpose Anti Compton spectrometer
Target chamber
4 π Ba F_2 spectrometer

Fig. 3.28(b).

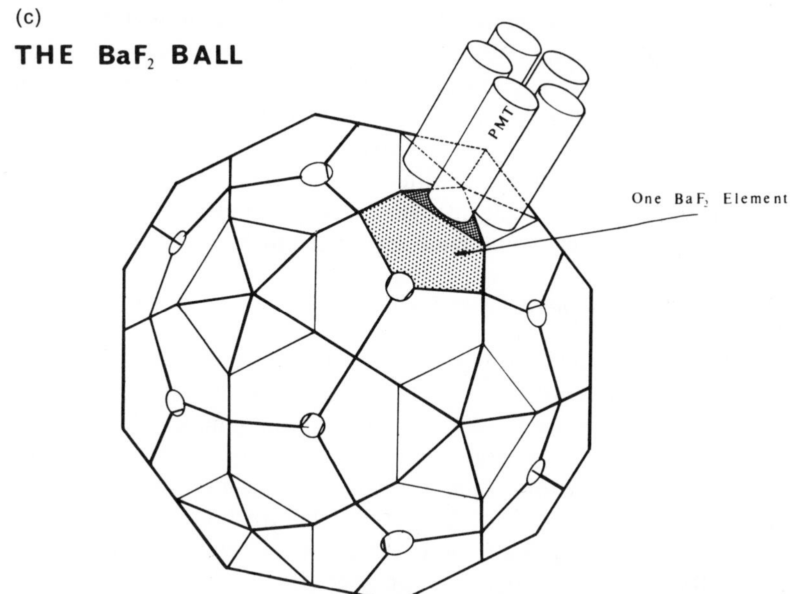

Fig. 3.28. The modular NORDBALL multi-detector assembly. Shown are the frame, 66 cm in diameter, which can accommodate 20 large detectors and 12 smaller ones (a). A cross-section through the centre of the ball (b). The black areas indicate the BaF_2 calorimeter and the grey ones the BGO CSS. This BaF_2 ball is shown in (c) consisting of 60 identical crystals (see shaded area). The light is collected at the triangle-shaped parts, where photomultipliers are mounted. The ball acts as an active collimator in front of the CSS with Ge detectors, which 'see' the target through the holes. (Herskind, 1985.)

Ge detector to optimize the rejection of backscattered γ rays from that Ge detector into the front detector. Conversely the front detector can be used as a low-energy detector with a superior Compton suppression by the back detector and the BGO shield. This additional feature can be of special interest, for instance in tracing low-energy M1 transitions to investiagate the magnetic properties of nuclear states under extreme conditions.

The Essa-30 (Polytessa) setup at Daresbury can be seen as a successor to the very successful Tessa-II spectrometer. This assembly is identical to the Nordball design, which allows the exchange of detectors between the two spectrometers. The basic design of the CSS is of the Liverpool type, as discussed in Section 3.2.1 (see Fig. 3.7). It is the largest high-resolution multi-detector spectrometer in use so far with superior background reduction.

The rapid development of detector systems as discussed in this section, combined with the access to a variety of heavy-ion beams, including radioactive beams, promises to prolong the renaissance in nuclear structure physics of the past ten years by another decennium. The emphasis may be put on the highly excited states in the spin region (40–70)ħ, which would then cover states up to the fission limit. A forseeable technical development would be the construction of a spectrometer covering the full 4π solid angle entirely with high-resolution Ge detectors with ~ 100 per cent absorption for 1 MeV γ rays. One could dream of a gigantic germanium crystal, segmented into very many small parts, which would remove the need for Compton veto material.

3.2.6 *Corrections of energy–energy correlations*

Both γ–γ energy correlation matrices displayed in Figs 6.27 and 6.28 (see Chapter 6) have been corrected for the background of uncorrelated events. The correction procedure is aimed at reducing the number of uncorrelated events not related to (full-energy) photo–photo events, which still amount to about 80 per cent of the total number of coincidences in Compton-suppressed data (see discussion in Section 3.2.1). A trivial correction for random events can easily be made by using a time gate in the time-difference spectrum.

The method introduced by Andersen *et al.* (1979) is based on the assumption that the number of uncorrelated events N_{ij}^{un} (excluding accidental ones) amounts approximately to the number of observed events N_{ij}^{exp} (the index i indicating small energy intervals in the spectrum of detector 1 and j in detector 2). It is thus assumed that all events are uncorrelated. This may be approximately true for bare Ge-detectors, but not for Compton suppressed spectrometers (see below). The probability of finding events in the interval i of detector 1 is given by $\Sigma_l N_{il}/\Sigma_{kl} N_{kl}$ and in the interval j of detector 2 by $\Sigma_k N_{kj}/\Sigma_{kl} N_{kl}$. Here the index l runs over the whole energy range of detector 2 and k over that of detector 1. The total number of calculated events is given by $\Sigma_{kl} N_{kl}$. The number of uncorrelated events (at the point i, j in the matrix) is given by the product of the two probabilities, multiplied by $\Sigma_{kl} N_{kl}$ as:

$$N_{ij}^{\text{un}} = (\Sigma_l N_{il} \times \Sigma_k N_{kj})/\Sigma_{kl} N_{kl}. \tag{3.2.53}$$

In first approximation one obtains the number of correlated (full energy) events $N_{ij}^{(1)}$ by subtracting N_{ij}^{un} from the observed number N_{ij}^{exp}

$$N_{ij}^{(1)} = N_{ij}^{\text{exp}} - N_{ij}^{\text{un}}. \tag{3.2.54}$$

Due to the underlying assumption of this method one may have subtracted too many counts in eqn (3.2.54). The background of truly uncorrelated events is certainly lower than N_{ij}^{un}. One may correct eqn (3.2.53) for the number of correlated events obtained from eqn (3.2.54) and repeat this in an iterative procedure (Herskind, 1980). The result for the nth step is:

$$N_{ij}^{(n+1)} = N_{ij}^{\text{exp}} - [\Sigma_l\{N_{il} - |N_{il}^{(n)}|\}\Sigma_k\{N_{kj} - |N_{kj}^{(n)}|\}]/\Sigma_{kl}\{N_{kl} - |N_{kl}^{(n)}|\} \tag{3.2.55}$$

The first-order result of eqn (3.2.54) will certainly yield false results for Compton suppressed spectrometers, because the number of uncorrelated events N_{ij}^{un} is significantly smaller than the number of observed events N_{ij}^{exp}. One may explicitly take into account the peak to total efficiency ratios of the used spectrometers ($\varepsilon_1 = \varepsilon_2 \approx 0.40$ averaged over energy) and replace N_{ij}^{un} in eqn (3.2.54) by $(1-\varepsilon_1\varepsilon_2)N_{ij}^{un}$. This method was followed for the matrix displayed in Fig. 6.27. The iteration procedure, discussed above, should be applied with care to avoid the introduction of spurious bumps or pockets in the matrix beyond the statistical errors. Note that the relative errors in the spectrum after subtraction is much larger than in the original one, which diminishes the significance of these possible irregularities.

An unfolding method has been proposed by Arciszewski (1984) using explicitly the approximated response function of the Compton suppression spectrometers (see also Arciszewski *et al.*, 1983). The response function should be deduced from measured spectra for radioactive sources in the same geometry as in the actual on-line experiments. The Compton edge is approximated assuming an exponential shape of the form $\exp\{\alpha x(x-\beta)\}$. Here α and β are adjustable parameters and $x = E_\gamma/E'_\gamma$. The background at an energy E_γ is attributed here entirely to Compton-scattered events of γ rays with energy E'_γ (thus $E_\gamma < E'_\gamma$). The background is in this way calculated as:

$$N^{un}(E_\gamma, E'_\gamma) = \varepsilon N(E'_\gamma) \exp\{\alpha x(x-\beta)\}. \qquad (3.2.56)$$

Here ε represents the energy-dependent detector efficiency $\varepsilon = c(E'_\gamma)^{1/2}$ with c as an adjustable parameter. The number of correlated events with energy E_γ is given by:

$$N(E_\gamma) = N^{exp}(E_\gamma) - \sum N^{un}(E_\gamma, E'_\gamma), \quad \text{with} \quad E'_\gamma > E_\gamma. \qquad (3.2.57)$$

The correction procedure has to start at the highest energy E'_γ (only photopeak–photopeak coincidences assumed). This method has been tested for the reaction ^{24}Mg + 44 MeV ^{16}O with very low background of continuum γ-radiation. The results were comparable with the iteration method discussed above.

The most spectacular result so far, using the γ–γ-energy correlation method, has been obtained at Daresbury with the discovery of superdeformed band structures in the correlation pattern at very high spins in ^{152}Dy, a nucleus which is nearly spherical at low spin. This will be discussed in Section 6.5.

3.3 Gamma-ray spectrometers for medium–high energy radiations

3.3.1 *Medium energy gamma-ray spectroscopy*

Electromagnetic (EM) transitions in the medium–high-energy region of $E_\gamma = 10$ to 100 MeV are of great interest in the study of new aspects of nuclear structures. High-energy γ rays are necessarily concerned with highly excited states and/or nuclear reactions through unbound (continuum) intermediate states. Studies of medium–high-energy γ rays reveal EM properties and EM interactions associated with high excitations of $E_{ex}=10$ to 100 MeV, high momentum transfers of $p=0.1$ to 0.5 GeV/c, and with fast reaction dynamics of $t=10^{-22}$ to 10^{-20}s. These new domains of nuclear structures and nuclear reactions show aspects quite different from those of low excitations and low meomentum transfers, which are associated with low-energy γ rays ($E_\gamma \lesssim 10$ MeV) from bound states. Medium–high-energy γ rays are excellent probes for such new domains since the EM transition operators are simple and well-known as mentioned in Chapter 1.

Various kinds of giant resonances appear in the medium-energy excitation region. They are excited by inelastic scattering and/or by charge–exchange reactions. Thus giant resonances with quantum numbers J^π and T, where J^π is the spin and parity and T is the isospin, can be studied exclusively by measuring medium-energy γ rays with the same J^π and T in coincidence with particles from inelastic scattering and/or charge–exchange reactions. EM strengths in the region far above the normal shell model states provide information about nucleon correlations and about effects of possible nucleon excitations such as isobars. Medium-energy γ rays following high momentum transfer reactions are strongly coupled to meson exchange currents.

Medium-energy γ rays can be studied by radiative capture reactions. The EM properties of highly excited particle states and those of deep-hole states are studied by measuring the γ rays following particle-transfer and particle-pickup reactions, respectively, and those of highly excited high-spin states by measuring the γ rays from compound nuclei produced by heavy-ion reactions. These new features associated with medium–high energy γ rays are discussed in Chapter 4.

Medium–high-energy γ spectroscopy differs from low-energy γ spectroscopy both from the point of view of the techniques as well as of the physics. The major technical features involved in medium–high-energy γ spectroscopy are as follows:

1. High efficiency γ detectors are required because the branching ratios of medium–high-energy γ radiations from highly excited unbound states are around 10^{-3} to 10^{-5}. The corresponding cross-sections are only of the order of 1 to 10 μb.

2. Large detectors are needed to get full absorption of medium–high-energy γ rays. Since the total absorption coefficient for 20 to 100 MeV γ rays is one order of magnitude smaller than that for 0.1 to 0.4 MeV γ rays, a long depth is needed for full absorption. Medium–high-energy γ rays are converted in detectors mainly to pair electrons. The range of the pair electrons is a few centimetres, but many bremsstrahlung photons are emitted along the electron

path. The positron annihilates by emitting two 511 keV γ rays. Thus the detector has to be large enough to absorb most of these photons and annihilation γ rays.

3. Energy resolutions of about 2 to 4 per cent are necessary for resolving γ transitions to discrete levels with typical level spacings of a few MeV. Thus the large detector has to be quite uniform and stable.

4. Large detectors used for in-beam measurements are exposed to a huge flux of low-energy γ rays, neutrons and charged particles. Their cross-sections are in the order of 1b. Thus the flux is about three to five orders of magnitude larger than the flux of the medium–high-energy γ rays. Charged particles are eliminated by a proper absorber. The time resolution has to be good enough to separate neutrons (and protons) by their time-of-flight. Since the detector has to accept high counting rates the electronic circuitry has to include some pile-up rejection device.

5. Backgrounds due to cosmic rays increase in proportion to the volume of the detector. Their signals are as large as those of medium-energy γ rays. Proper devices to reject cosmic rays are necessary.

Large NaI crystals activated with thallium have been shown to satisfy most of the above requirements for in-beam measurements of medium–high-energy γ rays. Cross-sections (the inverse of the mean absorption length) of NaI for γ-ray energies up to 100 MeV are shown in Fig. 3.29. It is noted that the cross-sections at energies around 10 MeV are due to the Compton and pair-production processes, while beyond 20 MeV the pair-production process becomes dominant. Large volume NaI(Tl) crystals with dimensions of about 30 cm $\phi \times 30$ cm are available. The NaI(Tl) scintillator shows adequate energy and time resolutions, and uniformity with proper treatment of the crystal surface. The characteristics of large NaI(Tl) detectors are described in the review article of Paul (1974). Most of the large NaI(Tl) crystals are surrounded by plastic scintillators, which are operated in anticoincidence with the NaI(Tl) detector. In this way charged cosmic rays and events accompanined by some bremsstrahlung photons escaping from the NaI crystal sre rejected (Kohler and Austin, 1963).

Large NaI detectors (POLYSCIN) made of polycrystalline scintillation material have been developed by HARSHAW. The POLYSCIN NaI(Tl) is composed of many small crystallites packed in close contact throughout the material. It is equivalent to a single crystal in optical and scintillation performance, but the size and the geometry are quite flexible. Consequently large NaI(Tl) detectors with proper geometries are produced by using POLYSCIN. A large γ-counter system consisting of many NaI(Tl) detector modules is feasible by using POLYSCIN. Details of large NaI(Tl) detectors

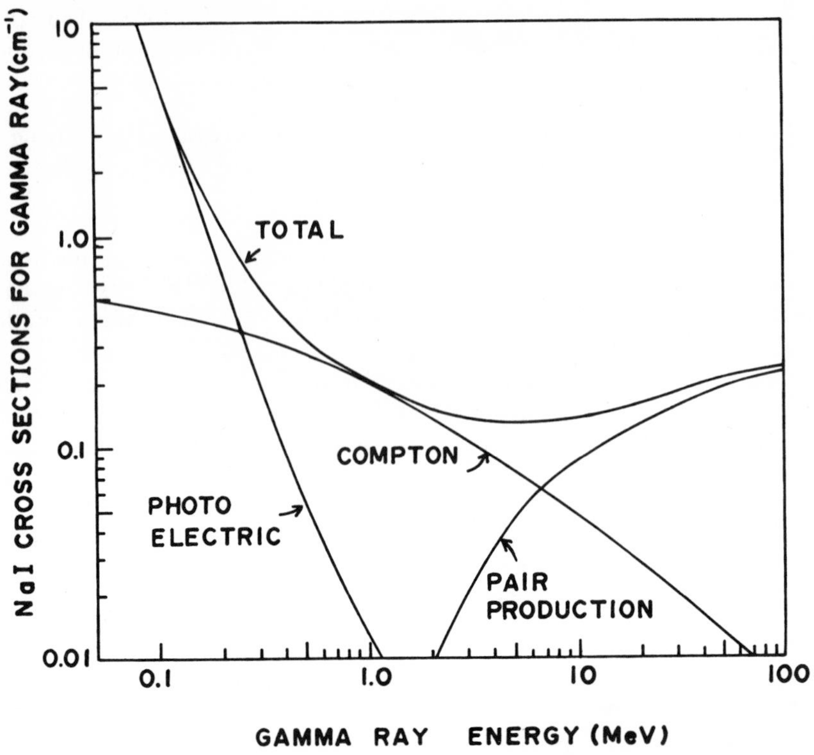

Fig. 3.29. Absorption cross-sections of NaI(Tl) for low, medium, and high energy γ rays.

are given in the following Section 3.3.2.; and those of multi-NaI(Tl) detectors in Section 3.3.3.

Bismuth germanate (BGO) crystals have recently been used for low-energy γ rays, particularly as anti-Compton counters, because of its high density and high Z (note that the photopeak efficiency is proportional to Z^5). These features are useful for detecting medium-energy γ rays as well. The effective volume of BGO is one order of magnitude larger than that of NaI of the same size since the absorption length is about a factor two shorter than that of the NaI (see Section 3.2.1). It is, however, still hard to build BGO crystals with an adequate energy resolution and a large effective volume equivalent to 10^5 to 10^4 cm^3 of NaI detectors. When these problems are solved the BGO crystal may be a nice compact detector for medium-energy γ rays.

Germanium detectors (Ge) with extremely high resolution of a few keV have been used for low energy γ rays. The density (5.4 g/cm^3) of Ge is 50 per cent larger than that of NaI (3.67 g/cm^3), but the radiation length for medium-energy γ rays is only about 25 per cent larger. Moreover a large-volume Ge crystal is hard to produce. The counting rate is limited to about 100 kHz, and the time resolution is worse than that of NaI because of the longer time constant of the semiconductor detector. The Ge detector is used for such special cases as γ transitions from isobaric analogue states in medium–heavy nuclei, where a good resolving power is necessary because of the small level spacing (see Section 4.5 and Ejiri *et al.*, 1968).

Medium-energy γ rays collimated into the central region of a large detector release a large fraction of their energies in the central region of the detector since the bremsstrahlung photons accompained by the pair electrons are emitted into a narrow cone along the electron path. This feature has been exploited in the Osaka detector HERMES, which is divided into a good central detector and an annular detector (Kishimoto *et al.*, 1982). The energy resolution is then determined mainly by that of the central detector which receives the largest fraction of the initial γ-ray energy. The spectrometer at Brookhaven (MKIV) is composed of a central Ge detector with diameter 7.6 cm and length 20.3 cm and a NaI (Tl) annulus with a wall thickness of 15.3 cm (Sandorfi and Collins, 1984). Adding the energies detected by the central Ge crystal and the NaI annulus, a net energy resolution of 0.3 to 0.5 per cent has been achieved for 20 MeV γ rays.

High-energy γ rays far beyond 100 MeV require very large detectors. Since the pair creation cross-section increases as the γ-ray energy increases (see Fig. 3.29), such γ rays are efficiently measured by using an external converter for the pair creation and a magnetic spectrometer for the electron momentum analysis (see Section 3.3.4).

3.3.2 *Single* NaI(Tl) *detectors*

A large NaI(Tl) detector surrounded by a plastic veto-counter has been used for medium energy γ rays by the Stanford group (Suffert *et al.*, 1968). Many groups have developed detectors composed basically of a configuration similar to the large NaI(Tl) with the plastic annular shield. The large NaI(Tl) crystals used by the Stanford group and the Stony Brook group (Diener *et al.*, 1970) are viewed by a single big photomultiplier (PM). The detectors used by the Oxford group (Kernel *et al.*, 1970), the Vancouver–Seattle group (Hasinoff *et al.*, 1974), the TUNL group (Weller *et al.*, 1976), the Ohio group (Kovash, 1978), and the Groningen group (Hofman *et al.*, 1985) have many small PM. Large NaI(Tl) detectors have been used also by others (Davidson *et al.*, 1971; Schaeffer *et al.*, 1971). A typical layout of the large NaI(Tl) detector system at Stony Brook is shown in Fig. 3.30, and typical examples of large NaI(Tl) detector components are given in Figs 3.31 and 3.32. Here the general

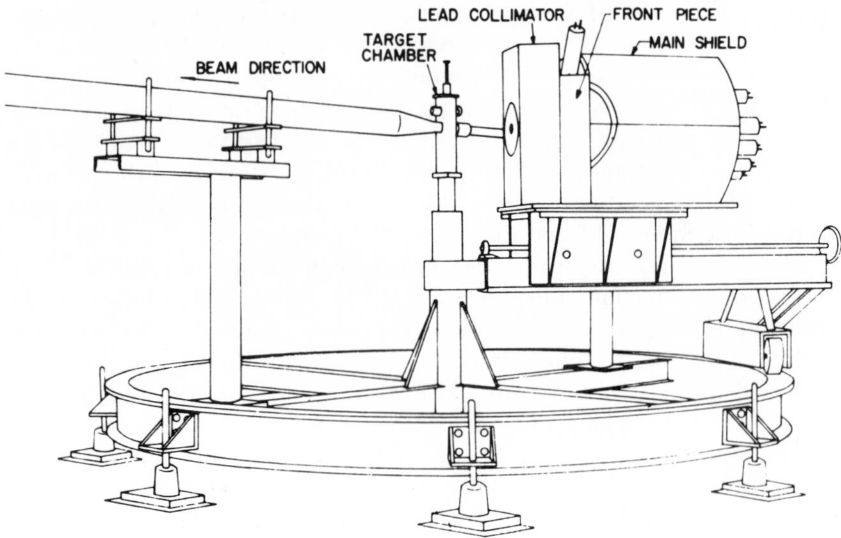

Fig. 3.30. Perspective view of the large NaI detector, the support table, and the beam line at Stony Brook. (Diener *et al.*, 1970.)

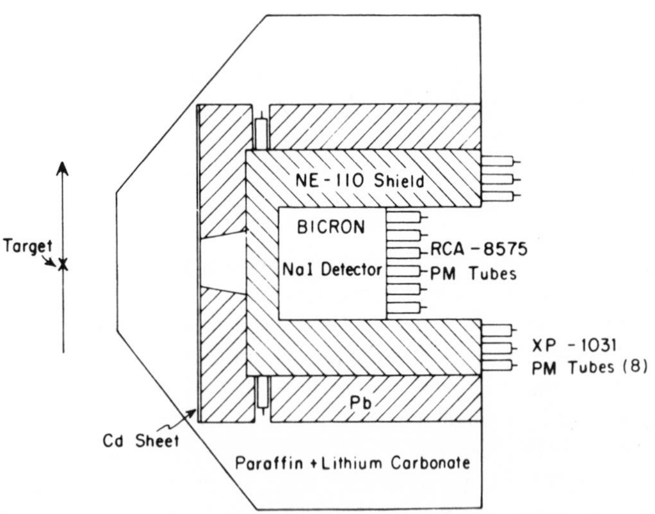

Fig. 3.31. Schematic view of the TUNL γ-ray detector. The shielding around the detector consists of 20.3 cm of paraffin-plus-lithium carbonate (50% mixture by weight), followed by an active plastic scintillator of 7.6 cm in thickness (Weller *et al.*, 1976; Collins *et al.*, 1982.)

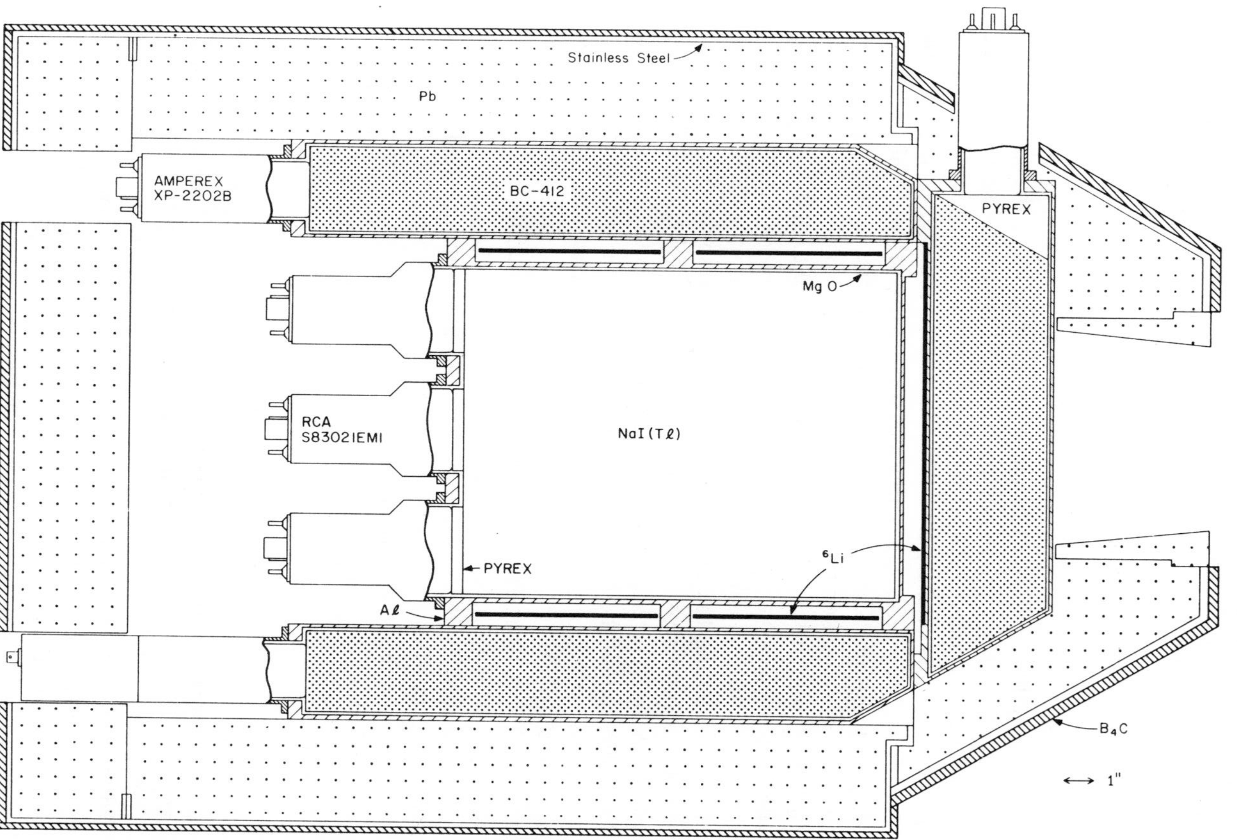

Fig. 3.32. Schematic view of the Ohio γ-ray detector. (Kovash, 1978.)

characteristics and the performance of large NaI detectors are briefly described (for details see the references given above).

The NaI(Tl) crystal has a mean absorption length of around 5 cm for 20 to 100 MeV γ rays. Thus 15 to 20 cm depth is needed for full absorption of such γ rays. Most of large NaI(Tl) detectors have a cylindrical shape with length 25 to 30 cm and about the same diameter to absorb most of the bremsstrahlung photons as well as the positron annihilation γ rays. The energy resolution of the large NaI(Tl) detector depends on the uniformity of the light output along the crystal length. A uniformity better than 1 per cent along the crystal depth is achieved by proper grading and processing of the crystal surface and the reflector. The uniformity check is made by observing the position dependence of the pulse heights of the γ rays from collimated radioactive sources. The 661 keV ^{137}Cs γ ray probes the surface region of the NaI, while the 2.61 MeV Th–C″ γ ray can probe the inner region as well as the surface region. The relative energy resolution improves with energy as $E_\gamma^{-1/4}$ (Heath *et al.*, 1979). Thus a resolution around $\Delta E = 4$ to 2 per cent for 20 to 100 MeV γ rays is obtained by using a NaI(Tl) with ΔE about 7 per cent for 1 MeV γ rays.

Plastic scintillators are most commonly used for veto-counters (active shields) surrounding the large NaI, since the time resolution is good and production of properly shaped scintillators is relatively easy. The NE-102 and NE-110 materials are used because of the long light-attenuation distance. The side and the front of the NaI detector are covered by the plastic scintillators. The charged cosmic rays penetrating the NaI detector can then be fully rejected by the plastic veto-counter. The energy release of the cosmic muons is about 2 MeV/cm. Bremsstrahlung photons, positron annihilation γ rays, and Compton scattered γ rays may escape from the NaI detector. Some of them may be absorbed by the plastic scintillator to give a finite signal from the scintillator, and others may still escape from the plastic counter without releasing any appreciable energy in the plastic shield. In this way the total NaI spectrum $T(E)$ can be decomposed as

$$T(E) = S(E) + f(E) + g(E), \tag{3.3.1}$$

where $S(E)$ is the NaI spectrum for fully absorbed events, $f(E)$ is that for events accompained by any signals from the plastic counter, and $g(E)$ corresponds to the cases where some energy escapes from the NaI detector without any veto-signals from the plastic counter. The difference $T(E) - f(E)$ is called the accepted spectrum (single subtracted spectrum), while $f(E)$ is the rejected one. The accepted spectrum is close to $S(E)$ provided $g(E)$ is very small. In practice the plastic scintillator is not thick enough to absorb all the 511 keV γ rays, and the back of the NaI detector is not covered by the plastic shield. Thus there is a finite contribution of $g(E)$. Assuming $g(E) \approx kf(E)$, with k being an energy-independent constant, one gets the full absorption spectrum by correcting for $g(E)$ by

$$S(E) = T(E) - (1+k)f(E). \qquad (3.3.2)$$

Here k is typically 0.2 to 0.3, $1+k$ is the effective subtraction coefficient and the spectrum is called the double subtracted spectrum. The energy resolution of the UCB–Seattle detector for 22.4 MeV γ rays is 3.4 per cent fwhm (full width half maximum) and 15 per cent fwtm (full width tenth maximum) in the case of the accepted (single subtracted) spectrum. It is improved to 3.0 per cent fwhm and 5 per cent fwtm by proper correction for $kf(E)$ (Hasinoff *et al.*, 1974).

The whole scintillator is shielded by lead bricks against background γ rays and bremsstrahlung photons following cosmic rays. Neutrons are slowed down by the plastic scintillator and are absorbed by the ^{6}LiH. Paraffin containing lithium carbonate can be used to shield the NaI.

High counting-rates in the large NaI detector demand a pile-up rejection circuit and a stabilizer for the photomultiplier (PM). The dc current generated by the high rate signals in the PM affects the voltage between the PM dynode stages, resulting in the rate-dependent PM gain. In order to stabilize the gain, a reference peak is inspected by a feed-back loop to regulate the high voltage for the PM (the details will be discussed later).

The BNL group has developed a large NaI(Tl) detector MKIII with a superior energy resolution using a new technique for grading the reflective surfaces of the NaI crystal and segmentation of the plastic anticoincidence shield. The characteristics of the detector are as follows (Sandorfi and Collins, 1984). The 24 cm $\phi \times$ 36 cm NaI(Tl) crystal was mapped by the 6.13 MeV γ rays from a collimated ^{244}Cm–^{13}C source. The 6.13 MeV γ ray can probe the inner part of the NaI crystal where most of the γ ray showers take place. A uniformity within ± 0.4 per cent was achieved. The 10 cm thick annular plastic scintillator surrounding the NaI(Tl) crystal was segmented into six optically isolated sections like those of the Osaka detector (see Section 3.3.3). It is noted that a single non-segmented annular plastic scintillator has a torus shape that hinders the collection of light by the PM at the back side. Thus the signal-to-noise ratio is worse than that for the segmented scintillator.

The BNL–MKIII NaI(Tl) is equipped with seven 10-stage RCA4900 PM's. The gain is stabilized by an active transistor-stabilized base. The last four dynodes are linked to a separate power supply to boost the current so as to stabilize the dynode voltage. Light from a green light-emitting-diode (LED) is guided to the NaI PM tube as shown in Fig. 3.33, and is used to stabilize the gain. A part of the LED light is introduced to a separate PM mounted on a small NaI, so that the LED light signal is always referred to the 1.274 MeV γ-ray signal from a radioactive ^{22}Na source in front of the small NaI crystal. The electronic circuitry used for the gain stabilizer is shown in Fig. 3.33. The resolutions obtained for the 22.4 MeV γ ray are 2.2 per cent fwhm and 6.4 per

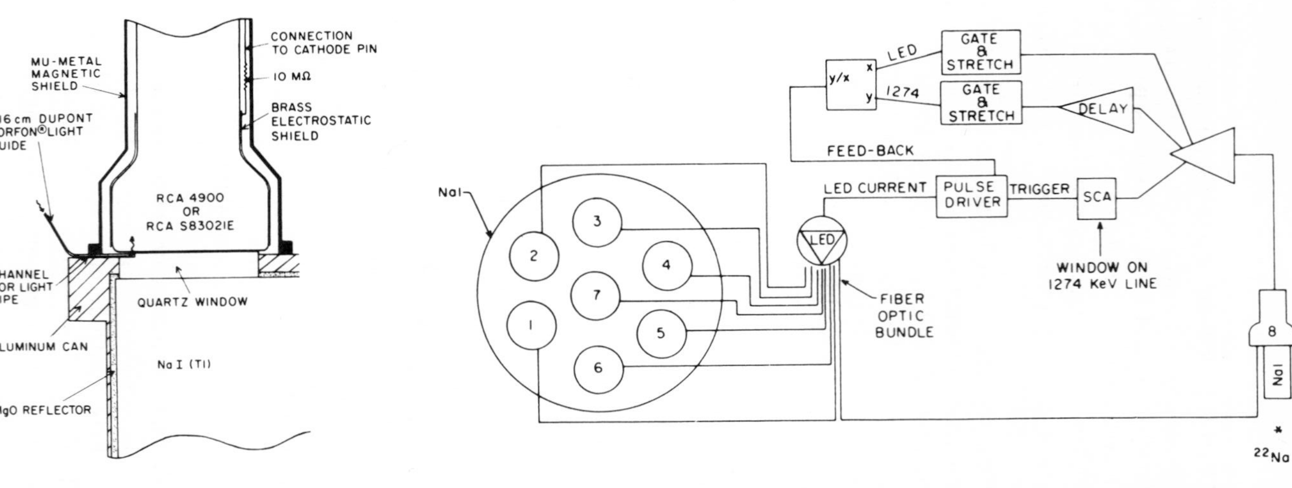

Fig. 3.33. Energy stabilizer for the BNL-MKIII detector. Left-hand side: The coupling of one of the 7.6-cm diameter PM tubes to the NaI crystal is shown here, together wtih the fibre-optic light pipe used to inject light from a light-emitting-diode (LED) into the photocathode. Right-hand side: The feedback and stabilization network for the LED pulser that is in turn used to stabilize the NaI gain (Sandorfi and Collins, 1984.)

cent fwtm in the accepted (single subtracted) spectrum, and are improved to 2.1 per cent fwhm and 4.4 per cent fwtm in the double-subtracted spectrum as shown in Fig. 3.34.

The Groningen group has recently developed a large NaI(Tl) detector (Hofman *et al.*, 1985). The NaI(Tl) crystal is equipped with seven PM's of 75 mm diameter and the plastic veto-shield is segmented into six parts, each viewed by two PM's of 52 mm diameter. The energy resolution of about 2 per cent fwhm is obtained for 22.6 MeV γ rays collimated into the central region of the NaI crystal, as shown in Fig. 3.35.

3.3.3 *Multi-*NaI(Tl) *detector systems for medium–high energy γ rays*

3.3.3a *Large detector with multi-NaI(Tl) crystals*

A large NaI(Tl) detector can be composed of several NaI(Tl) crystals which are optically isolated from each other. Energy and time signals from each NaI(Tl) section are processed separately, and those satisfying some conditions are stored on a magnetic tape. Proper on-line and off-line data processing of these signals leads to the true energy spectrum with an improved energy resolution and a good S/N (signal-to-noise) ratio. It is easy to build a large volume detector in the form of a multi-detector system. The multi-detector system may be composed of many units of standard NaI(Tl) modules. The development of computer data-taking systems has made it quite feasible to handle high-rate signals from many detector units. Electron showers generated by a medium–high-energy γ ray develop in the whole volume of a multi-detector ensemble, so that the energy loss of these electrons at the insensitive boundaries of the detector segments (modules) have to be negligibly small with respect to the total γ-ray energy. Consequently large detectors for medium-energy γ rays are made by segmenting them with thin optical insulators, while detectors for higher energy γ rays are made by assembling NaI(Tl) crystal modules covered with thin metal.

Localization of the impact point of the incident γ ray and mapping the energy release in the NaI(Tl) sections are possible by examining the energy signals from individual NaI(Tl) segments (modules). The impact point reveals the direction of the incident γ ray and the distribution of the energy release over the NaI segments reflects the linear polarization.

3.3.3b *Large detector with a segmented NaI(Tl) crystal*

A large NaI(Tl) detector segmented into several optically isolated sections has been developed for medium–high-energy γ rays at Osaka (Kishimoto *et al.*, 1982). The detector is called HERMES (high energy radiation measuring system). It is composed of a 27.7 cm $\phi \times 27.7$ cm NaI(Tl) crystal segmented

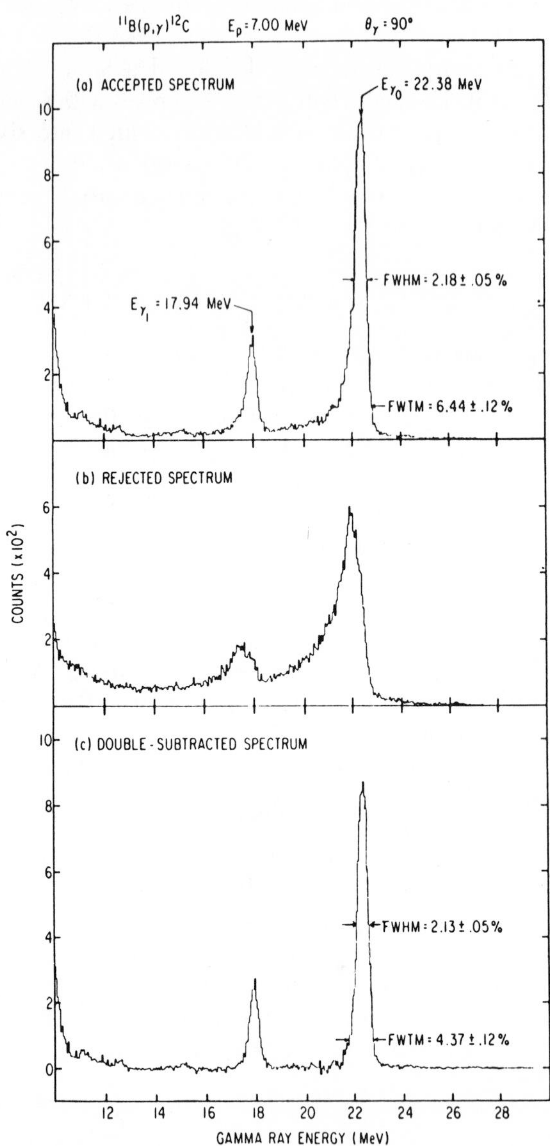

Fig. 3.34. Accepted, rejected, and double subtracted spectra of two high energy γ-ray lines from the ^{11}B(p, γ) reaction measured by BNL-MKIII. (Sandorfi and Collins, 1984.)

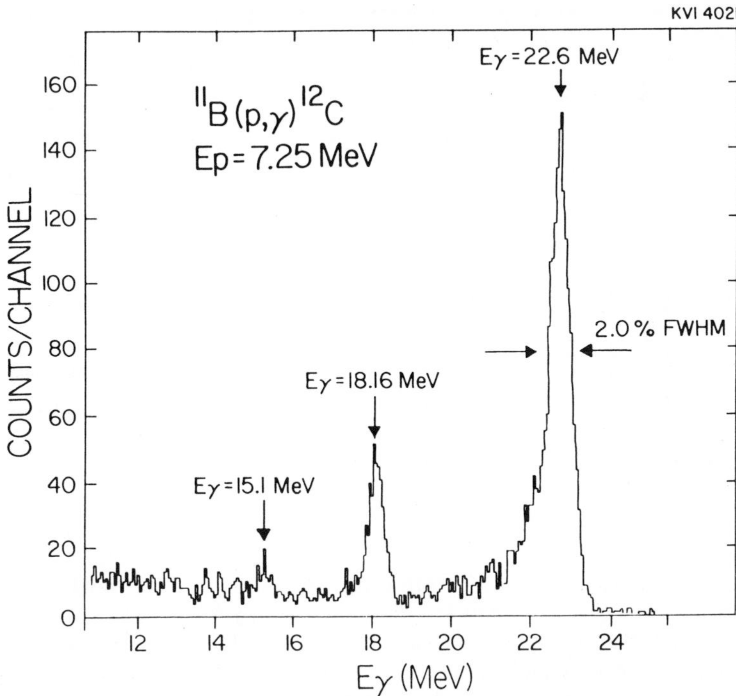

Fig. 3.35. Gamma-ray spectrum measured by the Groningen detector. (Hofman *et al.*, 1985.)

into five sections and a 10-cm-thick plastic annulus segmented into four sections. The layout of HERMES is shown in Fig. 3.36. The NaI(Tl) detector consists of a central 15.1 cm $\phi \times 27.7$ cm cylindrical NaI(Tl) crystal and an annular 27.7 cm $\phi \times 27.7$ cm NaI(Tl) crystal surrounding the central one. The annular crystal is optically divided into four segments.

The basic ideas of the segmented NaI(Tl) detector are the following. Gamma rays collimated into the central region of the NaI(Tl) detector lose most of their energies in the central crystal. Some fractions of the electron and bremsstrahlung showers escape from the central crystal and are absorbed by the annular NaI(Tl) crystal. The total energy is given by summing the signals from the central and annular NaI(Tl) segments. The net energy resolution is then written as

$$R_s = \{(1-f)^2 R_c^2 + f^2 R_a^2\}^{1/2}, \tag{3.3.3}$$

where R_c and R_a are the energy resolutions of the central and annular NaI(Tl) crystals, respectively, and f is the fraction of the total energy released in the annular NaI(Tl) detector. Since f is as small as a few per cent for 20 MeV γ

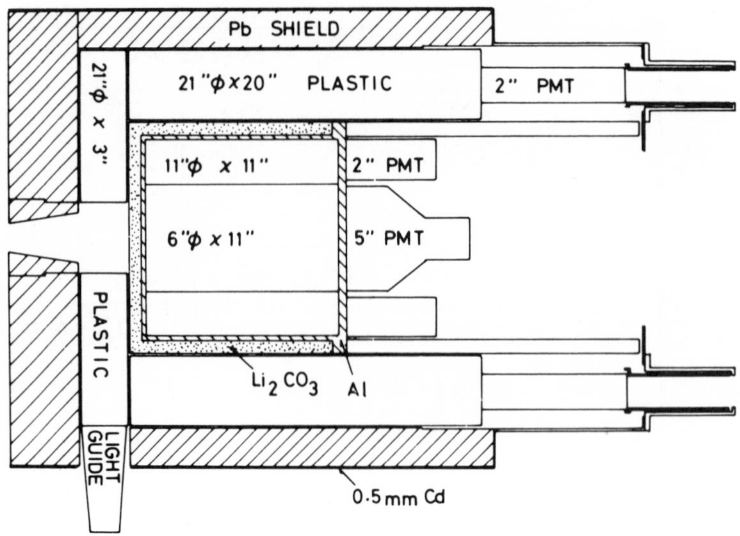

Fig. 3.36. Sectional layout of HERMES. The 27.7 cm $\phi \times$ 27.7 cm NaI(Tl) crystal is divided into two parts. Each part is optically isolated by 1.5 mm MgO layers. Eight 5 cm ϕ RCA 4523 PM's and the 12.6 cm ϕ RCA 4525 PM are attached to the annular 27.7 cm $\phi \times$ 27.7 cm crystal and the central 15.1 cm $\phi \times$ 27.7 cm crystal, respectively. The whole NaI assembly is packed in a 3 mm thick aluminum can. Four 5 cm ϕ Hamamatsu R329 PM's are attached to the front plastic scintillator disc and eight to the annular plastic scintillator, respectively. (Kishimoto *et al.*, 1982.)

rays, the net resolution R_s is approximately given by the resolution R_c of the central detector. The 15.1 cm ϕ cylinder is an adequate size to get a good resolution crystal with reasonable uniformity. The uniformity of the pulse height along the length of the central NaI(Tl) is achieved by changing the reflection coefficient of the NaI(Tl) surface. Actually the central crystal has a resolution of 7 per cent for 661 keV γ rays and a uniformity within 1.6 per cent (this should be improved), while the segments of the annular detector have resolutions of 8.5 per cent to 9 per cent.

Events are triggered by signals from the central NaI(Tl) crystal if they exceed a certain level of the discriminator. Then all the energy and timing signals from all NaI(Tl) crystals and plastic segments, being gated by the trigger pulses from the central crystal, are recorded on a magnetic tape. Consequently, the background rates are much reduced in proportion to the volume of the central crystal. The annular NaI(Tl) crystal, together with the annular plastic scintillator, serves as a good active shield for cosmic rays and background γ rays since events accompained by large energy releases in any of the segments of the annular NaI and plastic scintillators are rejected in the on-line or off-line data analysis. Segmentation of large NaI and plastic scintillators into small sections, being equipped with individual PM tubes, reduces the counting rate and the pile-up rate for each scintillator.

Neutrons are a serious problem in case of in-beam measurements of medium–high-energy γ rays with fairly low cross-sections. They have to be removed by means of TOF (time-of-flight) in the on-line or off-line analysis. Here a TDC (time-to-digital converter) is started by the fast pulse from the central NaI and is stopped by the signal associated with the beam pulse (RF signal in case of cyclotron beams). The layout of HERMES used for measuring (p, γ) reactions is shown in Fig. 3.37. The overall engrgy resolution of 2.7 per cent fwhm is obtained in the accepted spectrum for the 22.5 MeV γ ray, and the γ ray is well separated from neutrons in the TOF spectrum, as shown in Fig. 3.38.

HERMES has been used for studying radiative capture γ rays in the energy range $E_\gamma = 40$ to 100 MeV, with energy resolution is around 2 per cent. Details are given in Section 4.5. The spectrometer can be used for measuring the linear polarization of γ rays since the direction of the Compton scattering of the γ rays in the central crystal depends on the linear polarization. The Compton scattered γ rays are detected by either vertical or horizontal pairs of the four segments of the annular NaI(Tl) crystal. The annular NaI(Tl) crystal of

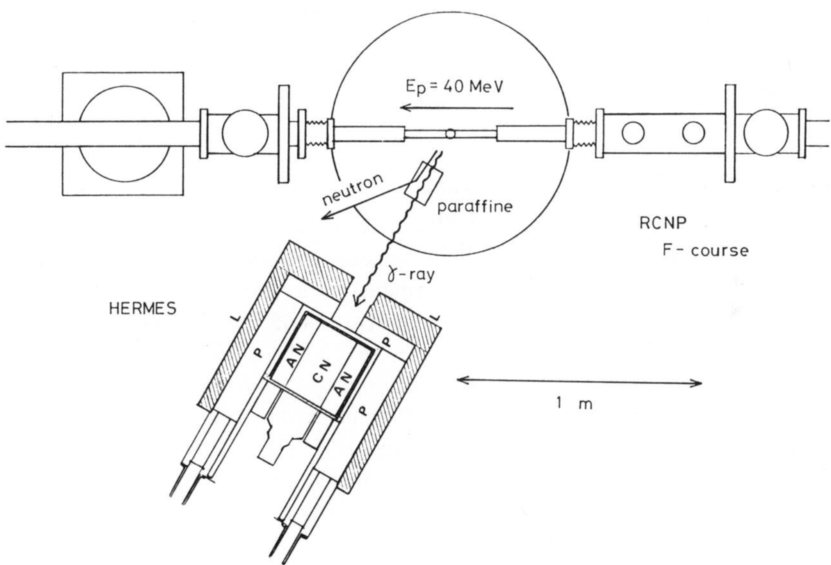

Fig. 3.37. An experimental arrangement for studying radiative capture reactions by means of HERMES. CN: central NaI detector, AN: annular NaI detector, P: plastic scintillator, and L: lead shield. (Ejiri *et al.*, 1984.)

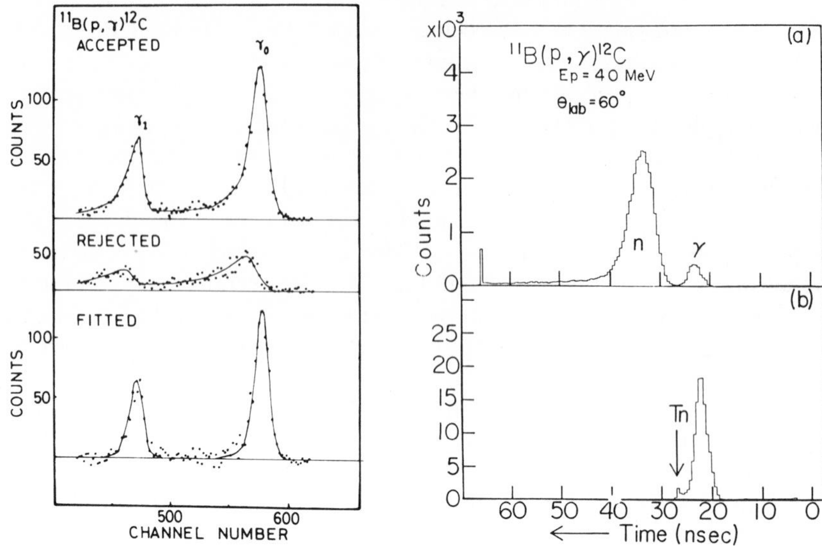

Fig. 3.38. The energy and time spectra measured by HERMES. Left-hand side: Energy spectra for 22.7 MeV γ rays from the $^{11}B(p, \gamma)$ reaction. Right-hand side: (a): Time spectra of neutrons and γ rays following the $^{11}B(p, \gamma)^{12}C$ reaction. (b): Time spectrum of γ rays selected by the threshold Tn. (Noumachi, 1983; Ejiri *et al.*, 1984.)

HERMES is also used as a Compton suppressor for medium–low-energy γ rays (see Section 4.2).

3.3.3c *A large array of NaI(Tl) modules*

A large close-packed array of NaI(Tl) modules is used to detect high energy γ rays. The modulation of the NaI(Tl) detector makes it possible to build a large-volume detector with arbitrary dimensions for high-energy γ rays and to locate the impact point of the γ rays. An array of 45 NaI(Tl) crystal modules has been used to search for the decay $\mu^+ \to e^+\gamma$ at LAMPF (Carrington *et al.*, 1979). Each module is 51 cm in length and hexagonal in cross-section with six 7.6 cm sides. It is encapsulated by a 50μm-thick stainless steel container, and is viewed at one end by a 7.6 cm ϕ RCA 8054 PM. The 45 modules are packed closely as shown in Fig. 3.39. Good events are such that a large fraction of the incident γ-ray energy is released in one NaI(Tl) module. Then the total γ-ray energy is obtained as

$$E_\gamma = \sum_{i=1}^{7} E_i, \tag{3.3.4}$$

where E_1 is the energy release in the NaI(Tl) module corresponding to

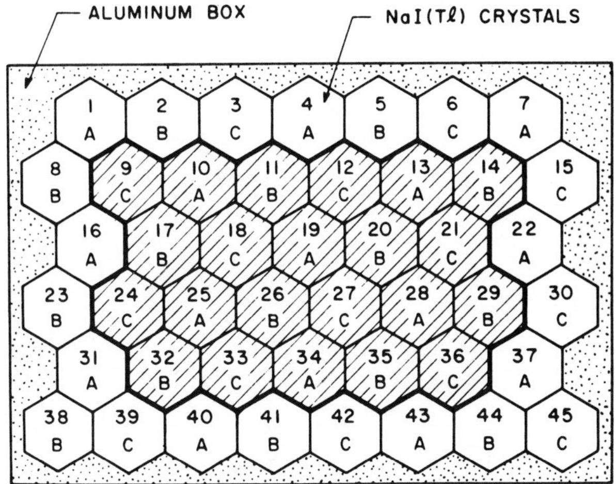

Fig. 3.39. The arrangement of many modules in a NaI(Tl) array. The fiducial aperture is indicated by the modules within the dashed area. The letters A, B and C identify the modules included in each of the three timing groups. (Carrington *et al.*, 1979.)

the γ ray impact point and $E_2, E_3, \ldots E_7$ are the energies released in the six NaI(Tl) modules surrounding the centre module. The impact point is given approximately as

$$X = \sum_{i=1}^{7} E_i X_i / E_\gamma, \qquad Y = \sum_{i=1}^{7} E_i Y_i / E_\gamma, \tag{3.3.5}$$

where (X_i, Y_i) is the axial coordinate of each module. The 22 modules within the fiducial aperture are surrounded by 23 outer-layer modules, as shown in Fig. 3.39, to avoid edge effects due to transverse energy leakage.

The energy resolution is around 7 per cent for medium-energy γ rays with $E_\gamma = 20$ to 60 MeV, and is improved to 5.5 per cent for high-energy γ rays with $E_\gamma = 129.4$ MeV. The time resolution for π^0-decay γ rays is about 1.9 ns fwhm. Typical energy and time spectra are shown in Fig. 3.40. The energy calibration of each NaI(Tl) module has to be achieved over a large dynamic range. Calibration γ rays used are as follows: The 0.90 and 1.8 MeV γ rays come from the ^{88}Y radioactive source; the 4.44 MeV γ ray from the ^{9}Be(α, n)^{12}C (4.44 MeV) reaction in the Pu–α–Be source; the 6.1 and 17.6 MeV γ rays from the ^{19}F(p, α)^{16}O (6.1 MeV) and ^{7}Li(p, γ)^{8}Be reactions, respectively, by using the auxiliary Van de Graaff accelerator; the 55.1 and 82.7 MeV γ rays from the decay of π^0 moving oppositely and along the NaI(Tl) modules, the π^0 being provided by the $\pi^- + \text{p} \rightarrow \text{n} + \pi^0$ reaction induced by π^- stopped at a liquid

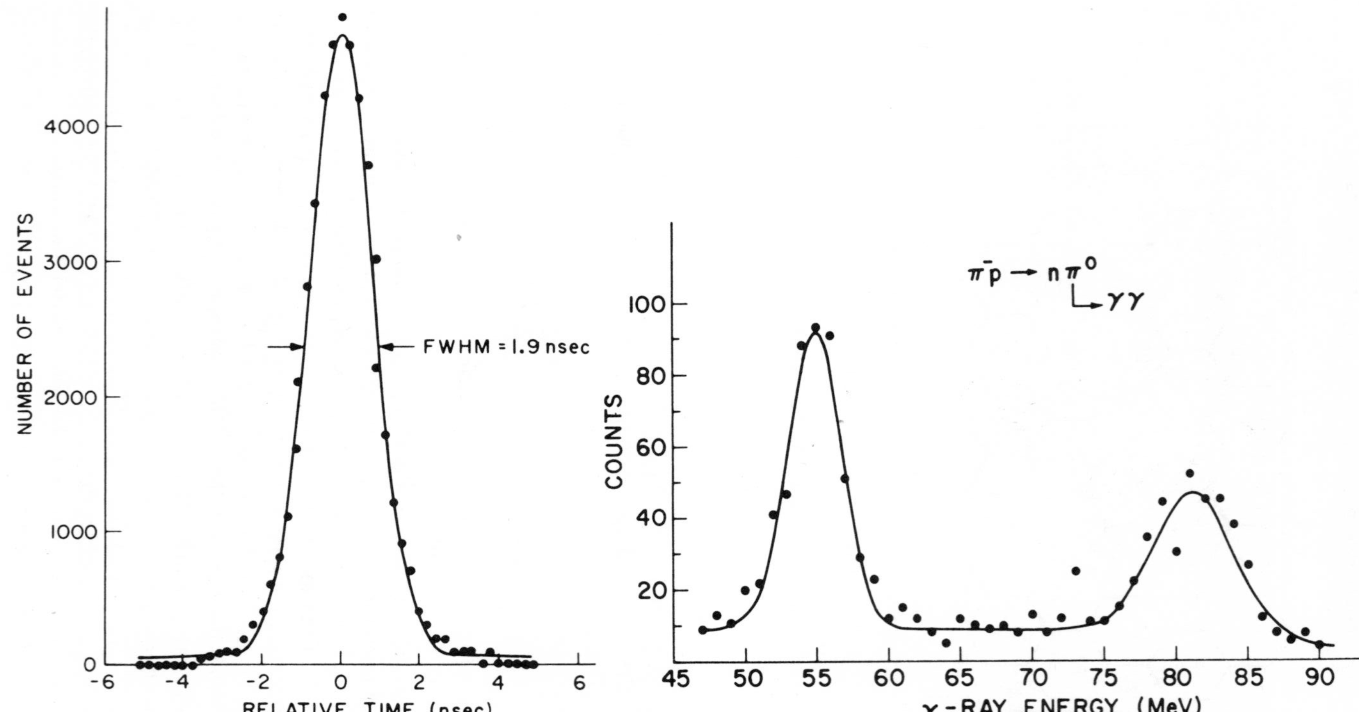

Fig. 3.40. Right-hand side: The calibration lines at 55.1 and 82.7 MeV obtained from the reaction $\pi^-p \rightarrow \pi^0 n$ by the detection of γ-ray pairs from the π^0 decay with a collinearity less than 5°. Left-hand side: The observed time spectrum for events of the type $\pi^-p \rightarrow \pi^0 n$ in which one π^0-decay γ ray is detected in the NaI(Tl) array and the other in the positron spectrometer. (Carrington *et al.*, 1979.)

hydrogen target; the 129.4 MeV γ ray is obtained from the similar $\pi^- + p \rightarrow n + \gamma$ reaction.

The plan view of the experimental apparatus used for studying the $\mu^+ \rightarrow e^+ + \gamma$ decay is shown in Fig. 3.41. The medium–high-energy γ ray was measured by the large arrary of the 45 NaI(Tl) modules in coincidence with the positron detected by the positron spectrometer.

Recently, a new NaI(Tl) array 'crystal box' has been built to search for $\mu^+ \rightarrow e^+\gamma$, $e^+\gamma\gamma$, and $e^+e^+e^-$ events (Bolton *et al.*, 1984). It consists of 396 NaI(Tl) crystals as shown in Fig. 3.42.

High-energy γ rays from $\psi(3095)$ and $\psi'(3684)$ decays have been observed by a large array of 19 NaI(Tl) modules, each consisting of a 50 cm long hexagonal NaI(Tl) crystal with six 7.5 cm sides (Biddick *et al.*, 1977). The overall energy resolution is 5 per cent at 0.2 GeV and is 2.5 per cent at 1.5 GeV. The experimental layout used for the ψ decays in the e^+e^- colliding beam

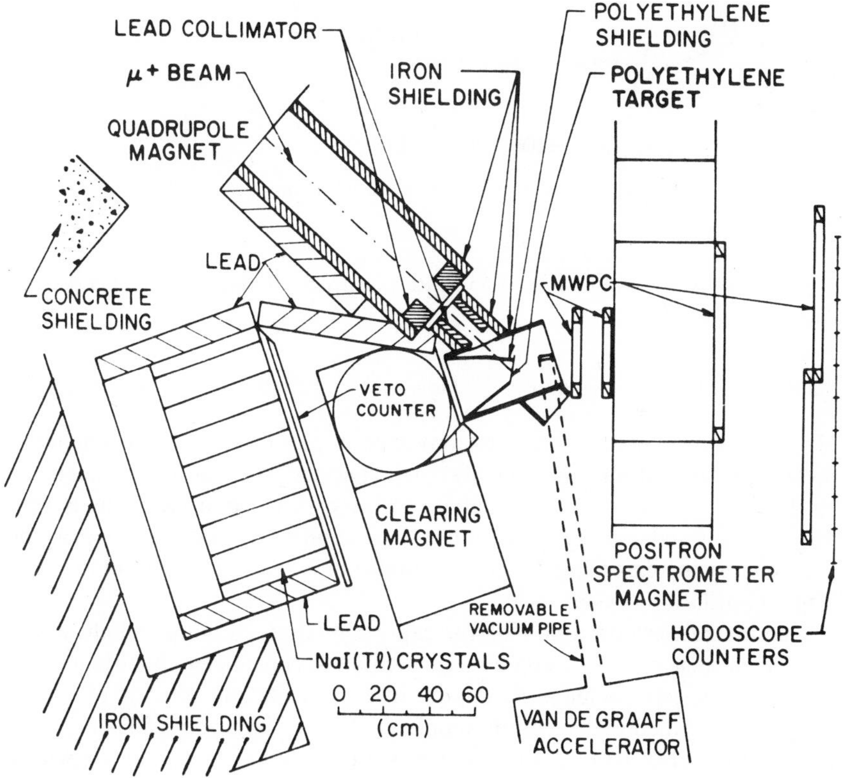

Fig. 3.41. A plan view of the experimental apparatus used in the search for $\mu^+ \rightarrow e^+\gamma$. (Carrington *et al.*, 1979.)

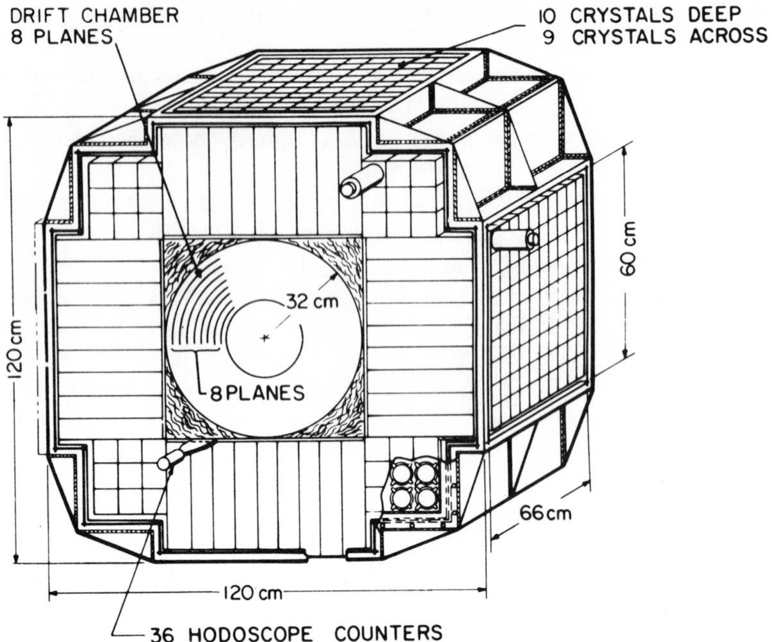

Fig. 3.42. A crystal box detector. (Bolton *et al.*, 1984.)

experiment is shown in Fig. 3.43, and examples of the energy spectra are shown in Fig. 3.44.

3.3.4 *Magnetic spectrometers for high-energy γ rays*

High-energy γ rays are externally converted to pair electrons by using heavy metal with large atomic number, and their energies are measured by magnetic spectrometers. The magnetic analysis of the external pair electrons is very attractive for high-energy γ rays since the pair conversion rate increases as the γ ray energy increases and the energy resolution is much better than that of a large scintillation crystal.

Gamma-ray spectra following the capture of stopped pions in light nuclei have been measured by using a magnetic pair spectrometer (Bistirlich *et al.*, 1972). The spectrometer is illustrated in Fig. 3.45(a). It consists of two C magnets combined with a common pole tip. It has a large gap of 33 cm with a maximum field of 10 KG. The converter used is a 0.11 mm gold foil with a conversion efficiency of 2.3 per cent for radiative capture γ rays. Six arrays of four gap spark chambers are arranged to measure electron trajectories. The energy resolution for the 129.4 MeV γ ray is about 2.0 MeV, and the overall

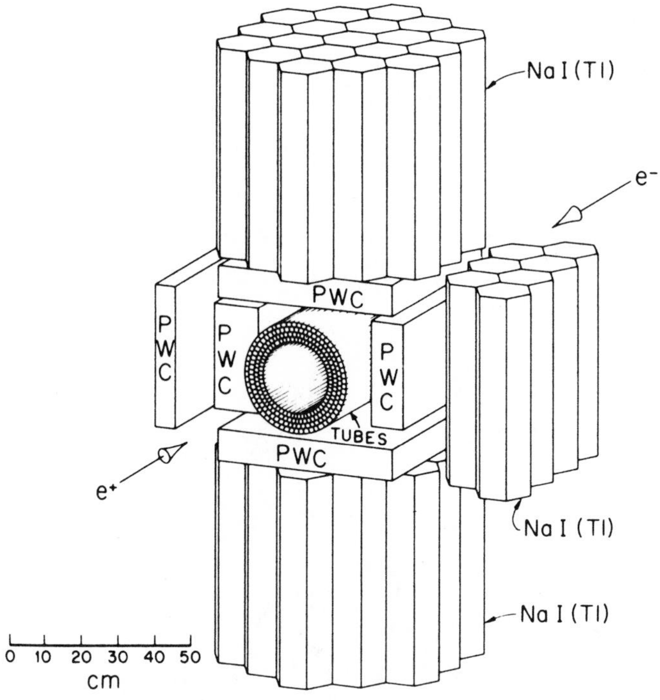

Fig. 3.43. A schematic view of the apparatus used for the inclusive γ ray measurement in the decay of ψ and ψ' particles. The tube counters provide azimuthal coordinates for charged-particle trajectories, while the proportional wire chambers (PWC) yield coordinates along the beam direction. The hexagonal NaI(Tl) modules are 20 radiation lengths thick. (Biddick *et al.*, 1977.)

efficiency, including the conversion rate and the solid angle with respect to 4π, is about 2.0×10^{-5} for 100 to 140 MeV γ rays. An energy spectrum for π^- capture in hydrogen is shown in Fig. 3.45(b). Major contributions to the resolution are the energy straggling in the converter and the spark chambers and the spatial resolution of the counter arrays.

An active converter such as a BGO scintillator can be used. A 4 mm BGO gives a conversion rate of 20 per cent for 100 MeV γ rays. Since the energy loss in the scintillator is corrected for by adding the signal from the BGO, it may have a better energy resolution.

Drift chambers or multi-wire proportional chambers are used to get good spatial resolution. A large efficiency pair spectrometer may be designed by using superconducting coils.

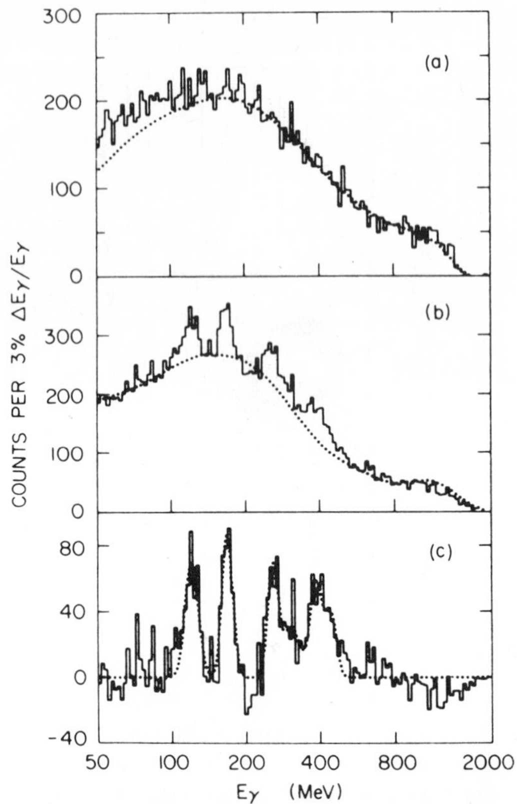

Fig. 3.44. The inclusive γ-ray distribution as a function of ln(E_γ). (a): for the ψ(3095) and (b): for the ψ'(3684) and (c): for the difference between the data in (b) and the continuum indicated by the dotted line in (b). (Biddick *et al.*, 1977.)

3.4 Electron spectrometers and electron transporters

3.4.1 *Electron spectroscopy associated with nuclear reactions*

Electromagnetic transitions between nuclear excited states are partly converted to internal electron emissions. The energy of the internal conversion electron and the conversion rate are written as

$$E_e(\alpha, E_\gamma) = E_\gamma - B_\alpha, \tag{3.4.1a}$$

$$T_e(\alpha, E_\gamma L^\beta) = \eta(\alpha, E_\gamma L^\beta) T_\gamma(E_\gamma L^\beta), \tag{3.4.1b}$$

where E_e, E_γ, and B_α are the electron energy, the transition energy and the electron binding energy, respectively. The symbol α stands for the electron orbit characterized by K, L, M, . . . and L^β is the transition multipolarity with $\beta = \mathrm{e}$ being the electric transition and $\beta = \mathrm{m}$ being the magnetic transition. The ratio of the conversion electron emission rate T_e to the γ emission rate T_γ is given by the internal conversion coefficient η. Gamma rays emitted from the nucleus are externally converted into electrons as well by using external converters placed outside the target.

Measurements of electrons following nuclear reactions are complementary to γ-ray measurements in the context of experimental and spectroscopic methods. Actually early high-resolution in-beam studies of collective states have been carried out by means of magnetic spectrometers (Hansen *et al.*, 1963; Sakai *et al.*, 1964; Diamond *et al.*, 1963; and Morinaga, 1964). The advent of the Ge semiconductor detector has made it possible to carry out high-resolution high-efficiency γ spectroscopy for both in-beam (on-line) and off-beam (off-line) measurements. On the other hand, measurement of electron spectra point by point by conventional magnetic spectrometers requires longer times and become less competitive than the Ge detector. Electrons

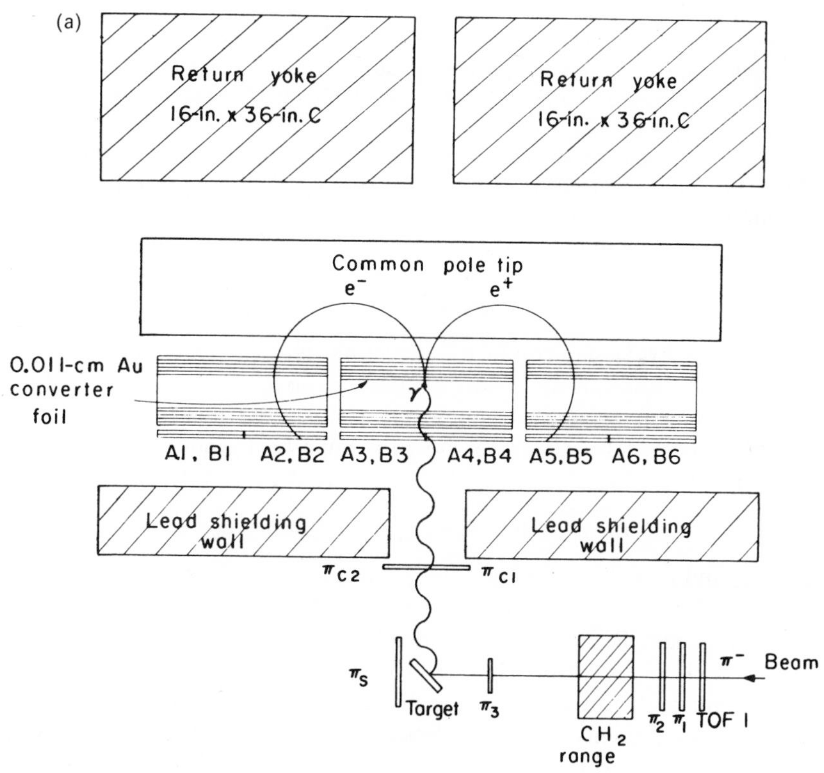

Fig. 3.45(a).

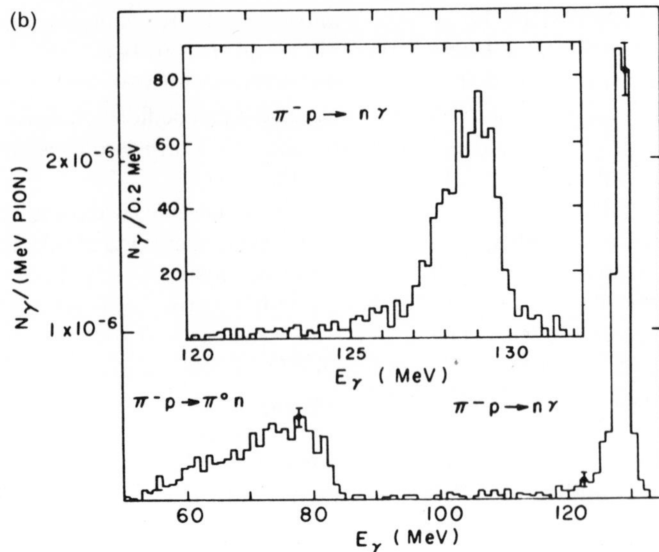

Fig. 3.45. (a) Experimental layout for the pair spectrometer used for (π^-, γ) reactions. The mirror system for photography of the spark chambers and the details of the magnet coils are omitted for clarity. The trigger for an event is made by $\pi_1 \times \pi_2 \times \pi_3 \times \bar{\pi}_s \times \bar{\pi}_c \times A_i \times B_i \times A_k \times B_k$; $i \neq k$, $k \pm 1$, where $\pi_i \times \ldots A_j \times \ldots B_k \times \ldots$ stand for coincidence with signals from counters π_i, A_j, B_k, . . . as indicated in the figure, while $\bar{\pi}_j$ stands for anti-coincidence. (Bistirlich *et al.*, 1972.) (b) Photon energy spectrum for π^- capture in hydrogen in 1 MeV bins. The insert shows the events from radiative capture leading to nγ, grouped in 0.2 MeV bins. (Bistirlich *et al.*, 1972.)

are measured by Si semiconductor detectors. Nuclear reactions, however, produce high multiplicity γ rays, many neutrons and many charged particles. Incident projectiles knock out atomic electrons from the target nuclei and these delta rays give rise to a huge background. Their energy extends up to

$$E(\delta) = 4\{(m_e/m_a)E_kE_a\}^{1/2} + 4(m_e/m_a)E_a, \tag{3.4.2}$$

where m_e and m_a are the electron and projectile masses, respectively, and E_k and E_a are the K-electron binding energy and the projectile energy, respectively (Klank and Ristinen, 1969). Therefore, the in-beam measurement of conversion electrons with fairly small conversion coefficients by means of a bare Si semiconductor detector is difficult. Despite these disadvantages of electron measurements, in-beam electron spectroscopy has many unique features and most of the experimental difficulties have been overcome by proper devices appropriate to in-beam electron measurements.

The unique features of electron spectroscopy associated with nuclear reactions are the following:

1. Conversion coefficients obtained by measuring both conversion electrons and γ rays are sensitive to the transition multipolarity, and are used for assignment of the multipolarity of the transition.

2. Since low energy and/or high multipole transitions have large conversion coefficients, they show up strongly with respect to many other transitions involved in nuclear reactions. E0 transitions appear strongly in conversion electron spectra, so that conversion electron measurement is a good way to study E0 transitions and weak transitions with large conversion coefficients.

3. Conversion electrons from spin-aligned states, which are easily obtained by nuclear reactions (Ejiri *et al.*, 1965, 1966), show angular distributions given by

$$W_e(\theta) = \Sigma A_\nu b_\nu P_\nu(\cos\theta), \tag{3.4.3}$$

where the particle parameter b_ν as well as angular distribution coefficients A_ν for the corresponding γ ray are sensitive to the transition multipolarity. Thus angular distributions are well suited for spin and multipole assignments. The b_ν coefficient for K conversion electrons differs from that for L conversion electrons, depending on the multipolarity. The K/L ratio at a given angle therefore depends on the multipolarity (Sakai *et al.*, 1966).

4. Conversion electrons following high-spin isomers, produced by nuclear reactions, are studied between beam bursts since isomers are likely to decay by high multipole transitions with large conversion coefficients.

5. Beta rays from short-lived radioactive nuclei produced by nuclear reactions are well studied by means of electron spectrometers used for in-beam measurements.

6. Many γ rays, charged particles and neutrons, which are emitted in most nuclear reactions, contribute seriously to the background for detectors near the target. Conversion electrons are easily transported far from the target by using magnetic transport devices. By setting a proper momentum window for the transport device a part of the electron spectrum of interest is transmitted and other parts including low-energy δ rays (atomic electrons ejected by projectiles) are eliminated.

7. High-resolution measurements over a wide energy range can be made by using magnetic spectrometers provided that the energy loss in the target is small.

8. By using double-focusing spectrometers or electron transport devices only the electrons from the beam spot on the target are selected, and all other

electrons due to Compton scattering, neutrons, etc. are eliminated. By shifting the target upstream from the source point by a small amount, the delayed electrons emitted from recoiling nuclei can be measured.

9. Full energy-loss peak efficiencies for most of electron detectors are close to 100 per cent in contrast to the peak efficiencies around 10^{-1} to 10^{-2} for γ detectors. On the other hand, conversion coefficients are of the order of $1 \sim 10^{-4}$. In order to compensate for the small conversion coefficient the transmissions of magnetic spectrometers and magnetic transport devices have to be very high.

The features 1 to 4 given above, which are unique to conversion electron spectroscopy, are associated with the finite mass of electrons. Because of the finite electron mass, in contrast to the mass-less γ ray, the phase space for the conversion electron is quite different from that for the γ ray. Then $\eta(\alpha, E_\gamma L^\beta)$ depends on the transition multipolarity and the transition energy. The features 6 to 8 given above, which are unique to electron measurements, are associated with the finite electric charge. In contrast to γ rays, electrons can be easily handled by all kinds of electromagnetic fields. Thus the conversion of γ rays into electrons by using external converters is worthwhile even though the conversion coefficients in a thin converter are small. Active converters such as a thin BGO plate and other heavy-mass detector plates may be used without greatly sacrificing the energy resolution provided that the energy loss in the converter is corrected for.

Methods for in-beam electron measurements are classified into two categories; one uses a magnetic spectrometer to transport electrons and to analyse the momentum, and another uses a magnetic field to transport electrons from a target to a semiconductor detector for energy analysis. The special features of various electron spectrometers are given in Table 3.4.1.

3.4.2 *Magnetic spectrometers for momentum analysis*

Magnetic spectrometers have several merits. The momentum resolution, $R = \delta P/P$, is as good as 10^{-3} to 10^{-2} if the energy loss in the target is relatively small. The detector can be as small as the beam spot provided that the magnification is of the order of one, and thus the backgrounds are rather small. The Compton electrons produced at the magnet poles by γ rays from the target and those from neutron capture on materials around the target are not focused on the detector point. Most magnetic spectrometers for light and heavy ions, as they are composed of huge magnets, have solid angles of the order of 10 msr. On the other hand, magnetic spectrometers for electrons require only small magnets because of the low magnetic rigidity of the electron. Therefore it is quite feasible to build spectrometers with large solid

Table 3.4.1 Some specifications of typical examples of electron spectrometers

	Magnetic Spectrometer		Magnetic electron transporter with Si detector			
	Orange-gap	Single-gap Multi-gap	Solenoid	Small sector magnet	Triple-focusing sector magnet	Mini-orange sector magnet
Resolution $\delta P/P$ or $\Delta E^{\dagger}$	10^{-2} to 10^{-3}	10^{-2} to 10^{-4}	ΔE(Si)	ΔE(Si)	ΔE(Si)	ΔE(Si)
Transmission ($\Omega/4\pi$)	0.01 to 0.1	0.02 to 0.003	0.5 to 0.05	0.5 to 0.05	0.01 to 0.1	0.15 to 0.02
Transmission width $\Delta P/P$	small	0.01	large	0.1	0.5 to 5	0.3 to 3
Maximum energy (MeV)	~ 1	~ 10	~ 2	$1 \sim 2$	5 to 15	~ 5
Dimension (m)	~ 1	~ 1	~ 1	~ 0.5	~ 0.5	~ 0.15
Background suppression[‡]	Y	Y	Y/N	Y	Y	Y
Angular distribution[‡]	N	Y	N	Y	Y	N
References	Hansen *et al.* (1963)	Sakai *et al.* (1964/1965)	Lindblad and Lindén (1975)	Cambi *et al.* (1972)	Ejiri *et al.* (1976)	van Klinken *et al.* (1972, 1975, 1978)
	Geiger *et al.* (1967) Moll (1964)	Ejiri *et al.* (1965) Ishihara (1972)	Konijn *et al.* (1975) Backe *et al.* (1978)	Halpern *et al.* (1974)	Nagai *et al.* (1982) Komma *et al.* (1978)	

[†] Momentum resolutions $\delta P/P$ are given for the magnetic spectrometers. Energy resolutions for the electron transporters with Si detectors are given by those of the Si detectors of $\Delta E(\mathrm{Si}) \approx 1 \sim 3$ keV for $E_e = 0.1 \sim 3$ MeV.

[‡] Yes and no are denoted by Y and N, respectively.

angles of the order of 100 msr (about 10 per cent of 4π) by introducing multi-gap magnets.

3.4.2a *Sector-type multi-gap spectrometer with a single detector*

A multi-gap orange-type spectrometer used by Hansen *et al.* (1963) is shown in Fig. 3.46. It consists of six gaps and the total solid angle is as large as 88 msr, corresponding to the transmission of $T \approx 7$ per cent. The momentum resolution is typically 1 per cent for a 1.5-mm-wide beam spot. Electrons from the target are detected by a 10 mm × 10 mm anthracene crystal 35 cm from the target. A Chalk River group has used the seven-gap magnetic spectrometer with $R \approx 0.5$ per cent and $T \approx 3$ per cent (Geiger and Graham, 1967), and a Pretoria group has used the six-gap magnetic spectrometer with $R \approx 1$ per cent and $T \approx 10$ per cent (Spoelstra and Rautenbach, 1968).

Iron-free (air-core) orange-type β-spectrometers have been used for in-beam electron studies. The spectrometer used at Jülich has a resolution of $R \approx 0.3$ per cent with the very high transmission of $T \approx 16$ per cent (Moll, 1964). Recently positrons produced by heavy-ion bombardment of medium–heavy atoms have been studied with the orange-type β spectrometer (Moll and Kankeleit, 1965; Kozhuharov *et al.*, 1979; and Clemente *et al.*, 1984). This spectrometer uses sixty current coils to produce a toroidal magnetic field. Heavy ions with about 6 MeV per nucleon are provided by the UNILAC at GSI. The positrons emitted between 40° and 70° with respect to the beam direction are focused through the magnetic field onto the hollow cylindrical plastic scintillator, as shown in Fig. 3.47. In the recent experiment by Clemente *et al.* (1984) the scintillator is surrounded by a position-sensitive proportional counter in the focal area. The accepted momentum range of $\Delta P/P = 14$ per cent is subdivided into six bins, and the total positron momentum is obtained by scanning the magnetic field. They found for the heavy projectile–target system sharp positron peaks at around 340 keV superimposed on continuum positron spectra.

3.4.2b *Sector-type single gap spectrometers*

Single gap double-focusing spectrometers are handy and useful for on-line studies of conversion electrons. They are set at backward angles with respect to the beam in order to avoid the intense δ rays at forward angles. An INS (Tokyo) group (Sakai *et al.*, 1964/1965; Ejiri *et al.*, 1965) used a spectrometer with the field parameters $\alpha = -1/2$ and $\beta = 1/8$. The parameters are defined by the field equation,

$$B(\rho, 0) = B_0\{1 + \alpha(\rho - \rho_0)/\rho_0 + \beta(\rho - \rho_0)^2/\rho_0^2\} \tag{3.4.4}$$

where ρ_0 is the radius of the central ray and B_0 is the field at $\rho = \rho_0$. The spectrometer is called S-RACE (single reaction conversion electron spectrometer). It has a transmission of $T \sim 0.15$ per cent with the resolution of $R \sim 0.3$

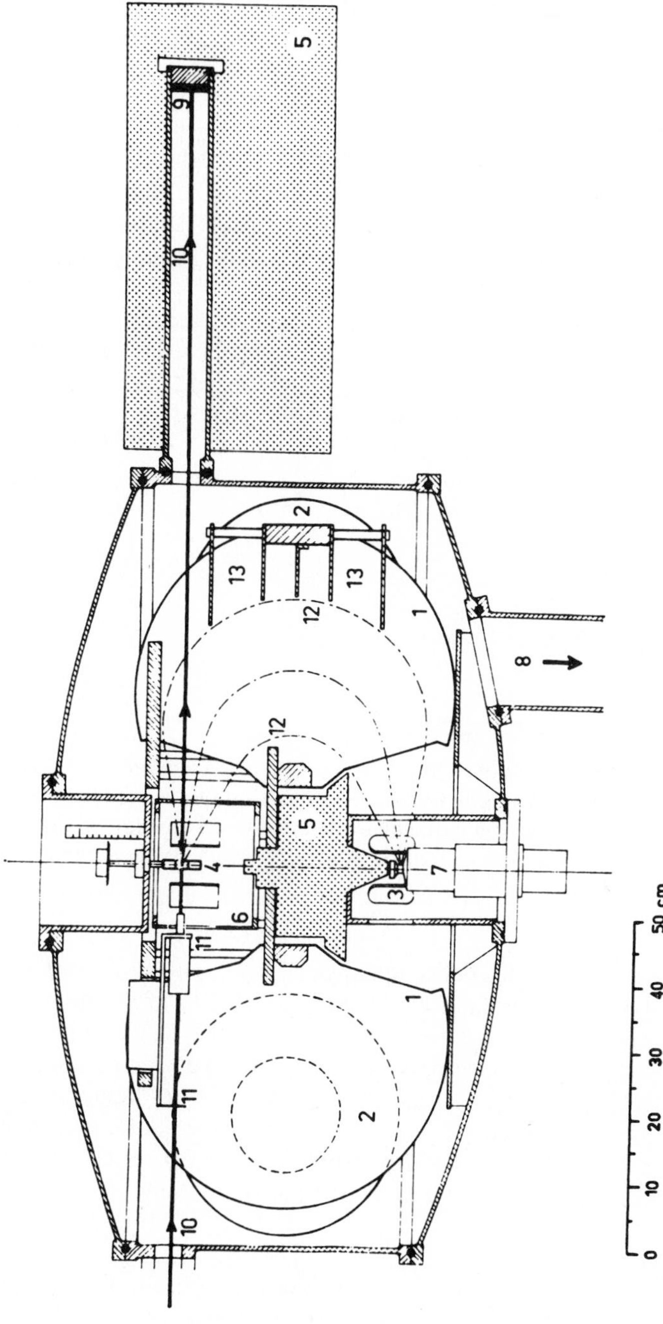

Fig. 3.46. The multi-gap magnetic electron spectrometer of the orange type. Pole plates (1), coils (2), anthracene crystal and exit slit (3), target (4), lead shield (5), apertures (6), photomutiplier (7), vacuum pumps (8), Faraday cup (9), beam (10), slits (11), electron path (12) and baffles (13). (Hansen *et al.*, 1963.)

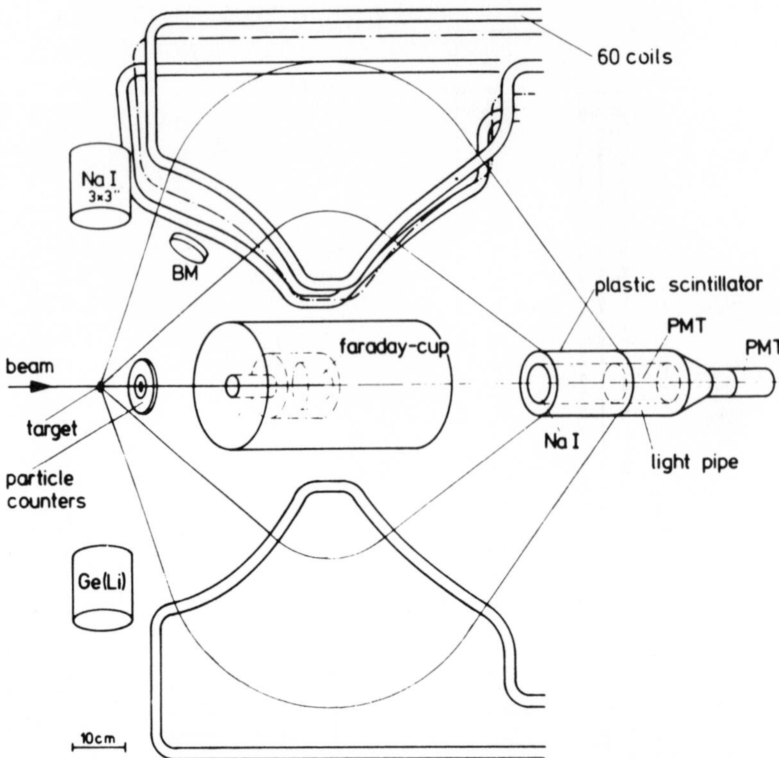

Fig. 3.47. Experimental set up of an iron-free orange-type β spectrometer. Positrons are focused by the toroidal magnetic field onto a cylindrical plastic scintillator which surrounds a NaI crystal. Scattered particles are detected by an annular parallel-plate avalanche counter with four concentric anodes. (Kozhuharov *et al.*, 1979.)

per cent. Spectrometers with a deflection angle of $\pi\sqrt{2}$, $\alpha = -1/2$ and $\beta = 3/8$ have been used by the Uppsala group (Antman *et al.*, 1970; Johansson *et al.*, 1973). Their transmissions and resolutions are $T \approx 0.2$ to 0.5 per cent at $R \approx 0.14$ to 0.3 per cent, respectively. The Bisgård type spectrometer with $T \approx 0.73$ per cent at $R \approx 0.45$ per cent has been used by a Manitoba group (Canty *et al.*, 1974). The air-core $\pi\sqrt{13/2}$ type spectrometer at Heidelberg has the very high resolution of $R \approx 0.022$ per cent with $T \approx 0.6$ per cent, but the practical resolution for on-line use is $R \approx 0.4$ per cent (Daniel *et al.*, 1970).

3.4.2c *Lens-type spectrometers*

The Siegbahn–Slätis type of iron-jacked spectrometer (Slätis and Siegbahn, 1949) has been used by a New Orleans group (Robert *et al.*, 1970). The transmission of $T \approx 0.7$ per cent and the resolution of $R \approx 2$ per cent are of the same order of magnitude as those of sector-type spectrometers.

3.4.2d *Sector-type multi-gap spectrometers with multi-detectors*

The multi-gap spectrometers discussed so far have one detector at one focusing point to collect all the electrons from all gaps. The electron yield therefore gives differential cross-sections averaged over the solid angles spanned by the multi-gaps. The INS (Tokyo) group has developed a multi-gap spectrometer with multi-detectors to measure the angular distributions of the conversion electrons. Figure 3.48 shows the side view of the spectrometer. It consists of five gaps to measure electrons emitted at 22.5°, 67.5°, 90°, 135° and 180° with respect to the beam, and they are detected individually by the five Si semi-conductor detectors. Each gap has the transmission of $T \approx 0.3$ per cent with a resolution of $R \approx 0.4$ per cent. The spectrometer is called M-RACE (multi-gap RACE) as the second version of the S-RACE. It is noted that the

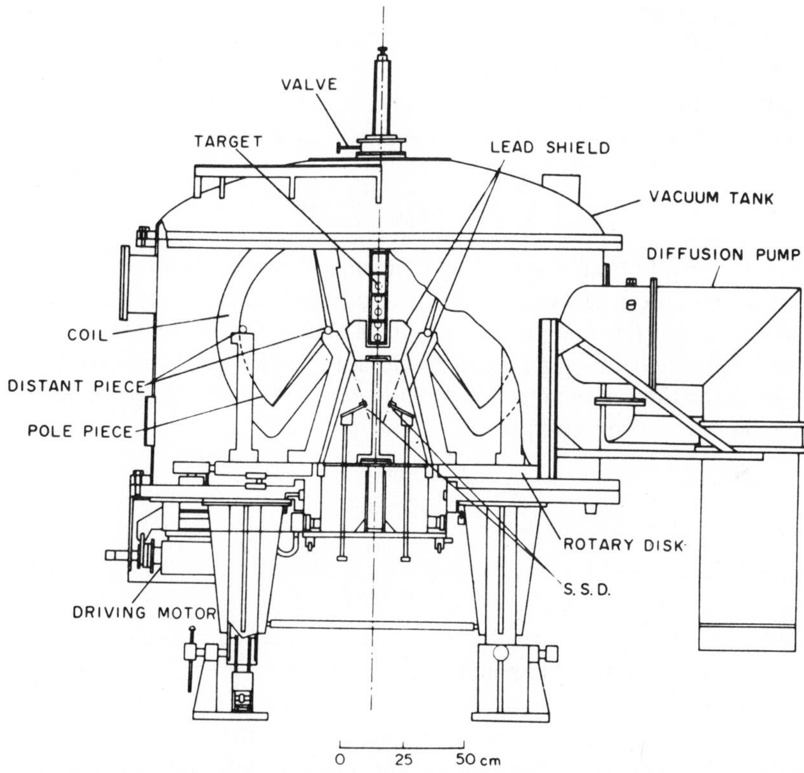

Fig. 3.48. Side view of the M-RACE (multi-gap reaction conversion electron spectrometer. (Ishihara, 1972.)

first anisotropy measurement of conversion electrons following statistical reactions was made by setting the S-RACE at 90° and 180° to the beam, and initiated the construction of the M-RACE for angular distributions (Ejiri *et al.*, 1965, 1966).

3.4.3 *Magnetic electron transport systems*

3.4.3a *Electron transport system with axially symmetric magnetic field*

Electrons emitted from a target are transported well by an axially symmetric magnetic field, and are detected by a semi-conductor detector. The magnetic field is used to transport only the electrons far from the target so as to reduce background due to γ rays and scattered particles. The electron energy is analysed by a cooled Si detector. The energy resolution is around $1 \sim 3$ keV, as in Ge γ detectors, unless the energy loss in the target exceeds the resolution.

A solenoidal coil with a homogeneous magnetic field is a simple transporter with a large solid angle. It has been used for in-beam studies of conversion electrons by many groups (Morinaga 1964; Burginyon and Greenberg 1966; Klank and Ristinen, 1969; Kotajima and Beringer, 1970; Konijn *et al.*, 1975; Lindblad and Lindén, 1975; and Backe *et al.*, 1978). Electrons emitted from a beam spot on the solenoid axis show helical motions in the uniform field. The distance of the electron orbit from the axis is given by

$$r = \frac{2Pc}{eH} \sin\left(\frac{ezH}{2Pc\cos\theta}\right)\sin\theta, \tag{3.4.5}$$

where P, θ, z, and H are the electron momentum, the angle with respect to the solenoid axis, the distance along the axis from the beam spot and the magnetic field strength. If P is given by $B\rho$ in units of Gauss·cm, $2Pc/eH$ is rewritten as $2B\rho/B$. The maximum momentum P_m for electrons transmitted by the solenoid and the maximum momentum P'_m covered fully by the Si-detector on the solenoid axis are given by

$$P_m = \frac{eHR}{2c\sin\theta}, \tag{3.4.6a}$$

$$P'_m = \frac{eHD}{4c\sin\theta}, \tag{3.4.6b}$$

where R is the inner radius of the solenoid and D is the diameter of the detector. Thus the transmission ($\Omega/4\pi$) for electrons up to $P = eHD/4c$ is 50 per cent, and decreases as P increases. The field at the target position is somewhat lower than the average field since the beam tube used to guide the accelerated beam is inserted there. The angle is then limited due to the magnetic mirror effect. Intense low energy δ electrons are rejected by inserting a small baffle on the solenoid axis between the target and the detector.

The solenoid-type spectrometer used by a Stockholm group is shown in Fig. 3.49 (Lindblad and Lindén, 1975). Examples of prompt and delayed conversion-electron spectra following the ^{209}Bi(α, xe) reaction are shown in Fig. 3.50. The energy resolution is 2.1 keV at 500 keV and the efficiency is about 5 per cent of 4π. A modulation system for the solenoid current has been introduced in order to smear out the fluctuation of the efficiency as a function of the electron energy. The solenoid type spectrometer developed by the Amsterdam group has a helical baffle to separate electrons from positrons (Konijn *et al.*, 1975). The absolute efficiency is 11 per cent of 4π at 481 keV and the Si detector has energy resolutions of 1.3 keV and 2.0 keV at 92 keV and 975 keV, respectively.

The Darmstadt group has developed an electron spectrometer consisting of a solenoid coil and a cooled Si detector for in-beam spectroscopy in three different modes (Backe *et al.*, 1978). The first mode is the broad-range electron transportation mode as discussed above. Electrons in a range from 80 keV to about 3 MeV are transmitted with the peak efficiency of around 5 per cent, as shown in Fig. 3.51. Low-energy δ rays are rejected by a tantalum disc on the axis between the detector and the target. The second mode is the lens-type spectrometer mode as shown in Fig. 3.52. Electrons within the momentum

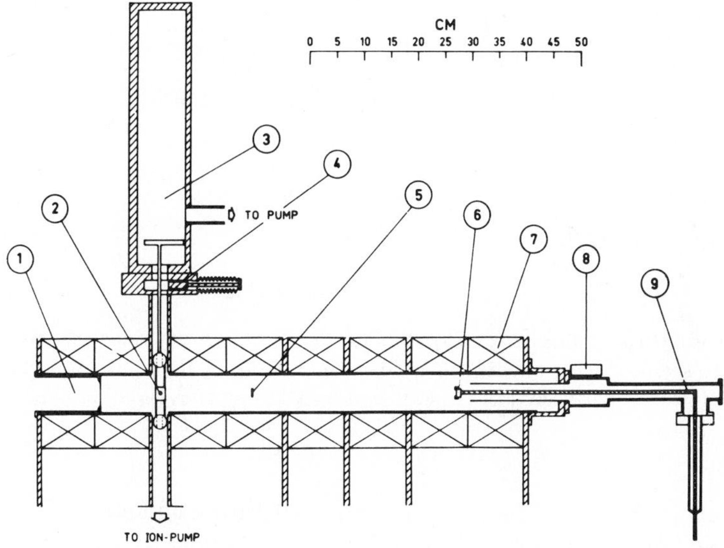

Fig. 3.49. Schematic drawing of a solenoid-type electron spectrometer. The direction of the beam from the accelerator is perpendicular to the paper. (1) position of photon detector, (2) target, (3) vacuum lock, (4) vent, (5) baffle, (6) Si(Li) detector, (7) coils, (8) pre-amplifier, and (9) cold finger. (Lindblad and Lindèn, 1975.)

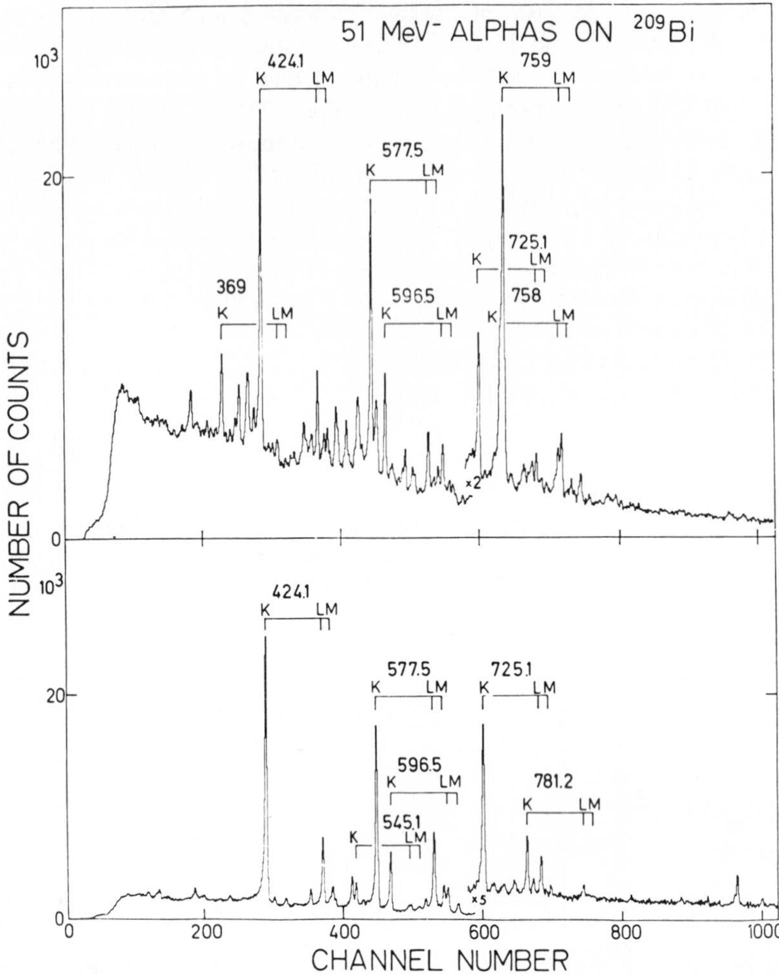

Fig. 3.50. Prompt (top) and delayed (bottom) conversion-electron spectra, obtained with the solenoid spectrometer displayed in Fig. 3.49 during the bombardment of ^{209}Bi with 51 MeV α-particles. The target thickness is 650 μg/cm^2. (Lindblad and Lindèn, 1975.)

window of $\Delta P/P = 12$ per cent are transported through four conical baffles with an efficiency of about 2 per cent of 4π. This mode is useful to select a momentum range of interest and to cut off other electrons. Positrons are rejected by two paddlewheel baffles. The third mode consists of the recoil shadow method, as shown in Fig. 3.53. It is used to reduce the very strong δ-electron background in the low-energy region.

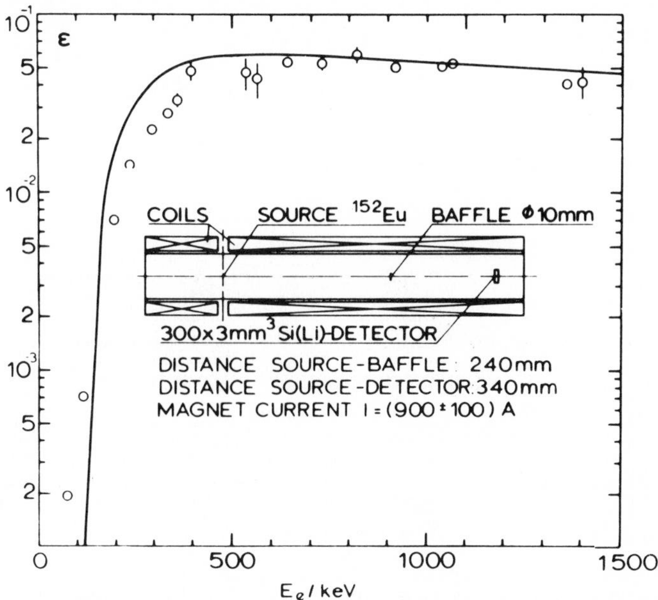

Fig. 3.51. Peak detection efficiency ε of electrons in the broad range mode of the Darmstadt solenoid as measured with an intensity-calibrated ^{152}Eu source and with a 1-mm thick 10-mm diameter tantalum baffle. To avoid transmission oscillations, the magnet current of 900 A is modulated by a triangular current signal of ± 100 A amplitude. The solid line is the calculated curve. (Backe *et al.*, 1978.)

The recoil shadow method is a unique feature of the Darmstadt spectrometer. It employs an aluminum diaphragm shaped like a semi-cylindrical tube as shown in Fig. 3.53. The target is set up-stream at a certain distance d from the solenoid axis so that prompt electrons such as δ rays are intercepted by the diaphragm acting as a shadow. On the other hand, delayed electrons are emitted from recoiling nuclei behind the target. Consequently they may well clear the shadow diaphragm provided that the recoil distance is larger than the distance d. This method is very useful for measuring low-energy conversion electrons below 100 keV since they are mostly delayed and the δ-ray background is serious. It is noted that the conversion coefficient in the low-energy region is so large that measurement of γ rays is less efficient than that of conversion electrons. The transmission of the delayed electrons depends on the flight distance of the recoiling nucleus from the target with respect to the target position d from the edge of the semi-cylindrical diaphragm. The flight distance is a linear function of the lifetime, and so the lifetime can be measured by observing the electron intensity as a function of the target position d. The lifetime measurement is demonstrated in Fig. 3.54 (Backe *et al.*, 1978).

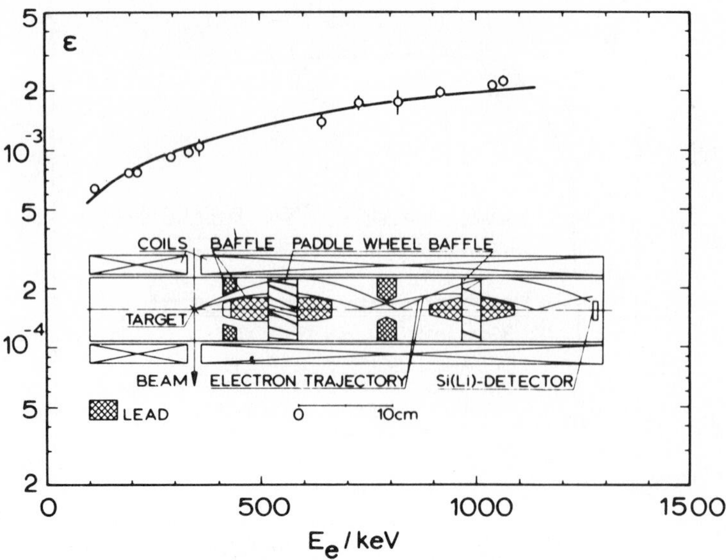

Fig. 3.52. Effective peak detection efficiency ε in the lens spectrometer mode of the Darmstadt solenoid as determined with an intensity-calibrated ^{152}Eu source. Due to the envelope baffle system also shown in this picture only a momentum band $\Delta P/P = 0.12$ can reach the Si(Li) detector with a 300 mm^2 area. The accepted solid angle is 2 per cent of 4π. The magnet current is linearly swept between 40 A and 240 A with a frequency of 6 Hz. (Backe *et al.*, 1978.)

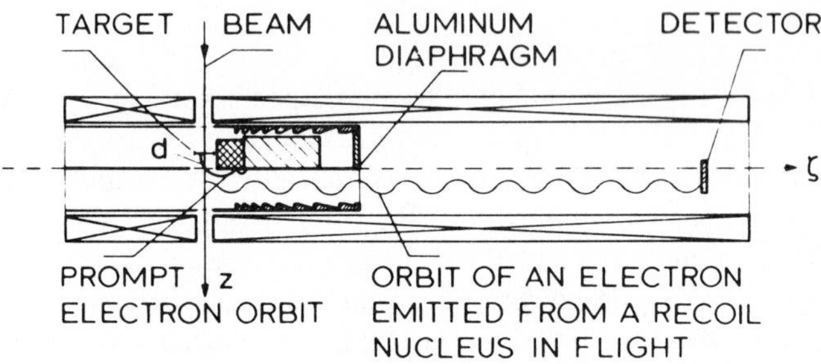

Fig. 3.53. The recoil shadow method with the third mode of the Darmstadt solenoid. Shown is a cut through the beam and solenoid symmetry axis of the electron transport system. The longitudinal baffle avoids detection of prompt electrons but allows very efficiently the transmission of delayed electrons emitted in flight. (Backe *et al.*, 1978.)

A lens-type spectrometer with a finite momentum window is used to transport electrons in a given momentum band at a fixed magnetic current, or to transport them over a wide range of momentum by sweeping the magnet current. The Texas group used the lens-type spectrometer with $T = 3.5$ per

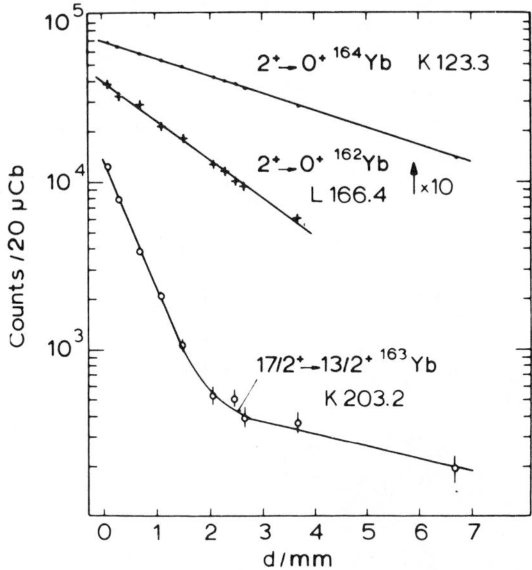

Fig. 3.54. Lifetime measurements on certain levels in ^{162,163,164}Yb with the recoil shadow method, as illustrated in Fig. 3.53, by variation of the target position d relative to the edge of the semicylindrical baffle. The results are $T_{1/2} = (971 \pm 31)$ps and $T_{1/2} = (439 \pm 37)$ps for the $2^+ \rightarrow 0^+$ transitions in ^{164}Yb and ^{162}Yb, respectively. For the 203.2 keV transition in ^{163}Yb the two delayed components are $T_{1/2}^{(1)} = (108 \pm 7)$ps and $T_{1/2}^{(2)} = (1.2 \pm 0.3)$ns. (Backe *et al.*, 1978.)

cent and $\Delta E/E = 14$ per cent to measure conversion electrons from (p, p') and (p, n) reactions (Picone *et al.*, 1972). Sweeping-current spectrometers have also been used by other groups (Avignone *et al.*, 1973; Westerberg *et al.*, 1975).

3.4.3b *Electron transport systems with a sector-type magnet*

A sector magnet is used to transport electrons from the target to a cooled Si detector for the energy analysis. In the sector-shaped magnetic field electrons are bent by the sector magnet so that the detector is easily hidden from the direct γ rays from the target. Positrons and charged particles are bent in other ways and do not reach the detector. Sector magnets produce strong magnetic fields, and thus they are built fairly compactly.

Cambi *et al.* (1972) used a compact magnet with uniform field to transport electrons in the momentum band $\Delta P/P = 7$ per cent. Even though the sector magnet is dispersive in the momentum, electrons in the large momentum band are well accepted by a 100 mm^2 Si detector. The efficiency is 0.2 per cent. A schematic view is given in Fig. 3.55. A simple sector magnet was used for measuring conversion electrons emitted backwards in coincidence with fission fragment (Halpern *et al.*, 1974).

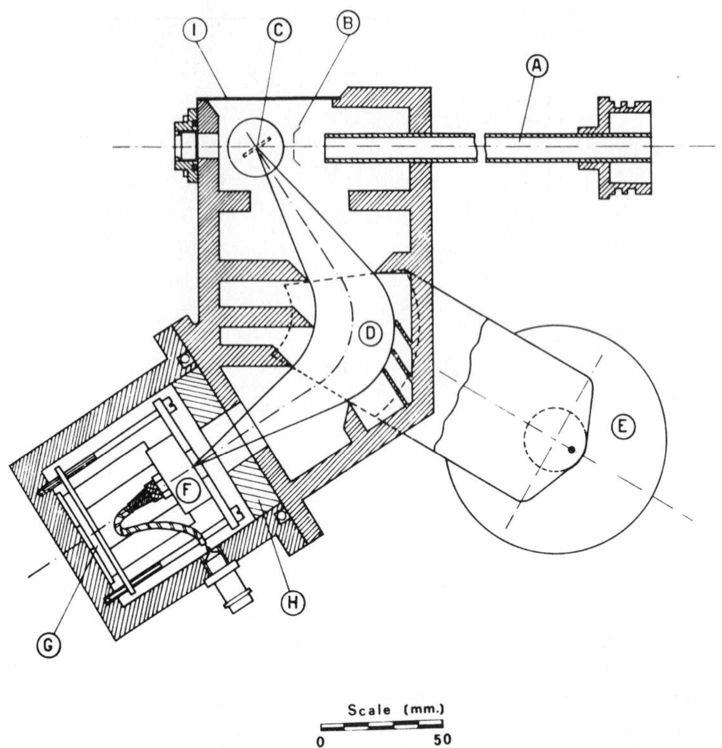

Fig. 3.55. Schematic view of a sector-type spectrometer, with uniform field. (A) Magnetic screening channel; (B) tantalum collimator, (C) target, (D) magnetic field region, (E) electromagnet, (F) Si(Li) detector, (G) thermocooler, (H) lead shield, (I) plexiglass window transparent for γ-rays. (Cambi *et al.*, 1972.)

Ejiri *et al.* (1976) developed an achromatic (broad range type) sector magnet for transporting electrons. The TESS (triple-focusing electron spectrum selector) is shown in Fig. 3.56. The magnet with uniform field is designed to give a triple-focusing image (double-focusing in vertical and horizontal coordinates, and momentum-focusing), so that electrons with a large momentum range are focused on a small Si detector with a moderate efficiency. The radial and axial images are given by using the transport matrices as

$$\begin{pmatrix} r_1 \\ r_1' \\ \Delta P/P \end{pmatrix} = \begin{pmatrix} a, b, c \\ d, e, f \\ 0, 0, 1 \end{pmatrix} \cdot \begin{pmatrix} r_0 \\ r_0' \\ \Delta P/P \end{pmatrix} \tag{3.4.7a}$$

$$\begin{pmatrix} z_1 \\ z_1' \end{pmatrix} = \begin{pmatrix} A & B \\ C & D \end{pmatrix} \cdot \begin{pmatrix} z_0 \\ z_0' \end{pmatrix} \tag{3.4.7b}$$

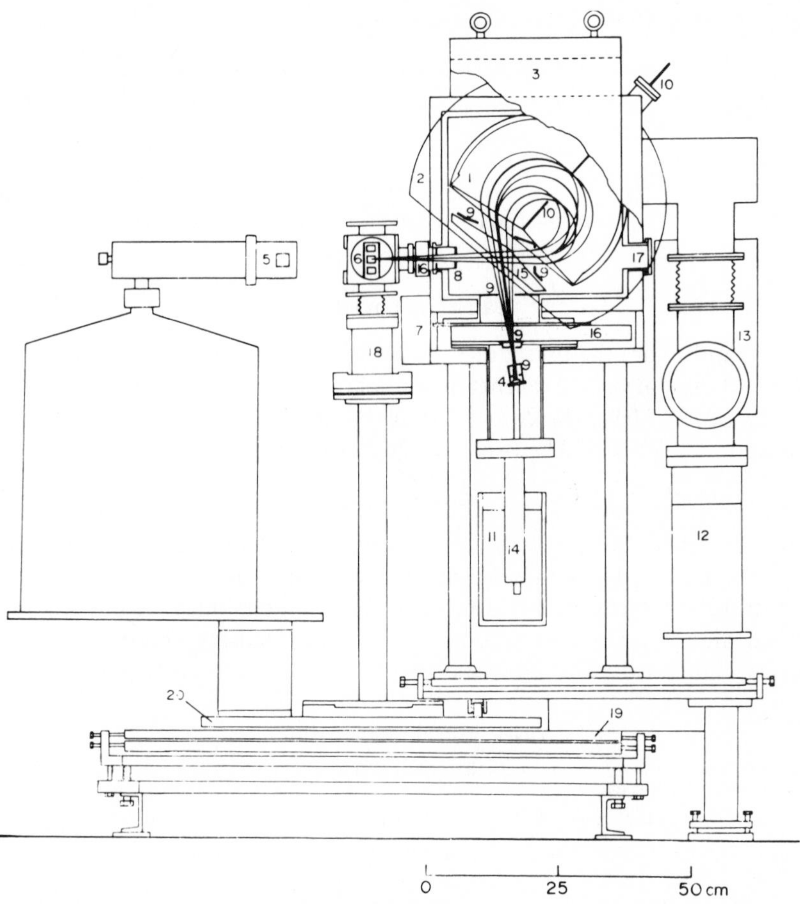

Fig. 3.56. Side view of the TESS (triple-focusing electron spectrum selector) and the Ge(Li) detector for inbeam electron gamma spectroscopy. Magnet poles (1), coils (2), yokes (3), Si(Li) detector (4), Ge(Li) detector (5), target (6), lead shield (7), entrance slits (8), baffles (9), momentum defining slits (10), liquid nitrogen (11), ion pump (12), sorption pump (13), cold finger (14), magnetic field clamp (15), valves (16), window (17), target adjuster (18), turntable for the TESS (19), and turntable for the Ge(Li) detector (20). The target can be moved up and down, and be rotated by using the mechanism (18). The window (17) is used to adjust a projectile beam and to measure electrons backward. Although the setting of the spectrometer in the picture is at 90° and 270° with respect to an incident beam, the TESS is usually set at $\theta=\pi-55°$ where $P_2(\cos\theta)=0$. (Ejiri *et al.*, 1976.)

where r and z are radial and axial coordinates, respectively, and r' and z' are their slopes. The deviation of the momentum from the central ray is given by $\Delta P/P$. The orbits are shown in Fig. 3.57. The radial, axial and momentum focusing points L_r', L_z', L_p', are obtained from the focusing conditions

$b=0$, $B=0$ and $c=0$ from the expressions

$$L'_r = \frac{L(\cos\phi + \sin\phi\tan u) + \sin\phi}{L\{\sin\phi - (\tan u + \tan u')\cos\phi - \tan u \tan u' \sin\phi\} - (\cos\phi + \tan u'\sin\phi)}, \tag{3.4.8}$$

$$L'_z = \frac{L\{1-(\tan u)\phi\} + \phi}{L\{\tan u + \tan u' - (\tan u \tan u')\phi\} - 1 + (\tan u')\phi}, \tag{3.4.9}$$

$$L'_p = \frac{\cos\phi - 1}{\tan u'(1-\cos\phi) + \sin\phi}. \tag{3.4.10}$$

Here L and L' are the object and the image distances, respectively, from the effective boundary of the magnet in units of the radius ρ of the central ray with momentum P. The entrance and exit angles are denoted by u and u', and the deflection angle by ϕ. The triple-focusing condition is

$$L'_r = L'_z = L'_p. \tag{3.4.11}$$

The symmetric magnet with $u=u'$ satisfies the triple focusing condition $L'_r = L'_z = L'_p = L$. The symmetric magnet with the intermediate radial focus is non-dispersive (achromatic). Dividing the symmetry magnet into the up-stream and down-stream sections, the double-focusing conditions at the intermediate focal plane are given by

$$L_{ir} = \frac{L(1 + \tan u \tan\frac{1}{2}\phi) + \tan\frac{1}{2}\phi}{L(\tan\frac{1}{2}\phi - \tan u) - 1} = 0, \tag{3.4.12a}$$

$$L_{iz} = \frac{L(1 - (\tan u)\frac{1}{2}\phi) + \frac{1}{2}\phi}{L\tan u - 1} = 0, \tag{3.4.12b}$$

where L_{ir} and L_{iz} are distances of the intermediate image from the intermediate plane. These equations correspond to eqns (3.4.8) and (3.4.9) with $u'=0$ and $\phi=\frac{1}{2}\phi$. The intermediate double focusing conditions are satisfied by

$$L = \frac{-\tan\frac{1}{2}\phi}{\tan u \tan\frac{1}{2}\phi + 1}, \tag{3.4.13}$$

$$L = \frac{\phi}{\phi\tan u - 2}. \tag{3.4.14}$$

The values for u and L are obtained for a given ϕ, and they obviously satisfy the triple-focusing condition. The magnifications for the symmetric magnet are $M_r = M_z = 1$.

The triple-focusing magnet with the double-focusing intermediate image has several unique features. Electrons in a large momentum range $\Delta P/P \approx 50$ per cent are selected at the intermediate slit, and are focused at the final image

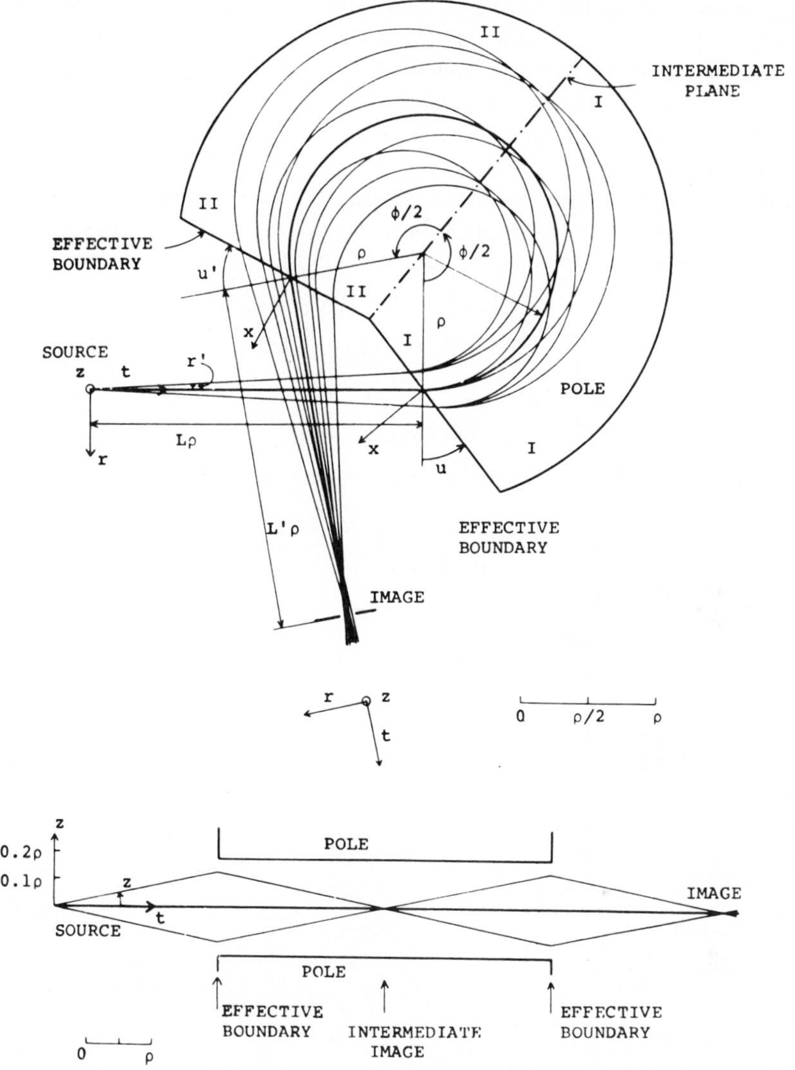

Fig. 3.57. Schematic picture of the electron orbits in the triple-focusing magnet as displayed in Fig. 3.56. (Ejiri *et al.*, 1976.)

point. In this way low-energy δ rays and high momentum particles are rejected. All kinds of Compton-scattered electrons and other backgrounds are reduced by the baffle with a narrow window at the intermediate image. It has a small and definite angular acceptance for emitted electrons, and can be rotated around the target in contrast with most of the solenoid-type transporters. This feature is important for measuring conversion electrons from spin-aligned states. The transmission is still kept as large as 1 per cent. This type of the electron transport system is very powerful for accurate in-beam measurements of conversion coefficients since the background is extremely small and angular distribution effects are well corrected for. An in-beam electron spectrum measured by TESS is shown in Fig. 3.58 together with the γ-ray spectrum.

Another triple-focusing electron transporter with an intermediate single (radial) focus is also used for in-beam measurements. The conditions for the symmetry magnet with the intermediate radial focusing image are

$$L_{ir} = \frac{L(1 + \tan u \tan \frac{1}{2}\phi) + \tan \frac{1}{2}\phi}{L(\tan \frac{1}{2}\phi - \tan u) - 1} = 0, \tag{3.4.15}$$

$$L_{iz} = \frac{L(1 - (\tan u)\frac{1}{2}\phi) + \frac{1}{2}\phi}{L \tan u - 1} = \infty. \tag{3.4.16}$$

The parameters L, u and ϕ are obtained from the condition

$$L = \frac{-\tan \frac{1}{2}\phi}{\tan u \tan \frac{1}{2}\phi + 1} = \frac{1}{\tan u}. \tag{3.4.17}$$

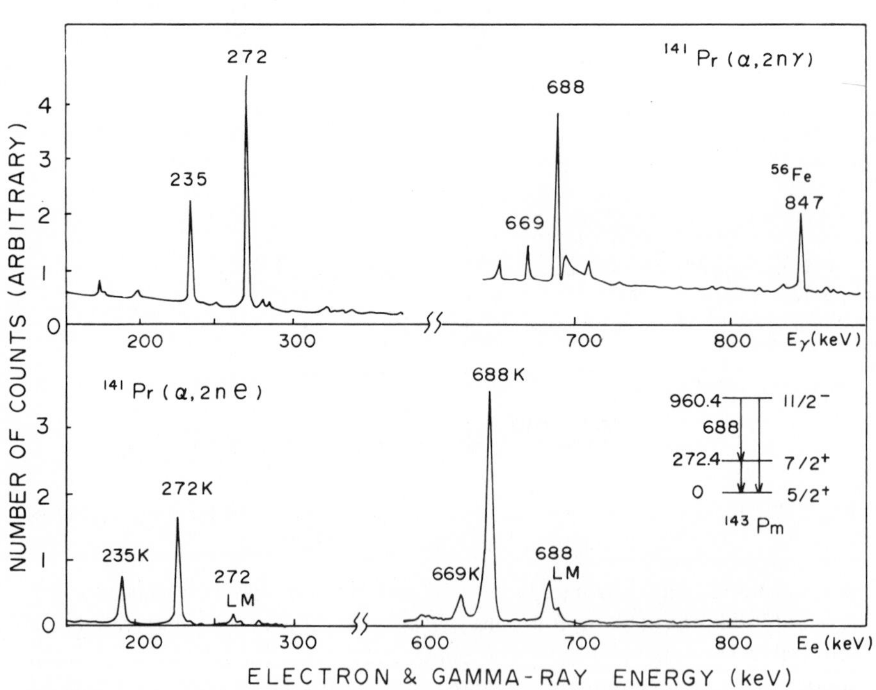

Fig. 3.58. Gamma-ray and conversion-electron spectra following the ^{141}Pr(α, 2n) reaction measured by TESS. (Ejiri *et al.*, 1976.)

The object distance L in this case becomes much shorter than that for the intermediate double focusing magnet, so that the solid angle and the momentum range become larger. An achromatic germinate nuclear electron selector (AGNES) with an intermediate radial focus has been developed for in-beam studies of conversion electrons (Nagai *et al.*, 1982). It consists of a pair of sector magnets with the uniform field as shown in Fig. 3.59. Conversion electrons emitted at 90° and 180° with respect to the beam axis are achromatically transported to the triple (radial, axial and momentum) focusing points, where they are energy-analysed individually by two Si detectors. The solid angle is as large as 50 msr for each magnet, and the momentum range is as large as $\Delta P/P = 57$ per cent. AGNES is useful for measuring anisotropies of conversion electrons from spin-aligned states and for measuring e^-, e^- or e^+, e^- pair electrons.

A very wide band electron transporter has been designed that incorporates a triple-focusing sector magnet (Komma, 1978). The ELSA (electron spectrum analyser) has achieved a very wide momentum range of $\Delta P/P = 490$ per cent ($P_{max}/P_{min} = 5.9$) with an adequate solid angle of 7.6 msr ($5° \times 5°$). Because of

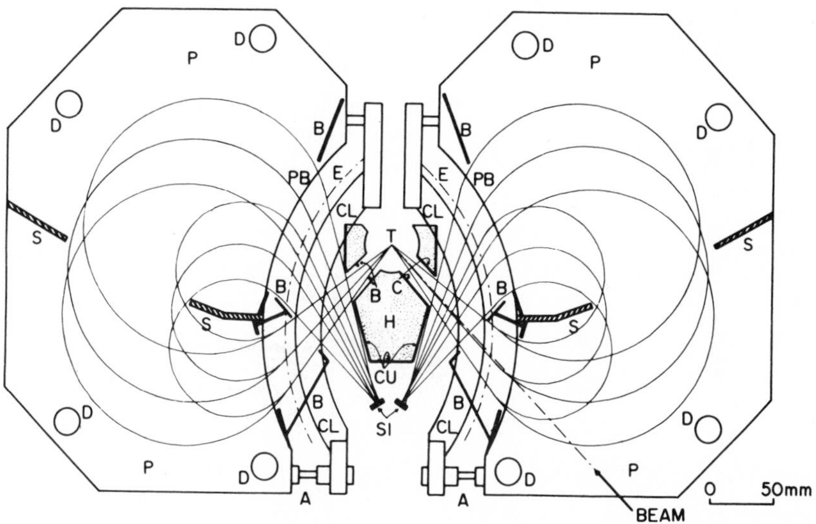

Fig. 3.59. Schematic view of AGNES and of the electron orbit trajectories. P: magnet poles, D: distance pieces, B: baffles to define entrance and exit electron orbits, S: slit at the intermediate radial focal plane to define the range of electron momentum, PB: pole boundaries, E: effective field boundaries, CL: field clamps, A: field clamp adjuster, H: heavy metal to shield the detector against γ-ray background, C: carbon liner, CU: copper plate to absorb Pb X-rays, T: target and SI: silicon semiconductor detectors. (Nagai *et al.*, 1982.)

the very large momentum band the optimization exceeds the limits as calculated within the linear theory. Thus the optimum shape of the magnet was obtained by using a ray-tracing program. The magnet and the electron trajectories are shown in Fig. 3.60.

A wide-range electron transporter has been designed by using the fringing field of a sector magnet (Watson *et al.*, 1967; Allan, 1971). A pair of pole tips has been used to transport conversion electrons emitted from a target to a detector along the fringing field (Gono and Sugihara, 1974). The pole tip and the electron trajectory are shown in Fig. 3.61.

Van Klinken *et al.* (1972, 1975, 1978) have developed mini-orange spectrometers for in- and off-beam conversion electron spectroscopies. Orange-type configurations of permanent magnets are used to transport electrons from a source to a detector as shown in Fig. 3.62. These spectrometers have the following unique features.

1. Conversion electrons are separated from positrons. The central absorber cuts off soft γ rays, X-rays, δ rays and other charged particles emitted from the target.

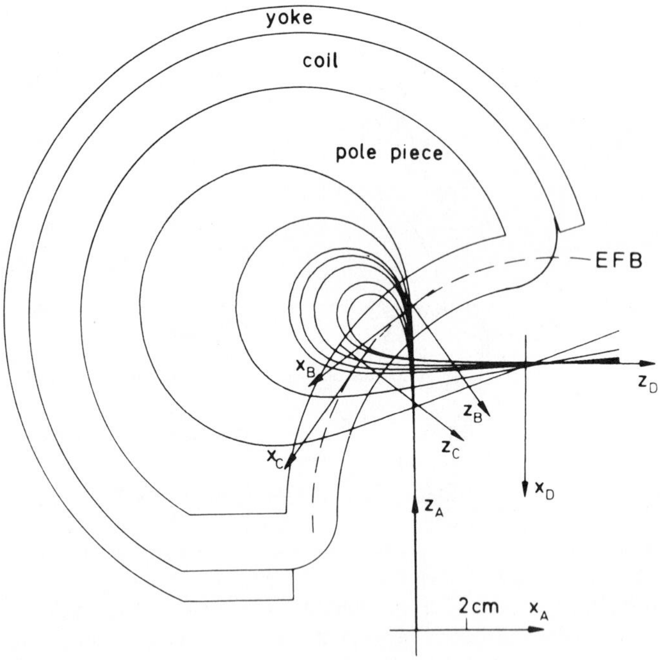

Fig. 3.60. Designed magnet of a wide-band, triple focusing spectrometer and particle trajectories with $\delta = \Delta P/P = 160\%$, 65%, 12%, 0%, -15%, -41%, -56%, EFB = effective field boundary. (Komma, 1978.)

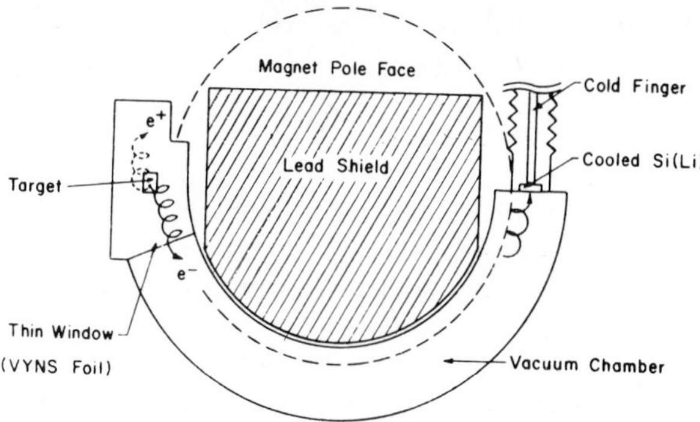

Fig. 3.61. Schematic top view of an inbeam conversion-electron spectrometer using a pair of magnetic pole tips. (Gono and Sugihara, 1974.)

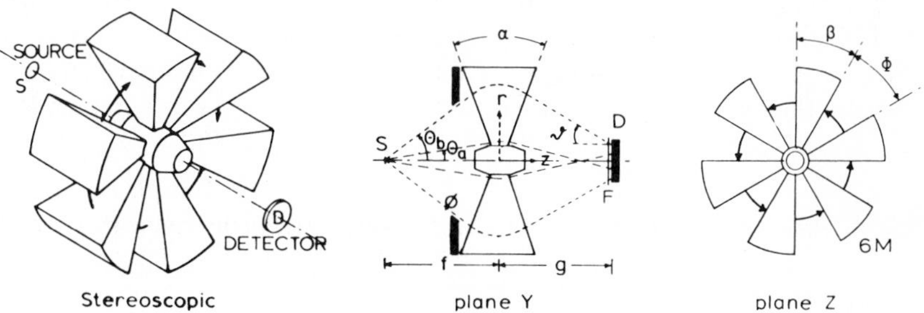

Fig. 3.62. Schematic view of a mini-orange spectrometer with an arrangement with six permanent magnets. Plane Y is through the system axis, and plane Z through the middle of the magnets as viewed from the source. The middle sketch shows a diaphragm with inner diameter of ~ 5 cm ϕ which is omitted in the stereoscopic picture. Various geometrical parameters are indicated. (van Klinken *et al.*, 1975.)

2. Since the magnet and the electron-orbit radius are comparable in size to the detector, electrons in a fairly large energy range from 0.1 to a few MeV are accepted by the detector. The low-energy electrons such as δ rays are blocked, and the maximum energy is determined by the detector thickness.
3. The absolute efficiency is as large as 7–10 per cent because of large acceptance angles of both θ and ϕ (see Fig. 3.62).
4. The size is small and thus very suitable for coincidence measurements with other detectors (Riezebos *et al.*, 1987). It is rather inexpensive and easy to operate.

The transmission for electrons of fixed energy is given by the following expression (van Klinken *et al.*, 1975),

$$T = \tfrac{1}{2}(\cos\theta_a - \cos\theta_b)(1-b)(1-k), \tag{3.4.18}$$

where θ_a and θ_b are effective boundary angles, and b is the blocking factor $b \sim \beta/\phi$ in the azimuthal angle, and k is the fraction of the low energy tail in the pulse-height spectrum. The inner angle θ_a depends on the diameter of the central shield with respect to the distance from the target, while the outer angle θ_b depends on the detector diameter and the field strength of the toroidal magnet. The factor $1-k$ gives the full energy loss (peak) efficiency. The transmission amounts to about 10 per cent with appropriate values of $\theta_a \sim 10°$, $\theta_b \sim 55°$, $b \sim 0.25$ and $k \sim 0.3$. By using various combinations of various kinds of magnets, one obtains electron filters with various momentum ranges and transmissions. The distances f between the target and the magnet and g between the magnet and the detector are changed easily, particularly the value of g during experiments. Narrow-range and broad-range transmission curves are obtained at adequate energy regions by using proper configurations of the magnets, as shown in Figs 3.63(a) and (b) respectively.

Recently four sets of mini-orange electron spectrometers have been used to measure the degree of longitudinal polarization of β-rays (see Chapter 7).

3.5 Measurements of dynamic and static electromagnetic moments

3.5.1 *Electromagnetic transition probabilities*

In-beam e–γ spectroscopy has been used extensively to study dynamic (transition) and static (diagonal) electromagnetic moments. Nuclear reactions induced by projectiles with a high momentum provide spin alignments of residual excited states and rapidly moving recoils via linear momentum transfers. Reactions induced by pulsed beams (i.e. from a cyclotron) produce nuclear excited states within certain time intervals. Multiple Coulomb excitation (MCE) predominantly populates states which are connected by collective (E2) transitions with the ground state. These characteristic features of nuclear reactions are fully exploited to measure the dynamic and static electromagnetic moments. This has opened up a new domain of experimental γ-ray spectroscopic studies. In this section we emphasize the relevant experimental methods of this spectroscopy. Some underlying theoretical notions are presented in Chapter 2 and we also refer to the standard works, for instance of Morinaga and Yamazaki (1976) and of Hamilton (1975), for many theoretical and experimental details.

The γ-ray transitions between nuclear excited states are of electric or magnetic character. The electric quadrupole (E2) and the magnetic dipole

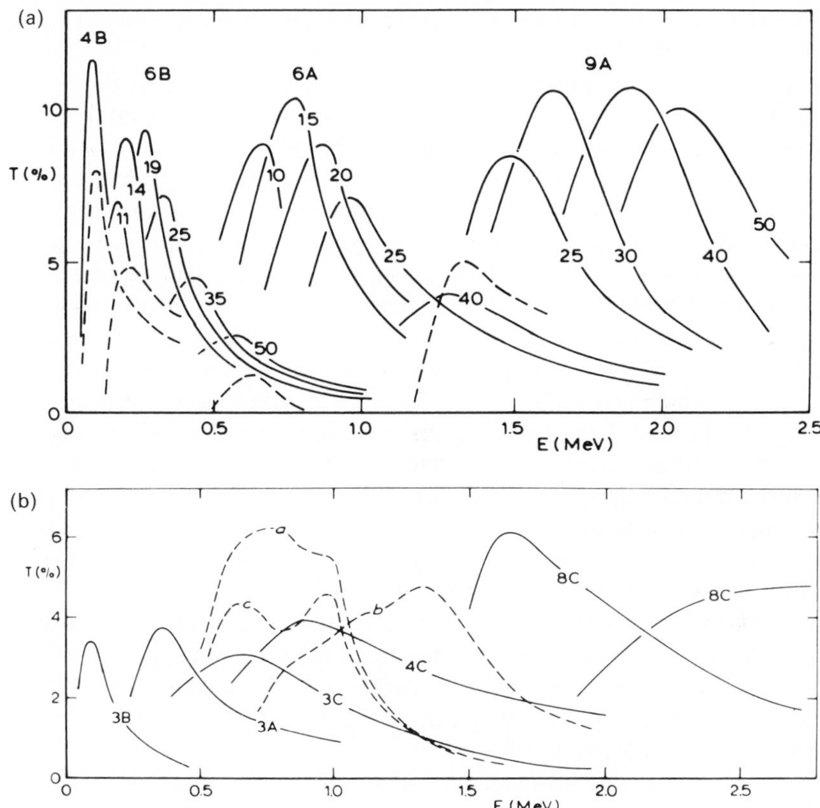

Fig. 3.63. (a) Transmission curves of the mini-orange spectrometer (see Fig. 3.62) labelled for the various configurations, together with some measured narrow-range transmission curves. Here f and g are the distances in mm from the magnet centre to the source and the detector, respectively. The magnet configurations are labelled as 4B, 6A and so on, where 4B stands for four B-type magnets, 6A for six A-type magnets, and so on. The A-type magnets are stronger than the B-type. 4B: $f=14$, $g=25$, 6B: $f=11$–50, $g=25$, 6A: $f=10$–40, $g=25$, 9A: $f=20$, $g=25$–50. The dashed lines are optimum curves. (van Klinken *et al.*, 1978.) (b) Some broad-range transmission curves obtained by using only 3 or 4 magnets (3B, 3A, 3C, 4C), by using different gap widths, by combining A and B magnets (a, b, c), or by using the circular C-magnets (8C). Dashed curves are for a configuration 4A + 4B. (van Klinken *et al.*, 1978.)

(M1) moments are most commonly observed in the in-beam studies of low-lying states. They appear as dynamic (transition) moments as well as static moments. Electric dipole (E1) transitions are mainly involved in the de-excitation of highly excited states, and higher multipole moments are involved in isomeric transitions. One has to establish the electromagnetic character of the transition by measuring the multipolarity λ and the parity of the γ ray. In most cases λ has the lowest value allowed by the spins and parities of the initial and final states associated with the transition. Mixed multipolarities, however, are involved particularly in the case when the lowest multipole transition is hindered (retarded) and/or a higher multipole is enhanced. The E2/M1 mixing is most commonly encountered in nuclei with collective nature. The mixing ratio δ of the intensities of the two multipolarities is expressed by eqn (2.1.7b) as discussed in Section 2.1.1. Multipolarities are obtained by measuring the angular correlations or angular distributions and linear polarizations of γ rays de-exciting spin-aligned (polarized) nuclear states as well as by measuring internal conversion electron coefficients.

The transition probability $T(\lambda)$ is directly obtained from the lifetime as $T(\lambda)=1/\tau(\lambda)$. Here $\tau(\lambda)$ represents the decay time constant of a particular transition. When that transition is the only one de-exciting a nuclear level then $\tau(\lambda)$ also represents the mean life of that level. Otherwise one has take into account the branching ratios connected with the decay of that level. Measurement of the lifetime $\tau(\lambda)$ of a nuclear level thus yields the transition probability under certain conditions.

Most commonly the transition rates are expressed as the reduced transition probabilities $B(\lambda)$, which are obtained from the branching ratio, the γ-ray energy, the total conversion electron coefficient and the lifetime. Expressions for the reduced transition probabilities are given in Chapters 2 and 6; i.e. by eqns (2.1.3), (6.3.84) and (6.3.85).

In this section we will focus our attention on transition moments employing: (a) electronic timing techniques, e.g. with pulsed beams, (b) the recoil-distance Doppler-shift technique, (c) the Doppler-shift attenuation method and (d) the method of multiple Coulomb excitation (MCE). A first indication of lifetimes in the picoseconds region can be obtained from a comparison of the spectra measured with a thin target with and without a backing that stops the recoiling excited nucleus. An example is given in Section 6.3.5, and the method will not be discussed here. In the next Section 3.5.2 the measurements of the magnetic and electric static moments will be discussed.

3.5.1a *Electronic timing techniques*

Lifetimes of nuclear states smaller than a few seconds can be measured with electronic techniques. The development of fast electronics and of detectors with a fast time response make it possible to measure lifetimes down to the sub-nanoseconds region. As discussed in Section 3.2.5 a small BaF_2 detector can have a time resolution as low as 112 picoseconds. The electronic clock determines the time difference between the arrival times of two signals, for instance those generated by two detectors. For a good time resolution one has to use detectors with a fast risetime and the pulse length must be small enough to allow high counting rates. The fastest timing is achieved with plastic scintillators and with BaF_2 crystals (time resolution of about 100 ps). Also NaI(Tl) is reasonably fast ($\leqslant 1$ ns) and small semiconductor detectors may achieve resolutions of a few ns, while the big germanium γ-ray detectors reach 5–10 ns, at best.

The electronic circuitry is designed to process the initial pulses of the detectors as rapidly as possible and subsequently to store the information in a computer or analyser. The time difference between two signals can be converted in a pulse amplitude by a time-to-amplitude converter (TAC). There is a linear relationship between that amplitude and the corresponding time difference. Subsequently the pulse-height spectra for time and also for energy can be digitized with an analogue to digital converter (ADC) and then processed with a computer. The time spectrum derived from the time difference between two signals can also be recorded directly with a computer by means of a time-to-digital converter (TDC) as part of a Camac system. The Camac system is a modern hardware interface between the computer and the particular electronical circuitry used for experiments. The use of Camac with ADC's and TDC's is now very common and it is even necessary in the fast processing of the many signals of the multi-detector assemblies.

The best way to employ the full power of the computer is to use the list mode, which stores all the data parameters in digitized form event by event on magnetic tape. In this way it is possible to play back the experiment in an off-line analysis and then to introduce certain constraints on the parameters, stored during the data taking. The constraints can be imposed on the time spectra as well as on the energy spectra. To reduce, for instance, the number of random coincidences between two detectors one sets a gate on the prompt peak in the time spectrum as well as on one of the accidental coincidence peaks when a pulsed beam is used. The corresponding energy spectra updated with those two gates are then subtracted from each other. Conversely time spectra can be obtained, which are related to a certain transition, by setting a gate in the energy spectrum on the corresponding γ-ray line. The background can be corrected for by also updating a time spectrum with the energy gate slightly shifted off the peak and then subtracting the two spectra.

The application of the coincidence method using the signals of a pulsed-beam and those generated by gammas or electrons is unique to in-beam e–γ spectroscopy. The beam signals are used to refer to the time of production of the excited state in the reaction, provided that the feeding time of that state is much smaller than its decay time. It is shown in Fig. 3.64 that the feeding time corresponds to the effective γ-transition time from the entry region (see Section 6.5.2) to the excited state, which is normally of the order of a picosecond. The (prompt) reaction must occur when one of the bursts of the pulsed beam hits the target. It is thus necessary to obtain a reference signal related to the burst structure of the beam. This reference signal may be derived

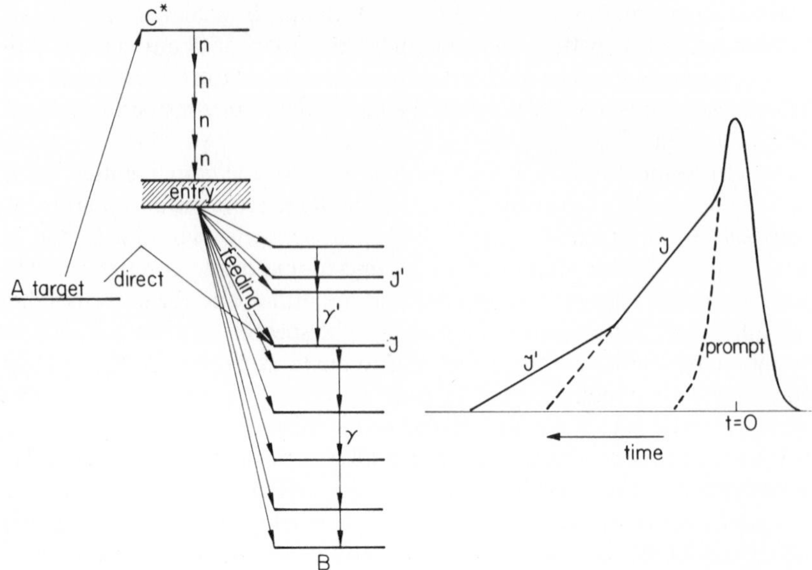

Fig. 3.64. A schematic illustration of the excitation of (isomeric) states in nucleus B via compound–nucleus formation C* and via a direct reaction (left). The time spectrum corresponding to the first case and for γ rays de-exciting the isomeric state J is given on the right. The effect of prompt feeding from entry states as well as of delayed feeding from another isomeric state J' is indicated.

from the phase of the radio frequency (RF) of the cyclotron oscillator or of the beam buncher, when a linear accelerator is used. These frequencies are mostly very high; in the case of a cyclotron in the range of mega cycles. Therefore it is recommended to use this signal to stop the time clock rather than to start it, in order to prevent large dead times of the counting system. The time calibration can be performed by introducing artificially known delays and recording the corresponding shift of the prompt peak or, more conveniently, by using a pulse generator or a RF signal with fixed frequency.

For a good time resolution it is necessary that the reference signal is stable, which is not always the case. In the operation of a cyclotron, for instance, phase shifts may occur which lead to shifts in the time spectrum. Also parasite beam bursts may be produced which do not necessarily come at fixed time intervals, thus leading to a broadening in the time spectra. A solution for those problems is to use an alternative reference signal produced by a detector or from a plastic scintillator mounted in the beam pipe, for instance in front of the Faraday cup. In this case the timing is entirely determined by the detectors. Also charged-particle and X-ray detectors as well as electron channeltrons or a sum spectrometer can be used. The sum spectrometer has the advantage of a high efficiency and a fast response of the NaI(Tl) crystal. Moreover it offers the additional possibility to perform energy and angular-momentum selection (see Section 3.2.3).

A time spectrum associated with a certain transition, observed in-beam, is often composed of more than one component corresponding to different decay constants (see Fig. 3.64). Most commonly there are two different types of components: i.e. a prompt peak due to the very fast feeding of the level and delayed components due to one or more isomeric states in the decay chain preceding the observed transition. In the schematical example of Fig. 3.64 the time spectrum is given for observed γ rays below the isomeric state indicated by J. To find the proper shape of the delayed component one has to subtract the prompt part, which can be obtained from a known prompt transition and proper normalization. Some examples are given in Fig. 3.65 for ^{152}Dy. The prompt 541 keV transition is used to correct the other spectra as indicated for the 525 keV transition. Different time spectra for ^{152}Dy are also given in Fig. 6.47.

To determine the proper lifetime of a nuclear level one has to correct for all delays involved in the de-excitation of the compound nucleus. In the case of equilibrium and non-equilibrium reactions the delay associated with the reaction mechanism and with the particle decays (four neutrons in the example of Fig. 3.64) is mostly negligible. Even the γ-ray feeding from the entry states is negligible with respect to the effect of nanoseconds isomers.

The analysis of decay curves of transitions following the decay of more than one isomer is complicated, due to the different time constants involved. Only when the time spectrum exhibits clearly distinguishable different slopes (as in Fig. 3.64) can the corresponding decay times be determined directly. In most cases, however, one has to determine first the time constant for the upper isomers and then use them with the measured corresponding intensities in the analysis of the lower-lying isomers. This procedure is analogous to the analysis of complex radioactive decay and is described in detail in several text books. Of course a computer least-squares fitting routine offers the possibility to obtain all the lifetimes of the isomeric levels involved in the decay, using all the experimental data as input simultaneously.

For the unambiguous localization of more than one isomeric state in a certain γ-ray decay chain one has to perform $E_{\gamma 1}$–$E_{\gamma 2}$-time coincidence measurements. The best method is to record five variables during the experiment; the two γ-ray energies, their time relationship, and the time relations of the two γ rays with respect to the beam burst. Subsequently one can determine the sequence and the position of the observed transitions by sorting the events, stored on magnetic tape, with different constraints on the various relations.

In the direct coincidence method the feeding ambiguity is removed because of the direct population of the level of interest (see Fig. 3.64) via the direct

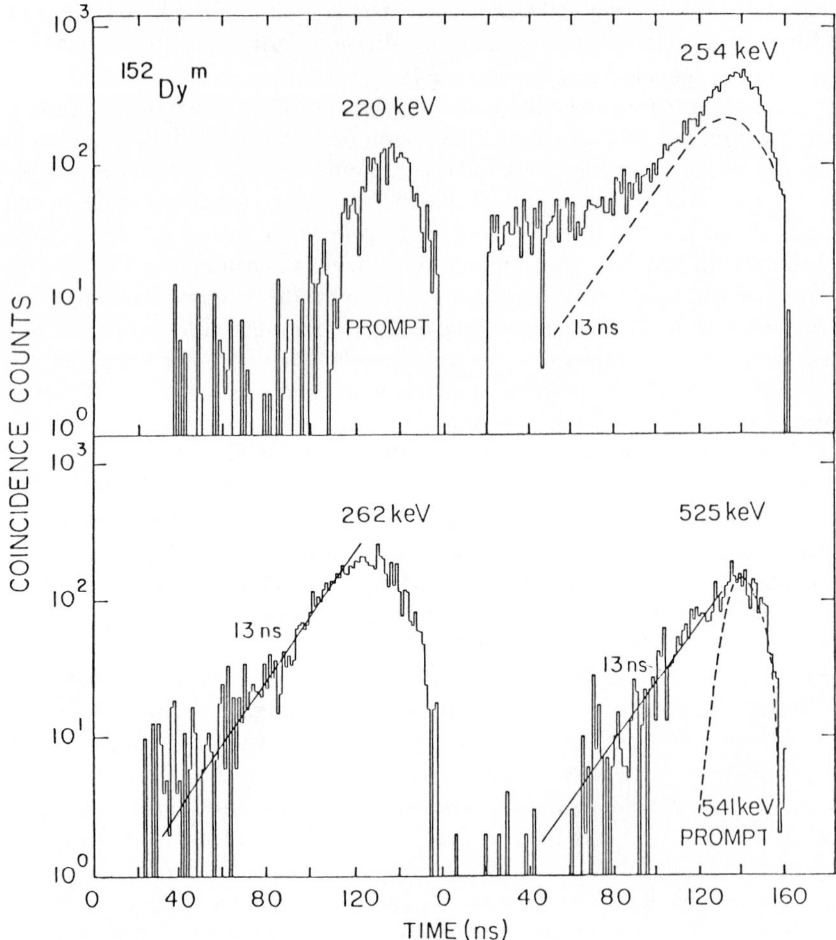

Fig. 3.65. Time spectra of four selected γ-rays in ^{152}Dy obtained with a multiplicity filter and 80 MeV ^{12}C beam to illustrate the 13 ns decay of the 254, 262 and 525 keV transitions. The complicated time spectrum of the 254 keV transition results from unresolved lines at 254 and 256 keV in the γ-spectrum. The prompt curves of the 221 and 541 keV γ-rays are shown for comparison. (Jansen *et al.*, 1979.)

reaction of the type $A(a, b)B^*(J) \to \gamma$. An example is the $(\alpha, \alpha' \gamma)$ reaction where the γ rays are measured in coincidence with the inelastically scattered alpha particles.

3.5.1b *The recoil-distance Doppler-shift technique*

Nuclear reactions provide excited nuclei with fairly large linear momenta and thus a time clock faster than one can obtain with the electronic technique.

This is particularly the case when heavy-ion beams are used. The in-beam γ spectroscopy employs only the Doppler shifted γ rays emitted by the rapidly moving recoils.

The velocity of the projectile-like fragments is approximately equal to that of the beam (v_a), of the order of (10^9 to 10^{10}) cm/s. A compound nucleus has a slower velocity, given by

$$v_0 = v_a m/(m + M) \tag{3.5.1}$$

where m is the mass of the beam particles and M that of the target nucleus. Thus v_0 is of the order of (10^8 to 10^9) cm/s. The products after direct reactions have velocities $< 10^{10}$ cm/s, strongly depending on the particular momentum transfer. The time scale $t = l/v$ depends on the measurable length l in the laboratory system. The present plungers to measure the flight path of an excited recoil nucleus can cover a range of $l = (10^{-1}$ to $10^{-3})$ cm, which corresponds, for a recoil velocity of 10^9 cm/s, to a time range of (10^{-10} to 10^{-12})s. This extends the nanoseconds region region of the electronic method to the picoseconds region with the plunger. The plunger thus adds a new (faster) clock to the electronic clock.

In compound nuclear reactions the full linear momentum of the projectile is transferred to the compound nucleus. This type of reaction leading to a complete fusion between target nucleus and projectile is widely used. When heavy-ion projectiles are used the velocity of the compound nucleus can become as large as ten per cent of the speed of light. The velocity of the compound nucleus depends on the mass numbers m, M, and on the energy of the beam E (in MeV), according to simple kinematics, as

$$v_0/c = 0.04635(mE)^{1/2}/(M + m) \tag{3.5.2}$$

This expression is valid for most cases in the non-relativistic limit. The velocity v_0 has still to be corrected for the average energy loss of the recoils in the target.

Most complete-fusion reactions are followed by the emission of a few light particles after which the final nucleus is formed as an evaporation product. The emitted particles are very slow and thus the velocity of the evaporation product is approximately the same as that of the compound nucleus. In the case of pre-equilibrium particle emission, however, the velocity of the particles can be high and the recoil velocity of the final product nucleus may be lower than that of the compound nucleus (see Section 4.2). Also a broadening of the recoil direction distribution takes place in this case. With heavy-ion beams one may also study projectile-like fragments, which have approximately the beam velocity and are well focused in the foreward direction.

In the case of Coulomb excitation the recoil velocity of the target nucleus in the laboratory system is given by

$$v_0(\mathrm{CE}) = 2v_{\mathrm{cm}} \cos \nu_2 \tag{3.5.3}$$

Here ϑ_2 is the scattering angle of the target nucleus in the laboratory system and v_{cm} its velocity in the centre-of-mass system as given by eqn (3.5.2). In the case of direct reactions the recoil velocity has to be calculated from the reaction kinematics, which depends strongly on the mass, velocity and direction of the ejectiles.

The energy E_γ of a γ ray, emitted by the moving recoil nucleus, will be Doppler shifted with respect to the transition energy $E_{\gamma 0}$. The shift depends on the velocity v of the nucleus and on the angle of emittance θ.

$$E_\gamma = E_{\gamma 0}\{1 + (v/c)\cos\theta\} \tag{3.5.4}$$

where the angle θ is measured between the direction of the emitted γ ray and that of the moving recoil. In heavy-ion induced fusion reactions the average direction of the recoiling ions is approximately along the beam axis. The angle θ is then defined by the direction of the γ ray with respect to the beam direction.

The results of recoil-distance measurements, displayed in Fig. 6.29 (see Chapter 6), have been carried out with an assembly consisting of a thin target and a stopper at adjustable distances behind the target to catch the recoiling nuclei. With the plunger a minimum distance between target and stopper of $d = 10$ μm can be realized with an accuracy of about 1.0 μm. One observes in the γ-ray spectra (e.g. taken at $\theta = 0°$) two components for each γ line, depending on the nuclear lifetimes involved in the transition. One component has the original transition energy $E_{\gamma 0}$ with intensity Y^s_γ, due to recoils being caught by the stopper (s-peak), and a Doppler-shifted component E_γ with intensity Y^m_γ, due to recoils moving between the target and the stopper (m-peak). For shorter lifetimes one observes more γ rays in flight and for longer lifetimes more γ rays from stopped nuclei. Thus for each transition one can deduce the lifetime from the measured ratio $R(d)$ of the intensities of those two components. An example is given in Fig. 6.30 and discussed in Section 6.3.5b.

$$R(d) = Y^s_\gamma/(Y^s_\gamma + Y^m_\gamma). \tag{3.5.5}$$

In practice the intensities are obtained from the integrations

$$Y^s_\gamma = Y(0)\int_{t_0}^{\infty} \exp(-t/\tau)\mathrm{d}t, \tag{3.5.6a}$$

$$Y^m_\gamma = Y(0)\int_{0}^{t_0} \exp(-t/\tau)\mathrm{d}t, \tag{3.5.6b}$$

where t_0 is the time when the stopper is reached and τ the nuclear lifetime. The time t_0 depends on the velocity of the recoils according to eqn (3.5.2). Also a finite stopping time in the stopper has to be taken into account.

A complication arises for the short-lived states, when they are shorter than the slowing-down time of the fast nuclei in the stopper foil. This slowing-down time may amount to about 2.5 ps and thus short-living states reaching the stopper (for short *d*) will decay during the slowing-down process. As a result no sharp s-peak can be observed in that case or part of the s-peak is smeared out over the region between the s and m peaks in the spectrum. Therefore one may follow a slightly different analysis procedure as follows. The total γ-ray intensity is measured for each transition as contained in the *m*-peak at very large distances (i.e. $d=5000$ μm), denoted by $Y_\gamma^m(\infty)$. The intensities of the s-peaks at distance d are then obtained as $Y_\gamma^s(d)=Y_\gamma^m(\infty)-Y_\gamma^m(d)$. The normalization is obtained from the summed intensities of the m- and truly s-peaks for long-living states. Corrections for (small) lineshape effects and for angular-distribution attenuation may also have to be applied. The decay curves are then constructed from

$$R(d)=1-Y_\gamma^m(d)/Y_\gamma^m(\infty). \tag{3.5.7}$$

A modification of the plunger technique enables the time scale to be extended to even shorter lifetimes (Emling *et al.*, 1984). The idea is to insert a gold foil in between the thin target and the stopper, as is illustrated in Fig. 3.66(a). The thickness of the retardation foil has to be adjusted such that the γ rays, emitted after the recoils have passed the foil, are significantly shifted in energy. In the example of Fig. 3.66 the reaction ^{25}Mg(^{136}Xe, 5n)^{156}Dy was used at a bombarding energy of 642 MeV. The velocity change in the 6 mg/cm^2 thick gold foil amounted to 20 per cent, causing a splitting of the γ lines of 1.6 per cent. The passing time through the foil of $\sim$0.1 ps is significantly smaller than the stopping time of a few picoseconds in gold. Part of the γ spectra, displayed in Fig. 3.66(b), were measured for three different flight times (t_F) from the target to the gold foil. It is seen that the intensity of the peak I_F, corresponding to recoils with the full velocity, changes drastically with respect to the retarded component I_R when the distance or flight time t_F increases. The main advantage of this method over that with a normal plunger is that the stopping time does not play a role any more, which shortens the measurable lifetimes to less than 1 ps.

The analysis involves again the construction of the decay curve $R(d)$ for each transition using the two observed intensities I_R and I_F as a function of d. The thickness of the gold foil implies a transient time of the recoil; in the above example of 0.13 ps, which is smaller than the lifetime of the highest spin states observed. A correction can be applied for nuclei decaying in the gold foil and thus not contributing to either one of the two peaks. In the analysis some knowledge or assumption is needed about the cascades of γ rays feeding each state of interest. The highest spin state analysed in the ^{156}Dy experiment was the $I^\pi=30^+$ yrast state resulting in a lifetime of $\tau(30^+)=0.16\pm0.08$ ps, and a feeding time of 0.39 ± 0.19 ps.

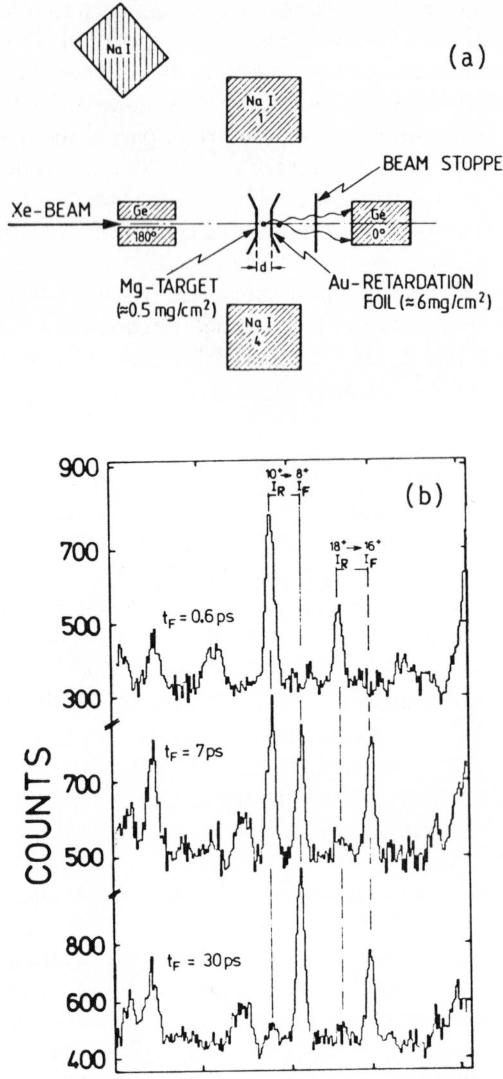

Fig. 3.66. Schematic view of the plunger setup, modified in such a way that a retardation foil can be inserted between the target and the stopper at adjustable distances d corresponding to a flight time t_F (a). Part of γ-ray spectra measured for three different flight times with the 0° detector in the reaction $^{25}Mg(^{136}Xe, 5n)^{156}Dy$ (b). Coincidences were required with at least two of the four NaI(Tl) detectors. (Emling *et al.*, 1984; Schwalm 1981.)

The measured nuclear lifetimes discussed so far yield the off-diagonal matrix elements of the electromagnetic transition operator. The diagonal matrix elements, needed to obtain the static (quadrupole) moments—see eqn (6.3.87)—can only be obtained by different methods. One of the most fruitful methods for high-spin states is multiple Coulomb excitation, as discussed below in Section 3.5.1d.

3.5.1c *The Doppler-shift-attenuation-method* (*DSAM*)

We showed in the previous subsection that the plunger clock time $1/v$ has a range of approximately (10^{-10} to 10^{-12})s. The limitation to $\sim 10^{-12}$ s is due to the fact that in the laboratory a minimum distance can be achieved of only $\sim 10\ \mu$m. A much faster time clock is provided by the slowing down process of excited nuclei in material. The energy loss ΔE of the recoil nucleus is related to the travelled distance Δx in the material via the stopping power $\mathrm{d}E/\mathrm{d}x$ by

$$\Delta x = (\mathrm{d}E/\mathrm{d}x)^{-1}\Delta E \qquad (3.5.8)$$

The minimum γ-ray energy shift still measurable is about $\Delta E_\gamma(\mathrm{min}) = 100$ eV, which is related to the shift in velocity and thus to a shift $\Delta E(\mathrm{min})$ in recoil energy. For a typical heavy-ion induced reaction such as $^{116}\mathrm{Cd}(^{40}\mathrm{Ar}, 4n)\,^{152}\mathrm{Dy}$ at 180 MeV, the recoil velocity is $v_0/c = 2.5$ per cent according to eqn (3.5.2) and the recoil energy is 47 MeV. The shift $\Delta E(\mathrm{min}) = 0.4$ MeV corresponds to $\Delta x = 7.5 \times 10^{-6}$ cm, according to eqn (3.5.8), using a stopping power of 5×10^4 MeV/cm. This measurable travel length $\Delta x = 7.5 \times 10^{-6}$ cm of the recoil nucleus in the material is thus at least two orders of magnitude smaller than with the plunger. The time clock $\Delta t = \Delta x/v$ can now reach times as short as $\sim 10^{-14}$ s and thus extends into the femto-seconds region. The DSAM covers in fact the time region of 10^{-12} to 10^{-15}.

For multiple Coulomb excitation (see Section 3.5.1d), or for heavy-ion induced fusion reactions (see 6.3.5b), using thick targets, the γ-ray lines are Doppler-broadened owing to the decay of the excited recoils during the slowing-down process. The lineshape depends on the lifetimes of the nuclear states involved and they may thus be extracted provided that the feeding through γ decay above the state of interest is taken into account or the feeding time is much shorter than the lifetime of that state. The lifetimes may be deduced either from the shift in the centroid of the γ lines or, more precisely, from the lineshapes. Another problem to overcome is the description of the slowing-down process, which relies on accurate stopping power data, if available. If no data are available, then rather accurate calculations can be performed. The best procedure, however, is to measure the stopping powers for the range of velocities of the actual experiment; see e.g. Grosse (1979). For not too low velocities the stopping power is dominated by the collisions of the ion with the electron clouds of the atoms in the slowing-down material. These electronic stopping powers are tabulated, for instance by Northcliff and Schilling (1970). At low velocities the energy loss is due largely to nuclear stopping, which is poorly known but can be estimated according to Blaugrund (1966).

The problem of the feeding γ rays is two-fold. First of all the measured lifetime of a particular state is influenced by the lifetimes of all higher-lying states in the cascade. Furthermore the delay in the high-lying states, which happen to have relatively long lifetimes, might result in the emission of the γ ray of interest when the nucleus has been slowed down significantly or even has come to rest. Then only small or no Doppler broadening is observed, which prevents an analysis of the line shape. Note that in any case the side feeding or any other feeding influences the slowing-down process and should be taken into account explicitly, because the measurable lifetimes are now very short.

A reduction of the feeding problem can be found by surrounding the target with a sum spectrometer covering a solid angle of $\sim 4\pi$. With the sum spectrometer one selects a certain energy region of excited states which feed the observed discrete (yrast) states. In this way one may observe the states which are fed via a narrow, nearly constant, feeding path in the broad quasi-continuum that is excited in the nuclear equilibrium reaction.

A much better selection and even elimination of the feeding can be obtained in multiple Coulomb excitation (see also the next subsection). Consider as a first example the schematic representation of the sum spectrometer used at GSI, as shown in Fig. 3.67. The sum spectrometer in the experiment consists of a NaI(Tl) annulus of 40 cm $\phi \times 30$ cm, separated into six segments. The axial hole of 8 cm ϕ contains the target in the centre and two Ge detectors (one with a hole through to introduce the beam). The total γ-ray energy of each de-excitation process is measured with the sum spectrometer in coincidence with the γ rays hitting the Ge detectors. In the off-line analysis one can impose conditions on the total-energy spectrum such that only a cascade is selected up to a maximum energy that corresponds to that of the state of interest. In this way one selects from all events in the experiment only those events in which each state of interest is excited as the nearly highest state. The

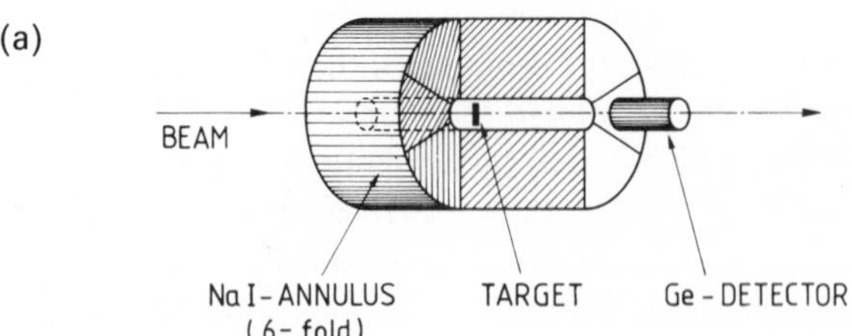

Fig. 3.67(a).

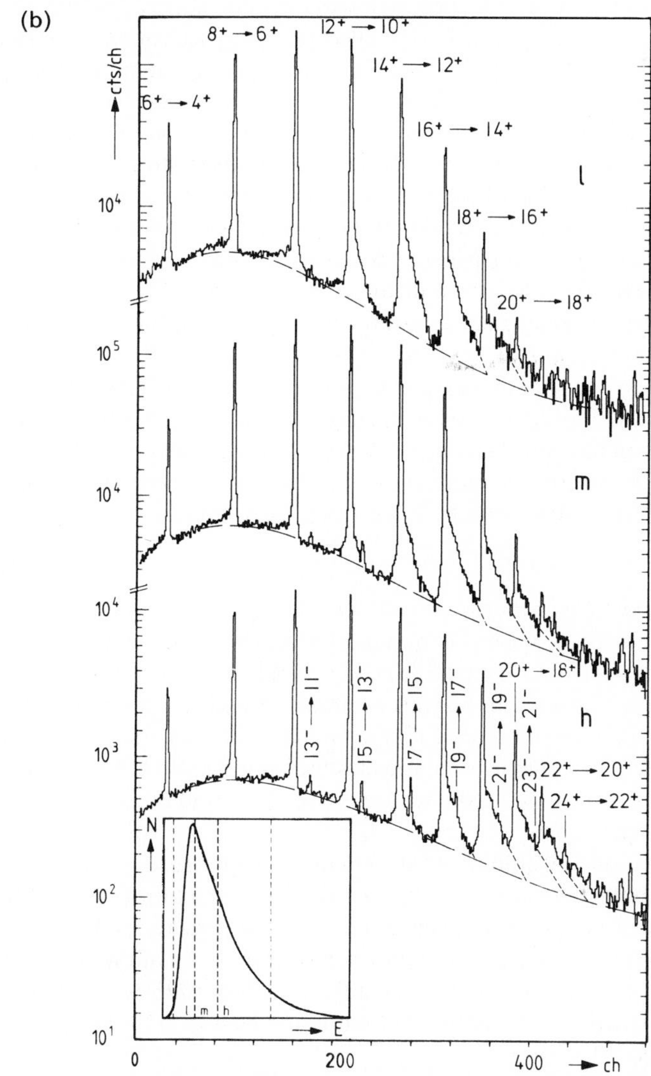

Fig. 3.67. The sum spectrometer is schematically drawn. The NaI(Tl) crystal is 40 cm $\phi \times 30$ cm. (a). Gamma-ray spectra obtained from MCE by bombardment of ^{238}U (thick target) with ^{208}Pb ions of 5.3 MeV/A. (b). By coincident detection of a γ-ray in one of the Ge detectors and the rest of the gamma cascade in a large solid-angle sum spectrometer and by setting a low, medium or high window on the sum energy, as indicated in the insert, the population pattern of the various states is influenced. The broken line indicates the decomposition of the spectra into Doppler-broadened full-energy peaks and background. (Schwalm, 1981.)

uncertainty in the energy selection is due to the finite energy resolution of the NaI(Tl) crystal (see Section 3.2.3). The effect of energy selection is illustrated in Fig. 3.67 for ^{238}U excited by ^{208}Pb projectiles. It is seen that for a given γ line a low energy gate causes broader line-shapes than a higher one. The sharp lines in the spectrum correspond to the unshifted transition energy and thus to γ rays emitted by recoils which have come to a complete stop. This is the case for low-lying states which live long enough. The shorter-living states, however, decay partly when the recoil is still moving. This causes the high-energy (Doppler shifted) tails of the peaks. A low-energy sum gate eliminates much of the feeding and thus also of the delay in the γ cascade. This diminishes the intensity of the stopped peaks with respect to the rapidly moving component in the tails.

The analysis of the lineshapes, employing precise knowledge of the slowing-down process, yields lifetimes of nuclear excited states smaller than 1 ps. Note that in the application of relativistic heavy-ion beams the time scale is enlarged by the relativistic factor $\{1-(v/c)^2\}^{1/2}$. So far only moderate beam energies with $v/c < 10$ per cent have been used, which do not need this correction.

3.5.1d *Multiple Coulomb excitation* (*MCE*).
Although the method of Coulomb excitation has been known for a long time (Huus and Zupancic, 1953), its full power could only be exploited in recent years when heavy-ion beams became available. We will focus our attention on some salient features in exciting high-spin states and on some experimental methods to deduce dynamic or static multipole moments. A detailed description of the process and related phenomena can be found in the work of Alder and Winther (1975).

The application of Coulomb excitation in γ-ray spectroscopy is encouraged by a few advantageous features. First of all, the interaction between the two colliding nuclei is assumed to be purely electromagnetic and is thus well understood. This assumption is valid when the nuclei remain well outside the range of their nuclear forces. One can adjust the projectile energy so that the Coulomb barrier is not exceeded. The Coulomb barrier in the centre-of-mass system can be calculated from

$$E_c = 1.44 Z_1 Z_2 / [1.16(A_1^{1/3} + A_2^{1/3} + 2)]. \qquad (3.5.9)$$

Here one obtains E_c in MeV when Z_1 and Z_2 are the atomic numbers and A_1 and A_2 the mass numbers for the projectile and the target, respectively. One may find in the literature different parameters for eqn (3.5.9) leading to slightly different values for E_c.

Secondly, MCE is a very clean process that produces much less background than ordinary fusion reactions. The bombarding energy for MCE is low enough for no nuclear reactions to take place in the target or in the surrounding material of the beam pipes. The refined γ spectroscopic techniques, like the plunger discussed above, can thus be applied using clean spectra, as they are free from all sorts of γ rays which are normally produced in nuclear reactions.

Thirdly, MCE provides an excitation mechanism of high-spin states complementary to (HI, xnγ) reactions. As discussed in Section 6.2 the (HI, xnγ) reactions populate yrast, yrare, and other low-lying levels in predominantly neutron-deficient isotopes, whereas MCE populates only levels connected with the ground state of the target nucleus via multiple transitions, mainly of the E2 type.

Nuclear states with fairly high spin can be excited by the multistep process in MCE, depending on the charge of the projectile. One can estimate, for instance, that with a ^{238}U target states can be populated up to spin $I=6$ with 73 MeV ^{16}O projectiles, up to $I=14$ with 185 MeV ^{40}Ar, up to $I=26$ with 605 MeV ^{132}Xe, up to $I=32$ with 1 GeV ^{208}Pb and up to $I=34$ with 1.15 GeV ^{238}U. The ^{208}Pb beam is particularly useful because the nucleus is doubly magic and the first excited ($I^{\pi}=3^{-}$) state is at 2.6 MeV. Projectile excitation is therefore unlikely, which simplifies the theoretical description of the MCE process.

One can also show that for relatively small v_0/c the magnetic type of excitation can be neglected compared with those of the electric type. The Coulomb interaction decreases drastically with increasing multipolarity although E1 transitions are usually also strongly hindered. This hindrance is caused by the small E1 effective charge (Z/A) of the projectile and of the small E1 strength in the low-lying states (see Chapter 5). Therefore members of the GSB as well as those of various side bands connected with the GSB in deformed nuclei are predominantly excited via consecutive E2 transitions.

A schematic excitation-decay process that can be initiated by MCE is given in Fig. 3.68(a). The figure shows that not only high-spin states can be reached, but also side bands of sometimes peculiar collective character. It is also shown in Fig. 3.68 that the excitation probability of states with a certain spin depends on the scattering angle. Let us define θ_1 as the scattering angle of the projectile in the centre of mass system and v_2 the one of the target nucleus in the laboratory frame of reference. If we assume that the excitation energy of a few MeV is much lower than the beam energy of a hundred MeV, then we have the relation $v_2=(\pi-\theta_1)/2$. Notice in Fig. 3.68(b) that for small θ_1 (large v_2) only low-spin states are excited, because the small scattering angle transfers small linear and angular moments. On the other hand, in the case of large momentum transfer for large θ_1 (small v_2) high-spin states are excited. In selecting the excitation of high-spin states one may thus detect exclusively the scattered projectiles at large scattering angles and/or detect the target nuclei at small angles. This is one of the reasons why in the techniques today one

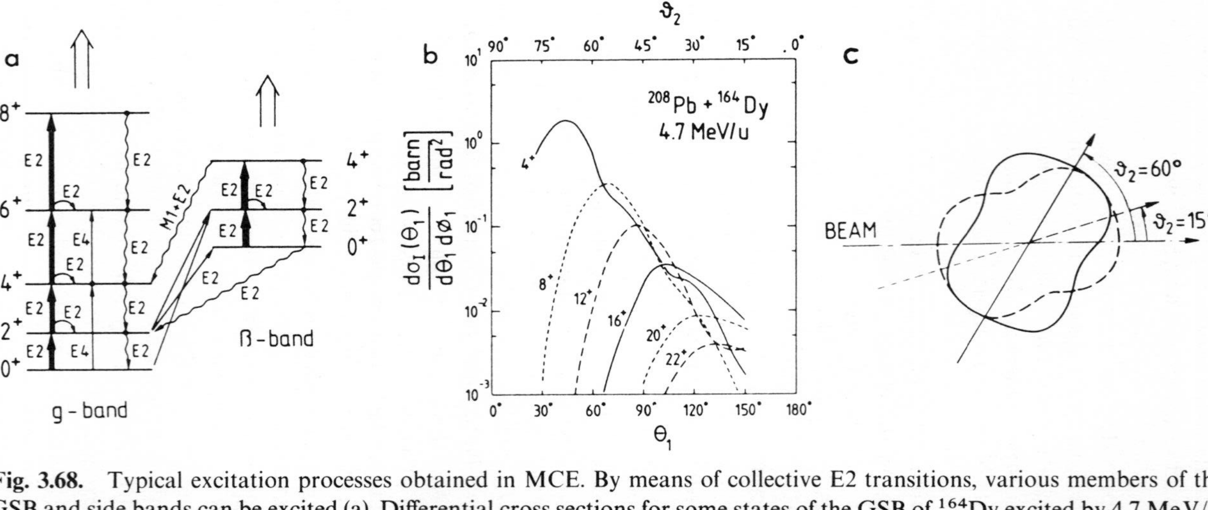

Fig. 3.68. Typical excitation processes obtained in MCE. By means of collective E2 transitions, various members of the GSB and side bands can be excited (a). Differential cross sections for some states of the GSB of ^{164}Dy excited by 4.7 MeV/A ^{208}Pb calculated as functions of the projectile (cm) scattering angle θ_1 (b). Schematic representation of the γ-ray angular correlation function for high-spin states at two different scattering angles of the target recoil in the laboratory system $v_2=(\pi-\theta_1)/2$ (c). (Schwalm, 1981.)

employs position-sensitive detectors for measuring scattering angles in addition to the γ-ray detectors. Another reason is to vary the reaction kinematics to obtain a large number of different excitation probabilities for the deduction of the many matrix elements involved in the MCE process. A third reason to fix the reaction kinematics is to overcome the severe Doppler broadening of the γ-ray lines in the spectra.

The necessary condition for the applicability of the semiclassical approach is expressed by $\eta \gg I$, where η is the Sommerfeld parameter

$$\eta = (0.158 Z_1 Z_2 \, m^{1/2})/E^{1/2}. \tag{3.5.10}$$

For the case of MCE of ^{164}Dy with ^{208}Pb projectiles of 4.7 MeV/A we have $\eta = 395$, indeed much larger than the highest observable spin of $I = 22$. A consequence of the classical trajectories being symmetric around $\nu_2 = (\pi - \theta_1)/2$ is the approximate symmetry of the γ-ray angular correlation function around the recoil axis of the target nuclei. This is illustrated in Fig. 3.68(c) for two different recoil angles ν_2.

Let us now discuss a particular example of experiments carried out at GSI, Darmstadt, with a detector arrangment as shown in Fig. 3.69. A thin ^{164}Dy target of $\leqslant 1$ mg/cm^2 was bombarded with 4.7 MeV/A ^{208}Pb projectiles. The decaying γ rays were measured by two Ge detectors in coincidence with recoiling ^{164}Dy nuclei. For that purpose large position-sensitive avalanche detectors were used, as shown in Fig. 3.69. The two Ge detectors were placed at an angle of 30° to the beam direction, which corresponds to $\nu_2 = 30°$ with the recoil axis and thus with the symmetry axis of the γ-ray angular correlation pattern. The angle ν_2 could be varied in the range $15° \leqslant \nu_2 \leqslant 58°$, corresponding to projectile cm scattering angles of $150° \geqslant \theta_1 \geqslant 60°$. The second particle detector was used to detect scattered projectiles in order to define uniquely the inelastic scattering process. In this way triple coincidences are recorded between the γ detectors, the recoil ion (in detector I) and the projectile (in II). For that reason also time-of-flight differences and/or angular correlations between the two coincident ions could be measured. More details of the set-up and analysis procedure, including the correction for the Doppler broadening are described by Schwalm (1981).

3.5.2 *Static electromagnetic moments*

3.5.2a *Magnetic dipole moments*

Magnetic moments can be measured because of the influence of a magnetic field on the orientation of the spin vector of nuclei. The magnetic moment $\boldsymbol{\mu}$ is a vector parallel or anti-parallel with the direction of the spin vector **I**. The observable nuclear magnetic moment μ is the z-component of $\boldsymbol{\mu}$ in the state of **I**, $m = I$ thus with the maximum value of the magnetic substate quantum number m. The magnetic moments are expressed in nuclear magnetons

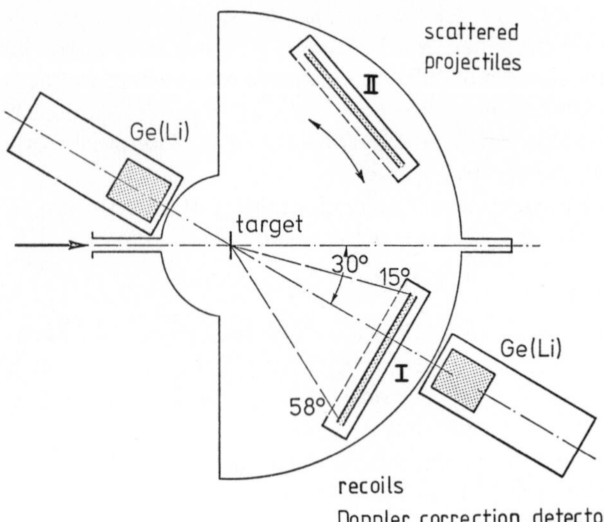

Fig. 3.69. Schematic drawing of a detector set-up used in MCE experiments to observe γ rays in coincidence with recoiling target nuclei and scattered projectiles. The position-sensitive avalanche detectors (I and II) correct for Doppler broadening of the line shapes and allow the impact parameter to vary in the MCE process. The angle $v_2 = 30°$ indicates a possible recoil direction chosen to position the Ge detectors. (Grosse, 1979.)

$\mu_N = e\hbar/(2m_p) = 5.0508 \times 10^{-27}$ JT^{-1}. The measured quantities are usually expressed as the dimensionless gyromagnetic factor (g-factor) as $g = \mu/(I\mu_N)$, as discussed in Section 6.3.5a.

For the measurements it is necessary to have the spins of the nuclei well aligned with respect to the magnetic field. Excited states produced by nuclear reactions are indeed well aligned in a plane perpendicular to the axis of momentum transfer. Compound nuclear reactions are associated with a large momentum transfer to the product nucleus and produce spin alignments in a plane perpendicular to the beam axis (Ejiri *et al.*, 1965, 1966). The effect of a magnetic field B on a nucleus with magnetic moment μ and spin I is a torque $\boldsymbol{\mu} \times \mathbf{B}$, which causes a precession with the Larmor frequency ω_L given by

$$\omega_L = \mu B/(\hbar I) = g\mu_N B/\hbar. \tag{3.5.11}$$

when B is expressed in Tesla (T), ω_L in s^{-1} and $\hbar = 1.0545 \times 10^{-34}$ Js, then μ is given in JT^{-1} and can thus be expressed in μ_N.

The frequency ω_L can be measured by observing the perturbed angular distribution of de-exciting γ rays or conversion electrons.

$$W(\theta, t) = \exp(-\lambda t)\ \Sigma_k A_k P_k\{\cos(\theta - \omega_L t)\}. \tag{3.5.12}$$

The exponential factor expresses the fact that the angular correlations are washed out with a decay time of $\tau = 1/\lambda$, the lifetime of the excited state. In the case that the relaxation time $\tau_r < \tau$ one has to insert τ_r rather than τ into eqn (3.5.12), as discussed below in the next subsection. This shows that the measurement is only meaningful during a time of the order of the lifetime τ. The frequency ω_L can be obtained from the measurements of the rotation of the angular distribution pattern over an angle $\Delta\theta = \omega_L t$. Thus the required magnetic field B can be estimated from eqn (3.5.11) as

$$B\tau = \Delta\theta\hbar/(g\mu_N), \tag{3.5.13}$$

where the nuclear lifetime is used as measuring time. For a g-factor of the order of one and a minimum observable $\Delta\theta = 0.1$ rad, a minimum $B\tau$ value of 2×10^{-9} Ts is required according to eqn (3.5.13). Consequently, excited states with lifetimes in the nanoseconds region require external fields of about one Tesla, which is the maximum achievable field with a simple conventional magnet. Such experiments are described in Section 3.5.2a1. Excited states with lifetimes in the picosecond region, however, require fields of the order of 1000 T, which cannot be provided by conventional magnets. In those cases internal (transient) fields are used as discussed in Section 3.5.2a2.

3.5.2.a1 *Magnetic-dipole-moment measurements with external magnetic fields* In the case of yrast traps, most lifetimes are in the nanosecond region. Yrast traps or isomeric states with lifetimes in this time region exhibit hindered M1, E2 transitions or higher multipole transitions like M2, E3. The magnetic moments of these states reveal their intrinsic structure, and this motivates most of the experiments. The lifetimes are sufficiently long to observe a significant time dependence of the spatial radiation pattern under influence of an external magnetic field, owing to the interaction with the nuclear magnetic dipole moment. The experimental method used is therefore called 'time dependent perturbed angular distribution' (TDPAD). The magnetic dipole moment μ, or the g-factor can be obtained from the measured Larmor-precession frequency ω_L, using eqn (3.5.11) or (6.3.67), as discussed in the introduction above and in Section 6.3.5.

A typical example of the experimental arrangement used in a $^{139}La(\alpha, 2n\gamma)^{141}Pr$ reaction is shown in Fig. 3.70 (Ejiri *et al.*, 1974). A common problem with the targets is that one may have to heat them in order to produce a favourable crystalline structure. In the above experiment the target had a closed-packed crystal structure of lanthanum at room temperature, which causes disturbing quadrupole fields (see below). Therefore a tantalum filament was mounted on the back of the target to anneal the foil at $\sim$900°K so as to change the crystal structure to face-centred cubic (at $\sim$500°K). Magnetic fields were applied from 1.2 to 1.9 T with a conventional magnet. The γ rays were measured with a Ge detector at angles of 0°, 45° and 90°. The

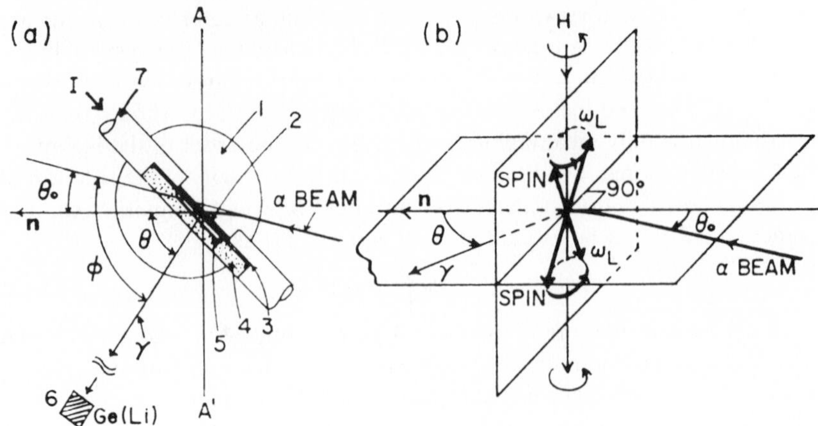

Fig. 3.70. Plane view (a) and schematic picture (b) of the experimental arrangement to measure magnetic-dipole moments; n: direction of the deflected α-beam. AA′: plane perpendicular to the beam axis (n). The spin of the compound state is aligned in the plane AA′. 1: magnet pole. 2: ^{139}La target. 3: Ta filament. 4: mica insulator. 5: Ta beam stopper. 6: Ge(Li) detector. 7: brass target holder. I: filament current. H: external magnetic field. (Ejiri *et al.*, 1974.)

measured time distribution of the ratio of the difference and sum of the γ ray yields at 0° and 90° are shown in Fig. 3.71. From the deduced Larmor frequency $\omega_L = 154 \pm 13$ MHz a g-factor of $g = 1.30 \pm 0.08$ has been deduced, taking into account a paramagnetic correction factor of $\beta(T = 900°\text{K}) = 1.34 \pm 0.05$ (see below). From this observation conclusions have been drawn about the magnetic dipole core polarization for the $1h_{11/2}$ state in ^{141}Pr (Ejiri *et al.*, 1974).

Core polarization effects have also been studied for $h_{11/2}$ states in ^{111,113}Sn (Brenn *et al.*, 1974). High magnetic fields are needed to observe the nuclear spin precession because of the small neutron g-factors. Therefore a Nb–Ti split solenoid superconducting magnet was used, as shown in Fig. 3.72. It is shown in the figure that the initial beam of 16 MeV α or 20 MeV $^7Li^{3+}$ is bent significantly by the strong magnetic field of about 6 T. This effect has to be taken into account in determining the angles in the angular distribution measurement. The spin precession for the $11/2^-$ state in ^{113}Sn is shown in Fig. 3.73. The measured Larmor frequency $\omega_L = 67.2 \pm 0.7$ MHz yields a g-factor of $g = -0.235 \pm 0.004$.

The main difficulty in those experiments is associated with relaxation effects, which tend to destroy the orientation of the nuclear spins. As the asymmetry in the spatial radiation pattern is directly related to this spin alignment, the deorientation will wash out the angular distribution. At the

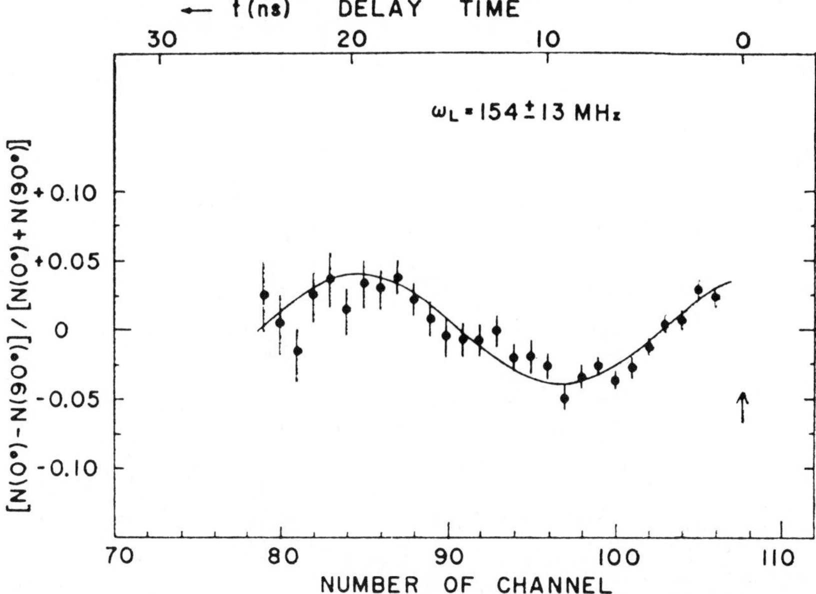

Fig. 3.71. A typical normalized ratio of the time distributions (measured at 0° and 90°) of the 972 keV gamma rays from the $1h_{11/2}$ state in ^{141}Pr. (Ejiri *et al.*, 1974.)

initial stage of heavy-ion fusion reactions the angualr momentum can be assumed to be perpendicular to the beam direction (complete spin alignment). Subsequent evaporation of light particles (e. g. neutrons) and the emission of γ rays will cause some dealignment. This effect is considered to be small, particularly in the case of high-spin states. When the decay is delayed by an isomeric state, however, the time of interaction between the nuclear and atomic electromagnetic fields (hyperfine interactions) may be so long that the orientation of the nuclear spins is considerably disturbed.The angular distribution may thus become more or less isotropic, depending on the lifetime of the isomeric state and on the hyperfine intereaction strength.

The asymmetry of the anuglar distribution is expressed by the a_2 coefficient of the Legendre polynomial expansion of the angular emission pattern. An exponential function for this a_2 coefficient seems to represent a reasonable time dependence.

$$a_2(t)=a_2(0)\exp(-t/\tau_r) \tag{3.5.14}$$

where τ_r indicates the relaxation time. In a measurement of magnetic moments in some Gd isotopes Häusser *et al.* (1979) have determined τ_r as a function of the temperature of the target material tin. Above the melting point of tin (505°K) they obtained nearly constant values $\tau_r = 80 \pm 10$ ns for ^{144}Gd

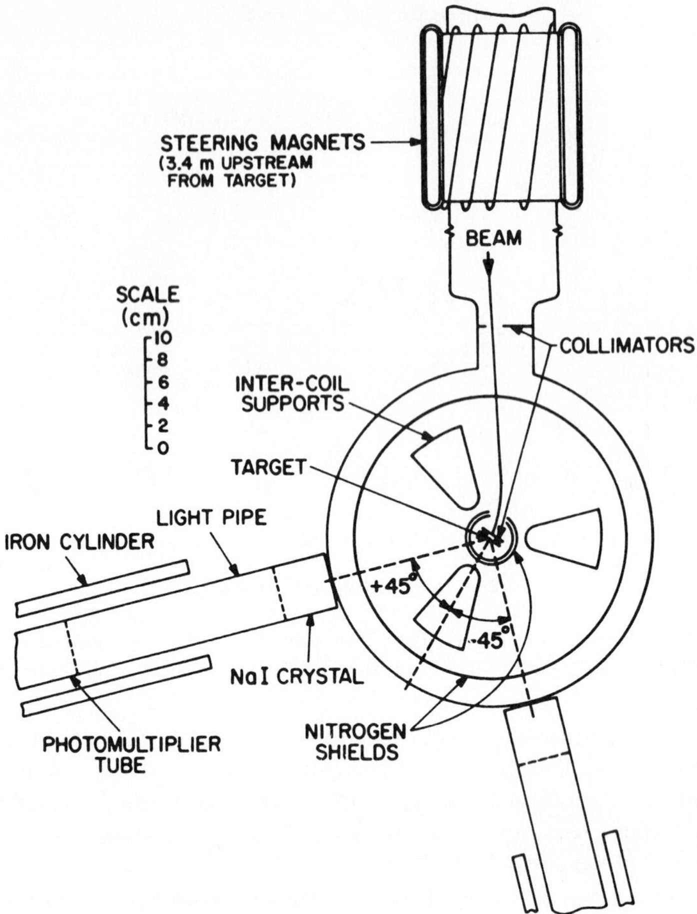

Fig. 3.72. Experimental set-up of a superconducting Nb–Ti split solenoid, which achieves fields of ~6 T, for measuring g-factors. (Brenn *et al.*, 1974.)

and $\tau_r = 750 \pm 200$ ns for ^{147}Gd, inversely proportional to g^2. They suggested that here magnetic relaxation effects may dominate, whereas below the melting point τ_r becomes much smaller, probably due to additional effects from electric quadrupole interactions (see below).

The orientation effects in the rare earth region are mainly caused by the paramagnetism of the lanthanides. The unfilled 4f atomic shell generates large internal magnetic fields acting on the magnetic moment of the nucleus. These paramagnetic ions, embedded into a solid, experience thermal fluctuations and the rapidly fluctuating field, acting on the nucleus, tends to disorient the nuclear spins. Moreover, the paramagnetism enhances the external field

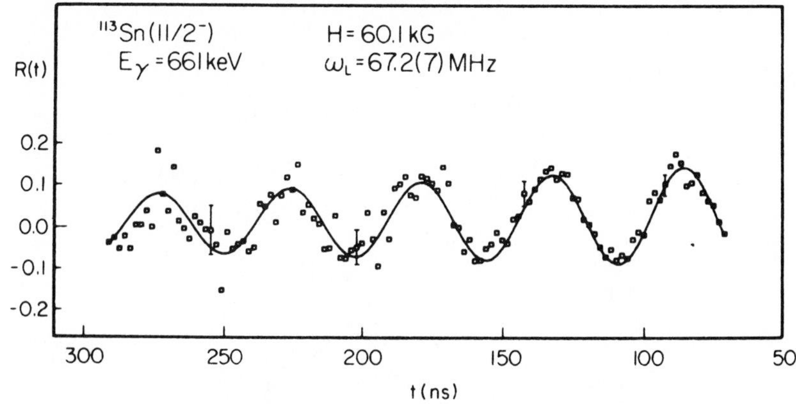

Fig. 3.73. The spin precession pattern measured with a NaI(Tl) detector for the 661 keV transition de-exciting the $h_{11/2}$ state in ^{113}Sn. The applied field of the superconducting solenoid was 6 *T*. (Brenn *et al.*, 1974.)

owing to the induction of unequally populated electronic substates. The gadolinium isotopes, however, form an exception. Since the Gd^{3+} ions have an $^8f_{7/2}$ electronic (nearly spherical) ground state there are no large orbital contributions to the hyperfine fields. It is thus not surprising that the measured g-factors in the light rare earth region are restricted to Gd isotopes, with the exception of ^{152}Dy, as discussed in Section 6.4.3c. Another cause of the disorientation effect can be the quadrupole interaction (QI), as discussed in Section 3.5.2b. A common way to avoid this is to heat up the target so as to change, for instance, a hexagonal close-packed crystal into a face-centred cubic structure, as in the example of ^{141}Pr, discussed above.

From the measured Larmor frequency ω_L one obtains the quantity $g\beta(T)$, using eqn (3.5.11), where $\beta(T)$ represents the paramagnetic correction factor. This correction has normally the value of 1, but it may be very different for rare earth nuclei. Correction factors $\beta(T)$ have been measured for Gd stopped in metallic Sn, Sm or Pb, using as probes the 135 ns 10^+ isomer in ^{144}Gd and the 530 ns $49/2^+$ isomer in ^{147}Gd. Assuming a small Zeeman splitting, relative to the thermal energy, one can approximate $\beta(T)$ by

$$\beta(T) = 1 + g_J \mu_B (J+1) B_0 / (3KT). \tag{3.5.15}$$

Here $g_j = 1.9913$ is the Landé factor, $J = 7/2$ for the Gd^{3+} ground state, μ_B the Bohr magneton and B_0 the internal magnetic field at $T = 0$°K.

The measured $\beta(T)$ values are plotted in Fig. 3.74. This figure shows that a significant correction has to be applied at room temperature, which diminishes to only a few per cent when the experiment is carried out at 1000°K. The work of Häusser *et al.* (1979) also yielded values for the internal hyperfine field

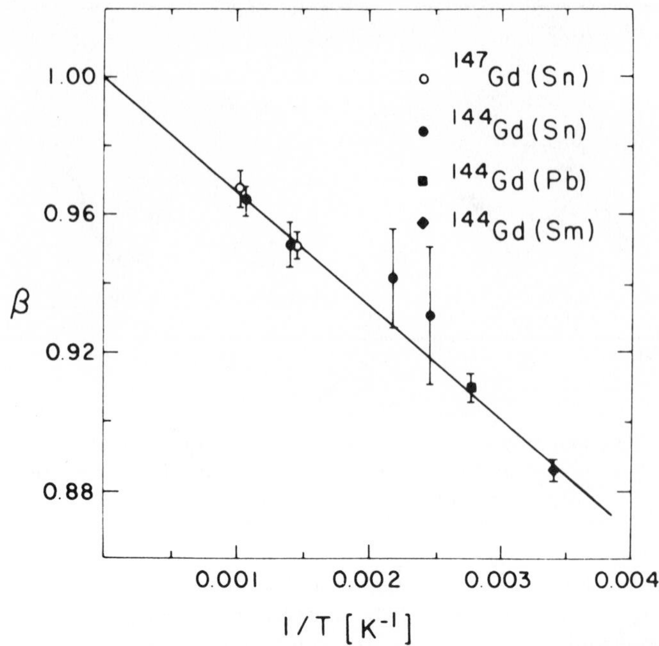

Fig. 3.74. The paramagnetic correction factor $\beta(T)$ for Gd recoils in three metallic environments as a function of the inverse of the host temperature T, expressed in K^{-1}. The solid line, corresponding to $\beta(T) = 1 - 33/T$, is consistent with all data points. (Häusser *et al.*, 1979.)

of $B = -16.5$ Tesla for all stopping materials. This value differes from that for free Gd^{3+} ions ($B^{free} = -34 \pm 2$ T), indicating a substantial contribution of the conduction electrons for each host material.

3.5.2.a2 *Magnetic-dipole-moment measurements by means of the transient-field technique* Nuclear collective states, decaying mainly by fast E2 transitions, normally have short lifetimes, of the order of picoseconds. Such short lifetimes are also involved in the decay of quasi-continuum states (see Section 6.5). In those cases very high magnetic fields are required in order to observe any measurable precession. The transient field technique is the only possibility known so far to measure magnetic dipole moments of nuclear excited states with lifetimes in the picoseconds region. A quantitative description of the transient field cannot be given, although one understands its atomic origin.

It is believed that the dominant contribution to the transient magnetic field arises from the ion–solid interaction between the electrons of the moving recoils and the polarized electrons of the ferromagnetic solid medium. The recoil velocity must be large enough; i. e. $v \geqslant 2v_0$, where $v_0 = c/137$ is the Bohr velocity. The action of the transient field on the nuclear magnetic moment can be described by that of the classical magnetic field, pointing in the the same direction as the external field used to magnetize the ferromagnetic medium. This remarkable effect of a small external field of $\sim$0.03 Tesla is a transient field of $\sim$4000 Tesla for Dy ions traversing iron with a speed corresponding to $v/c = 5$ per cent. This field is strong enough to cause a significant Larmor precession within the very short decay time of the nuclear states, of the order of a picosecond.

The main problem up to now is the field calibration at high recoil velocities. This is mainly due to the lack of states with known g-factors, which can be populated cleanly without excessive feeding through higher lying states. A series of calibration experiments with recoiling ions in the range $Z = 62$–86 for magnetized iron and gadolinium have been carried out by Andrews *et al.* (1982). Their conclusion was that the transient field in Gd is 30–45 per cent larger than in Fe and also the precession effects are a factor of 2 to 2.5 larger than in Fe. Moreover, since the Coulomb barrier for Gd is higher than for Fe, the beam causes much less or no nuclear reactions in Gd as compared to Fe, which leads to a much smaller background. A difficulty with Gd, however, is to prepare samples with a constant and high magnetization.

In the excitation process of nuclear states via (HI, xn) fusion reactions the reliable determination of g factors is often hampered because of the complex feeding patterns. Also in this respect Coulomb excitation offers a much less complicated situation. To reach states with sufficiently high spin in MCE one has to use heavy projectiles like ^{208}Pb. The experimental arrangement used at GSI to measure g-factors of rotational yrast states in ^{158}Dy using MCE with 4.7 MeV/A ^{208}Pb projectiles is shown in Fig. 3.75. The rotation of the angular correlation pattern was measured by the six Ge detectors placed symmetrically around the beam direction. The Dy recoils of scattered Pb projectiles were measured in coincidence with the γ rays in the two position-sensitive parallel-plate detectors. The target assembly consisted of a thin ^{158}Dy target, a magnetized Fe foil and a very thin Al layer. The Al layer prevented the possible pick-up of polarized electrons of the recoiling Dy ions at the exit from the ferromagnetic foil. Recoiling Dy ions were distinguished from recoiling Fe ions and from Pb projectiles by means of the kinematical correlation between their scattering angles and their flight times with respect to the beam burst.

We can distinguish two different ways of detecting the heavy ions. First, consider the case when the scattered Pb ions are detected in the forward direction ($20° \leqslant \nu_p \leqslant 27°$). This corresponds to Dy recoils moving almost perpendicular to the beam and thus being stopped in the Fe medium. The coincident γ rays thus stem from low-spin states, since the projectiles are only weakly affected and transfer only a small amount of angular momentum. For

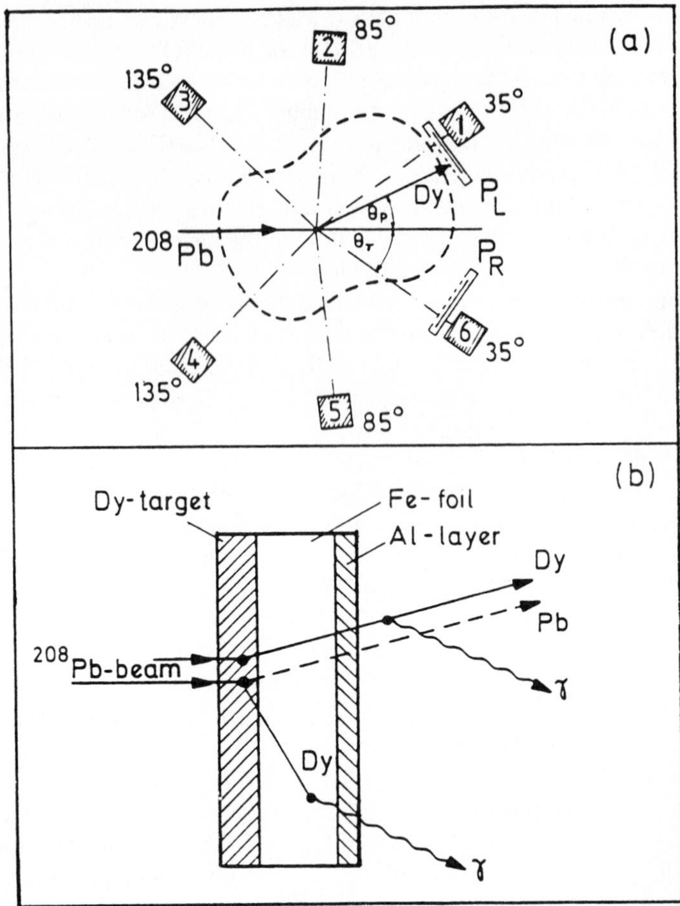

Fig. 3.75. A schematic representation of the experimental arrangement (a) to measure g-factors in ^{158}Dy using MCE with 4.7 MeV/A ^{208}Pb projectiles. The dashed curve represents an angular correlation pattern for a stretched E2 transition measured in coincidence with Dy ions recoiling at an angle θ_p (a). The target arrangement is given in (b). (Grosse, 1979; Seiler-Clark *et al.*, 1983.)

the deduction of g-factors, using eqn (3.5.11), one has also to take into account the static field, opposite to the transient field (see Schwalm, 1981). Secondly, one may detect Dy recoils in the forward detector in coincidence with the γ rays. In this case the Pb projectiles are deflected much more than in the first situation, leading to the excitation of high-spin states. The transit time of excited Dy nuclei through the Fe foil was 0.4 ps, still shorter than the shortest lifetime involved $\{\tau(18^+) = 0.8 \text{ ps}\}$.

3.5.2b *Measurement of electric quadrupole moments*

The relation between the observed static quadrupole moment $Q(I)$ for a state with spin I and the intrinsic quadrupole moment eQ_0 is given by eqn (6.3.71) in Section 6.3.5. The shape of the nuclear state with spin I is characterized by the shape parameters $a_{\lambda\mu}$ as defined by eqn (6.3.29). In the most common case of quadrupole deformation with $\gamma=0°$ (see Section 6.3), only one parameter determines the magnitude and sign of the deformation, namely a_{20} (also denoted by β_2; see eqn 6.3.30). The relation between β_2 and eQ_0 can be derived for the simple picture of a rotating ellipsoid as:

$$eQ_0=3/(5\pi)^{1/2}\{Z(r_0A^{1/3})^2\beta_2\}e, \qquad (3.5.16)$$

where Z is the proton number and r_0 is the nuclear radius parameter ($r_0=1.25$ fm). It appears that a positive sign for eQ_0 and β_2 implies a prolate nuclear shape and the negative sign an oblate shape. The sign and magnitude of the measured quadrupole moment $Q(I)$ thus determines directly the shape and size of the deformation, assuming $\gamma=0°$. Note that the signs of $Q(I)$ and Q_0 are opposite when $I(I+1)>3K^2$, according to eqn (6.3.71).

Quadrupole moments for low-spin states, i. e. 2^+ states with lifetimes in the picosecond region or with even shorter lifetimes, have been measured using the reorientation effect. This is not possible for the long-lived high-spin yrast traps discussed in Section 6.4. In that case the quadrupole moment may be determined by measuring the quadrupole interaction (***QI***) when the excited nucleus moves with a certain velocity through an electric field gradient (EFG). This EFG is most commonly supplied by the surrounding atoms of a crystal lattice of non-cubic symmetry. The ***QI*** causes a time-dependent change in the shape of the spatial emission pattern of the delayed γ-rays. The method of measurement is exactly the same as discussed above in Section 3.5.2a1, namely the TDPAD.

A particular difficulty arises from the fact that the energy splitting of the nuclear magnetic substates decrease qudratically with spin. This can be understood by noticing that the splitting in energy is given by the product of an electric-field gradient (EFG) and Q(I). The Q(I) is given by eqn (6.3.71), but using the magnetic substate quantum number m rather than K. Equation (6.3.71) then shows that the energy difference between states with m and $m\pm1$ is proportional to I^{-2}. As a consequence the ***QI*** becomes weaker for high-spin states. That makes the determination of quadrupole moments difficult in the domain of high angular momentum. It is also important to have such a long relaxation time τ_r before the spin alignment is relaxed, yielding a measurable angular asymmetry during the slowing-down of the ion in the target. As is discussed above for measurement of g-factors, Gd^{3+} keeps the magnetic hyperfine fields weak enough to reach spin relaxation times up to $\tau_r=1$ μs.

Häusser *et al.* (1980) performed special experiments to determine the EFG at temperatures above the Curie point. They used a 6 MeV ^{4}He beam to

Coulomb excite a single Gd crystal at $T = 332°$K and measured the perturbed angular distribution for the $2^+ \rightarrow 0^+$ γ-rays in ^{156,158,160}Gd. From the known quadrupole moments an EFG was deduced as $(3.43 \pm 0.14)10^{17}$ V/cm^2. This value can be compared with the known EFG of $(3.41 \pm 0.14)10^{17}$ V/cm^2 at 4°K. The temperature dependence in the ferromagnetic regime was thus found to be very weak. Also in the paramagnetic region only a small temperature dependence was found for the EFG. From the established EFG values quadrupole moments have been deduced for the three lowest yrast traps in ^{147}Gd, as listed in Table 6.4.1 (see Chapter 6). It should be noted that the TDPAD method discussed above does not yield the sign of the quadrupole moment. It leaves the shape of the nucleus uncertain since the sign of the deformation parameter β_2 is not determined.

Both the sign and the magnitude of the quadrupole moments of yrast traps can be obtained from the combined method of the TDPAD and the tilted multifoil technique. This method has been applied to isomers in ^{144,147}Gd by Dafni *et al.* (1984, 1985). The multifoil technique has been developed by Hass *et al.* (1984). The reaction products are polarized owing to the surface interaction at the exit from each foil of a set which is tilted with respect to the beam direction. In the experiment of Dafni *et al.* (1984, 1985) the Gd isomers have been excited by a pulsed ^{28}Si beam, hitting a thin Sn target. That target and the following stack of carbon foils were positioned at an oblique angle with their normals at 60° to the beam axis (see Fig. 3. 76). The stack consisted of 18–24 carbon foils, each 4–8 μg/cm^2 thick and separated by 150 to 450 μm. Electronic polarization was induced along an axis (tilt axis), perpendicular to the normal of the foils and the beam direction, corresponding to a rolling

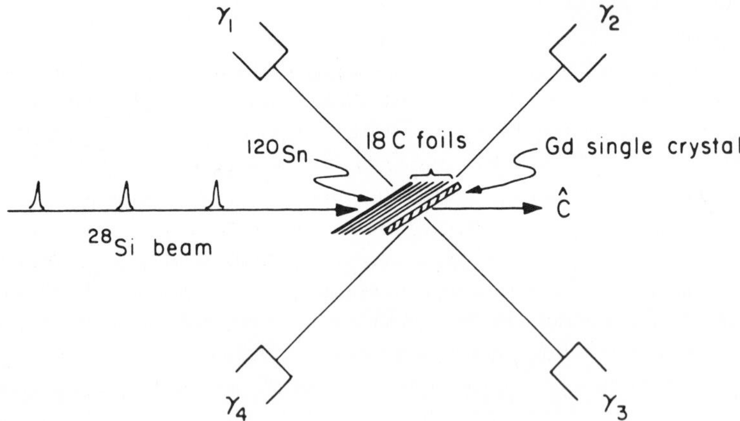

Fig. 3.76. The arrangement used in the tilted-foil TDPAD measurements. Gamma rays are measured in the four detectors indicated. The nuclear polarization vector is pointing out of the page for this orientation of the foils. (Dafni *et al.*, 1984, 1985.)

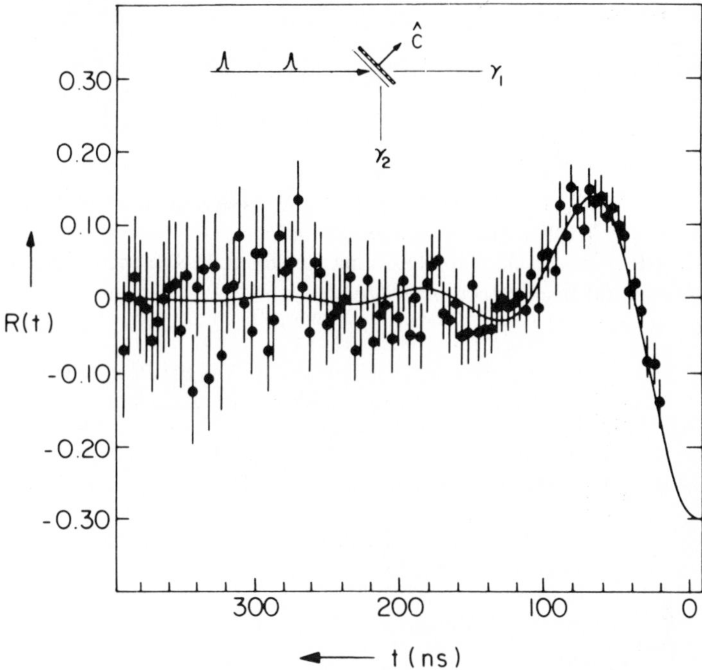

Fig. 3.77. The ratio function for the γ-rays measured at 0° (γ_1) and 90° (γ_2) from the aligned 10^+ isomeric state in ^{144}Gd. (Dafni *et al.*, 1985.)

motion of the ions on the surface. This polarization of the electrons was subsequently transferred to the nuclei via the hyperfine interaction in vacuum during the flight time between successive foils. The (spin) polarized nuclei were stopped in a single hexagonal Gd host crystal that was oriented with the *c*-axis along the beam direction. In that crystal, a spin precession was induced owing to the QI with internal EFG, with a sense of rotation depending on the sign of Q(I) and of the EFG. The modulation pattern can be measured by means of the TDPAD method as discussed above. Such a pattern is shown in Fig. 3.77 for the aligned 10^+ isomeric state in ^{144}Gd. The negative sign was found for the deformation, which established the oblate nature of the isomer (see also Fig. 6.54 for ^{147}Gd, discussed in Section 6.4).

It has been shown in this section that precision measurements of the electric quadrupole moments and magnetic dipole moments yield valuable information on various details of the nuclear structure. One may expect that further improvements of the technology might extend the measurements in the direction of higher spin states.

3.6 References

Albrecht, R. *et al.* (1980). Annual Report, Max Planck Institut für Kernphysik, Heidelberg, p. 41.

Alder, K. and Winther, A. (1975). *Electromagnetic Excitations.* North-Holland, Amsterdam.

Allan, C. J. (1971). *Nucl. Instr. Meth.* **91,** 117.

Andersen, O., Garrett, J. D., Hagemann, G. B., Herskind, B., Hillis, D. L. and Riedinger, L. L. (1979). *Phys. Rev. Lett.* **43,** 687.

Andrews, H. R., Häusser, O., Ward, D., Taras, P., Nicole, R., Keinonen, J., Skensved, P. and Haas, B. (1982). *Nucl. Phys.* **A383,** 509.

Antman, S., Grunditz, Y., Johansson, A., Nyman, B., Pettersson, H., Svahn, B. and Siegbahn, K. (1970). *Nucl. Instr. Meth.* **82,** 13.

Arciszewski, H. F. R., Aarts, H. J. M., Kamermans, R., Van der Poel, C. J., Holzmann, R., Van Hove, M. A., Vervier, J., Huyse, M., Lhersonneau, G., Janssens, R. V. F. and de Voigt, M. J. A. (1983). *Nucl. Phys.* **A401,** 531.

Arciszewski, H. F. R. (1984). Thesis, University of Utrecht, The Netherlands.

Avignone III, F. T., Pinkerton, J. E. and Trueblood, J. H. (1973). *Nucl. Instr. Meth.* **107,** 453.

Backe, H., Richter, L., Willwater, R., Kankeleit, E., Kuphal, E., Nakayama, Y. and Martin, B. (1978). *Z. Phys.* **A285,** 159.

Biddick *et al.* (1977). *Phys. Rev. Lett.* **38,** 1324.

Bistirlich, J. A., Crowe, K. M., Parsons, A. S. L., Skarek, P. and Truöl, P. (1972). *Phys. Rev.* **C5,** 1867.

Blaugrund, A. E. (1966). *Nucl. Phys.* **88,** 591.

Bolton, R. D. *et al.* (1984). *Phys. Rev. Lett.* **53,** 1415.

Brenn, R., Bhattachjee, S. K., Sprouse, G. D. and Young, L. E. (1974). *Phys. Rev.* **C10,** 1414.

Bron, K., Kosse, B., Poortema, K., Schaafsma, W., de Voigt, M. J. A. and van der Werf, S. Y. (1985). *Statistica Neerlandica* **39,** 261.

Burginyon, G. A. and Greenberg, J. S. (1966). *Nucl. Instr. Meth.* **41,** 109.

Cambi, A., Fazzini, T. F., Giannatiempo, A. and Maurenzig, P. R. (1972). *Nucl. Instr. Meth.* **103,** 331.

Canty, M. J., Debenham, P. H., Dohan, D. A., Skalsey, M. and Watson, J. W. (1974). *Nucl. Instr. Meth.* **122,** 547.

Carrington, R. L. *et al.* (1979). *Nucl. Instr. Meth.* **163,** 203.

Clemente, M., Berdermann, E., Kienle, P., Tsertos, H., Wagner, W., Kozhuharov, C., Bosch, F. and Koening, W. (1984). *Phys. Lett.* **137B,** 41.

Collins, M. T., Manglos, S., Roberson, N. R., Sandorfi, A. M. and Weller, H. R. (1982). *Phys. Rev.* **C26,** 332.

Dafni, E., Bendohan, J., Broude, C., Goldring, G., Hass, M., Naim, E., Rafailovich, M. H., Chasman, C., Kistner, O. C. and Vajda, S. (1984). *Phys. Rev. Lett.* **26,** 2473; and (1985). *Nucl. Phys.* **A443,** 135.

Daniel, H., Jahn, P., Kuntze, M. and Martin, B. (1970). *Nucl. Instr. Meth.* **82,** 29.

Davidson, W. F., Black, J. L. and Najam, M. R. (1971). *Nucl. Phys.* **A168,** 399.

Diamond, R. M., Elbek, B. and Stephens, F. S. (1963). *Nucl. Phys.* **43,** 560.

Diener, E. M., Amann, J. F., Blatt, S. L. and Paul, P. (1970). *Nucl. Instr. Meth.* **83,** 115.

Ejiri, H., Ishihara, M., Sakai, M., Katori, K. and Inamura, T. (1965). *Phys. Lett.* **18,** 314.

Ejiri, H., Ishihara, M., Sakai, M., Katori, K. and Inamura, T. (1966). *Nucl. Phys.* **89,** 641.

Ejiri, H., Richard, P., Ferguson, S., Heffner, R. and Perry, D. (1968). *Phys. Rev. Lett.* **21,** 373.

Ejiri, H., Shibata, T. and Takeda, M. (1974). *Nucl. Phys.* **A221,** 211.

Ejiri, H., Shibata, T., Nagai, Y. and Nakayama, S. (1976). *Nucl. Instr. Meth.* **134,** 107.

Ejiri, H., Shibata, T., Nagai, Y., Kishimoto, T., Ohsumi, H., Kamikubota, N. and Satoh, T. (1984). *AIP Conf. Proc. No.* 125, *Capture Gamma–ray Spectroscopy and Related Topics*, Tennessee (ed. S. Raman), p. 582.

Emling, H., Grosse, E., Kulessa, R., Schwalm, D. and Wollersheim, H. J. (1984). *Nucl. Phys.* **A419,** 187.

Emling, H., Azgui, F., Daether, M., Doebereiner, S., Grein, H., Michel, C. and Wollersheim, H. J. (1986). *Nucl. Instr. Meth.*, **A249,** 320.

Endt, P. M. and van der Leun, C. (1978). *Nucl. Phys.* **A310,** 1.

Geiger, J. S. and Graham, R. L. (1967). *Can. J. Phys.* **45,** 2281.

Gono, Y. and Sugihara, T. (1974). *Phys. Rev.* **C10,** 2460.

Grosse, E. (1979). *Proc. Symp. on High-Spin Phenomena in Nuclei*, Argonne, p. 223.

Habs, D., Stephens, F. S. and Diamond, R. M. (1979). Internal Report Lawrence Berkeley Laboratory, LBL-8945.

Hagemann, G. B., Broda, R., Herskind, B., Ishihara, M., Ogaza, S. and Ryde, H. (1975). *Nucl. Phys.* **A245,** 166.

Hageman, D. C. J. M. (1981). Thesis University of Groningen, The Netherlands.

Halpern, I., Heffner, R., Pedersen, J., Sletten, G., Swanson, H. and Vandenbosch, R. (1974). Univ. Washington Annual Report, p. 153.

Hamilton, W. D. (1975). *The Electromagnetic Interaction in Nuclear Spectroscopy.* North-Holland, Amsterdam.

Hansen, G. B., Elbek, B., Hagemann, K. A. and Hornyak, W. F. (1963). *Nucl. Phys.* **47,** 529.

Hasinoff, M. D., Lim, S. T., Measday, D. F. and Mulligan, T. J. (1974). *Nucl. Instr. Meth.* **117,** 375.

Hass, M., Dafni, E., Bertschat, H. H., Broude, C., Davidovsky, F. D., Goldring, G. and Lesser, P. M. S. (1984). *Nucl. Phys.* **A414,** 316.

Häusser, O., Taras, P., Trautmann, W., Ward, D., Alexander, T. K., Andrews, H. R., Hass, B. and Horn, D. (1979). *Phys. Rev. Lett.* **42,** 1451.

Häusser, O., Andrews, H. R., Mahnke, H. E., Sharpey-Schafer, J. F., Swanson, M. L., Taras, P., Ward, D. and Alexander, T. K. (1980). *Phys. Rev. Lett.* **44,** 132.

Heath, R. L., Hofstadter, R. and Hughes, E. B. (1979). *Nucl. Instr. Meth.* **162,** 431.

Herskind, B. (1980). *J. Phys. (Paris) Colloq.* **41–10,** 106 (Int. Conf. on Nuclear Behaviour at High Angular Momentum, Strasbourg).

Herskind, B. (1985). *Proc. Int. Conf. on Nucleon–Nucleon Collisions*, Visby, Gotland, Sweden.

Hofman, H., Harakeh, M., Hesmondhalgh, S. K. B., Poelhekken, T. D., Slatius, L. and Smid, I. (1985). Annual Report, KVI, University of Groningen, The Netherlands.

Huus, T. and Zupancic, C. (1953). *K. Dan. Vidensk, Selsk. Mat. Phys. Medd.* **28,** No. 1, 1.

Ishihara, M. (1972). *Nucl. Phys.* **A179,** 223.

Jansen, J. F. W., de Voigt, M. J. A., Sujkowski, Z. and Chmielewska, D. (1979). *Nucl. Phys.* **A321,** 365.

Johansson, A., Nyman, B., Svahn, B., Bergman, D. and Pettersson, H. (1973). *Nucl. Instr. Meth.* **108,** 225.

Johnson, N. R., Baktash, C. and Lee, I. Y. (1984). ORNL, Oak Ridge, private communication.

Kernel, G., Mason, W. W. and Tanner, N. W. (1970). *Nucl. Instr. Meth.* **89,** 1.

Kishimoto, T., Shibata, T., Sasao, M., Noumachi, M. and Ejiri, H. (1982). *Nucl. Instr. Meth.* **198,** 269.

Klank, B. and Ristinen, R. A. (1969). *Proc. Int. Conf. on Radioactivity in Nuclear Spectroscopy*, Nashville (ed. J. H. Hamilton and J. C. Manthuruthil), p. 207. Gordon Breach, N.Y.

Klinken, J. van and Wisshak, K. (1972). *Nucl. Instr. Meth.* **98,** 1.

Klinken, J. van, Feenstra, S. J., Wisshak, K. and Faust, H. (1975). *Nucl. Instr. Meth.* **130,** 427.

Klinken, J. van, Feenstra, S. J. and Dumont, G. (1978). *Nucl. Instr. Meth.* **151,** 443.

Kohl, W., Kolb, D. and Giese, I. (1978). *Z. Phys.* **A285,** 17.

Kohler, D. and Austin, S. M. (1963). *Bull Am. Phys. Soc.* **8,** 290.

Komma, M. (1978). *Nucl. Instr. Meth.* **154,** 271.

Konijn, J., Goudsmit, P. F. A. and Lingeman, E. W. A. (1973). *Nucl. Instr. Meth.* **109,** 83.

Konijn, J., Posthumus, W. L., Goudsmit, P. F. A., Schiebaan, C., Geerke, H. P., Maarkveld, J. L., Andringa, J. H. S. and Evers, G. J. (1975). *Nucl. Instr. Meth.* **129,** 167.

Kotajima, K. and Beringer, R. (1970). *Rev. Sci. Instr.* **41,** 632.

Kovash, M. A. (1978). Thesis, Ohio State Univ., USA.

Kozhuharov, C., Kienle, P., Berdermann, E., Bokemeyer, H., Greenberg, J. S., Nakayama, Y., Vincent, P., Backe, H., Handschug, L. and Kankeleit, E. (1979). *Phys. Rev. Lett.* **42,** 376.

Lieder, R. M., Jäger, H., Neskasis, A., Venkova, T. and Michel, C. (1984). *Nucl. Instr. Meth.* **220,** 363.

Lindblad, Th. and Lindén, C. G. (1975). *Nucl. Instr. Meth.* **126,** 397.

Moll, E. (1964). Thesis, München.

Moll, E. and Kankeleit, E. (1965). *Nucleonika* **7,** 180.

Morinaga, H. (1964). Private communication.

Morinaga, H. and Yamazaki, T. (1976). *In-beam Gamma-ray Spectroscopy.* North-Holland, Amsterdam.

Nagai, Y., Ejiri, H., Shibata, T., Okada, K., Nakayama, S., Suzuki, H., Ohsumi, H., Adachi, Y. and Sakai, H. (1982). *Nucl. Instr. Meth.* **204,** 101.

Nolan, P. J., Gifford, D. W. and Twin, P. J. (1985). *Nucl. Instr. Meth.* **A236,** 95.

Northcliff, L. C. and Schilling, R. F. (1970). *Nucl. Data Tables* **A7,** 256.

Noumachi, M. (1983). Ph.D. Thesis, Osaka University.

Ockels, W. J. (1978). Thesis, University of Groningen, The Netherlands, and (1978) *Z. Phys.* **A286,** 181.

Paul, P. (1974). In *Nuclear Spectroscopy and Reactions, part A* (ed. J. Cerny), p. 345. Academic Press, New York.

Penninga, H. (1986). Thesis, Free University, Amsterdam, The Netherlands.

Picone, J. M., Fitch, J. F., Baker, J. W., Hoffmann, G. W. and Moore, C. F. (1972). *Nucl. Instr. Meth.* **105,** 377.

Riezebos, H. J., Balanda, A., Dudek, J., van Klinken, J., Nazarewicz, W., Sujkowski, Z. and de Voigt, M. J. A. (1987). *Phys. Lett.* **B183.**

Robert, K. Q., Linn, J. R. and Durham, F. E. (1970). *Nucl. Inst. Meth.* **79,** 251.

Sakai, M., Yamazaki, T. and Ejiri, H. (1964/1965). *Phys. Lett.* **12,** 29; *Nucl. Phys.* **74,** 81.

Sakai, M., Yamazaki, T. and Ejiri, H. (1966). *Proc. Internal Conversion Processes* (ed. J. H. Hamilton), p. 197. Academic Press, New York.

Sakai, M., Yamazaki, T., Ejiri, H. and Ishihara, M. (1966). Institute for Nuclear Study Report, INST-99.

Saladin, J. X., Baktash, C., Lee, I. Y., Benenson, W., Blue, R., Ronningen, R. and Sorensen, R. A. (1980). Proposal MEHRACDATEMS, University of Pittsburg.

Sandorfi, A. M. and Collins, M. T. (1984). *Nucl. Instr. Meth.* **222,** 479.

Schaeffer, M., Suffert, M. and Magnac-Valette, D. (1971). *Nucl. Phys.* **A175,** 217.

Schwalm, D. (1981). *Nuclear Structure* (eds. K. Abrahams, K. Allaart and A. E. C. Dieperink). Plenum Press, New York.

Seiler-Clark, G., Pelte, D., Emling, H., Balanda, A., Grein, H., Grosse, E., Kulessa, R., Schwalm, D., Wollersheim, H. J., Hass, M., Kumbartzki, G. J. and Speidel, K. H. (1983). *Nucl. Phys.* **A399,** 211.

Simon, R. R. (1984). GSI, Darmstadt, private communication.

Slätis, H. and Siegbahn, K. (1949). *Phys. Rev.* **75,** 1955.

Spoelstra, B. and Rautenbach, W. L. (1968). *Nucl. Instr. Meth.* **66,** 336.

Stephens, F. S., Lark, N. L. and Diamond, R. M. (1965). *Nucl. Phys.* **63,** 82.

Stephens, F. S. (1982). *Proc. Conf. on High Angular Momentum Properties of Nuclei*. p. 479. Oak Ridge, Tennessee.

Suffert, M., Feldman, W., Mahieux, T. and Hanna, S. S. (1968). *Nucl. Instr. Meth.* **63,** 1.

Twin, P. J., Nolan, P. J., Aryaeinejad, R., Love, D. J. G., Nelson, A. H. and Kirwan, A. (1983). *Proc. Int. Conf. on Heavy Ion and Nuclear Physics*, Catania, Italy, March 1983.

Vivien, J. P. (1983). Internal Report CRP/PN 83-25, Strasbourg, France.

Watson, R. L., Rasmussen, J. O. and Bowman, H. R. (1967). *Rev. Sci. Instr.* **38,** 105.

Weller, H. R., Blue, R. A., Roberson, N. R., Rickel, D. G., Maripuu, S., Cameron, C. P., Ledford, R. D. and Tilley, D. R. (1976). *Phys. Rev.* **C13,** 922.

Werf, S. Y. van der (1978). *Nucl. Instr. Meth.* **153,** 221.

Westerberg, L., Edvardson, L. O., Madueme, G. Ch., and Thun, J. E. (1975). *Nucl. Instr. Meth.* **128,** 61.

4

GAMMA-RAY SPECTROSCOPY AND NUCLEAR REACTIONS

4.1 Nuclear reaction modes and gamma-ray spectroscopy

4.1.1 *Nuclear reactions and gamma-decay schemes*

Nuclear reactions induced by hadron projectiles are essentially governed by the strong interaction. The electro-magnetic interaction, which is much weaker than the strong interaction, is involved at various stages of the nuclear reaction. The gamma rays associated with nuclear reactions are shown schematically in Fig. 4.1.

The nuclear reaction is normally a fast dynamic process. The time scale is in the order of 10^{-22} to 10^{-20} s for direct and pre-equilibrium reactions, and extends to around 10^{-18} s for equilibrium (compound nucleus) reactions. The former time scale corresponds to a typical nuclear period of $R{\cdot}M/P_f$ where R, M and P_f are the nuclear radius, the nucleon mass and the fermi momentum, respectively. The latter one corresponds to a typical level width of around 1 keV for a compound state. This time scale is still much shorter than the time scale of γ transitions of around $10^{-15} \sim 10^{-9}$ s. Consequently, γ-ray spectroscopy is mostly concerned with effectively bound excited states produced after all the prompt nuclear reaction processes which are induced by the strong interaction have taken place (see B–γ in Fig. 4.1). Gamma-ray spectroscopy, however, yields decisive and detailed information on the residual bound states such as the identification of reaction products (mass and charge), excitation energies, linear and angular momenta, parities and other properties, as discussed in Section 4.2. These quantities necessarily reflect directly or indirectly the reaction dynamics. Gamma transitions are partially converted to internal conversion electrons, leaving holes in inner electron shells, and the hole is filled by emission of an X-ray characteristic of the atomic number Z. This X-ray may be used to identify the atomic number Z of the residual nucleus. If the reaction feeds the ground state of a β-unstable nucleus, β–γ spectroscopy may also be used to identify the residual nucleus.

Gamma-ray spectroscopy can be extended to highly excited unbound states of the reaction product (UB–γ in Fig. 4.1). Here the γ-decay branch is as small as 10^{-3} to 10^{-5} because of the competition with particle decay channels available in the high excitation-energy region. Nevertheless, high-intensity accelerated beams, combined with high-efficiency γ-ray and particle detectors,

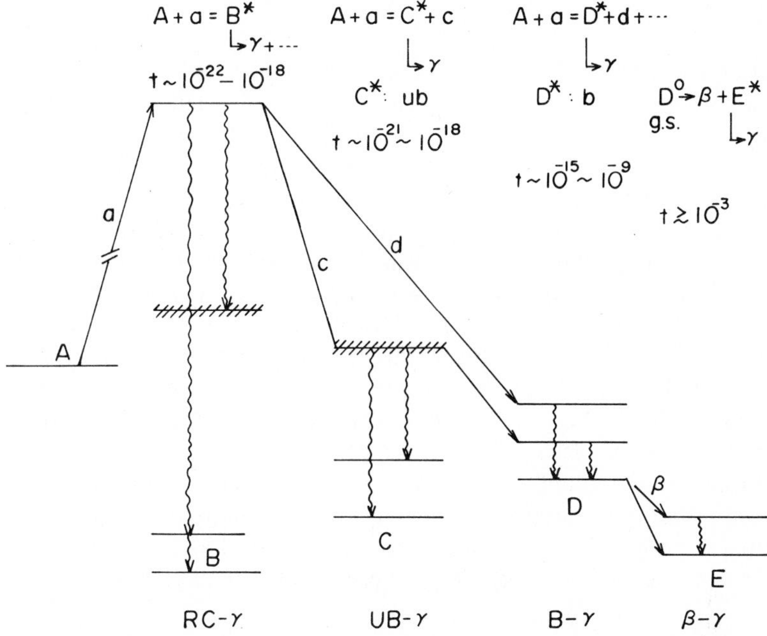

Fig. 4.1. Schematic diagram of nuclear reactions and γ decays. A: target nucleus, a: projectile, t: typical time-scale in units of sc. RC-γ: radiative capture γ decay, UB-γ: unbound (ub) state γ decay, B-γ: bound (b) state γ decay, and β-γ: γ decay following β decay. Note that a sharp resonance with $\Gamma_t < 1$ keV gives a longer time scale than 10^{-18} s for the RC-γ and UB-γ processes, and a long-lived isomer gives a longer time scale than 10^{-9} s for the B-γ process.

make it feasible to use γ-ray spectroscopy for studying the prompt reaction process, which necessarily involves particle unstable (unbound) intermediate states.

The radiative capture reaction is a special case where the electromagnetic interaction is involved directly in the first dynamic stage of the nuclear reaction (RC–γ in Fig. 4.1.). Then the γ-ray spectroscopy elucidates straight forwardly the electromagnetic properties of highly excited states. In this way various modes of collective motion associated with isospin, spin, and space coordinates have been studied by the radiative capture reaction, as discussed in Chapter 5.

In this section, we present briefly some general types of nuclear reactions related to γ-ray spectroscopy. Details of the reaction processes studied by the relevant e–γ measurements are described in the following sections of this chapter.

4.1.2 *Nuclear reaction modes and gamma decay*

Nuclear reaction modes associated with γ-ray spectroscopy are classified schematically into four types as illustrated in Fig. 4.2. The major features of these reaction modes in view of the γ-ray spectroscopy are as follows.

4.1.2a *Equilibrium and pre-equilibrium reactions (EQ-PEQ)*

A projectile with energy E_a interacts strongly with nucleons and/or nucleon clusters in nuclei, and initiates a nuclear cascade. Particle–hole pairs (excitons) are produced by the internuclear cascade process. If the projectile energy is high enough, some nucleons and/or nucleon clusters may be emitted in the course of the cascade process before thermal equilibrium is reached, namely at the pre-equilibrium (PEQ) stage. The PEQ stage is followed by the thermal equilibrium (EQ) phase, where the energy and momentum are distributed over the whole nucleus. In this case many nucleons are involved in the excitation. Equilibrated nuclei with excitation energy above the particle emission threshold energy cool down by evaporating particles. The nucleons emitted at the PEQ stage are characterized by a large kinetic energy and large linear and angular momenta, while those at the EQ stage have a relatively

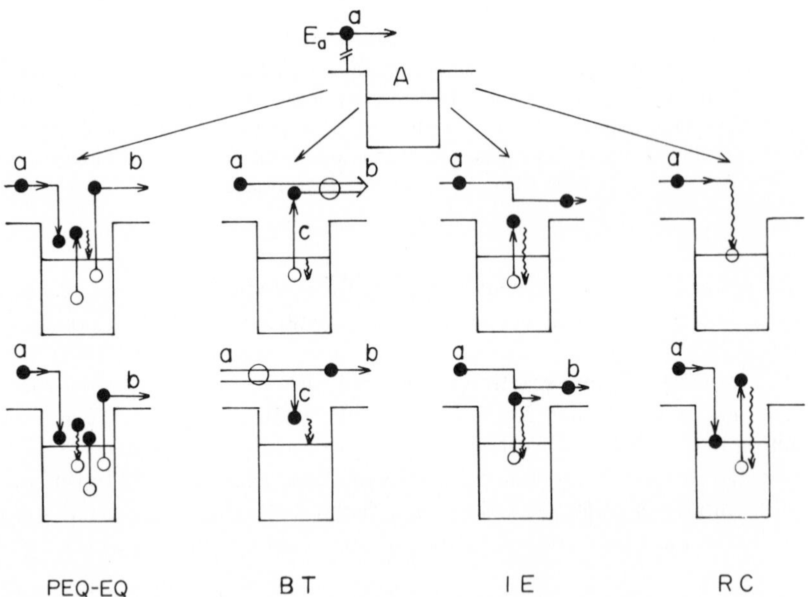

Fig. 4.2. Schematic diagrams of nuclear reaction modes for $A+a \rightarrow X+b+\ldots$ and associated γ radiations. a: projectile, E_a: its kinetic energy, A: target nucleus. PEQ–EQ: pre-equilibrium and equilibrium processes, BT: break-up and transfer processes, IE: inelastic process, and RC: radiative capture process.

small energy and small linear and angular momenta. A small number of fast particles (neutrons and protons) are emitted in the PEQ reaction, while many slow nucleons (mostly neutrons from medium and heavy nuclei) are evaporated in the EQ reaction. The PEQ process followed by the EQ process in heavy-ion reactions corresponds to incomplete fusion, while the pure EQ process without the preceding PEQ stage corresponds to the complete fusion.

If these PEQ and EQ reactions leave residual nuclei in excited states, they de-excite by emitting γ rays and/or conversion electrons. The PEQ and EQ processes may then be investigated by spectroscopic studies of the e–γ rays associated with the residual nuclei, as described in Section 4.2.

The PEQ and EQ reactions, in combination with γ-ray spectroscopy, are often used for studying nuclear structure. By selecting properly the projectile (mass and charge), the incident energy and the target nucleus, one may populate a particular region of residual states in a wide range of the N–Z plane, E–I plane, and other nuclear coordinates such as K, T, etc. Here N, Z, E, I, K and T stand for neutron number, proton (atomic) number, excitation energy, angular momentum, its projection on the symmetry axis if it exists, and isospin. As a result, in-beam γ-ray spectroscopy associated with the EQ (compound) reaction has been used extensively for investigation of nuclear properties in a wide range of the N, Z, E, I, K and T space.

4.1.2b *Break-up and transfer reactions* (*BT*)

The internal properties of the projectile are mostly lost through the internuclear cascade of the PEQ–EQ processes, although gross (external) properties such as energy, linear and angular momenta, and to some extent also isospin are preserved in the reaction. There are several modes of nuclear reaction where some memory of the projectile is preserved. One is the break-up and transfer reactions (BT) where nucleons (nuclear clusters) are transferred between the projectile and the target nuclei (see BT in Fig. 4.2). Thus the γ rays associated with the target-like nucleus (fragment) and those with the projectile-like nucleus (fragment) are used to study the reaction mechanism.

The particle pick-up process leaves the residual nucleus in a hole state, while the particle transfer process leaves it in a particle state. They are identified by the γ-decay to the hole state and that from the particle state, respectively. The mass, energy and momentum of the transferred particle are found by measuring the γ rays from the residual nucleus, as described in the following sections. The angular momentum transfer evaluated from the γ-ray multiplicity gives the impact parameter involved in break-up and transfer reactions. In the case that a heavy-ion projectile breaks up into two fragments as $\mathrm{a} \to \mathrm{b} + \mathrm{c}$, a heavier fragment c may be transferred (absorbed) into the target nucleus. This is a massive–transfer reaction. Break-up and transfer reactions studied by γ-ray spectroscopy are discussed in Section 4.3.

Observation of the out-going particle b together with the kinematical condition of the transfer reaction determines the energy and momentum of the transferred particle c. This makes it possible to perform γ-ray spectroscopy following nuclear reactions induced by the transferred particle c with proper mass and angular momentum windows.

4.1.2c *Inelastic scattering* (*IE*)

Inelastic scattering (IE) transfers only energy and momentum into the target nucleus. Thus the initial doorway of the IE is a particle–hole state, and γ decays relevant to the particle–hole at the first initial stage occur with a certain probability.

There are various types of collective motions which are composed of a coherent superposition of particle–hole excitations, and these are strongly excited by inelastic scattering. One type of collective motion in an unbound excitation region shows a giant resonance (GR) for that type of the excitation process. Here the excitation of the GR is due to the strong interaction with rather complex properties, while the γ decay from the GR is due to the electromagnetic interaction with simple selection rules. Thus one can get clear information on the multipolarity of the GR by studying its γ decay.

Deep inelastic processes of heavy ions produce highly excited nuclei with many particle–hole excitations. In these reactions, energy, angular-momentum and mass are transferred between the projectile and the target. The excitation energy and the angular momentum are studied by measuring the γ-ray multiplicity and the summed γ-ray energy.

Inelastic scattering is discussed from the point of view of the relevant γ spectroscopy in Section 4.4.

4.1.2d *Radiative capture reactions* (*RC*)

Electromagnetic interaction in the first stage of the reaction may induce a γ radiation while the projectile is captured into the target nucleus (RC in Fig. 4.2). The γ-radiation strength itself gives the electromagnetic strength distribution of the capture state in a highly excited unbound region. Radiative capture into an excited orbit gives the nucleon (cluster) property of the excited state. If the incident particle is captured into a resonance state, the radiative capture γ ray gives directly the electromagnetic transition probability of the resonance state. Many single particle (cluster) resonances and isobaric analogue states have been studied by the radiative capture reaction.

Because of a strong coupling of the J-multipole γ-radiation and the J-mode giant resonance, the radiative capture reaction is used to study giant resonances in the high excitation region. Here the giant resonance based on highly excited and/or high-spin states may be studied by observing the radiative capture into highly excited high spin states. The radiative capture reaction is described in Section 4.5.

4.1.3 *Nuclear reactions and impact parameters*

The major reaction modes depend largely on the projectile mass (m_a), the projectile energy (E_a), and the impact parameter (b). Low-energy projectiles are likely to form compound nuclei through the EQ process, while the PEQ and the direct reactions (break-up, transfer, inelastic scattering, etc) become important as the projectile energy per nucleon exceeds 10 MeV. The wave length $\lambda\!\!\!^{-}$ of the projectile is given by $\lambda\!\!\!^{-} \approx 4.5/\{m_a \cdot \sqrt{(E/m_a)}\}$ fm, where E/m_a is the projectile energy per nucleon in units of MeV. Projectiles with $\lambda\!\!\!^{-}$ much smaller than the nuclear radius are treated semiclassically, and the reaction modes are classified by the geometrical (impact) parameter b. Since most of the heavy ions have small wavelengths, the reaction modes are characterized by the impact parameter b and accordingly by the input angular momentum l_a. As illustrated schematically in Fig. 4.3, they are divided into three regions (see also Section 6.2).

1. The distant collision region with a large value of $b > R_A + R_a$ or $E < B_c$, where R_a, R_A, and B_c are the projectile radius, the target radius and the Coulomb barrier. In this case the projectile cannot reach the nuclear interaction region, and only the long-range Coulomb force causes Rutherford scattering or Coulomb excitation.

2. The grazing–collision region. The direct and pre-equilibrium processes dominate this reaction mode. If the interaction with the target is weak, simple break-up and transfer reactions and inelastic scatterings take place, while the deep inelastic scattering and the pre-equilibrium process take place if the interaction is strong. (Here "weak and strong" refer to the interaction strengths, and not to the fundamental forces.)

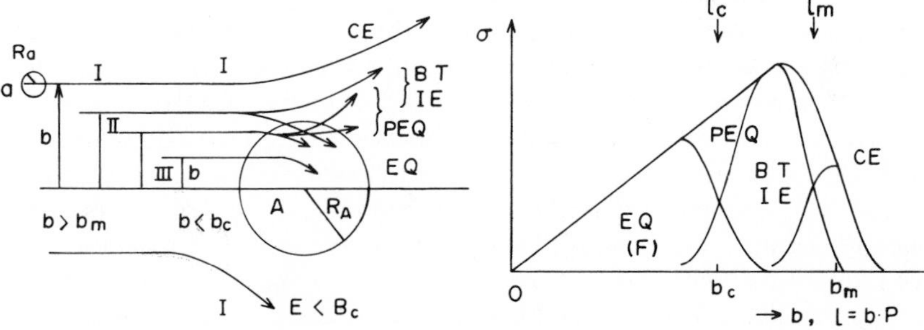

Fig. 4.3. Reaction modes shown schematically as a function of the impact parameter b. b_m: maximum value for nuclear reactions, b_c: critical value for the equilibrium (EQ) compound process, l_m and l_c: corresponding angular momenta. See text and captions of Fig. 4.1 and 4.2 for other symbols. These reaction schemes apply to low-energy heavy ions with $E/m_a \lesssim 20$ MeV.

3. The compound (fusion) reaction region. The projectile and target composite system is then equilibrated to form a (compound) system in thermal equilibrium. This process is characterized by critical and maximum impact parameters, b_c and b_m, respectively, corresponding to the critical and the maximum angular momenta, l_c and l_m, respectively (see Section 6.2).

Classification of the reaction mode by the impact parameter is not strict, even for heavy ions. Low-energy light ions with $A \lesssim 2$ and $E \lesssim 10-20$ MeV have large wave lengths of the order of $\lambda \approx 1$ fm (which is comparable with the nuclear radius) and a long mean free path of the order of the nuclear radius. They even show some resonances at particular energies, so that their reactions can hardly be treated geometrically.

Medium-energy light and heavy ions with $E_a/m_a \gtrsim \varepsilon_F$, where ε_F is the Fermi energy of around 30 MeV, do not dissipate their whole energy and momentum at the first collision and some fraction of the projectile energy and momentum can still escape from the composite system, even in collisions with small b. Therefore the impact parameter does not specify the reaction modes of these medium-energy projectiles, but various types of reaction modes may occur with certain probabilities, even for a collision with a given impact parameter. Noting that the multiplicity of γ rays determines the angular momentum and accordingly the impact parameter, one can study the reaction modes as a function of the impact parameter.

4.2 Gamma-ray spectroscopy for equilibrium and pre-equilibrium reactions: statistical and quasi-statistical processes

4.2.1 *Gamma rays following equilibrium and pre-equilibrium reactions*

A projectile interacting strongly with a target nucleus A produces a highly excited composite nucleus C*, and particles are emitted in the pre-equilibrium (PEQ) and/or equilibrium (EQ) phases of the reaction. Such particle decays are characterized as quasi-statistical and statistical decays. In most cases the residual nucleus is left in an excited state (B*), which is effectively below the particle emission threshold, so that it de-excites to the ground state B by emitting γ rays. The reaction process is expressed as A(a, xy . . .)B*(γ)B or A(a, xy . . . , γ)B, where a is the projectile and x, y, . . . are the decaying particles. Studies of the γ rays following the A(a, xy . . .)B* (γ)B reaction reveal the properties of the residual nucleus B* such as its mass, charge, excitation energy, angular momentum and its direction, linear momentum, and so on. The conservation of these quantities throughout the reaction process leads directly to the corresponding quantities removed by the particles x, y, z, . . . in the reaction. Since many kinds of particles are involved in the statistical and quasi-statistical particle decays, observation of all of them at all angles is very difficult. On the other hand γ-ray spectroscopy associated with the reaction residues is a simple and useful way of obtaining the gross properties of these decaying systems.

4.2.1a *Reaction channels and transfers of mass, charge and energy*
Mass, charge and energy transfers in the EQ and PEQ reactions are schematically illustrated in Fig. 4.4. High-energy-resolution γ detectors (Ge) and/or conversion–electron spectrometers are used to resolve most of the discrete γ rays characteristic of the residual nuclei, even when a great number of γ rays from many kinds of residual nuclei are emitted. Conservation of nucleon number and charge in the reaction gives uniquely the total charge ΔZ and the mass ΔM of the decaying (out-going) particles,

$$\Delta Z = \Sigma Z_i = Z_a + Z_A - Z_B. \tag{4.2.1a}$$

$$\Delta M = \Sigma M_i = \Delta Z + N_a + N_A - N_B, \tag{4.2.1b}$$

where the suffixes a, A, B and i refer to the projectile, the target nucleus, the residual nucleus and the i-th emitted particle, respectively, and N and Z are the proton and neutron numbers.

The composite nucleus C* is normally located on the neutron deficient side in the N–Z plane as shown in Fig. 4.4b. Some fast protons and fast neutrons may be emitted at the PEQ stage. Particles evaporated at the EQ stage of medium and heavy nuclei are mostly slow neutrons because charged-particle

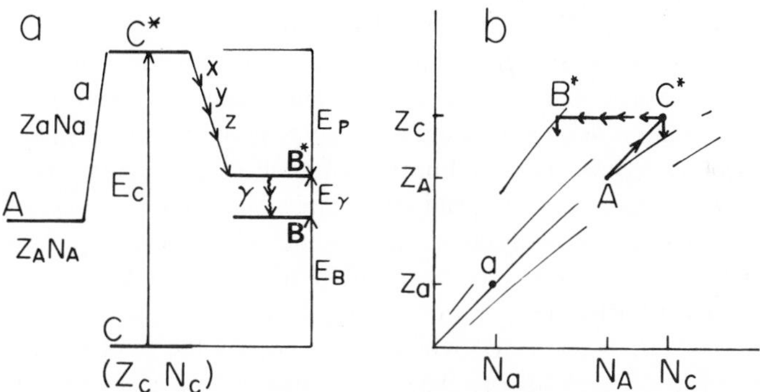

Fig. 4.4. Schematic diagrams of the A(a, xyz . . .)B* reaction through the compound (thermal equilibrium) nucleus C*. Z_i and N_i are the atomic and neutron numbers of the nucleus i. a: Level scheme of the A(a, xy. . .)B* reaction. E_C is the excitation energy of C, and E_p and E_B are the kinetic and binding energies carried away by emitted particles, E_γ is the average excitation energy of B*. The E_γ is carried away by the decaying γ rays. b: Nuclear chart showing the projectile a, the target nucleus A, the compound nucleus C* and the residual nucleus B*. The arrows indicate the incoming projectile and decaying nucleons.

evaporation is suppressed by the Coulomb barrier, so that residual nuclei well on the neutron-deficient side are produced. As the nucleus becomes more neutron-deficient, the neutron binding energy becomes larger while the proton binding energy becomes smaller, and then proton emission may start to compete with neutron emission.

The cross-section $\sigma(\mathrm{B})$ for the reaction channel A(a, xy . . .)B* is obtained from the cross-section $\sigma(\gamma_{\mathrm{B}}^{i})$ for the discrete γ ray γ_{B}^{i} as

$$\sigma(\mathrm{B})=\frac{4\pi}{W(\theta)}\cdot\frac{\mathrm{d}\sigma(\gamma_{\mathrm{B}}^{i})}{\mathrm{d}\Omega}\cdot\frac{1+\alpha_i}{b_i}, \tag{4.2.2}$$

where b_i is the branching ratio of the i-th γ transition with respect to the total production of B*, α_i is the coefficient for the internal conversion electron, and $W(\theta)$ represents the γ-ray angular distribution. The quantity b_i is obtained from the γ-decay scheme. As discussed later the normalized angular distribution is generally given by $W(\theta)=1+A_2P_2(\cos\theta)+A_4P_4(\cos\theta)+\ldots$ with $A_4\ll 1$. Then the γ-ray yield at $\theta_0\approx 55°$, where $P_2(\cos\theta)=0$, is commonly used because $W(\theta_0)\approx 1$ and so the correction for $W(\theta)$ is not necessary.

The excitation energy decreases by emitting particles until the particle and γ decay probabilities become comparable. Then the nucleus B* starts to de-excite by emitting statistical γ rays with low multipolarities, followed by spin-stretched γ decays along the yrast and near-yrast bands (see Section 4.2.2d). The yrast level, defined as the level with the lowest energy for a given spin, is separated in energy from other (non-yrast) levels in even–even nuclei. As a result, the single yrast line in the low excitation region accumulates most of γ-ray flows, and the branching ratio b_i for such yrast transition approaches unity. These γ rays are used to evaluate $\sigma(\mathrm{B})$ since the uncertainty in b_i is small. In the case of odd-mass nuclei there are several strong γ-ray flows along low-lying collective bands. Thus a number of such strong γ rays are used to evaluate the total yield of the residual nucleus. Such strong γ-decay flows in the near-yrast region of medium and heavy nuclei have been used for in-beam γ-ray spectroscopy, initiated by Morinaga and Gugelot (1963).

The major EQ–PEQ reaction channels for medium and heavy nuclei are expressed as A(a, xn yp . . .)B* where x, y, . . . , are the numbers of emitted neutrons, protons, and so on. The mass and charge distributions of the reaction products B* then give the multiplicity distribution $M(x, y, \ldots)$ for the emitted particles, which reflects directly the PEQ and EQ processes. In fact alpha-particles are also emitted from highly excited high-spin states in heavy nuclei, and they cannot be distinguished from the 2n and 2p emissions by measuring only γ rays from the residual nuclei.

The total kinetic energy E_{p} carried away by the emitted particles with energies $E_{\mathrm{p}}(j)$ in the reaction channel is evaluated from the energy conservation law by

$$E_c = \sum_i \{E_p(j) + E_B(j)\} + \sum_i E_\gamma(i)$$

$$= E_p + E_B + E_\gamma \tag{4.2.3a}$$

and

$$E_p = \sum_j E_p(j), \quad E_B = \sum_j E_B(j), \quad E_\gamma = \sum_i E_\gamma(i), \tag{4.2.3b}$$

where E_γ is the excitation energy of B* at the entry region, E_C is the excitation energy of C*, E_B is the sum of the binding energies $E_B(j)$ of the emitted particles in the reaction channel (see Fig. 4.4).

The total excitation energy E_C is divided into three parts, E_p, E_B and E_γ as given by eqn (4.2.3a). The balance of E_p and E_B then reflects the relative contributions of the PEQ and EQ processes in the reaction channel. The PEQ process releases a large amount of energy in the form of the kinetic energy E_p, while the EQ process releases a large amount of energy in the form of the binding energy E_B by emitting a large number of slow particles.

The excited nuclei following nuclear reactions decay by emitting mostly γ rays, and to some extent by internal conversion electrons, particularly in case of heavy nuclei. The internal conversion electron produces a K or higher-shell electron hole, and thus is followed by the K or higher-shell X-ray. The X-ray, which is characteristic of the residual atom, is used to identify the atomic number Z_B of the residual nucleus. K X-rays are quite useful for selecting residual isotopes with high Z because of the large coefficients of internal conversion electrons and the large energies of the K X-rays.

In the case of the (a, ypxn) reaction the K X-ray of the residual isotopes with the atomic number Z_B gives the charge transfer $\Delta Z = Z_a + Z_A - Z_B = y$. Thus the yield of the K X-rays is written as

$$N_K = \sum_x \langle M_K^x \rangle \sigma(\text{a, ypxn}), \tag{4.2.4}$$

where $\langle M_K^x \rangle$ is the multiplicity of the K X-rays in the (a, ypxn) reaction and σ(a, ypxn) is the cross-section.

The K X-ray yield and multiplicity for the (^{6}Li, xn) fusion reaction were measured by Karwowski *et al.* (1979). The multiplicity was obtained from the ratio of the X–γ coincidence yield $N_{X\gamma}$ to the singles γ-ray yield N_γ as $\langle M_K^x \rangle = N_{X\gamma}/(N_\gamma \cdot \eta_X)$, where the γ ray specifies the reaction channel (neutron multiplicity x) and η_X is the X-ray detection efficiency. The result shows that the K X-ray measurement gives reliable total cross-sections for the evaporation residues. It is not necessary to know all the details of the γ-ray branching ratios for individual reaction residues. The multiplicity gives the average conversion coefficient of the γ-ray cascades, which is used to extract the multipolarity (see Section 6.5.5). K X-rays have also been used for identifying the reaction channels of break-up reactions induced by heavy ions (see Section 4.3).

4.2.1b *Angular momentum transfer and angular momentum alignment*
Conservation of the angular momentum $\mathbf{I}\hbar$ and its z-component m in the A(a, xyz . . .)B* reaction leads to the simple relations,

$$\Delta\mathbf{I}\hbar = (\mathbf{I}_a - \mathbf{I}_B)\hbar \tag{4.2.5a}$$

and

$$\Delta m\hbar = (m_a - m_B)\hbar, \tag{4.2.5b}$$

where $\Delta\mathbf{I}\hbar$ and $\Delta m\hbar$ are the angular momentum and its z-component carried away by the emitted particles, and the target spin is neglected for simplicity. Assuming that the projectile spin is negligible, the orbital angular momentum brought in by the projectile is $I_a\hbar = P_a \cdot b$, with b being the effective impact parameter for the reaction channel. Since the orbital angular momentum ($I_a\hbar$) is perpendicular to the beam axis, one gets $m_a\hbar = 0$ by choosing the beam direction as the z-axis. The angular and linear momenta involved in the reaction is shown in Fig. 4.5. The target and projectile spins, which are oriented at random, have little effect on $I_a\hbar$ and $m_a\hbar$ since their spin values are usually much less than the orbital angular momentum of the projectile.

The angular momentum of the residual nucleus can be obtained by studying the γ-ray multiplicity M_γ (see Sections 3.1 and 6.5). The γ-de-excitation flow in medium and heavy nuclei is divided into two parts,

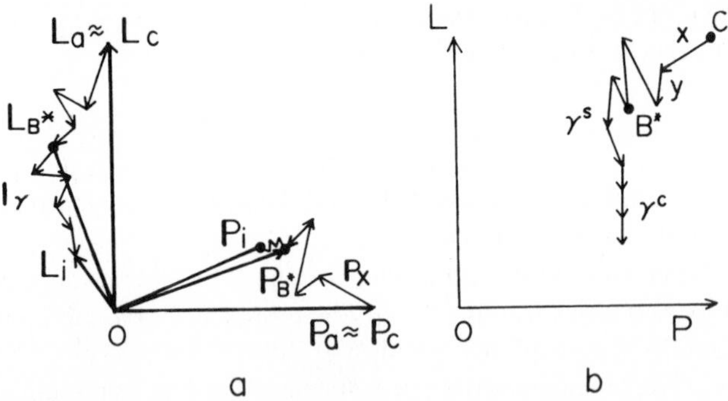

Fig. 4.5. Angular and linear momenta associated with the A(a, xy. . .)B reactions. L_α and P_α are the angular and linear momenta, respectively, where α = a, x, B* γ and i stand for the projectile, emitted particles, the residual nucleus, γ ray and the excited level i in the nucleus B. (a) linear and angular momentum vectors. (b) the reaction scheme in the L–P plane.

one is the statistical decay and the other is the collective decay along collective bands. (Grover and Gilat, 1967; Newton *et al.*, 1970.) The angular momentum I_B may then be related to M_γ by

$$I_B = M_\gamma^s \Delta J^s + M_\gamma^c \Delta J^c, \tag{4.2.5c}$$

where M_γ^s and M_γ^c are the γ-ray multiplicities for the statistical and collective de-excitations, respectively, and ΔJ^s and ΔJ^c are the average spin changes for the corresponding γ decays. Since the γ decays along the collective bands are mostly spin-stretched transitions of $I \to I - J$ where J is the multipolarity, ΔJ^c is approximately given by the average multipolarity. In the case of even–even deformed nuclei one gets $\Delta J^c = J = 2$ because of the E2-dominant collective transition. The statistical transitions are mostly of the dipole type, and ΔJ^s is small in the low-spin region because the statistical transitions $I \to I+1$, I, and $I-1$ are all possible. It approaches $\Delta J_\gamma^s \approx 1$ in the high-spin region, where the γ transition $I \to I-1$ with $J_\gamma = 1$ is more favoured than others because of the larger level density for the lower-spin state. The γ-ray energy for statistical decays is of the order of 2 to 4 MeV, while that for most collective transitions is less than one MeV. Thus the quantities M_γ^s and M_γ^c may be measured separately by selecting the γ rays in the proper energy region. Details of the γ-ray multiplicity measurements are described for high-spin states in Sections 6.2 and 6.5.

Alignment of the angular momentum I_B is measured by observing the angular distributions of known multipole γ transitions. The angular distribution is written

$$W(\theta) = \sum_{v=0}^{L} a_{2v} P_{2v}(\cos\theta), \tag{4.2.6}$$

where $P_{2v}(\cos\theta)$ is the Legendre polynomial. The coefficient a_{2v} is written $a_{2v} = \alpha_{2v} A_{2v}$, where A_{2v} is the angular distribution coefficient for the completely plane-aligned initial spin ($m_i \approx 0$) and α_{2v} is the attenuation factor due to spin de-alignment. The spin alignment is disturbed by the particle decays feeding the state I_B and by the γ decays feeding the initial state I_i. The $m_B\hbar$ in eqn (4.2.5b) is evaluated by correcting the observed attenuation factor α_{2v} for the state I_i for the spin dealignment due to the preceding γ decays ($I_B \to \ldots \to I_i$) feeding the state I_i. Then one finally obtains the spin dealignment $\Delta m\hbar$, which corresponds to the z-component of the angular momentum carried away by the emitted particles.

The spin dealignment throughout the PEQ–EQ particle decays and statistical-collective γ decays is fairly small and thus spins of excited states in the residual nucleus are well aligned as was first shown by Ejiri *et al.* (1965/1966). The spin alignment plays an essential role in spectroscopic studies of the nuclear properties such as magnetic moments, spins, parities, and γ-ray multipolarities.

Since the $I_B\hbar$ and $m_B\hbar$ for the reaction residue B* are obtained from the γ-ray measurements as discussed above, the $\Delta\mathbf{I}\hbar$ and $\Delta m\hbar$ for the emitted particles are evaluated by using eqn (4.2.5). They are sensitive to the PEQ and EQ processes. Fast particles emitted in the forward direction during the PEQ phase carry off large angular momenta along the beam direction, leaving the spin alignment practically unaltered. Particles emitted in random directions remove little angular momenta on the average, but disturb the spin alignment (see Fig. 4.5).

4.2.1c *Linear momentum transfer*

Linear momentum transfers in PEQ–EQ reactions are considered in a similar way to angular momentum transfers, as shown in Fig. 4.5. The conservation of the linear momentum leads to

$$\Delta\mathbf{P} = \mathbf{P}_a - \mathbf{P}_B. \tag{4.2.7}$$

Here the projectile momentum $\mathbf{P}_a$ is aligned along the beam axis (z-axis) and the momentum $\Delta\mathbf{P}$ removed by the emitted particles is obtained from the momentum $\mathbf{P}_B = M_B \cdot \mathbf{V}_B$ of the residual nucleus.

The mass M_B is known from discrete γ rays characteristic of B*, and $\mathbf{V}_B$ is known from the Doppler shifts of the γ rays, provided that either the γ-decay lifetime is much shorter than the stopping time of the recoiling nucleus in the target material or the recoiling nucleus is emitted outside the target material without much energy loss. The former condition is satisfied in the case of low multipole (E1, M1, E2) γ decays with $E_\gamma \gtrsim 1 \sim 3$ MeV in light residual nuclei, while the latter case occurs when residual nuclei are produced by heavy-ion reactions with very thin targets.

Particles evaporated from compound nuclei are emitted nearly isotropically, so that the sum of the linear momenta vanishes, and one gets $\Delta\mathbf{P} = 0$. Thus the average momentum $\mathbf{P}_B$ of the residual nuclei produced by a compound reaction is approximately given by the projectile momentum $\mathbf{P}_A$. Using this simple relation the Doppler effect on the γ rays emitted from the recoiling nuclei has been used for lifetime measurements. The stopping time of the recoiling nucleus in the target material is typically in the order of 10^{-12} to 10^{-16} s, depending on the energy and the mass. The γ-decay lifetime in that region is measured by observing the Doppler attenuation. The lifetime in the region of 10^{-9} to 10^{-13} s is measured by means of the recoil distance (plunger) method. Here the recoiling nuclei with large momenta, introduced by heavy projectiles, are ejected outside the thin target foil, and some are stopped in a stopper foil placed close behind the target foil. Comparison of the yield of the Doppler-shifted γ rays emitted from the recoiling nuclei with that of the unshifted ones emitted from those caught in the stopper gives the lifetime in the region of the flight time (see Section 3.5.1 for details of the Plunger method and the Doppler-shift attenuation method).

The linear momenta of the recoiling nuclei are directly measured from the range in either the target material or the stopper material. A layer of thin foils is conveniently used for this purpose. The residual nucleus stopped at the foil is detected by measuring γ rays following the radioactive decay of the residual nucleus, provided that the residual nucleus is radioactive in an adequate range of half-lives.

The momentum $\Delta\mathbf{P}$ removed by the emitted particles reflects straightforwardly the reaction dynamics involved. It is written as $\Delta\mathbf{P}=\sum_i \mathbf{P}_i$, where $\mathbf{P}_i$ is the linear momentum of the i-th emitted particle. Fast particles emitted foward in the PEQ phase take away large amounts of linear momentum $\mathbf{P}_i$ in the longitudinal (beam) direction. The average momentum of the residual nucleus is given by $P_B \approx P_a - \Delta P_z$, where ΔP_z is the projection of $\Delta\mathbf{P}$ on the beam axis. Thus the particle decay in the PEQ phase decreases the momentum P_B of the residual nucleus just as the γ decays along the collective band decreases the angular momentum.

The statistical particle decays are nearly isotropic, and show a forward-backward symmetry in the c.m. (centre of mass) frame. Consequently the sum $\Sigma\mathbf{P}_i$ is nearly zero in the c.m. frame. Then ΔP_z in the laboratory frame is given by $\Delta P_z \approx \Delta m \cdot V_{c.m.}$, where $V_{c.m.}$ is the c.m. velocity of the compound nucleus and Δm is the total mass of the emitted particles. Then P_B is given by $P_B = P_a M_B/(M_a + M_A)$.

4.2.2 *Gamma-ray spectroscopy and equilibrium reactions: statistical and fusion processes*

4.2.2a *Reaction channels and population of levels*

The equilibrium (EQ) process of the A(a, xyz . . .) reaction takes place firstly by formation of a compound nucleus C* by equilibration of the incident energy and momentum, and secondly by sequential emission (evaporation) of particles x, y, z . . . . The excitation energy of the compound nucleus C* is given by $E_c = E_a + B_a$, where E_a is the projectile energy in the c.m. system and B_a is the binding energy of the projectile in the nucleus C. Since C* is a thermally equilibrated system, a nuclear temperature T_c is defined as $T_c = \sqrt{(E_c/a)}$, where a is the nuclear level density parameter. The compound nucleus C* cools by evaporating particles, x, y, . . . , which are treated in the framework of a statistical model.

The cross-section for the formation of the compound nucleus with spin I_c is given by

$$\sigma_c(I_c) = \pi\left(\frac{\hbar}{P_a}\right)^2 \sum_{l=|I_c-S|}^{I_c+S} \sum_{S=|I_t-s_a|}^{I_t+s_a} \frac{(2I_c+1)}{(2s_a+1)(2I_t+1)} T(l_a E_a), \qquad (4.2.8a)$$

where P_a is the projectile momentum, s_a the projectile spin, I_t the target spin, S

the channel spin, $l_a\hbar$ the orbital angular momentum of the projectile, and $T(l_aE_a)$ the transmission coefficient of the projectile with l_a and E_a into the target nucleus. In the case $l_a \gg s_a$ and I_t, eqn (4.2.8a) reduces to the simple form,

$$\sigma_c(I_c) \approx \pi\left(\frac{\hbar}{P_a}\right)^2 (2I_c+1)\sum_{l_a} T(l_a E_a), \tag{4.2.8b}$$

with $l_a \approx I_c$. The transmission coefficient $T(l_aE_a)$ is obtained from the optical potential for the projectile. A sharp cut-off model gives $T(l_a E_a)=1$ for $l_a \lesssim l_{cr}$ and $T(l_a E_a)=0$ for $l_a > l_{cr}$, where l_{cr} is the critical angular momentum for the EQ (complete fusion) process. This is commonly referred to a black-body absorption with the critical radius $R_{cr}=l_{cr}\hbar/P_a$. In this case the total cross-section for the EQ process (complete fusion reaction) is written as

$$\sigma_c(t) \approx \sum_{I_c}^{l_{cr}} \sigma_c(I_c) \approx \pi\left(\frac{\hbar}{P_a}\right)^2 (l_{cr}+\tfrac{1}{2})^2. \tag{4.2.8c}$$

The probability of particle decay from the compound state is written in the statistical model

$$P_x(E_cI_c, E_1I_1) = C_1^{-1}\rho(E_1I_1) \sum_{s_1=|I_1-s_x|}^{I_1+s_x} \sum_{l_x=|I_c-S_1|}^{I_c+S_1} T\{l_xE_p(x)\}, \tag{4.2.9}$$

where $\rho(E_1I_1)$ is the level density of the final state with excitation energy E_1 and spin I_1, and S_1 is the channel spin for the decay of the particle x (Hauser and Feshbach, 1952; Ericson, 1960). $E_P(x)$, l_x and s_x are the energy, orbital angular momentum and spin of the emitted particle x, respectively, and $T\{l_xE_P(x)\}$ is the transmission factor. The sequential decay of the compound nucleus C* is expressed as

$$C^*(E_cI_c) \xrightarrow{x_1} C_1^*(E_1I_1) \xrightarrow{x_2} C_2^*(E_2I_2) \rightarrow \ldots B^*(E_BI_B), \tag{4.2.10}$$

where C_j^* is the j-th intermediate nucleus with the excitation energy E_j and spin I_j, and x_j is the particle decaying from C_{j-1}^* to C_j^*. Then the cross-section for populating the final nucleus with energy E_B and spin I_B is

$$\sigma_\alpha(E_BI_B) = \sum_{I_c} \sigma_c(I_c)\cdot \sum_{E,I} \prod_j P_j(E_jI_j, E_{j-1}I_{j-1}), \tag{4.2.11}$$

where α denotes the reaction channel A(a, $x_1x_2 \ldots x_j \ldots$)B* and P_j is the relative probability (branching ratio) for emitting the particle x_j. The sum over E and I extends over all possible E and I in the intermediate states.

The cross-section for feeding the final nucleus B* is obtained as

$$\sigma(B^*) = \sum_{I_BE_B} \sigma(E_BI_B).$$

The cross-sections $\sigma(E_B I_B)$ and $\sigma(B^*)$ are just the values obtained from yields of γ rays in the nucleus B^*. In the cases where the final nucleus B^* is fed by several reaction channels (for example (a, pxn), {a, d(x−1)n}, {a, t(x−2)n} channels) the summation has to be made over these possible channels as well.

The compound (EQ) process for medium momentum projectiles with $P_a \gtrsim$ a few hundred MeV/c on medium heavy target nuclei is fairly simple. The compound nuclear formation is given by eqn (4.2.8b) because $l_a = P_a R_c/\hbar \gg s_a$ and I_t. Charged particles are hardly ever evaporated because of the Coulomb barrier unless the Q-value and the angular momentum consideration greatly favour charged particle emission. Thus the major reaction channel is the neutron evaporation process expressed as the (a, xn) reaction.

The level density in eqn (4.2.9) is written as (Bohr and Mottelson, 1969; Ericson, 1960; Lang, 1966)

$$\rho(E\,I) = C(2I+1)\frac{1}{\tilde{E}^2}\,\mathrm{exps}\,\{2\sqrt{(a\tilde{E})}\}, \tag{4.2.12}$$

with

$$\tilde{E} = E - I(I+1)\hbar^2/(2\mathscr{I}_r), \tag{4.2.13}$$

where $\tilde{E}$ is the effective excitation energy corrected for the rotational energy and $\mathscr{I}_r$ is the effective moment of inertia. The level density parameter a is given by $a = A/a'$, where A is the mass number and $a' = 10$ to 15 MeV. The temperature of the compound nucleus produced by medium energy (20 to 200 MeV) projectiles on medium heavy nuclei is then $T = \sqrt{(E_c/a)} = 1$ to 4 MeV.

The kinetic energy $E_p(j)$ of the j-th particle x_j is given by

$$E_p(j) = E_{j-1} - E_B(j) - E_j, \tag{4.2.14}$$

where $E_B(j)$ is the binding energy of x_j in the nucleus C_j^*. Noting that $E_p(j)$ is of the order of the nuclear temperature, one has $E_p(j) \ll E_{j-1} - E_B(j)$, and the energy spectrum of the emitted (evaporated) particle x_j is given by

$$s\{E_P(j)\} \propto C E_P(j) \exp\left\{-\frac{E_p(j)}{T_j}\right\} T\{l_{x_j}, E_P(j)\}. \tag{4.2.15}$$

The average energy is given by $\bar{E}_p \sim 2T$. The orbital angular momentum is limited by the centrifugal potential to $l_j\hbar \lesssim 2\sqrt{(M_j T)}\cdot R$, where M_j is the mass of the particle x and R is the compound nuclear radius. In the case of medium heavy nuclei with $R \sim 6$ fm and $E_j \approx 50$ MeV, l_j for nucleon emission $(M_j = 1)$ is limited to $l_j \lesssim 2$, and increases very slowly as E_j increases.

It should be remarked that one can deduce the properties of excited states in the entry region of the residual nucleus B^* from the intensities of decaying γ rays by using the statistical model for particle decays feeding such excited states (Hansen *et al.*, 1963; Sakai *et al.*, 1965).

4.2.2b *Reaction channels for reactions induced by light projectiles*

The projectile energy for the EQ process has to be restricted to rather low values of $E_a/M_a \lesssim 20$ to 15 MeV, depending on the projectile mass M_a. Thus the excitation energy of the compound nucleus and that of intermediate states are at most 80 to 50 MeV for light projectiles (light ions) such as p, d, ^{3}He, and α. The nuclear temperature is as low as $T \lesssim 0.5$ to 2 MeV, and accordingly the average kinetic energy of the evaporating neutrons is only a few MeV and their partial waves are mainly s and p waves. Therefore the major reaction channel (neutron multiplicity x in case of (a, xn) reactions) is well specified for a given projectile energy in accordance with energy conservation—eqn (4.2.3). The particle decay process for alpha-induced EQ reactions is schematically illustrated in Fig. 4.6.

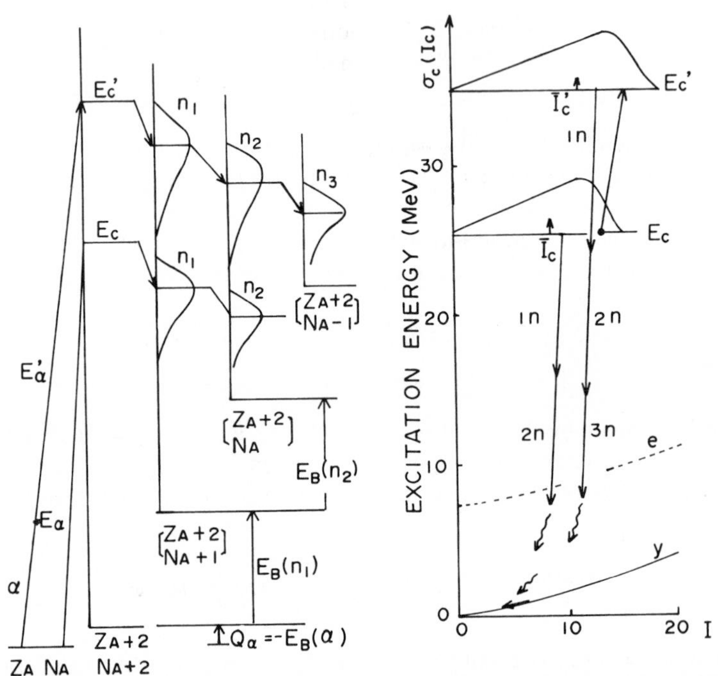

Fig. 4.6. Reaction scheme for the A(α, xn)B reaction through the compound (equilibrium) nucleus C. Left-hand side: Level schemes involved in the reaction. E_a and E_c are the projectile energy and the excitation energy of the compound nucleus and $E_B(n_i)$ and $E_B(\alpha)$ are the binding energy of the i-th decaying neutron (n_i) and the projectile. Right-hand side: Reaction process in a plane of the excitation energy and the angular momentum I. $\sigma_c(I_c)$ and $\bar{I}_c$ are the cross-section for formation of the compound nucleus with the angular momentum I_c and the average angular momentum, and e and y stand for the entry and yrast lines.

The compound nucleus produced by alpha-particle bombardment has an excitation energy of $E_c \approx E_\alpha$. It loses excitation energy of about $B_n + 2T \approx 10$ MeV by every neutron evaporated, where $B_n = \bar{E}_B(j)$ is the average binding energy of neutrons. The average excitation energy E_γ of the residual nucleus is about 5 MeV. Then the proper choice of $E_\alpha \approx 10x + 5$ MeV for a given x leads exclusively to the (α, xn) reaction channel. The angular momentum changes little due to the neutron evaporation because of the small l_n and the small spin dependence of the level density in the present region of $I_c = 10$ to 15. Increase of the projectile energy results in an increase of the angular momentum I_c. Thus the dependence of the γ-ray intensity on the projectile energy is used to evaluate the spin of the residual states.

The excitation functions of (p, 2n) and (α, xn) reactions have been used to study the spins of vibrational levels (Sakai *et al.*, 1965; Ejiri *et al.*, 1966, 1968; Jett and Lind, 1970) and the spins of rotational levels (Hansen *et al.*, 1963; Johnson *et al.*, 1972). Excitation functions for yrast transitions following the ^{162}Dy(α, 4nγ)^{162}Er and the ^{168}Er(α, 4nγ)^{168}Yb reactions are shown in Fig. 4.7. The intensities of the γ rays from high-spin states increase more rapidly than those from low-spin states as the incident alpha-particle energy increases. The excitation functions, together with the γ-ray intensity and the γ–γ coincidence yield, were used to assign the spins of levels in the backbending region (Johnson *et al.*, 1972).

EQ reactions near threshold energies have unique features. In the case of an A(p, n)B* reaction in the vicinity of the threshold energy of $E_B \approx E_c - B_n$ the transmission coefficient $T(l_n E_n)$ for neutrons feeding a residual state with

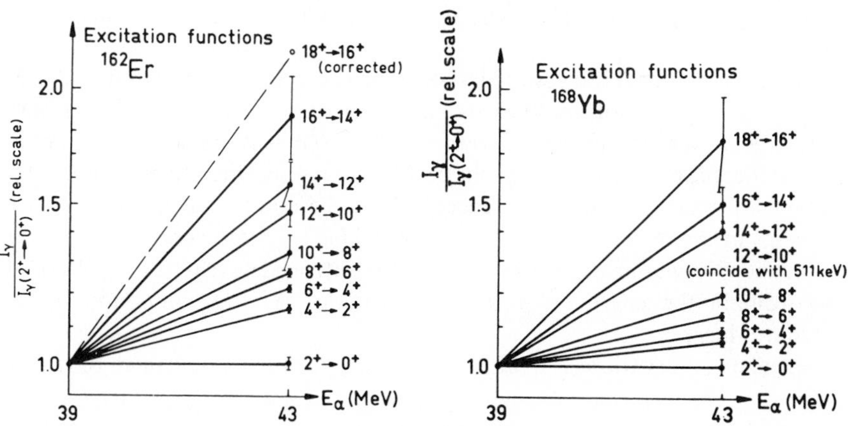

Fig. 4.7. Excitation functions for yrast transitions in ^{162}Er and ^{168}Yb populated by the ^{162}Dy(α, 4nγ)^{162}Er and ^{168}Er(α, 4nγ)^{168}Yb reactions. The relative yields normalized at $E_\alpha = 39$ MeV are plotted. (Johnson *et al.*, 1972.)

energy E_B and spin I_B depends strongly on l_n and E_n since the neutron energy given by $E_n = E_c - B_n - E_B$ is very small. Therefore the excitation functions for γ rays from such residual states are sensitive to E_B and I_B, and so they may be used to determine E_B and I_B. Since the excitation energies, spins and parities of the isobaric analogue resonances (IAR) are known from those of the parent nucleus, (p, n) reactions through the IAR's is very useful. The residual states populated by low energy neutrons from IAR's are studied by measuring the de-excition γ rays (Lieb and Hausmann, 1969; Mitarai and Minehara, 1983).

The A(α, x)B* reaction with the projectile energy near the Coulomb barrier is a very powerful tool for γ-ray spectroscopy of low-spin levels well above the yrast line (Brentano *et al.*, 1984; Zell *et al.*, 1976; Kajrys *et al.*, 1983). The cross-section $\sigma_f(I_B E_B)$ for feeding the residual state with energy E_B and spin I_B is obtained by comparing the yields of feeding and decaying γ rays of that state. The dependence of σ_f on the projectile energy is used to determine the spin I_B. Actually $\sigma_f(E_B\ I_B)$ for a given spin I_B decreases as E_B increases, and the slope $d\sigma_f(E_B\ I_B)/dE_\alpha$ increases as E_B increases. Feeding cross-sections for states with various spins are shown as a function of the bombarding energy near the threshold in Fig. 4.8.

4.2.2c *Reaction channels for heavy projectiles and angular momentum transfers*

The EQ process for heavy projectiles (heavy ions) shows unique features because of the large angular momenta introduced by the heavy projectiles. The projectile energy per nucleon for the EQ (complete fusion) process is in the range of $3 \lesssim E_a/M_a \lesssim 8$ (MeV). The lower limit corresponds to the Coulomb barrier and the upper one is the limiting energy above which the PEQ or incomplete fusion process becomes important as discussed in Section 4.2.4.

Angular momenta introduced by heavy projectiles are given by $l_a\hbar \approx$ (2 to 3) $M_a\hbar$, and may amount to 40 to 100 $\hbar$. The angular momenta of the residual nuclei, however, are limited by fission to $I_c \lesssim 70\,\hbar$ as discussed in Chapter 6. The projectile energy E_a for a given E_a/M_a increases as the projectile mass M_a increases, while the excitation energy E_c of the compound nucleus always remains in the region $E_c = 60$ to 100 MeV. This is because compound nuclei formed by heavier projectiles become more neutron-deficient (see Fig. 4.4) and accordingly the energy required for formation of the compound nuclei becomes larger.

Energy conservation (eqn 4.2.3) for the A(a, xn)B reaction can be rewritten as

$$E_c = x(\bar{B}_n + \bar{E}_n) + \bar{E}_\gamma(x), \qquad (4.2.16)$$

where $\bar{B}_n$ and $\bar{E}_n$ are the average binding energy and the kinetic energy for the evaporated neutrons, respectively, and $\bar{E}_\gamma(x)$ is the average excitation energy

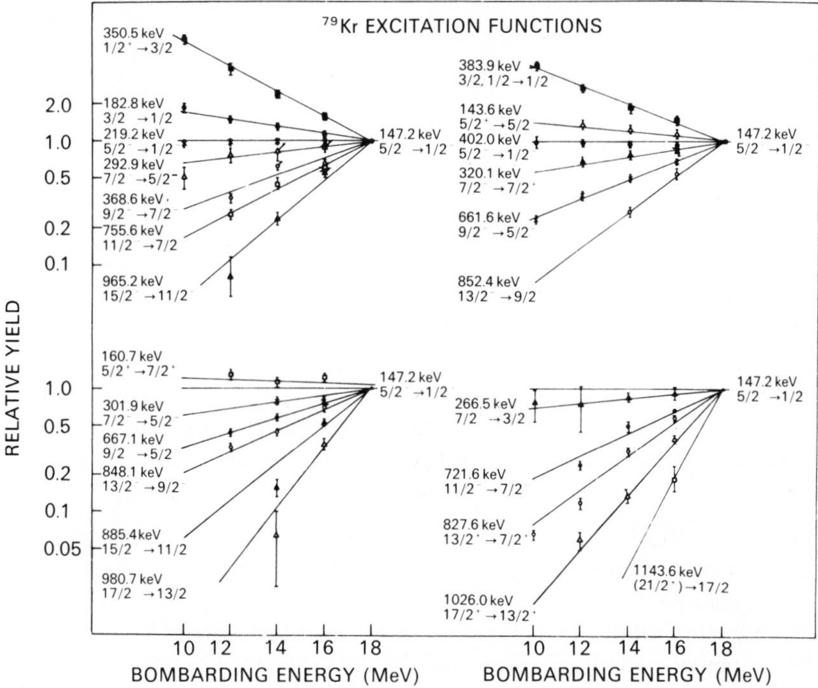

Fig. 4.8. Some examples of excitation functions for transitions in ^{79}Kr populated by the ^{76}Se(α, nγ)^{79}Kr reactions. All yields are normalized to the yield of the 147.2 keV transition. The solid lines are eye-guides. (Kajrys *et al.*, 1983.)

of the residual nucleus. The $\bar{E}_n$ is given by $\bar{E}_n \approx \frac{2}{3}\cdot 2T \approx \frac{4}{3}\sqrt{(E_c/a)}$, where T is the nuclear temperature of the compound state. The levels populated in the residual nucleus are those between the entry line and the yrast line. The exctitation energy of the yrast line is fairly large. The average excitation energy is given by $\bar{E}_\gamma(x) \approx (\bar{I}(x))^2 \cdot \hbar^2/2\mathscr{I}$, where $\mathscr{I}$ is the effective moment of inertia for the levels populated in the residual nucleus. Then the average angular momentum is obtained as

$$\bar{I}(x) \approx \left[\left\{E_c - x\left(\bar{B}_n + \frac{4}{3}\sqrt{\frac{E_c}{a}}\right)\right\}\cdot\frac{2\mathscr{I}}{\hbar^2}\right]^{1/2}. \tag{4.2.17}$$

The quantity $\bar{I}(x)$ increases as E_c increases, while it decreases as x increases, and accordingly $\bar{E}_\gamma(x)$ follows the similar trend as the $\bar{I}(x)$. These features are illustrated in Fig. 4.9. Then $\bar{I}(x)$ and $\bar{E}_\gamma(x)$ are assigned to the proper (a, xn) reaction channel (the proper neutron multiplicity x). Here the window of the excitation energy for the given x is as large as $\bar{B}_n + \frac{4}{3}T$, and accordingly the width of the angular momentum for given x is large. In fact they are not

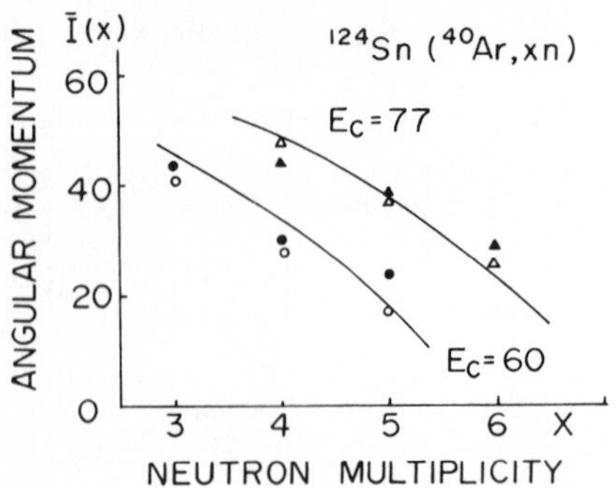

Fig. 4.9. Average angular momentum $\bar{I}(x)$ as a function of the neutron multiplicity x for the ^{124}Sn(^{40}Ar, xn) reaction. The solid lines are evaluations with $\bar{B}_n = 8.4$ MeV, $a = A/7$ MeV and $\hbar^2/2\mathscr{I} = 0.0125$ MeV (see text). The open and closed circles are statistical model calculations and experimental values given by Hillis *et al.* (1979).

sharply defined, and the angular momentum distribution for given x overlaps partly with those for the neighbouring channels with $x \pm 1$.

Statistical model calculations for heavy-ion induced EQ (fusion) reactions have been carried out by many authors (Grover and Gilat, 1967a; Pülhofer, 1977; Cormier *et al.*, 1977; Hillis *et al.*, 1979; Hageman 1981, and refs therein). Optical models are used to evaluate transmission coefficients of the projectile and those of decaying particles. Calculation codes such as ALICE (Blann and Plasil, 1973; Plasil and Blann, 1975) and GROGI2 developed by Gilat (1970) are used to evaluate the angular momentum and the excitation energy of the evaporation residue (see Fig. 6.8). An example of a statistical model calculations is shown in Fig. 4.10. The channel cross-section $\sigma(x)$ for the A(a, xn) reaction is divided among neighbouring channels with the larger x associated with lower $\bar{I}(x)$ and $\bar{E}_\gamma(x)$, as the kinetic energy E_n extends beyond the average value of $2T$—see eqn (4.2.17).

The angular momentum $\bar{I}(x)$ and the excitation energy $\bar{E}_\gamma(x)$ are evaluated from the γ-ray multiplicity M_γ and the summed energy of all cascade γ rays, respectively. The M_γ distribution is obtained from the ratio of the singles to the multifold coincidence yields of a discrete γ ray associated with a certain reaction channel. The average multiplicity $\bar{M}_\gamma$ and higher moments $\langle (M_\gamma - \bar{M}_\gamma)^n \rangle$ are used to evaluate the average angular momentum and the shape of the angular momentum distribution (Hagemann *et al.*, 1975). Details of this procedure are described in Section 3.2 and applications to high-spin states are given in Section 6.5.4.

In deformed nuclei with well-developed collective bands the γ-decay starts with statistical transitions, followed by stretched E2 transitions along collective bands. Thus $\bar{I}$ is related to $\bar{M}_\gamma$ by

$$\bar{I} \approx 2(\bar{M}_\gamma - k), \tag{4.2.18}$$

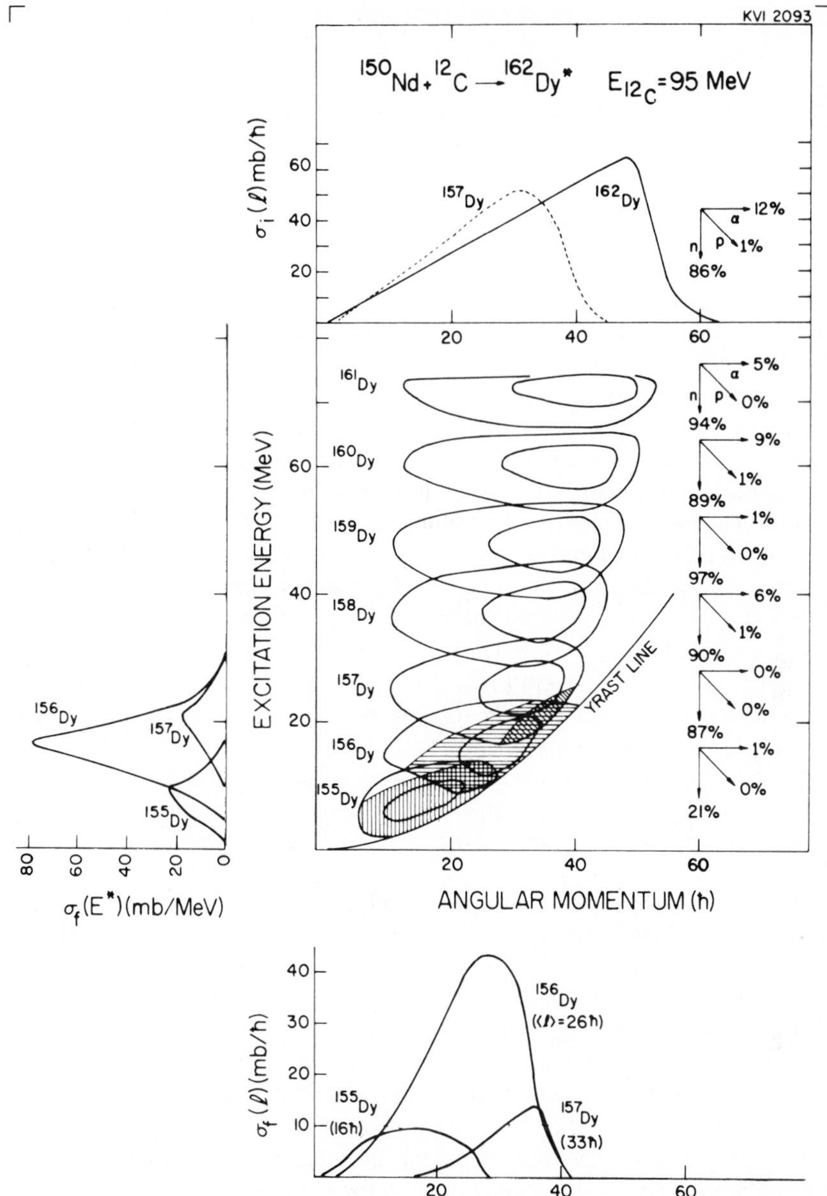

Fig. 4.10. Statistical model calculation with the GROGI code for the decay of the ^{162}Dy compound system with an excitation energy of 82 MeV formed by the $^{150}Nd + ^{12}C$ reaction. Top: Initial population $\sigma_i(l)$ with $l = I_c$ for the compound nucleus. At the right-hand side, the relative probabilities of n, p and α decays are shown. Middle: Population probability contours for the intermediate nuclei in the E–I^* plane. Shaded areas indicate the region where γ decays start. Bottom and left-hand side: Projections of these entry populations on the excitation energy and angular momentum axes, respectively. (Hageman, 1981.)

where $k \approx 4$ is used to correct for the statistical decays (Britt *et al.*, 1977; Hillis *et al.*, 1979). The relation eqn (4.2.18) is consistent with the total fusion cross-section σ_F (Herskind, 1976) given by

$$\sigma_F = \pi \lambda\!\!\!^{-2} l_{\max}^2 = \tfrac{9}{4}\pi \lambda\!\!\!^{-2} \langle \bar{I} \rangle^2$$
$$= \tfrac{9}{4}\pi \lambda\!\!\!^{-2} (\langle \bar{M}_\gamma \rangle - 4)^2, \qquad (4.2.19)$$

where $\langle \bar{I} \rangle$ and $\langle \bar{M}_\gamma \rangle$ are the channel-weighted average values given by $\langle \bar{I} \rangle = \Sigma \bar{I} \sigma(x)/\Sigma \sigma(x)$ and $\langle \bar{M}_\gamma \rangle = \Sigma \bar{M}_\gamma \sigma(x)/\Sigma \sigma(x)$, respectively. Here σ_F stands for the total cross-section of the (a, $xn\gamma$) reaction contributing.to the γ-ray multiplicity. Angular correlation data (Deleplanque *et al.*, 1978) for quasi-continuum (unresolved) γ rays following heavy-ion EQ reactions support the picture of γ decay via statistical dipole and collective quadrupole transitions.

The multiplicity distributions of γ rays following compound nucleus decay of Er nuclei have been studied by Hillis *et al.* (1979), and they are shown for individual reaction channels in Fig. 4.11(a). The M_γ values are plotted as a function of the neutron multiplicity x for various bombarding energies in Fig. 4.11(b). Noting the simple linear dependence of the $\bar{I}$ on $\bar{M}_\gamma$, the angular momentum (i.e. γ-ray multiplicity) is indeed divided among the reaction channels with larger $\bar{M}_\gamma$ (and $\bar{I}$) for fewer neutron multiplicity x, as shown in Fig. 4.11(a). $\bar{M}_\gamma$ and accordingly $\bar{I}$ for most of E_α, except for the data at the highest E_a, decrease monotonically with increasing x—see Fig. 4.11(b)—as expected from eqn (4.2.17). The data at $E_a = 236$ MeV include some PEQ process as discussed in Section 4.2.4. The $\bar{I}$-values as derived from the $\bar{M}_\gamma$ are in accord with results of the statistical model calculations as indicated in Fig. 4.9.

Most of the heavy-ion induced reactions discussed so far are mass-asymmetric systems with $M_a \ll M_A$, where M_a and M_A are the projectile and target masses. Angular momentum transfers and neutron multiplicities (reaction channels) for mass-symmetric systems with $M_a \approx M_A$ have been studied by measuring γ rays (Haas *et al.*, 1985). In the energy regime near the Coulomb barrier the angular momentum transfer is much larger than the value expected from the sharp cut-off model, and the reaction channel with the small neutron multiplicity is enhanced over the statistical model estimate. These features are explained by taking into account the potential-barrier fluctuation related to the ground-state shape vibration.

The population of the reaction channels (evaporation residues) is sensitive to the γ-decay width in competition with the particle-emission width at the entry region. The γ-decay probability for the multipolarity J is written as

$$\Gamma_\gamma(EI, E_f I_f) = k(EIJ)(E - E_f)^{2J+1} \rho_f/\rho, \qquad (4.2.20)$$

where E and I are the energy and the spin of the initial state in the entry region, respectively, and ρ and ρ_f are the level densities of the initial and final

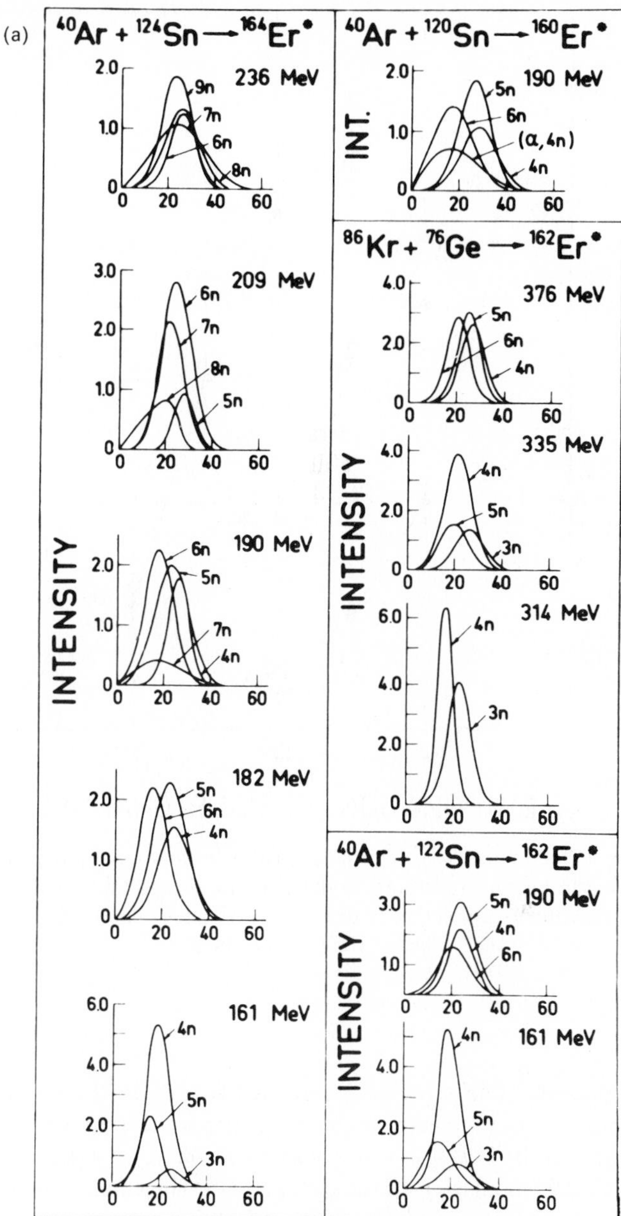

Fig. 4.11(a). Multiplicity distributions obtained from a fit of the measured folded distributions to a skewed-Gaussian multiplicity distribution. (Hillis *et al.*, 1979.)

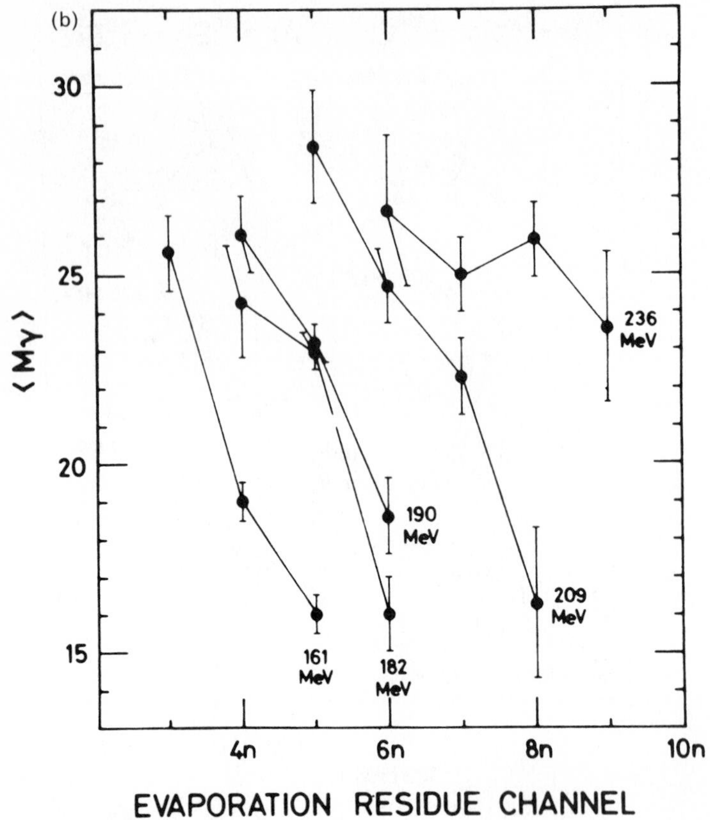

Fig. 4.11(b). Plot of the $\bar{M}_{\gamma}$, determined from the analysis of the discrete lines corresponding to known transitions in the evaporation residues for the $^{124}Sn(^{40}Ar, xn)^{164-x}Er$ reaction as a function of evaporation-residue channel and incident ^{40}Ar energy. (Hillis *et al.*, 1979.)

states respectively. The strength parameter $k(EIJ)$ in the entry region is interesting in view of the electromagnetic properties of states with large E and I. The parameter for low-spin states is known from (n, γ) data, and is used in statistical model calculations (Hillis *et al.*, 1979; Hageman, 1981).

Heavy-ion EQ reactions have been used for γ-ray spectroscopy in the high-spin region (see Chapter 6). In particular, studies of γ rays as functions of E_{γ}, M_{γ} and the bombarding energy E_{a} are useful because of the simple dependence of the heavy-ion EQ process on these quantities.

4.2.2d *Gamma de-excitation processes and population of low-lying levels*
Statistical particle decays of EQ (compound) nuclei populate in principle all kinds of levels in the entry region of the I–E plane, where I and E are the angular momentum and the excitation energy, respectively.

Since γ decay follows simple selection rules, the population of the lower levels reflects the γ de-excitation process and the level structure. In the highly excited entry region with high-level density statistical (dipole) transitions dominate the γ-decay process unless there are competing non-statistical (collective) transitions. Giant resonances with enhanced (collective) γ widths are above the particle threshold and thus the branching ratio is very small as discussed in Section 4.4. The branching ratio of collective E2 transitions along collective (rotational) bands to statistical EJ transitions is given by

$$T_c(\mathrm{E2})/T_s(\mathrm{E}J)=h_J E_{co}^5/E_{st}^{2J+1}, \tag{4.2.21}$$

where E_{co} and E_{st} are collective E2 and statistical EJ transition energies, in units of MeV, respectively. The factor h_J is proportional to the ratio of the enhancement of the collective transition to the reduction of the non-collective transition expressed in Weisskopf units. The E_{st} depends on the nuclear temperature T, and thus on the excitation energy above the yrast line, while E_{co} is given by $E_{co}\approx 2I\hbar^2/\mathscr{I}$. Consequently the highly excited residual nucleus de-excites first by emitting a number of statistical γ rays to the region near the yrast line, and then by emitting collective E2 γ rays. The decay flow following (a, xnγ) reactions is schematically illustrated in Fig. 4.12.

Gamma decays in axially symmetrical deformed nuclei follow the K-selection rule, where K is the projection of the angular momentum on the symmetry axis. A K-band de-excitation model (KBD), in which the K-selection rule is explicitly taken into account, has been proposed (Ejiri and Halpern, 1968/1970; Ferguson *et al.*, 1972; Ejiri, 1981). The density of levels with given K is proportional to $\exp(-K^2/2k_0^2)$, while the Coriolis interaction energy, which causes mixing of the K quantum number, is given by $\langle H_c\rangle\approx b\sqrt{(I^2-K^2)}$. The mixing interaction, especially for high K states, is still small in the region of $I\leqslant 20$. Then γ decays from such high K-levels follow the K-selection rule. In the medium-spin region most of the high-K levels de-excite within the bands till the γ flow reaches the band head with $I=K\approx 5\sim 12$, as shown in Fig. 4.12. Such band heads are likely to be isomeric states because the γ-decay to the yrast level ($K=0$) is forbidden. Some high K states may decay to levels in the γ band with $K=2$ and to other non-yrast levels with $K=0$, which may eventually decay to the yrast band.

The population of the yrast levels reflects the γ-de-excitation process above the yrast line. The intensities $Y(I\rightarrow I-2)$ of yrast (ground-state rotational) transitions following (α, 4n) reactions on deformed nuclei are shown as a function of spin I in Fig. 4.13. The difference between $Y(I\rightarrow I-2)$ and

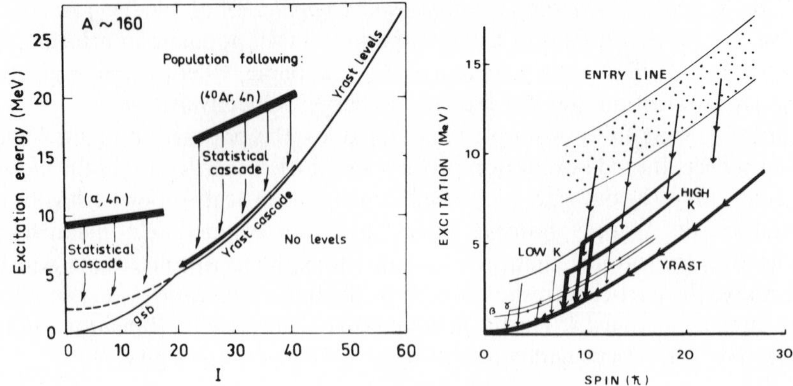

Fig. 4.12. Left-hand side: Estimated spin-ranges and γ decays following different (a, xn) reactions. (Newton *et al.*, 1970.) Right-hand side: Schematic picture of the K-band de-excitation model. The dotted region is the entry region (line) populated by the (particle, xnyp) reaction. 'High K' and 'Low K' stand for many bands based on two quasi-particle states with large K quantum numbers and those with small K ($K \lesssim 5$) quantum numbers, respectively. Low lying β and γ bands are also indicated. (Ejiri, 1981.)

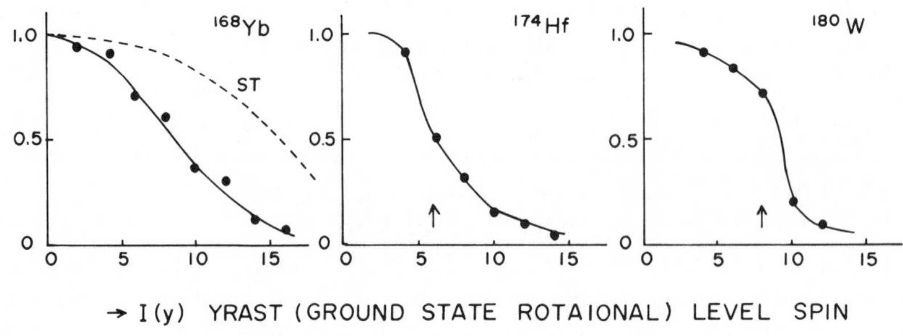

Fig. 4.13. The intensities of yrast (ground-state rotational) transitions following (α, 4n) reactions at $E_\alpha = 40$ MeV. The data points are taken from Ferguson *et al.* (1972). The dashed line is calculated for statistical γ decays to yrast levels, and the solid lines are eye-guides. The arrows indicate the feeding from K-isomers, the 6^+ isomer with $K=6$ in ^{174}Hf and the 8^- isomer with $K=8$ in ^{180}W.

$Y(I+2 \to I)$ gives the yrast population (side-feeding intensity). The intensities of yrast transitions are strongly shifted towards the low-spin region compared with those expected from a simple statistical population of the yrast levels (Shepherd *et al.*, 1966; Williamson *et al.*, 1968). The angular momentum dissipation is mainly due to collective (non-statistical) transitions along high-K excited bands. Isomers with high K accumulate a large fraction of the

γ-decay flow from high K levels, and greatly affect the population of the yrast levels (see Figs 4.12 and 4.13). These features are well reproduced by the KBD-model (Ferguson *et al.*, 1972; Ejiri *et al.*, 1972b; Dracoulis *et al.*, 1977; Kishimoto *et al.*, 1979).

It is interesting to see how the K–isomer ratio depends on the reaction. The K–isomer ratio is defined as $P_K(K)/P_0(I)$, where $P_K(I)$ is the population of the K–isomer with spin I and $P_0(I)$ is that of the yrast ($K=0$) level with the same spin I. The K–isomer ratio for the 8^- state in ^{180}W has been studied for various projectiles and energies (Ejiri *et al.*, 1978a). The ratio first increases as the angular momentum introduced by the projectile increases and then saturates as shown in Fig. 4.14. These features support the KBD model. The ratio for a given input angular momentum is smaller for heavier projectiles, suggesting that heavier projectiles tend to excite levels with low K. A similar effect has been observed for the K–isomer in ^{182}Os populated by (α, xn) and (p, xn) reactions (Ishihara *et al.*, 1972).

4.2.3 *Alignment of angular momentum and linear momentum in statistical reactions*

4.2.3a *Spin alignment in statistical reactions*

In the EQ reaction the projectile a introduces into the compound nucleus C a large linear momentum along the beam axis and a large angular momentum aligned in the plane perpendicular to the beam axis. The linear and angular momenta of the compound nucleus are well preserved through particle and γ decays (Ejiri *et al.*, 1965/1966; Diamond *et al.*, 1966; Halpern *et al.*, 1968; Williamson *et al.*, 1965, 1968; Newton *et al.*, 1967). The angular momenta

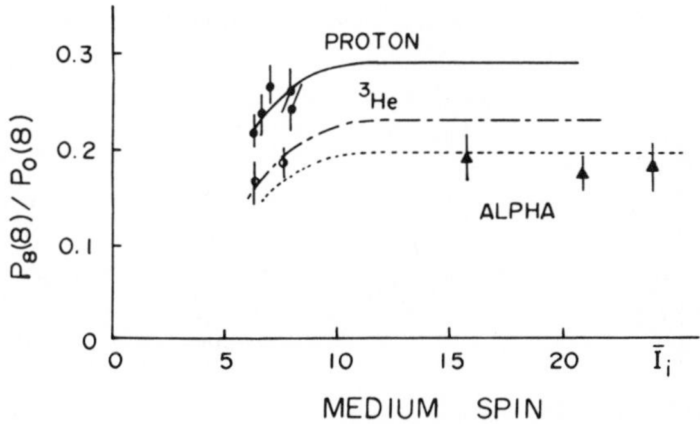

Fig. 4.14. The isomer ratios for the 1832.5 keV 8^- isomer with $K=8$ in ^{180}W populated by (p, xn), (^{3}He, xn) and (α, xn) reactions as a function of the medium spin $\bar{I}_i$ introduced by the projectiles. The solid line, the dot-dash line and the dotted line are calculated in terms of the KBD model with $K_0=4.4$, 3.9 and 3.8, where K_0 is the parameter in the K-level density given by $\exp(-K^2/2K_0^2)$. (Ejiri *et al.*, 1978a.)

associated with the EQ reaction and the γ decays are schematically illustrated in Fig. 4.15.

The degree of spin alignment of the residual state with spin I is obtained from the angular distribution of a decay γ ray,

$$W(\theta) = \Sigma \alpha_\nu A_\nu^0 P_\nu(\cos\theta), \tag{4.2.22}$$

where A_ν^0 is the coefficient of the Legendre polynomial P_ν for the γ ray from the completely aligned state and the α_ν is the attenuation coefficient due to the spin dealignment. They are given for the γ transition $I \to I'$ as follows (Blin-Stoyle and Grace, 1957; Ferguson, 1965; Biedenharn and Rose, 1953; Biedenharn, 1960).

$$A_\nu^0(I\,I'; J) = (2J+1)(2I+1)^{1/2}(-)^{I-I'+1}(J1J-1|\nu 0)W(JJII;\nu I'), \tag{4.2.23}$$

$$\alpha_\nu(I) = \sum_m (-)^{-m}(ImI-m|\nu 0)P_m(I)/(I0I0|\nu 0), \tag{4.2.24}$$

where J is the multipolarity (angular momentum) of the γ transition, $P_m(I)$ is the population of the magnetic substate m (the projection of I on the beam axis), and W is a Racah coefficient. The compound nucleus produced by a spinless projectile on a spinless target has a complete plane alignment of $P_m(I) = \delta_{m0}$. The attenuation factor reflects all the spin history of the particle and γ-ray de-excitation processes populating the state with spin I. An attenuation factor due to a transition $I_1 \to I_2$ is written in general as

$$\alpha_\nu(I_1, I_2; j) = \{(2I_1+1)(2I_2+1)\}^{1/2}(-)^{I_1+I_2-j-\nu}W(I_1I_1I_2I_2;\nu j), \tag{4.2.25}$$

where j is the angular momentum of the emitted particle or γ ray. In eqns

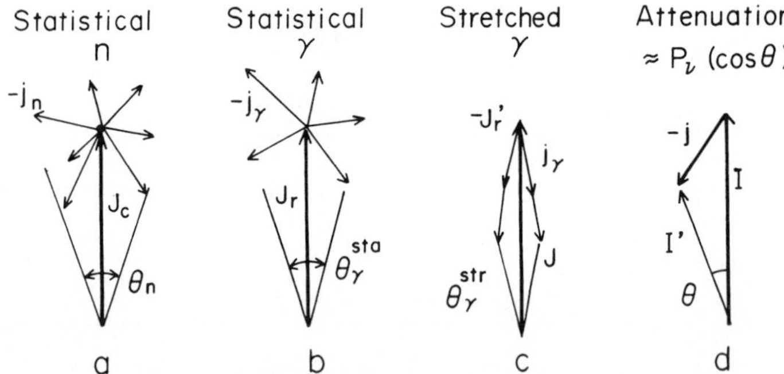

Fig. 4.15. The angular momenta associated with the (a, $xn\gamma$) reaction. (a) Statistical neutron evaporation. (b) Statistical γ decays, (c) Stretched γ decays along collective bands. (d) Angular momenta associated with attenuation factors for angular distribution coefficients by a transition of $I \to I'$, j being the angular momentum of the transition.

(4.2.22–4.2.25) ν is restricted to $\nu \leqslant \min(2J, 2I)$, and to even numbers in the γ decays from bound (definite parity) states (see Section 2.1).

The attenuation factor is indeed close to one in the high-spin region, as shown in Fig. 4.16. Thus the spin-dealignment is small in such cases. The attenuation factors are reduced to simple forms in the case of I_1, $I_2 \gg j$ (Rasmussen and Sugihara, 1966; Ejiri *et al.*, 1972a). Using the limiting expression for the Racah coefficient in eqn (4.2.25) (Rose, 1957) the attenuation factor is reduced to a simple form,

$$\alpha_\nu(I_1, I_2; j) \approx P_\nu(\cos\theta) \tag{4.2.26}$$

where θ is the angle between the angular momentum vectors $\mathbf{I}_1$ and $\mathbf{I}_2$ with triangle relation $\mathbf{I}_1 = \mathbf{I}_2 + \mathbf{j}$—see Fig. 4.15(d). The θ decreases with the increasing I_1 and with decreasing j, and accordingly the α_ν approaches unity. Stretched transitions with $I_1 \to I_2 = I_1 \pm j$ have small angles θ, resulting in little disturbance of the alignment—see Fig. 4.15(c). Using $\cos\theta = \{I_1(I_1+1) + I_2(I_2+1) - j(j+1)\}/2\{I_1(I_1+1)I_2(I_2+1)\}^{1/2}$, one gets for I_1, $I_2 \gg j$

$$\alpha_2(I_1, I_2 = I_1 \pm j; j) \simeq 1 - \frac{3}{2} \cdot \frac{j}{\left(I_1 \pm \frac{j}{2}\right)\left(I_1 \pm \frac{j}{2} + 1\right)}, \tag{4.2.27}$$

$$\alpha_2(I_1, I_1; j) \approx 1 - \frac{3}{2} \cdot \frac{j^2 + j}{I_1(I_1+1)}. \tag{4.2.28}$$

Let us consider for simplicity a schematic decay chain $I_i \to \ldots (N_n) \ldots (N_\gamma) \ldots (N'_\gamma) \ldots \to I$, where N_n, N_γ and N'_γ are the numbers of statistical neutrons, statistical γ rays and collective (stretched) γ rays, respectively. Then the attenuation coefficient is written as

$$\alpha_\nu(I_i \to I) = \prod_i^{N_n} \alpha_\nu(\text{sta. } i) \prod_j^{N_\gamma} \alpha_\nu(\text{sta. } k) \prod_{k'}^{N'_\gamma} \alpha_\nu(\text{str. } k'), \tag{4.2.29}$$

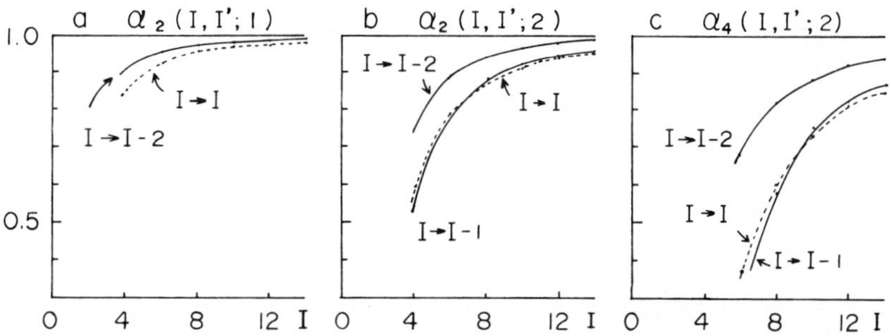

Fig. 4.16. The attenuation factors for angular distribution coefficients of Legendre polynomials $P_\nu(\cos\theta)$. The $\alpha_\nu(I, I'; j)$ is the factor due to the transition $I \to I'$ by the particle (γ-ray) with the angular momentum (multipolarity) j.

where $\alpha_v(\text{sta. } i)$, $\alpha_v(\text{sta. } k)$ and $\alpha_v(\text{str. } k')$ are attenuation coefficients for the i-th statistical neutron, the k-th statistical γ ray and the k'-th stretched γ ray, respectively. Since the statistical process does not greatly change the spin values, the spins of intermediate states remain as large as I_i, while the angular momenta for the evaporating neutrons and statistical γ rays are much smaller than I_i, i.e. $j \ll I_1$ in eqns (4.2.27) and (4.2.28). Thus the attenuation coefficient due to the statistical decay is rather close to one. The stretched transition removes angular momenta in a stretched way, resulting in little disturbance of the spin alignment—see eqn (4.2.27)—and consequently, the spin alignment is fairly well preserved throughout the neutron and γ de-excitation processes. The attenuation coefficient for the level with spin I populated by the EQ A(a, xnγ)B reaction is given by

$$\alpha_v(I) = \sum_{I_c} \rho(I_c)\alpha_v(I_c \rightarrow I) / \sum \rho(I_c), \tag{4.2.30}$$

where $\rho(I_c)$ is the spin distribution of the compound nucleus. The $\alpha_v(I_y)$ for yrast levels populated by the $(\alpha, \text{xn}\gamma)$ reaction are calculated by using the schematic decay model as described by eqn (4.2.29). They reproduce well the experimental values obtained from angular distributions of the yrast γ transitions, as shown in Fig. 4.17. The $\alpha_2(I_y)$ increases as I_y increases since the spin dealignment in the high spin region is small as seen from eqns (4.2.27) and (4.2.28).

So far we discussed the spin dealignment due to decaying particles and photons. The target spins and projectile spins are oriented randomly unless polarized targets and polarized projectiles are used. The spin dealignment due

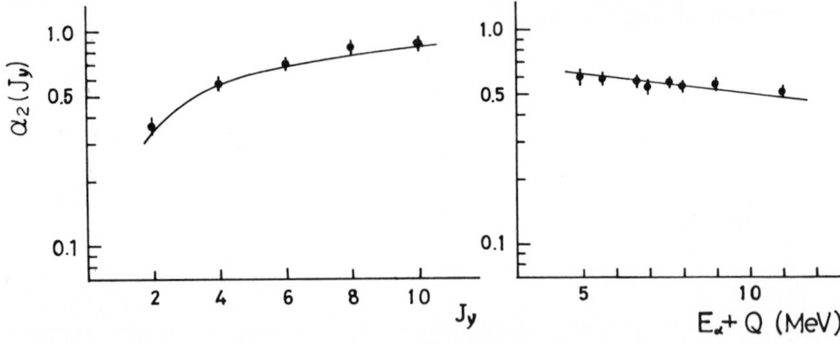

Fig. 4.17. Left-hand side: The attenuation factors $\alpha_2(I_i)$ as a function of the spin I_i for the yrast (ground-band) levels of ^{152}Gd populated by the $(\alpha, 4\text{n})$ reaction at $E_\alpha = 43$ MeV. The solid line represents the calculated value using the KBD model. Right-hand side: The attenuation factors as a function of the excitation energy for the 4^+ yrast (ground band) levels of Nd and Sm isotopes populated by $(\alpha, 2\text{n}\gamma)$ reactions at $E_\alpha = 22$ MeV. The solid line represents the calculated value. (Ejiri *et al.*, 1972a.)

to these random spins introduces an additional attenuation factor. In the case of a proton ($j_p = 1/2$) induced reaction on a target with spin I_t, the attenuation factor for the compound state with spin I_c is given by

$$\alpha_\nu(I_t, I_c; j_p) = (2j_p+1)(2I_c+1)^{1/2}(-)^{I_c - I_t + \frac{1}{2}}\, C(j_p j_p \nu; \tfrac{1}{2}\tfrac{-1}{2}) \times W(j_p j_p I_c I_c; \nu I_t). \tag{4.2.31}$$

The Racah coefficient corresponds to the coefficient for the decay of a particle with spin I_t from the state with spin I_c. Thus $\alpha_\nu(I_t, I_c; j_p)$ is approximately given by $P_\nu(\cos\theta)$, θ being the angle between $\mathbf{j}_p$ and $\mathbf{I}_c$ in the triangle relation $\mathbf{j}_p + \mathbf{I}_t = \mathbf{I}_c$.

The attenuation coefficients for a state with a Gaussian distribution of the substate population are given by (Rassmussen and Sugihara, 1966)

$$\alpha_2(I) = 1 - \frac{3\sigma^2}{I(I+1)}, \tag{4.2.32a}$$

$$\alpha_4(I) = 1 - \frac{10\sigma^2}{I(I+1)} + \frac{35}{3} \times \frac{\sigma^4}{I^2(I+1)^2}, \tag{4.2.32b}$$

where σ is the width parameter for the Gaussian distribution given by $P_J(m) = N\exp(-m^2/2\sigma^2)$.

4.2.3b *Spin alignment and gamma-ray angular distributions*

The angular distribution of a γ transition from a state with spin I populated by a statistical (EQ) reaction is obtained by summing the angular distributions for all possible paths which contribute to the feeding of that state. The distribution for the transition $I \to I'$ with the multipolarity J is expressed as

$$W(\theta) = \sum_\nu \sum_{I_c} P(I_c)\alpha_\nu(I_c) \sum_{n I_r} \rho(\mathrm{n})\alpha(I_c \to I_r; n) \times \sum_\gamma \rho(\gamma)\alpha(I_r \to I; \gamma) A_\nu^0(I, I'; J) P_\nu(\cos\theta), \tag{4.2.33}$$

where $A_\nu^0(I, I'; J)$ is the coefficient for the completely aligned state as given in eqn (4.2.23). The $\alpha(I_c)$, $\alpha(I_c \to I_r; \mathrm{n})$ and $\alpha(I_r \to I; \gamma)$ are the attenuation coefficients for the formation of the compound nucleus I_c, for the particle decays from the compound nucleus I_c to the residual nucleus I_r, and for the γ decays from the residual nucleus I_r to the initial state I for the present γ transition of $I \to I' + J$, respectively. The summations in eqn (4.2.33) are extended over all possible particle decays (n), the residual states I_r and the γ decays (γ), while $\rho(\mathrm{n})$ and $\rho(\gamma)$ stand for the weighting factors for the neutron and γ decays, respectively.

The spin alignment and the γ-ray angular distributions in statistical (EQ) reactions have several features which are useful in nuclear spectroscopy (Ejiri *et al.*, 1965/1966; 1972a):

1. The spin alignment is fairly good. The degree of spin alignment, being expressed by the attenuation coefficient for the γ-ray angular distribution, is of the order of $\alpha_2(J) \approx 1 - 3N_S/\bar{I}_c^2$, where N_S is the number of statistical decays and $\bar{I}_c$ is the average spin of the compound nuclei.

2. The spin alignment is not sensitive to how the state is fed. It changes slowly as a function of the spin I of the state and the number of statistical γ rays feeding the state. The alignment is either obtained from the γ-ray angular distribution of the known spin sequence or evaluated from a statistical model calculation.

3. The angular distributions of stretched transitions along a collective band, $I_1, \rightarrow I_1 - j, \ldots, I_n, I_n - j$, are independent of the spin I. This is due to the relation

$$\prod_{i=1}^{n-1} \alpha_\nu(I_i, I_i - j; j) A_\nu(I_n, I_n - j; j) = A_\nu(I_1, I_1 - j; j). \tag{4.2.34}$$

Thus a change of the angular distribution indicates the contribution of non-stretched side-feeding processes.

4. The spin alignment of isomeric states may be used for magnetic moment measurements provided that the alignment is kept finite after slowing down in the target (catcher foil), as discussed in Section 3.5. In particular, spin-aligned states produced by pulsed beams or naturally bunched beams are well suited for such measurements (Yamazaki and Ewan, 1967; Yamazaki and Matthias, 1968).

4.2.3c *Linear momentum in statistical reactions*

The linear momentum **P** is the counter parts of the angular momentum **I**. Corresponding to the spin alignment with $P_J(m) \sim \delta_{m0}$, the linear momentum is aligned along the beam axis (z) as $P_P(P_z) \approx \delta_{P_z P}$, where m and P_z are projections of **I** and **P**, respectively, and $P = |\mathbf{P}|$. The orientation of the linear momentum **P** introduced by a projectile is disturbed slightly by evaporated particles since they carry away small momentum **P**(i) at random direction. Gamma decays disturb little the linear momentum **P** because γ-ray linear momenta are much smaller than P. Thus evaluation of the final linear momentum is straightforward, as it is independent of the γ-de-excitation process. In particular, the P_z of the residual state remains almost the same as the projectile momentum P because the sum of the z-components of **P**(i) for all evaporated particles (i = 1, 2, . . . , x) vanishes.

Large momenta or velocities of excited nuclei produced by EQ reactions are used for in-beam (on-line) study of the Doppler shift of γ rays as discussed in Section 3.5. Radioactive nuclei produced by (a, xy . . . γ) reactions with large P_a may be ejected from thin targets and be analyzed by on-line mass separators for β–γ spectroscopic studies.

4.2.4 *Gamma-ray spectroscopy and pre-equilibrium reactions*

4.2.4a *The PEQ reaction channel and the particle multiplicity*

In nuclear reactions induced by medium-energy projectiles some fast particles are emitted before the thermal equilibration, i.e. at the pre-equilibrium (PEQ) phase. Thereafter slow nucleons are evaporated in the equilibrium (EQ) phase. Then eqn (4.2.3a) is rewritten as

$$E_c = \sum_i^{x_p} \{E_B(i) + E_p(i)\} + \sum_j^{x_e} \{E_B(j) + E_p(j)\} + E_\gamma. \qquad (4.2.35)$$

where x_p and x_e are the numbers of the emitted particles in the PEQ and EQ phases, respectively.

In the case of the A(a, xn)B reaction eqn (4.2.35) is rewritten as

$$E_c = x_p \bar{E}_n(p) + 2x_e k\bar{T}_e + E_\gamma + x\bar{B}_n, \qquad (4.2.36)$$

where $\bar{B}_n$ is the average binding energy of the emitted neutrons, x is the total neutron multiplicity given by $x = x_p + x_e$, $2k\bar{T}_e$ is the average kinetic energy of the evaporated neutrons, and $\bar{E}_n(p)$ is the average kinetic energy of the fast particles emitted in the PEQ phase. $\bar{T}_e$ is the effective nuclear temperature at the EQ phase. Note that kT in this section (4.2) corresponds to T used conventionally in other sections. The total and average kinetic energies of the emitted particles are given by

$$E_p(x) = x_p \bar{E}_n(p) + 2x_e k\bar{T}_e = E_c - E_\gamma - x\bar{B}_n, \qquad (4.2.37a)$$

$$\bar{E}_n(x) = \frac{E_c - E_\gamma}{x} - \bar{B}_n. \qquad (4.2.37b)$$

The energy excess beyond the threshold energy is given by

$$\Delta E(x) = E_p(x) + E_\gamma = E_c - x\bar{B}_n = E_a^{cm} + Q(x), \qquad (4.2.38)$$

where $Q(x)$ is the Q-value for the (a, xn) reaction channel, and E_a^{cm} is the projectile energy in the c.m. system. The quantities $\Delta E(x)$ and $\bar{E}_n(x)$ are plotted as a function of the neutron multiplicity x in Fig. 4.18.

The PEQ-dominant process has a large $\bar{E}_n(x)$ and $E_p(x)$, and accordingly a small x, while the EQ-dominant process has a small $\bar{E}_n(x)$ and $E_p(x)$, and accordingly a large x. Actually the $E_p(i)$ for the PEQ particle extends far

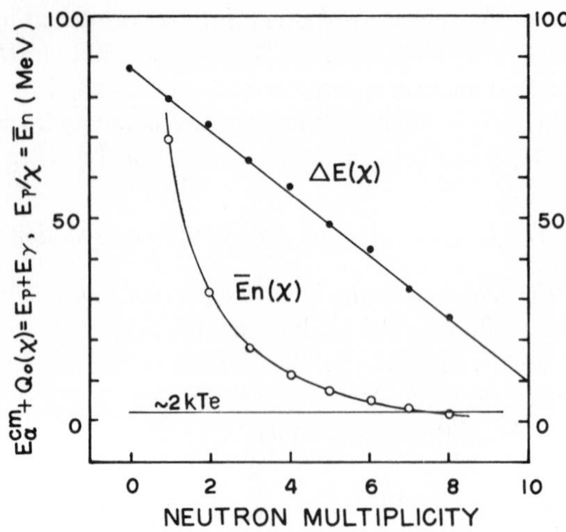

Fig. 4.18. The average neutron energies $\bar{E}_n(x)$ and energy excesses $\Delta E(x)$ above the reaction threshold energies for the ^{164}Dy(α, $xn\gamma$) reactions at $E_\alpha^{cm} = 88$ MeV. $\Delta E(x) = E_\alpha^{cm} + Q_0(x) = E_p + E_\gamma$ and $\bar{E}_n(x) = E_p/x$, where E_p is the total kinetic energy (sum of energies) of emitted particles, and E_γ is that of emitted γ rays (see text). The value $2kT_e$ gives the mean energy of the evaporation neutrons. (Ejiri *et al.*, 1980.)

beyond $2k\bar{T}_e$. Consequently x in the PEQ process for given E_c is not definite but is distributed over a wide range, in contrast to the EQ process.

The gamma-ray spectra following PEQ and EQ reactions contain very many lines because there are many reaction channels, spread over a wide range of particle multiplicity, as demonstrated by the γ-ray spectrum shown in Fig. 4.19. Here γ rays in five reaction channels of (p, xn) with $x = 2$, 3, 4, 5, and 6, and those in four reaction channels of (p, pxn) with $x = 2$, 3, 4, and 5 are clearly seen. The multiplicity distribution $\sigma(x)$ for the A(a, xn)B reaction, which is obtained from the γ-ray yields, is converted to the kinetic energy distribution $\sigma(E_p)$ by using eqn (4.2.37a). Figure 4.20 shows the $\sigma(E_p)$ distribution for the Dy(α, xn) reaction at $E_\alpha = 90$ MeV. The observed distributions are much broader than expected for the EQ process, and are shifted towards large E_p(small x). These features indicate large contributions of the PEQ process.

4.2.4b *The properties of PEQ decay particles*

The energy spectra of particles from PEQ–EQ reactions are composed of the fast PEQ component and the slow EQ (evaporation) component, as shown in Fig. 4.21. These two components are effectively given by $S_p(E) = E\exp(-E/kT_p)$ and $S_e(E) = E\exp(-E/kT_e)$, respectively. By analogy with the nuclear temperature T_e for the EQ phase, one may assign the effective

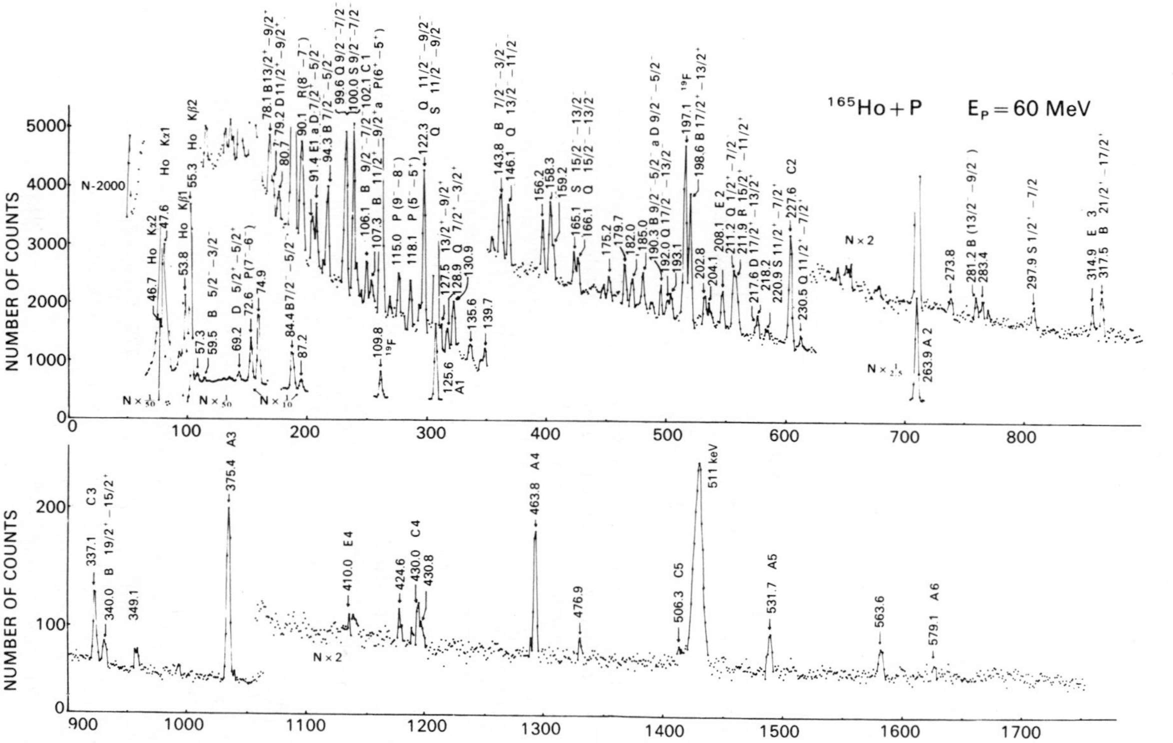

Fig. 4.19. The energy spectrum of γ-rays observed at $\theta_{\text{lab}}=40°$. Gamma-ray energies are given in keV. The Ai, Bi, Ci, Di and Ei stand for the rotational γ-rays following the (p, 6n), (p, 5n), (p, 4n), (p, 3n) and (p, 2n) reactions, respectively. The numbers $i=1, 2, 3, \ldots$ indicate, respectively, the transitions $2^+ \to 0^+$, $4^+ \to 2^+$, $6^+ \to 4^+$, and so on. P, Q, R and S stand for the γ rays following (p, p5n), (p, p4n), (p, p3n) and (p, p2n) reactions, respectively. (Ejiri *et al.*, 1978b.)

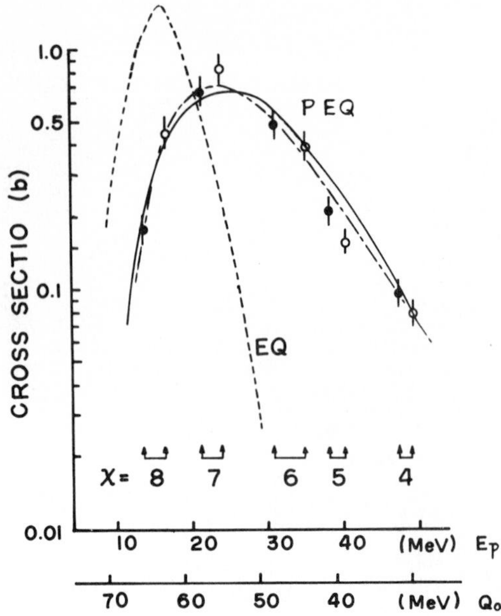

Fig. 4.20. The neutron multiplicity (x) distributions as functions of the energy sum $E_T = \sum E_n$ for emitted neutrons and of the ground-state Q value (Q_0). The closed and open circles are the cross-sections for the ^{162}Dy and ^{164}Dy targets, respectively. The solid and the dot-dash lines are the PEQ–EQ calculations in terms of the two-phase (PEQ and EQ) model and the exciton model, respectively. The dotted line is the EQ calculation. (Ejiri *et al.*, 1980; Maeda *et al.*, 1983.)

quasi-temperature T_p to the PEQ phase. Actually energy spectra of particles as well as multiplicity distributions (the E_p distribution) following medium-energy light-ion reactions are fitted well by the two-phase model, as illustrated in Figs 4.20 and 4.21 (Ejiri *et al.*, 1978b, 1980; Ejiri, 1981; de Voigt *et al.*, 1982; Sakai *et al.*, 1979, 1980; Maeda *et al.*, 1983). Exclusive neutron spectra for (α, 2n) reactions at $E_\alpha \approx 30$ MeV show some PEQ contributions (Fields *et al.*, 1981).

The PEQ process has been extensively studied in terms of exciton models (Griffin, 1966; Cline and Blann, 1971; Blann, 1975; Agassi *et al.*, 1975; Gadioli *et al.*, 1976; Yoshida, 1977; Machner, 1981). The exciton model calculations agree well with the observations as shown in Figs 4.20 and 4.21. The PEQ features show up conspicuously by selecting the low multiplicity (small x) channel by coincidence measurements with γ-rays characterizing the small x channel.

Excitation functions for (a, xyz . . . γ) reactions have been studied by many authors (Kurz *et al.*, 1971; Demeyer *et al.*, 1970; Jahn *et al.*, 1973; Djaloeis *et*

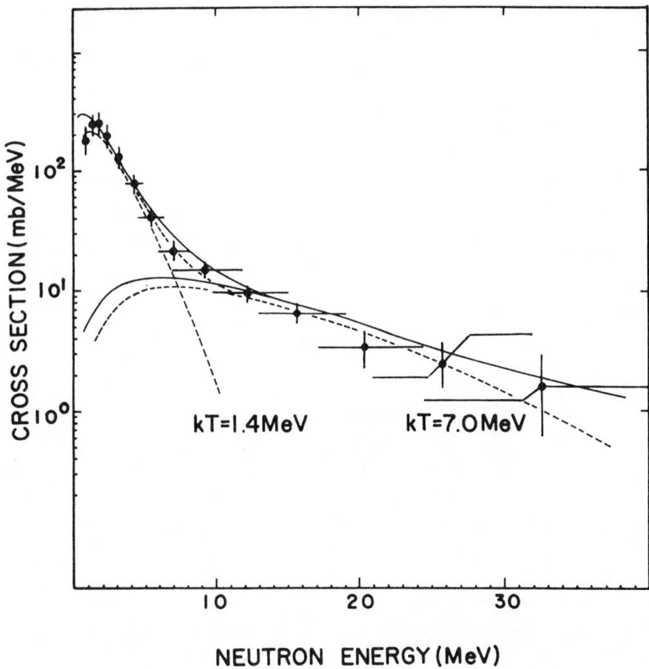

Fig. 4.21. The energy spectrum of neutrons following the ^{158}Gd (α, $xn\gamma$) reaction at $E_\alpha = 70$ MeV. The dotted lines are the results of the two-phase (PEQ–EQ) calculations. The solid lines are the results of the exciton model calculations. (Ejiri, 1981.)

al., 1975). Figure 4.22 shows the excitation functions for the ^{197}Au(d,$xn\gamma$) and ^{197}Au(d, p$xn\gamma$) reactions. The excitation curve for a given x has a peak at E_a where the compound (EQ) process dominates the reaction channel, and a long tail at the higher bombarding energy because of the PEQ contribution. These features are in accord with exciton model calculations.

Fast charged particles as well as fast neutrons are emitted in the PEQ phase, and consequently the evaporation of slow charged particles in the EQ phase of medium heavy nuclei is suppressed. Protons from (α,pxn) reactions, being measured in coincidence with the γ rays associated with the (α,pxn) channel, show PEQ features depending on the neutron multiplicity x (Ejiri, 1982; Lieder *et al.*, 1981). The proton spectrum from the ^{165}Ho(α,p4n) reaction has a high energy peak and a lower energy tail as shown in Fig. 4.23 (Shibata *et al.*, 1985). The peak corresponds to one fast PEQ proton followed by four evaporation neutrons in the EQ phase, while the lower tail corresponds to the case where two PEQ particles (one proton and one neutron) are followed by three evaporation neutrons. The excitation energy after proton

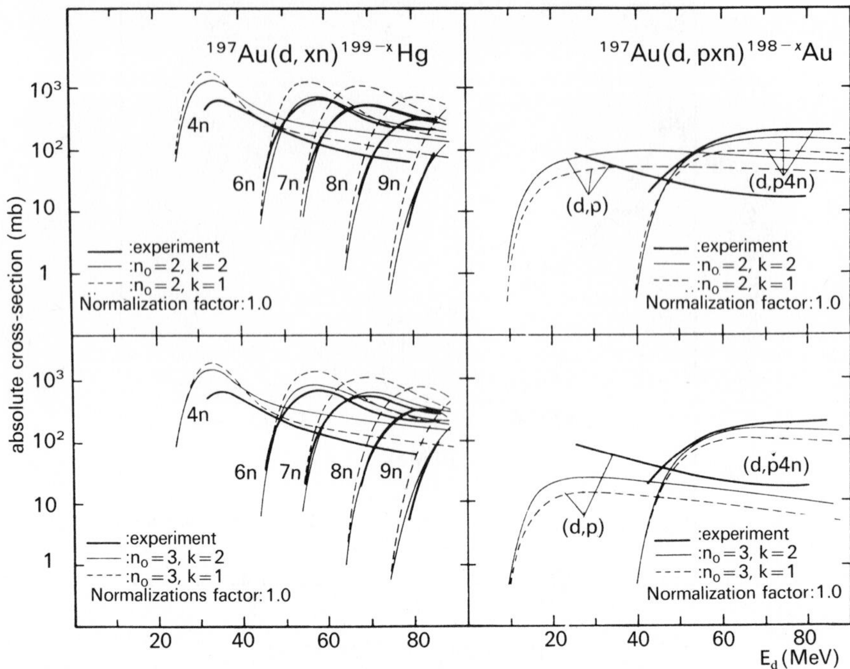

Fig. 4.22. Excitation functions of the ^{197}Au(d, $xn\gamma$) and ^{197}Au(d, pxn) reactions. n_0 is the initial exciton number and K is the multiplier of the mean free path in the hybrid-exciton model. The experimental data (thick lines) are compared with the results of the calculations. (Jahn *et al.*, 1973.)

emission is given by $E'_c = E_c - B_p - E_p$, where B_p is the proton binding energy, so it decreases as E_p increases. Thus the proton energy spectrum (the proton yield as a function of E_p) for the (α, p4n) reaction may be related to the excitation function of the (t, 4n) reaction plotted against $E'_c = E_t + B_t$, where B_t and E_t are the triton binding energy and the triton energy, respectively.

Particles following heavy-ion induced reactions with $E_a/A \geqslant 6$ to 10 MeV show a fast component due to the PEQ phase (Westerberg *et al.*, 1978). Neutron spectra following (^{12}C, αxn) and (^{12}C, xn) reaction channels are displayed in Fig. 4.24. The fast component is pronounced in low x (neutron multiplicity) channels at forward angles.

The angular distributions of neutrons following A(a, xn)B reactions with $E_a \gtrsim 40$ MeV show anisotropic distributions characteristic of the PEQ process (Ejiri *et al.*, 1978b; de Voigt *et al.*, 1982; Sakai *et al.*, 1979, 1980). The fast neutrons following (a, xn) reactions with small x are greatly enhanced at forward angles, as shown in Fig. 4.25. The enhanced fast neutrons at forward angles for all the (a, xn) reaction channels over a wide range of x indicate the

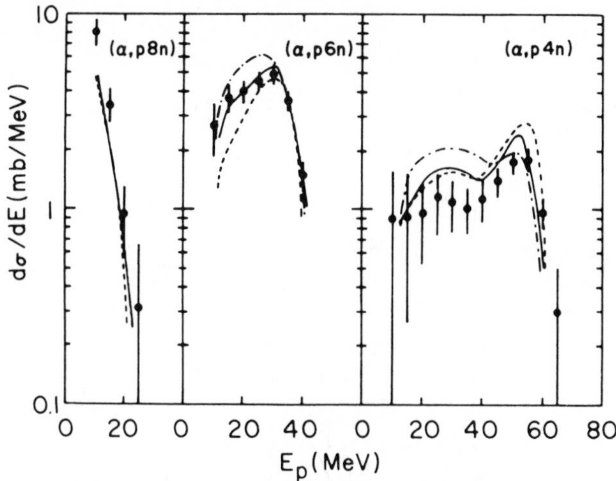

Fig. 4.23. The angle-integrated proton spectra for individual reaction channels of the ^{165}Ho(α, pxn) reactions at $E_\alpha = 109$ MeV. Dashed, solid and dot-dashed lines are calculations with initial excitons of (4, 0) with C = 10, (4, 1) with C = 14 and (5, 1) with C = 20, respectively, where C is the reduction factor for the spreading width. (Shibata *et al.*, 1985.)

existence of some PEQ process in all the reaction channels. The number x_p of pre-equilibrium particles in (a, xn) reactions is obtained by analysing the neutron multiplicity distribution, and the angular distribution in terms of the PEQ–EQ phase model as well as the exciton model. As shown in Fig. 4.26(a), the number of neutrons emitted in the pre-equilibrium phase increases as the neutron multiplicity x decreases and also as the total available (kinetic) energy E_p in eqn (4.2.37) increases. It is noted that only one or two neutrons are emitted in the PEQ phase produced by light projectiles and that the number tends to saturate as the projectile energy increases.

The production of various nuclei following proton-induced reactions has been studied by measuring prompt and delayed γ rays for medium energy protons of 80 to 160 MeV (Sadler *et al.*, 1977). The pre-equilibrium particle number is between one and two, in accord with the values for lower energy projectiles. These observations suggest a simple PEQ–EQ reaction process, in which a few fast nucleons emitted at the PEQ phase take away so much energy that the remaining nucleus moves to the EQ phase.

The total number of neutrons, i.e. the neutron multiplicity x in (a,xn) reactions, is obtained from eqn (4.2.37b) as $x = (E_c - E_\gamma)/(\bar{E}_n + \bar{B}_n)$. The multiplicities for (p,xn) and (α,xn) reactions are plotted in Fig. 4.26(b) as a function of the effective excitation energy $E^* = E_c - E_y$, where E_y is the correction for the yrast energy. In the EQ process with $\bar{E}_n \ll \bar{B}_n$, x increases

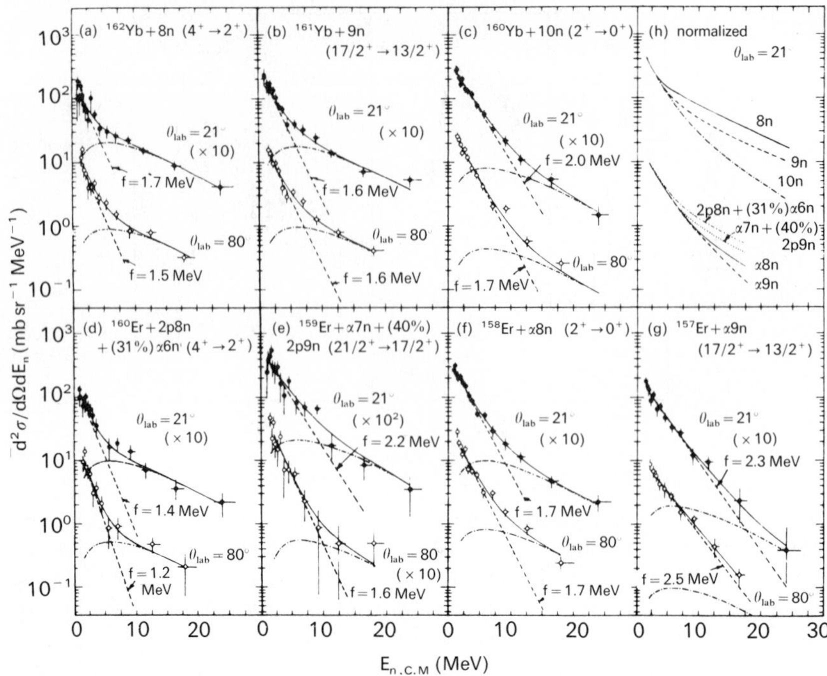

Fig. 4.24. Spectra of the neutrons recorded by time-of-flight in coincidence with the indicated γ transitions, observed in the Ge(Li) detector, which characterize the various exit channels for 152 MeV ^{12}C on ^{158}Gd. The cross-sections give the yield per neutron in the various channels (not the cross-section for forming the xn product). The spectra have been separated into an equilibrium component (dashed curves) and a pre-equilibrium component (dash-dot curves). The strength of the PEQ component decreases with increasing x and with increasing angle. The solid curves are sums of each pair of dashed and dash-dot curves. In (h) the normalized spectra for the indicated channels at 21° are shown for comparison. The indicated temperatures t are from fits to Boltzmann distributions. (Westerberg *et al.*, 1978.)

almost linearly as E^* increases—see the line x_e in Fig. 4.26(b)—, while the data tend to saturate as E^* increases beyond 60 to 80 MeV because $\bar{E}_n$ starts to increase in the PEQ region. A similar feature of the average neutron multiplicity is seen also in the ^{124}Sn(^{40}Ar, xn) reaction beyond $E^* \approx 80$ MeV (Hillis *et al.*, 1979).

Average neutron energies for A(a, xn)B reactions are given by

$$\bar{E}_n = \sum_x x \bar{E}_n(x)\sigma(x) / \sum_x x\sigma(x), \tag{4.2.39}$$

where the average energy $\bar{E}_n(x)$ for a certain x is given by eqn (4.2.37b) and the

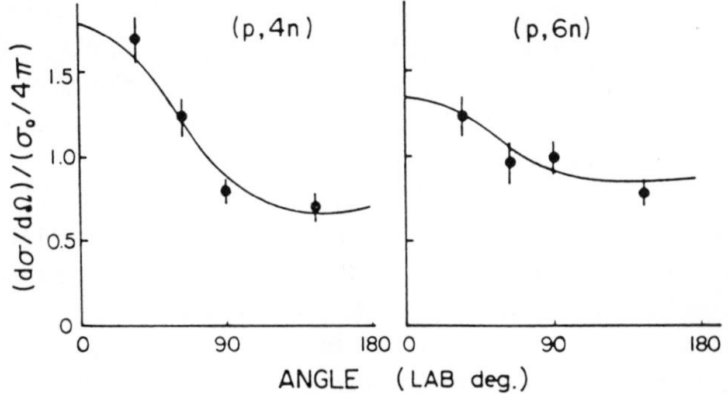

Fig. 4.25. Angular distributions of neutrons following the ^{165}Ho(p, xn)$^{166-x}$Er reactions with $x=4$ and 6. The closed circles are the data points for all neutrons above 1.1 MeV measured in coincidence with discrete γ rays from the ^{162}Er and ^{160}Er. The solid lines are the calculations in terms of the two-phase (PEQ and EQ) model, (Ejiri *et al.*, 1978b.)

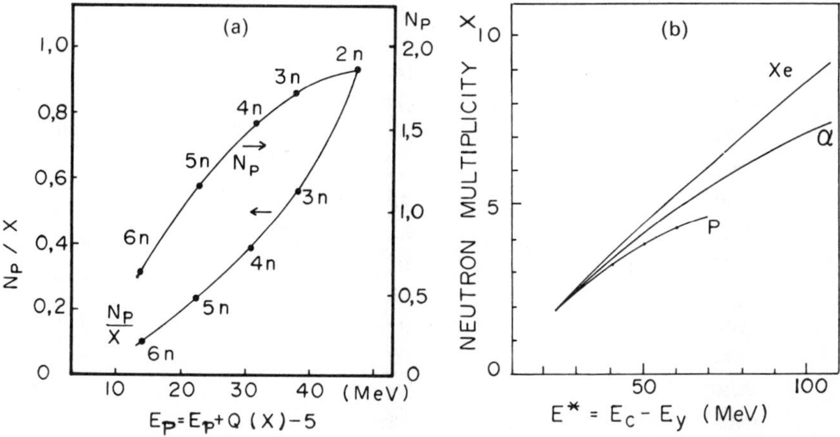

Fig. 4.26. (a) The number N_p of neutrons emitted in the pre-equilibrium phase and the ratio N_p/x for the (p, xn) reactions as functions of the energy sum E_p of the emitted neutrons. E_p is given by $E_p = E_i + Q(x) - 5$ MeV, where E_p, $Q(x)$ and 5 MeV are the incident proton energy, the (p, xn) reaction Q-value and the average excitation energy E_γ for the residual nucleus, respectively—see eqn (4.2.3) in text. N_p corresponds to x_p in eqns (4.2.35) and (4.2.36) in text (Ejiri *et al.*, 1978b). (b) Neutron multiplicities x for the (p, xn) and (α, xn) reactions. The expected value for the EQ (compound) process is denoted by x_e, and E^* is the excitation energy of the compound nucleus above the yrast level (E_y). p and α denote the data of (p, xn) and (α, xn) reactions, respectively.

multiplicity distribution $\sigma(x)$ is obtained from the yields of characteristic γ rays. The quantity $\bar{E}_n(x)$ is plotted as a function of the effective excitation energy E^* for various projectiles in Fig. 4.27. The energy excess $\bar{E}_n - 2k\bar{T}_e$ beyond the value $2k\bar{T}_e$ expected for the EQ neutron evaporation is due to the PEQ contribution. This excess increases with increasing E^* and with decreasing projectile mass m_a. The projectile mass dependence reflects the average kinetic energy of the excitons in the PEQ phase for a given excitation energy. A small number of high-energy excitons are produced by light projectiles, while a large number of low energy excitons are created by heavy projectiles.

4.2.4c *Angular momentum transfer and spin alignment in the PEQ process*
Since particles in the PEQ phase are emitted preferentially at forward (beam) direction some fractions of the linear and angular momenta brought in by the projectile are removed by such particles. The average angular momenta I_γ of the residual nuclei produced by A(a, xn)B reactions have been studied over a

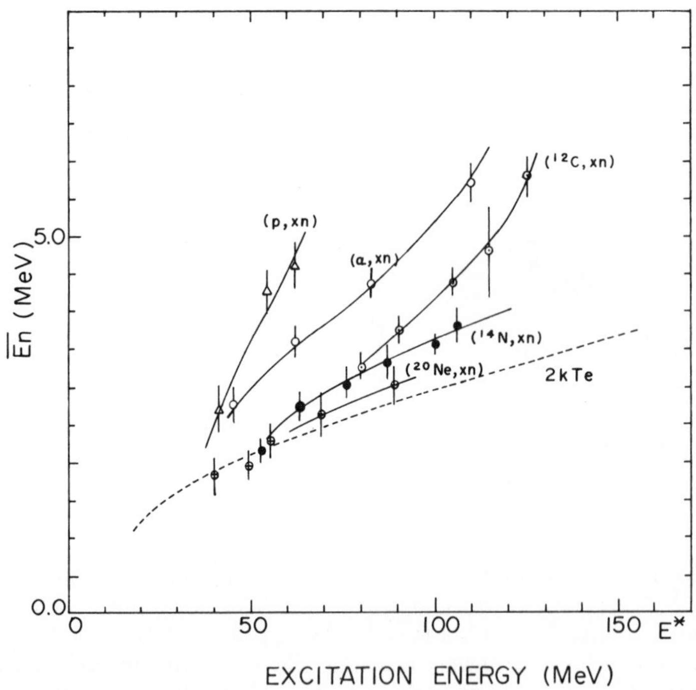

Fig. 4.27. The average neutron energies for (particle, $xn\gamma$) reactions as a function of the effective excitation energy E^*. The solid line is the value for the equilibrium evaporation process. The effective excitation energy is measured from the yrast level with the mean spin-value of the compound nucleus. (Ejiri, 1981.)

wide region of the projectile energy E_a by measuring the γ-ray multiplicity. I_γ is found to increase first and then saturate with increasing E_a, and even decreases as the E_a increases beyond 60 to 100 MeV, depending on the projectile mass (Ejiri *et al.*, 1978b, 1980; Ejiri, 1981; Nagai *et al.*, 1979; Ogawa *et al.*, 1978; Ockels *et al.*, 1978). The angular momenta associated with the $^{164}\mathrm{Dy}(\alpha, \mathrm{xn}\gamma)^{168-x}\mathrm{Er}$ reactions are plotted as a function of E_α in Fig. 4.28. The I_γ of the residual nucleus is smaller than the I_c introduced by the projectile at $E_\alpha \gtrsim 90$ MeV, and the missing angular momentum is considered to be carried off by the PEQ neutrons.

The angular momenta associated with PEQ (a, *x*n) reactions are studied by measuring both neutrons and γ rays (Ejiri *et al.*, 1978b, 1980; de Voigt *et al.*, 1979, 1982). The momentum of the PEQ neutron is decomposed as $\mathbf{P} = \mathbf{P}_{\|} + \mathbf{P}_{\perp}$, where $\mathbf{P}_{\|}$ is parallel to the beam axis and $\mathbf{P}_{\perp}$ is perpendicular to it. Since $\mathbf{P}_{\perp}$ vanishes on the average, the angular momentum change due to

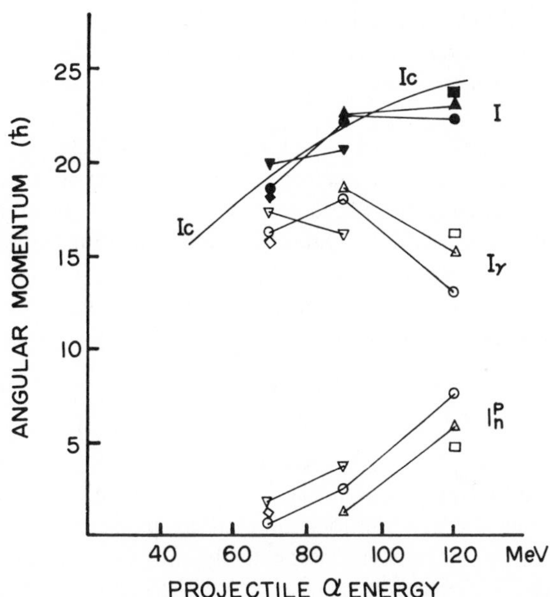

Fig. 4.28. The angular momenta for the $^{164}\mathrm{Dy}(\alpha, x\mathrm{n}\gamma)\mathrm{Er}$ reactions. The angular momenta (I_c) introduced by alpha-particle projectiles are given by the solid line. The angular momenta removed by pre-equilibrium neutrons and those carried away by γ-de-excitations are denoted by l_n^p and I_γ, respectively. The sum $I = l_n^p + l_n^e + I_\gamma$, where l_n^e is for equilibrium neutrons, is given by closed symbols. The inverse triangles, diamonds, circles, triangles and squares are the values for the neutron multiplicities $x = 4$, 5, 6, 8, and 10, respectively. (Ejiri *et al.*, 1980.)

the PEQ neutron with mass m_n is given by

$$l_n^p \hbar = R^e P_{\|} \approx \tfrac{2}{9} R \sqrt{(2m_n \bar{E}_n^p)} A_1^p N_p, \tag{4.2.40}$$

where N_p is the number of PEQ neutrons, A_1^p the coefficient of the Legendre polynomial $P_1(\cos\theta)$ of the angular distribution, $\bar{E}_n^p$ the average kinetic energy of the PEQ neutron and R^e the effective nuclear radius in the PEQ phase. The angular momentum l_n^p increases as E_α increases (see Fig. 4.28). For the EQ phase l_n^e is evaluated from the statistical model with a spin-dependent level density. The sum $I = l_n^p + l_n^e + I_\gamma$ is the total angular momentum removed by the PEQ and EQ neutrons and γ rays and it agrees with the input angular momentum I_c evaluated by $I_c \approx \frac{2}{3} I_{max} \approx \frac{2}{3}\{\Sigma\sigma(xn\gamma p)/\pi\lambda^2\}^{1/2}$, as shown in Fig. 4.28.

The spin alignment in the PEQ process depends largely on the way the PEQ particles are emitted since they carry off a large amount of angular momentum $l_n^p \approx 4$ to 8. The spin alignment for A(a, xn)B reactions has been studied over a wide range of projectile energies and for various reaction channels (Ejiri, 1981; Kishimoto *et al.*, 1980). The spin alignments of low-lying yrast levels in the reaction residues are shown in Fig. 4.29. They are fairly good, being independent of the projectile energy and the reaction channel. Thus the PEQ process preserves the spin alignment as much as the EQ process.

The PEQ particles emitted in forward direction carry off the angular momentum perpendicular to the beam axis. Thus the spin of the residual nucleus is still aligned in the plane perpendicular to the beam axis as in the initial stage. In this sense the PEQ neutron emission along the beam direction is analogous to the stretched transition of $I \rightarrow I - l_n^p \rightarrow \ldots$ in the angular momentum space. The spin alignment expressed by the attenuation factor $\alpha_2(I)$ for the $P_2(\cos\theta)$ coefficient of the decaying γ ray is given by

$$\alpha_2 \approx 1 - 2N_e k\bar{T}_e \frac{8m_n R^2}{9\hbar^2 I_e^2} - 2N_p k\bar{T}_p \cdot \frac{8m_n R^2}{9\hbar^2 I_p^2}, \tag{4.2.41}$$

where m_n, I, N and $\bar{T}$ are the neutron mass, the average nuclear spin (angular momentum), the average number of emitted particles, and the effective temperature (energy parameter), respectively, and the suffixes e and p stand for the EQ and PEQ phases, respectively. N_e, N_p, $\bar{T}_e$ and $\bar{T}_p$ increase as the E_α increases, but I_e^2 and I_p^2 increase as well, so that α_2 stays fairly constant over wide ranges of E_α and x. The attenuation factor for the yrast level is calculated by using the PEQ–EQ model for the particle decays and the K-band de-excitation model (see Section 4.2.2) for the γ decays. The calculations reproduce well the observed values.

The PEQ aspects of heavy-ion induced (a, xn) reactions are seen in the γ-ray multiplicity M_γ at higher bombarding energies (Hillis *et al.*, 1979; Sarantites *et*

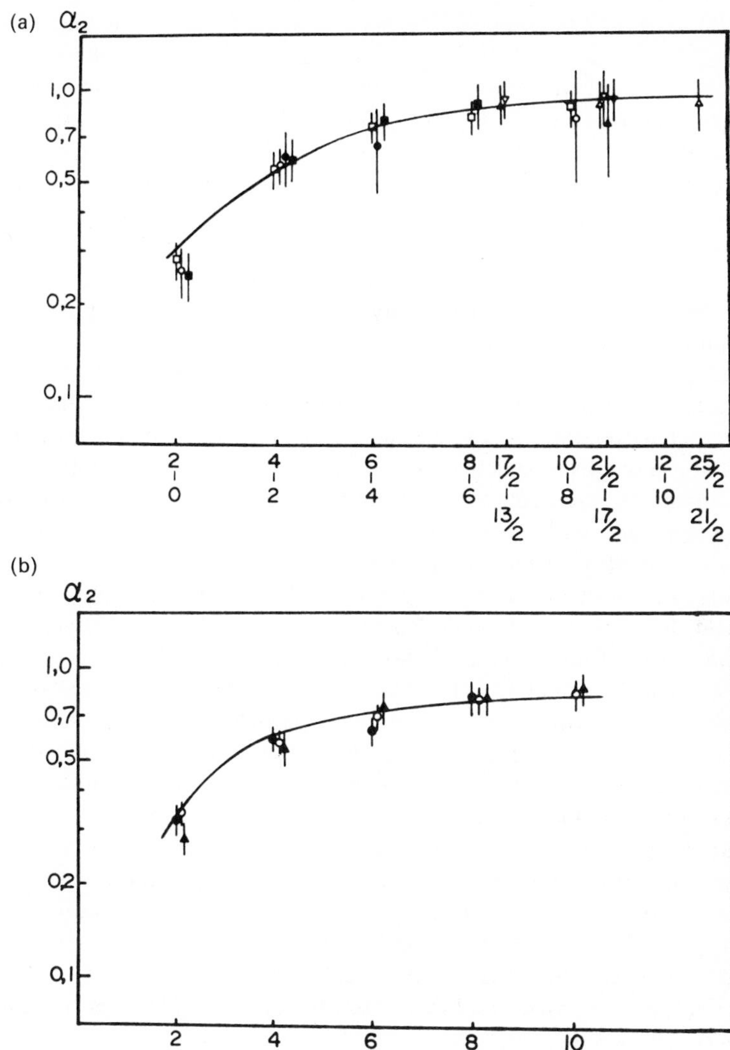

Fig. 4.29. (a) The degree of spin alignment for the yrast levels versus the yrast-level spin. The closed and open symbols are for the ^{162}Dy(α, $xn\gamma$) and ^{164}Dy(α, $xn\gamma$) reactions, respectively. The circles, triangles, squares, and inverse triangles stand for the values for the (α, 6nγ), (α, 7nγ), (α, 8nγ) and (α, 9nγ) reactions, respectively. The solid line is the KBD model calculation. (b): Comparison of the spin alignment for the (α, $xn\gamma$) reactions at three different projectile energies. Closed circles, open circles and closed triangles stand for the ^{148}Nd(α, 2nγ)^{150}Sm at $E_\alpha = 22$ MeV, ^{154}Sm(α, 4nγ)^{154}Gd at $E_\alpha = 43$ MeV, and ^{162}Dy(α, 8nγ)^{158}Er at $E_\alpha = 110$ MeV, respectively. The solid line is the KBD model calculation. (Ejiri 1981; Kishimoto *et al.*, 1980.)

al., 1978). In the EQ region M_γ increases with decreasing x, as discussed in Section 4.2.2, while at higher bombarding energies well above the Coulomb barrier, the M_γ distributions for neighbouring xn evaporation residues overlap, as shown in Fig. 4.11. The saturation of the M_γ as the x decreases, together with the constant value of the average γ-ray energy, indicates that the average kinetic energy of the emitted neutrons must increase with decreasing neutron multiplicity x. These features are expected from the fast neutron emission at the PEQ phase.

4.2.4d *Recoil ranges and linear momentum transfer*

Linear momentum transfers, which are counterparts of angular momentum transfers, are also sensitive to the PEQ–EQ processes. The average momentum transfers for fissile heavy nuclei have been studied by measuring the angular correlations of the fission fragments. On the other hand, recoil ranges of radioactive reaction products in the stopping material are used to study transferred momenta over a wide mass range of residual nuclei. Here β–γ spectroscopy with radioactive reaction products provides a sensitive method for identifying the reaction channels.

The recoil ranges in the stopping material have been studied for reaction products following light and heavy-ion bombardments (Miyano and Nakahara, 1973; Chan *et al.*, 1983; Jastrzębski *et al.*, 1984; Batsch *et al.*, 1985). The mass distribution $\sigma(A_r)$ of the reaction products is related directly to the multiplicity distribution $\sigma(x)$ of the emitted particles by $A_r = A_c - x$, with A_c being the mass of the composite (compound) nucleus. The recoil momenta (momentum transfer) attain the maximum value at the lightest residual nucleus, corresponding to the maximum number x of the removed mass, and decrease with increasing A_r. These features are interpreted as the maximum momentum transfer in the EQ process with the maximum multiplicity x of the evaporated particles, and as the removal of the linear momentum by the forward particles in the PEQ phase with the smaller x. The maximum and average momentum transfers for the ^{59}Co(^{4}He, x) reaction first increase as the projectile energy increases, and then saturate beyond 30 MeV per nucleon, as shown in Fig. 4.30 (Jastrzębski *et al.*, 1984).

The recoil ranges of heavy α-active nuclei produced by (p, xn) reactions on ^{209}Bi show saturation beyond $E_p = 22$ to 28 MeV, corresponding to the tail region of the excitation function (Miyano and Nakahara, 1973). The saturation indicates that momentum removed by direct or PEQ neutrons indeed increases as the projectile momentum increases, resulting in a constant momentum transfer to the residual nucleus. The saturation of the transferred linear momentum with the increasing projectile energy is similar to the saturation of the angular momentum transfer shown in Fig. 4.28. These saturations correspond to the saturation of the particle multiplicity, which reflects the saturation of the energy transfer to the compound nucleus in the EQ phase—see Fig. 4.26(b).

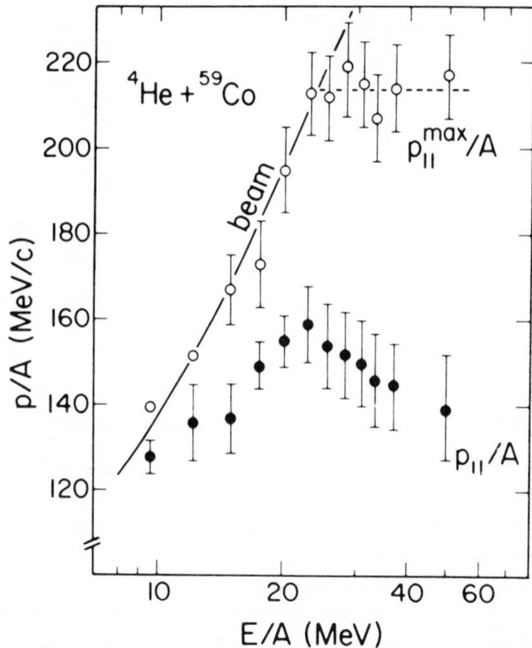

Fig. 4.30. The average (solid circles) and maximum (open circles) longitudinal momentum transfer per projectile nucleon as a function of alpha-particle bombarding energy. The solid line represents the incident projectile momentum per nucleon. The error bars in $P_{\parallel}/A$ reflect the uncertainty in the averaging procedure. (Jastrzębski *et al.*, 1984.)

4.2.4e *PEQ and EQ de-excitation charts*

It is interesting to see the PEQ–EQ reaction process in coordinates of the excitation energy E and the angular momentum I. The (α, 6nγ) reaction process for a deformed nucleus is shown schematically in Fig. 4.31(a). The highly excited high-spin nucleus decays first by emitting non-statistical neutrons in the PEQ phase, and later it cools down by statistically evaporating neutrons. If the excitation energy falls below the entry line, γ decays are switched on (see Section 6.5.2) and the nucleus cools down further first by emitting γ rays statistically to the yrast region and finally it de-excites to the ground state by nonstatistical γ decays along collective bands. Here non-statistical neutrons are to some extent spin-stretched transitions analogous to the nonstatistical (collective) γ rays.

In the exciton model the PEQ–EQ decay process is described in terms of the exciton number m and the excitation energy E as shown in Fig. 4.31(b). The first doorway stage of the reaction consists of a number of excitons (particles and holes). Then m increases through the internucleon collisions

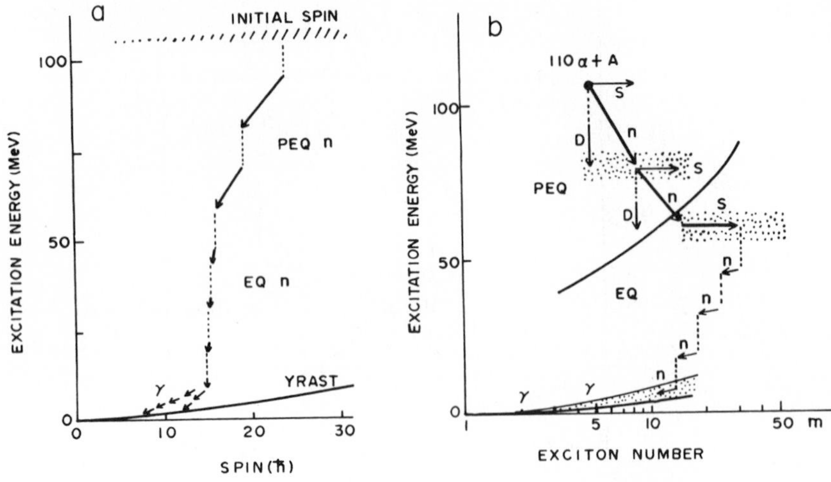

Fig. 4.31. (a) The de-excitation scheme for the (α, $xn\gamma$) reaction induced by 120 MeV alpha-particles is illustrated in the I–E space, where I is the angular momentum and E is the excitation energy. The dotted lines show the energy decrease corresponding to the neutron binding energy. (b). The de-excitation scheme of the (α, $xn\gamma$) reaction induced by 110 MeV alpha-particles is illustrated in the m–E space, where m is the exciton number and E is the excitation energy. S and D stand for the spreading and the neutron escaping, respectively. The dotted lines show the energy decrease corresponding to the neutron binding energy. (Ejiri, 1981.)

(spreading width $\Gamma^{\downarrow}$), while E decreases through particle emission in the PEQ phase. If E becomes so low or m becomes so large that the particle hardly ever escapes at the PEQ phase, then the nucleus moves to the thermal equilibrium (EQ) phase through the spreading process. The EQ nucleus cools down first by evaporating particles, and then by emitting statistical γ rays followed by nonstatistical (collective) γ rays.

The PEQ phase consists of several excitons with large linear and angular momenta, which are rather uniformly aligned as introduced by the projectile. Thus it is a sort of collective motions of excitons. Such a regular motion in the PEQ phase damps with the spreading width $\Gamma^{\downarrow}$ through the internucleon collisions acting like friction. In this way the energy and momentum of the uniform motion are converted to the random motion in the thermally equilibrated nucleus.

In short, there are four phases of the de-excitation process, i.e. the PEQ phase with high energy E and small exciton number m, the EQ phase with medium E and large m, the effectively-bound warm phase with dominant statistical γ decays, and the cold phase with small m and dominant non-statistical (collective) γ decays. The phase transition regions are rather broad in all the E, m, and I coordinate spaces since the nucleus consists of a finite number of nucleons. Such vague boundaries, characteristic of nuclear phenomena, are in contrast with sharp phase boundaries in the solid state with practically infinite number of atoms.

4.3 Break-up and transfer reactions and gamma-ray spectroscopy

4.3.1 *Break-up and transfer reactions*

Direct and quasi-direct nuclear reactions occur at a fairly early (fast) stage of the collision process. The reaction (interaction) time is about $R/v \approx 10^{-22}$ s, where R is the nuclear radius and v is the projectile velocity. Accordingly, the transfers of energy, of linear and angular momenta, and of nucleons are rather small. These features contrast with the large energy- and momentum-transfers involved in slow processes such as compound and precompound reactions (see Section 4.2).

In the fast processes, the properties of projectile and target nuclei are preserved to some extent even after the reaction. Such direct and quasi-direct reactions involve projectile break-up, nucleon transfer (pick-up), projectile and/or target fragmentation processes and so on. The direct and quasi-direct processes are dominant in nuclear reactions induced by medium- and high-energy projectiles with $E/A \gg E_f$, where $E_f \approx 30$ MeV is the Fermi energy. The interaction time is short and the transferred energy is much smaller than the projectile energy. The direct and quasi-direct processes also occur in peripheral (surface) reactions induced by lower energy projectiles since the interactions between projectile and target nuclei are weak at the surface.

The reaction scheme is illustrated in Fig. 4.32. The break-up reactions of a complex projectile $\mathrm{a = b + c}$ are written as

$$\text{EB:} \quad \mathrm{a + A \rightarrow b + c + A}, \tag{4.3.1}$$

$$\text{IEB:} \quad \mathrm{\rightarrow b + c + A^*}, \tag{4.3.2}$$

$$\text{BF:} \quad \mathrm{\rightarrow b + B}, \tag{4.3.3}$$

$$\text{with} \quad \mathrm{B = A + c}.$$

EB is the elastic break-up process and IEB is the inelastic break-up process leaving the target in an excited state (A*). If the excitation energy exceeds a particle threshold energy the excited nucleus A* will decay by emitting the particle. BF denotes the break-up fusion process. A strong interaction of c, which is a break-up partner of b, with the target A leads to a fused nucleus of $\mathrm{B = A + c}$. The inverse process to BF is the transfer fusion process,

$$\text{TF:} \ \mathrm{a + A \rightarrow b + C},$$

$$\text{with} \quad \mathrm{A = C + c,\ b = a + c}. \tag{4.3.4}$$

The BF and TF transfer a nucleon or a nuclear fragment (cluster), in BF from projectile to target and in TF from target to projectile. The break-up, transfer and fusion processes may thus occur on both the projectile and target nuclei. The target and/or projectile fragmentation process is generally written as

$$\text{F:} \qquad \mathrm{a+A \to a'+a''+\ \ldots\ +A'+A''+\ \ldots,} \tag{4.3.5}$$

where a′ and A′ still keep some memory (mass, momentum, composite nucleons, etc) of a and A, respectively. Here a′ or A′ can be larger than a or A if the break-up partner is fused (transferred) into a or A.

Gamma-ray spectroscopy associated with the direct and quasi-direct reactions is used to study the nuclear structure of the reaction products as well as the dynamics of the reaction process. Gamma spectroscopy offers several unique features for studying the reaction mechanism. Coincidence measurement of a break-up fragment b with discrete γ rays characteristic of the reaction product gives a unique assignment of the break-up (transfer) process involved in the reaction. EB, IEB, BF, and TF processes lead to residual nuclei A, A*, B, and C, respectively, which are identified by their characteristic γ rays. Note that A is not accompanied by γ rays. Observation of all fragments in the F process is not required for identification since discrete γ rays identify uniquely a′ and A′ (see Fig. 4.32).

Linear- and angular-momentum transfers, which are rather small in the direct and quasi-direct reactions, are studied by measuring Doppler shifts of discrete γ rays and relative intensities of γ rays from high-spin levels. The break-up fusion process of low-energy heavy ions is a peripheral process with an impact parameter b close to the nuclear radius of the target, so that the angular momentum window is fairly narrow. The transferred angular momentum is expected to be aligned approximately in a plane perpendicular to the axis of the momentum transfer. Furthermore the distortion of incoming

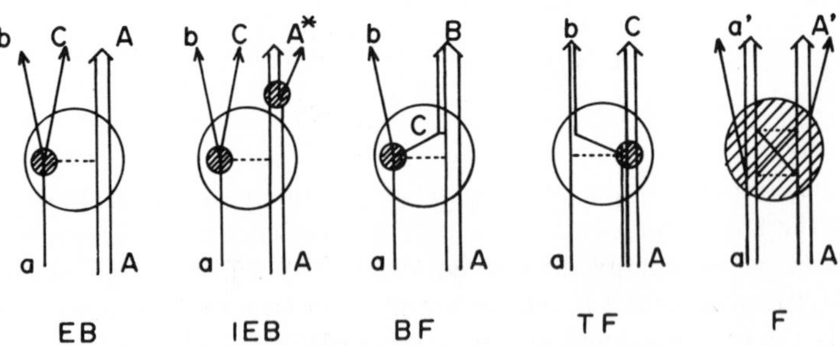

Fig. 4.32. Schematic diagram of break-up and particle transfer reactions. EB; elastic break-up, IEB; inelastic break-up, BF; break-up fusion into the target nucleus, TF; transfer fusion into the projectile, and F; projectile and/or target fragmentation.

and outgoing waves in the nuclear and Coulomb fields gives rise to a spin (angular momentum) polarization, which would be washed out if the reaction would be a non-direct (statistical) process. The spin polarization of the reaction product is very useful for understanding the reaction mechanism.

4.3.2 *Gamma-rays following break-up reactions in light nuclei*

Weakly-bound light nuclei such as d, ^{3}He, ^{6}Li, etc. tend to break-up into two or three composite particles through a weak collision at the nuclear surface even at fairly low energy. The break-up process for tightly bound nuclei (alpha, etc) becomes important at incident energies above $E/A \approx 30$ MeV. Experimental studies have mostly been performed by measuring break-up particles (mainly charged particles) (Matsuoka *et al.*, 1978/1980; Pampus *et al.*, 1978, and refs. therein). A review of break-up process with light projectiles is given by de Meijer and Kamermans (1985). Theoretical studies of elastic, inelastic and break-up fusion processes have been carried out in terms of the DWBA (Udagawa *et al.*, 1982; Li *et al.*, 1984; Shyam *et al.*, 1984).

Recently γ-ray spectroscopy has been applied to study the individual break-up process. The (^{3}He, d or p, xpynγ) reactions on ^{165}Ho and ^{166,167}Er were studied by measuring the break-up d or p in coincidence with discrete γ-rays characteristic of the residual nucleus (Ejiri *et al.*, 1982; Motobayashi *et al.*, 1984). Such γ rays identify the reaction channels and the yields give the cross-sections without observation of x protons and y neutrons at all angles. The neutron multiplicity distributions $\sigma(x)$ for the ^{165}Ho(^{3}He, dxn)$^{166-x}$Er, has been obtained from the yields of $^{166-x}$Er γ rays, and they are shown in Fig. 4.33. The observed data are well reproduced by the exciton model calculation for the neutrons following the break-up fusion process of the ^{165}Ho(^{3}He, d) reaction. The total cross-sections for elastic, inelastic and break-up fusion for deuterons at various angles are obtained by summing the cross-sections $d\sigma/d\Omega$ over all the multiplicities x. They are well reproduced by the DWBA calculation as shown in Table 4.3.1.

The angular momentum transfer $l_R\hbar$ depends on the reaction process. In the break-up process at forward angles θ_i one gets $l_i\cos\theta_i \approx l_i$. Thus the transferred angular momentum is

$$l_R\hbar \approx l_a - (l_b\hbar + l_c\hbar \cdot k), \tag{4.3.6}$$

where $l_a\hbar$, $l_b\hbar$ and $l_c\hbar$ are the angular momenta of the projectile a, the break-up particle b and the other break-up partner c, respectively. The coefficient k is 1 for simple EB and IEB and 0 for a simple BF with c being fused into the target. If the break-up particles are emitted in the forward direction, the angular momentum transfers may be expressed using the impact parameter $\tilde{b}$

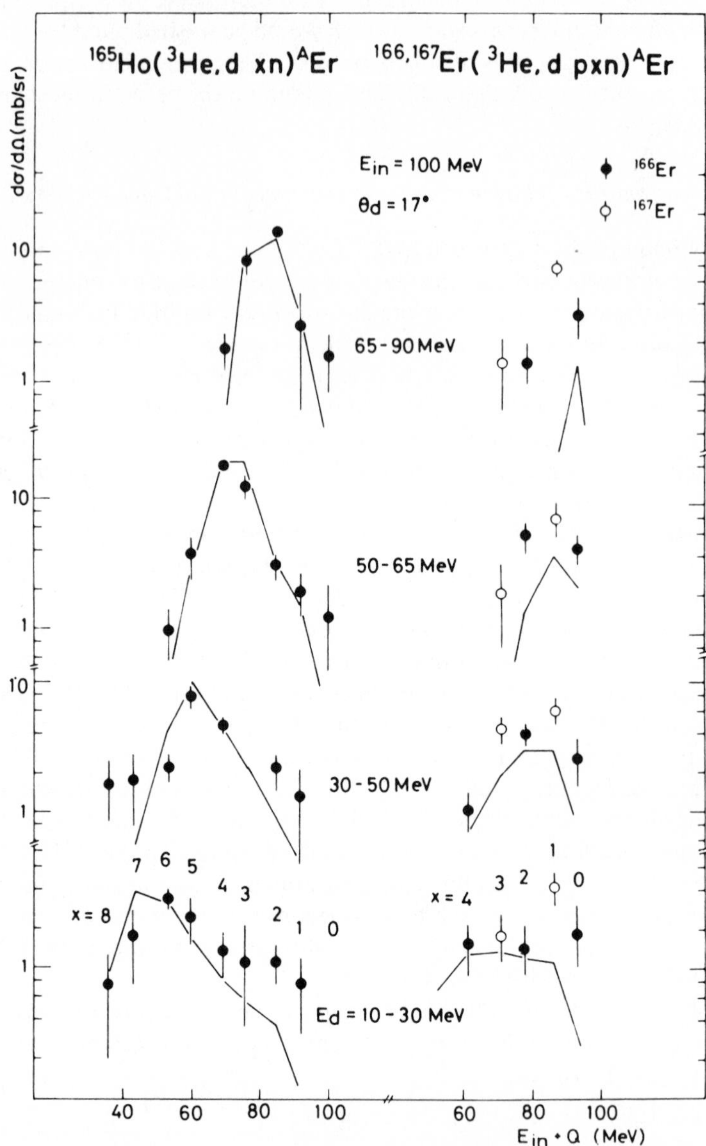

Fig. 4.33. Coincidence cross sections of the ^{166,167}Er(^{3}He, dp$x$n) and ^{165}Ho(^{3}He, dxn) reactions as a function of $E_{in}+Q$ for various energy windows of deuterons. Here E_{in} is the ^{3}He energy in the c.m. system and Q is the reaction Q-value. The neutron multiplicity x is also labelled. Lines indicate the exciton-model calculations assuming the break-up fusion process. (Motobayashi *et al.*, 1984.)

Table 4.3.1 Cross-sections for the elastic break-up, the inelastic break-up and the break-up fusion processes for the ^{165}Ho(^{3}He, d) reaction at E(^{3}He) = 100 MeV. (Expt: Motobayashi *et al.*, 1984; Theory: Shyam *et al.*, 1984)

$\frac{d\sigma}{d\Omega}\left(\frac{mb}{sr}\right)$ at θ(d)	θ(d) = 17°		θ(d) = 35°	
	Expt.	Theory	Expt.	Theory.
EB Elastic break-up	62 ± 12	60.17	12 ± 4	8.98
IEB Inelastic break-up	23 ± 5	130.80 (IEB + BF)	4 ± 2	31.72 (IEB + BF)
BF Break-up fusion	119 ± 10		35 ± 4	
R Ratio (IEB + BF)/EB	2.29	2.17	3.25	3.53

as follows

$$\text{EB} \qquad l_R\hbar \approx (P_a - P_b - P_c)\tilde{b} \approx 0, \tag{4.3.7}$$

$$\text{IEB} \qquad l_R\hbar \approx (P_a - P_b - P_c)\tilde{b} \ll P_a\tilde{b}, \tag{4.3.8}$$

$$\text{BF} \qquad l_R\hbar \approx (P_a - P_b)\tilde{b} \approx (M_c/M_a)P_a\tilde{b}, \tag{4.3.9}$$

where P_i is the linear momentum of the particle i and M_j the mass number of the particle j. For the (^{3}He, d) reaction with $E(^3\text{He}) = 100$ MeV, the transferred angular momentum $l_R\hbar$ for the IEB (^{3}He, dpn) reaction is only 3ħ, while for the BF (^{3}He, d4n) reaction it is about 11ħ. Here the impact parameter is assumed to be given by the radius of the target nucleus. (Li *et al.*, 1984). These values of l_R are clearly demonstrated by the relative intensities of cascade γ rays along yrast (rotational) levels. The γ ray intensities for the (^{3}He, dpn) reaction increases as the spin (I) of the rotational level decreases from $I = 8$ to 2, while those for the (^{3}He, d4n) reaction are fairly constant below $I = 8$ as shown in Fig. 4.34. These features are in accord with the angular momentum transfer of ~3ħ and ~10ħ with a narrow l-window (Motobayashi *et al.*, 1984; Shyam *et al.*, 1984).

4.3.3 *Break-up fusion process in heavy-ion reactions and γ-ray spectroscopy for high-spin nuclei*

4.3.3a *Break-up fusion of heavy ions (massive transfer)*

Break-up (projectile fragmentation) fusion reactions of heavy projectiles (heavy ions, HI) have unique features from the point of view of γ ray spectroscopy of high-spin states. The gamma-ray intensities of cascade transitions along the yrast band and the γ-ray multiplicities provide the angular momentum transfer, which is crucial for understanding the HI reaction mechanism. The strongly absorptive nature of HI in a nucleus may

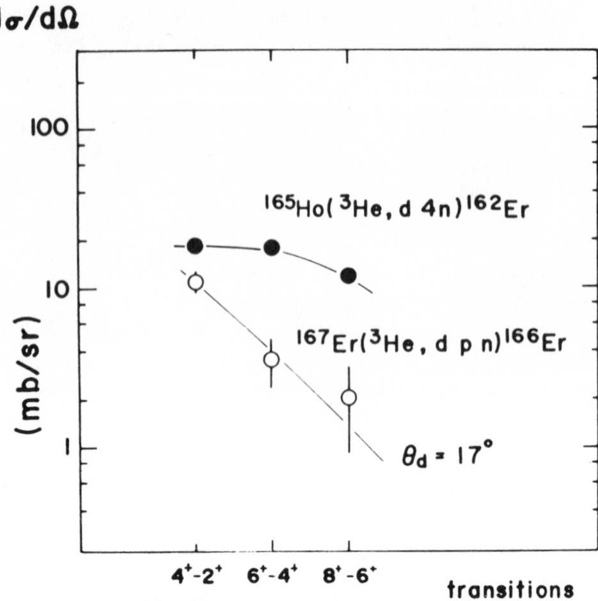

Fig. 4.34. Intensities of the ground-band γ-ray transitions for the ^{165}Ho(^{3}He, d4n) and ^{167}Er(^{3}He, dpn) reactions. The deuterons were detected at 17°. The solid lines are eye-guides. (Motobayashi *et al.*, 1984.)

give rise to a polarization of the transferred angular momentum. The polarization is detected by measuring circular polarization of γ rays.

The break-up particles of a HI projectile are massive fragments, and one of them may be transferred to the target nucleus. This process is therefore called a massive transfer reaction, or a partial fusion process in an incomplete fusion reaction. The HI reaction process greatly depends on the impact parameter as shown in Fig. 4.35. For the break-up ($a \to b + c$) and break-up fusion ($c + A \to B$) processes, due to a weak collision at the surface an optimal condition may be chosen such that the transferred particle (c) on the average is at rest with respect to the projectile (Siemens *et al.*, 1971). Then the relative velocity of b with respect to B is $V_{Bb} = V_{Aa}(M_A/M_B)$ and the angular momentum transfer to B is

$$J_B\hbar \approx l_a\hbar M_c/M_a. \tag{4.3.10}$$

Here the impact parameter $\tilde{b}$ has to be around the optimal value b_0 for the break-up fusion process, and then the angular momentum J_B is well-defined with a narrow window as

$$J_B\hbar = b_0 P_a M_c/M_a, \tag{4.3.11}$$

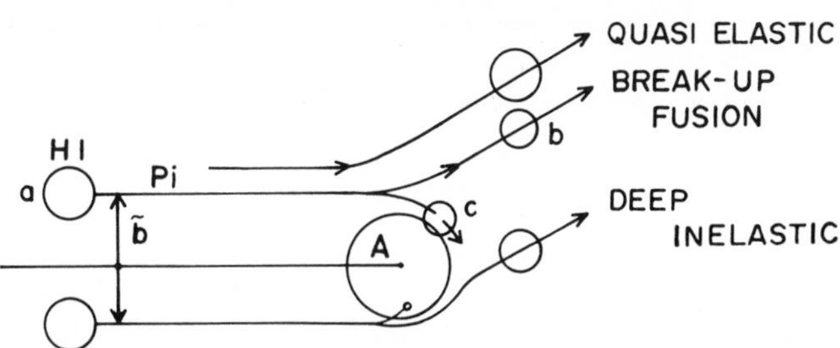

Fig. 4.35. Schematic trajectories of HI reaction processes. In the break-up fusion process the break-up partner (projectile fragment (c) is fused (transferred) with the target nucleus (A).

where b_0 is around the nuclear radius for the peripheral reaction. The localization of the l-window is very useful for the spectroscopic studies of high-spin states. The transferred angular momentum is oriented in the plane perpendicular to the momentum of c, and is even polarized provided that the break-up fusion process is a near-side process as the quasi-elastic scattering (see Fig. 4.35). Break-up and incomplete fusion reactions have been discussed in review articles (Siemssen, 1983, 1984).

4.3.3b *Gamma-rays following break-up fusion reactions and the localization of angular momentum transfer*

Localization of the angular momentum transfer has been shown in relative intensities of cascade γ rays along yrast bands in deformed nuclei. The γ rays from the ^{159}Tb (^{14}N, αxn)$^{169-x}$Yb reaction with 95 MeV ^{14}N were observed in coincidence with fast and slow alpha-particles as shown in Fig. 4.36 (Inamura *et al.*, 1977). The relative intensities of yrast E2 γ rays in coincidence with fast (break-up) alphas in the forward direction are fairly constant up to J (yrast)$=10$, and fall off rapidly thereafter. On the other hand the γ rays in coincidence with slow alphas in the backward direction decrease monotonically as the yrast level spin increases. These observations indicate a localization of the entry point (starting point at which γ cascades start) at $\langle J_\gamma \rangle = 12$ to 14, depending on the angular momentum carried away by quasi-continuum γ transitions, in case of the break-up fusion process. This localization is in contrast to a rather wide angular momentum range in the compound reaction. The average angular momentum of the entrance channel is given by

$$\langle J_a \rangle = \langle J_\gamma \rangle + \langle J_n \rangle + \langle J_\alpha \rangle. \tag{4.3.12}$$

Using $\langle J_{\mathrm{n}} \rangle = 6\hbar$ for the evaporating neutrons, $\langle J_\alpha \rangle \approx 21\hbar$ for the break-up alpha and $\langle J_\gamma \rangle \approx 13\hbar$, one gets $\langle J_a \rangle \approx 40\hbar$. The angular momentum transfer by the break-up fusion (massive transfer) is $\langle l_t \rangle = \langle J_\gamma \rangle + \langle J_{\mathrm{n}} \rangle \approx 19\hbar$. The

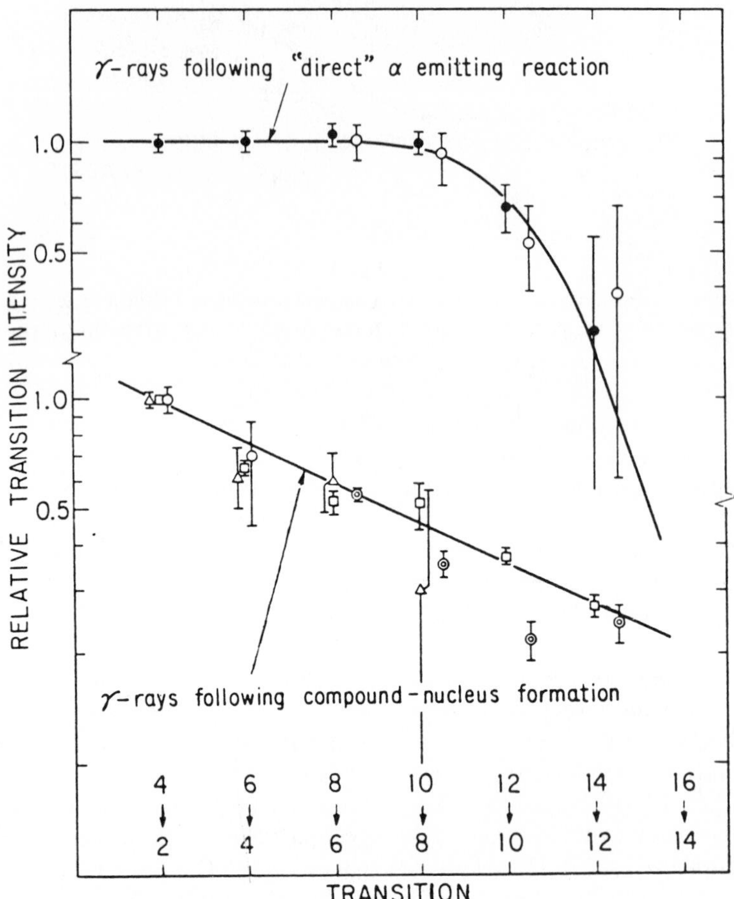

Fig. 4.36. Gamma-transition intensities relative to the $4^+ \to 2^+$ ($17/2^+ \to 13/2^+$ for ^{165}Yb) transition in the de-excitation of the ^{159}Tb + 95 MeV ^{14}N reaction products. The upper drawing shows the data obtained in coincidence with direct α-particles: ● for ^{166}Yb and ○ for ^{165}Yb; the solid line was obtained for ^{166}Yb. The lower drawing shows the data on the residues of compound–nucleus formation: □ for ^{168}Hf, ○ for ^{166}Yb, ⊚ for ^{165}Yb and △ for ^{164}Yb; for ^{165}Yb the normalization was made at 0.54 on account of the spin concerned, and the line is drawn to guide the eye. (Inamura *et al.*, 1977.)

value for $\langle J_a \rangle$ is just above the critical angular momentum for the compound nuclear formation.

The energy and angular distributions of the break-up particles, being observed in coincidence with discrete γ rays characteristic of the break-up fusion reaction, are in accord with the localization of the angular momentum transfer (Gerschel *et al.*, 1982, Siemssen, 1983).

The angular momentum of the entry point in the break-up (partial) fusion reaction is obtained by measuring the γ ray multiplicity. The multiplicity M_γ for the γ rays in coincidence with fast forward particles associated with partial fusion reactions are found to be fairly constant for all the yrast (rotational) transitions (Yamada *et al.*, 1981; Hageman *et al.*, 1981; Inamura, 1984). This implies a localization of the transferred angular momentum as given by $\langle l_t \rangle = 2(M_\gamma - 4) + \langle J_n \rangle$ (see Section 4.2). For example, the γ-ray multiplicities in coincidence with a fast forward p, d or t from the ^{154}Sm(^{14}N, p, d, or t, xn)^{158}Er reactions with 167 MeV ^{14}N are constant ($M_\gamma \approx 31$) for all the yrast transitions up to states with 26ħ. This fact leads to $\langle l_t \rangle \approx 63\hbar$, in accord with the critical angular momentum associated with the fusion of the break-up partner (C) into ^{154}Sm (see Section 6.2). The relation between the transferred mass and the transferred angular momentum, as given by eqn 4.3.10 is clearly demonstrated in the incomplete fusion of 153 MeV ^{16}O on ^{154}Sm (Geoffroy *et al.*, 1979). The γ-ray multiplicity was measured in coincidence with those projectile fragments emitted in the forward direction with velocities near that of the projectile. The transferred angular momentum J_B is found to be proportional to the mass of the captured particle c. The entrance channel spins deduced for the capture of ^{12}C, ^{8}Be and ^{4}He are 52, 60 and 74ħ, respectively, and thus increase as the captured mass decreases.

Spin alignment and spin polarization for residual nuclei are obtained from the γ-ray angular distribution and the circular polarization. The in-plane out-of-plane anisotropy, $W\{\theta_\gamma = (\pi/2),\ \psi_\gamma = (\pi/2\}/W\{\theta_\gamma = 0,\ \psi_\gamma = (\pi/2)\}$, for the yrast (E2) transitions following the ^{159}Tb(^{14}N, α, 5n)^{164}Yb reaction at $E(^{14}\mathrm{N}) = 115$ MeV is around 1.5 for the fast (direct) forward alphas, while it is nearly unity for the slow alphas (Inamura *et al.*, 1984). Thus the spin is well aligned in a plane perpendicular to the reaction plane in the first process. Assuming the spin alignment is due to the spin polarization, the evaluated spin polarization is as large as $P_{zz} = 0.35$. Actually, a circular polarization of the γ rays has been observed in coincidence with the direct forward alphas (Ieki *et al.*, 1985). The large negative spin polarization for the break-up alpha particle with near-beam velocity is observed as shown in Fig. 4.37. It is consistent with the break-up fusion process as schematically illustrated in Fig. 4.35.

4.3.3c *K X-rays following incomplete fusion reactions*

K X-rays are quite useful for identifying the reaction channels of incomplete fusion reactions with heavy nuclei (heavy ions) (see Section 4.2.2). Here the K X-rays are caused by the internal K-electron conversion of the γ cascade in the residual nuclei. The incomplete or break-up fusion reaction of the heavy-ion projectile a is written as $\mathrm{a} + \mathrm{A} = \mathrm{b} + \mathrm{B}^*$, where B* is the fused (compound)

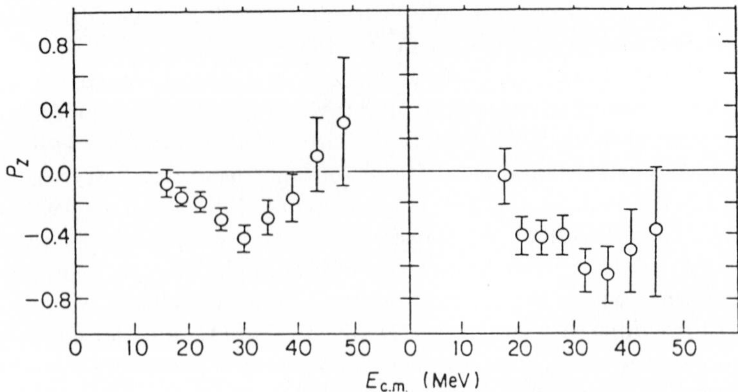

Fig. 4.37. The spin polarization versus emitted alpha-particle energy for ^{14}N-induced reactions at $\theta_{lab} = 30°$ on ^{159}Tb (left-hand side) and on ^{181}Ta (right-hand side). The $E(^{14}\text{N})$ is 115 MeV and the $E(\alpha)$ corresponding to the ^{14}N beam velocity is 33 MeV (Ieki *et al.*, 1985.)

nucleus of A and the break-up partner c. The compound nucleus B* with high atomic number Z_B decays by evaporating mostly neutrons. Then the cross-section for the incomplete fusion reaction is given as

$$\sigma(\text{a, bB*}) = \sum_x \sigma(\text{a, bxn}). \tag{4.3.13}$$

Since all the neutron-evaporation residues have the same atomic number of Z_B, the K X-ray yield gives the sum of the residual isotopes with the same atomic number Z_B. Thus the K X-ray of the Z_B atom selects the incomplete fusion channel of (a, b) with $Z_b = Z_A + Z_a - Z_B$, where A, a and b stand for the target, projectile and projectile fragment, respectively.

Incomplete fusion reactions are not necessarily binary processes, but some charged particle Y may be emitted as well. Then the K X-ray of the residual nucleus with Z_R gives the missing charge $Z_Y = Z_A + Z_a - Z_b - Z_R$.

K X-rays following the $^{14}\text{N} + {}^{197}\text{Au}$ reaction have been measured in coincidence with projectile-like fragments (PLF) at the KVI, Groningen (Wilschut *et al.*, 1983). K X-ray spectra following the $^{197}\text{Au}(^{14}\text{N, bXxn})$ reaction are shown in Fig. 4.38. The X-ray spectra, gated by the particle b, show reaction channels with all atomic numbers Z_R ranging from Z_A to $Z_A + Z_a - Z_b$. The exclusive K X-ray yield for the A(a, bYxn) reaction is written as

$$I(Z_R \cdot \text{b}) = c \sum_x \langle M_K \rangle \sigma(\text{a, bYxn}), \tag{4.3.14}$$

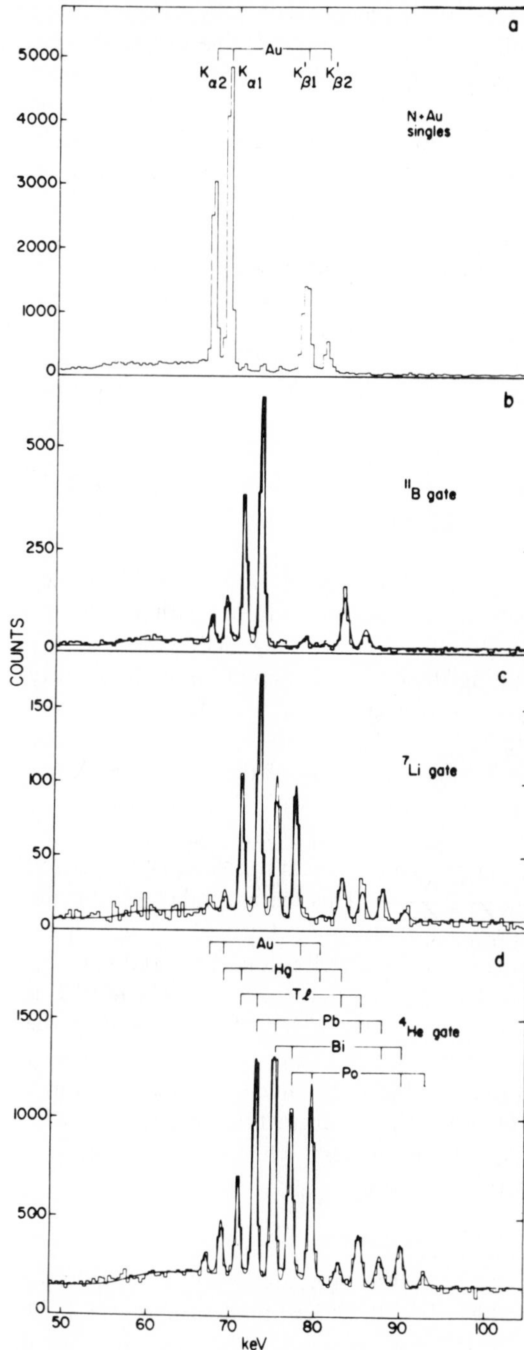

Fig. 4.38. Some examples of X-ray spectra. (a) Singles X-ray spectrum. The characteristic K X-rays are mainly due to target ionization. (b, c and d): Coincident X-ray spectra showing the contributions of various reaction channels. The lines through the data points are obtained from the fitting procedure that was used to determine the elemental yields. (Wilschut *et al.*, 1983.)

where $I(Z_R \cdot \mathrm{b})$ is the Z_R K X-ray yield in coincidence with b, $Z_R = Z_A + Z_a - Z_b - Z_Y$, and $\langle M_K \rangle$ is the multiplicity of the K X-rays. The multiplicity is obtained from the ratio of the particle-X–X triple coincidence rate to the particle-X double coincidence rate. It is found that for ejectiles with $Z_b \geqslant 4$ the charge–binary (incomplete fusion) reaction channel with $Z_R = Z_A + Z_a - Z_b$ is major, while for the lighter ejectiles K X-rays of $Z_R = Z_A + Z_a - Z_b - 2$ are very strong, suggesting emission of some charged particle Y with $Z_Y = 2$ such as alpha-particles (Wilschut *et al.*, 1983).

The exclusive K X-rays of Z_R isotopes in coincidence with the PLF b for the ^{159}Tb(^{14}N, bYxn) reaction have been studied at $E(^{14}\mathrm{N}) = 115$ and 168 MeV (Balster *et al.*, 1984). Since charged particles are unlikely to be emitted from a fused system with high Z_R, the charged particles b and Y are considered to be mainly due to the sequential decay of the excited PLF into b and Y. Then the decay probability is given as

$$P(\mathrm{E} \rightarrow \mathrm{b} + \mathrm{Y}) = \frac{I(Z_R \cdot \mathrm{b})}{I(Z_R \cdot \mathrm{b}) + \sum_{b'} I(Z_R \cdot \mathrm{b}')}. \qquad (4.3.15)$$

Here the K X-ray multiplicity is assumed to depend only on the charge Z_R (Chmielewska *et al.*, 1981). The obtained values are in accord with the particle (b)–particle (Y) correlation data (Bhowmik *et al.*, 1982). Thus the K X-ray method is quite useful for obtaining the decay probability in a simple way without requiring measurements of the missing particles Y all over the angles.

Siemssen *et al.* (1985) have studied the ^{14}N + ^{159}Tb reaction at the higher energies of $E(^{14}\mathrm{N}) = 236$ and 309 MeV by means of the particle K X-ray coincidence method. Charge non-binary channels with $Z_Y \neq 0$ increase as E_a increases, and dominate at higher energies. As the ejectile mass M_b (and the charge Z_b) become smaller, the charge of the residual nuclei Z_R is distributed in a wider range, beyond $Z_R = Z_a + Z_A - Z_b$. This feature indicates that the missing charge Z_Y becomes larger for smaller Z_b. Particularly the alpha-particle cross-sections have a much broader distribution as a function of Z_R. These features are shown in Fig. 4.39. Noting that the energy of the PLF is roughly given by the mass ratio as $E_b = (M_b/M_a)E_a$, the ejectile with the smaller mass M_b carries the smaller part of the energy and accordingly transfers the larger energy and mass to the target nucleus. This results in a larger mass distribution of the final nuclei. It is argued that the ejectiles with $Z_b > 2$ are mainly due to a quasi-elastic process, while the alpha-particles ($Z_b = 2$) originate from more damped reactions.

4.3.3d *Models for incomplete fusion and break-up fusion reactions*

The break-up fusion process has been analysed microscopically in terms of the DWBA calculation (Kerman and McVoy, 1979; Kishimoto and Kubo, 1979; Udagawa and Tamura, 1980, 1981; Udagawa *et al.*, 1982). The differential

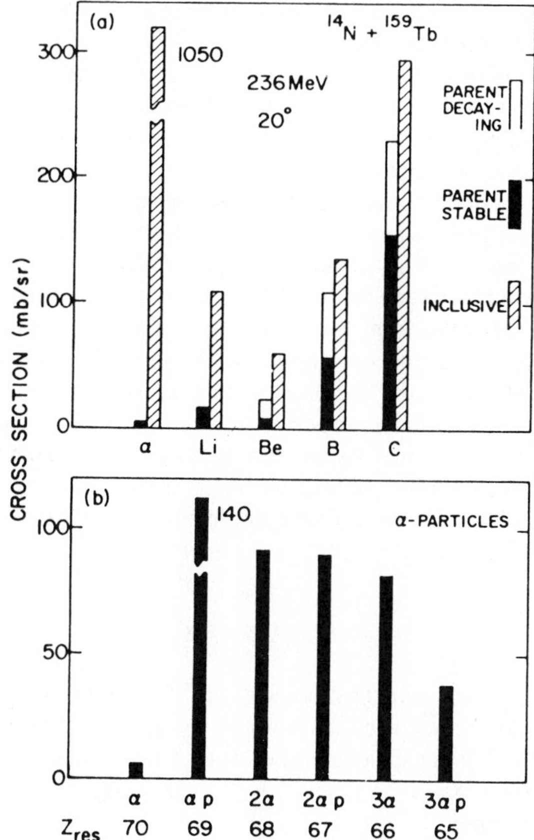

Fig. 4.39. (a) Cross-sections of primary fragments emitted in a particle-stable (black bars) or a particle-unstable (open bars) state. For Li the particle-unstable fraction could not be determined reliably. In addition the inclusive differential cross sections (shaded bars) measured at the same angle (20°) are also shown. (b) Alpha-particle cross-sections, corrected for multiplicities, for the different channels. (Siemssen *et al.*, 1985.)

cross-section for the break-up fusion, $a+A \rightarrow b+B$ with a composite particle $a=b+c$ and $B=A+c$, is written in terms of the DWBA as follows

$$\frac{d^2\sigma}{dE_b d\Omega_b} = \frac{m_a m_b}{(2\pi\hbar^2)^2} \cdot \frac{k_b}{k_a} \sum_l A_l \sum_m |\beta_{lm}(\mathbf{K}_b)|^2, \tag{4.3.16}$$

where m_i and k_i are the mass and momentum of the particle i and $\beta_{lm}(\mathbf{K}_b)$ is the amplitude of the break-up process. Here b is emitted with a momentum $\mathbf{K}_b$ while c is fused into A with a relative angular momentum lm. The elastic break-up process is given by $A_l = 1$ for all l. The additional contribution due to the break-up fusion process is given by $A_l = P_l/4$, where P_l is the penetrability of c into A (Kerman and McVoy, 1979). The coefficient is reformulated as $A_l = P_l/|s_l|^2$, where s_l is the elastic scattering s-matrix, and good agreement with observed data is obtained (Udagawa and Tamura, 1980, 1981).

The break-up fusion process is essentially a surface phenomenon since massive PLF's are preferentially ejected from the surface region. On the other hand, light PLF's such as alpha-particles may also be ejected from the inner region of the nucleus. Exact finite range DWBA calculations for the ^{181}Ta(^{14}N, α) reaction shows in fact that the alpha-particle is emitted from the inner region rather than from the peripheral region (Udagawa *et al.*, 1982).

In contrast to the microscopic DWBA for the break-up fusion model the sum rule model treats the incomplete fusion reactions macroscopically. Wilczyński *et al.* (1982) have proposed the sum rule model by extending the concept of the critical angular momentum to the incomplete fusion reaction of the A(a, b)B type. All HI reactions within a critical impact parameter are assumed to be either complete or incomplete (massive transfer) fusion. The relative weight $P(i)$ of each exit channel i is proportional to the exponential factor,

$$P(i) \propto \exp\{(Q^i_{gg} - \Delta Q^i_c)/T\}, \tag{4.3.17}$$

where Q^i_{gg} is the ground state Q-value with both b and B being in their ground states and ΔQ^i_c is the change in the Coulomb energy from the initial channel to the final one and T is the nuclear temperature. The angular momentum of the transferred particle X = a − b is given by $l_x = M_x l_a / M_a$. The i channel cross-section then becomes

$$\sigma(i) = \pi \lambda\!\!\!^{-2} \sum_{l=0}^{l_m} (2l+1) \frac{T_l(i)P(i)}{\sum_j T_l(j)P(j)}, \tag{4.3.18}$$

where $T_l(i)$ is written as $\{1 + \exp(l - l^i_c)/\Delta\}^{-1}$ in terms of the critical angular momentum l^i_c for the A + X system, and Δ is the width parameter.

The relative probabilities for the main reaction channels in the ^{12}C + ^{160}Gd system and the excitation functions are shown in Fig. 4.40. The excitation functions for the ^{160}Gd(^{12}C, α) and ^{160}Gd(^{12}C, 2α) reactions are obtained by the particle $-\gamma$ coincidence measurement, and they agree well with the model calculation (Wilczyński *et al.*, 1980). The cross-sections of the binary reactions for the ^{14}N + ^{159}Tb system have been measured for a large number of exit channels by means of the particle $-\gamma$ coincidence method (Wilczyński *et al.*, 1982). They agree well with the sum rule prediction as shown in Fig. 4.41.

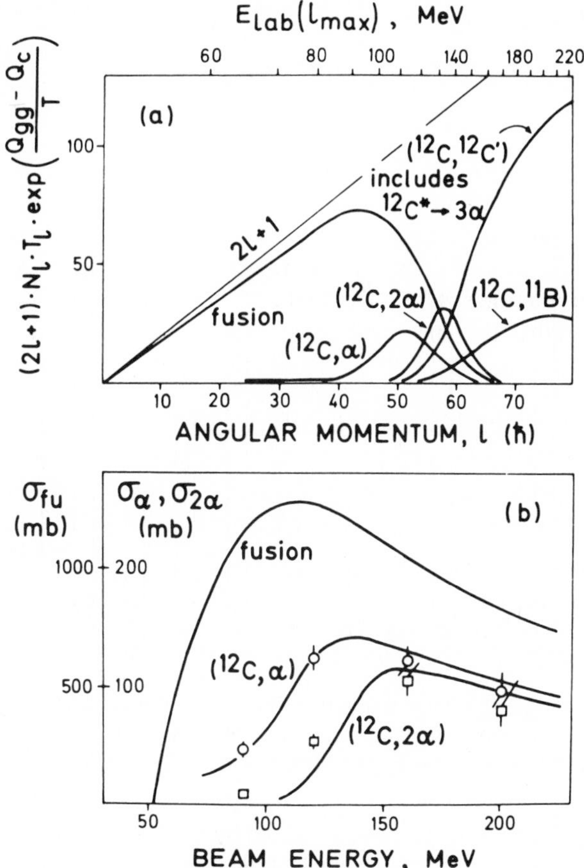

Fig. 4.40. (a) The reaction probabilities for the main reaction channels in the $^{12}C + {}^{160}Gd$ system ($T = 3$ MeV, $\Delta_l = 2\hbar$, $q_C = 0.06\ \text{fm}^{-1}$). (b) The excitation functions for complete-fusion and two main incomplete-fusion channels in $^{12}C + {}^{160}Gd$ collisions calculated with the same set of parameters. (Wilczyński *et al.*, 1980.)

4.3.4 *Particle transfer reactions studied by means of β–γ spectroscopy*

Particle γ-ray coincidence measurements for particle transfer reactions are useful for studying the properties of particles and holes in residual nuclei. The gamma rays to low-lying high-spin states are used to select high-spin states in the high excitation region produced by transfer reactions and the γ rays to low-lying particle (hole) states to select the particle (hole) states in that region.

High-spin proton states in ^{209}Bi have been studied by means the $^{208}Pb(\alpha, t\gamma)$ reaction (Ohsumi 1986). The $1j_{15/2}$ proton strength distribution

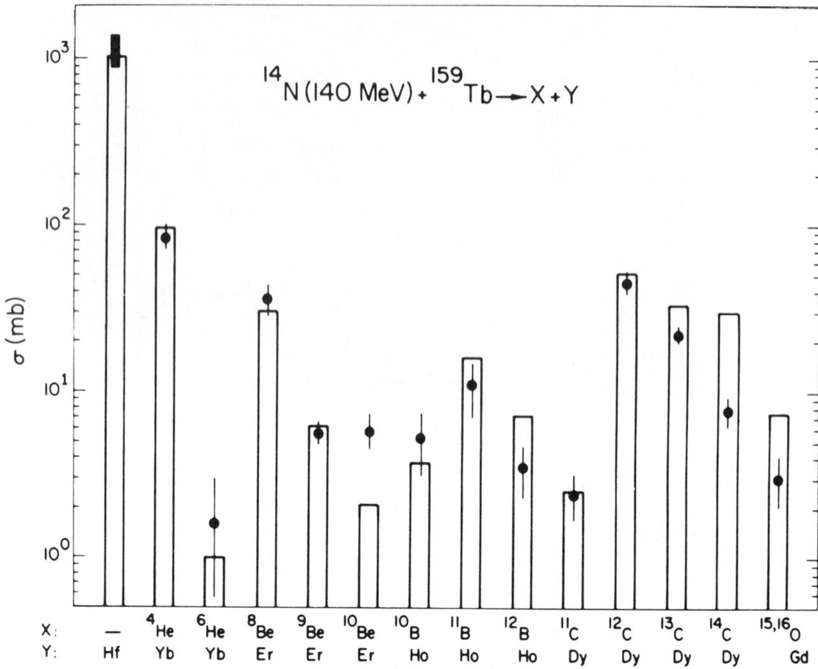

Fig. 4.41. Comparison of experimental cross-sections (circles) with predictions of the sum-rule model (rectangular bars). The range of possible magnitudes of the fusion cross section is indicated by the thick black bar. (Wilczyński *et al.*, 1982.)

was obtained from the (α, t) spectrum gated by the E1 γ decay to the low-lying $1i_{13/2}$ proton state. High-spin particle–hole states in ^{116}Sn were selectively populated by the ^{115}In(α, t) reaction in coincidence with γ-rays from rotational levels (Blasi *et al.*, 1983). High-spin 1p–1h states in ^{116}Sn have been studied by ($\vec{p}$, p′), (e, e′), (^{3}He, d) and (α, t) reactions (Van der Werf *et al.*, 1986).

Single-hole states and their γ decays have been studied by the (^{3}He, αγ) reaction (Sakai *et al.*, 1985). The 2.4 MeV $1g_{9/2}$ neutron–hole state in ^{101}Pd shows a sharp peak in the (^{3}He, α) spectrum, while the broad bump of the deep $1g_{9/2}$ neutron–hole is seen in the 3.5–4.5 MeV excitation region. The sharp hole state decays to the low-lying $1g_{7/2}$ state by a M1 transition, while the broad bump decays statistically to many states owing to the spreading of the hole state into many complicated states. These features are demonstrated by a strong γ ray from the 261 keV $1g_{7/2}$ state in coincidence with the sharp 2.4 MeV peak and by many weak γ rays in coincidence with the broad bump as shown in Fig. 4.42.

Polarization in HI-induced reactions gives crucial information on the reaction mechanism. Polarization of the radioactive ejectile (^{12}B) in transfer

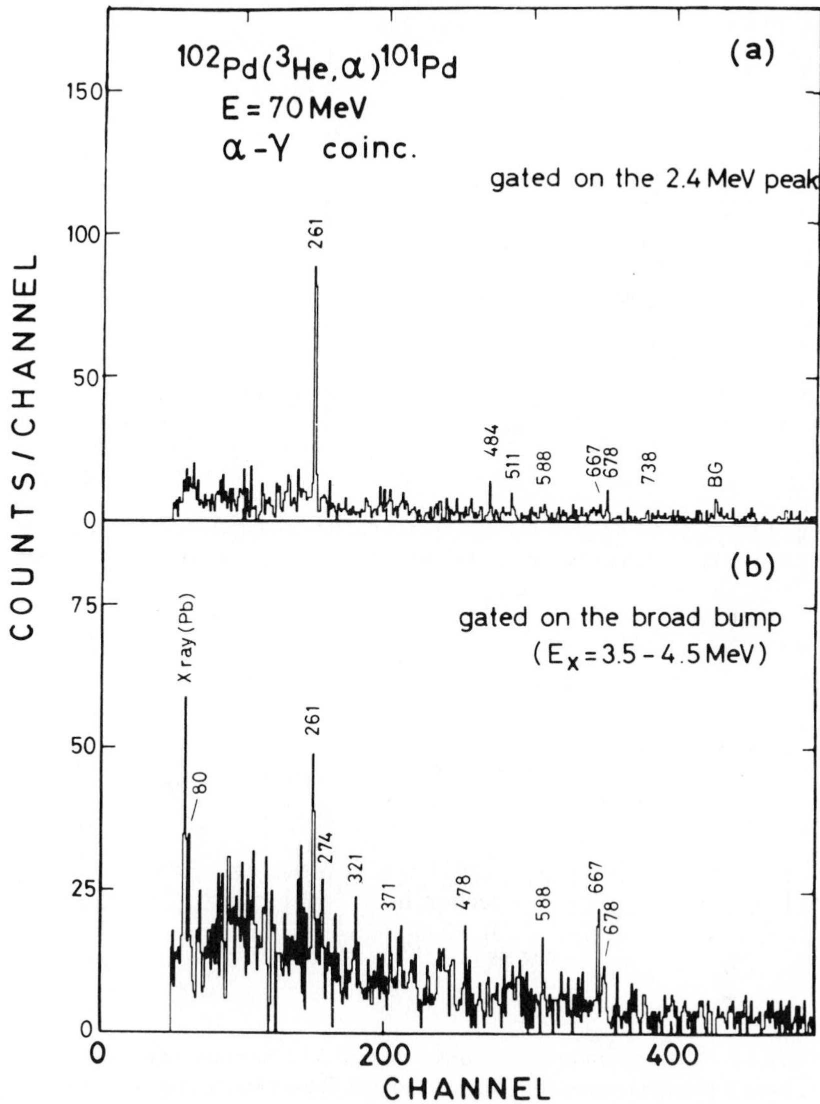

Fig. 4.42. Gamma-ray spectra gated by the 2.4 MeV peak and by the $E_x = 3.5$–4.5 MeV region (Sakai *et al.*, 1985.)

reactions has been studied by measuring the asymmetry of the β rays from the ejectile as shown in Fig. 4.43 (Sugimoto *et al.*, 1977). The observed data for the ^{100}Mo(^{14}N, ^{12}B)^{102}Ru reaction at 90 MeV are consistent with the simple transfer process.

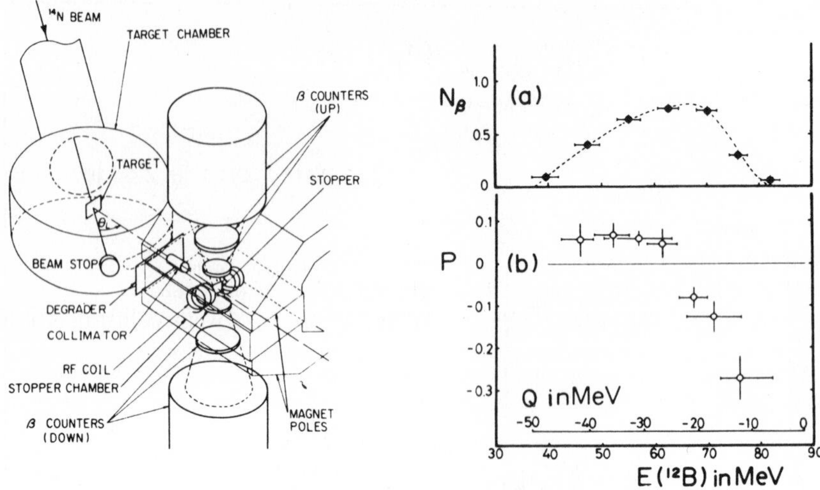

Fig. 4.43. Left-hand side: Schematic drawing of the experimental set-up. Right-hand side: Energy spectrum (a) on arbitrary scale and polarization (b) of ^{12}B in the ^{100}Mo(^{14}N, ^{12}B)^{102}Ru induced by 90 MeV ^{14}N at 20°, as a function of ^{12}B kinetic energy *E* and *Q* value. (Sugimoto *et al.*, 1977.)

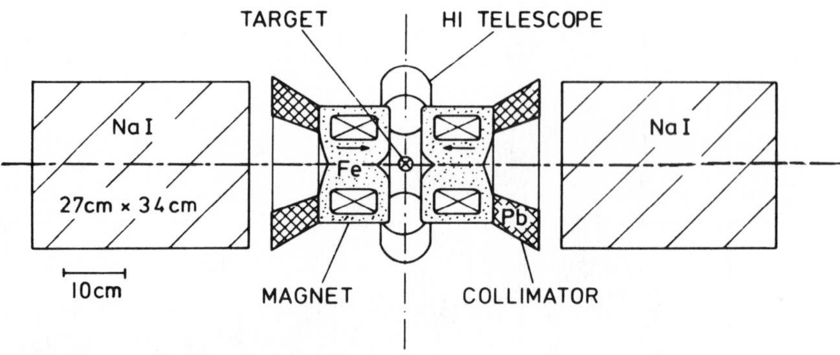

Fig. 4.44. The detector arrangement for the circular polarization measurement. The beam direction points into the plane of the figure (Trautmann *et al.*, 1981.)

The polarization of excited projectile-like fragments in the reaction ^{16}O + ^{58}Ni at 100 MeV has been studied by measuring the circular polarization of the de-exciting γ rays (Trautmann *et al.*, 1981). The circular polarization has been measured by means of the transmission method as illustrated in Fig. 4.44. The observed polarizations are large upto 100 per cent. The sign and the magnitude of the polarization are interpreted in terms of the matching conditions of the kinematics in the direct heavy-ion reaction.

4.3.5 *Gamma rays following relativistic heavy ion reactions*

The extension of γ-ray spectroscopy for relativistic heavy-ion reactions in the GeV region provides new information on both dynamic and static properties of nuclei. The central collision of an energetic heavy-ion with a target nucleus may produce many nuclear fragments with large velocities. Such a strong collision process makes the observation of discrete γ-rays difficult. On the other hand a peripheral collision, in which the transferred energies and momenta are fairly small, may produce a target fragment with small velocity and a projectile fragment with a velocity around the beam velocity. The reaction scheme is illustrated in Fig. 4.45(a). This type of weak collision is used

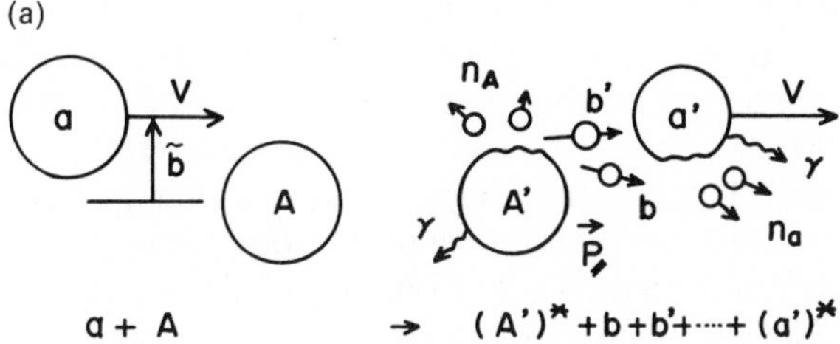

Fig. 4.45(a). Schematic picture of a peripheral HI collision producing a target fragment A′ and a projectile fargment a′.

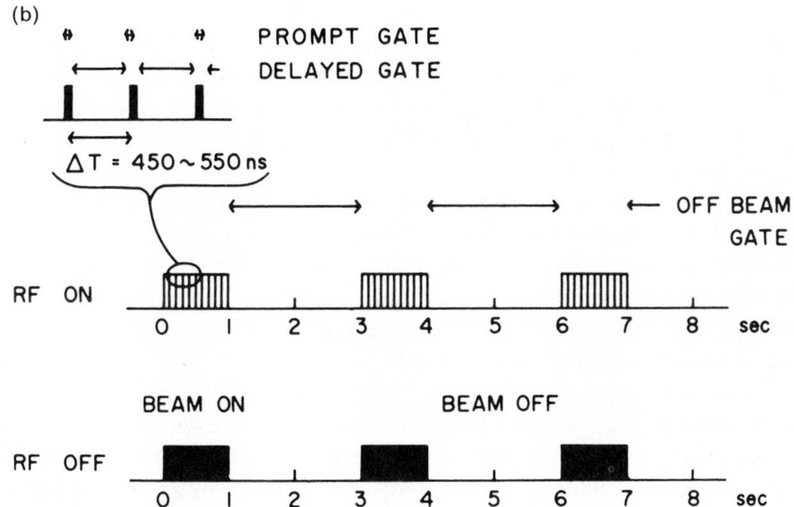

Fig. 4.45(b) The time structure of the Bevatron beam and gate positions for prompt, delayed, and off-beam spectra. (Shibata *et al.*, 1978.)

for discrete γ-ray spectroscopy to study the reaction mechanism and the structure of fragment nuclei. This spectroscopy includes identification of fragment nuclei, linear and angular momentum transfers, spin alignment and spin polarization. The first observation of discrete γ rays following relativistic HI was made by the Osaka–LBL–Tokyo collaboration (Shibata *et al.*, 1978). A brief summary of the experiment is given below.

Relativistic beams of alphas and ^{12}C in a 1.6–12.6 GeV region were provided from the Bevatron, and discrete γ rays following collisions of these relativistic HI with s–d shell nuclei were measured by a Ge (Li) detector. Application of discrete γ-ray spectroscopy to such relativistic beams is experimentally hard since the beam intensity is only of the order of 10 to 100 pico A, the beam spot is as large as a few centimetres in diameter, and all kinds of discrete and continuum γ-rays are emitted. The synchrotron beam has a time structure as shown in Fig. 4.45(b). The r.f. power for the accelerator resonator is turned off for the coincidence measurements. The beam is also extracted with the r.f. power turned on, so that the microscopic beam structure is used. This gives a prompt spectrum within the microscopic beam pulse and a delayed one between the microscopic beam pulses. An off-beam spectrum between the beam spills gives γ rays following radio-active decays. These three different types of spectra are used to identify reaction processes. A typical γ-ray spectrum is shown in Fig. 4.46. It is interesting to find many discrete γ rays superposed on a large continuum, which are due to surface collisions with rather weak disturbance of the nucleus (see Fig. 4.45(a)).

The reaction mechanisms revealed by observation of discrete γ rays are as follows.

(i) Various kinds of the target-fragment nuclei are produced. The production cross-section decreases monotonically as the minimum Q-value Q(min) decreases. Here Q(min) is defined as the minimum excitation energy of the target to form the given fragment. This may be attributed to the decrease of the available phase-space volume.

(ii) No residual nuclei with mass number greater than that of the target are observed. This fact shows that neither compound nucleus formation nor break-up fusion (transfer of projectile fragments into the target nucleus) have measurable cross-sections. This is due to the large momentum-mismatch between the fermi momentum of the target nucleon and the nucleon momentum of the relativistic projectile.

(iii) Gamma rays for light (s–d shell) target nuclei with $T_z=0$, or $\frac{1}{2}$ are mostly from product nuclei with $T_z=0, \frac{1}{2}$ or 1, and those with $T_z=-\frac{1}{2}$ are not produced. On the other hand γ rays from medium–heavy nuclei are mostly from products of neutron removal reactions. These observations indicate excitation of the target (fragments) by the abrasion and de-excitation by nucleon evaporation, i.e. protons from light nuclei because of the lower

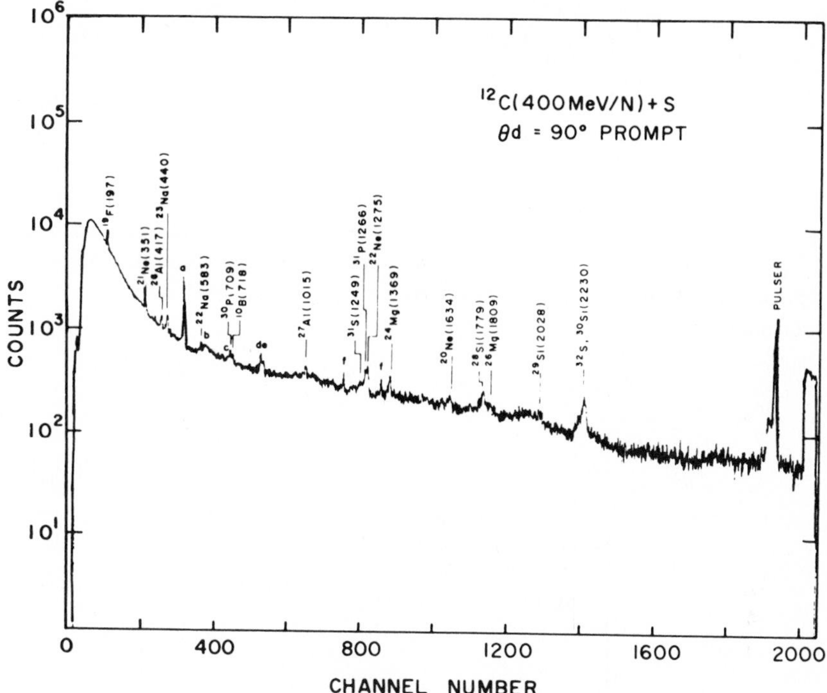

Fig. 4.46. The γ-ray spectrum following the ${}^{nat}S({}^{12}C, X)Y$ reaction at $E({}^{12}C)=4.8$ GeV. The γ rays denoted by letters (a) to (f) are as follows: (a) 511 KeV, (b) 596 keV ${}^{74}Ge(n, n'\gamma)$, (c) 690 keV ${}^{72}Ge(n, n'\gamma)$, (d) 835 keV ${}^{54}Mn$, (e) 846 keV ${}^{27}Al$, (f) 1117 and 1332 keV ${}^{60}Co$. (Shibata *et al.*, 1978.)

binding energy and neutrons from medium heavy nuclei because of the suppression of proton emission by the Coulomb barrier (see Fig. 4.45(a)).

(iv) In going from light to heavier projectiles an increasing number of nucleons or alpha-particles are knocked out in the fast cascade (abrasion) as shown in Fig. 4.47.

(v) The Doppler effects on discrete γ rays from short-lived states are fairly small. The Doppler shift gives the momentum P along the beam direction and the Doppler broading gives the spread $\sigma(P)$. They are found to be $P<50$ MeV/c and $\sigma(P)\lesssim 250$ MeV/c. The cross-sections for given fragments produced by ${}^{12}C$ are only twenty per cent larger than those produced by alpha-particles, and this indicates the peripheral nature of the reaction.

(vi) Projectile and target fragments produced by relativistic HI offer good possibilities for spectroscopic studies of exotic nuclei. The product nuclei span

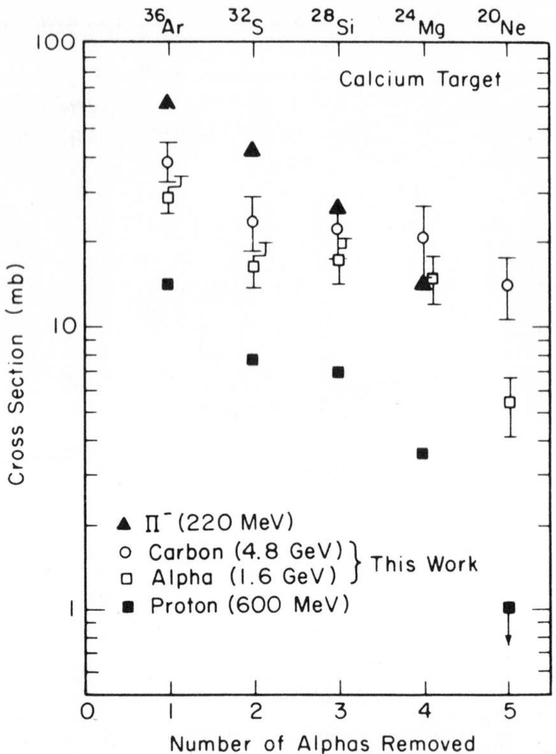

Fig. 4.47. Comparison with pion and proton data for alpha-particle nuclei: Closed triangles for 220 MeV π^-, open circles for 4.8 GeV ^{12}C, open squares for 1.6 GeV α and closed squares for 600 MeV protons. (Shibata *et al.*, 1978 and refs. therein for experimental data.)

a quite wide region in the N–Z plane of the nuclear chart around $N/Z \approx N_i/Z_i$, where N_i and Z_i are the proton and neutron numbers of the target or projectile nucleus, respectively. Thus nuclei far from stability, particularly in the neutron-rich side, are produced in contrast with the neutron-deficient nuclei produced by compound (fusion) reactions (see Section 4.2). Projectile fragments provide beams of unstable nuclei at the original beam direction with a velocity similar to the projectile. It should be noted that such new beams may be useful for astrophysics and in some other fields. Heavy-ion accelerators combined with high-quality mass-separators may soon provide beams of unstable nuclei for detailed γ–e spectroscopic studies.

It is noted that the cross-section of electromagnetic (Coulomb) excitation increases as the projectile velocity increases. Thus relativistic HI is used to study electromagnetic properties (giant resonances, etc) of target nuclei, projectile nuclei and even projectile fragments.

4.4 Gamma-ray spectroscopy associated with inelastic scattering

4.4.1 *Inelastic scattering and gamma-ray spectroscopy*

The inelastic scattering of medium-energy hadrons has been used extensively to excite particle–hole states and collective states in nuclei. Inelastic scattering gives rise to one-particle (1p) one-hole (1h) excitations as shown in Fig. 4.48. Here the 1p–1h state is the initial doorway of the inelastic reaction process. The collective state is a special case of the 1p–1h excitation, and it consists of a coherent superposition of 1p–1h pairs coupled to a definite spin and parity. The wave function is expressed as

$$|C_\alpha\rangle = \Sigma C^{\alpha}_{ii'} [u^{+}_{i'} u_i]_\alpha , \tag{4.4.1}$$

where α is the quantum number characterizing the collective motion, and u_i^+ and u_i are quasi-particle creation and annihilation operators, respectively. The collective motion with the quantum number α is induced by an α-type particle–hole interaction such as $\mathbf{H}_\alpha = \chi_\alpha \cdot \mathbf{T}^+_\alpha \mathbf{T}_\alpha$, where $\mathbf{T}^+_\alpha$ is the creation operator of the α-mode phonon (collective motion) (Bohr and Mottelson, 1975). An attractive interaction with $\chi_\alpha < 0$ pushes the collective state down in

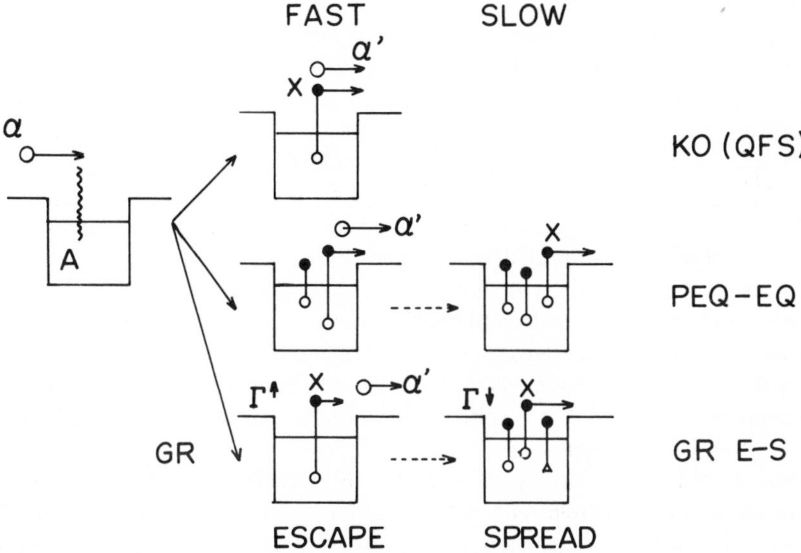

Fig. 4.48. Schematic doorways associated with the A(α, α'x)B reaction. FAST and SLOW mean the fast and slow processes, respectively. GR is a special coherent sum of 1p–1h states with a definite spin and parity. KO(QFS), PEQ, EQ, GR E–S stand for the knock-out (quasi-free scattering) process, the pre-equilibrium process, the equilibrium process, and the giant resonance escape–spreading processes, respectively. (Ejiri, 1984.)

excitation energy, while a repulsive interaction with $\chi_\alpha > 0$ pushes the collective state up. Actually there are many types of residual interactions with quantum numbers $\alpha = T, S, L, J$ where $T = 0$ and 1 stand for isoscalar and isovector modes, $S = 0$ and 1 for electric and magnetic modes, L for the orbital angular momentum, and J for the total angular momentum. Accordingly, many types of collective states appear in a wide region of excitation. Collective states in spin and/or isospin modes are discussed in Chapter 5.

Giant resonances (GR) are some kind of collective states, appearing in the high-excitation region as strong resonances in inelastic scattering. The resonance width is as great as a few MeV because of the large damping width of the particle–hole states. Damping is in fact the mechanism that spreads the particle–hole states through the nuclear interaction with the other states in the same excitation region. Studies of GR's have mostly been made by measuring single spectra of inelastically scattered particles as a function of angle. The angular momentum of the GR is obtained from the angular distribution, and other properties such as T, S, etc. are assigned by using appropriate probes for the inelastic scattering. Over the past decade various kinds of GR's have been investigated by using inelastic scattering of medium-energy light ions (see, for example, Bertrand, 1976, 1980, 1981; Bertsch, 1979; Hanna, 1983; van der Woude 1987; and references therein).

There are several problems in studying only the singles spectrum of the inelastic scattering (Bertrand, 1980). One GR often overlaps with another GR since there are often two or three broad GR's with similar excitation energies. They are excited by inelastic scattering, some strongly at one angle and others at some different angle. In other words, the inelastic scattering excites not only one type of GR but also other types at the same time, since the wave length of the incident particle is as short as the nuclear radius and accordingly many angular momenta L are involved in the excitation. This feature is in contrast with the γ transition, where the lowest L is predominant (see Section 2.1). The GR is likely to be located on a huge continuum background which is due either to a quasi-free scattering or to a multiple scattering. The subtraction of the continuum background is rather difficult. Extraction of the resonance strength requires a precise DWBA analysis of the inelastic scattering and it is rather model-dependent.

Gamma-ray spectroscopy associated with inelastic scattering, which deals with coincidence (correlation) measurements of γ rays with the inelastically scattered particle, is one step more advanced than the singles-spectrum measurement. It gives vital information on properties of GR's. A direct γ decay of a GR gives straightforwardly the GR strength, as it is rather independent of the reaction model. Good selection rules of the γ decay owing to the long wave-length property of the γ ray lead to a unique determination of the multipole strength of the GR. The large background underneath the GR may be removed by recording coincidences with direct γ rays since neither quasi-free scattering nor multiple scattering are followed by such direct γ decay. An important mechanism of damping of the GR is a coupling of the GR with other modes (ψ_i) of excitations such as surface vibrations and particle–hole excitations. It can be written as $|\mathrm{GR}\rangle = \alpha|\mathrm{GR}\rangle + \Sigma\beta_i|\mathrm{GR} \times \psi_i\rangle$, and the coupling may be investigated by looking for the direct γ-decay to the excited state ψ_i.

The experimental studies of direct γ decays of GR's require special devices. The γ-branching ratio is only of the order of one part of 10^4 to 10^5 since particle decay owing to the strong interaction is much more probable than is electromagnetic decay. Thus, extremely weak γ rays have to be selected among a huge flux of particles, mostly neutrons in the case of medium and heavy nuclei.

Since GR's in the unbound excitation region decay mainly by emitting one or two particles, their particle natures are investigated by measuring such particles in coincidence with the inelastic scattering. Gamma rays following particle emission of the GR are also used for studying the particle decay. Here γ rays following an (a, a$'$xγ) reaction are measured in coincidence with a$'$. Discrete γ rays characteristic of the residual nucleus give uniquely the final nucleus and hence information about the emitted particle x. The level- and decay-schemes relevant to γ spectroscopy associated with inelastic scattering are schematically illustrated in Fig. 4.49.

4.4.2 *The direct gamma-decay of giant resonances*

Gamma transitions from unbound states such as GR's in a highly excited continuum (unbound) region have in principle a range of multipolarities since many spins and parities are involved in the overlapping resonances as well as in the underlying continuum components. In practice, however, they are limited to the low multipole E1, M1 and E2 transitions since the wave length of the γ ray in the $E(\mathrm{GR}) = 10$ to 20 MeV region is still much longer than the nuclear radius.

The gamma decay strengths are conventionally described in units of the classical sum rule limits. The sum rules for electric multipole excitations in a nucleus with mass number A are written as (Bohr and Mottelson, 1975).

$$S(\mathrm{E1}) = \frac{9}{4\pi}\cdot\frac{\hbar^2}{2M}\cdot\frac{NZ}{A}\,\mathrm{e}^2, \tag{4.4.2a}$$

$$S(\mathrm{E}\lambda) = \frac{\lambda(2\lambda+1)^2}{4\pi}\,\frac{\hbar^2}{2M}\,Ze^2\langle r^{2\lambda-2}\rangle_{\mathrm{prot}} \quad \lambda \geqslant 2, \tag{4.4.2b}$$

where N and Z are the numbers of neutrons and protons, respectively. The centre-of-mass degree of freedom for the E1–γ is corrected for by using the effective charge N/A, and that for the higher Eλ, which is

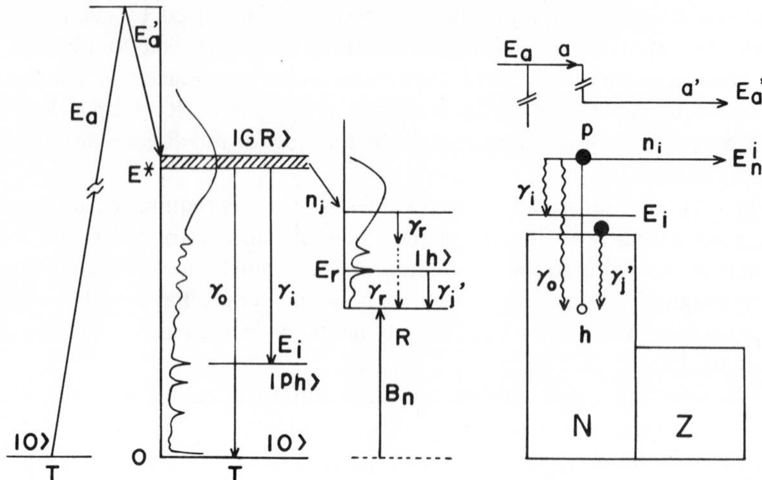

Fig. 4.49. Level and γ-transition schemes associated with the inelastic scattering (a, a′). p and h are a particle and a hole produced by the inelastic scattering. E^*, the excitation energy. γ_o, the direct γ decay to the ground state. γ_i, the γ decay to an excited state $\psi_i = |\text{ph}\rangle$, n_i, neutron decay from the ith particle state. γ_j', the γ decay from the j-th hole state. T and R are the target nucleus and the residual nucleus after one neutron emission, respectively.

in the order $ZA^{-\lambda}$, is neglected. For medium–heavy nuclei with $Z \approx 0.4A$ one gets $S(\text{E1}) \approx 3.5\ \text{e}^2\text{fm}^2$ MeV, $S(\text{E2}) \approx 50\ A^{5/3}\text{e}^2\text{fm}^4$ MeV, and $S(\text{E3}) \approx 200\ A^{7/3}\text{e}^2\text{fm}^6$ MeV. Noting that the reduced γ-de-excitation probability is smaller than that of the γ-excitation probability by a factor $2\lambda + 1$, the sum rule values for the γ-decay probabilities are

$$T(\text{E1}) \approx 1.86 \cdot 10^{15} \cdot E^2, \tag{4.4.3a}$$

$$T(\text{E2}) \approx 1.22 \cdot 10^{10}\ A^{5/3} E^4, \tag{4.4.3b}$$

$$T(\text{E3}) \approx 1.62 \cdot 10^4\ A^{7/3} E^6, \tag{4.4.3c}$$

where E is the γ-ray energy in units of MeV. Therefore the major multipolarities in the 10 to 15 MeV excitation region are E1 and E2. The E3 component may become appreciable near the E3 GR. Conversion electrons and electron pairs are quite useful for selecting E0 strength. Thus the E0 GR with 0^+, which has no γ-decay branch to the ground state 0^+ state, is exclusively studied by electron (pair) spectroscopy.

One type of GR with the quantum number α is in practice selected by a proper choice of the projectile and scattering angle, combined with measurement of γ rays. Inelastic scattering of the alpha particle with $T=0$ excites predominantly the isoscalar electric GR (Bohr and Mottelson, 1975; Lewis and Bertrand, 1972; Pitthan and Walcher, 1971), and only weakly the isovector ones. In fact the isovector E1 GR is weakly excited by Coulomb excitation with charged particles. Inelastic scattering at very forward angles (less than a few degrees) enhances the E0 and M1 excitations. It should be noted that inelastic scatterings of electrons excite both the isoscalar and isovector GR's (Fukuda and Torizuka, 1972).

The isoscalar E2 GR (giant quadrupole resonance GQR) is well studied by γ spectroscopy following inelastic hadron scatterings, which strongly excites the isoscalar Eλ GR's with $\lambda \geqslant 2$. Note that the E1 GR is dominantly of the isovector type, and it is studied by radiative capture processes as discussed in Section 4.5.4. Direct γ decays from the GQR in ^{208}Pb have been studied at ORNL (Bertrand *et al.*, 1984; Beene *et al.*, 1985). The GQR was excited by inelastic scattering of ^{17}O. Here the choice of the loosely bound ^{17}O projectile makes the γ-ray spectrum beyond the neutron threshold of 4.1 MeV free from the interference from the projectile excitation.

The γ detector has to be highly efficient and selective for the direct γ decay since the decay branch is as small as about 10^{-4} and thus the γ ray has to be selected from 10^4 times as many neutrons and low-energy γ rays.

An ORNL group studied γ rays from GQR in ^{208}Pb by using the ORNL spin spectrometer (Beene *et al.*, 1985). High efficiency was achieved by having the spherical shell of NaI divided into 72 independent modules. The selection of the direct γ decay was accomplished as follows: The total γ-ray energy is obtained by summing all prompt pulses from the NaI modules as $H = \Sigma h_i$, where h_i is the pulse height in detector i. The value H has to agree with the excitation energy given by the inelastic scattering. Most of the neutrons following the inelastic scattering are slow, and their signals are delayed by the time-of-flight. Thus they do not come in the prompt time window for the γ rays. Since the direction of the γ ray is identified by the location $\mathbf{r}_i$ of the NaI segment i with respect to the target centre, the quantity defined as $V = |\Sigma h_i \mathbf{r}_i|/H$ is used to get the multiplicity of the γ rays. The single γ decay to the ground state gives $V \sim 1$, while the cascade γ decays give $V \approx 0$ as shown in Fig. 4.50. A single γ ray is not always absorbed by only one NaI segment but rather by a cluster consisting of the five or six nearest neighbouring ones. Thus the cluster sum of the pulse heights and the cluster multiplicities are realistic quantities to define the γ-ray energy and the multiplicity.

The γ-ray energy spectrum and the ratio to the ground state decay are shown in Fig. 4.51. The ground state decays from the 10.6 MeV GQR and the 13.4 MeV GDR (isovector E1 GR) are clearly seen. The energy levels and GR's in ^{208}Pb are shown in Fig. 4.52. The observed branching ratio of the ground state decay from the GQR is $\Gamma(\gamma_0)/\Gamma(t) = (3.27 \pm 0.45) \cdot 10^{-4}$. Using the observed total width of $\Gamma(t) = 2$ MeV, the γ decay probability for the GQR is

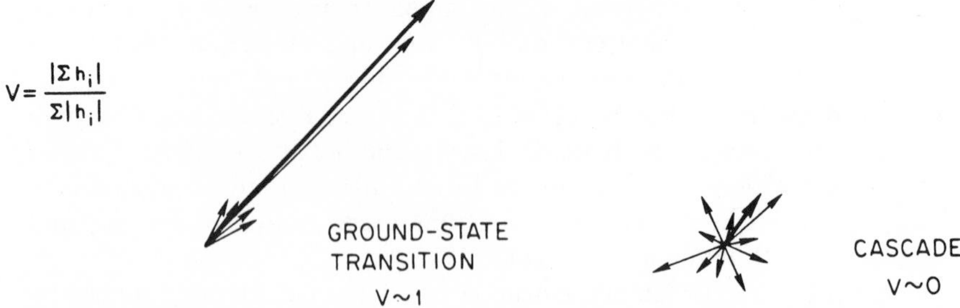

Fig. 4.50. Schematic representation of the method for selecting ground-state γ-ray transitions. The heavy arrow is the vector sum of the transitions in typical single events for the ground-state transition (left-hand side) or cascade decay to the ground state (right-hand side). The magnitude of the parameter V distinguishes between the two. (Bertrand *et al.*, 1984.)

found to be

$$\Gamma(\gamma_0) = 654 \pm 91 \text{ eV}, \tag{4.4.4a}$$

$$B(\mathrm{E2}) = (5.81 \pm 0.81)\cdot 10^3 \text{ e}^2\text{fm}^4. \tag{4.4.4b}$$

The energy weighted sum for the γ excitation is obtained as $E_\gamma \cdot B(\mathrm{E2})\cdot 5$, where 5 is the spin factor for the 2^+ excitation. This can be compared with the theoretical value of eqn (4.4.2b). The observed value amounts to 85 ± 12 per cent of the theoretical value, indicating that the GQR indeed absorbs most of the E2 strength. The cascade γ decays via low-lying excited states are selected by requiring a cluster multiplicity of two, provided that the low-lying states decay directly to the ground state. The GQR is found to decay to the 4.97 MeV 3^- and to 1^- states in the 5 to 7 MeV excitation region, but not to the 2.61 MeV 3^- state. These observations suggest weak coupling of the GQR with the 2.61 MeV 3^- surface vibration, and strong coupling with other 3^- and 1^- particle–hole states mentioned above. The γ decay matrix element for the transition from the GQR to the low-lying p–h state is given by

$$M(J_i \to J_f) = \Sigma a_{ii'} \langle [u_{j'}^+ u_j]_{J_f} \| T_i \| [u_{i'}^+ u_i]_{J_i} \rangle \, \delta_{i'j'} + \Sigma a_{ii'} \langle [u_{j'}^+ u_j]_{J_f} \| T_{i'} \| [u_{i'}^+ u_i]_{J_i} \rangle \, \delta_{ij}, \tag{4.4.5}$$

where the GQR is assumed to be written as $|\mathrm{GQR}\rangle = \Sigma a_{ii'} [u_{i'}^+ u_i]_{J_i}$ with $J_i = 2$ and the low-lying state is given by $[u_{j'}^+ u_j]_{J_f}$. Thus one may study microscopic p–h configurations from the particle ($T_{i'}$) or hole (T_i) transitions to a particle or hole component of low-lying p–h states.

Giant monopole resonances (GMR) have been located a few MeV above the GQR (Marty *et al.*, 1975; Sasao and Torizuka, 1977; Harakeh *et al.*, 1977; Youngblood *et al.*, 1977; Willis *et al.*, 1980). It is found that the three broad GR's, namely the GDR, GMR, and GQR, coexist. Electromagnetic decays of

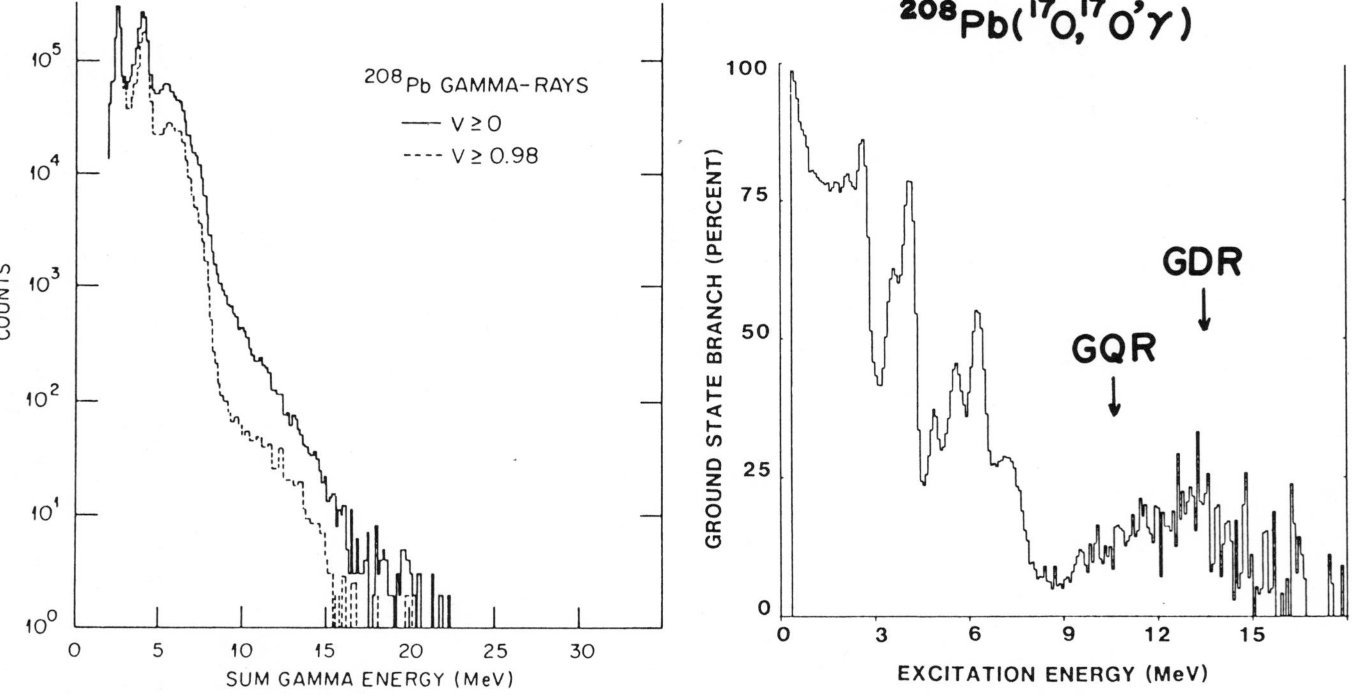

Fig. 4.51. Left hand side: Gamma-ray spectra from ^{208}Pb for $V > 0.98$ (only ground-state γ rays) and for $V \geqslant 0$ (all γ rays). Right hand side: Ratio of ground-state γ-decay events to total γ-ray yield as a function of ^{208}Pb excitation energy. (Beene *et al.*, 1985.)

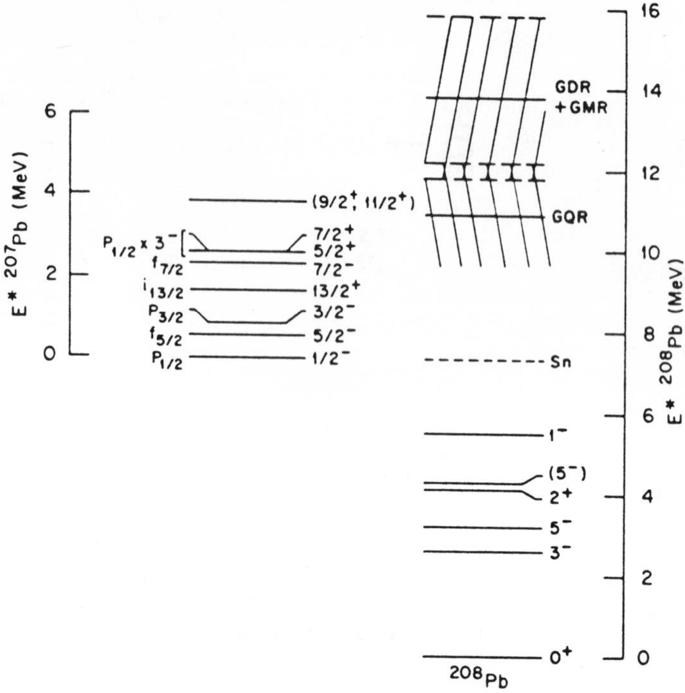

Fig. 4.52. Selected levels in ^{208}Pb and ^{207}Pb. The configuration labels on the ^{207}Pb states refer to neutron hole states. (Beene *et al.*, 1985.)

the GMR are either by the nuclear-pair creation mainly for light nuclei or by the conversion electron emission. The E0 sum rule is (Bohr and Mottelson, 1975)

$$S(\mathrm{E0}) = \frac{2\hbar^2}{M} Ze^2 \langle r^2 \rangle_{\mathrm{prot}}. \tag{4.4.6}$$

Conversion-electrons or nuclear electron pair in coincidence with the inelastic scattering at very forward angles are used to select the E0 strength of the GMR since other GR's and continuum regions hardly ever decay by emitting such electrons.

4.4.3 *Particle and gamma-ray spectroscopy associated with inelastic scattering and giant resonances*

The inelastic excitation of unbound p–h states is followed most frequently by particle emission, either at the pre-equilibrium stage or at the final equilibrium stage (see Fig. 4.48). Particles emitted at the pre-equilibrium stage of the internucleon cascade (exciton) development are fast particles and retain a memory of the initial doorway p–h configuration. Particle emission at the equilibrium stage is a slow process and low-energy particles (mostly neutrons in medium heavy nuclei) are evaporated. Here the memory of the initial p–h configuration is almost lost. In any case, γ rays follow the particle emission unless the particle decays directly to the ground state of the residual nucleus.

The GR decay width is composed of the escape width $\Gamma^{\uparrow}$ and the spreading width $\Gamma^{\downarrow}$. The direct neutron decay of the initial p–h pair corresponds to the escape of one neutron from the neutron particle–hole (1p–1h) pair, while the internucleon cascade forming (2p–2h), (3p–3h) states and so on, corresponds to the spreading process. Thus the direct decay is of particular importance since it reflects the microscopic features of the GR which is composed of a coherent sum of special 1p–1h pairs. The escape particle gives the particle component, and the hole state in the residual nucleus reflects the hole component. Charged particles such as p, α, etc. may be directly emitted from GR's in light nuclei (Wagner, 1979/1980; Knöpfle *et al.*, 1978, 1979, 1981; van der Woude 1987, and references therein). They give the proton and the alpha-cluster components. Since most of GR's lie only 5 to 10 MeV above the particle threshold energies, charged particle emission is greatly suppressed in medium heavy nuclei because of the Coulomb barrier. The decay particles are then exclusively neutrons, which are experimentally hard to measure. Furthermore the direct escape process is a minor process for GR's in medium heavy nuclei since the spreading process is enhanced by the high level density of 2p–2h, etc.

Direct neutron decays following inelastic scattering have been studied in Osaka by means of multi-neutron detector arrays combined with multi position-sensitive detectors for charged particles. The energy and angular correlations for decay neutrons and inelastically scattered alphas were studied for the GQR's in ^{92}Zr and ^{119}Sn (Ejiri, 1984; Ohsumi *et al.*, 1985; Okada *et al.*, 1982). These nuclei are simple semi-magic nuclei with low neutron threshold energies. The decaying neutron energy (E_n) and the inelastically scattered alpha energy (E'_α) are related simply by (see Fig. 4.49)

$$E_{\mathrm{n}} = E^* - E_r - B_{\mathrm{n}} = E_{\mathrm{a}} - E'_{\mathrm{a}} - E_r - B_{\mathrm{n}} \tag{4.4.7a}$$

$$E'_{\mathrm{a}} = E_{\mathrm{a}} - E^* = E_{\mathrm{a}} - E_{\mathrm{n}} - E_r - B_{\mathrm{n}}, \tag{4.4.7b}$$

where E^* is the excitation energy given by the inelastic scattering and E_r is the excitation energy of the residual (hole) state populated by the neutron decay. The incident energy (E_{a}) and the neutron binding energy (B_{n}) are fixed and the recoil energy of the residual nucleus is neglected. Therefore the neutron energy spectrum gated by the inelastic alpha with the fixed energy E'_{a} reflects the hole state distribution (E_r). Similarly the a′ energy spectrum gated by the decaying neutron with the fixed energy E_n is related to the hole states energy E_r (see eqn 4.4.7b).

The singles ^{92}Zr(α, α') spectrum in the GQR region and the coincidence spectrum with the 3 to 4 MeV fast neutron decay from the GQR are shown in Fig. 4.53. The neutron energy spectrum measured in coincidence with the inelastic alphas exciting the GQR in ^{119}Sn is shown in Fig. 4.54. Here the neutrons were detected at the anti-recoil direction of the (α, α') reaction to be free from the fast non-GR neutrons that are emitted predominantly in the recoil and forward directions. The inelastic alpha particles were detected in the direction where the angular distribution, characteristic of the E2 excitation, shows a peak. As shown in Fig. 4.54, fast neutron groups feeding the hole states in the residual nuclei are clearly seen. These fast neutron peaks are assigned to the direct escape neutrons from the GQR's on the basis of the full energy and angular correlations for n and α'.

The fast neutron reactions to low-lying hole states greatly exceed the Hauser–Feshbach statistical model calculation and thus they are attributed to the direct escape process. Comparison of the (α α'n) data with the (p, d) data

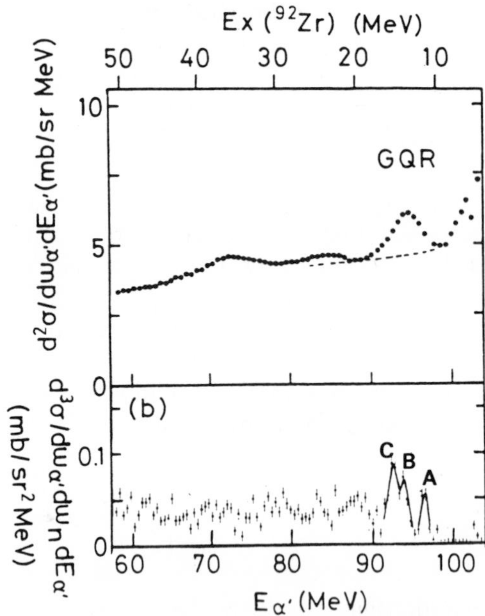

Fig. 4.53. Top: Singles energy spectrum of the inelastic alpha-particle scattering on ^{92}Zr near the GQR peak of $\theta(\alpha')=15.9°$. Bottom: Coincidence α' spectrum gated by fast neutrons of $E_n=3-4$ MeV at the anti-recoil direction of $\theta(\alpha')=110°$. A and B correspond to the neutrons feeding the $2d_{5/2}$ and $1g_{9/2}$ hole states in ^{91}Zr, respectively. (Ohsumi *et al.*, 1985.)

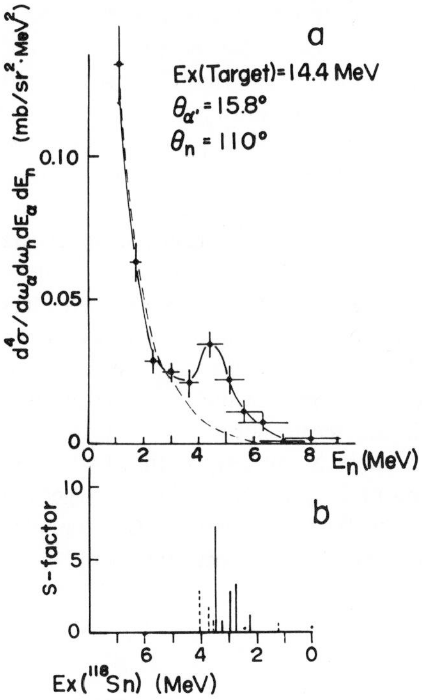

Fig. 4.54. Top: Energy spectrum of neutrons gated by inelastic α' feeding the GQR in ^{119}Sn. The dashed line shows the spectrum for the continum region of $E^* = 20$ MeV, which is well above the E(GQR) = 14 MeV. Bottom: Spectroscopic factors of the single-neutron holes obtained from the (p, d) reaction. Solid, dashed and dotted lines stand for $l_n = 4$, 2 and 1 transfers, respectively. (Okada *et al.*, 1982.)

reveals a large fraction of the $1g_{7/2}$ hole component in the GQR of ^{119}Sn, and three components of the $2f_{5/2}$, $1g_{9/2}$ and $2p_{1/2}$ holes in the GQR of ^{92}Zr. The direct escape probabilities are only 10 to 20 per cent of the total decay widths in medium–heavy nuclei and therefore precise measurements sensitive to the direct escape are required to distinguish it from major statistical decays. In fact, the major decay process of GR's in Ni and Pb is statistical (Collins *et al.*, 1979; Steuer *et al.*, 1981).

Gamma-ray spectroscopy following ($\alpha\ \alpha'$x) reactions requires in principle a triple correlation measurement for the γ-ray (γ'), the decay particle (x) and the inelastically scattered particle (α'). The γ-spectroscopy can resolve all the residual states which are hardly resolved by the neutron measurements and thus it provides details on the hole states constituting the initial doorway and the GR in the inelastic scattering. So far the γ rays in residual nuclei have been measured in coincidence only with the inelastically scattered particle (Eyrich *et al.*, 1979; Ejiri, 1984; Ohsumi *et al.*, 1985). Here the decay particle x remains unobserved, as is also the case with in-beam γ-ray spectroscopy following (a, xnγ) reactions. The gamma-ray spectra following (a, a′nγ) reactions, where neutrons are unobserved, may show strong peaks corresponding to the γ decay from hole states. However these γ rays do not necessarily mean that the hole states are populated directly after neutron emission as they may be fed by γ decay from higher excited states as well.

Discrete γ rays, characteristic of the residual nuclei, can be used to identify the reaction channel and hence the neutron multiplicity for the (a, a′xnγ) reaction. Inelastic alpha-particle and γ-ray coincidence spectra for the ^{118}Sn(α, α'xnγ) reactions are shown in Fig. 4.55. The γ-ray spectrum gated by the α' exciting the 18 to 25 MeV region of ^{118}Sn shows γ lines of ^{117}Sn as well as those of ^{116}Sn. Thus ^{118}Sn with such low excitation energy, being excited by the inelastic alpha scattering, decays by emitting either one fast neutron at the pre-equilibrium stage or by evaporating two slow neutrons from the equilibrium compound nucleus. The large fraction of pre-equilibrium neutrons is illustrated by the tails at the low energy side in the α' energy spectra gated by discrete γ rays following the (α, α'n) and (α, α'2n)

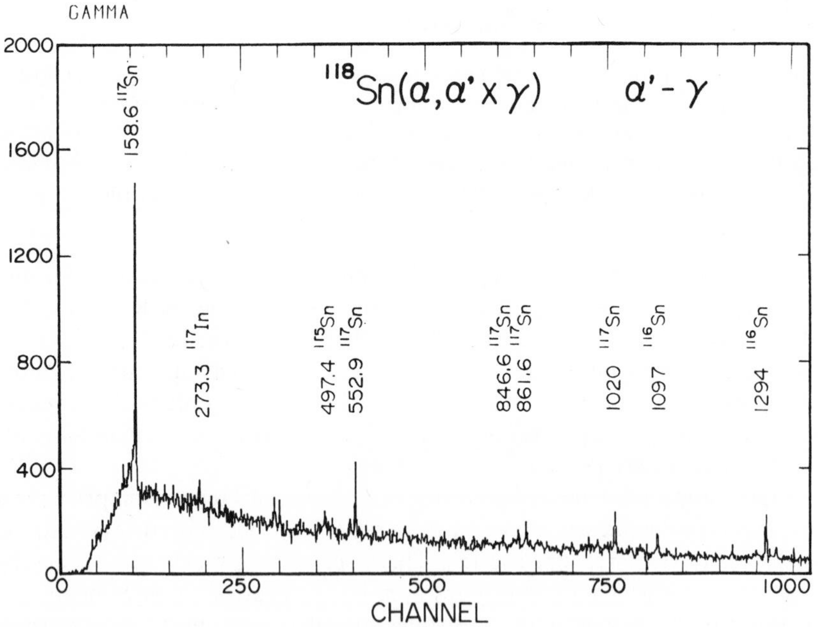

Fig. 4.55. Gamma-ray spectrum of the ^{118}Sn(α, α'xnγ) reaction in coincidence with the α' exciting the 24–32 MeV region of ^{118}Sn. (Ohsumi *et al.*, 1984.)

reactions (see Fig. 4.56). These features indeed reflect the simple 1p–1h exciton configuration of the initial doorway for the inelastic scattering.

4.4.4 *Gamma-ray spectroscopy for particle–hole states produced by inelastic scattering through isobaric analogue resonances*

Inelastic scattering excites particle–hole (1p–1h) states in general, and their configurations are studied by investigating their γ decay. Gamma-ray spectroscopy following inelastic proton scattering through isobaric analogue resonances (IAR) is useful for studying the microscopic p–h structure. Since the IAR is given by $|\mathrm{IAR}\rangle=(N)^{-1/2}\ [a_{\mathrm{p}}^{+}(j)+\Sigma\{a_{\mathrm{p}}^{+}(j')a_{\mathrm{n}}(j')\}_0\ a_{\mathrm{n}}^{+}(j)]$, the 1p–1h states with $\{\alpha_{\mathrm{n}}^{+}(j)\alpha_{\mathrm{n}}(j')\}_J$ are populated by inelastic scattering through the IAR (Fig. 4.57). The IAR formed by the j-orbit proton $\{a_{\mathrm{p}}^{+}(j)\}$ constitutes the initial doorway. Thus the 1p component in the 1p–1h states is given by the j-orbit neutron $\{\alpha_{\mathrm{n}}^{+}(j)\}$. Gamma decays from various types of 1p–1h states to the ground and excited states can be investigated. Inelastic scattering through IAR's on a doubly-magic nucleus populates 1p in a empty $(\mathrm{N}+1)\hbar\omega$ shell and 1h in an occupied $\mathrm{N}\hbar\omega$ shell. Thus typical E1 transitions from the $(\mathrm{N}+1)\hbar\omega$ shell to the $\mathrm{N}\hbar\omega$ shell are studied. Many E1 γ-decays from the $6\,\hbar\omega$ shell to the $5\,\hbar\omega$ shell in $^{208}\mathrm{Pb}$ have been observed, as shown in Fig. 4.58 (Cramer *et al.*, 1968).

4.4.5 *Charge exchange reactions and inelastic electron scattering*

Charge exchange reactions are regarded as special inelastic scattering processes accompanied by rotation in an isospin space. They are expressed in terms of isospin coordinates as

$$a(t, t_z)+A(T, T_z)=a'(t', t_z\pm 1)+A'(T', T_z\mp 1), \tag{4.4.8}$$

where (t, t_z) and (T, T_z) are the isospins and their z-components for the projectile and the target, respectively. The reactions most frequently used are (p, n), (^{3}He, t), ($\pi^{\pm}$, π^0) and so on. Excitation of the isobaric analogue state corresponds to the elastic scattering, and excitation of the charge exchange E2 GR corresponds to the inelastic excitation of the isovector E2 GR. Therefore the particle- and γ-spectroscopic techniques associated with inelastic scattering discussed so far (subsections 1 to 4) are also applicable to the spectroscopy associated with charge exchange reactions.

Charge exchange reactions such as (p, n), (n, p), ($\pi^{\pm}$, π^0) etc. involve neutral particles (neutron, π^0) in either the initial or exit channel. The efficient detection of such neutral particles is hard. Some charge exchange reactions with charged particles in both incident and exit channels are (^{3}He, t), (^{6}He, ^{6}Li), (^{16}O, ^{16}N), and so on. Their cross-sections are fairly small so that γ

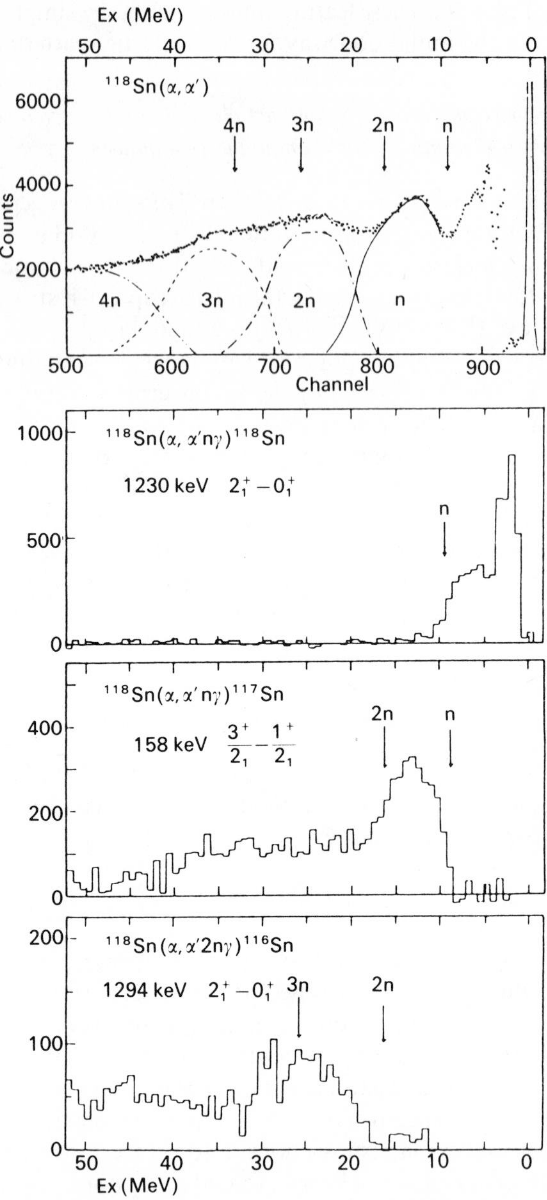

Fig. 4.56. Top: Singles α' energy spectrum for the ^{118}Sn(α, α') reaction at $E(\alpha) = 109$ MeV. Lower three figures: Coincidence α' spectra gated by discrete γ rays. The dotted lines in the singles α' spectrum are the statistical model expectations for the coincidence α' spectra gated by individual γ rays of the (α, $\alpha'x$n) reaction channels. The arrows indicate the threshold energies of the reaction channels. (Ohsumi *et al.*, 1984.)

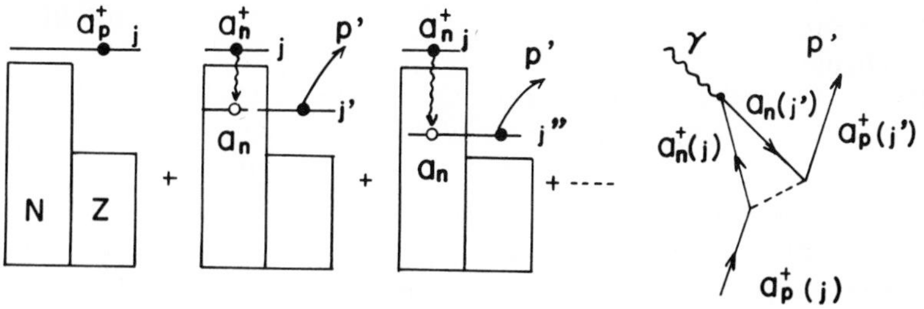

Fig. 4.57. Schematic diagram of an A(p, p′γ)B reaction through an isobaric analogue resonance. $a_p^+(j)$ is the incident proton and $a_p^+(j')$ is the inelastically scattered proton. $a_n^+(j)$ and $a_n(j')$ are the neutron particle and the neutron hole left after the inelastic scattering, and the γ decay occurs from $a_n^+(j)$ to $a_n(j')$. The configuration of the analogue state is shown in the left-hand side.

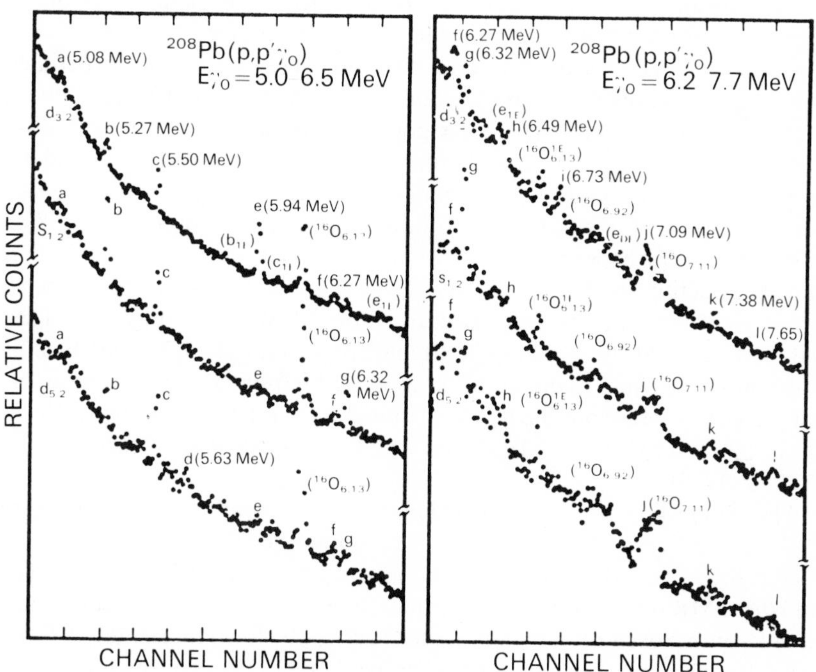

Fig. 4.58. High-energy portion of the gamma-ray spectra observed with a 20.7 cm^3 Ge(Li) detector at proton bombarding energies corresponding to analog resonances with $d_{3/2}$, $s_{1/2}$ and $d_{5/2}$ in ^{209}Bi. The peaks labelled a–k are identified as second-escape peaks from ground-state transitions in ^{208}Pb. The γ-transition energy is given in the parentheses. (Cramer *et al.*, 1968.)

spectroscopy following these charge exchange reactions is hard from an experimental point of view.

It should, however, be emphasized that the experimental difficulties of γ-ray measurements in coincidence with inelastic scattering can be overcome by using 4π γ spectrometers. Thus γ-ray spectroscopy with a large solid angle γ counter surrounding the target chamber, combined with a time-of-flight measurement for neutrons or a magnetic spectrometer for charged particles, can be a very powerful tool for such coincidence measurement in the near future.

Inelastic electron scattering has been used for studying GR's (Fukuda and Torizuka, 1972; Nagao and Torizuka, 1973; Dolbilkin *et al.*, 1982). The interaction involved is the electromagnetic interaction which is weak and well-known. Thus the electromagnetic transition strength is unambiguously obtained, while inelastic scattering calculations involving the strong interaction are rather model-dependent. The inelastic electron scattering provides a broader range of momentum transfer. Since high intensity, high duty-factor electron beams are available, microscopic features of giant multipole resonances have been investigated by coincidence measurements of decaying particles with inelastically scattered electrons. The coincidence experiment using the (e, e′x) reaction is good compared with the (a, a′x), (γ, x) and (x, γ) reactions. The (e, e′x) provides the x-particle (cluster) structure of the various types of GR's in a wide momentum range. Furthermore the coincidence spectrum is free from the radiation tail, which dominates the singles spectrum. These unique features of the (e, e′x) have been discussed by Hanna (1984).

A ^{12}C (e, e′p) experiment has been carried out by using the Stanford superconducting accelerator (Hanna, 1984). The singles spectrum, the coincidence spectrum, and the angular correlation are shown in Fig. 4.59. These data indicate that the main dipole strength at around 23 MeV consists of the non-spin-flip $p_{3/2}^{-1}\, d_{5/2}$ configuration and the high-lying one at around 24.5 MeV consists of the spin-flip $p_{3/2}^{-1}\, d_{3/2}$ configuration. These are consistent with the (γ, p_0) data (Allas *et al.*, 1964) and with the p–h calculation (Donnelly and Walker, 1970).

The Illinois group studied coincidence spectra from the (e, e′n) reaction on ^{208}Pb (Bolme *et al.*, 1984). The (e, e′) reaction in the 10 to 15 MeV excitation region shows broad resonances composed of the GDR (E1), the GQR (E2) and the GMR (E0). The coincidence data show a finite non-statistical neutron decay from an excitation region of the GQR to the low-lying states of ^{207}Pb, in accordance with the analogous experiment of the (α, α'n) on medium–heavy nuclei (Ejiri *et al.*, 1984; Ohsumi *et al.*, 1985; Okada *et al.*, 1982). The (e, e′n) spectrum, after removing the E1 contribution as obtained from the (γ, n) cross-section, was analysed in terms of the E0 and E2 strengths. They were found to be qualitatively consistent with the values derived from the (α, α') reaction (Morsch *et al.*, 1979; Hanna, 1984).

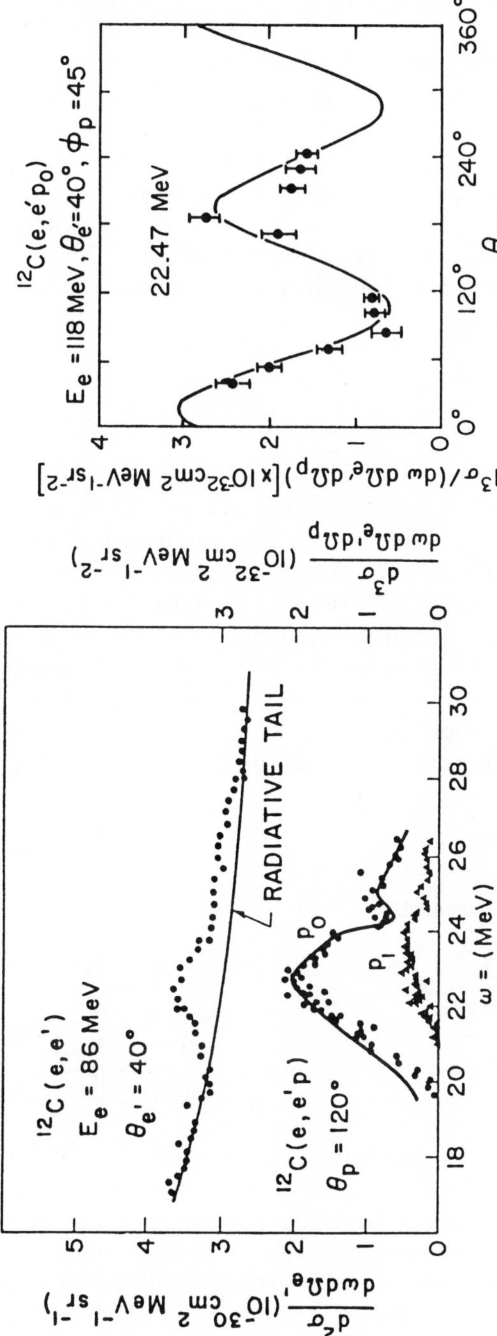

Fig. 4.59. Left-hand side: Singles spectrum from the $^{12}C(e, e')$ reaction and coincidence spectra from the $^{12}C(e, e'p_0$ or $p_1)$ reaction. The solid line superimposed on the p_0 points is the (γ, p_0) curve. Right-hand side: Angular correlation data for the $^{12}C(e, e'p_0)$ at $\theta_{e'} = 40°$. The solid line is a fit to the theoretical curve for a single 1^- level. (Hanna, 1984).

4.4.6 *Gamma rays following deep–inelastic scattering of heavy ions*

Deep-inelastic process in heavy-ion inelastic scattering has been studied by measuring de-excition γ rays following deep–inelastic scattering. The γ-ray circular polarization gives the sign of the spin polarization of the residual nucleus, and hence the sign of the deflection angle for the inelastic particle. The circular polarization of the de-excitation γ rays following bombardment of natural Ag targets with 284 MeV and 303 MeV ^{40}Ar has been measured by means of Compton polarization (Trautmann *et al.*, 1977). The experimental arrangement is shown in Fig. 4.60. The circular polarization was obtained from the asymmetry of Compton scattered γ rays from magnetized iron. The measured asymmetry of 0.65 ± 0.15 per cent for deep–inelastic events indicates a finite polarization of the product nuclei in the direction of the reaction normal $\mathbf{k}_i \times \mathbf{k}_f$, and hence the negative classical angle for the deep inelastic process, as illustrated in Fig. 4.61.

The gamma rays from highly-excited nuclei produced in the $^{16}O + {}^{58}Ni$ deep–inelastic reactions have been studied (Trautmann *et al.*, 1983). The circular polarization P_γ of continuum γ rays has been measured with the transmission method. Gamma rays have been measured by means of a large NaI detector both in-plane ($\theta_\gamma = 90°$) and out-of-plane ($\theta_\gamma = 0°$). The measured polarization and the out-of-plane/in-plane intensity ratio are shown in Fig. 4.62. The polarization is large around $E_\gamma \sim 2$ MeV and decreases as E_γ increases, while the intensity ratio $I(0°)/I(90°)$ is small around $E_\gamma \sim 2$ MeV and approaches unity (isotropic distribution) as E_γ increases. These features demonstrate the existence of a strong quadrupole component in the statistical γ transitions above $E_\gamma \sim 2$ MeV.

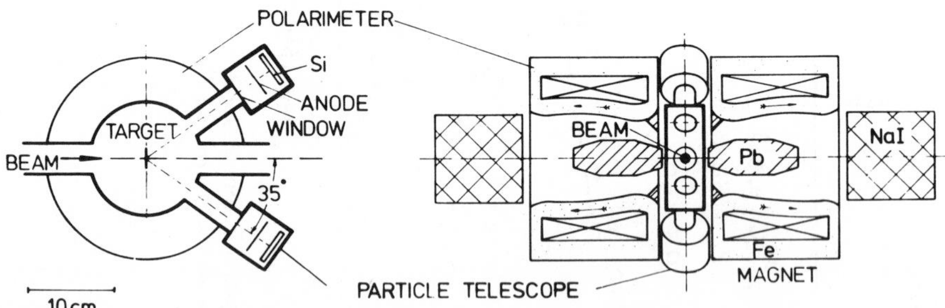

Fig. 4.60. Detector arrangement for circular polarization measurement. The left-hand side of the figure is a section of the apparatus along the scattering plane. The right-hand side is a section containing the scattering normally viewed along the beam direction. The arrows in the magnets indicate the direction of magnetization. (Trautmann *et al.*, 1977.)

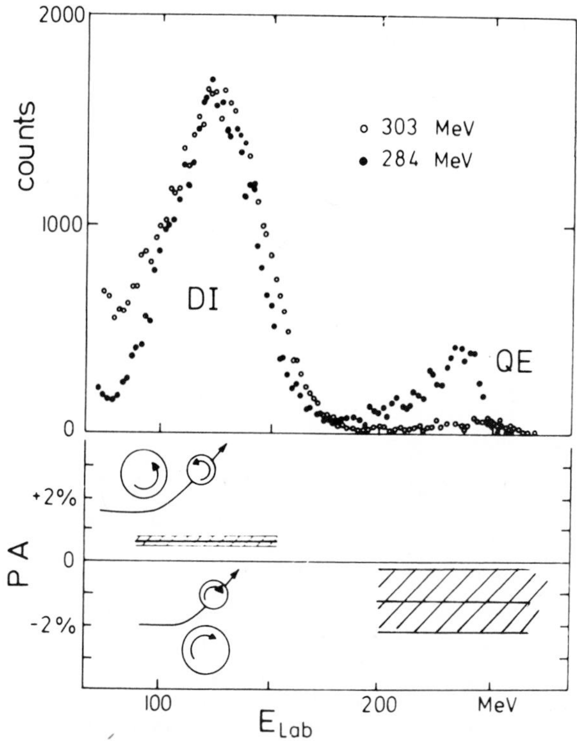

Fig. 4.61. Coincident energy spectra for ejectiles with $11 \leqslant Z \leqslant 21$ (upper part) and count-rate asymmetry PA (lower part). The deep inelastic (DI) group is clearly separated from the quasielastic (QE) group. The counts below about 50 MeV are due to light particles. The sense of rotation of the fragments corresponding to positive and negative PA and, accordingly, to negative and positive deflection angles is indicated. (Trautmann *et al.*, 1977.)

4.5 Gamma-ray spectroscopy associated with radiative capture reactions

4.5.1 *Radiative capture reactions*

An incident particle, interacting with a nucleus, can undergo a radiative transition from its scattering state ψ_i to a lower-lying state ψ_f. When the final state ψ_f is a bound state it is called a radiative capture process into a bound nuclear potential. The idea of radiative capture can be extended to transitions to unbound final states provided that they live much longer than a typical transit nuclear time of $R/v = 10^{-23}$ to 10^{-22} s. The radiative capture process is a one-step direct process induced by an electromagnetic interaction between the incident particle and the target nucleus. The reaction process is schematically illustrated in Figs 4.63 and 4.64. The reduced transition matrix element

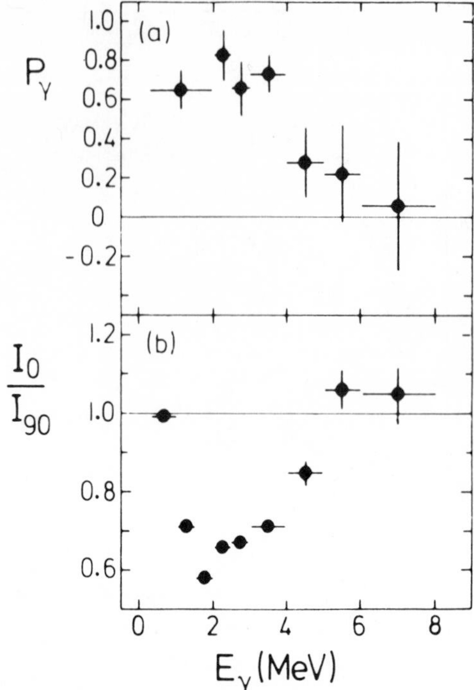

Fig. 4.62. (a) Gamma-ray circular polarization P_γ with respect to the scattering normal $\mathbf{k}_i \times \mathbf{k}_f$ and (b) the ratio I_0/I_{90} of the intensities measured in the direction of the scattering normal and perpendicular to it for γ-radiation from heavy fragments of $^{16}O + ^{58}Ni$ deep-inelastic reactions. (Trautmann *et al.*, 1983.)

is given by

$$T_D = \langle \psi_f \| g_{em} \mathbf{T}_L^P \| \psi_i \rangle, \tag{4.5.1}$$

where $\mathbf{T}_L^P$ is the electric ($P=1$) or magnetic ($P=0$) transition operator with the multipolarity L, and g_{em} is the electromagnetic coupling constant (see Section 2.1).

The incident particle interacts with the target nucleus (nucleon) through the strong (nuclear) interaction as well as the electromagnetic interaction. The strong interaction gives rise to a nuclear excitation ψ_I, while the incident particle is captured to the lower orbit ψ_f (see Fig. 4.63) and the excited state ψ_I de-excites by emitting a γ ray. This is a kind of two-step process where the nuclear excitation in the first step is followed by the γ de-excitation as the second step. One of most important intermediate states for such γ de-excitation is the giant resonance (GR) assóciated with the γ radiation. The GR is expressed as a coherent superposition of particle–hole excitations with

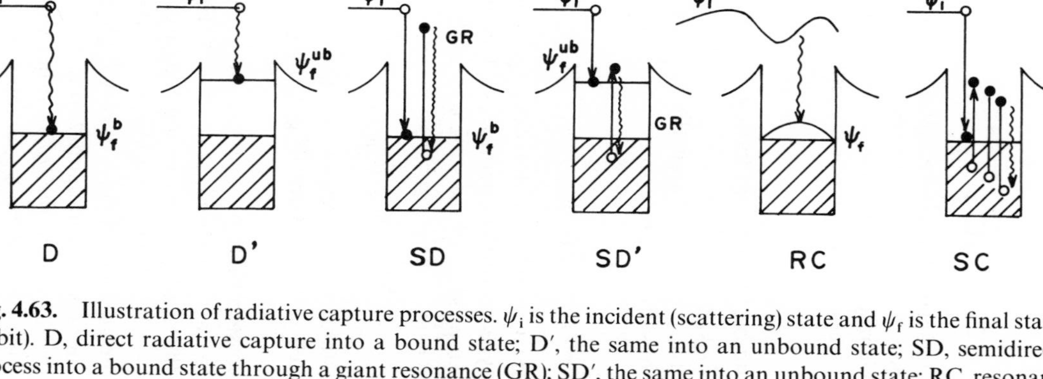

Fig. 4.63. Illustration of radiative capture processes. ψ_i is the incident (scattering) state and ψ_f is the final state (orbit). D, direct radiative capture into a bound state; D′, the same into an unbound state; SD, semidirect process into a bound state through a giant resonance (GR); SD′, the same into an unbound state; RC, resonant compound process, and SC, statistical compound process.

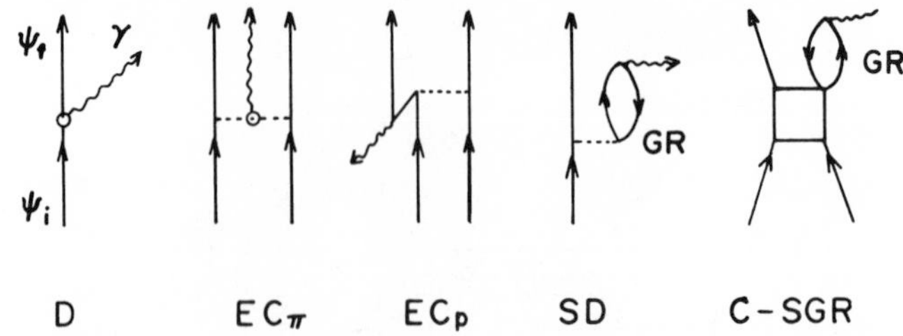

Fig. 4.64. Schematic diagram of the radiative capture processes. D, direct (one body) process, EC_π, an exchange current (two-body) process for pion current; EC_p, the same for pair current; SD, semidirect process through a giant resonance (GR); and C-SGR, compound process for the statistical giant resonance decay.

the same quantum number as the γ radiation, and it absorbs most of the γ-transition strength.

The radiative capture process through the GR was first introduced by Brown (1964) as the semi-direct (SD) process. The transition matrix element is given by

$$T = \langle \psi_f \| g_{em} \mathbf{T}_\alpha \| \psi^\alpha_{GR} \rangle \langle \psi^\alpha_{GR} \| \mathbf{H}_\alpha \| \psi_i \rangle, \tag{4.5.2}$$

where $\mathbf{H}_\alpha$ is the nuclear (strong) interaction operator to excite the α-mode GR (ψ^α_{GR}), and $\mathbf{T}_\alpha$ is the α-type transition operator (see Chapter 1 and Section 2.1). Here ψ^α_{GR} is written as a product of the state ψ_f and the giant resonance $\tilde{\psi}^\alpha_{GR}$ built on the target nucleus,

$$\psi^\alpha_{GR} = [\psi_f \cdot \tilde{\psi}^\alpha_{GR}]_{J_f}. \tag{4.5.3}$$

If the ψ_f is an excited state, the ψ^α_{GR} responsible for the SD process is the GR coupled with the excited state ψ_f. Then ψ^α_{GR} is the α-mode GR built on the state ψ_f.

The idea of the two-step (SD) process can be extended to a multi-step process, and finally to a compound process (C). Here multiple nuclear interactions between the incident particle and the target nucleons produce a compound nucleus ψ_c, which decays to a lower state ψ_f by γ radiation. If the excitation energy of the compound state is very high, there are many lower states ψ_f and then ψ_c decays predominantly to a state ψ_f such that ψ_c has a good overlap with the GR built on the ψ_f. In other words, the final state ψ_f, produced by annihilating the GR on ψ_c, accepts the strong γ-de-excitation strength. This mechanism is a statistical giant resonance process (SGR) in the high excitation region (Figs 4.63 and 4.64). The transition matrix element is

$$T_c = \langle \psi_f | g_{em} \mathbf{T}_\alpha \| \psi^\alpha_{GR} \rangle k \langle \psi_c \| \mathbf{H}_c \| \psi_i \rangle, \tag{4.5.4}$$

where $\mathbf{H}_c$ is the nuclear interaction to form a compound nucleus and k is the coefficient of the component of $\psi^{\alpha}_{\mathrm{GR}}$ in ψ_{c}. Then ψ_{c} is given by the sum of the non-GR and GR components,

$$\psi_{\mathrm{c}} \approx \psi'_{\mathrm{c}} + k \cdot [\psi_{\mathrm{f}} \cdot \tilde{\psi}^{\alpha}_{\mathrm{GR}}]_{J_{\mathrm{c}}}. \tag{4.5.5}$$

A special type of compound states is a resonance state, where the incident particle is captured into a particular resonance state (orbit) ψ_{R} in the compound nucleus, and then decays to the lower state ψ_{f} by γ radiation (see Fig. 4.63. RC). Then the transition matrix element is written as

$$T = \langle \psi_{\mathrm{f}} \| g_{\mathrm{em}} \mathbf{T}_{\alpha} \| \psi_{\mathrm{R}} \rangle \langle \psi_{\mathrm{R}} \| \mathbf{H}_{\mathrm{R}} \| \psi_{\mathrm{i}} \rangle, \tag{4.5.6}$$

where $\mathbf{H}_{\mathrm{R}}$ is the nuclear interaction relevant to the resonant capture.

Electromagnetic interactions with mesons in nuclei are treated as exchange current terms which interfere coherently with the nucleonic current terms. The exchange current processes are shown in Fig. 4.64. They become important at the medium energy (E_{γ} = 40 to 200 MeV) region since the large momentum transfer reduces the direct (one-body) process and enhances relatively the exchange process involving two strongly-interacting nucleons.

The γ-ray spectroscopy associated with the radiative capture process has two unique features. Firstly one can study the single-particle matrix element, $\langle \psi_{\mathrm{f}} \| g_{\mathrm{em}} \mathbf{T}_{\alpha} \| \psi_{\mathrm{i}} \rangle$. The electromagnetic interaction $g_{\mathrm{em}} \mathbf{T}_{\alpha}$ is weak but well-known and is quite selective of the particular mode α, the spin and parity of which are well determined from e–γ spectroscopy (see Chapter 2). The wavefunction of initial state ψ_{i} is obtained from the optical potential. The matrix element gives the projection of ψ_{i} on ψ_{f} through the operator $\mathbf{T}_{\alpha}$, and it is proportional to the spectroscopic factor $CS_{\mathrm{i}'}$ of the single particle $\psi_{\mathrm{i}'}$ connected directly to ψ_{i}. The radiative capture process of p, d, t, α and HI to the final state ψ_{f} gives the spectroscopic components of p, d, t, α and HI in ψ_{f}.

In the case of a semi-direct (SD) process with matrix element $\langle \psi_{\mathrm{GR}} \| \mathbf{H}_{\alpha} \| \psi_{\mathrm{i}} \rangle$, the nuclear interaction which excites the α-mode GR is written $\mathbf{H}_{\alpha} = \chi_{\alpha} \mathbf{T}'_{\alpha} \cdot \mathbf{T}'_{\alpha}$. Thus the matrix element is rewritten

$$\langle \psi^{\alpha}_{\mathrm{GR}} \| \mathbf{H}_{\alpha} \| \psi_{\mathrm{i}} \rangle \propto \chi_{\alpha} \langle \psi_{\mathrm{f}} \| \mathbf{T}'_{\alpha} \| \psi_{\mathrm{i}} \rangle \langle \psi_{\mathrm{GR}} \| \mathbf{T}'_{\alpha} \| 0 \rangle. \tag{4.5.7}$$

The $\mathbf{T}'_{\alpha}$ relevant to excitation of the α-mode GR has to be similar to the $\mathbf{T}_{\alpha}$ for the γ decay of the GR, and therefore the matrix element $\langle \psi_{\mathrm{f}} \| \mathbf{T}'_{\alpha} \| \psi_{\mathrm{i}} \rangle$ contains similar information to $\langle \psi_{\mathrm{f}} \| \mathbf{T}_{\alpha} \| \psi_{\mathrm{i}} \rangle$ except for the radial form factor.

The second unique feature of radiative capture spectroscopy is the study of the GR through the SD and SGR processes. The matrix elements involved are $\langle \psi_{\mathrm{f}} \| g_{\mathrm{em}} \mathbf{T}_{\alpha} \| \psi^{\alpha}_{\mathrm{GR}} \rangle$ and $\langle \psi^{\alpha}_{\mathrm{GR}} \| \mathbf{H}_{\alpha} \| \psi_{\mathrm{i}} \rangle$, where ψ_{GR} represents the GR based on the ψ_{f}, which is not necessarily the ground state. Thus the GR's built on such excited states as particle–hole states, vibrational states, and even a GR can be investigated by studying radiative capture reactions to these states. Since the compound nucleus produced by HI bombardment has a very high spin and a high excitation-energy, the GR based on highly excited high-spin states can be studied.

4.5.2 *Experiments on radiative-capture γ rays*

The spectroscopy of radiative-capture γ rays is beset by several experimental difficulties. Since the radiative capture reaction necessarily involves the electromagnetic decay of an unbound (scattering) state, the transition rate is as small as 10^{-4} to 10^{-5} compared with the particle decay rate. Let us consider for simplicity the direct E1 decay of a proton orbiting in a nuclear potential, and evaluate the transition rate in the classical limit. The emission rate of the electromagnetic power for the E1 radiation is

$$W = \frac{(e/2)^2 R^2 \omega^4}{3c^3/4}, \tag{4.5.8}$$

where ω is the angular velocity of the orbiting proton, R is the radius, and the effective charge corrected for the centre-of-mass motion is assumed to be $e/2$. Then the emission rate is written by using the velocity $v = R\omega$, giving

$$T_\gamma = W/E_\gamma = W/\hbar\omega = \frac{e^2}{\hbar c}\cdot\left(\frac{1}{3}\right)\cdot\left(\frac{v}{c}\right)^2\cdot\frac{v}{R}. \tag{4.5.9}$$

The particle decay rate is given by $T_p \approx v/R$ if the proton emission is not suppressed by the potential. Then one gets

$$T_\gamma/T_p \approx \frac{e^2}{\hbar c}\cdot\left(\frac{1}{3}\right)\cdot\left(\frac{v}{c}\right)^2. \tag{4.5.10}$$

The value for v/c is estimated to be $(\hbar\omega/\hbar c)R \approx E_\gamma/40$ for $R \approx 5$ fm. Thus one gets $v/c \approx 1/4$ for $E_\gamma = 10$ MeV, and the wavelength $\lambda\!\!\!^{-}$ is still much larger than the nuclear size R. Because of the factor $(v/c)^2 \approx (R/\lambda\!\!\!^{-})^2 \approx 6{\cdot}10^{-2}$ and the factor $e^2/\hbar c \approx 8{\cdot}10^{-3}$, the ratio T_γ/T_p becomes as small as 10^{-3}. This order of magnitude agrees with the quantum-mechanical evaluation of the T_γ and T_p.

This is why the radiative capture process is a very rare process and it has to be selected among a huge flux of nucleon decays. The peak efficiency of the detectors for medium energy γ rays is small, so this situation is similar to the γ decay of GR's in an unbound region excited by inelastic scattering (see Section 4.4).

The advent of high-efficiency high-resolution γ detectors and of high-intensity high-quality incident beams have stimulated much γ-ray spectroscopy of radiative capture reactions. In particular, high-quality polarized beams are very powerful probes for studying the spin configuration of the nuclear states involved in radiative processes and for multipole assignments.

Reviews on the radiative capture reactions have been published by several authors (Hanna, 1979; Paul, 1974, 1977; Weller and Roberson, 1980 and refs. therein). Most of the experimental and theoretical details are found in these reviews. In what follows, the essential features of radiative capture γ-ray spectroscopy are briefly illustrated by some experimental data.

4.5.3 *Angular distributions and analysing powers*

The angular distributions of radiative capture γ-rays provide important information about the γ-ray multipolarity and about the spin and parity of the states involved in the transitions, as discussed in Chapter 2. The initial state ψ_i for the radiative process is the scattering (unbound) state, and thus in general several spins and parities are involved, except for the resonant capture reaction. The multipolarity of the radiation is not unique for a given final state, and one usually has to deal with mixed multipoles, mixed channel spins and mixed initial orbits. Polarized proton, deuteron, and helion beams are now available in many accelerator laboratories. These measurements of analysing powers provide useful information on the radiative process.

The theory of angular distributions and analysing powers for radiative γ rays has been formulated by many authors and the relevant coefficients have been tabulated. (Devons and Goldfarb, 1957; Carr and Baglin, 1971; Maute, *et al.*, 1974; Laszewski and Holt, 1977; Seyler and Weller, 1979; Weller and Roberson, 1980, and refs therein.) The reaction scheme is given in Fig. 4.65. The differential cross-section for a radiative nucleon capture reaction, A(a, γ)B, is written

$$\frac{d\sigma}{d\Omega} = \sum_k A_k P_k (\cos\theta) + (\mathbf{p}\cdot\mathbf{n}) \sum_k B_k P_k^1 (\cos\theta)$$

$$= A_0 \left\{ 1 + \sum_k a_k P_k (\cos\theta) + (\mathbf{p}\cdot\mathbf{n}) \sum_k b_k P_k^1 (\cos\theta) \right\}, \quad (4.5.11)$$

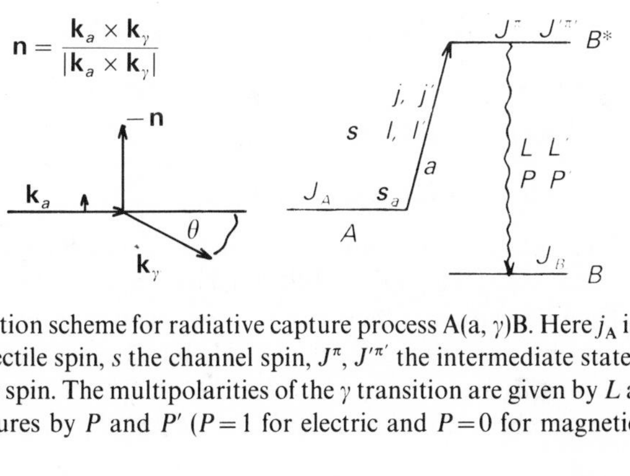

$$\mathbf{n} = \frac{\mathbf{k}_a \times \mathbf{k}_\gamma}{|\mathbf{k}_a \times \mathbf{k}_\gamma|}$$

Fig. 4.65. Reaction scheme for radiative capture process A(a, γ)B. Here j_A is the target spin, s_a the projectile spin, s the channel spin, J^π, $J'^{\pi'}$ the intermediate state spins, and J_B the final-state spin. The multipolarities of the γ transition are given by L and L', and the natures by P and P' ($P=1$ for electric and $P=0$ for magnetic).

where $a_k = A_k/A_0$ and $b_k = B_k/A_0$ are the coefficients of the Legendre polynomial expansions, $\mathbf{p}$ is the nucleon beam polarization, and $\mathbf{n} = (\mathbf{k}_a \times \mathbf{k}_\gamma)/|\mathbf{k}_a \times \mathbf{k}_\gamma|$ is the unit vector normal to the reaction plane. The total cross-section is given by $\sigma_t = 4\pi A_0$. The analysing power $A(\theta)$ is defined as

$$A(\theta) = \frac{1}{p}\left\{\left(\frac{d\sigma}{d\Omega}\right)_+ - \left(\frac{d\sigma}{d\Omega}\right)_-\right\} \Big/ \left\{\left(\frac{d\sigma}{d\Omega}\right)_+ + \left(\frac{d\sigma}{d\Omega}\right)_-\right\}, \tag{4.5.12}$$

where $(d\sigma/d\Omega)_\pm$ are the differential cross-sections for the polarized (spin-up and spin-down) beams. This expression may be rewritten as

$$A(\theta) = \frac{\sum b_k P_k^1(\cos\theta)}{1 + \sum a_k P_k(\cos\theta)}. \tag{4.5.13}$$

The total cross-section and the coefficients a_k and b_k can be written in terms of l–s and j–j coupling schemes (Laszewski and Holt, 1977; Weller and Roberson, 1980; and refs therein. In what follows use is made of equations from these references). The coefficients in the l–s scheme are

$$4\pi A_0 = 4\pi g^2 \Sigma (2J+1)|T|^2 \tag{4.5.14a}$$

$$g^2 = (\lambda^2/4)\{(2J_A+1)(2s_a+1)\}^{-1} \tag{4.5.14b}$$

$$\begin{aligned}
a_k = {} & A_0^{-1} \cdot g^2 \sum_{tt'} (-)^{s-J_B+1}\{1+(-)^{L+P+L'+P'+k}\}/2 \\
& \times \{(2l+1)(2l'+1)(2L+1)(2L'+1)\}^{1/2}\,(2J+1)(2J'+1) \\
& \times (l0\,l'0|k0)\,W(lJ\,l'J';\,sk) \\
& \times (L1\,L'-1|k0)\,W(LJ\,L'J';\,J_Bk)\mathrm{Re}(TT'^*),
\end{aligned} \tag{4.5.15a}$$

$$\begin{aligned}
b_k = {} & A_0^{-1} g^2 3\{s_a(2s_a+1)\}^{1/2}(2k+1)^{1/2}\{(s_a+1)k(k+1)\}^{-1/2} \\
& \times \sum_{tt'} \{1+(-)^{L+P+L'+P'+k}\}2^{-1}\cdot(2J+1)(2J'+1) \\
& \times \{(2s+1)(2s'+1)(2l+1)(2l'+1)(2L+1)(2L'+1)\}^{1/2} \\
& \times (-)^{J_A-s_a+J_B-J-s+l}(l0\,l'0|k0)\,W(s_a s s_a s';\,J_A 1) \\
& \times (L1L'-1|k0)\,W(LJL'J';\,J_B k) \\
& \times X(lsJ;\,l's'J';\,k1k)\,\mathrm{Re}(iTT'^*),
\end{aligned} \tag{4.5.15b}$$

where s, l, J, L, and J_B are the spins for the incident channel, the incident orbit, the intermediate state, the γ-ray and the final (residual) state, respectively (see Fig. 4.65). Then the nature of the γ-transition is given by $P = 1$ for the electric transition and by $P = 0$ for the magnetic one. They are restricted by the triangle relations, $\mathbf{J}_A + \mathbf{s}_a = \mathbf{s}$, $\mathbf{l} + \mathbf{s} = \mathbf{J}$ and $\mathbf{L} + \mathbf{J}_B = \mathbf{J}$. The t and t' stand for the quantum numbers s, l, J, L, P and s', l', J', L', P', respectively and the summation in eqn 4.5.14 extends over all possible t and t'. The reduced transition matrix element is

$$T = \langle PLJ_{\mathrm{B}}J^{\pi} \| \mathbf{T}_L^P \| l(J_{\mathrm{A}}s_{\mathrm{a}})sJ^{\pi} \rangle \tag{4.5.16}$$

The coefficients a_k and b_k in the j–j coupling scheme of $\mathbf{j} = \mathbf{l} + \mathbf{s}_{\mathrm{a}}$ and $\mathbf{j} + \mathbf{J}_{\mathrm{A}} = \mathbf{J}$ are given as

$$\begin{aligned} a_k = A_0^{-1} g^2 \sum_{tt'} (-)^{J_{\mathrm{A}} - J_{\mathrm{B}} + 1 - s_a + k + j - j'} (2J+1)(2J'+1) \\ \times \{(2j+1)(2j'+1)(2l+1)(2l'+1)(2L+1)(2L'+1)\}^{1/2} \\ \times (l0l'0|k0)\, W(ljl'j';\, s_a k)\{1 + (-)^{L+P+L'+P'+k}\}/2 \\ \times W(jJj'J';\, J_A k)\, W(LJL'J';\, J_B k)(L1L'-1|k0)\mathrm{Re}(TT'^*) \end{aligned} \tag{4.5.17a}$$

$$\begin{aligned} b_k = A_0^{-1} g^2 3\{s_a(2s_a+1)\}^{1/2}(2k+1)^{1/2}\{(s_a+1)k(k+1)\}^{-1/2} \\ \times \sum_{tt'} \{(2j+1)(2j'+1)(2l+1)(2l'+1)(2L+1)(2L'+1)\}^{1/2} \\ \times (2J+1)(2J'+1)(-)^{J_A - J_B + 1 + l' - j'}\{1 + (-)^{L+P+L'+P'+k}\}/2 \\ \times (l0l'0|k0)\, W(jJj'J';\, J_A k)(L1L'-1|k0) \\ \times W(LJL'J';\, J_B k)\cdot X(ls_a j;\, l's_a j';\, k1k)\mathrm{Re}(iTT'^*) \end{aligned} \tag{4.5.17b}$$

In general the summation extends over all relevant t and t' quantum numbers and the Legendre expansion goes over the integral values of k restricted by $l + l' + k = \text{even}$ and by $\mathrm{Max}(|l - l'|,\, |L - L'|;\, |J - J'|) \leqslant k \leqslant \mathrm{Min}\,(l + l',\, L + L',\, J + J')$.

In practice, the quantum numbers associated with radiative capture reactions are estimated from the analysis of the reaction. In this way physical quantities such as the transition matrix elements, the relative phases, and the spins and orbits of the states involved in the radiation are obtained from measured a_k and b_k coefficients. Extraction of these physical quantities is not unique if the number of unknown parameters is larger than that of the observed ones. Nevertheless the physical quantities essential for the radiative process are investigated by means of the radiative capture γ-ray spectroscopy using various types of polarized and unpolarized projectiles (proton, neutron, etc.), supplemented by considerations (calculations) of the nuclear structure and reaction mechanism.

Some useful features relevant to radiative capture processes are as follows.

1. Finite a_k coefficients for the $P_k\,(\cos\theta)$ term with k = odd indicate parity mixing of the radiation. A finite a_1 is due to E1–M1 or E1–E2 mixing, and a finite a_3 is due to El–E2 mixing provided that the main radiation is E1.

2. Finite b_k coefficients for the P_k^1 ($\cos\theta$) term indicate channel mixing of incident projectiles; $b_2, b_4, \ldots$ are due to mixing of incident waves of l and $l \pm 2$, and $b_1, b_3, \ldots$ are due to that of l and $l \pm 1$.

3. The a_k and b_k are related by (Devons and Goldfarb, 1957)

$$a_k = \sum_{tt'} C_{tt'} \operatorname{Re}(T_t T_{t'}'^{*}). \tag{4.5.18a}$$

$$b_k = \{3/(2s_a + 2)\} \sum_{tt'} C_{tt'} \operatorname{Re}(iT_t T_{t'}'^{*})$$
$$\{j(j+1) - l(l+1) - j'(j'+1) + l'(l'+1)\}/\{k(k+1)\}. \tag{4.5.18b}$$

4. The direct electric interaction of an incident projectile with a target nucleus has to be corrected for the centre of mass motion, which is spurious. Thus the effective charge (Moszkowski, 1965; Buck and Pilt, 1977; see Section 2.1) used for the $\mathrm{a} + \mathrm{A} \rightarrow \mathrm{B} + \gamma$ reaction is

$$q^{\mathrm{eff}} = (M_{\mathrm{A}}/M_{\mathrm{B}})^L q_{\mathrm{a}} + (-)^L (M_{\mathrm{A}}/M_{\mathrm{B}})^L q_{\mathrm{A}}, \tag{4.5.19}$$

where q_i and M_i are the charge and mass for the particle (nucleus) i, respectively. This gives eN/A and $-eZ/A$ for single proton and neutron E1 transitions, respectively, where Z, N, and A refer to the proton, neutron and mass numbers of the compound nucleus B. For higher multipole radiations the correction term is very small if $M_{\mathrm{a}} \ll M_{\mathrm{B}}$, so that a neutron with $q_{\mathrm{a}} = 0$ has a negligible charge q^{eff} for E2, E3 etc, resulting in little contribution to the direct neutron capture radiation. Consequently, finite E2 or E3 radiation induced by neutron capture leads to a finite SD or C process associated with the GR excitation (Arthur *et al.*, 1974; and Wender *et al.*, 1978).

5. The analysing power at $\theta = 90^\circ$ has a special meaning, (Saporetti and Guidotti, 1979), which can be seen as follows. It is written as

$$A(90^\circ) = \frac{1}{p}(b_1 - \tfrac{3}{2}b_3 + \tfrac{15}{8}b_5 + \ldots)/(1 - \tfrac{1}{2}a_2 + \tfrac{3}{8}a_4 - \tfrac{5}{16}a_6 + \ldots). \tag{4.5.20}$$

A finite value of $A(90^\circ)$ excludes pure E1 γ radiation, but indicates mixing of E1–M1 and/or E1–E2, and so on. As discussed before a non-zero $A(90^\circ)$ for radiative neutron capture reveals the existence of non-E1 GR's.

6. The cross-section and the angular distribution for the odd-A target, which consists of a particle with spin J_{A} and a even core, are simply related to those for the spin-zero (core) target if the particle is assumed to be a spectator. The direct and semi-direct (DSD) meet this condition since the GR excited by the SD process does not depend on whether the particle is there or not. Then the T matrix and the cross-section are expressed as follows (Weller and

Roberson, 1980),

$$\langle LJ_BJ\|\mathbf{T}\|(ls_a)jJ_AJ\rangle = \sum_{j'}\langle Lj'j\|\mathbf{T}\|(ls_a)j0j\rangle\sqrt{(S_j)}$$
$$\{(2J_B+1)(2j+1)\}^{1/2}\,W(J_Aj'JL;J_Bj), \quad (4.5.21a)$$

where $S_{j'}$ is the spectroscopic factor for the final state (J_B) composed of the target particle (J_A) coupled with the captured nucleon (j'). The angular distribution and the analysing power for the odd target with spin J_A are the same as those for the even core without the particle, and the cross-section is given as

$$\frac{d\sigma}{d\Omega}(J_A\rightarrow J_B)=\frac{2J_B+1}{2J_A+1}\sum_{j'}\frac{S_{j'}}{(2j'+1)}\frac{d\sigma}{d\Omega}(0^+\rightarrow j'), \quad (4.5.21b)$$

where $d\sigma/d\Omega(0^+\rightarrow j')$ is the differential cross-section (including the analysing power) for the radiative capture on the 0^+ target core into the final j' orbit. (Blatt *et al.*, 1984; Weller and Roberson, 1980).

4.5.4 *Radiative capture reactions through resonances and isobaric analogue states*

4.5.4a. *Single resonances*

The radiative capture reaction through a resonance with definite quantum numbers (spin and parity of J_R^{π}, and other quantum numbers) is given by a Breit–Wigner form,

$$\frac{d\sigma}{d\Omega}=\frac{2J+1}{(2s_a+1)(2J_A+1)}\pi\lambda_a^2\cdot\frac{\Gamma_p\cdot\Gamma_\gamma}{(E-E_R)^2+\Gamma_t^2/4}$$
$$\cdot\{1+\sum a_kP_k(\cos\theta)+(\mathbf{p}\cdot\mathbf{n})\sum b_kP_k^1(\cos\theta)\}, \quad (4.5.22)$$

where Γ_t, Γ_p and Γ_γ are the total width, the partial proton width and the partial γ width for the resonance, respectively, and J and E are the resonance spin and the energy. The angular distribution and the analysing power are used to assign the resonance spin and parity, the γ-ray multipolarity and the single nucleon orbit (s_a, l, j) constituting the resonance. In particular, a finite analysing power indicates a mixing of another single-nucleon orbit j_a' into the main resonance orbit j_a. Odd-k terms in eqn 4.5.22 do not exist unless there are other overlapping resonances with different parity.

The γ decay width Γ_γ is obtained from the measured cross-section (eqn 4.5.22) provided that the proton width Γ_P and the total width Γ_t are known from other experiments. In fact Γ_P and Γ_t are obtained from the analysis of the resonant elastic scattering. At very low energy the elastic channel dominates the total resonance, so that $\Gamma_P/\Gamma_t\approx 1$. Then the thick target yield or an integral of the excitation function over the resonance gives Γ_γ. Thus the resonant capture reaction provides a unique method for obtaining the γ-decay width Γ_γ in a range of 0.1 to 100 eV. This region, corresponding to the life-time of 10^{-14} to 10^{-17} s, is hard to reach by other methods. The resonance theories are well developed (Blatt and Weisskopf, 1954).

4.5.4b *Isobaric analogue resonances and γ-ray spectroscopy*

Isobaric analogue resonances (IAR) have been used for both nuclear structure studies and nuclear reaction studies from various points of view. The IAR is a collective isospin vibration or isospin giant resonance and it appears as a sharp resonance in a region of high excitation over a wide range of nuclei (Fox and Robson, 1966; Anderson *et al.*, 1969; and references therin). Gamma-ray spectroscopy with IAR has very important and unique features (Fujita 1963, 1967; Ejiri *et al.*, 1968/1969; Wilkinson, 1966. See Section 5).

IAR γ-ray spectroscopy provides γ-matrix elements for single particle transitions (particularly E1 γ transitions) since IAR's are clean and sharp resonances with single-particle nature in the region of high excitation energies.

The isobaric analogue state ($|\mathrm{IA}\rangle$) is expressed as

$$|\mathrm{IA}\rangle = \frac{1}{\sqrt{2T}} T_- |\mathrm{i}\rangle, \tag{4.5.23}$$

where T_- is the isospin lowering operator and T is the isospin of the initial state $|\mathrm{i}\rangle$. The wave function is expressed in terms of proton creation $\{a_j^+(p)\}$ and neutron annihilation $\{a_j(\mathrm{n})\}$ operators. We take for simplicity a single neutron state $a_{J'}^+(n)|0\rangle$ as the initial state $|\mathrm{i}\rangle$ and a single proton state $a_{J_\mathrm{f}}^+(p)|0\rangle$ as the final state, where $|0\rangle$ is the even–even core nucleus. Then the states relevant to the γ decay and the transition matrix element are expressed as

$$|\mathrm{i}\rangle = a_{J'}^+(\mathrm{n})|0\rangle, \tag{4.5.24a}$$

$$\begin{aligned} |\mathrm{IA}\rangle &= (2T)^{-1/2}[\Sigma\{a_\delta^+(\mathrm{p})a_\delta(\mathrm{n})\}_0|\mathrm{i}\rangle] \\ &= (2T)^{-1/2}[a_{J'}^+(\mathrm{p}) + a_{J'}^+(\mathrm{n})\Sigma\{a_\delta^+(\mathrm{p})a_\delta(\mathrm{n})\}_0]|0\rangle, \end{aligned} \tag{4.5.24b}$$

$$|\mathrm{f}\rangle = a_{J_f}^+(\mathrm{p})|0\rangle, \tag{4.5.24c}$$

$$T = \langle \mathrm{f} \| \mathbf{M}_\gamma^L \| \mathrm{IA}\rangle, \tag{4.5.24d}$$

$$\begin{aligned} \mathbf{M}_\gamma^L = {} & \Sigma\langle\delta\| q_\mathrm{P} m_\gamma \|\delta'\rangle a_\delta^+(\mathrm{p})a_{\delta'}(\mathrm{p}) \\ & + \Sigma\langle\varepsilon\| q_\mathrm{n} m_\gamma \|\varepsilon'\rangle a_\varepsilon^+(\mathrm{n})a_{\varepsilon'}(\mathrm{n}), \end{aligned} \tag{4.5.24e}$$

where $\mathbf{M}_\gamma^L$ is the L-multipole γ transition operator, and q_P and q_n are proton and neutron effective charges. In case of an E1 transition, $q_\mathrm{P} = Ne/A$, $q_\mathrm{n} = -Ze/A$ and $m_\gamma = r\mathbf{Y}_1$. The summations in eqns 4.5.24b and 4.5.24e extend over all possible particle–hole pairs of $\delta\delta$, $\delta\delta'$ and $\varepsilon\varepsilon'$. As shown in Fig. 4.66

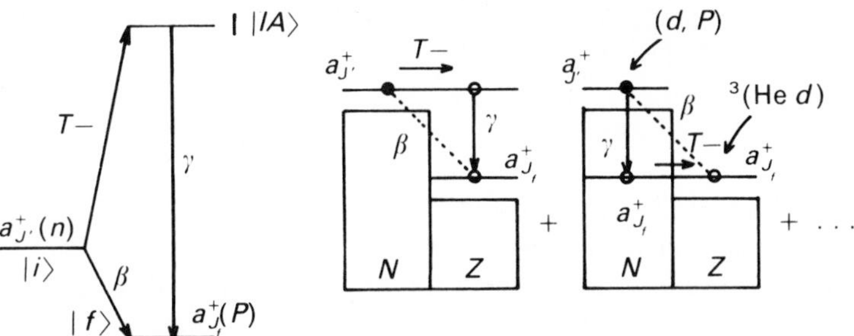

Fig. 4.66. The reaction scheme for radiative proton capture process through an isobaric analogue resonance (IA), and comparison with the analogue β decay, the (d,p) reaction and the (^{3}He, d) reaction. The IAS γ decay proceeds as $a^+_{J'}(\mathrm{p})\rightarrow a^+_{J_\mathrm{f}}(\mathrm{p})$ and $a^+_{J'}(\mathrm{n})\rightarrow a^+_{J_\mathrm{f}}(\mathrm{n})$, the analogue β decays as $a^+_{J'}(\mathrm{n})\rightarrow a^+_{J_\mathrm{f}}(\mathrm{p})$, the (d, p) reaction on the core ($|0\rangle$) as $a^+_{J'}(\mathrm{n})+|0\rangle$, and the (^{3}He, d) reaction on the core as $a^+_{J_\mathrm{f}}(\mathrm{p})+|0\rangle$.

and in eqn 4.5.24, the γ-transition amplitude is a coherent sum of the proton and neutron transitions. Thus the E1 matrix element is (Ejiri *et al.*, 1968/1969)

$$T(\mathrm{E1}) = \frac{1}{\sqrt{2T}}\left\langle a^+_{J_\mathrm{f}}(\mathrm{p})\left\|\frac{Ne}{A}\mathbf{m}'_\gamma\right\|a^+_{J'}(\mathrm{p})\right\rangle + \left\langle a^+_{J_\mathrm{f}}(\mathrm{n})\left\|\frac{Ze}{A}\mathbf{m}'_\gamma\right\|a^+_{J'}(\mathrm{n})\right\rangle$$

$$= \frac{1}{\sqrt{2T}}\langle a^+_{J_\mathrm{f}}\,\|\,e\mathbf{m}'_\gamma\,\|\,a^+_{J'}\rangle. \tag{4.5.25}$$

The first forbidden β decay operator consists of $g_V\mathbf{r}$, $g_\mathrm{V}\boldsymbol{\alpha}$, $g_\mathrm{A}\boldsymbol{\sigma}\cdot\mathbf{r}$, $g_\mathrm{A}(\boldsymbol{\sigma}\times\mathbf{r})_1$, $g_\mathrm{A}(\boldsymbol{\sigma}\times\mathbf{r})_2$, and $g_\mathrm{A}\gamma_5$, where g_V and g_A are vector and axial vector coupling constants, and $\boldsymbol{\alpha}$ and γ_5 are the Dirac matrix operators. Among these the term $g_V\mathbf{m}_\mathrm{B}\equiv g_V\mathbf{r}$ is analogous to the E1 γ operator of $e\mathbf{m}_\gamma\equiv e\mathbf{r}$. They are related as (Fujita 1963; Ejiri *et al.*, 1968/1969)

$$\langle \mathrm{f}|g_V\mathbf{m}_\mathrm{B}|\mathrm{i}\rangle = \frac{g_V}{e}\langle \mathrm{f}|[e\mathbf{m}_\gamma, \mathbf{T}_-]|\mathrm{i}\rangle$$

$$\approx \frac{g_V}{e}\cdot\sqrt{(2T)}\langle \mathrm{f}|e\mathbf{m}_\gamma|\mathrm{IA}\rangle, \tag{4.5.26}$$

since $\langle \mathrm{f}|T_-\mathbf{m}_\gamma|\mathrm{i}\rangle\approx 0$ for a low-lying state ($|\mathrm{f}\rangle$) in medium heavy nuclei. Therefore the term $\langle g_V\mathbf{r}\rangle$ is obtained from the IAS E1 γ decay, and is used to extract other β-decay terms. On a basis of a single (quasi) particle model, the radiative capture probability is proportional to the proton spectroscopic factor $S_p(J')$ for the IAR and to $S_p(J_\mathrm{f})$ for the final state. $S_p(J')$ for the IAR is related to the neutron spectroscopic factor $S_n(J')$ for the parent state by $S_p(J')=S_n(J')/2T$. The factors $S_n(J')$ and $S_p(J_f)$ are just the spectroscopic factors involved in the (d, p) and (^{3}He, d) reactions on the core nucleus ($|0\rangle$), respectively (see Fig. 4.66).

4.5.4c *Interference of two overlapping resonances, isobaric analogue resonance and giant resonances*

A resonant capture process does not always proceed through a single resonance but is often modified by a broad resonance near the single resonance. The IAR provides a very good example of overlapping resonances since the IAR is located just in the region of the GDR (giant dipole resonance). The radiative capture reactions through IAR and GDR have been studied by Ejiri and Bondorf (1968) and the reaction scheme is given in Fig. 4.67. The T-matrix is written as a coherent superposition of the two resonant amplitudes. It is obtained as follows (Ejiri and Bondorf, 1968; Terasawa and Fujita, 1974).

$$T=\frac{i\tilde{g}^{\mathrm{p}}\,\tilde{g}^{\gamma}}{E-\tilde{E}+\frac{1}{2}\mathrm{i}\tilde{\Gamma}}\delta_{JJ'}+\frac{ig^{\mathrm{p}}\cdot g^{\gamma}}{E-E_G+\frac{1}{2}\mathrm{i}\Gamma}, \tag{4.5.27}$$

where the first and the second terms stand for the IAR and GDR, respectively. The proton and γ widths are given by $\Gamma_{\mathrm{p}}=|g^{\mathrm{p}}|^2$ and $\Gamma_\gamma=|g^\gamma|^2$, respectively. For simplicity the GDR is expressed by a coherent sum of forward-type

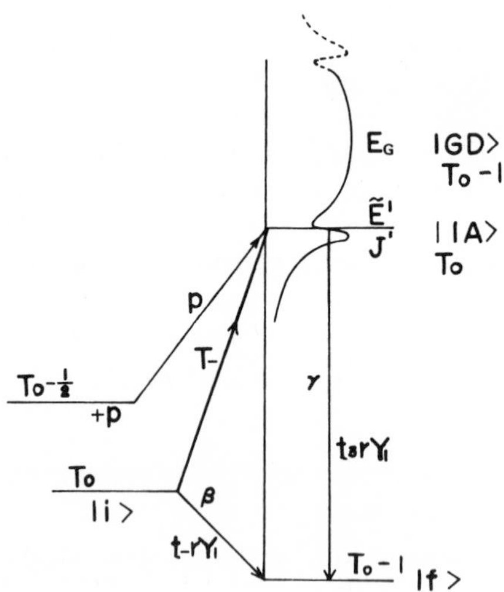

Fig. 4.67. Schematic level diagram of the radiative capture process through an isobaric analogue resonance ($|\mathrm{IA}\rangle$) and a giant dipole resonance ($|\mathrm{GD}\rangle$). (Ejiri and Bondorf, 1968.)

particle–hole excitation as follows (Brown and Bolsterli, 1959),

$$|\mathrm{GD}\rangle = \{Q^{+}\cdot a^{+}_{J_{\mathrm{f}}}(\mathrm{p})\}_{J}|0\rangle, \tag{4.5.28a}$$

$$Q^{+} = \Sigma C^{\mathrm{p}}_{\delta\delta'}\{a^{+}_{\delta}(\mathrm{p})a_{\delta'}(\mathrm{p})\}_{1} + \Sigma C^{n}_{\varepsilon\varepsilon'}\{a^{+}_{\varepsilon}(n)a_{\varepsilon'}(n)\}_{1} \tag{4.5.28b}$$

$$C^{\mathrm{p}}_{\delta\delta'} = D^{\mathrm{p}}_{\delta\delta'}/(\Sigma D^{2}_{\delta\delta'})^{1/2}. \tag{4.5.28b}$$

$$D^{\mathrm{p}}_{\delta\delta'} = \langle a_{\delta}(\mathrm{p})\|\mathbf{m}_{\gamma}\|a^{+}_{\delta'}(\mathrm{p})\rangle, \text{ and similarly for } C^{n}_{\varepsilon\varepsilon'}. \tag{4.5.28d}$$

Using eqns (4.5.24) and (4.5.28), eqn (4.5.27) is reduced to a simple form,

$$T = \mathrm{i}g^{\mathrm{p}}_{\mathrm{sp}}g^{\gamma}_{\mathrm{sp}}\frac{D_{JJ_{\mathrm{f}}}}{q_{\mathrm{p}}}\left[\left(\frac{g^{\mathrm{eff}}_{\mathrm{IA}}/g_{\mathrm{IA}})/2T}{E-\tilde{E}+\frac{1}{2}\mathrm{i}\tilde{\Gamma}}\right)\delta_{JJ'} + \frac{q_{\mathrm{p}}\{3/(2J+1)\}^{1/2}}{E-E_{G}+\frac{1}{2}\mathrm{i}\Gamma}\right], \tag{4.5.29}$$

where g^{eff}/g is the renormalization factor of the single-particle E1 operator in a nucleus (see Section 5), and $g^{\mathrm{p}}_{\mathrm{sp}}$ and g^{γ}_{sp} are the single-particle amplitudes for the proton $\{a^{+}_{J}(\mathrm{p})\}$ capture and the γ decay, respectively. The g^{p} and g^{γ} for the IAR are given by $g^{\mathrm{p}}_{\mathrm{sp}}/\sqrt{2T}$ and $g^{\gamma}_{\mathrm{sp}}\cdot(g^{\mathrm{eff}}/g)/\sqrt{2T}$, respectively. Thus the T matrix is reduced by the factor $2T$, which is the number of p–h configurations constituting the IAR. On the other hand the g^{p} for the GDR is reduced by the factor $C_{JJ_{\mathrm{f}}} = D_{JJ_{\mathrm{f}}}/(\Sigma_{vv'}D^{2}_{vv'})^{1/2}$, which is the amplitude of the initial doorway, while the g^{γ} is enhanced by the factor $(\Sigma_{vv'}D^{2}_{vv'})^{1/2}$ due to the collectivity. The two factors cancel each other, resulting in the same single particle matrix element $D_{JJ_{\mathrm{f}}} = \langle 0|a_{J}(\mathrm{p})\|\mathbf{m}_{\gamma}\|a^{+}_{J_{\mathrm{f}}}(\mathrm{p})|0\rangle$ as the IA process.

The phase difference $\Delta\phi = \phi_{J'} - \tilde{\phi}_{J'}$ between the IAR and GDR amplitudes at the IAR resonance energy ($E = \tilde{E}$) is simply given by the resonance phase since the phases for the initial doorways for both the IAR and GDR resonances are considered to be the same. Thus it is written as

$$\Delta\phi = \tan^{-1}\{2(\tilde{E} - E_{G})/\Gamma\}. \tag{4.5.30}$$

The interference pattern below the GDR is therefore opposite to the pattern above the GDR as shown in Fig. 4.67.

4.5.4d *Radiative capture processes through IAR's and GDR's in medium heavy nuclei*

Experimental studies of (p, γ) reactions through IAR's and GDR's in medium heavy nuclei have been carried out for the neutron–magic nucleus ^{140}Ce (Ejiri *et al.*, 1968/1969). A Ge(Li) detector with high-energy resolution was used to resolve γ rays feeding individual low-lying states, even though the peak efficiency for capture γ rays of around 10 MeV was extremely small. The reaction scheme is shown in Fig. 4.68. The differential cross-section at

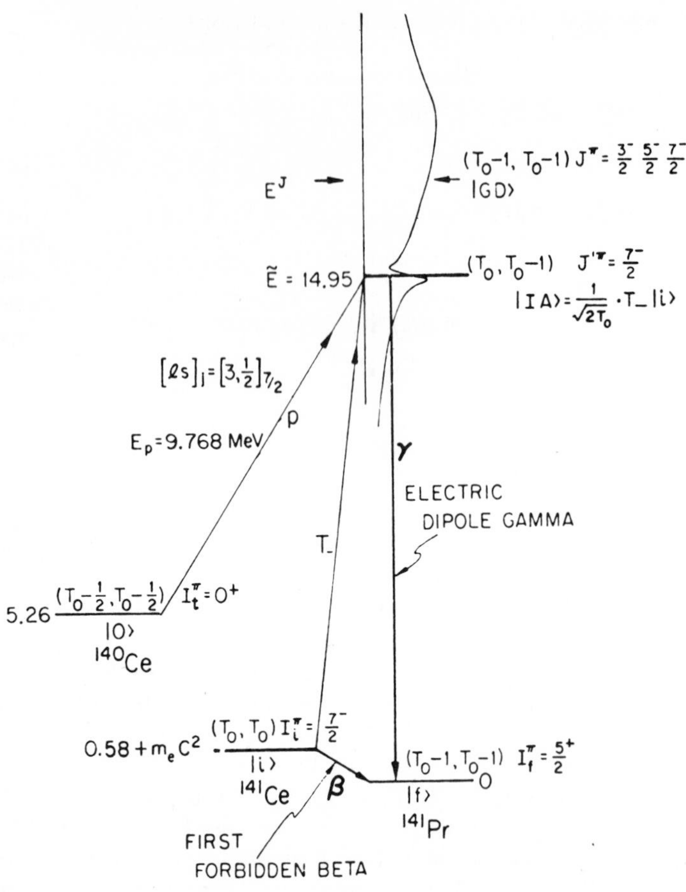

Fig. 4.68. Schematic diagram of an isobaric analogue state and the electric dipole γ and the analogous first-forbidden β transitions. The initial state $|\mathrm{i}\rangle$, the analogue state $\{1/\sqrt{(2T_0)}\}T_-|\mathrm{i}\rangle$ and the final state $|\mathrm{f}\rangle$ have configurations $(2\mathrm{f}_{7/2})_\mathrm{n}|0\rangle$, $\{1/\sqrt{(2T_0)}\}\{(2\mathrm{f}_{7/2})_\mathrm{p}+(2\mathrm{f}_{7/2})_\mathrm{n}\Sigma(\mathrm{a}_\sigma^+\mathrm{a}_\sigma)_0\}|0\rangle$ and $(2\mathrm{d}_{5/2})_\mathrm{p}|0\rangle$, respectively, where $|0\rangle$ is the ^{140}Ce core $T_0=12.5$). (Ejiri *et al.*, 1968/1969.)

$\theta_0 = 55°$, where $P_2(\cos\theta_0)\approx 0$, is written as

$$(\mathrm{d}\sigma/\mathrm{d}\Omega)_{\theta_0} = \tfrac{1}{8}\lambda\!\!\!^{-2}[|A^I_{J'}|^2 + \sum_J |A^D_J|^2 + \sum |A^C_J|^2 + 2\mathrm{Re}(A^I_{J'}\cdot A^{D}_{J'}{}^*)], \tag{4.5.31}$$

where $A^I_{J'}$, A^D_J and A^C_J are amplitudes for the IAR with the spin J', for the GDR, and for the slow compound processes, respectively. In the case of a sharp resonance overlapping with a broad resonance, the amplitude of the broad one is approximately given by a linear function of E.

Thus, by using the relative phases ϕ_J and the $\phi_{J'}$, the amplitudes are

$$A_J^D = (2J+1)^{1/2} e^{i\phi_J}[a_J\{1+\varepsilon_J(E-\tilde{E})\}]^{1/2} \tag{4.5.32}$$

$$A_{J'}^I = \mathrm{i}(2J'+1)^{1/2}(\tilde{\Gamma}_{\mathrm{P}}\tilde{\Gamma}_\gamma)^{1/2} e^{i\phi_{J'}}\{(E-\tilde{E})+\mathrm{i}\tilde{\Gamma}/2\}^{-1}. \tag{4.5.33}$$

and the differential cross-section is

$$\left(\frac{\mathrm{d}\sigma}{\mathrm{d}\Omega}\right)_{\theta_0} = \frac{\lambda\!\!\!^{-2}}{8}\sum_J (2J+1)[(a_J(1+\varepsilon_J(E-\tilde{E}))+\frac{\tilde{\Gamma}_{\mathrm{P}}\tilde{\Gamma}_\gamma}{(E-\tilde{E})^2+(\tilde{\Gamma}/2)^2}\delta_{JJ'}$$
$$+\frac{(a_J(1+\varepsilon_J(E-\tilde{E}))\tilde{\Gamma}_{\mathrm{P}}\tilde{\Gamma}_\gamma)^{1/2}}{(E-\tilde{E})^2+(\tilde{\Gamma}/2)^2}(\tilde{\Gamma}\cos\Delta\phi-2(E-\tilde{E})\sin\Delta\phi)\,\delta_{JJ'}. \tag{4.5.34}$$

The excitation function is fitted by eqn (4.5.34) as shown in Fig. 4.69. The IAR parameters $\tilde{\Gamma}_{\mathrm{P}}$ and $\tilde{\Gamma}_\gamma$ obtained from the (p, p) reaction were used for the fitting. The negative phase difference $\Delta\phi = -0.2\pi$ is consistent with eqn (4.5.30). The γ-decay width derived for the E1 γ decay of the $2f_{7/2}$ IAR to the $2d_{5/2}$ state is $\tilde{\Gamma}_\gamma = (3.6)^2\cdot\mathrm{eV}$. The E1 transition matrix element is derived from the observed width by the relation,

$$\Gamma_\gamma = \frac{16}{9}\pi\left(\frac{E_\gamma}{\hbar c}\right)^3 e^2|M_{\mathrm{IA}}^\gamma|^2. \tag{4.5.35}$$

One gets $|eM_{\mathrm{IA}}^\gamma| = (0.0063 \pm 0.015)$ efm. Comparison with the single quasiparticle estimate leads to a renormalization factor, $g^{\mathrm{eff}}/g = 0.26 \pm 0.05$. The (p, γ) reaction through the IAR resonances above the GDR in ^{209}Bi shows the IAR–GDR interference pattern with a positive phase difference as indicated by eqn (4.5.30) (Snover *et al.*, 1971). Radiative capture reactions through IAR's have been performed on many medium-heavy nuclei (Ejiri *et al.*, 1969; Hasinoff *et al.*, 1972; Paul, 1972). It is interesting to note that similar data have been obtained by inverse reactions like (e, e′p) (Shoda *et al.*, 1971). Those data show uniform quenching of single quasiparticle E1 transitions (see Chapter 5 for the details).

The beta-decay matrix elements are obtained from Γ_γ for the IAR as follows (Ejiri *et al.*, 1968/1969). The first forbidden β operator with $\Delta J = 1$ is expressed on the basis of the ξ-approximation ($\xi \equiv \alpha Z/2R = 12.5 \gg E_\beta(\max)/m_e c^2$) as

$$g_v \mathbf{T}_\beta = g_v\langle \mathbf{r}\rangle + g_v \mathrm{i}\langle \boldsymbol{\alpha}\rangle/\xi - g_{\mathrm{A}}\mathrm{i}\langle \boldsymbol{\sigma}\times\mathbf{r}\rangle. \tag{4.5.36}$$

The partial matrix element $\langle r\rangle$ is derived from the IAR-γ matrix element as $g_v\langle \mathbf{r}\rangle = \sqrt{(2T_0)}eM_\gamma\, g_v/e = (0.32 \pm 0.08)g_v f_m$. Combining with the total β decay matrix element $|g\,T_\beta|$, obtained from the β decay rate, one gets $|-\mathrm{i}\langle\boldsymbol{\alpha}\rangle/\xi\langle\mathbf{r}\rangle - 1.2\mathrm{i}\langle\boldsymbol{\sigma}\times\mathbf{r}\rangle/\langle\mathbf{r}\rangle - 1| = 0.48 \pm 0.07$. This is consistent with the value 0.5 obtained from the angular distribution of the oriented

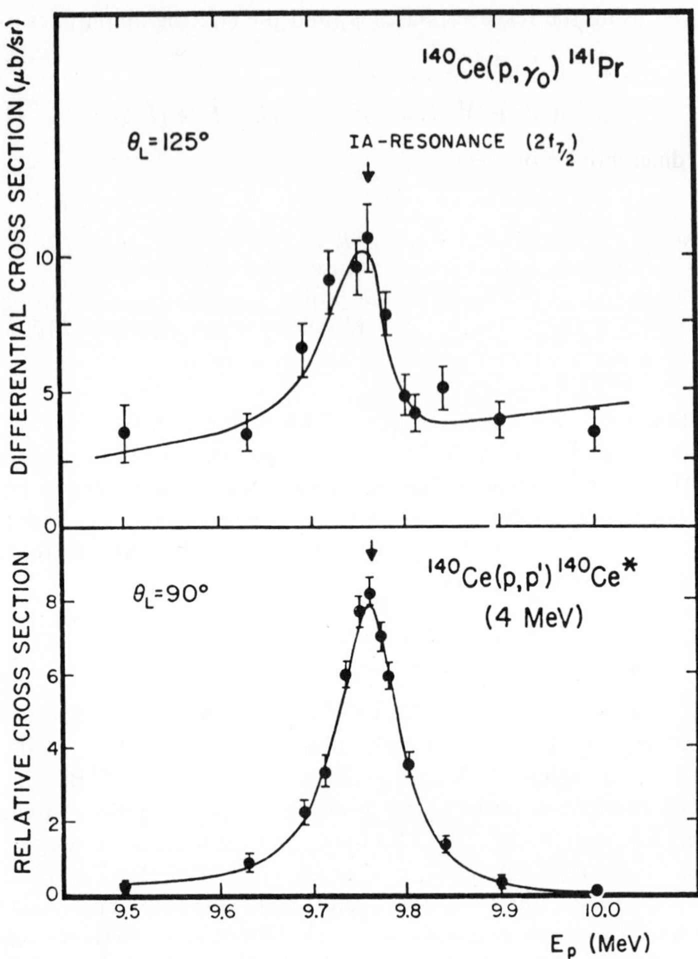

Fig. 4.69. Excitation function of the ^{140}Ce(p, γ)^{141}Pr reaction through the $2f_{7/2}$ IA resonance for the ground-state transition (upper part) and that of the inelastic reaction leading to the 4^- particle–hole state at 4 MeV (lower part). The solid lines are calculated ones. For the fit to the (p, p′) excitation function, a slight interference effect $\sqrt{\Gamma_{p'}}(E-\tilde{E})$ is also included. (Ejiri *et al.*, 1968/1969.)

^{141}Ce nucleus (Hoppes *et al.*, 1961). Use of Fujita's value $\Lambda = 2.4$ obtained on the basis of the CVC theory and the Ahrens–Feenberg approximation (Fujita, 1962, 1967; Ahrens and Feenberg, 1952) leads to $\mathrm{i}\langle \boldsymbol{\sigma} \times \mathbf{r} \rangle / \langle \mathbf{r} \rangle = 0.81 \pm 0.1$. This method is used generally to extract the important matrix elements $\langle \boldsymbol{\sigma} \times \mathbf{r} \rangle_1$, which are otherwise difficult to obtain.

4.5.5 *Direct and semi-direct processes and giant resonances*

4.5.5a *Direct and semi-direct models*

The gamma rays following the radiative capture of a low-energy ($E_a = 5$ to 20 MeV) projectile (a) are mostly in the energy range of $E_a + B_a \approx 10$ to 30 MeV, where B_a is the binding energy of the projectile in the final nucleus. This is the energy region where various types of low-multipole giant resonances (GR's) exist. The GR of the mode T_α is of particular importance for the γ radiation of the same mode T_α because of the coherent contribution of all p–h excitations involved in the GR. The radiative capture process through the GR has been described in terms of the direct-semidirect (DSD) model (Brown, 1964). The semi-direct (SD) capture, in which the projectile excites first the T_α-mode GR which subsequently decays by the T_α-mode γ transition, is predominant in the GR region. Let us for simplicity discuss the radiative capture of a particle a by a target with spin 0 through an intermediate state with spin J. Then the transition matrix in eqn (4.5.14) is written (Brown, 1964; Weller and Roberson, 1980)

$$T = e\varepsilon_L \left(\frac{8k_\gamma}{m_a v_a}\right)^{1/2} \left(\frac{2J_f+1}{2J+1}\right)^{1/2} B_L \cdot t, \tag{4.5.37a}$$

where

$$B_L = k_\gamma^L \left(\frac{L+1}{L}\right)^{1/2} \frac{(2L+1)^{1/2}}{(2L+1)!!}, \tag{4.5.37b}$$

$$t = (C^2 S)^{1/2} \mathrm{i}^{l-l_f-L} (j_f \tfrac{1}{2} L0 | j\tfrac{1}{2}) I, \tag{4.5.37c}$$

and

$$I = \langle \chi_0(\xi)\chi_f(r) | g_a^L \mathbf{m}^L(r) | \chi_0(\xi)\chi_a^+(r) \rangle + \frac{\langle \chi_0(\xi)\chi_f(r) | \mathbf{h}^L(r,\xi) | \chi_0(\xi)\chi_a^+(r) \rangle}{E - E_R + (\Gamma/2)i}, \tag{4.5.37d}$$

where χ_a^+ and χ_f are the scattering state of the incident particle and the captured final orbit, respectively, $g_a^L \mathbf{m}^L$ the γ operator for the projectile and $\mathbf{h}_r^L$ the interaction coupling with the GR, and $\chi_0(\xi)$ is the target state. The first term in eqn 4. 5.37d stands for the direct capture and the second one for the semi-direct capture through the GR with energy E_R and width Γ.

The SD term for the GDR is evaluated by using a schematic interaction model as follows (Brown, 1964). The matrix element of the SD term—the second term of eqn (4.5.37d) is written

$$\{\chi_0(\xi)\chi_f(r) | (\mathbf{V} - \bar{\mathbf{V}}) \mathbf{D}(\xi) | \chi_0(\xi)\chi_a^+(r)\}, \tag{4.5.38}$$

where $\bar{\mathbf{V}}$ is the average Hartree potential, $\mathbf{V}$ is the interaction between the projectile and target nucleon, and $\mathbf{D}(\xi)$ the dipole operator for the target nucleons. $\mathbf{D}(\xi)$ is simply given by the sum $\sum_i \mathbf{D}(\xi_i)$ over all nucleons involved in the GDR. The interaction is approximately expressed by the schematic interaction $\mathbf{V} = \lambda \mathbf{D}(r)\mathbf{D}(\xi)$. The matrix element then reduces to a simple form, $\lambda \mathbf{D}_{\mathrm{af}} \cdot \Sigma \mathbf{D}^2_{\mathrm{ii'}}$, where $\mathbf{D}_{\mathrm{af}} = \langle \chi_{\mathrm{f}} | \mathbf{D}(r) | \chi_{\mathrm{a}}^+ \rangle$ and $\lambda \Sigma \mathbf{D}^2_{\mathrm{ii'}}$ is the energy shift ΔE of the GDR from the average unperturbed p–h energy. Thus eqn 4.5.37d becomes

$$I \approx (q^{\mathrm{eff}}(E)/q_0) \langle \chi_{\mathrm{f}} | \mathbf{D}(r) | \chi_{\mathrm{a}}^+ \rangle, \tag{4.5.39a}$$

with

$$q_0^{\mathrm{eff}}(E)/q_0 \approx \{1 + \Delta E/(E - E_{\mathrm{GD}} + \tfrac{1}{2}\mathrm{i}\Gamma_D)\}. \tag{4.5.39b}$$

The SD effect is represented by the effective charge for the direct process. Since $\Delta E/\Gamma \approx 2$ to 3, the cross-section is an order of magnitude larger than that for the direct process. The SD term contributes constructively to the direct term below the GDR and destructively above it, as in the case of the IAR–GDR interference (see Section 4.5.2).

The interaction matrix is conveniently expressed by the coupling interaction function $\mathbf{h}^L(r)$ (form factor) as $\langle \chi_{\mathrm{f}}(r) | q^L \mathbf{h}(r) | \chi_{\mathrm{a}}^+(r) \rangle$, where q^L is the electro-magnetic coupling constant (charge in case of electric transition). Then the transition matrix is given by

$$I = \langle \chi_{\mathrm{f}}(r) | q_{\mathrm{a}}^L \mathbf{m}^L + \frac{q^L \mathbf{h}(r)}{E - E_R + (\mathrm{i}\Gamma/2)} | \chi_{\mathrm{a}}^+(r) \rangle. \tag{4.5.40a}$$

Thus the effective coupling constant (charge) is

$$(q_{\mathrm{a}}^L)^{\mathrm{eff}} = q_{\mathrm{a}}^L + \frac{\langle q^L \mathbf{h}(r) \rangle / \langle \mathbf{m}^L \rangle}{E - E_R + (\mathrm{i}\Gamma/2)}. \tag{4.5.40b}$$

where $(q_{\mathrm{a}}^L)^{\mathrm{eff}}$ is the effective L-multipole charge corrected for the SD process. The cross-section depends on the form factor as well as on the optical potential used to calculate the scattering state $\chi^+(r)$.

The DSD model is used to extract the direct and GR amplitudes and the multipolarities in the 10 to 30 MeV excitation region, where major electromagnetic transitions are restricted to the lowest orders, E1 and E2. Excitation functions and angular distributions are used to extract properties of both electromagnetic radiation and particle orbits associated with the radiative capture process. In particular, high-precision data on the angular distribution and the analysing power for the radiative capture of polarized projectiles yield a_k and b_k coefficients of the differential cross-section (eqn 4.5.11), which are used to obtain amplitudes (including phases) of the partial waves involved in the reaction.

The cross-section for the DSD radiative capture to a single particle orbit χ_j is written as

$$\sigma(p,\gamma) = C \cdot (q_{\mathrm{a}}^L)^2 \cdot S_j |M_j|^2 \left| 1 + \frac{h'}{E - E_R + (\mathrm{i}\Gamma/2)} \right|^2, \tag{4.5.41}$$

where S_j is the spectroscopic factor of the single j nucleon for the final state χ_f, and M_j is the single-particle matrix element given by $\langle \chi_j | \mathbf{m}_L | \chi_a^+ \rangle$. Therefore the radiative capture process is used to study the single-particle properties of the final state χ_f. Here χ_f can be any excited state. Thus the radiative process is a kind of projection from a high excitation scattering state χ_a^+ to the lower excitation state through a simple electromagnetic operator $\mathbf{m}^L$. The giant resonance (SD process) may show-up in principle for any final state χ_f in the resonance-like shape of the cross-section as a function of the projectile energy E (see 4.5.41). If the incident energy is high enough, the resonance shape may be seen as a broad peak at $E_\gamma \approx E_R$ in the γ-ray spectrum for a given incident energy. The integrated cross-section for a given final state over the giant resonance built on the final state leads to the spectroscopic strength for the final state. It gives a single nucleon spectroscopic factor in case of a nucleon radiative capture, and a cluster spectroscopic factor in case of a cluster radiative capture. In fact the DSD process for a cluster (complex projectile) is not dominant and one has to consider the compound process as well.

4.5.5b *Radiative capture reactions and giant dipole resonance*

The giant dipole resonance (GDR) is one of the strongest resonances and it exhausts most of the E1 sum rule, the lowest multipole E1 radiation being dominant over a wide excitation region around the GDR. Thus the DSD process for the E1 transition has been extensively studied by using unpolarized and polarized protons and unpolarized neutrons. The excitation function of the radiative capture reaction $^{208}\text{Pb}(\text{n}, \gamma)^{209}\text{Pb}$ is shown in Fig. 4.70, and the data are well reproduced by the DSD model. The excitation function depends to some extent on the form factor in eqn (4.5.40a). Various authors used different types of form factors (Brown, 1964; Clement *et al.*, 1965; Zimanyi *et al.*, 1970; Potokar *et al.*, 1977). The importance of using a complex coupling interaction in the DSD calculation of (n, γ) reactions has been emphasized (Potokar *et al.*, 1977).

A pure resonance model (PRM) has been proposed by Dietrich and Kerman (1979). The model is based on the same assumptions as in the DSD model, but it projects the GDR explicitly out of the continuum. The PRM process is easily understood in the inverse process of the (γ, p) reaction, where a photon excites a single GDR (a single door-way), which de-excites by emitting a proton in a scattering state. The PRM is found to be very insensitive to the form factor.

The microscopic properties of radiative capture reactions have been studied particularly by using polarized protons (Glavish *et al.*, 1972; Hanna *et al.*, 1972; Cameron *et al.*, 1976; Turner *et al.*, 1978; Weller *et al.*, 1978). In the low excitation regions of light nuclei the radiation is assumed to be mainly E1 in character and the number of partial waves involved is small. Thus the analysis is made model-independently to deduce microscopic information on

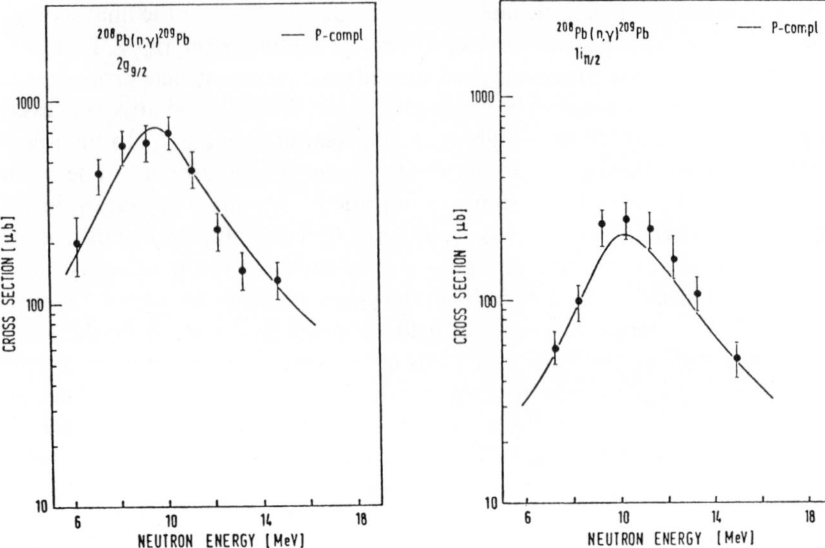

Fig. 4.70. Theoretical excitation functions for the capture of neutrons to the single-particle states $2g_{9/2}$ (left-hand side) and $1i_{11/2}$ (right-hand side) corresponding to the 90° incident angle compared with the experimental data. The values of free parameters are $V_1 = 60$ MeV and $W_1 = 145$ MeV. (Potokar *et al.*, 1977.)

the GDR. The $^{15}N(\vec{p}, \gamma_0)^{16}O$ reaction in the GDR region was studied by Hanna *et al.* (1972) and the cross-section and coefficients a_i and b_i are shown in Fig. 4.71. Since the target spin is $1/2^-$ and the final state spin is 0^+, $s_{1/2}$ and $d_{3/2}$ partial waves contribute to the E1 radiative capture through the intermediate state $J^\pi = 1^-$. They are written as $s_{1/2}\exp(i\phi_s)$ and $d_{3/2}\exp(i\phi_d)$, where $s_{1/2}$ and $d_{3/2}$ are the real amplitudes in the T matrix and ϕ_s and ϕ_d are their phases. The $s_{1/2}$, $d_{3/2}$ and $\phi_s - \phi_d$ are deduced from the observed a_2 and b_2 coefficients and the normalization condition $s_{1/2}^2 + d_{3/2}^2 = 1$. They are plotted in Fig. 4.71. Here the solution I is found to be consistent with the (γ, n_0) data. It is interesting to note that the $s_{1/2}$ and $d_{3/2}$ real amplitudes stay constant over the GDR region.

There are several T matrix elements contributing to the radiative capture on medium heavy nuclei. Although there are in general more unknown than known, the major T matrix elements associated with non spin–flip may be extracted from the measured a_i and b_i coefficients. The angular distributions and analyzing powers for the radiative capture of polarized protons into the $1f_{7/2}$ shell are shown in Fig. 4.72. They give the major amplitudes $d_{5/2}$ and $g_{9/2}$, and the relative phase of $\phi_g - \phi_d$ (Cameron *et al.*, 1976). These quantities are well analysed in terms of the DSD model (Weller *et al.*, 1978).

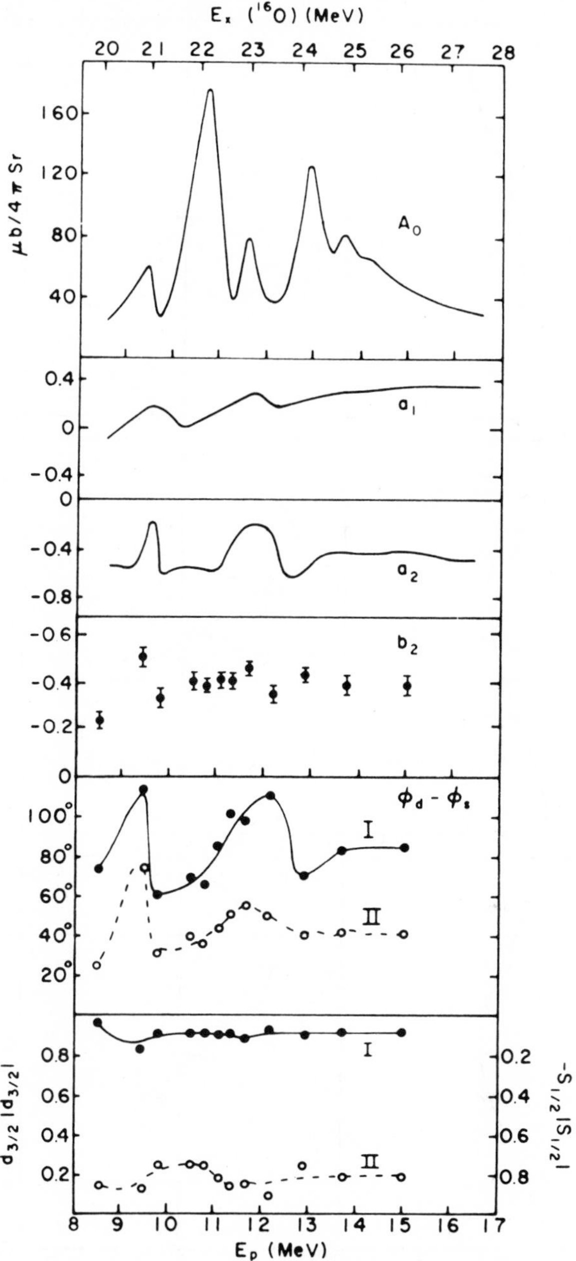

Fig. 4.71. Total yield (A_0 and angular distribution coefficients (a_1, a_2, b_2) for $^{15}N(\vec{p}, \gamma_0)^{16}O$. Two solutions (I and II) for the phases and the amplitudes are obtained from fitting to the data. (Hanna *et al.*, 1972.)

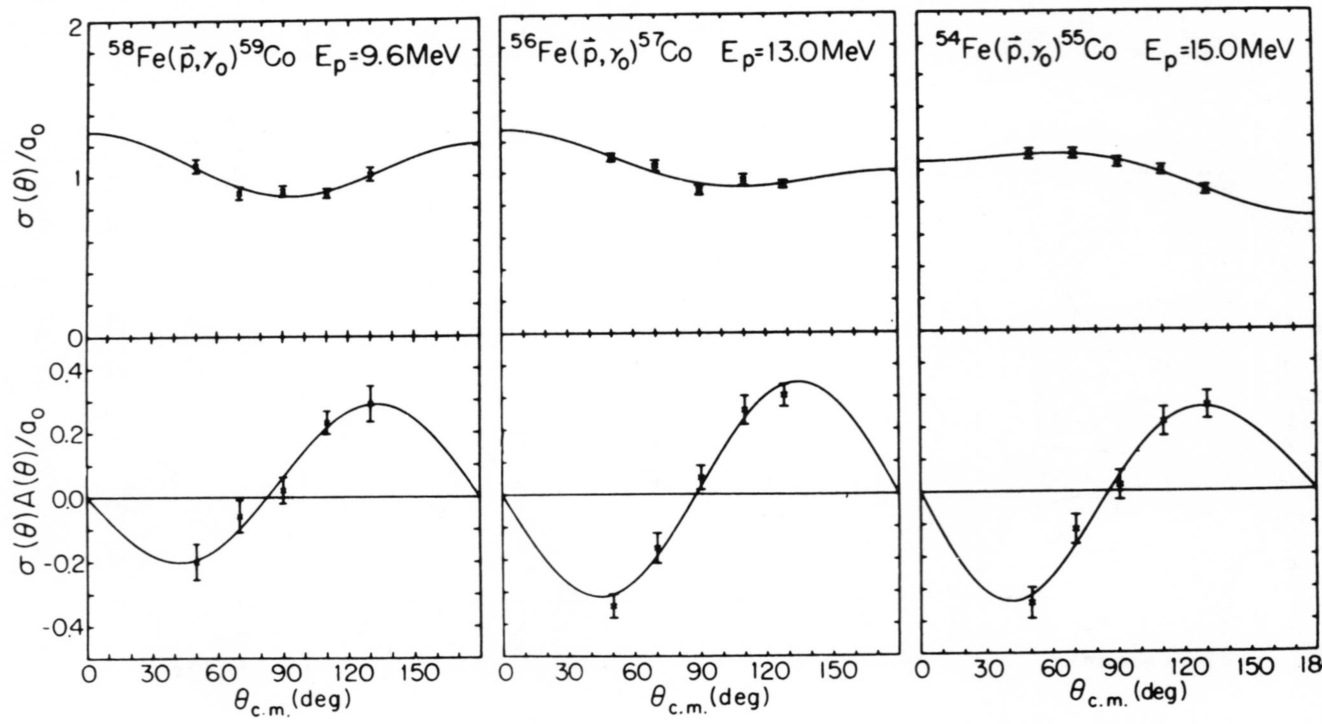

Fig. 4.72. Typical data for $(\vec{p}, \gamma_0)$ reactions on Fe isotopes at three energies for the quantities $\sigma(\theta)/a_0$ and $\sigma(\theta)A(\theta)/a_0$. The error bars represent the statistical errors associated with the data points. The solid curves are the result of fitting the data. (Cameron *et al.*, 1976.)

4.5.5c *Radiative capture reactions and giant quadrupole resonances*
Giant quadrupole resonances (GQR) and giant octupole resonances (GOR) have been studied by means of inelastic scattering and radiative capture reactions. Even though the E2 and E3 transitions are much weaker than the E1 transitions, they are readily studied by observing interference with the dominant E1 transition in the radiative capture process. Isovector and isoscalar GQR's have been studied by means of (p, γ), (n, γ) and (α, γ) reactions. A large E2 strength was found in the 20 to 27 MeV excitation region of ^{16}O by studying the polarized proton capture reaction $^{15}N(\vec{p}, \gamma_0)^{16}O$ (Hanna *et al.*, 1974). In this case the reaction amplitudes contributing to the E1 and E2 transitions are $s_{1/2}\exp(i\phi_s)$, $d_{3/2}(i\phi_d)$, $p_{3/2}\exp(i\phi_P)$, and $f_{5/2}\exp(i\phi_f)$. Since one phase is arbitrarily chosen, seven quantities are obtained from the seven observables A_0, a_1, a_2, a_3, b_1, b_2 and b_3. The results are shown in Fig. 4.73. The major E1 process shows several intermediate structures, while the E2 process interfering with the E1 process shows a broad resonance around 24 MeV excitation energy. The E2 strength amounts to 30 per cent of the E2 sum rule. On the other hand $(\vec{p}, \gamma_0)$ reactions on ^{14}C and ^{13}C are well described by the DSD model with only the direct E2 process (Snover *et al.*, 1976; Turner *et al.*, 1979). These observations indicate no resonance-like concentration of the E2 strength in the p_0 channel of these nuclei.

The neutron capture γ-rays, which have no direct E2 contribution (see Section 4.5.2), are suitable to select the resonant (SD) E2 strength. In fact the E2–E1 interference term $\{a_1 P_1(\cos\theta)$ term$\}$ observed in the $^{14}C(p, \gamma_0)^{15}N$ (Weller *et al.*, 1976; Harakeh *et al.*, 1975) disappears in the $^{14}N(n, \gamma_0)^{15}N$ reaction (Wender *et al.*, 1980). This clearly indicates no E2 resonant strength in ^{15}N. Isoscalar GQR's in medium nuclei are seen in (n, γ_0) reactions (Likar *et al.* 1978; Wender *et al.*, 1978). Jensen *et al.*, (1979) has found the E2 resonance strength of 1 to 3 per cent of the isoscalar E2 sum rule in the n_0 channel for ^{41}Ca by means of polarized neutron radiative capture. The polarized neutron beam was obtained by means of the $^2H(\vec{d}, n)\,^3He$ reaction. The large b_1 and a_1 coefficients, which arise from the interference with the E2 resonance, are seen in Fig. 4.74, and this corresponds to the direct n_0 decay of the GQR excited by inelastic scattering (see Section 4.4).

The isovector GQR's (Pitthan, 1980) with excitation energies of about 130 MeV/$A^{1/3}$ are studied by (n, γ) reactions with medium energy neutrons (Drake *et al.*, 1981). An experimental set-up for measuring the (n, γ) angular distributions is shown in Fig. 4.75 (Arthur *et al.*, 1975). The isovector GQR in ^{41}Ca was observed around 32 MeV in excitation energy by means of the (n, γ) reaction with $E_n = 20$ to 28 MeV (Bergqvist *et al.*, 1984). The data and the DSD analysis are shown in Fig. 4.76. The E2 strength extracted from the data amounts to one-third of the isovector E2 sum rule, which corresponds to the full strength of the $T_<$ (lower isospin) component selected by the neutron channel.

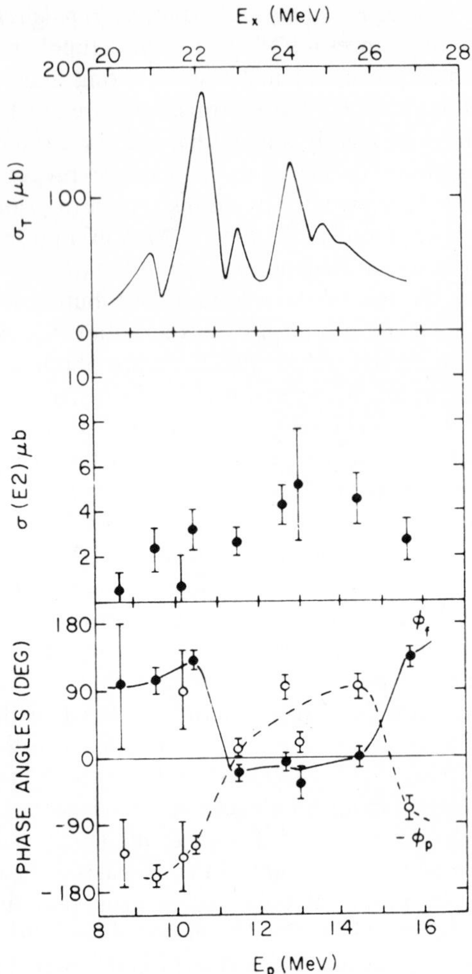

Fig. 4.73. Top: total cross section curve of $^{15}N(p, \gamma)^{16}O$. Middle: plot of total E2 cross-sections derived from the analyses. Bottom: plot of the phases of the partial-wave amplitudes for proton capture leading to E2 radiation. (Hanna *et al.*, 1974 and refs. therein.)

The isoscalar GQR's are readily excited by radiative alpha capture reactions. In fact, the $^{12}C(\alpha, \gamma_0)^{16}O$ reaction has revealed a large fraction of the isoscalar GQR strength around 15 MeV excitation energy in ^{16}O (Snover *et al.*, 1974). E1 and E2 radiative α-capture reactions on SD-shell nuclei have been analysed in terms of the direct, SD, and compound processes (Shikazono and Terasawa, 1975). The experimental data are consistent with the existence

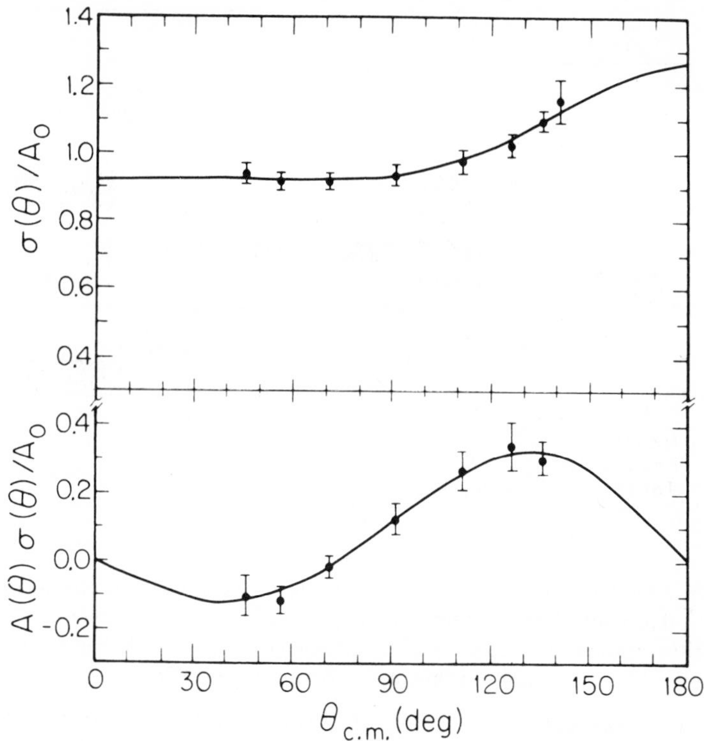

Fig. 4.74. The cross-section and analysing powers obtained for the reaction ^{40}Ca(n_{pol}, γ)^{41}Ca at $E_n = 10.0$ MeV. The solid lines are the curves generated by fitting the data with the *T*-matrix amplitudes and phases (Jensen *et al.*, 1979.)

of the isoscalar GQR as shown in Fig. 4.77. Some of the α capture reactions proceed via a statistical process, without E1–E2 interference (Meyer-Schützmeister *et al.*, 1968), while a finite non-statistical component has been reported in (α, γ) reactions (Meyer-Schützmeister *et al.*, 1978; Kuhlmann *et al.*, 1979).

4.5.5d *Magnetic dipole resonance and higher order electric giant resonances*
The magnetic dipole resonance is essentially due to a spin–flip transition within one $\hbar\omega$ shell. Thus it lies in a rather low excitation region, which is hard to reach by means of radiative capture reactions. Doubly closed (*l–s* closed) shell nuclei, where the spin–flip partner is occupied, are supposed to have no M1 strength. Consequently observation of M1 transitions in doubly closed shell nuclei provides the M1 strength distribution in the high excitation region and the core-breaking contribution. The M1 strength in the doubly-closed ^{16}O has been clearly demonstrated by the E1–E2–M1 interferences in the

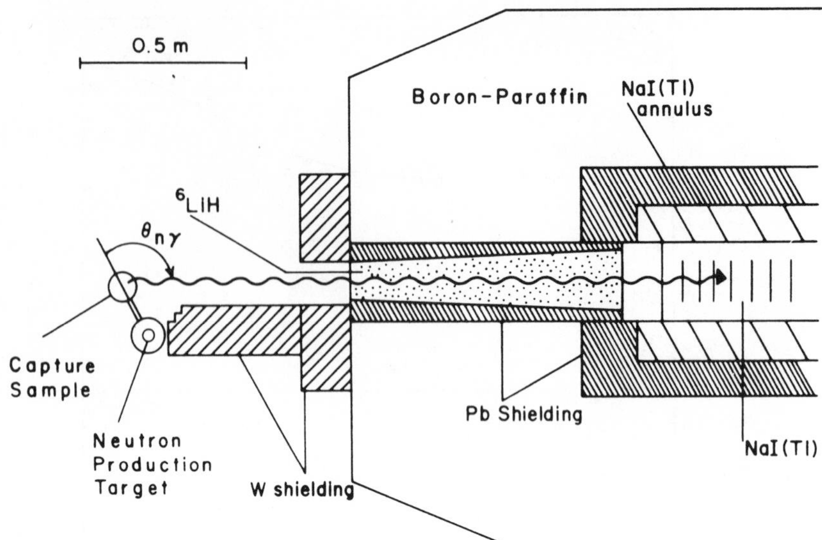

Fig. 4.75. Experimental arrangement for the (n, γ) reaction with neutrons produced by the t + d reaction. A section through the apparatus taken in the plane perpendicular to the triton beam is shown. The angle $\theta_{n\gamma}$ is varied by displacing the capture target left or right in this plane. (Arthur *et al.*, 1975.)

$^{15}N(p, \gamma_0)$ ^{16}O reaction (Snover *et al.*, 1979). The data at 16.22, 17.14 and 18.8 MeV in excitation energy of ^{16}O deviate strongly from the best fit in terms of only the E1 and E2 radiations, and require the M1 component admixture, as shown in Fig. 4.78. The total M1 strength in the 16 to 20 MeV excitation region is as large as $B(M1\downarrow) \gtrsim 0.24\mu_0^2$.

Searching for M1 strength in a high-excitation region is interesting in view of the possible M1 strength spread by the tensor interaction (see Chapter 5). High multipole electric giant resonances have been studied by inelastic scattering. Calculations of radiative capture reactions through E1, E2, and E3 resonances show the sensitivity of polarized neutron capture to the E3 resonance (Saporetti and Guidotti, 1979). The E3 strength in the high-excitation region of ^{90}Zr is indicated by a finite a_5 coefficients in the (p, γ) angular distribution (Dietrich *et al.*, 1977).

4.5.6 *Giant resonances built on excited states*

4.5.6a *Giant resonance built on highly excited states*

A giant resonance (GR) is a collective state expressed as $\chi'(\xi) = Q^+\chi(\xi)$ where Q^+ is the creation operator of the collective phonon. The core state $\chi(\xi)$ may be any excited state as well as the ground state (Brink, 1955). Therefore all excited states may possibly have a GR built on them. This idea can be

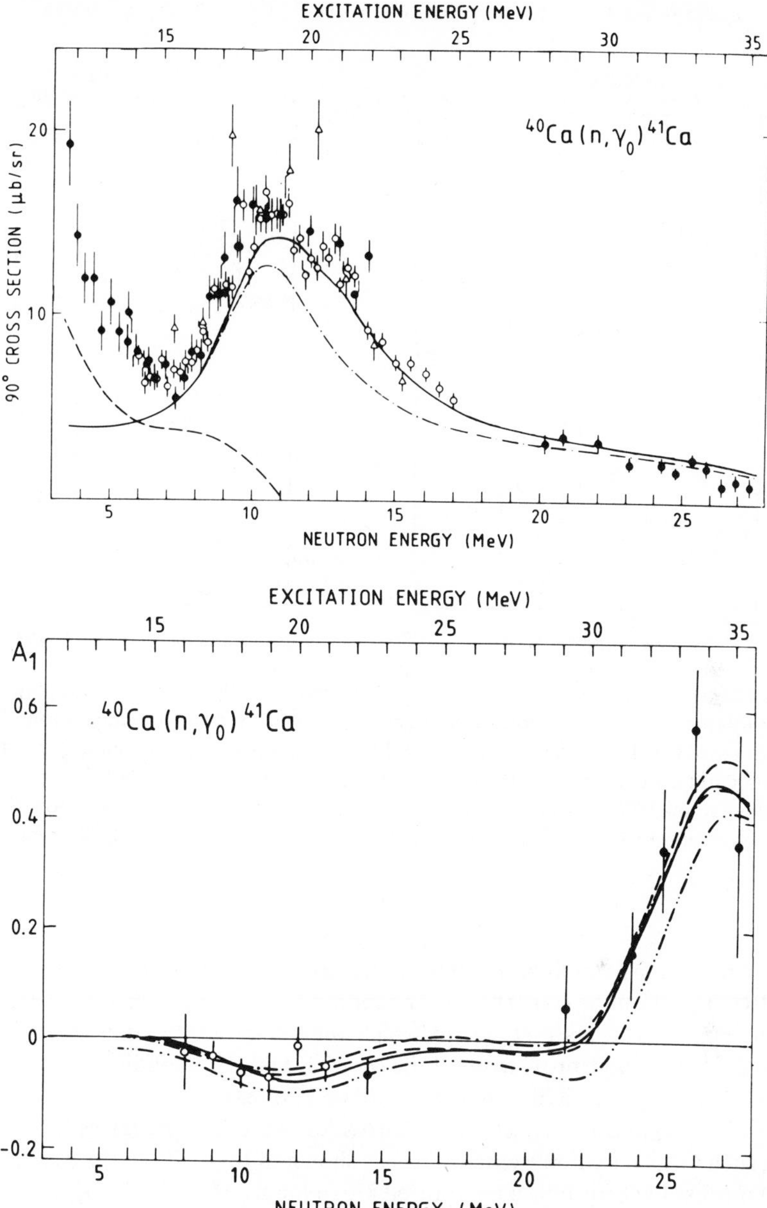

Fig. 4.76. Top: the 90° cross-section of the $^{40}Ca(n, \gamma_0)^{41}Ca$ reaction. The full and dot-dashed curves represent cross-sections calculated on the basis of the DSD model. The dashed curve is the compound-nucleus cross section. Bottom: forward–backward asymmetries calculated from the DSD model using various optical-model potentials and experimental data points. (Bergqvist *et al.*, 1984; see the text of the ref. for the optical potential used and the refs. therein for the data points.)

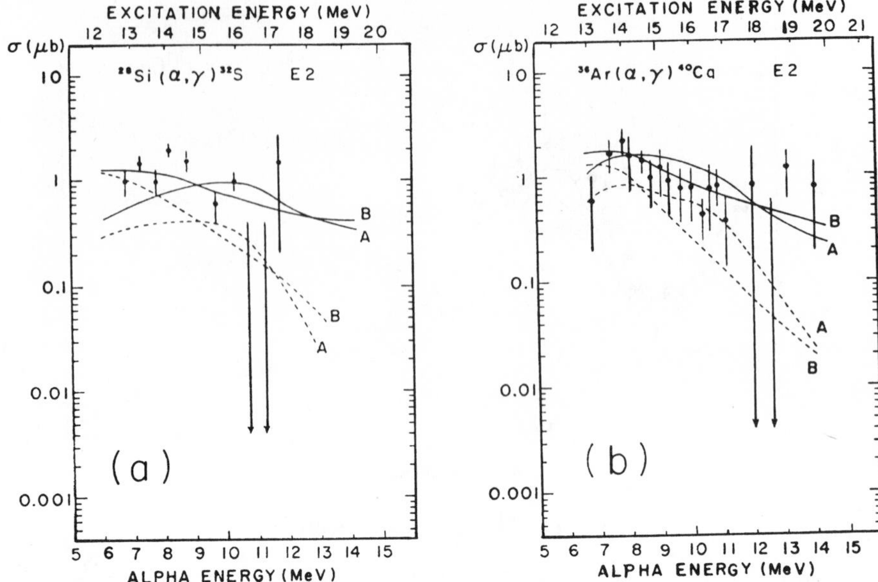

Fig. 4.77. Comparison of the calculated cross-sections with the experimental ones for E2 radiative capture. The optical parameter set II was used for the calculations. The solid and dashed lines represent the total (compound + direct and semidirect) cross section and the compound cross-section, respectively. The letters A and B correspond to the assumed Lorentzian and rectangular shapes for the total γ-absorption cross-section, respectively. (a) For ^{28}Si(α, γ_0)^{32}S. (b) For ^{36}Ar(α, γ_0)^{40}Ca. (Shikazono and Terasawa, 1975.)

extended to the annihilation (Q) of the GR as given by $Q\chi'(\xi)$. Here the GR(Q^+) built on the state $Q\chi'(\xi)$ corresponds to the $\chi'(\xi)$. Then highly excited states may find a state $Q\chi'(\xi)$ to which they de-excite by emitting the Q^+ phonon (GR-γ ray), provided that the $Q\chi'(\xi)$ is real. This condition is satisfied when the $\chi'(\xi)$ is a highly excited compound state.

Radiative capture γ-spectroscopy provides two useful processes for studying such GR's based on excited states as shown in Fig. 4.79. In the case of the SD process the projectile a is captured into a highly excited orbit χ'_a, while the target χ_A is excited to the GR, which de-excites by emitting the GR-γ ray. The intermediate state is given by $\chi'_a Q^+ \chi_A$, where the GR built on χ_A is coupled with the excited state χ'_a. This is approximately equivalent to the state $Q^+ \chi'_a \chi_A$, i.e. the GR built on the excited state $\chi'_a \chi_A$. Here the GR(Q^+) consists of a superposition of many 1p–1h pairs and it does not make much difference whether a is a participant in $Q^+ \chi'_a \chi_A$ or a spectator in $\chi'_a Q^+ \chi_A$. In the case of the compound process the highly excited compound state χ_c consists of many

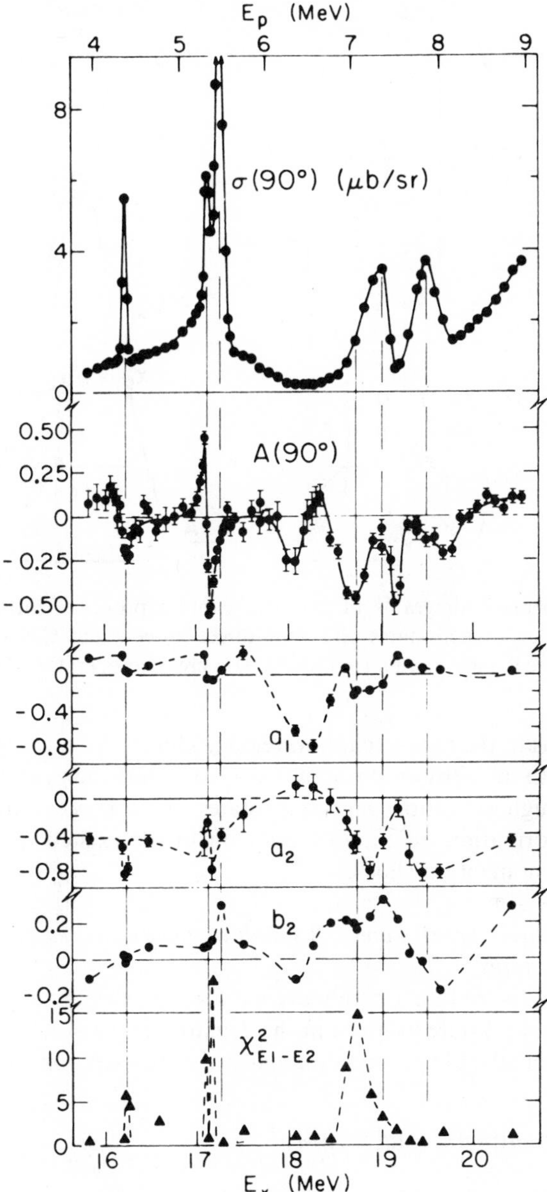

Fig. 4.78. Excitation curves for the $^{15}N(p_{pol}, \gamma_0)^{16}O$ reaction. From top to bottom: $\sigma(90^\circ)$, $A(90^\circ)$, and reduced χ^2 for the angular distribution fits assuming only E1 and E2 radiations. a_1, a_2 and b_2 are coefficients of the angular distribution; the curves eye guides. The vertical solid and dashed lines indicate M1 and E1 resonances, respectively. (Snover *et al.*, 1979.)

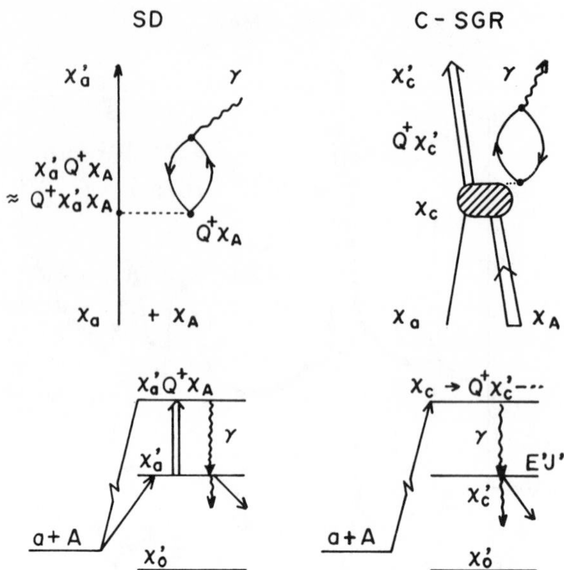

Fig. 4.79. Schematic diagrams of radiative capture processes through giant resonances built on excited states. SD; semi-direct process and C-SGR; compound statistical giant resonance process. Q^+: creation operator of GR.

p–h excitations in the high-excitation region. Thus it has a component $Q^+\chi_c'$ which decays to the χ_c' by emitting the GR-γ ray. Here the χ_c and accordingly also χ_c' are high-excitation high-spin states. This process thus provides important information on the GR built on highly-excited (many particle) high-angular momentum states.

4.5.6b *Semi-direct process and GR's built on highly excited particle (hole) states*

The DSD process feeding a highly excited state χ_a' is essentially the same as that for the ground state as given in the T matrix (eqns (4.5.39–5.41)). In the GR region where the SD process dominates, the cross-section—eqn (4.5.41)—reduces to

$$\sigma(\mathrm{p},\gamma)=C\cdot(q_a^L)^2S_j|M_j|^2\frac{h'^2}{(E_\gamma-E_{\mathrm{GR}})^2+(\Gamma^2/4)}, \qquad (4.5.42)$$

where the γ-ray energy is given by $E_\gamma=E_a+B_a-E_f$, and E_a, B_a, and E_f are the incident projectile energy, the binding energy, and the excitation energy of the final state. Thus the excitation functions (the yield curves as a function of the incident energy E_a) for various final states, when plotted as a function of E_γ, should show a similar resonance peak centred at $E_\gamma\approx E_{\mathrm{GR}}$ with intensity proportional to the spectroscopic factor. These features are nicely demonstrated in the $^{27}Al(p,\gamma)$ reaction (Dowell, *et al.*, 1983). Gamma-rays feeding various excited states are clearly observed as shown in Fig. 4.80a. The cross-section for each γ transition has a broad bump at $E_\gamma = 20$ MeV, corresponding to the GDR energy of the ground state. The widths, however, increase from $\Gamma = 4.5$ to 12 MeV as E_f increases from 0 to 14.4 MeV. The integrated inverse cross-section, $I = \int (2J_f + 1)\sigma(\gamma, p_0)dE_\gamma$, follows closely the spectroscopic strength $\Sigma(2J_f + 1)S$ as shown in Fig. 4.80b. Giant dipole resonances built on excited states in other s–d shell nuclei have also been seen (Dowell, 1985).

Radiative capture γ-rays to high-lying particle–hole states in lighter nuclei have been studied (Kovash *et al.*, 1979; Arnold, 1979; Manglos *et al.*, 1981; Weller *et al.*, 1982). Strong γ-transitions to simple particle–hole states (19 MeV $\{(p\frac{3}{2})^{-1}(d\frac{5}{2})\}_{4^-}$ state in ^{12}C, etc) are seen and they may be explained in terms of direct E1 and E2 capture processes with proper final wave functions.

4.5.6c *Giant resonances built on highly-excited high-spin states*

The cross-section for statistical γ-emission from a compound nucleus is written as

$$\sigma(a, \gamma) = \sigma_c \cdot \frac{\Gamma_\gamma(E_\gamma)}{\Gamma_t}, \tag{4.5.43}$$

where σ_c is the compound nucleus formation cross-section and Γ_γ/Γ_t is the branching ratio of the γ-emission. The γ-decay width is given by

$$\Gamma_\gamma(E_\gamma) = Kf(E_\gamma) \cdot \exp\{(E - E_\gamma)/T\}, \tag{4.5.44}$$

where $f(E_\gamma)$ is the γ-transition strength function and $\exp\{(E - E_\gamma)/T\}$ gives the level density factor with effective nuclear temperature T and excitation energy $E - E_\gamma$. On the basis of the GDR strength distribution for the highly excited compound nucleus, $f(E_\gamma)$ is given by a Lorentzian-type GDR form,

$$f(E_\gamma) = K'\{(E_\gamma^2 - E_{GR}^2)^2 + E_\gamma^2 \Gamma_{GR}^2\}^{-1}. \tag{4.5.45}$$

Therefore the cross-section is rewritten as

$$\sigma(a, \gamma) = C \cdot \exp(-E_\gamma/T) \cdot f(E_\gamma). \tag{4.5.46}$$

The GDR-type strength function has been found in γ-ray spectra following compound nuclei produced by heavy-ion bombardment (Newton *et al.*, 1981). The γ-ray spectrum does not fall off simply as $\exp(-E_\gamma/T)$ but shows an excess around $E_\gamma \approx E_{GR}$. This is clearly demonstrated in the γ-ray spectra corrected for the level density factor as shown in Fig. 4.81. The GDR γ-rays have to be emitted in competition with the neutron evaporation. The branching ratio is proportional to the level density ratio as given by

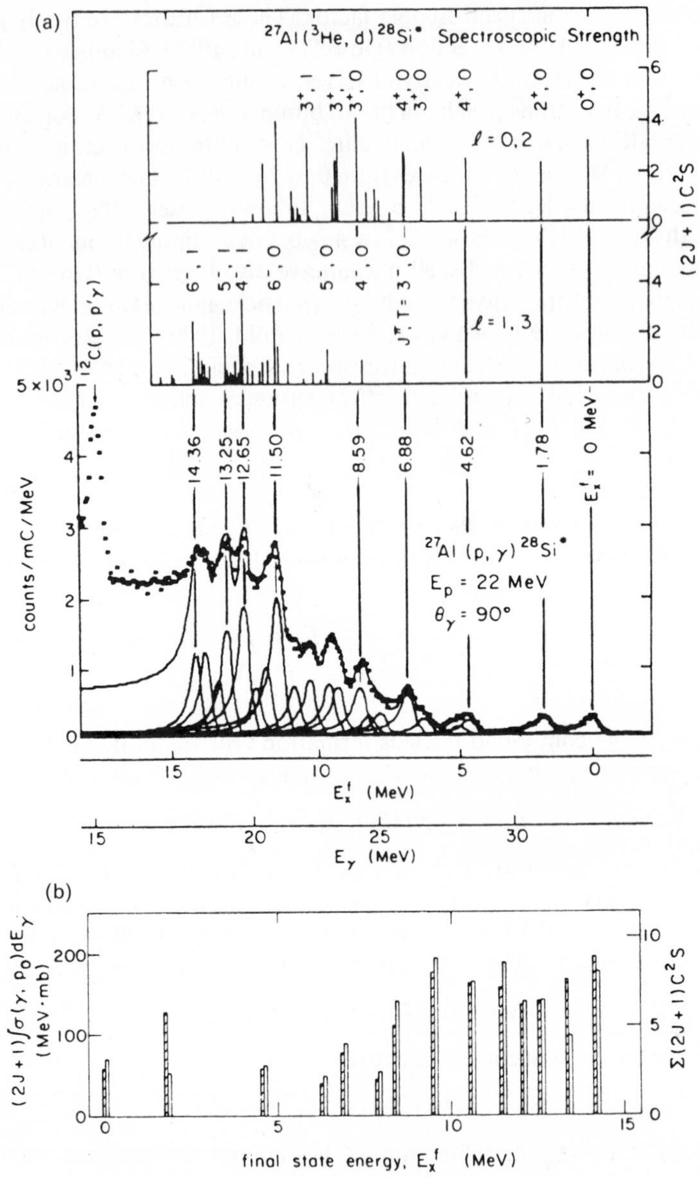

Fig. 4.80. (a) ^{27}Al(p, γ)^{28}Si γ-ray spectrum with line-shape decomposition measured at BNL and the ^{27}Al(^{3}He, d)^{28}Si single-proton spectroscopic factor $(2J+1)C^2S$. The vertical lines indicate the correspondence of proton capture and proton stripping for some of the strong γ transitions. (b) Hatched bars and left scale; integrated strengths $(2J+1)\int\sigma(\gamma, p_0)dE_\gamma$. Open bars and right scale: single-proton spectroscopic strength $\Sigma(2J+1)C^2S$ (Dowell *et al.*, 1983.)

$\Gamma_{\gamma}/\Gamma_{n} \propto \exp\{B_n/T') - (E_{\gamma}/T)\}$, where B_n is the neutron binding energy. Since $E_{\gamma} - E_n \gg T$ for the GDR γ rays, the ratio $\Gamma_{\gamma}/\Gamma_{n}$ increases as the T increases. Thus the GDR γ rays are mainly emitted from the compound nucleus with the highest temperature since $T \propto \sqrt{E_x}$, where E_x is the excitation energy (Newton *et al.*, 1981).

The GDR shown in the statistical compound decay has opened up the possibility of studying collective (resonance) motions of excited nuclei over a wide range of excitation energy and angular momentum. Experimentally it requires careful and sensitive measurements of 15 to 30 MeV γ rays. The γ-decay branch itself is as small as 10^{-5} to 10^{-4} as in other unbound states. Thus background due to pile-up signals and to the huge flux of neutrons and γ rays from light element impurities have to be carefully corrected for. Measurement of the γ-ray multiplicity by means of a sum-spectrometer is useful to select high-spin states (Newton *et al.*, 1981).

In s–d shell nuclei considerable broadening of GDR's were observed in the radiative capture of ^{3}He as shown in Fig. 4.82 (Snover, 1985). The shape of a highly excited (high-spin) state can be studied by investigating the splitting of the GDR build on that state. GDR's in Sn and Er isotopes with $E \sim 60$ MeV and $I\hbar \sim 60\,\hbar$ have been studied by Gaardhøje *et al.* (1984). The γ-ray spectra for $^{16}\text{O} + {}^{92}\text{Mo} \rightarrow {}^{108}\text{Sn}^*$ and $^{16}\text{O} + {}^{150}\text{Nd} \rightarrow {}^{166}\text{Er}^*$ reactions are shown in Fig. 4.83. The spectrum for $^{108}\text{Sn}^*$ shows a single GDR resonance in consistent with a spherical shape. The spectrum for ^{166}Er shows a broad bump, which can be decomposed in two GDR components with intensities of 60 per cent and 40 per cent for the lower and higher components, respectively. This indicates a predominantly oblate shape at such high excitation region of $T \sim 1.6$ MeV in contrast to the prolate shape of the ground state (note that the prolate shape gives 33 per cent and 67 per cent for the two GDR components).

4.5.7 *Medium-energy gamma rays*

The radiative capture of medium-energy light ions gives medium-energy γ-rays in the range 40 to 200 MeV. The momentum transfer involved is as large as 1 to 0.3 GeV/c. Thus the study of medium-energy γ-rays is quite important for investigating the electromagnetic properties of highly-excited nuclei far above the giant resonances and radiation mechanisms associated with large momentum transfer. The E1 radiation strength is closely related to the photo-absorption (γ, x) cross-section. The dipole sum rule for photo-absorption is given by (Levinger and Bethe, 1950)

$$S(\text{E1}) \equiv \int \sigma_{\gamma}^{\text{E1}}(E_{\gamma})\text{d}E_{\gamma} = \sigma_0(1+\kappa) \qquad (4.5.47\text{a})$$

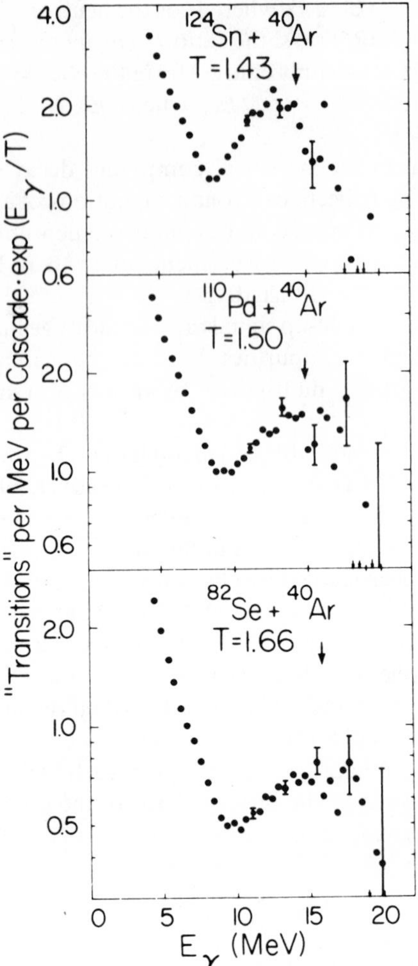

Fig. 4.81. The background subtracted γ-ray spectra multiplied by $\exp(E_\gamma/T_e)$ for 170 MeV ^{40}Ar-induced reactions. The arrows indicate $E_\gamma = 78/A^{1/3}$ MeV, the centroid of the ground-state GDR. (Newton *et al.*, 1981.)

where

$$\sigma_0 = 2\pi^2 e^2 \frac{\hbar}{mc}\frac{NZ}{A}, \tag{4.5.47b}$$

and

$$\kappa = \frac{mA}{e^2\hbar^2 NZ}\langle \psi_0 | [\mathbf{D}_z, [\mathbf{V}, \mathbf{D}_z]] | \psi_0 \rangle. \tag{4.5.47c}$$

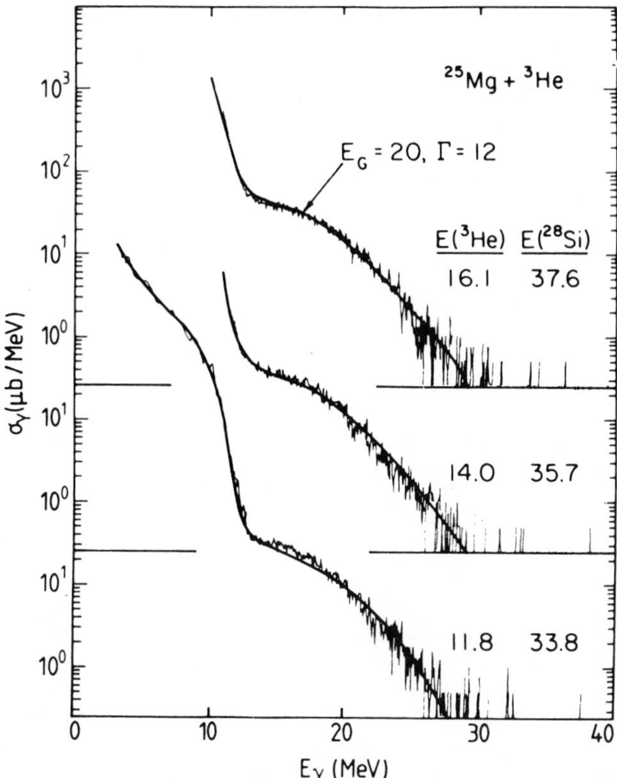

Fig. 4.82. The gamma-ray production cross-section for ^{3}He plus ^{25}Mg, indicating a broad GDR width of $\Gamma \approx 12$ MeV at the GDR energy of $E_G = 20$ MeV. The smooth curves are statistical model calculations with the GDR parameters indicated. (Snover, 1984, 1985.)

In these expressions σ_0 is the classical Thomas–Reiche–Kuhn sum rule and $D_z = \Sigma \frac{1}{2} e\{1 + \tau_z(i)\} z_i$ is the dipole operator. The enhancement factor κ is a measure of the contribution of the nucleon–nucleon interaction **V** which does not commute with the dipole operator. It reflects the velocity dependent term or the exchange force (current) term. Experimentally the enhancement factor has been found to be as large as $\kappa \approx 0.5$ to 1.0 (Ahrens *et al.*, 1975; Leprêtre *et al.*, 1981, and references therein). Theoretically it is interpreted as partly due to the exchange current which is related to the δg_l and partly due to the tensor correlation which plays an essential role in the medium-energy region far beyond the GDR (Fujita and Hirata, 1971; Arima *et al.*, 1973; Brown and Rho, 1980).

The radiative proton capture (p,γ) is the inverse process of the photo-absorption (γ,p_0) into the particular p_0 channel. Here the initial and final

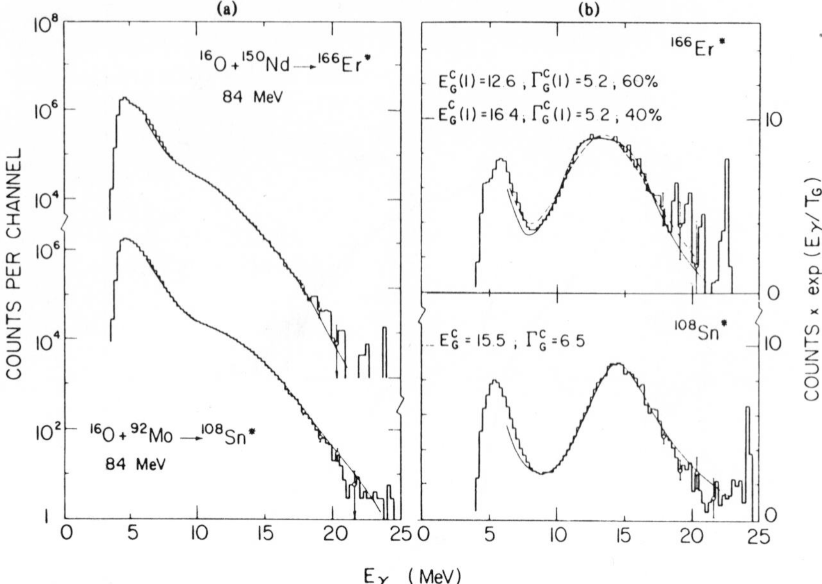

Fig. 4.83. (a) The measured gamma-ray spectra from the decay of $^{166}Er^*$(top) at $E^* = 61.5$ MeV and $^{108}Sn^*$ (bottom) at $E^* = 61.2$ MeV, plotted as a function of the detected gamma-ray energy. The $^{166}Er^*$ spectrum has been fitted with a statistical model calculation using an electric dipole strength function given by a double Lorentzian shape with the parameter values indicated in (b). (b) Same spectra multiplied by $\exp(E_\gamma/T_G)$, with $T_G(Sn) = 1.6$ MeV and $T_G(Er) = 1.4$ MeV, and normalized to approximately the same height at $E_\gamma = 14$–15 MeV. The ordinate is in arbitrary units. The quality of the fits and the different structures of the GDR's in the two systems is emphasized in this way. (Gaardhøje *et al.*, 1984.)

states are well known and one may hope to clarify if the exchange current (tensor correlation) effect is important there. A magnetic transition strength in the medium excitation region has been discussed as a possible origin of the missing spin–isospin strength in the GR region associated with the spin–isospin mode (Gaarde, 1983; Ejiri, 1983).

Experimental studies of (p,γ) reactions in the medium energy (30 to 100 MeV) region have mostly been made on light nuclei. The general trends of the angular distributions and some of analysing powers for γ rays below 40 MeV may be explained in terms of the direct E1–E2 process (Weller *et al.*, 1982). Extensive studies of ($\vec{p}$, γ) reactions on ^{11}B and ^{12}C in the region of $E_\gamma = 40$–100 MeV have been carried out for medium energy (40–80 MeV) polarized protons at Indiana and Osaka. Medium-energy γ-rays following ($\vec{p}$,γ) reactions on B and C have uniform features characteristic of the medium-energy region and they are summarized as follows (Nomachi *et al.*,

1985; Ejiri *et al.*, 1985): The angular distribution shows a uniform bell-shape with enhancement at forward angles, as shown in Fig. 4.84. The shape is independent of the final state. The coefficients a_1 and a_2 of the Legendre polynomials $P_1(\cos\theta)$ and $P_2(\cos\theta)$ increase as the γ-ray energy (the incident energy) increases. On the other hand the analysing power, as a function of the angle θ_γ, depends on the spin–orbit coupling of the final state, as shown in Fig. 4.85. The b_2 coefficients of the term $(\mathbf{p}\cdot\mathbf{n})P_2^1(\cos\theta)$ (eqn 4.5.11) for the final orbits of $j_> = l+\frac{1}{2}$ and $j_< = l-\frac{1}{2}$ show the simple features $b_2(j_>) \lesssim 0 < b_2(j_<)$ and $|b_2(j_>)| < |b_2(j_<)|$. Both the $|b_2(j_>)|$ and $|b_2(j_<)|$ decrease as the proton (γ-ray) energy increases.

The uniform features indicate a uniform E1 and E2 interference with a simple phase relation. Where the major transitions are E1 with amplitude $A(\mathrm{E1}) = R_1\exp(i\delta_1)$; and E2, with amplitude $A(\mathrm{E2}) = R_2\exp(i\delta_2)$, the a_1 coefficient for the radiative capture into the final (j,l) orbit is

$$a_1 = C(jl)(R_1^2+R_2^2)^{-1}R_1R_2\cos(\delta_2-\delta_1). \tag{4.5.48}$$

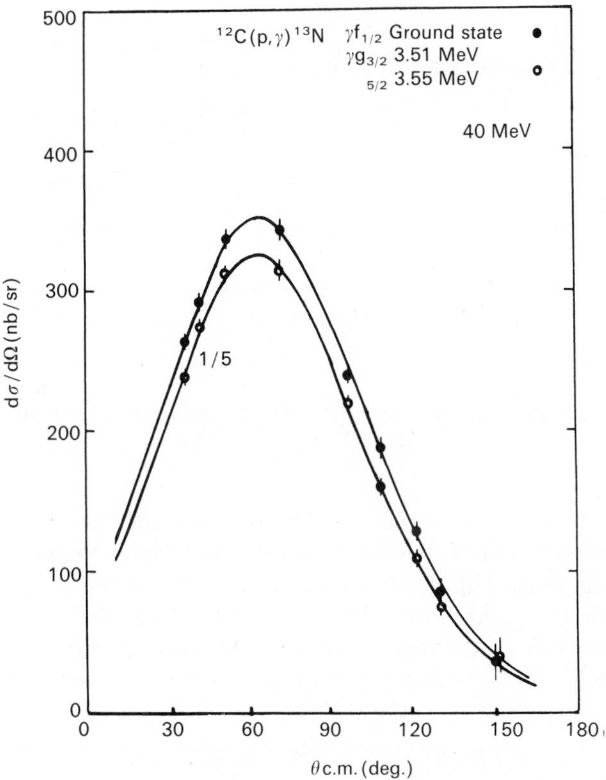

Fig. 4.84. The angular distributions of the $^{12}C(p,\gamma)$ reaction at $E_p = 40$ MeV. (Ejiri *et al.*, 1985.)

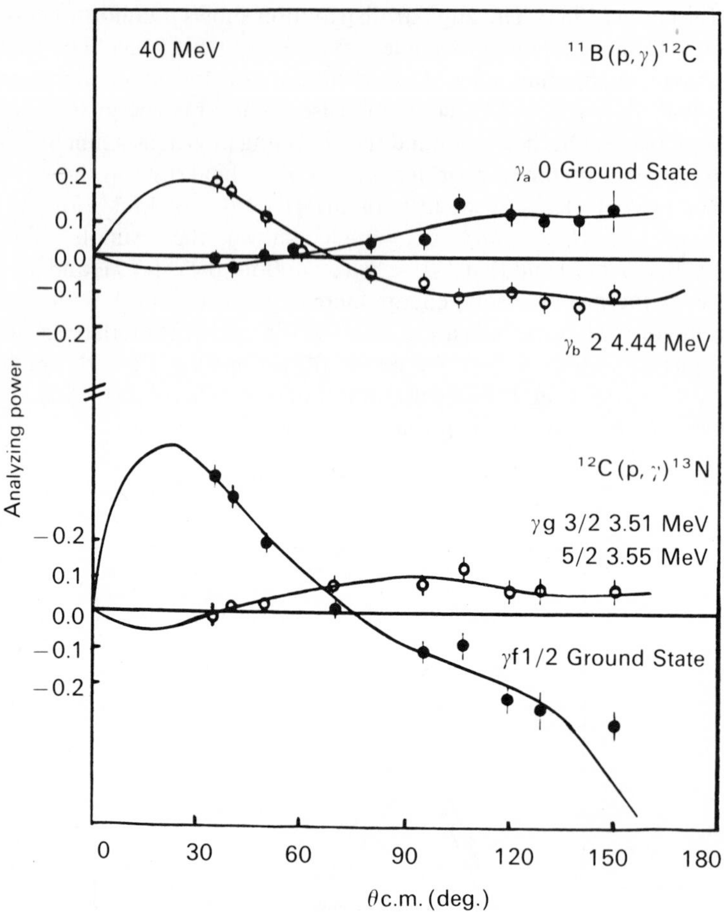

Fig. 4.85. The analyzing powers f the ^{11}B(p, γ) and ^{12}C(p, γ) reactions at $E_p = 40$ MeV. (Ejiri *et al.*, 1985.)

Assuming the most favourable (stretched) transitions $(j+2, l+2) \rightarrow (jl)$ for the E2 transition and $(j+1, l+1) \rightarrow (jl)$ for the E1 case, the two amplitudes interfere regularly. This leads to the uniform behavior of the a_1, being independent of the final (jl). The b_2 coefficient for the analysing power is due to the two interfering radiations with same parity. Introducing an E1 amplitude $A'(\text{E1}) = R'\exp(i\delta')$ for the other stretched transition of $(j-1, l-1) \rightarrow (jl)$, the b_2 coefficient is

$$b_2 = B_2(jl)(R_1^2 + R_1'^2 + R_2^2)^{-1} R_1 R_1' (-\sin(\delta_1 - \delta_1')). \qquad (4.5.49)$$

The phase is approximately given by $\delta = (l\pi)/2 + \delta(jl)$, where $l\pi/2$ refers to the incoming proton phase and $\delta(jl)$ is the phase shift. The phase of the matrix element is common as long as the most favourable (stretched) transition is concerned, and thus falls off as the difference $\delta_1-\delta'_1$. Then b_2 is reduced to

$$b_2 = B_2(jl)(R_1^2+R_1'^2+R_2^2)^{-1} R_1 R_1' \sin(\delta_{l+1}-\delta_{l-1}). \qquad (4.5.50)$$

The phase shift difference is mainly due to the centrifugal potential $V_{cf}l(l+1)$. The repulsive interaction ($V_{cf}>0$) leads to $\delta_{l+1}-\delta_{l-1}<0$. Noting that $R_1R_1'>0$ for the major stretched transition and $B_2(jl)$ is positive for $j=l+\frac{1}{2}$ and negative for $j=l-\frac{1}{2}$, one gets $b_2<0$ for $j=l+\frac{1}{2}$ and $b_2>0$ for $j=l-\frac{1}{2}$. The spin–orbit interaction gives a phase shift. It contributes constructively with the phase shift due to the centrifugal potential for $j_< = l-\frac{1}{2}$, and destructively for $j_> = l+\frac{1}{2}$. Therefore the absolute value of the phase difference $\delta_1-\delta'_1$ for $j_>$ is smaller than the value for $j_<$. The decrease of the absolute value of the b_2 coefficient as E_γ increases is attributed to the decrease of the phase difference and the increase of the E2 admixture (Ejiri *et al.*, 1985).

The analysing power for the electric radiative capture reaction can be schematically understood in terms of a geometrical consideration. Since the electric gamma transition does not flip spin, initial protons with $j^{\rm i}_> = l_{\rm i}+\frac{1}{2}$ contribute to the radiative capture into the final orbit of $j_> = l+\frac{1}{2}$, and protons with $j^{\rm i}_< = l_{\rm i}-\frac{1}{2}$ to the capture into the $j_< = l-\frac{1}{2}$ orbit. Thus the incident spin-up proton captured into the $j_> = l+\frac{1}{2}$ final orbit must interact with the right-hand side of the nucleus because the orbital angular momentum is oriented upwards. On the other hand, the incident spin-down proton must interact with the left-hand side to have the downward orbital angular momentum. For $j_< = l-\frac{1}{2}$, the right–left relation is simply reversed. These situations are schematically illustrated in Fig. 4.86. The analysing power, which is the difference between the spin-up and spin-down cross-sections, is thus given by $\mathrm{A}_R-\mathrm{A}_L$ for $j_>$, while it is given by $\mathrm{A}'_L-\mathrm{A}'_R$ for $j_<$. Therefore the b_2 coefficient, which arises from the interference between the proton waves of $l+1$ and $l-1$, changes sign for the change from the $j_>$ state to the $j_<$ state, in the case that $\mathrm{A}_R \sim \mathrm{A}'_R$ and $\mathrm{A}_L \sim \mathrm{A}'_L$. This is so in the geometrical consideration of the proton–nuclear interaction provided that the effect of the spin–orbit interaction is much smaller than the effect of the centrifugal potential. With increasing proton energy, both the impact parameter and the phase shift difference between the $l+1$ and $l-1$ waves becomes small, resulting in the reduction of the difference between the b_2 values for the $j_<$ and $j_>$ final states.

A simple relation for the b_k coefficient is obtained for direct radiative capture reactions by assuming spin-independent distortion (Seyler and Weller, 1984). In this case the radial matrix element of the transition amplitude is the same for both $j_> = l+\frac{1}{2}$ and $j_< = l-\frac{1}{2}$ incident waves and similarly for the captured orbits. The ratio of the b_k coefficients for the $j_{\rm f} = \mathrm{j}_>$

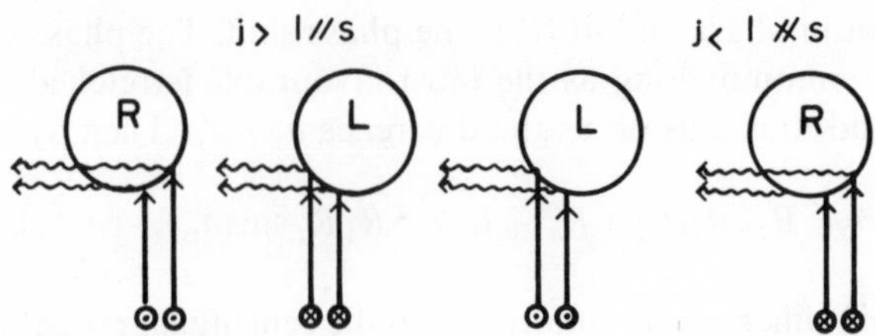

Fig. 4.86. Schematic diagrams of the analysing power for the E1 radiative capture process. R(L) indicates the case of interaction of the incident proton with the right- (left-) hand side of the nucleus. The asymmetry for the $j_>$ $(=l+\frac{1}{2})$ is R–L, while it is L–R for the $j_<$ $(=l-\frac{1}{2})$.

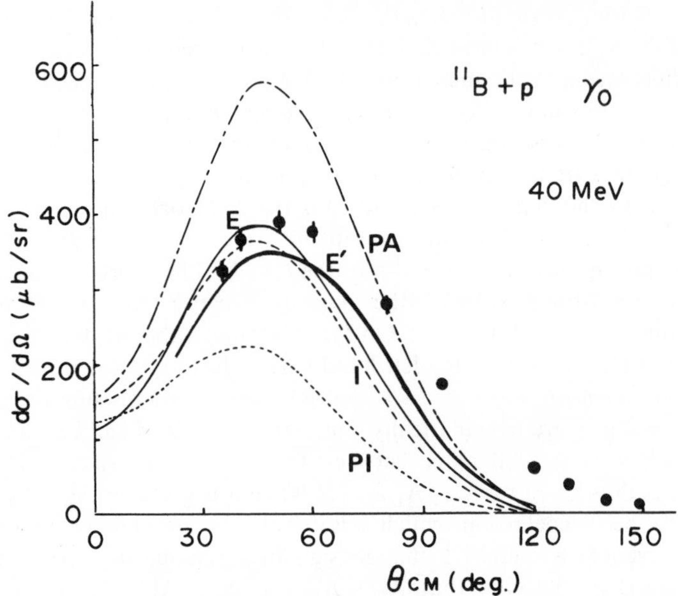

Fig. 4.87. The angular distributions for the $^{11}B(p, \gamma_0)^{12}C$ at $E_p = 40$ MeV (Ejiri *et al.*, 1985). *I*, *PA* and *E* are calculations for the impulse term, the impulse term + the pion current term, the impulse term + the pair current term, and the full exchange + impulse terms $(I+PI+PA)$ (Ohtsubo and Honmura, 1985). *E′* is calculated including the exchange current. (Tsai *et al.*, 1979).

and $j_<$ is simply given by

$$b_k(j_<)/b_k(j_>) = -(l_f+1)/l_f. \tag{4.5.51}$$

Theoretical calculations, including exchange-current contributions, have been made for radiative reactions on 1p shell nuclei. It has been shown that

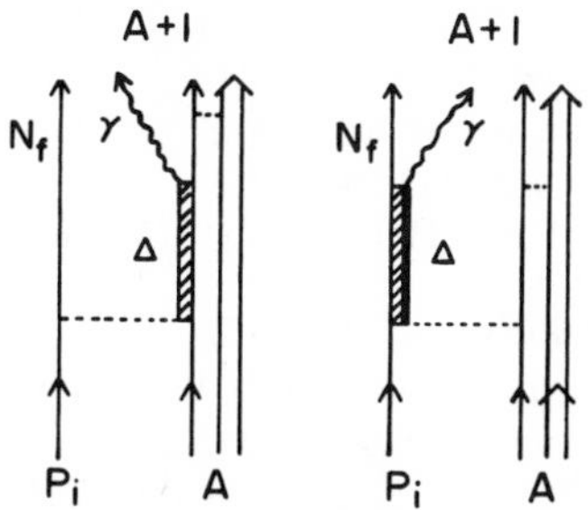

Fig. 4.88. Radiative capture process through the Δ(1232) isobar. P_i is the incident proton and N_f is the nucleon in the final orbit.

the combined contributions of convection-plus-exchange currents are larger by factors 2 to 3 than the convection current contribution and are consistent with the observed data (see Fig. 4.87) (Tsai and Londergan, 1979; Gari and Hebach, 1978). This shows the importance of the two-body process for medium energy (p, γ) reactions involving large momentum transfer. On the other hand, a full account of the exchange-current contribution in the framework of the Siegert theorem gives differential cross-sections twice as large as a model calculation for the pion exchange current (Ohtsubo and Hommura, 1985). This suggests the use of an improper wave function and/or the importance of heavier meson currents. The exchange-current effect is an important open problem.

In the higher energy region of $E_\gamma \gtrsim 100$ MeV the radiative capture process via the Δ(1232) isobar plays an essential role. The reaction scheme is illustrated in Fig. 4.88. It is found that the experimental data for the inverse (γ, p_0) reaction on ^{16}O do not show a very rapid fall-off as E_γ increases beyond 100 MeV, which would be expected from the simple direct process calculation. This indicates a major contribution of the two-step process through the Δ isobar (Londergan *et al.*, 1976; Matthews *et al.*, 1977). Here the incident proton interacts with one of the target nucleons, and excites the Δ isobar. The Δ isobar then de-excites by emitting a medium-energy γ ray, while the incident proton is captured in the final nucleon orbit f.

4.6 References

Agassi, E., Weidenmüller, H. A. and Mantzouranis, G. (1975). *Phys. Rep.* **C22,** 145.
Ahrens, T. and Feenberg, E. (1952). *Phys. Rev.* **86,** 64.
Ahrens, T., Borchert, H., Czock, K. H., Eppler, H. B., Gimm, H., Gundrum, H., Krönig, M., Riehn, P., Sitarama, G., Zieger, A. and Ziegler, B. (1975). *Nucl. Phys.* **A251,** 479.
Allas, R. G., Hanna, S. S., Meyer-Schützmeister, L. and Segel, R. E. (1964). *Nucl. Phys.* **58,** 122.

Anderson, J. D., Bloom, S. D., Cerny, J. and True, W. W. (1969). *Proc. Conf. Nucl. Isospin, Asilomar*, Academic Press, New York.

Arima, A., Brown, G. E., Hyuga, H. and Ichimura, M. (1973). *Nucl. Phys.* **A205,** 27.

Arnold, L. G. (1979). *Phys. Rev. Lett.* **42,** 1253.

Arthur, E. D., Drake, D. M. and Halpern, I. (1975). *Phys. Rev. Lett.* **35,** 914.

Balster, G. J., Bhowmik, R. K., Goldhoorn, P. B., Siemssen, R. H., Sujkowski, Z. and Wilschut, H. W. (1984). *Phys. Lett.* **143B,** 79.

Batsch, T., Blachot, J., Chen, Q., Crancon, J., Fatyga, M., Gizon, A., Jastrzebski, J., Kurcewicz, W., Pieńkowski, L., Singh, P. P., Ward, T. and Zychor, I. (1985). *Proc. 4th Int. Conf. Nuclear Reaction Mechanisms, Varenna* (ed. Gadioli, E.).Università Degli Studi Di Milano, Supplemento no. 46, p. 247.

Beene, J. R., Bertrand, F. E. and Halbert, M. L. (1985). *Proc. Gamma-Ray Spectroscopy and Related Topics—1984*, AIP Conf. Proc., New York, No. 125, p. 623.

Bergqvist, I., Drake, D. M. and McDaniels, D. K. (1972). *Nucl. Phys.* **A191,** 641.

Bergqvist, I., Zorro, R., Håkansson, A., Lindholm, A., Nilsson, L., Olsson, N. and Likar, A. (1984). *Nucl. Phys.* **A419,** 509.

Bertsch, G. F. (1979). *Nature* **280,** 639.

Bertrand, F. E. (1976). *Ann. Rev. nucl. Sci.* **26,** 457.

Bertrand, F. E. (1980). *Proc. Topical Conf. on Giant Multipole Resonances, Oak Ridge 1979*, Harwood Academic, New York.

Bertrand, F. E. (1981). *Nucl. Phys.* **A354,** 129C.

Bertrand, F. E., Beene, J. R. and Sjoreen, T. P. (1984). *J. Phys.* Coll. C4, Suppl. 3, **45,** C4–99.

Bhowmik, R. K., Driel, J. van, Siemssen, R. H., Balster, G. J., Goldhoorn, P. B., Gonggrijp, S., Iwasaki, Y., Janssens, R. V. F., Sakai, H., Siwek-Wilczyńska, K., Sterrenburg, W. A. and Wilczyński, J. (1982). *Nucl. Phys.* **A390,** 117.

Biedenharn, L. C. and Rose, M. E. (1953). *Revs mod. Phys.* **25,** 729.

Biedenharn, L. C., 1960, in *Nuclear Spectrscopy* (ed. F. Ajzenberg-Selove), p. 732. Academic Press, New York.

Blann, M. and Plasil, F. (1973). ALICE, Atomic Energy Commission Report No COO-3494-10.

Blann, M. (1975). *Ann. Rev. nucl. Sci.* **25,** 123.

Blann, M., Mignerey, A. and Scobel, W. (1976). *Nucleonika* **21,** 335.

Blasi, N., Dijk, J. H. van, Jansen, J. F. W., Riezebos, H., Voigt, M. J. A. de., van der Werf, S. Y. and Wilschut, H. W. (1983). KVI Annual Report, p. 15.

Blatt, J. M. and Weisskopf, V. F. (1954). *Theoretical Nuclear Physics*, John Wiley, New York.

Blatt, S. L., Hausman, H. J., Arnold, L. G., Seyler, R. G., Boyd, R. N., Donoghue, T. R., Koncz, P., Kovash, M. A., Bacher, A. D. and Foster, C. C. (1984). *Phys. Rev.* **C30,** 423.

Blin-Stoyle, R. J. and Grace, M. A. (1957). in *Handbuch der Physik*, (ed. S. Flugge), p. 42. Springer, Berlin.

Bohr, A. and Mottelson, B. R. (1969/1975). *Nuclear Structure*, Vol. I and Vol. II. Benjamin, New York.

Bolme, G. O., Cardman, L. S., Doerfler, R. W., Koester, Jr. L. J., Miller, B. L., Papanicolas, C. N., Rothaas, H. and Williamson, S. E. (1984). Private communication.

Breit, G., *Handbuch der Physik* (ed. S. Flügge), Vol. XLI, p. 1. Springer, Berlin.

Brentano, P. von, Dewald, A., Gelberg, A., Lieberz, W., Reinhardt, R., Panqueva, J. and Zell, K. O. (1984). *Proc. In-Beam Nuclear Spectroscopy, Hungary*, (ed. Zs. Dombradi and T. Fénuyes), p. 189. Akadémia Kiadó, Budapest.

Brink, D. M., Ph. D. Thesis, Oxford Univ. (1955). (unpublished).

Britt, *et al.* (1977). *Phys. Rev. Lett.* **39,** 1458.

Brown, G. E. and Bolsterli, M. (1959). *Phys. Rev. Lett.* **3,** 472.

Brown, G. E. (1964). *Nucl. Phys.* **57,** 339.

Brown, G. E. and Rho, M. (1980). *Nucl. Phys.* **A338,** 269.

Buck, B. and Pilt, A. A. (1977). *Nucl. Phys.* **A280,** 133.

Cameron, C. P., Roberson, N. R., Rickel, D. G., Ledford, R. D., Weller, H. R., Blue, R. A. and Tilley, D. R. (1976). *Phys. Rev.* **C14,** 553.

Carr, R. W. and Baglin, J. E. E. (1971). *Nucl. Data Tables* **10,** 143.

Chan, Y., Murphy, M., Stokstad, R. G., Tserruya, I., Wald, S. and Budzanowski, A. (1983). *Phys. Rev.* **C27,** 447.

Chmielewska, D., Sujkowski, Z., Janssens, R. V. F. and Voigt, M. J. A. de. (1981). *Nucl. Phys.* **A366,** 142.

Clement, C. F., Lane, A. M. and Rook, J. R. (1965). *Nucl. Phys.* **66,** 273, 293.

Cline, C. K. and Blann, M. (1971). *Nucl. Phys.* **A172,** 225.

Collins, M. T., Chang, C. C., Tabor, S. L., Wagner, G. J. and Wu, J. R. (1979). *Phys. Rev. Lett.* **42,** 1440.

Collins, M. T., Chang, C. C. and Tabor, S. L. (1981). *Phys. Rev.* **C24,** 387.

Cormier, T. M., Cosman, E. R., Lazzarini, A. J., Wegner, H. E., Garrett, J. D. and Pulhofer, F. (1977). *Phys. Rev.* **C15,** 654.

Cramer, J. G., Brentano, P. von, Phillips, G. W., Ejiri, H., Ferguson, S. M. and Braithwaite, W. J. (1968). *Phys. Rev. Lett.* **21,** 297.

Deleplanque, M. A., Byrski, Th., Diamond, R. M., Hubel, H., Stephens, F. S., Herskind, B. and Bauer, R. (1978). *Phys. Rev. Lett.* **41,** 1105.

Demeyer, A., Cery, R., Chevarier, N., Chevarier, A., Tousset, J. and Duc, T. M. (1970). *J. Phys.* **31,** 847.

Devons, S. and Goldfarb, L. J. B. (1957). *Handbuch der Physik*, (ed. S. Flügge), Vol. **42,** p. 362. Springer, Berlin.

Dewald, A., Reinhardt, R., Panqueva, J., Zell, K. O. and Brentano, P. von. (1984). *Z. Phys.* **A375,** 77.

Diamond, R. M., Mattias, E., Newton, J. O. and Stephens, F. S. (1966). *Phys. Rev. Lett.* **16,** 1205.

Dietrich, F. S., Heikkinen, D. W., Snover, K. A. and Ebisawa, K. (1977). *Phys. Rev. Lett.* **38,** 156.

Dietrich, F. S. and Kerman, A. K. (1979). *Phys. Rev. Lett.* **43,** 114.

Djaloeis, A., Jahn, P., Probst, H.-J. and Mayer-Böricke, C. (1975). *Nucl. Phys.* **A250,** 149.

Dolbilkin, B. S., Ohsawa, S., Torizuka, Y., Saito, T., Mizuno, Y. and Saito, K. (1982). *Phys. Rev.* **C25,** 2255.

Donnelly, T. W. and Walker, G. E. (1970). *Ann. Phys.* **60,** 209.

Dowell, D. H., Feldman, G., Snover, K. A., Sandorfi, A. M. and Collins, M. T. (1983). *Phys. Rev. Lett.* **50,** 1191.

Dowell, D. H. (1985). *Capture Gamma-Ray Spectroscopy and Related Topics, Knoxville*, 1984, Conference Proc. No. 125. (ed. S. Raman), p. 597. American Inst. Phys.

Dracoulis, G. D., Ferguson, S. M., Newton, J. O. and Slocombe, M. G. (1977). *Nucl. Phys.* **A279,** 251.

Drake, D. M., Joly, S., Nilsson, L., Wender, S. A., Aniol, K., Halpern, I. and Storm, D. (1981). *Phys. Rev. Lett.* **47,** 1581.

Ejiri, H., Ishihara, M., Sakai, M., Katori, K. and Inamura, T. (1965). *J. Phys. Soc. Japan* **24,** 1189.

Ejiri, H., Ishihara, M., Sakai, M., Inamura, T. and Katori, K. (1965/1966). *Phys. Lett.* **18,** 314; *Nucl. Phys.* **89,** 641.

Ejiri, H., Ishihara, M., Sakai, M., Inamura, T. and Katori, K. (1966). *J. Phys. Soc. Japan* **21,** 1021.

Ejiri, H. and Bondorf, J. (1968). *Phys. Lett.* **28B,** 304.

Ejiri, H., Richard, P., Ferguson, S., Heffner, R. and Perry, D. (1968/1969). *Phys. Rev. Lett.* **21,** 373; *Nucl. Phys.* **A128,** 388.

Ejiri, H. and Halpern, I. (1968/1970). *Bull Am. phys. Soc.* **13,** 700; *Proc. Int. Symp. Nuclear Structure*, Joutsa, Finland.

Ejiri, H., Ferguson, S. M., Halpern, I. and Heffner, R. (1969). *Phys. Lett.* **29B,** 111.

Ejiri, H., Shibata, T., Shimizu, A. and Yagi, K. (1972a). *J. Phys. Soc. Japan* **33,** 1515.

Ejiri, H., Hagemann, G. B. and Hammer, T. (1972b). *Proc. Int. Conf. Statistical Properties of Nuclei, Albany* (ed. J. B. Garg), p. 483. Plenum Press, New York.

Ejiri, H., Shibata, T., Itahashi, T., Nagai, Y., Sakai, H., Hoshi, M., Nakayama, S., Kishimoto, T. and Maeda, K. (1978a). *Phys. Lett.* **78B,** 192.

Ejiri, H., Itahashi, T., Nagai, Y., Shibata, T., Nakayama, S., Kishimoto, T., Maeda, K., Sakai, H. and Hoshi, M. (1978b). *Nucl. Phys.* **A305,** 167.

Ejiri, H., Nagai, Y., Sakai, H., Shibata, T., Kishimoto, T. and Maeda, K. (1980). *J. Phys. Soc. Japan*, **49,** 2103.

Ejiri, H. (1981). *Physica Scripta*, **24,** 130.

Ejiri, H. (1982). *Proc. Int. Conf. Nuclear Reaction Mechanisms, Varenna, Italy, June 1982* (ed. E. Gadioli), Suppl. No. 28, p. 287. Ricerca Scientifica ed Educazione Permanente.

Ejiri, H. (1983). *Nucl. Phys.* **A396,** 181C.

Ejiri, H. (1984). *J. Physique*, Coll. C4, *Suppl.* 3, **45,** C4–135.

Ejiri, H. *et al.* (1985). *Capture Gamma-Ray Spectroscopy and Related Topics 1984, Knoxville* (ed. S. Raman), p. 582. AIP Conf. Proc. 125.

Ericson, T. (1960). *Phil. Mag. Suppl. adv. Phys.* **9,** 425.

Eyrich, W., Hofmann, A., Scheib, U., Schneider, S. and Vogler, F. (1979). *Phys. Rev. Lett.* **43,** 1369.

Ferguson, A. J. (1965). *Angular Correlation Methods in Gamma-Ray Spectroscopy*. North-Holland, Amsterdam.

Ferguson, S. M., Ejiri, H. and Halpern, I. (1972). *Nucl. Phys.* **A188,** 1.

Fields, C. A., de Boer, F. W. H., Prull, D. E., Ristinen, R. A., Samuelson, L. E., Smith, P. A. and Sugarbaker, E. (1981). *Nucl. Phys.* **A366,** 38.

Fox, J. D. and Robson, D. (1966). *Proc. Conf. Isobaric Spin and Nuclear Physics* Academic Press, New York.

Fujita, J. I. (1962). *Phys. Rev.* **126,** 202.

Fujita, J. I. (1963). Brookhaven National Laboratory Report No. BNL 887C–39, p. 340.

Fujita, J. I. (1964). *Progr. theor. Phys.* **32,** 438.

Fujita, J. I. (1967). *Phys. Lett.* **24B,** 123.

Fujita, J. I. and Hirata, M. (1971). *Phys. Lett.* **37B,** 237.
Fukuda, S. and Torizuka, Y. (1972). *Phys. Rev. Lett.* **29,** 1109.
Gaarde, C. (1983). *Nucl. Phys.* **A396,** 127C.
Gaardhøje, J. J., Ellegaard, C., Herskind, B. and Steadman, S. G. (1984). *Phys. Rev. Lett.* **53,** 148.
Gadioli, E., Gadioli-Erba, E., Sajo-Bohus, L. and Tagliaferri, G. (1976). *Rev. nuovo Cim.* **6,** 1.
Gari, M. and Hebach, H. (1978). *Phys. Rev.* **C18,** 1071.
Geoffroy, K. A., Sarantites, D. G., Halbert, M. L., Hensley, D. C., Dayras, R. A. and Barker, J. H. (1979). *Phys. Rev. Lett.* **43,** 1303.
Gerschel, C. (1982). *Nucl. Phys.* **A387,** 297C.
Gilat, J. (1970). GROGI2, BNL 50246 (T-580), Brookhaven National Lab. New York; *ibid. Phys. Rev.* **C1,** 1432.
Glavish, H. F., Hanna, S. S., Avida, R., Boyd, R. N., Chang, C. C. and Diener, E. (1972). *Phys. Rev. Lett.* **28,** 766.
Griffin, J. J. (1966). *Phys. Rev. Lett.* **17,** 478.
Grover, J. R. and Gilat, J. (1967a). *Phys. Rev.* **157,** 802; *ibid.* 814.
Grover, J. R. and Gilat, J. (1967b). *Phys. Rev.* **157,** 823.
Haas, B. *et al.* (1985). *Phys. Rev. Lett.* **54,** 398.
Hagemann, G. B., Broda, R., Herskind, B., Ishihara, M., Ogaza, S. and Ryde, H. (1975). *Nucl. Phys.* **A245,** 166.
Hageman, D. C. J. M. (1981). Ph.D. Thesis, Rijksuniversiteit, Groningen.
Hageman, D. C. J. M., Janssens, R. V. F., Lukasiak, J., Ockels, W. J., Sujkowski, Z. and Voigt, M. J. A. de (1981). *Physica Scripta* **24,** 145.
Halpern, I., Shepherd, B. J. and Williamson, C. F. (1968). *Phys. Rev.* **169,** 805.
Hanna, S. S. (1979). *Proc. Int. Conf. Nucl. Phys. with Electromagnetic Interactions,* (ed. H. Arenhövel and D. Drechsel), p. 288. Springer, Berlin.
Hanna, S. S. (1983). *Comments nucl. part. Phys.* **11,** 79.
Hanna, S. S. (1984). *Proc. Int. Symp. Nuclear Spectroscopy and Nuclear interactions, Osaka.* (ed. H. Ejiri and T. Fukuda), p. 299. World Scientific.
Hanna, S. S., Glavish, H. F., Diener, E. M., Calarco, J. R., Chang, C. C., Avida, R. and Boyd, R. N. (1972). *Phys. Lett.* **B40,** 631.
Hanna, S. S., Glavish, H. F., Avida, R., Calaco, J. R., Kuhlmann, E. and LaCanna, R. (1974). *Phys. Rev. Lett.* **32,** 114.
Hansen, G. B., Elbek, B., Hagemann, K. A. and Hornyak, W. F. (1963). *Nucl. Phys.* **47,** 529.
Harakeh, M. H., Paul, P., Kuan, H. M. and Warburton, E. K. (1975). *Phys. Rev.* **C12,** 1410.
Harakeh, M. N., Borg, K. van der, Ishimatsu, T., Morsch, H. P., van der Woude, A. and Bertrand, F. E. (1977). *Phys. Rev. Lett.* **38,** 676.
Hasinoff, M., Fisher, G. A., Kurjan, P. and Hanna, S. S. (1972). *Nucl. Phys.* **A195,** 78.
Hauser, W. and Feshbach, H. (1952). *Phys. Rev.* **87,** 366.
Herskind, B. (1976). in *Proc. Symp. Macroscopic Features of Heavy Ion Collisions,* Argonne, Illinois, Vol. 1, p. 385. ANL Report, ANL/PHY-76-2.
Hillis, D. L., Garrett, J. D., Christensen, O., Fernandez, B., Hagemann, G. B., Herskind, B., Back, B. B. and Folkmann, F. (1979). *Nucl. Phys.* **A325,** 216.
Hommura, T. and Ohtsubo, H. (1985). Private communication.
Hoppes, D. D., Ambler, E., Hayward, R. W. and Kaeser, R. S. (1961). *Phys. Rev. Lett.* **6,** 115.

Ieki, K., Ishihara, M., Inamura, T., Kohmoto, S., Utsunomiya, H., Sueki, K. and Kudo, H. (1985). *Phys. Lett.* **150B,** 83.

Inamura, T., Ishihara, M., Fukuda, T., Shimoda, T. and Hiura, H. (1977). *Phys. Lett.* **68B,** 51.

Inamura, T. (1984). Lecture at the 16th Masurian Summer School on Nucl. Phys., Mikolajki, Poland, IPCR-Cyclotron Report, 50.

Ishihara, M., Yoshikawa, N., Kawakami, H., Sakai, M. and Ishii, K. (1972). *Nucl. Phys.* **A179,** 769.

Jahn, P., Probst, H. J., Djaloeis, A., Davidson, W. F. and Mayer-Böricke, C. (1973). *Nucl. Phys.* **A209,** 333.

Jastrzębski, J., Singh, P. P., Mróz, T., Karwowski, H. J., Vigdor, S. E. and Fatyga, M. (1984) *Phys. Lett.* **136B,** 153.

Jensen, M., Tilley, D. R., Weller, H. R., Roberson, N. R., Wender, S. A. and Clegg, T. B. (1979). *Phys. Rev. Lett.* **43,** 609.

Jett, J. H. and Lind, D. A. (1970). *Nucl. Phys.* **A155,** 182.

Johnson, A., Ryde, H. and Hjorth, S. A. (1972). *Nucl. Phys.* **A179,** 753.

Kajrys, G., Landsberger, S., Lecomte, R. and Monaro, S. (1983). *Phys. Rev.* **C27,** 1983.

Karwowski, H. J., Vigdor, S. E., Jacobs, W. W., Kailas, S., Singh, P. P., Soga, F. and Ploughe, W. D. (1979). *Phys. Rev. Lett.* **42,** 1732.

Kerman, A. K. and McVoy, K. (1979). *Ann. Phys. N.Y.* **122,** 197.

Kishimoto, T., Nakayama, S., Takahashi, N. and Ejiri, H. (1979). *J. Phys. Soc. Japan* **46,** 6.

Kishimoto, T., Sasao, M., Suzuki, H., Sakai, H. and Ejiri, H. (1980). *Nucl. Phys.* **A337,** 493.

Kishimoto, T. and Kubo, K. I. (1979). *Proc. Symp. High Spin Phenomena in Nuclei*, ANL-PHY-79-4, 535.

Knopfle, K. T., Wagner, G. J., Paul, P., Breuer, H., Mayer-Böricke, C., Rogge, M. and Turek, P. (1978). *Phys. Lett.* **74B,** 191.

Knopfle, K. T. (1979). *Lecture Notes in Physics*, **108,** 311.

Knopfle, K. T., Riedesel, H., Schindler, K., Wagner, G. J., Mayer-Böricke, C., Oelert, W., Rogge, M. and Turek, P. (1981). *Phys. Rev. Lett.* **46,** 1372.

Kovash, M. A., Blatt, S. L., Boyd, R. N., Donoghue, T. R., Hausman, H. J. and Bacher, A. D. (1979). *Phys. Rev. Lett.* **42,** 700.

Kuhlmann, E., Calarco, J. R., Cheng, V. K. C., Mavis, D. G., Hall, J. R. and Hanna, S. S. (1979). *Phys. Rev.* **C20,** 5.

Kurz, H. E., Jasper, E. W., Fischer, K. and Hermes, F. (1971). *Nucl. Phys.* **A168,** 129.

Lang, D. W. (1966). *Nucl. Phys.* **77,** 545.

Laszewski, R. M. and Holt, R. J. (1977). *Atomic Data and Nuclear Data Table*, **19,** 305.

Leprêtre, A., Beil, H., Beigère, R., Carlos, P., Fagot, J., Miniac, A. de and Veyssière, A. (1981). *Nucl. Phys.* **A367,** 237.

Levinger, J. S. and Bethe, H. (1950). *Phys. Rev.* **78,** 115.

Lewis, M. B. and Bertrand, F. E. (1972). *Nucl. Phys.* **A196,** 337.

Li, X. H., Udagawa, T. and Tamura, T. (1984). *Phys. Rev.* **C30,** 1349.

Lieder, R. M., Bochev, B., Kutsarova, T., Beuscher, H., Haenni, D. R., Morek, T., Muller-Verggian, M., Neskakis, A., Mayer-Böricke, C. and Didelez, J. P. (1981). *Physica Scripta* **24,** 123.

Lieb, K. P. and Hausmann, T. (1969). *Phys. Rev.* **186,** 1229.

Likar, A. A., Lindholm, A., Nilsson, L., Bergqvist, I. and Palsson, B. (1978). *Nucl. Phys.* **A298,** 217.

Londergan J. T., Nixon, G. D. and Walker, G. E. (1976). *Phys. Lett.* **65B,** 427.

Machner, H. (1981). *Z. Phys.* **A302,** 125.

Maeda, K., Shibata, T., Ejiri, H. and Sakai, H. (1983). *Phys. Rev.* **C28,** 635.

Manglos, S., Roberson, N. R. and Weller, H. R. (1981). *Phys. Rev.* **C24,** 2378.

Marty, N., Morlet, M., Willis, A., Comparat, V. and Frascaria, R. (1975). *Proc. Int. Symp. Highly Excited States in Nuclei*, Jülich, (ed. A. Faessler, C. Mayer-Böricke and P. Turek), Vol. 1, p. 17.

Matsuoka, N., Shimizu, A., Hosono, K., Saito, T., Kondo, M., Sakaguchi, H., Toba, Y., Goto, A. and Ohtani, F. (1978/1980). *Nucl. Phys.* **A311,** 173; *ibid.* **A377,** 269.

Matthews, J. L., Bertozzi, W., Leitch, M. J., Peridier, C. A., Roberts, B. L., Sargent, C. P., Turchinetz, W., Findlay, D. J. S. and Owens, R. O. (1977). *Phys. Rev. Lett.* **38,** 8.

Maute, R. E., D'Amato, D. P. and Blatt, S. L. (1974). *Atomic Data and Nuclear Data Tables* **13,** 499.

Meijer, R. J. de and Kamermans, R. (1985). *Rev. mod. Phys.* **57,** 147.

Meyer-Schützmeister, L., Vager, Z., Segel, R. E. and Singh, P. P. (1968). *Nucl. Phys.* **A108,** 180.

Meyer-Schützmeister, L., Segel, R. E., Raghunathan, K., Debevac, P. T., Wharton, W. R., Rutledge, L. L. and Ophel, T. R. (1978). *Phys. Rev.* **C17,** 56.

Mitarai, S. and Minehara, E. (1983). *Nucl. Phys.* **A406,** 55.

Miyano, K. and Nakahara, H. (1973). *J. Phys. Soc. Japan*, **35,** 953.

Morinaga, H. and Gugelot, P.C. (1963). *Nucl. Phys.* **46,** 210.

Morsch, H. P., Rogge, M., Turek, P., Sükösd, C. and Mayer-Böricke, C. (1979). *Phys. Rev.* **C20,** 1600.

Moszkowski, S. A. (1965). *Alpha-, Beta- and Gamma-Ray Spectroscopy*, (ed. K. Siegbahn), p. 863. North-Holland, Amsterdam.

Motobayashi, T., Ejiri, H., Shibata, T., Okada, K., Sasao, M., Maeda, K. and Suzuki, H. (1984). *Nucl. Phys.* **A413,** 290.

Nagai, Y., Shibata, T., Sakai, H., Kishimoto, T. and Ejiri, H. (1979). *J. Phys. Soc. Japan*, **46,** 1025.

Nagao, M. and Torizuka, Y. (1973). *Phys. Rev. Lett.* **30,** 1068.

Newton, J. O., Stephens, F. S., Diamond, R. M., Kotajima, K. and Mattias, E. (1967). *Nucl. Phys.* **A95,** 357.

Newton, J. O., Stephens, F. S., Diamond, R. M., Kelley, W. H. and Ward, D. (1970). *Nucl. Phys.* **A141,** 631.

Newton, J. O., Herskind, B., Diamond, R. M., Dines, E. L., Draper, J. E., Linderberger, K. H., Schück, C., Shih, S. and Stephens, F. S. (1981). *Phys. Rev. Lett.* **46,** 1383.

Nomachi, M., Shibata, T., Okada, K., Motobayashi, T., Ohtani, F. and Ejiri, H. (1985). *Phys. Rev.* **C31,** 242.

Ockels, W. J., Voigt, M. J. A. de, and Sujkowski, Z. (1978). *Phys. Lett.* **76B,** 401.

Ogawa, M., Kleinheinz, P., Lunardi, S., Schult, O. W. B. and Fenze, M. (1978). *Z. Phys.* **A284,** 271.

Ohsumi, H., Ejiri, H., Shibata, T., Nagai, Y., Okada, K., Motobayashi, T., Shimizu, A., Maeda, K. and Nomachi, M. (1984). *Proc. Int. Symp. Nuclear Spectroscopy and Nuclear Interactions*, Osaka (ed. H. Ejiri and T. Fukuda), p. 317. World Scientific.

Ohsumi, H., Ejiri, H., Shibata, T., Nagai, Y., Okada, K. and Motobayashi, T. (1985). *Phys. Rev.* **C32,** 1789.

Ohsumi, H. (1986). Private communication.

Ohtsubo, H. and Hommura, T. (1985). Private communication.

Okada, K., Ejiri, H., Shibata, T., Nagai, Y., Motobayashi, T., Ohsumi, H., Noumachi, M., Shimizu, A. and Maeda, K. (1982). *Phys. Rev. Lett.* **48,** 1382.
Pampus, J., Bisplinghoff, J., Ernst, J. H., Mayer-Kuckuk, T., Rama Rao, J., Baur, G., Rosel, R. and Trautmann, D. (1978). *Nucl. Phys.* **A311,** 141.
Paul, P. (1972). *Proc. Int. Conf. Nuclear Structure Studies Using Electron Scattering and Photo Reactions*, Sendai (ed. K. Shoda and H. Ui), p. 343.
Paul, P. (1974). *Nuclear Spectroscopy and Reactions, Part A*, (ed. J. Cerny), p. 345. Academic Press, New York.
Paul, P. (1977). *Elementary Modes of Excitation in Nuclei* (ed. A. Bohr and R. A. Broglia), p. 495. North-Holland, Amsterdam.
Pitthan, R. and Walcher, Th. (1971). *Phys. Rev. Lett.* **B36,** 563.
Pitthan, R. (1979). *Proc. Giant Multipole Resonances*, Oak Ridge, (ed. F. E. Bertrand), p. 161. Harwood Academic, New York (1980).
Plasil, F. and Blann, M. (1975). *Phys. Rev.* **C11,** 508.
Potokar, M. (1973). *Phys. Lett.* **46B,** 346;
Potokar, M., Likar, A., Budnar, M. and Cvelbar, F. (1977). *Nucl. Phys.* **A277,** 29.
Pühlhofer, F. (1977). *Nucl. Phys.* **A280,** 267.
Rasmussen, J. O. and Sugihara, T. T. (1966). *Phys. Rev.* **151,** 992.
Rose, M. E. (1957). *Elementary Theory of Angular Momentum.* John Wiley, New York.
Sadler, M., Jastrzebski, J., Nadasen, A., Singh, P. P., Rutledge, Jr., L. L., Chen, T. and Segel, R. E. (1977). *Phys. Rev. Lett.* **38,** 950.
Sakai, H., Ejiri, H., Shibata, T., Nagai, Y. and Okada, K. (1979). *Phys. Rev.* **C20,** 464.
Sakai, H., Shimizu, A., Ejiri, H., Shibata, T., Okada, K., Kishimoto, T. and Maeda, K. (1980). *Phys. Rev.* **C22,** 1002.
Sakai, H., Bhowmik, R. K., Brandenburg, S., Dijk, J. H. van, Drentje, A. G., Harakeh, M. N., Iwasaki, Y., Siemssen, R. H., van der Werf, S. Y. and van der Woude, A. (1985). *Nucl. Phys.* **A441,** 640.
Sakai, M., Yamazaki, T. and Ejiri, H. (1965). *Nucl. Phys.* **74,** 81.
Saporetti, F. and Guidotti, R. (1979). *Nucl. Phys.* **A330,** 53.
Sarantites, D. G., Westerberg, L., Halbert, M. L., Dayras, R. A., Hensley, D. C. and Barker, J. H. (1978). *Phys. Rev.* **C18,** 774.
Sasao, M. and Torizuka, Y. (1977). *Phys. Rev.* **C15,** 217.
Seyler, R. G. and Weller, H. R. (1979). *Atomic Data and Nuclear Data Tables* **23,** 99.
Seyler, R. G. and Weller, H. R. (1979). *Phys. Rev.* **C20,** 453.
Seyler, R. G. and Weller, H. R. (1984). *Phys. Rev.* **C30,** 1146.
Shepherd, B. J., Williamson, C. F. and Halpern, I. (1966). *Phys. Rev. Lett.* **17,** 806.
Shibata, T., Ejiri, H., Chiba, J., Nagamiya, S., Nakai, K., Anholt, R., Bowman, H., Ingersoll, J. G., Rauscher, E. A. and Rasmussen, J. O. (1978). *Nucl. Phys.* **A308,** 513.
Shibata, T., Maeda, K., Okada, K., Ejiri, H., Sakai, H. and Shimizu, A. (1985). *Nucl. Phys.* **A441,** 445.
Shikazono, N. and Terasawa, T. (1975). *Nucl. Phys.* **A250,** 260.
Shoda, K., Suzuki, A., Sugawara, M., Saito T., Miyase, H. and Oikawa, S. (1971). *Phys. Rev.* **C3,** 1999, 2006.
Shyam, R., Bauer, G., Rösel, F. and Trautmann, D. (1984). *Phys. Rev.* **C30,** 1109.
Siemens, P. J., Bondorf, J. P., Gross, D. H. E. and Dickmann, F. (1971). *Phys. Lett.* **36B,** 24.
Siemssen, R. H. (1983). *Nucl. Phys.* **A400,** 245C.
Siemssen, R. H. (1984). *Nuclear Structure and Heavy-Ion Dynamics*, LXXXVII Corso, Soc. Italiana di Fisica, Bologna.

Siemssen, R. H., Balster, G. J., Wilschut, H. W., Bond, P. D., Crouzen, P. C. N., Goldhoorn, P. B., Shukui, HAN. and Sujkowski, Z. (1985). *Phys. Lett.* **161B,** 261.
Snover, K. A., Amann, J. F., Hering, W. and Paul, P. (1971). *Phys. Lett.* **37B,** 29.
Snover, K. A. Adelberger, E. G. and Brown, D. R. (1974). *Phys. Rev. Lett.* **32,** 1061.
Snover, K. A., Bussoletti, J. E., Ebisawa, K., Trainor, T. A. and McDonald, A. B. (1976). *Phys. Rev. Lett.* **37,** 273.
Snover, K. A., Ikossi, P. G. and Trainor, T. A. (1979). *Phys. Rev. Lett.* **43,** 117.
Snover, K. A. (1984/1985). *Comments on nucl. part. Phys.* **12** (1984) 243; *Proc. Capture Gamma-Ray Spectroscopy and Related Topics, Knoxville,* 1984 (ed. S. Raman), p. 660. AIP Conf. Proc. No. 125.
Steuer, H., Eyrich, W., Hofmann, A., Ortner, H., Scheib, U., Stamminger, R. and Steuer, D. (1981). *Phys. Rev. Lett.* **47,** 1702.
Sugimoto, K., Takahashi, N., Mizobuchi, A., Nojiri, Y., Minamisono, T., Ishihara, M., Tanaka, K. and Kamitsubo, H. (1977). *Phys. Rev. Lett.* **39,** 323.
Terasawa, T. and Fujita, J. I. (1974). *Nucl. Phys.* **A223,** 492.
Tsai, S. F. and Londergan, J. T. (1979). *Phys. Rev. Lett.* **43,** 576.
Trautmann, W., Boer, J. de, Dünnweber, W., Graw, G., Kopp, R., Lautenbach, C., Puchta, H. and Lynen, U. (1977). *Phys. Rev. Lett.* **39,** 1062.
Trautmann, W., Dahme, W., Dönnweber, W., Hering, W., Lauterbach, C., Puchta, H., Köhn, W. and Wurm, J. P. (1981). *Phys. Rev. Lett.* **46,** 1188.
Trautmann, W., Puchta, H., Dünnweber, W., Hering, W. and Lautenbach, C. (1983). *Phys. Lett.* **123B,** 177.
Turner, J. D., Cameron, C. P., Roberson, N. R., Weller, H. R. and Tilley, D. R. (1978). *Phys. Rev.* **C17,** 1853.
Turner, J. R., Roberson, N. R., Wender, S. A., Weller, H. R. and Tilley, D. R. (1979). *Phys. Rev.* **C21,** 525.
Udagawa, T. and Tamura, T. (1980). *Phys. Rev. Lett.* **45,** 1311.
Udagawa, T. and Tamura, T. (1981). *Phys. Rev.* **C24,** 1348.
Udagawa, T., Price, D. and Tamura, T. (1982). *Phys. Lett.* **118B,** 45.
Voigt, M. J. A. de, Ockels, W. J., Sujkowski, Z., Zglinski, A. and Mooibroek, J. (1979). *Nucl. Phys.* **A323,** 317.
Voigt, M. J. A. de, Maeda, K., Ejiri, H., Shibata, T., Okada, K., Motobayashi, T., Sasao, M., Kishimoto, T., Suzuki, H., Sakai, H. and Shimizu, A. (1982). *Nucl. Phys.* **A379,** 160.
Wagner, G. J. (1979/1980). *Lecture Notes in Physics* **92,** 269; *Proc. Topical Conf. Giant Multipole Resonances, Oak Ridge 1979,* (ed. F. E. Bertrand), p. 251. Harwood Academic, New York.
Weller, H. R., Blue, R. A., Roberson, N. R., Rickel, D. G., Maripuu, S., Cameron, C. P., Ledford, R. D. and Tilley, D. R. (1976). *Phys. Rev.* **C13,** 922.
Weller, H. R., Roberson, N. R. and Cotanch, S. R. (1978). *Phys. Rev.* **C18,** 65.
Weller, H. R. and Roberson, N. R. (1980). *Rev. mod. Phys.* **52,** 699.
Weller, H. R., Hasan, H., Manglos, S., Mitev, G., Roberson, N. R., Blatt, S. L., Hausman, H. J., Seyler, R. G., Boyd, R. N., Donoghue, T. R., Kovash, M. A., Bacher, A. D. and Foster, C. C. (1982). *Phys. Rev.* **25,** 2921.
Wender, S. A., Roberson, N. R., Potokar, M., Weller, H. R. and Tilley, D. R. (1978). *Phys. Rev. Lett.* **41,** 1217.
Werf, S. Y. van der, *et al.* (1986). *Phys. Lett.* **166B,** 372.
Westerberg, L., Sarantites, D. G., Hensley, D. C., Dayras, R. A., Halbert, M. L. and Barker, J. H. (1978). *Phys. Rev.* **C18,** 796.

Wilczyński, J. (1973). *Nucl. Phys.* **A216,** 386.

Wilczyński, J., Siwek-Wilczyńska, K., van Driel, J., Gongrijp, S., Hageman, D. C. J. M., Janssens, R. V. F., Lukasiak, J. and Siemssen, R. H. (1980). *Phys. Rev. Lett.* **45,** 606.

Wilczyński, J., Siwek-Wilczyńska, K., van Driel, J., Gonggrijp, S., Hageman, D. C. J. M., Janssens, R. V. F., Lukasiak, J., Siemssen, R. H. and Van der Werf, S. Y. (1982). *Nucl. Phys.* **A373,** 109.

Wilkinson, D. H. (1966). *Proc. Conf. Isobaric Spin in Nucl. Phys.* (ed. J. D. Fox and D. Robson), p. 612. Academic Press, New York.

Williamson, C. F. and Shepherd, B. J. (1965). *Bull. Am. phys. Soc.* **10,** 428.

Williamson, C. F., Ferguson, S. M., Shepherd, B. J. and Halpern I. (1968). *Phys. Rev.* **174,** 1544.

Willis, A., Morlet, M., Marty, N., Frascaria, R., Djalali, C. and Comparat, V. (1980). *Nucl. Phys.* **A344,** 137.

Wilschut, H. W., Bhowmik, R. K., Goldhoorn, P. B., Jansen, J. F. W., Siemssen, R. H., Siwek-Wilczyńska, K., Sujkowski, Z. and Wilczyński, J. (1983). *Phys. Lett.* **123B,** 173.

Woude, A. van der. (1987). *Progress in Particle and Nuclear Physics* (ed. A. Faessler).

Yamada, H., Maguire, C. F., Hamilton, J. H., Ramayya, A. V., Hensley, D. C., Halbert, M. L., Robinson, R. L., Bertrand, F. E. and Woodward, R. (1981). *Phys. Rev.* **C24,** 2565.

Yamazaki, T. and Ewan, G. T. (1967). *Phys. Lett.* **24B,** 278.

Yamazaki, T. (1967). *Nucl. Data Tables* **A3,** 1.

Yamazaki, T. and Matthias, E. (1968). *Phys. Rev.* **175,** 1476.

Yoshida, S. (1977). *Proc. Int. Symp. Heavy Ion and Pre-equilibrium Reactions*, IPCR Cyclotron Progress Report Suppl. **6,** 359.

Youngblood, D. H., Rozsa, C. M., Moss, J. M., Brown, D. R. and Bronson, J. D. (1977). *Phys. Rev. Lett.* **39,** 1188.

Zell, K. O., Heits, B., Gast, W., Schuh, H.-W. and Brentano, P. von. (1976). *Z. Phys.* **A279,** 373.

Zimanyi, J., Halpern, I. and Madsen, V. A. (1970). *Phys. Lett.* **33B,** 205.

5

ELECTROMAGNETIC PROPERTIES IN SPIN AND ISOSPIN MODES

5.1 Electromagnetic transitions in spin and isospin modes

5.1.1 *General spin and isospin aspects of electromagnetic transitions*

Electromagnetic (EM) interactions in nuclei respond to the EM moments of nucleons (hadrons) in nuclei, such as electric charges, electric currents and magnetic moments. Accordingly, they also respond to the isospins (τ), the spins (σ), and the orbital motions of the constituent nucleons (hadrons). The spin–isospin ($\sigma\tau$) multipole moments are quite unique to the nucleon ensemble since nuclear forces (strong interactions) include large spin and isospin components. Such $\sigma\tau$ type interactions are mediated by π, ρ and other mesons. Electron–gamma spestroscopy is therefore used to probe the spin–isospin motions of nucleons, spin–isospin interactions and the roles of mesons associated with spin–isospin variables.

One of the most important aspects of the EM properties in spin–isospin modes is the renormalization of the EM multipole moments due to relevant nuclear medium effects. The renormalization of such EM moments has recently been reviewed by one of the authors and his collaborator (Ejiri and Fujita, 1978). We follow to a large extent that review, particularly in the discussion of the renormalization, and include also new data and recent developments.

EM transitions in $\sigma\tau$ modes are closely related to spin–isospin polarizations, spin–isospin giant resonances, and appropriate meson degrees of freedom. Recently much progress on such spin–isospin nuclear phenomena has been made experimentally as well as theoretically.

The EM transition operator associated with the $\sigma\tau$ variables is expressed on the basis of the long wave-length approximation as

$$\mathbf{T}_{TLSJ} = g_{TLSJ}\mathbf{t}^{T}\cdot(i^{L}r^{L}\mathbf{Y}_{L}\times\boldsymbol{\sigma}^{S})_{J}, \tag{5.1.1}$$

where $\mathbf{t}=(t_{+}, t_{-}, \sqrt{2}t_{3})$ and g_{TLSJ} is the EM coupling constant. Here the transition operator given by eqn. (1.1.7) in Chapter 1 is modified to a simple form like eqn (5.1.1) for convenience. $T=1$ and 0 give isovector and isoscalar modes, and $S=1$ and 0 the magnetic, and electric modes, respectively. L and J are the orbital and total angular-momenta, and the magnetic and electric γ transitions correspond to the axial-vector and vector β transitions. One can therefore study effectively the nuclear properties of various spin, isospin, and multipole modes by investigating the EM moments, EM transitions and analogous β transitions for the appropriate $\sigma\tau$ modes. Some typical examples of $\mathbf{T}_{TLSJ}$ with spin and/or isospin operators are as follows

$\mathbf{T}_{1101}=g_{1101}t_3[ir\mathbf{Y}_1]_1$	Isovector E1 γ transition
$\mathbf{T}_{1101}=g_{1101}t_\pm[ir\mathbf{Y}_1]_1$	First forbidden vector β transition
$\mathbf{T}_{1011}=g_{1011}t_3[\boldsymbol{\sigma}]_1$	Isovector M1 γ transition
$\mathbf{T}_{0011}=g_{0011}[\boldsymbol{\sigma}]_1$	Isoscalar M1 γ transition
$\mathbf{T}_{1011}=g_{1011}t_\pm[\boldsymbol{\sigma}]_1$	GT β transition
$\mathbf{T}_{1112}=g_{1112}t_3[ir\mathbf{Y}_1\times\boldsymbol{\sigma}]_2$	Isovector M2 γ transition
$\mathbf{T}_{0112}=g_{0112}[ir\mathbf{Y}_1\times\boldsymbol{\sigma}]_2$	Isoscalar M2 γ transition
$\mathbf{T}_{111J}=g_{111J}t_\pm[ir\mathbf{Y}_1\times\boldsymbol{\sigma}]_J$	First forbidden axial-vector β transition

The single-particle E1 transitions with $\mathbf{T}_{1101}=g_{1101}t_3[ir\mathbf{Y}_1]_1$ are essentially particle–hole de-excitations of high-lying states over one $\hbar\omega$ shell. There are no good single-particle states in the 1 $\hbar\omega$ excitation region of medium and heavy nuclei. E1 transitions between low-lying states near the Fermi surface are j-forbidden because $\Delta J \geqslant 2$ (see Section 2.3 and Fig. 2.13) and therefore most of the E1 transition rates between low-lying states in medium and heavy nuclei are much smaller than the single-particle (Weisskopf) unit. Thus, quantitative evaluation of the effective charge is rather difficult. E1 matrix elements in medium and heavy nuclei have been obtained from γ transitions through isobaric analogue states (IAS) (Ejiri *et al.*, 1968/1969; see Section 4.5). The observed values are found to be uniformly smaller than the single (quasi)-particle estimates by a reduction (hindrance) factor ~ 3.

M1 transitions and M1 moments with $\mathbf{T}_{1011}=g_{1011}\mathbf{t}\boldsymbol{\sigma}$ and $\mathbf{T}_{0011}=g_{0011}\boldsymbol{\sigma}$ have been extensively studied both experimentally and theoretically, and they are known to be quenched (retarded) much with respect to single-particle estimates.

Magnetic quadrupole transitions with $\mathbf{T}_{1112}$ and $\mathbf{T}_{0112}$ have been surveyed by Kurath and Lawson (1967), and most of them have been found to be retarded. They analysed the transitions in terms of the K selection rule and isobaric spin effects. An extensive survey of experimental data has been carried out by Letessier and Foucher (1969). Detailed studies of single quasi-particle M2 transitions have been made systematically by means of inbeam conversion–electron gamma spectroscopy (Ejiri *et al.*, 1972, 1973a, 1975; Brenn *et al.*, 1974) and they found that the single quasi-particle M2 transition matrix elements are smaller by factors $3\sim 4$.

Higher multipole transitions with $\mathbf{T}_{1L1J}$ and $\mathbf{T}_{0L1J}$ have, in general, not been extensively studied except for M4 transitions. Because of the large spin–orbit splitting, there are many stretched M4 transitions between single (quasi) particle states and they were surveyed by Goldhaber and Sunyar (1951). Bohr and Mottelson (1969) pointed out that the observed M4 transition rates were significantly less than the single-particle estimate due to an appreciable readjustment of the collective field. The M4 transition matrix elements are indeed retarded, like the M2 ones.

It is interesting to compare these $\sigma\tau$-mode γ transition rates with the analogous β transitions. Fermi β matrix elements in medium and heavy nuclei are known to be hindered strongly by 1 to 3 orders of magnitude. Gammow–Teller (GT) β matrix elements in such nuclei are reduced by factors 0.2 to 0.1 (Ikeda *et al.*, 1963; Kisslinger and Sorensen, 1963). The vector β decays with $t_{\pm}r\mathbf{Y}_1$ in the first forbidden β transitions, which are analogous to the IAS E1 γ decays, are retarded like the corresponding γ ones. Axial vector β decays with $\mathbf{T}_{1L1J} = g_{1L1J}\mathbf{t}i^L[r^L\mathbf{Y}_L \times \boldsymbol{\sigma}]_J$ with $J = L+1$ are obtained from the *ft* values of the unique *L*-th forbidden β transitions. The unique first-forbidden transitions have been extensively surveyed (Rose and Osborn, 1954; King and Peaslee, 1954; Lipnik and Sunier, 1964) and the matrix elements were found to be reduced by a similar factor 0.3 to the vector transitions. Analyses of non-unique first-forbidden beta transitions in the Pb region have given renormalizatin factors $g_v^{\text{eff}}/g_v \simeq 0.3$ to 0.6 and $g_{\text{A}}^{\text{eff}}/g_{\text{A}} \approx 0.7$ to 0.5 for the vector ($S = 0$) and axial vector ($S = 1$) transitions, respectively (Bohr and Mottelson, 1969/1975; Damgaad and Winter, 1964).

To summarise, the single-particle matrix elements for the $\sigma\tau$ mode γ operators and the analogous β ones are mostly reduced in medium and heavy nuclei by renormalization factors of about 0.3 compared with the simple single (quasi)-particle estimations.

5.1.2 *Giant resonances and nuclear core polarization in spin–isospin modes*

The uniform reduction of the $\sigma\tau$ γ- and β-transition rates between single (quasi)-particle (valence nucleon) states suggests some destructive effects of the nuclear medium such as spin–isospin core polarizations. The core polarization of the $\mathbf{T}_\alpha$ mode with $\alpha = \text{TLSJ}$ consists of a coherent superposition of the $\mathbf{T}_\alpha$ mode particle–hole (p–h) excitations of the nucleons in the nuclear core. Such a superposition is a kind of $\mathbf{T}_\alpha$-mode collective state (see Section 2.4). Thus the single-particle state $|\text{f}\rangle$ with a polarized core is expressed schematically as a sum of the unperturbed single-particle state $|\text{f}\rangle_0$ with a large amplitude and the polarized core $|\text{C}_\alpha\rangle$ (collective state) with a small amplitude ($|\varepsilon| \ll 1$). The state $|\text{f}\rangle$ shows single-particle behaviour in most of the nuclear phenomena except in the particular transition of the $\mathbf{T}_\alpha$ mode. For the latter transition the matrix element $\langle\text{f}|\mathbf{T}_\alpha|\text{i}\rangle$ is schematically given by a sum of the single-particle matrix element $M_{\text{sp}} = \sqrt{(1-\varepsilon^2)}\ {}_0\langle\text{f}|\mathbf{T}_\alpha|\text{i}\rangle$ and the matrix element $\varepsilon M_{\text{c}} = \varepsilon\langle\text{C}_\alpha|\mathbf{T}_\alpha|\text{i}\rangle$ for the polarized core (collective state). The latter is as important as the former (M_{sp}) even though the mixing amplitude ε is small, since the collective state matrix element M_{c} is very large due to the coherent contributions of many p–h terms. When the phase of the polarization (admixture of the collective state) is the same as that of the single-particle configuration the transition is much enhanced because of the constructive effect of M_{sp} and M_c, and when it is opposite the transition is much hindered because of the cancellation (destructive) effect. Constructive polarization effects (enhancement) in isoscalar electric transitions (E2, E3, etc.) and the associated collective states are described by Bohr and Mottelson (1969/1975). The retardation phenomena of the EM transitions in the spin–isospin mode are related to the destructive effect of the spin–isospin polarization.

The polarization interaction is symbolically expressed in terms of a scalar product $\mathbf{H}_I \sim \chi_{TLSJ} \mathbf{T}_{TLSJ} \cdot \mathbf{T}_{TLSJ}$. The Majorana interaction $\mathbf{H}_M = \chi_M(1+\boldsymbol{\sigma\sigma})(1+\boldsymbol{\tau\tau})\mathrm{f}(r_1 r_2)$ is dominant in exchange interactions (see, for example; Austin, 1972). Therefore, one may expect that the polarization strengths χ_α for the isospin mode $(\mathbf{H}_I \sim \chi_{1L0J} \cdot \mathbf{T}_{1L0J} \cdot \mathbf{T}_{1L0J})$, the spin mode $(\mathbf{H}_I \sim \chi_{0L1J} \mathbf{T}_{0L1J} \cdot \mathbf{T}_{0L1J})$ and the isospin spin mode $(\mathbf{H}_I \sim \chi_{1L1J} \mathbf{T}_{1L1J} \cdot \mathbf{T}_{1L1J})$ are about the same. The phase of the core polarization (sign of the mixing coefficient ε) depends on the sign of the interaction and the relative energies of the unperturbed particle–hole states with respect to those of the single-particle state.

The polarization interaction gives rise to the collective state in the high or low excitation region by pushing up or down one state among the unperturbed particle–hole states, depending on whether the interaction is repulsive or attractive. Since the Majorana interaction between particle–hole states is repulsive ($\chi_M > 0$), the collective states of the isospin $(\mathbf{t}r^L\mathbf{Y}_L \times \boldsymbol{\sigma})$ modes are expected to occur in the high excitation region, and the core polarization effect for a low-lying state below the unperturbed particle–hole states has a negative (destructive) phase.

The collective state in the $\mathbf{T}_\alpha$ mode is a kind of $\mathbf{T}_\alpha$ mode vibration. If it occurs in the high excitation region, it appears as a giant resonance (GR) in the $\mathbf{T}_\alpha$ mode excitation. The retardation of the isovector E1 transition is qualitatively explained by the destructive effect due to the admixture of the E1 GR with negative phase.

The idea of the E1 γ GR and E1 core polarization effect can be extended to other spin and isospin modes of various multipolarities on the basis of the Majorana type exchange interaction with $\chi_M > 0$. Then GR's with the GT $(t_\pm \boldsymbol{\sigma})$, the first forbidden vector $(t_\pm r\mathbf{Y}_1)$, the M2 $(t_3 r\mathbf{Y}_1 \times \boldsymbol{\sigma}, r\mathbf{Y}_1 \times \boldsymbol{\sigma})$, the first forbidden axial vector $(t_\pm r\mathbf{Y}_1 \times \boldsymbol{\sigma})$ and other higher multipole spin–isospin modes are expected to occur in the higher excitation region. Thus they may contribute destructively to the single-particle transitions between low-lying states. The core polarizations and the GR are schematically illustrated in Fig. 5.1. Such $\sigma\tau$ polarization effects on β decays have been discussed by several authors (Überall, 1965; Morita, 1967; Ejiri, 1972).

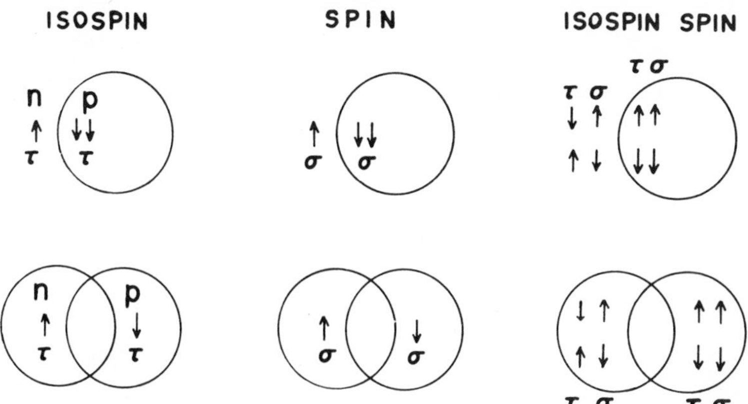

Fig. 5.1. Schematic pictures of isospin, spin and spin–isospin core polarizations (upper pictures) and isospin, spin and spin–isospin giant resonances (lower pictures). (Ejiri and Fujita, 1978.)

The existence of various kinds of GR for photo-nuclear reactions has been predicted by Danos and Steinwedel (1951) and Glassgold *et al.* (1959). GR's in the modes $t_3[r\mathbf{Y}_1 \times \boldsymbol{\sigma}^S]_J$ with $(T, S) = (0, 1)$ and $(1, 1)$ are expected to lie near the E1 GR with $(T, S) = (1, 0)$, and the retardation factors for those modes are about the same if the exchange interaction of the Majorana-type is predominant in $\mathbf{H}_\text{I}$.

The GT GR associated with the GT β decay was predicted long ago (Ikeda *et al.*, 1963; Yamada, 1965), and has now been found experimentally (Doering *et al.*, 1975). The core polarizations and possible giant resonances for first forbidden β decays and analogous γ decays (E1, M2) have been studied (Ejiri *et al.*, 1968a; Ejiri, 1971, 1972, 1973b). The M2 core polarizations and M2 gamma giant resonances were studied by many authors (Wild 1956; Überall 1966; Ejiri *et al.*, 1973a; Grecksch *et al.*, 1975; Lindgren *et al.*, 1975). The spin–isospin polarization effects on unique forbidden β transitions, muon absorption and magnetic γ transitions were also studied (Bohr and Mottelson, 1969/1975; Ejiri, 1972; Überall, 1965; Morita, 1967). The fact that all these β and γ transitions are uniformly retarded suggests the existence of possible giant resonances which absorb a considerable amount of the transition strength from low-lying single particle states. In fact giant resonances corresponding to the isovector M2-γ and first forbidden β transitions have been observed (Horen *et al.*, 1980, 1981; Bainum *et al.*, 1980; Gaarde *et al.*, 1981).

The transitions and possible GR's are shown schematically in Fig. 5.2.

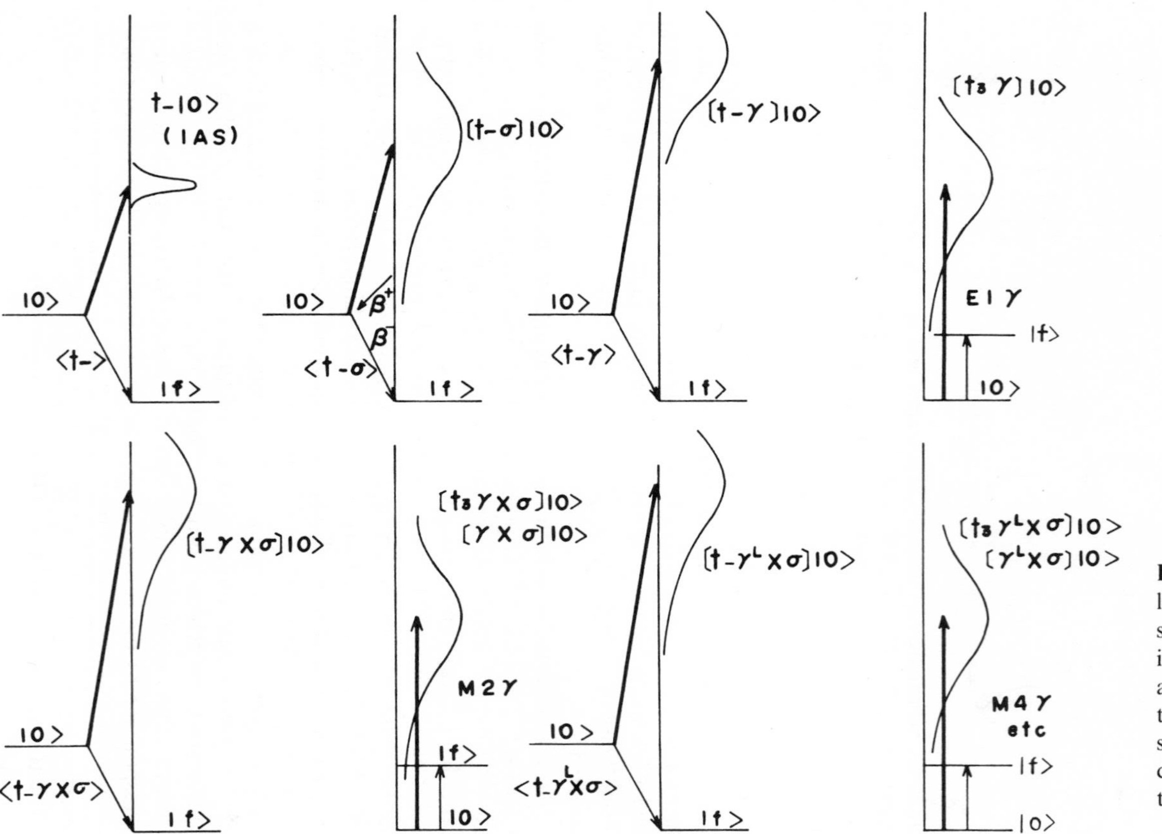

Fig. 5.2. Transition and level schemes showing schematically the (possible) giant resonances absorbing a large fraction of the transition strengths, and the reduced single particle transitions. (Ejiri and Fujita, 1978.)

5.1.3 *Effective coupling constant and susceptibility*

The core polarization, which strongly affects the single-particle $\mathbf{T}_\alpha$ moment, is induced firstly by the valence nucleon ($|f\rangle_0$) outside the core through the $\mathbf{T}_\alpha$-mode polarization interaction. In this case the contribution to the $\mathbf{T}_\alpha$ transition from the polarized core is proportional to the single-particle transition matrix element, and is given by $-\kappa_0\langle\mathbf{T}_\alpha\rangle_0$, where κ is the susceptibility (polarizability) for the $\mathbf{T}_\alpha$ mode (note $\kappa>0$ for a negative polarization in the $\sigma\tau$ mode). Taking into account subsequent contributions of the polarization induced by the polarized core, the transition matrix element is

$$\langle\mathbf{T}_\alpha\rangle = {}_0\langle\mathbf{T}_\alpha\rangle_0 - \kappa_0\langle\mathbf{T}_\alpha\rangle_0 + \kappa^2{}_0\langle\mathbf{T}_\alpha\rangle_0 - \ldots . \tag{5.1.2}$$

An effective coupling constant g_α^{eff} may be introduced to represent the renormalization due to the core polarization (Migdal, 1944/1945, 1968; Ejiri and Fujita, 1978) by

$$\langle\mathbf{T}_\alpha\rangle = (g_\alpha^{\text{eff}}/g_\alpha)\cdot{}_0\langle\mathbf{T}_\alpha\rangle_0, \tag{5.1.3}$$

$$g_\alpha^{\text{eff}} = \frac{1}{1+\kappa_\alpha}g_\alpha. \tag{5.1.4}$$

Here the susceptibility κ is schematically written as proportional to the polarization interaction strength χ_α and the square of the transition matrix element for the polarized core (collective states), and inversely proportional to the average energy required to polarize the core (to excite the particle–hole states). The spin and isospin susceptibilities correspond to the electric and magnetic susceptibilities, respectively.

When the polarization effect is so small that the susceptibility $|\kappa|\ll 1$, one may use a simple first-order perturbation approach giving $\langle\mathbf{T}_\alpha\rangle = {}_0\langle\mathbf{T}_\alpha\rangle_0 - \kappa_0\langle\mathbf{T}_\alpha\rangle_0$. The isospin, spin and spin–isospin core polarizations greatly modify the single-particle transitions since for such modes κ is of the order of 1. Therefore the simple perturbation expression $g_\alpha^{\text{eff}} = g_\alpha(1-\kappa)$ cannot be used, but instead the expression given by eqn (5.1.4). The theoretical expressions for the effective coupling constant and the susceptibility are given in Section 5.2 and the experimental data are analysed in terms of the g^{eff} and κ in Section 5.3.

5.2 The renormalization of the spin–isospin operators in nuclei

The EM transition moments and analogous β ones in spin–isospin ($\sigma\tau$) modes in nuclei are renormalized by a factor that is expressed in terms of the nuclear spin–isospin susceptibilities, as described briefly in Section 5.1. In the following sub-sections (5.2.1 and 5.2.2) the theoretical aspects of the renormalization in the $\sigma\tau$ mode and the $\sigma\tau$ susceptibility are briefly described (see Ejiri and Fujita, 1978 for details).

5.2.1 *The renormalization of electromagnetic transitions*

In the field of electromagnetism the notions of 'electric susceptibility' and 'dielectric constant' are quite familiar. The electric polarization P of the dielectric is known to be roughly proportional to the electric field E' at the point in dielectrics,

$$P = \varepsilon_0 \chi E', \tag{5.2.1}$$

where ε_0 is the dielectric constant of the vacuum and χ is the electric susceptibility. It should be noted that E' is not equal to the applied field E_{ext} due to the external true charge, but includes the contribution from the overall and local polarizations of the dielectric. In the case of an insulator placed between a pair of condenser plates, the field E' is given by

$$E' = E_{\text{ext}} + E_1 + E_2, \tag{5.2.2}$$

where $E_1 = -P/\varepsilon_0$ is the electric field due to the surface polarization charge induced at the surface of the insulator and E_2 is the remaining field called the 'local field correction'. The electric displacement D and the dielectric constant ε are related by

$$D = \varepsilon_0 E_{\text{ext}} = \varepsilon E = \varepsilon_0 E + P, \tag{5.2.3}$$

where E is the average electric field, i.e. $E_{\text{ext}} + E_1$.

If we neglect all the effects of the interactions among the electrons by substituting $E' \approx E_{\text{ext}}$ into eqn (5.2.1), we obtain from eqn (5.2.3)

$$(\varepsilon/\varepsilon_0)^{-1} = 1 - \chi. \tag{5.2.4}$$

To improve the accuracy, we use the relation, $E' \approx E_{\text{ext}} + E_1$ in eqn (5.2.1) by neglecting only the local field correction E_2. Then we obtain, instead of eqn (5.2.4),

$$\varepsilon/\varepsilon_0 = 1 + \chi. \tag{5.2.5}$$

This latter approximation is called the random phase approximation (RPA). If we further include the effect of local field correction, we obtain the Lorentz–Lorenz formula,

$$\varepsilon/\varepsilon_0 = \{1 + \chi(1-\gamma)\}/(1 - \chi\gamma), \tag{5.2.6}$$

where γ is the strength of local field correction (Noziéres and Pines, 1958).

Similar arguments can be given for magnetic susceptibility. The idea described here for the electromagnetism can be applied to nuclear physics (Mottelson, 1967) and the Lorentz–Lorenz formula was used in the discussion of pion–nuclear interactions (Ericson and Ericson, 1966; Ericson *et al.*, 1973).

Nuclear EM matrix elements have long been studied by means of a shell model with configuration mixing. One of the most thoroughly-studied quantities is the magnetic-dipole moment. Since the so-called Schmidt lines (or Landé lines) cannot explain the experimental data quantitatively, configuration mixing effects have been studied first by using first-order perturbation theory for the residual interaction of the δ-function type (Blin-Stoyle, 1953; Arima and Horie, 1954; Noya *et al.*, 1958). Analyses with configuration mixing have also been made for M1 transitions (Arima *et al.*, 1957; Sorensen, 1963; Shibata, 1967; Hsieh and Ogata, 1970; Freed and Kisslinger, 1968; Hirata, 1970) and for M4 transitions (Horie and Oda, 1964; Maripuu, 1970).

In the case of the collective model, more specifically the so-called the Bohr–Mottelson model (Bohr and Mottelson, 1953), the core polarization is explicitly taken into account as a core deformation or surface oscillation of the ellipsoidal type. The first-order configuration-mixing model was also applied to the study of static E2 moments (Horie and Arima, 1955). The beta transition matrix elements are reduced to some extent by a similar core polarization effect (Towner *et al.*, 1971, 1979; Morita *et al.*, 1971, 1976).

Migdal (1944/1945) has used the sum rule $\int\{\sigma(w)/w^2\}dw$ to predict the resonance energy of the dipole vibration, where $\sigma(w)$ represents the E1 photo-absorption cross-section as a function of the energy w. When a constant electric field is applied to a nucleus, protons and neutrons are displaced in opposite directions in the centre-of-mass frame of the nucleus. The symmetry energy term in the semi-empirical mass formula is proportional to $(N-Z)^2$ and is reinterpreted as $(\rho_n - \rho_p)^2$, where ρ_n and ρ_p are the neutron and proton densities, respectively. The centres-of-mass of the protons and neutrons cannot be separated too much because of the strong restoring forces. This is the so-called 'dipole vibration'. More quantitative models have been given later by Goldhaber–Teller (1948) and Steinwedel–Jensen (1950), and the E1 GR was nicely explained (Baldwin and Klaiber, 1947). Such type of a collective vibration is a sort of isospin dipole ($t_3\mathbf{Y}_1$) vibration.

The single-particle transition is generally accompanied by a polarization of the nuclear medium (the nuclear core) and the transition strength is enhanced or reduced (hindered). The core polarization effect on single (quasi)-particle transitions can also be expressed as the effect of the coupling to the phonon (collective state) of the same type as the transition.

The isoscalar EJ phonon states (E2, E3, etc) are discrete states in the low-excitation region. The enhancement of the EJ transitions has been extensively analysed in terms of the in-phase mixing of the EJ phonon (Bohr and Mottelson, 1975). On the other hand spin–isospin phonons appear as resonances (GR) in the high excitation region. Thus in order to explain the retardation of the spin–isospin transition by the out-of-phase mixing of the GR, the energy, the width and the coupling with low-lying states have to be known (see Section 5.1).

Experimental data, as described in Sections 5.1 and 5.3, show the need of an additional retardation factor due to the core polarization effect and so on.

Thus even if we take into account the configuration mixing in the valence nucleons in terms of the pairing reduction factors (Kisslinger and Sorensen, 1963; Sakai and Yoshida, 1964; Fujita *et al.*, 1964) we cannot explain the absolute magnitude of nuclear EM and analogous β matrix elements.

Various theoretical methods have been proposed in order to treat the core polarization effect; (a) the configuration mixing method (Arima and Horie, 1954; Spector, 1960); (b) the configuration mixing method in the pairing model (Hamamoto, 1965); (c) the commutator method (Fujita and Ikeda, 1965b; Ejiri *et al.*, 1968a); (d) the random phase approximation (Bohr and Mottelson, 1975; Halbleib and Sorensen, 1969; Ejiri, 1971, 1973b; Überall, 1966; Liu and Brown, 1976); (e) the method of Migdal (Migdal, 1964); and (f) the method of the nuclear response function (Ring *et al.*, 1973; Bertsch and Tsai, 1975). These methods are closely related to each other, and are more or less overlapping. Similar theoretical consideration have been extensively made in relation to the pion optical potential by Ericson *et al.* (1966).

In the next section we follow the scheme;

(1) Brown and Bolsterli's schematic model (Brown and Bolsterli, 1959; Brown *et al.*, 1966) for giant resonances;

(2) the commutator method as developed by Fujita and Ikeda (1965b);

(3) the RPA method used for the forbidden matrix elements by Ejiri (1971, 1973b) and for the M1 matrix element by Mottelson (1967).

This is done simply because in these treatments the retardation factor is expressed as $1/(1+\kappa)$, as in the theory of Migdal (1964), and it corresponds exactly to the familiar form for the dielectric constant in RPA in the case of electromagnetism, as stated in Section 5.1.

5.2.2 *Spin–isospin polarization and spin–isospin susceptibility*

5.2.2a *General remarks*

There are several ways of defining the renormalization factor of the form $1/(1+\kappa)$, where κ is the susceptibility. The concept of 'polarizability' of nuclei was first introduced by Migdal (1944) when discussing photo-abosorption cross-section, which is proportional to

$$\left\langle \mathrm{i} \left| D_z \frac{1}{E_\mathrm{i} - \mathbf{H}} \mathrm{D}_z \right| \mathrm{i} \right\rangle, \tag{5.2.7}$$

where D_z is the dipole operator $\frac{1}{2}\Sigma_j \tau_z^j z_j$, $\mathbf{H}$ the total nuclear Hamiltonian, and E_i the energy of the state $|\mathrm{i}\rangle$.

The quantity (5.2.7) corresponding to the polarizability is related to the sum rule $\int\{\sigma(w)/w^2\}\mathrm{d}w$, where $\sigma(w)$ stands for the E1 photo-absorption cross-section at energy w. If the excitation function $\sigma(w)$ has a giant resonance peak, the peak energy can be estimated from the quantity

$$[\int \sigma(w)\mathrm{d}w / \int \{\sigma(w)/w^2\}\mathrm{d}w]^{1/2}. \tag{5.2.8}$$

Various types of sum rules have been extensively studied by Levinger (1960) and his collaborators for nuclear photo-reactions. For our purpose a more convenient expression for the polarizability is

$${}_0\left\langle \mathrm{i} \middle| D_z \frac{1}{E_i^0 - \mathbf{H}^0} \mathrm{D}_z \middle| \mathrm{i} \right\rangle_0, \tag{5.2.9}$$

where $|\mathrm{i}\rangle_0$ is an unperturbed eigen-state of the model Hamiltonian $\mathbf{H}^0$, and E_i^0 is the eigenvalue.

5.2.2b *Schematic model for giant resonances*

Using a schematic model (Brown and Bolsterli, 1959; Brown *et al.*, 1966) the single-body transition operator **m** is written in the form

$$\mathbf{m} = \sum_{\mathrm{rs}} \langle \mathrm{r} | \mathbf{m} | \mathrm{s} \rangle c_\mathrm{r}^+ c_\mathrm{s}, \tag{5.2.10}$$

where c^+ and c represent the creation and annihilation operators, respectively. A set of n states is defined by

$$|\phi_j\rangle = \frac{1}{\sqrt{N_j}} c_\mathrm{r}^+ c_\mathrm{s} |\mathrm{i}\rangle \quad (j = 1, 2, \ldots n), \tag{5.2.11}$$

where j represents a pair of states (r, s), $|\mathrm{i}\rangle$ the true eigenstate of the total Hamiltonian **H**, and N_j the normalization factor. N_j is written

$$N_j = \langle \mathrm{i} | c_\mathrm{s}^+ c_\mathrm{r} c_\mathrm{r}^+ c_\mathrm{s} | \mathrm{i} \rangle. \tag{5.2.12}$$

The model Hamiltonian **H** is given by

$$\mathbf{H} = \mathbf{H}^0 + \mathbf{H}^1, \tag{5.2.13}$$

with

$$\mathbf{H}^0 = \bar{E}^0 \sum_{j=1}^{n} |\phi_j\rangle\langle\phi_j|, \tag{5.2.14}$$

and

$$\mathbf{H}^1 = G\left(\sum_{j=1}^{n} c_j \sqrt{N_j} |\phi_j\rangle \right)\left(\sum_{i=1}^{n} \langle\phi_i| \sqrt{N_i} c_i^* \right), \tag{5.2.15}$$

where $c_j = \langle \mathrm{r} | \mathbf{m} | \mathrm{s} \rangle$. Then one of the eigenstates of **H** is given by

$$\mathbf{m}|\mathrm{i}\rangle = \sum_{j=1}^{n} c_j \sqrt{N_j} |\phi_j\rangle. \tag{5.2.16}$$

The eigenvalue is written

$$E=\bar{E}^0+NG \tag{5.2.17}$$

$$N\equiv\langle \mathrm{i}|\mathbf{m}^+\mathbf{m}|\mathrm{i}\rangle=\sum_{j=1}^{n}|c_j|^2N_j, \tag{5.2.18}$$

$$NG=\langle \mathrm{i}|\mathbf{m}^+\mathbf{H}^1\mathbf{m}|\mathrm{i}\rangle/\langle \mathrm{i}|\mathbf{m}^+\mathbf{m}|\mathrm{i}\rangle. \tag{5.2.19}$$

All the other (n − 1) eigenstates are degenerate, having energy $\bar{E}^0$. The state $\mathbf{m}|\mathrm{i}\rangle$, corresponding to the m-mode collective vibration (m-mode GR), absorbs all the transition strength, since

$$\sum_{\mathrm{f}}|\langle \mathrm{f}|\mathbf{m}|\mathrm{i}\rangle|^2=\langle \mathrm{i}|\mathbf{m}^+\mathbf{m}|\mathrm{i}\rangle=|\langle \mathrm{i}|(\mathbf{m}^+/\sqrt{N})\mathbf{m}|\mathrm{i}\rangle|^2. \tag{5.2.20}$$

Thus the matrix elements for all other $n-1$ states vanish. The interaction is expressed as

$$\mathbf{H}^1=\mathbf{P}\mathbf{H}^1\mathbf{P}, \tag{5.2.21}$$

where $\mathbf{P}\equiv 1-\mathbf{Q}$ is the projection operator defined by

$$\mathbf{P}\equiv\mathbf{m}|\mathrm{i}\rangle\langle \mathrm{i}|\mathbf{m}^+/N. \tag{5.2.22}$$

The matrix element of $\mathbf{H}^1$ is written in a separable form,

$$\langle\phi_k|\mathbf{H}^1|\phi_j\rangle=G(c_k\sqrt{N_k})(\sqrt{N_j}c_j^*). \tag{5.2.23}$$

Now let us relax the condition that all unperturbed states are degenerate—see eqn (5.2.14)—and introduce a non-degenerate state $|\mathrm{f}\rangle_0\equiv|\phi_1\rangle$ with energy $\bar{E}^0-\delta$. The Hamiltonian is written

$$\mathbf{H}^0=(\bar{E}^0-\delta)|\phi_1\rangle\langle\phi_1|+\bar{E}^0\left(\sum_{j=2}^{n}|\phi_j\rangle\langle\phi_j|\right). \tag{5.2.24}$$

Then the matrix element for $|\mathrm{f}\rangle$ has a finite value

$$\langle \mathrm{f}|\mathbf{m}|\mathrm{i}\rangle\approx{}_0\langle \mathrm{f}|\mathbf{m}|\mathrm{i}\rangle/(1+NG/\delta). \tag{5.2.25}$$

Thus the susceptibility (or polarizability) is given by the ratio of the energy shift NG for the states $\mathrm{m}|\mathrm{i}\rangle$ to the deviation of the unperturbed energy of $|\mathrm{f}\rangle_0$ from the average energy of all the other unperturbed states $|\phi_j\rangle$, namely NG/δ.

5.2.2c *The commutator method*

The commutator method has been proposed and applied mainly to β-decay matrix elements (Fujita and Ikeda, 1965b; Fujita *et al.*, 1967; Ejiri *et al.*, 1968a). It can be applied to the general transition operator **m**. We start from the identity,

$$\langle \mathrm{f}|\mathbf{m}|\mathrm{i}\rangle = \frac{\langle \mathrm{f}|\mathbf{Q}[\mathbf{H},\mathbf{m}]|\mathrm{i}\rangle}{E_\mathrm{f}-E_\mathrm{i}-\Delta}, \tag{5.2.26}$$

with

$$\Delta = \frac{\langle \mathrm{i}|\mathbf{m}^+[\mathbf{H},\mathbf{m}]|\mathrm{i}\rangle}{\langle \mathrm{i}|\mathbf{m}^+\mathbf{m}|\mathrm{i}\rangle}, \tag{5.2.27}$$

provided that $E_\mathrm{f}-E_\mathrm{i}-\Delta\neq 0$. Although eqn (5.2.26) is exact for the true wave functions $|\mathrm{i}\rangle$ and $|\mathrm{f}\rangle$, both sides of eqn (5.2.26) give different values if the true wave function $|\mathrm{f}\rangle$ is replaced by the model wave function $|\mathrm{f}\rangle_0$. The right-hand side of eqn (5.2.26) then becomes

$$\frac{{}_0\langle \mathrm{f}|\mathbf{Q}[\mathbf{H},\mathbf{m}]|\mathrm{i}\rangle}{E_\mathrm{f}-E_\mathrm{i}-\Delta} = \frac{{}_0\langle \mathrm{f}|\mathbf{m}|\mathrm{i}\rangle + S}{1+\kappa'_\mathrm{f}}, \tag{5.2.28}$$

with

$$\Delta^0 = \langle \mathrm{i}|\mathbf{m}^+[\mathbf{H}^0,\mathbf{m}]|\mathrm{i}\rangle/\langle \mathrm{i}|\mathbf{m}^+\mathbf{m}|\mathrm{i}\rangle, \tag{5.2.29a}$$

$$\Delta^1 = \langle \mathrm{i}|\mathbf{m}^+[\mathbf{H}^1,\mathbf{m}]|\mathrm{i}\rangle/\langle \mathrm{i}|\mathbf{m}^+\mathbf{m}|\mathrm{i}\rangle, \tag{5.2.29b}$$

$$S = \frac{{}_0\langle \mathrm{f}|\mathbf{Q}[\mathbf{H}^1,\mathbf{m}]|\mathrm{i}\rangle}{E_\mathrm{f}-E_\mathrm{i}-\Delta^0}, \tag{5.2.29c}$$

and

$$\kappa'_\mathrm{f} = \frac{\Delta^1 + E^0_\mathrm{f} - E_\mathrm{f}}{E_\mathrm{i}+\Delta^0 - E^0_\mathrm{f}}. \tag{5.2.29d}$$

Where $\Delta = \Delta^0 + \Delta^1$ and we assumed that $\mathbf{H}^0|\mathrm{i}\rangle = E_\mathrm{i}|\mathrm{i}\rangle$ in eqn (5.2.29d).

It is interesting to compare the approximate expression (5.2.28) with eqn (5.2.25): The denominator is essentially the same as $1+NG/\delta$ since $E^0_\mathrm{f}\approx E_\mathrm{f}$, $\Delta^1\approx NG$ and $E_\mathrm{i}-E^0_\mathrm{f}+\Delta^0\approx\delta$. The correction term S in the numerator arises because of the deviation of $\mathbf{H}^1$ from the separable one in eqn (5.2.23). One of the merits in starting from the right-hand side of eqn (5.2.26) is that the relationship between $\langle \mathrm{f}|\mathbf{m}|\mathrm{i}\rangle$ and the symmetry-breaking part of $\mathbf{H}$ is clear; if $[\mathbf{H},\mathbf{m}]\approx 0$, $\langle \mathrm{f}|\mathbf{m}|\mathrm{i}\rangle\approx 0$ unless $E_\mathrm{f}=E_\mathrm{i}+\Delta$. However, it must be remembered that $|\mathrm{i}\rangle$ in eqn (5.2.28) is the true wave function, and the initial and final states are not treated symmetrically. More detailed arguments are given in Fujita (1968) and Arita *et al.* (1976).

In order to treat the initial and final states symmetrically, we must consider the state $\mathbf{m}^+|\mathrm{f}\rangle$ as well as $\mathbf{m}|\mathrm{i}\rangle$. Here $\mathbf{m}|\mathrm{i}\rangle$ corresponds to the forward correlation and $\mathbf{m}^+|\mathrm{f}\rangle$ to the backward correlation, as shown in Fig. 5.3. Then the matrix element is (Arita *et al.*, 1976)

$$\langle \mathrm{f}|\mathbf{m}|\mathrm{i}\rangle \approx \frac{{}_0\langle \mathrm{f}|\mathbf{m}|\mathrm{i}\rangle_0}{1+\bar{\kappa}^- + \bar{\kappa}^+ + F\langle \mathrm{f}|\mathbf{m}|\mathrm{i}\rangle}, \tag{5.2.30}$$

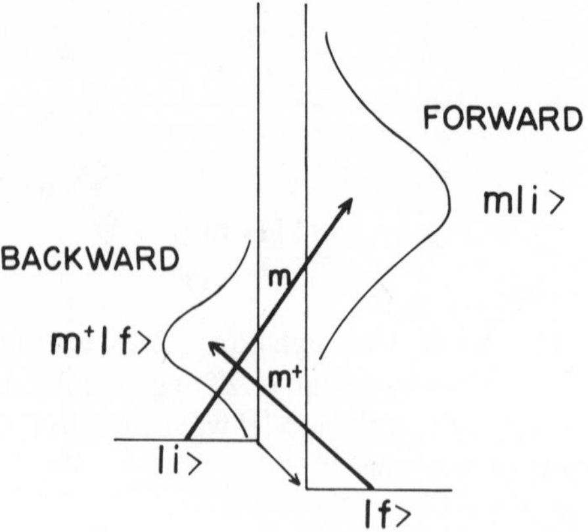

Fig. 5.3. Schematic picture showing collective states (giant resonances) associated with forward and backward correlations. (Ejiri and Fujita, 1978.)

where

$$\bar{\kappa}^- = \frac{1}{\bar{E}_f^0 - E_f^0} \frac{\langle i|\mathbf{m}^+(\mathbf{H}^1 - \delta_f)\mathbf{m}|i\rangle}{\langle i|\mathbf{m}^+\mathbf{m}|i\rangle}, \tag{5.2.31}$$

and similarly for $\bar{\kappa}^+$, and

$$F = -\bar{\kappa}^-\bar{\kappa}^+ \frac{\langle i|\mathbf{m}^+\mathbf{qmqm}^+|f\rangle}{\langle i|\mathbf{m}^+\mathbf{qm}|i\rangle \langle f|\mathbf{mqm}^+|f\rangle}. \tag{5.2.32}$$

where $\mathbf{q} = 1 - |f\rangle_{00}\langle f|$, $\bar{E}_f^0 = \langle i|\mathbf{m}^+\mathbf{qH}^0\mathbf{m}|i\rangle / \langle i|\mathbf{m}^+\mathbf{qm}|i\rangle$ and $\delta_f = E_f - E_f^0 = \langle f|\mathbf{H}^1|f\rangle_0 / \langle f|f\rangle_0$. Since F can be neglected in the usual physical situation the formula (5.2.30) agrees in practice with the one obtained by the usual RPA with separable interactions, as shown in Section 5.2.2d. The susceptibility (polarizability) is then given by $\kappa = \bar{\kappa}^- + \bar{\kappa}^+$, where $\bar{\kappa}^-$ corresponds to the susceptibility for the forward correlation (polarizability of the final state) and $\bar{\kappa}^+$ to that for the backward one (polarizability of the initial state).

5.2.2d *The random phase approximation for the separable polarization interaction*

Since the core polarization effect on the $\sigma\tau$ transition moment is a kind of collective effect, arising from admixture of the $\sigma\tau$ collective vibration, it is treated theoretically with the random phase approximation (RPA) as used for various types of collective phenomena (Takagi, 1959; Marumori, 1960; Arvieu and Veneroni, 1960; Baranger, 1960). The RPA analysis for the coupling of the low-lying single-particle state with the $\sigma\tau$ collective vibration has been carried out by Bohr and Mottelson (1975), and other investigators (Halbleib and Sørensen, 1969; Ejiri, 1971, 1972).

The core polarization interaction $\mathbf{H}_\alpha^P$ for the $\mathbf{T}_\alpha$ mode transition, where $\alpha = TLSJ$, can be written as the separable $\mathbf{T}_\alpha$ phonon interaction (Bohr and Mottelson, 1975):

$$\mathbf{H}_\alpha^P = \chi_\alpha \mathbf{T}_\alpha \cdot \mathbf{T}_\alpha. \tag{5.2.33}$$

The $\mathbf{T}_\alpha$ is written as $\mathbf{T}_\alpha = \mathbf{t}^T \cdot \mathbf{M}_\alpha$, where the isospin operator is $t^1 = (t_-, t_+, t_3\sqrt{2})$ for the isovector ($T=1$) mode β^-, β^+ and γ transitions, respectively, and $t^0 = 1/\sqrt{2}$ for the isoscalar mode. The Hamiltonian $\mathbf{H}$ is divided into two parts of the polarization interaction namely $\mathbf{H}_\alpha^P$ and the residue $\mathbf{H}_0$:

$$\mathbf{H} = \mathbf{H}_0 + \mathbf{H}_\alpha^P. \tag{5.2.34}$$

The idea of eqn (5.2.34) is to take into account explicitly the polarization interaction $\mathbf{H}_\alpha^P$, which is most important for the $\mathbf{T}_\alpha$ mode transition because the particle–hole excitations induced by the $\mathbf{H}_\alpha^P$ contribute coherently to the $\langle \mathbf{T}_\alpha \rangle$. On the other hand those ε_j induced by other residual interactions contribute incoherently only as the second-order term $(1 - \Sigma \varepsilon_j^2)^{1/2}$.

The unperturbed Hamiltonian is given for simplicity by a sum of the shell-model potential and the pairing interaction. Let us first consider the $t_-(\beta^-)$ mode for a doubly-even nucleus $|\mathrm{i}\rangle \equiv |0\rangle$ and follow the RPA derivation given by Ejiri (1971). The $t_+(\beta^+)$ and $t_3(\gamma)$ modes can be easily obtained in the same way as the $t_-(\beta^-)$ mode. The final states are then given by

$$|\mathrm{f}_n\rangle = \mathbf{D}_n^+ |0\rangle,$$

$$\mathbf{D}_n^+ = \sum_{iI} [b_n(Ii)\mathbf{B}^+(Ii)_\mu^J + c_n(Ii)C(Ii)_\mu^J], \tag{5.2.35}$$

$$\mathbf{B}^+(iI)_\mu^J = \sum_{mM} (imIM|J\mu)\alpha_{im}^+ \alpha_{IM}^+, \tag{5.2.36}$$

$$\mathbf{C}(\mathrm{iI})_\mu^J = \sum_{mM} (imIM|Ju)\,(-)^{i-m}\alpha_{i-m}(-)^{I-M}\alpha_{I-M}, \tag{5.2.37}$$

where α_J^+ and α_j^+ are creation operators for protons (capital J) and neutrons (small j), respectively, and α_J and α_j are the corresponding annihilation operators. The particle–hole excitation is schematically illustrated in Fig. 5.4. The wave function for the n-th state is obtained by solving the Schrödinger equation

$$[\mathbf{H}_0 + \mathbf{H}_\alpha^P, \mathbf{D}_n^+] = E_n \mathbf{D}_n^+. \tag{5.2.38}$$

The dispersion equation for a simple semi-magic nucleus is written

$$\chi_\alpha \left(\frac{\Sigma G_{iI}^2 V_i^2 U_I^2}{E(Ii) - E_n} + \frac{-\Sigma G_{Ii}^2 U_i^2 V_I^2}{-E(Ii) - E_n} \right) = -1, \tag{5.2.39}$$

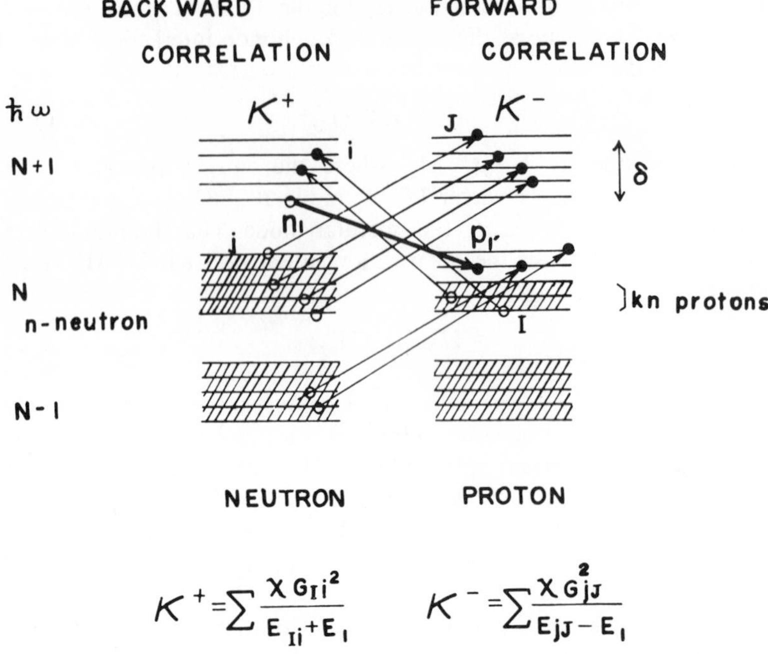

Fig. 5.4. Level structure in a harmonic oscillator model with the closed $N\hbar\omega$ neutron shell (n neutrons) and the non-closed $N\hbar\omega$ proton shell (kn protons with $k<1$). The main single (quasi)-particle transition is $n_1 \to p_1$. Many coherent particle–hole excitations due to core polarization are expressed as $j \to J$ (forward correlation) and $I \to i$ (backward correlation), respectively. (Ejiri and Fujita, 1978.)

where $E(Ii)$ is the unperturbed energy of the quasi-particle hole pair (Ii), and U^2 and V^2 are the vacancy and occupation probabilities, respectively. The interaction matrix element is

$$G_{iI} = \frac{1}{\sqrt{(2J+1)}} \langle \mathrm{i} \| \mathbf{M}_\alpha \| I \rangle. \tag{5.2.40}$$

The coefficients for the eigenfunction are given by

$$b_n(Ii) = \frac{\sqrt{(\chi_\alpha)} G_{Ii}}{E(Ii) - E_n} (-U_I V_i B), \tag{5.2.41a}$$

$$c_n(Ii) = \frac{\sqrt{(\chi_\alpha)} G_{Ii}}{-E(Ii) - E_n} (U_i V_I B), \tag{5.2.41b}$$

where B is obtained from the normalization condition. The eigenvalue E_n is obtained from the dispersion equation—eqn (5.2.39). The highest-energy state

(n=max) is a collective state since the amplitudes for the (iI) configurations are all in-phase with respect to the transition matrix element G_{iI}. The other state $|f_k\rangle$ consists mainly of the unperturbed state $\alpha_{IK}^{+}\alpha_{ik}^{+}$ because the denominator $E(I_k i_k)-E_k$ is very small. The matrix element for the lowest state $|f_1\rangle$ consists mostly of the unperturbed component with a positive amplitude $-\chi_\alpha G_{I_1 i_1}^2/\{E(I_1 i_1)-E_1\}$ (note $\chi_\alpha>0$ and $E(I_1 i_1)<E_1$) and all other terms with negative amplitudes $-\chi_\alpha G_{Ii}^2/\{E(Ii)-E_1\}$. Thus all configurations induced by the polarization interaction contribute coherently to reduce the unperturbed matrix element. The transition matrix element for the $|f_1\rangle$ is given by

$$\langle f_1 \| \mathbf{T}_\alpha \| 0\rangle \approx \langle I_1 \| \mathbf{T}_\alpha \| i_1\rangle_{\mathrm{sp}} U_{I_1} V_{i_1} \frac{1}{1+\kappa}, \tag{5.2.42}$$

where $\langle \mathbf{I}_1 \| \mathrm{T}_\alpha \| \mathrm{i}_1\rangle_{\mathrm{sp}}$ is the single-particle matrix element and $U_{I_1} V_{i_1}$ is the pairing reduction factor. The effective coupling constant g_α^{eff} in the α-mode and the relevant susceptibility κ are (see eqn 5.1.4 and Fig. 5.4)

$$g_\alpha^{\mathrm{eff}}/g_\alpha = 1/(1+\kappa), \tag{5.2.43}$$

$$\kappa = \sum_{\substack{I\neq I_1\\ i\neq i_1}} \left[\frac{\chi V_i^2 U_I^2 G_{Ii}^2}{E(Ii)-E_1} - \frac{\chi U_i^2 V_I^2 G_{Ii}^2}{-E(Ii)-E_1}\right]. \tag{5.2.44}$$

where g_α is the coupling constant for a free nucleon. In eqns (5.2.42 – 5.2.44) all first-order terms proportional to the ε_j which contribute coherently, are included. The transition to the lowest (and low-lying) state can be rewritten by using the unperturbed single quasi-particle matrix element $\langle I_1 \| \mathbf{T}_\alpha \| i_1\rangle_{\mathrm{p}} = \langle I_1 \| \mathbf{T}_\alpha \| i_1\rangle_{\mathrm{sp}} U_{I_1} V_{i_1}$ (see Section 5.3) and the effective coupling constant as follows:

$$\langle f \| \mathbf{T}_\alpha \| i\rangle \approx \frac{g_\alpha^{\mathrm{eff}}}{g_\alpha} \langle I_1 \| T_\alpha \| i_1\rangle_{\mathrm{p}} \tag{5.2.45a}$$

$$\frac{g_\alpha^{\mathrm{eff}}}{g_\alpha} = \frac{1}{1+\kappa} = \frac{1}{1+\kappa^- + \kappa^+} \tag{5.2.45b}$$

$$\kappa^\pm = \sum_\lambda (2\lambda+1)\kappa_\lambda^\pm \tag{5.2.45c}$$

$$\kappa_\lambda^\pm = \sum_\nu (2\nu+1)\kappa_{\lambda\nu}^\pm \tag{5.2.45d}$$

$$\kappa_{Ii}^- = \frac{h_{Ii}}{E(Ii)-E_1} V_i^2 U_I^2, \quad \kappa_{Ii}^+ = \frac{h_{Ii}}{E(Ii)+E_1} V_I^2 U_i^2, \tag{5.2.45e}$$

$$h_{Ii} = \chi_\alpha G_{Ii}^2/\{(2i+1)(2I+1)\}. \tag{5.2.45f}$$

Here κ^+ and κ^- are the susceptibilities due to the forward and backward correlations, respectively. The $\kappa_{\lambda\nu}$ and κ_λ are the susceptibilities for the $(\lambda\nu)$ pair excitation and the λ shell orbit. The susceptibility κ_{Ii}^- for the $i\to I$ polarization is proportional to the interaction matrix element h_{Ii} and inversely proportional to the energy required to excite (polarize) the (Ii) pair. The effective coupling constant and the susceptibility for odd A nuclei are simply obtained if one replaces E_1 in the eqns (5.2.44) and (5.2.45) by $E_1 - E(i_1)$, where $E(i_1)$ is the unperturbed energy of the initial state $|i_1\rangle$.

The present expressions for g_α^{eff} and κ are useful for transitions between low-lying quasi-particle states from which the GR in the α-mode is well separated in energy. Actually the GR in the spin–isospin mode is pushed up to a high excitation region because of the large repulsive spin–isospin interaction. Thus the concept of the g^{eff} can be applied in general for the relevant γ and β decays. The present expressions for $g_{\text{a}}^{\text{eff}}$ and κ are essentially consistent with eqn (5.2.30) derived from the commutator method in the sense that both take into account all the coherent contributions coming effectively from the spin–isospin polarization interaction.

Since the core polarization effect on the low-lying transitions is an uniform effect, the susceptibility and the GR energy may be evaluated by using average values $\bar{G}^2$ and $\bar{E}$ for the G_{iI}^2 and the E_{iI}, respectively. The dispersion equation eqn (5.2.39) is rewritten in the form

$$-\frac{x_a n_f \bar{G}_f^2}{\bar{E}_f - E^{\text{GD}}} - \frac{x_a n_b \bar{G}_b^2}{\bar{E}_b + E^{\text{GD}}} = 1, \tag{5.2.46}$$

$$k = \frac{x_a n_f \bar{G}_f^2}{\bar{E}_f - E_1} + \frac{x_a n_b \bar{G}_b^2}{\bar{E}_b + E_1}, \tag{5.2.47}$$

where $\bar{E}_f$ and $\bar{E}_b$ are average particle–hole energies for the forward and backward correlations, respectively. The average matrix elements for the forward and backward correlations are defined by $n_f \bar{G}_f^2 = \Sigma V_i^2 U_I^2 G_{iI}^2$ and $n_b \bar{G}_b^2 = \Sigma U_i^2 V_I^2 G_{Ii}^2$, respectively.

The GR energy E^{GD} and the susceptibility for γ transitions are written in terms of $\bar{E} = \bar{E}_f = \bar{E}_b$, and $n\bar{G}^2 = n_f \bar{G}_f^2 = n_b \bar{G}_b^2$ (Ejiri, 1973b) as

$$\begin{aligned} E^{\text{GD}} &\approx (2\chi_\alpha n\bar{G}^2 \bar{E} + \bar{E}^2)^{1/2} \\ &\approx \{1 + \kappa - (E_\gamma/\bar{E})^2 \kappa\}^{1/2} \cdot \bar{E} \end{aligned} \tag{5.2.48}$$

$$\kappa \approx \chi_\alpha n\bar{G}^2 2\bar{E}(\bar{E}^2 - E_\gamma^2)^{-1}. \tag{5.2.49}$$

Thus the GR energy (centre of the distribution of the transition strength) can be obtained from the measured κ or g^{eff}. The E1 excitation is mainly a one

$\hbar\omega$-jump with $\bar{E} \approx \hbar\omega = 41\, A^{-1/3} \gg E_\gamma$. Then one gets

$$\kappa \approx 2\chi_\alpha n\bar{G}^2/\hbar\omega, \tag{5.2.50}$$

$$E^{\mathrm{GD}} \approx (2\chi_\alpha n\bar{G}^2\hbar\omega + (\hbar\omega)^2)^{1/2}, \tag{5.2.51}$$

$$\approx \hbar\omega\sqrt{(1+\kappa)}. \tag{5.2.52}$$

The quantity $\chi n\bar{G}^2$ can be obtained by measuring either the κ or E^{GD}. The observed value $E^{GD}(\mathrm{E1}) \approx 2\hbar\omega$ for the E1 giant resonance energy gives $\chi n\bar{G}^2 = \frac{3}{2}\hbar\omega$ and $\kappa = 3$. If one assumes that the exchange interaction is effectively given by the Majorana type interaction $\mathbf{H}_M = \chi_M(1+\boldsymbol{\tau}\boldsymbol{\tau})(1+\boldsymbol{\sigma}\boldsymbol{\sigma})(r\mathbf{Y}_1 r\mathbf{Y}_1)$, the susceptibilities and the GR energies for the isovector ($\mathbf{t}\boldsymbol{\sigma}r\mathbf{Y}_1$) and isoscalar ($\boldsymbol{\sigma}r\mathbf{Y}_1$) modes are also given by $\kappa \approx 3$ and $E^{\mathrm{GD}} = 2\hbar\omega$ as in the E1 case ($\mathbf{t}r\mathbf{Y}_1$). (See Sections 5.3.3 and 5.3.4 for detailed discussions of β and γ decays with $\Delta\pi = \text{yes}$.)

5.3 The effective coupling constants for electromagnetic transitions in spin–isospin modes

5.3.1 *Effective coupling (renormalization) constants*

The spin–isospin EM (γ) and analogous β transition moments for a single quasi-particle (valence nucleon) are greatly modified in nuclei because of the spin–isospin core polarization effect (the effect of the spin–isospin giant resonance). The effect is represented by the renormalization factor $g_\alpha^{\mathrm{eff}}/g_\alpha$—see Section 5.2 and eqns (5.2.43) and (5.2.45). Since this is essentially the effect of the core on the single-particle transition, the effect manifests itself clearly in fairly good single quasi-particle transitions. Thus we discuss in this section favoured (normal) single quasi-particle transitions in simple shell-model nuclei with N (and/or Z) $\approx$ magic number. The effective coupling constants for individual transitions $\mathbf{T}_{TLSJ} = g_{TLSJ}\mathbf{t}^T[\mathrm{i}^L r^L \mathbf{Y}_L \times \boldsymbol{\sigma}^S]_J$ with $L = 0, 1, 2, \ldots$ are analysed, where $T = 1$ is for β^+, β^- and isovector γ transitions and $T = 0$ is for isoscalar γ transitions. The value for g_α^{eff}, where $\alpha = TLSJ$, is obtained experimentally as

$$g_\alpha^{\mathrm{eff}}/g_\alpha = \langle \mathbf{T}_\alpha \rangle_{\mathrm{exp}} / {}_0\langle \mathbf{T}_\alpha \rangle_0, \tag{5.3.1}$$

where ${}_0\langle \mathbf{T}_\alpha \rangle_0$ is the unperturbed matrix element in the model space $\mathbf{H}_0$. The $\mathbf{H}_0$ is a simple model Hamiltonian without the polarization interaction $\mathbf{H}_I = \chi_\alpha \mathbf{T}_\alpha \cdot \mathbf{T}_\alpha$. The idea behind eqn (5.3.1) is that the actual transition matrix element with coupling constant g_α in a free space is given by the simple (single quasi-particle) model matrix element by introducing the effective coupling constant g_α^{eff}, which stands for the core polarization and other possible effects.

We define for simplicity the model space as

$$\mathbf{H}_0 = \mathbf{H}_{\rm sp} + \mathbf{H}_{\rm p} + \mathbf{H}_{\rm QQ} + \mathbf{H}_{\rm OO}, \tag{5.3.2}$$

where $\mathbf{H}_{\rm sp}$, $\mathbf{H}_{\rm p}$, $\mathbf{H}_{\rm QQ}$ and $\mathbf{H}_{\rm OO}$ are Hamiltonians for the single-particle shell model, the pairing interaction, the quadrupole interaction, and the octuple interaction, respectively. Actually the $\mathbf{H}_{\rm QQ}$ and $\mathbf{H}_{\rm OO}$, which are important for E2 and E3 transitions, have little effect on the spin–isospin transitions in nearly-magic and semi-magic nuclei. The interaction which is most important for the T_α mode is the polarization interaction $\mathbf{H}_{\rm I}$ relevant to the α-mode since it introduces coherence in the α-mode excitation. Therefore the matrix element in the model space $\mathbf{H}_0$ is approximately given by the single quasi-particle matrix element $\langle \mathbf{T}_\alpha \rangle_{\rm sqp}$ as

$${}_0\langle \mathbf{T}_\alpha \rangle_0 \approx \langle \mathbf{T}_\alpha \rangle_{\rm sqp} = \mathrm{C}_0(j_{\rm i})\mathrm{C}_0(j_{\rm f})P\langle \mathbf{T}_\alpha \rangle_{\rm sp}, \tag{5.3.3}$$

where $\mathrm{C}_0(j_{\rm i})$ and $\mathrm{C}_0(j_{\rm f})$ are the single quasi-particle amplitudes for the initial and final states, respectively, $\langle \mathbf{T}_\alpha \rangle_{\rm sp}$ is the single-particle matrix element, and P is the pairing reduction factor.

The pairing reduction factor for the γ transition $j_{\rm i} \to j_{\rm f}$ between odd (unpaired) quasi-particles in $S=0$ (electric) and $S=1$ (magnetic) modes is given by $P = U_{\rm i}U_{\rm f} - (-)^s V_{\rm i} V_{\rm f}$ and that for the β transition by $P = U_{\rm i}U_{\rm f}$—see eqn (2.3.6) and Kisslinger and Sorensen, 1963), where U_k^2 and V_k^2 are the vacancy and occupation probabilities, respectively.

The single-particle matrix elements for the electric and magnetic J-pole stretched transitions with $J_{\rm i} - J_{\rm f} = \pm J$ are simply expressed by (Bohr and Mottelson, 1969)

$$\langle j_{\rm f} \| T(EJ) \| j_{\rm i} \rangle_{\rm sp} = \mathrm{e}(-)^{j_{\rm i} - j_{\rm f} + J}\ \mathrm{i}^{l_{\rm i} - l_{\rm f} + J} \sqrt{\frac{(2J+1)(2j_{\rm i}+1)}{4\pi}} \times \langle j_{\rm f} | r^J | j_{\rm i} \rangle \langle j_{\rm i} \tfrac{1}{2} J 0 | j_{\rm f} \tfrac{1}{2} \rangle, \tag{5.3.4}$$

$$\langle j_{\rm f} \| T(MJ) \| j_{\rm i} \rangle_{\rm sp} = -\frac{\mathrm{e}\hbar}{2Mc}\left(\mu_s - \frac{1}{J+1} g_l\right) \cdot 2 \cdot J \sqrt{\left(\frac{2J+1}{4\pi}\right)} \sqrt{(2j_{\rm i}+1)} \times \langle j_{\rm f} | r^{J-1} | j_{\rm i} \rangle \langle j_{\rm i} \tfrac{1}{2} J 0 | j_{\rm f} \tfrac{1}{2} \rangle. \tag{5.3.5}$$

The unique first-forbidden matrix element is

$$\langle j_{\rm f} \| t_{\pm} ir[\mathbf{Y}_1 \times \boldsymbol{\sigma}]_2 \| j_{\rm i} \rangle_{\rm sp} = \sqrt{\left(\frac{3}{16\pi}\right)} \cdot \langle r \rangle \cdot \sqrt{\left\{\frac{4(2j_{\rm i}+1)(2j_{\rm f}+1)(j_{\rm i}+j_{\rm f}+1)}{(j_{\rm i}+j_{\rm f})(j_{\rm i}+j_{\rm f}+2)}\right\}}. \tag{5.3.6}$$

The single-particle matrix elements $\langle j_{\rm f} \| \mathrm{i}^L [\mathbf{Y}_L \times \boldsymbol{\sigma}^S]_J \| j_{\rm i} \rangle_{\rm sp}$ are tabulated by Lipnik and Sunier (1964), Lipnik (1966), and Delabaye and Lipnik (1966). The radial single-particle matrix element $\langle r \rangle$ is expressed by using a harmonic oscillator potential as $\langle nl|r|n+1\ l-1\rangle = -\sqrt{(\hbar/M\omega_0)}\sqrt{n}$, and $\langle nl|r|nl+1\rangle = \sqrt{(\hbar/M\omega_0)}\ \sqrt{(n+l+1/2)}$, where $\hbar\omega_0 \simeq 41A^{-1/3}$ MeV and M is the nucleon mass. Values for $\langle r^2\rangle$ and $\langle r^3\rangle$ are given by Yoshida (1961, 1962).

It should be noted here that the values $|C_0(j)V_j|^2$ and $|C_0(j)U_j|^2$ are obtained experimentally from the spectroscopic factors in stripping and pick-up reactions. The U and V factors are also calculated from the unperturbed Hamiltonian $\mathbf{H}_0' = \mathbf{H}_{\text{shell}} + \mathbf{H}_{\text{pair}}$, and the single quasi-particle amplitude is obtained by adding proper residual interactions such as quadrupole ($\mathbf{H}_{QQ}$) and octuple ($\mathbf{H}_{OO}$) interactions. In the present case C_0 is close to unity and thus the errors in the calculation for C_0 are small.

In most cases the renormalization factors—eqn (5.3.1)—are deduced from the experimental values for the transition probability $|\langle T_\alpha\rangle|^2_{\text{exp}}$. Thus one gets only the absolute value $|g_\alpha^{\text{eff}}/g_\alpha|$. The theoretical consideration of the core polarization effect gives $0 < g_\alpha^{\text{eff}}/g_\alpha < 1$ for the transition between low-lying states. Therefore in the following discussions the quantities $g_\alpha^{\text{eff}}/g_\alpha$ are given assuming they have positive values.

5.3.2 *Electric dipole transitions*

Some single-particle E1 matrix elements in medium and heavy nuclei have been obtained for the first time from (p, γ) reactions through isobaric analogue resonances (IAR) by Ejiri *et al.* (1968/1969) for ^{141}Pr ($N = 82$). The γ-ray spectrum is shown in Fig. 5.5. The radiative capture process is a very rare process, but the IAR γ peak is seen clearly in the highest energy region. The IAR E1 matrix element has been extracted by taking into account the interference of the IAR γ process with the E1 giant resonance (Ejiri and Bondorf, 1968b). Details of the (p, γ) reaction mechanism are given in Section 4.5.4. Snover *et al.* (1971) measured IAR γ rays in lead region ($N \approx 126$), and their excitation function for the ^{208}Pb (p,γ) reaction is shown in Fig. 5.6. Richard *et al.* (1969), Paul *et al.* (1972) and Ejiri *et al.* (1969) studied IAR E1 γ rays in nuclei with $N \approx 50$, and Hashinoff *et al.* (1972) measured IAR E1 γ rays in ^{142}Nd. The inverse (γ, p) reactions were investigated extensively by means of the (e, e′p) process through IAR (Shoda *et al.*, 1971). Both favoured (normal) and unfavoured transitions are studied by these methods. The transition scheme for ^{93}Tc is shown as an example in Fig. 5.7. We discuss in the present work mostly the single quasi-particle E1 matrix elements for favoured transitions measured through IAR, since they provide fairly precise values for single quasi-particle transitions.

The first-forbidden β transition has a vector component $\langle t_\pm ir\mathbf{Y}_1\rangle$, which is analogous to the E1 transition from the IAS (see Fig. 5.7 and Section 4.5). Discoveries of well-defined isobaric analogue states (IAS) (Anderson and Wong, 1961) and sharp resonances of IAR (Fox *et al.*, 1964) have provided an experimental method to get the $\langle t_- ir\mathbf{Y}_1\rangle$ operator by measuring the

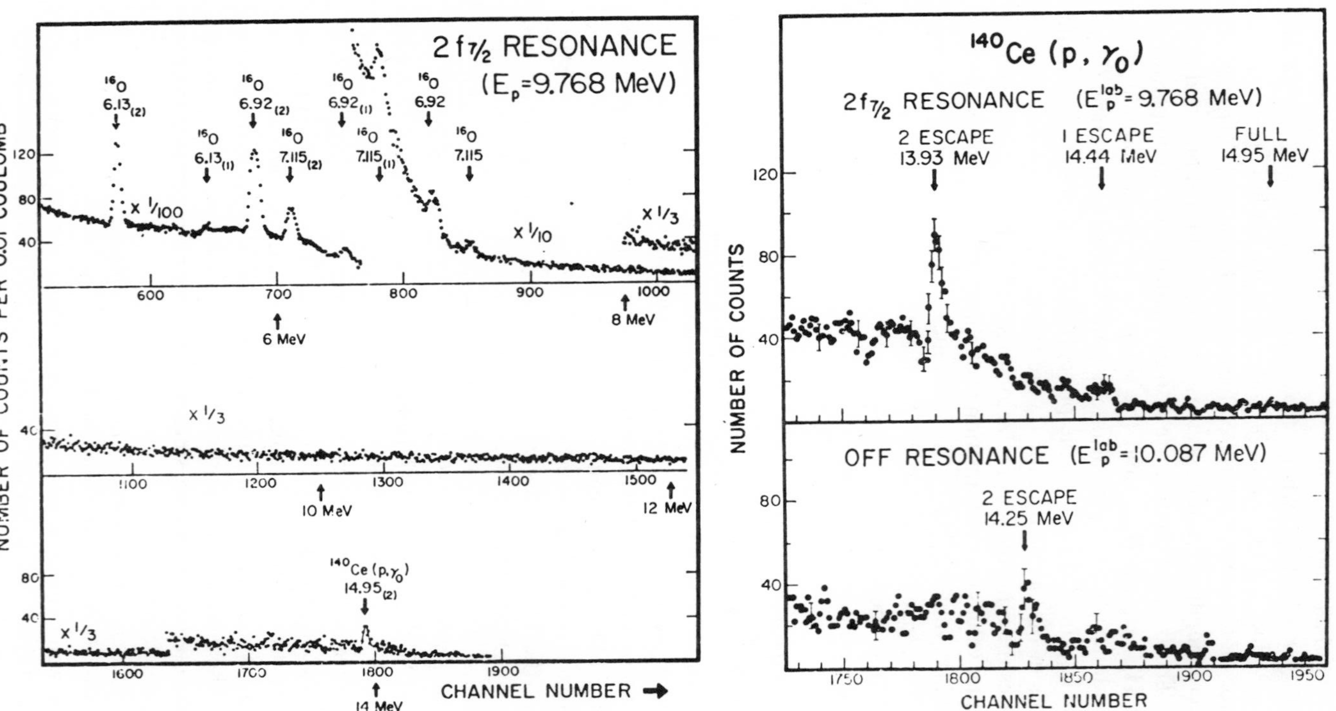

Fig. 5.5. Left hand side: Energy spectrum of γ rays from the ^{140}Ce(p, γ)^{141}Pr reaction at $E_p = 9.768$ MeV 2f$_{7/2}$ IA resonance in ^{141}Pr at $\theta_{lab} = 90°$. The double- and single-escape peaks are labelled $E_{(2)}$ and $E_{(1)}$, respectively. Right-hand side: Expanded energy spectra of the ground state transition (γ_0) at $E_p = 9.768$ MeV (2f$_{7/2}$ resonance) and at $E_p = 10.087$ MeV (off-resonance). The overall energy resolution is about 40 keV, which is due to both the Ge detector and the target thickness. (Ejiri *et al.*, 1968/1969.)

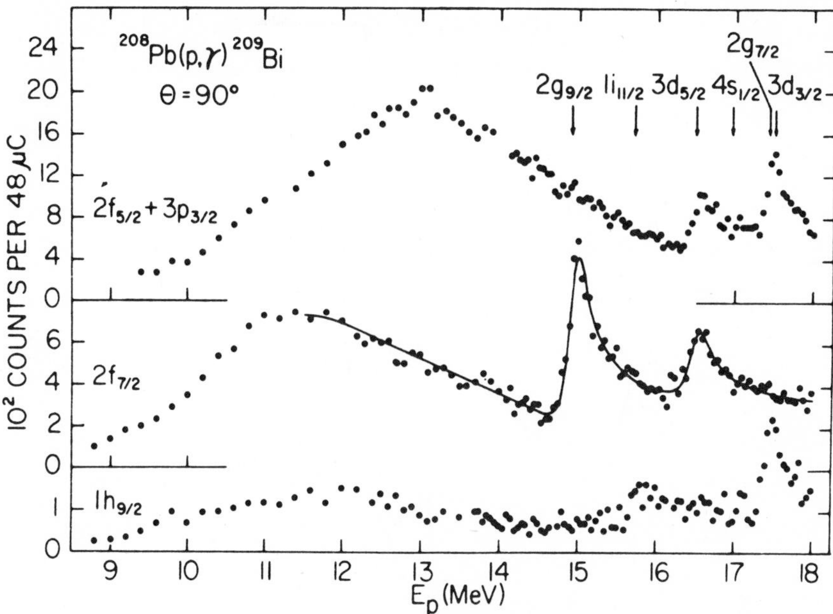

Fig. 5.6. The ^{208}Pb(p, γ)^{209}Bi excitation curves at $\theta = 90°$ for transitions to the $1h_{9/2}$, $2f_{7/2}$ and $2f_{5/2} + 3p_{3/2}$ final states. The absolute normalization for the ordinate is given by 0.94 ± 0.19 μb/sr per 100 counts/48 μC. The energies and configurations of the known single-particle analogue resonances are indicated in the upper right corner. The solid curve represents a least-squares fit to the data. (Snover *et al.*, 1971.)

analogous IAS E1 transition as suggested by Fujita (1963/1967). Many good data for the $\langle t_{\pm} r \mathbf{Y}_1 \rangle$ have been obtained by this method.

The E1 γ-transition operator $T_\gamma(\mathrm{E1}) = ig_\gamma(\mathrm{E1}) r\mathbf{Y}_1$ for the IAS and the corresponding β-transition operator $T_\beta(\mathrm{V1}) = g_v i t_{\pm} r \mathbf{Y}_1$ are related by—Fujita, 1963/1967, Ejiri *et al.*, 1968/1969; see also eqn (4.5.26)—

$$\langle \mathrm{f}|\mathbf{T}_\beta(\mathrm{V1})|\mathrm{i}\rangle \approx (g_v/\mathrm{e})\sqrt{2T_i})\langle \mathrm{f}|\mathbf{T}_\gamma(\mathrm{E1})|\mathrm{IAR}\rangle. \tag{5.3.7}$$

Here we assume that $|\mathrm{IAR}\rangle$ is a well-defined resonance given by $|\mathrm{IAR}\rangle \approx |\mathrm{IAS}\rangle \equiv \mathbf{T}_-|\mathrm{i}\rangle/\sqrt{(2T_\mathrm{i})}$, where T_i is the isospin of the initial state.

The renormalization factors are related by

$$\frac{g^{\mathrm{eff}}_{\beta v}(\mathrm{V1})}{g_v} = \frac{g^{\mathrm{eff}}_\gamma(\mathrm{E1, IA})}{\mathrm{e}/\sqrt{(2T_\mathrm{i})}}, \tag{5.3.8}$$

where the denominators correspond to the coupling constants for the free nucleon transitions.

The observed effective coupling constants for simple neutron–magic nuclei are shown in Table 5.3.1, and they are plotted in Fig. 5.8 as a function of k (the

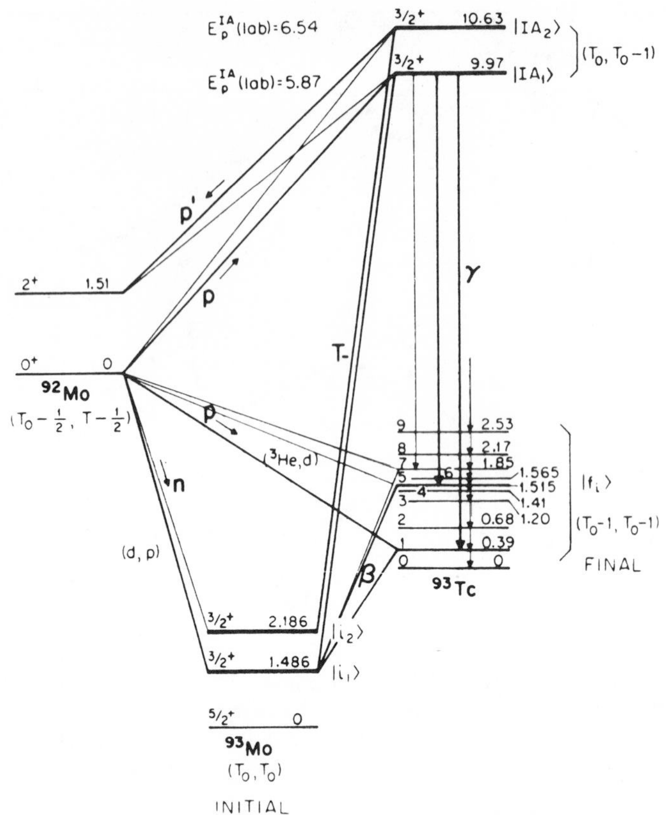

Fig. 5.7. The levels and transition schemes relating to the ^{92}Mo(p, γ)^{93}Tc reaction proceeding through the IAS. The level energies are given in units of MeV. The heavy lines show strong transitions. The beta-transitions corresponding to the analogue E1 γ decays from IAS are actually not observed. (Ejiri *et al.*, 1969.)

occupation probability for protons in the unfilled major shell). The observed values are indeed reduced uniformly by factors of 0.25 to 0.4. The observed values for the nearly doubly-magic nucleus (^{209}Bi, $k \approx 1$) are $g^{\text{eff}}/g \approx 0.4$, while those for other nuclei ^{141}Pr, ^{142}Nd, ^{89}Y with $k < 1$ are $g^{\text{eff}}/g \approx 0.3$. The k-dependence can be explained by the blocking effect of the backward correlation —κ^+ in eqns (5.2.30) and (5.2.45)—in the doubly-magic nuclei with $k \approx 1$. The observed renormalization factors are in accord with calculations based on the schematic model as shown in Fig. 5.8. (See Section 5.3.4c for a detailed description of the schematic model.) Numerical calculations in terms of the RPA are made using eqn (5.2.45) and the results are shown in Table 5.3.1. The interaction strength is chosen so as to reproduce the observed E1 giant

Table 5.3.1 Effective coupling constants for favoured (normal) E1 transitions from IAR

N	Z	Transition	$(g^{\mathrm{eff}}/g)_{\mathrm{exp}}$	κ^-	κ^+	$(g^{\mathrm{eff}}/g)_{\mathrm{cal}}$	Ref.
83	58	$2f_{7/2} \rightarrow 2d_{5/2}$	0.26 ± 0.05	1.84	1.33	0.24	†
127	82	$2g_{9/2} \rightarrow 2f_{7/2}$	~ 0.49	1.56	0.06	0.38	‡
127	82	$3d_{5/2} \rightarrow 2f_{7/2}$	~ 0.41	1.39	0.08	0.40	‡
50	39	$3s_{1/2} \rightarrow 2p_{1/2}$	~ 0.37	1.26	1.23	0.29	§

† Ejiri *et al.* (1968/1969); ‡ Snover *et al.* (1971); § Paul (1972).

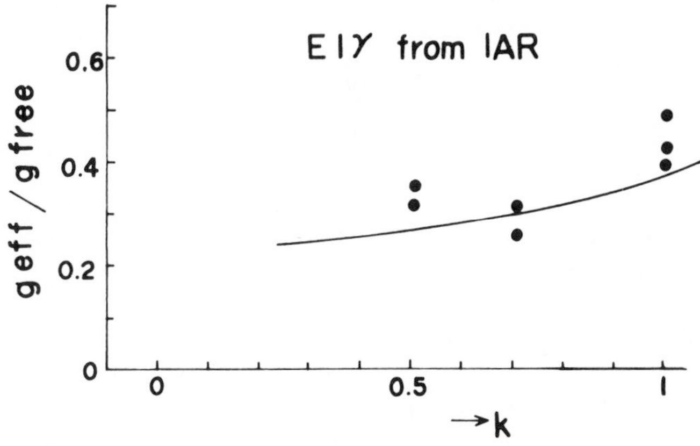

Fig. 5.8. The effective coupling constants for E1 transitions from IAR in neutron–magic nuclei. The order of magnitude is estimated by using a simplified schematic model—eqn (5.2.47)—as shown by the solid line. The k is the occupation probability of protons in the unfilled major shell (see Section 5.2.2). (Ejiri and Fujita, 1978.)

resonance energies. The agreement between the calculated and the observed effective coupling constants g^{eff} indicates that the uniform reduction (renormalization) of the isovector E1 mode and the observed E1 GR energies (E^{GD}) may be understood as mostly due to the same (effective) isospin interaction in the framework of the RPA. This, however, does not exclude the existence of any other effects which contribute to both the g^{eff} and E^{GD}.

So far we have discussed the favoured transitions with large overlap integrals for both radial and angular wave functions, but we should also take into account the effects of the other interactions for unfavoured transitions. Spin–flip E1 transitions are unfavoured because the E1 operator has no spin term, and in this case a spin-dependent interaction becomes important for spin–flip transitions. The spin–flip E1 transition is effectively written by introducing a spin–flip coupling term of the $\boldsymbol{\sigma} \times r\mathbf{Y}_1$ field with the $r\mathbf{Y}_1$ field (Fujita *et al.*, 1970; Ejiri, 1973b). It is

$$g\langle \mathrm{f} \| \mathbf{t}ir\mathbf{Y}_1 \| \mathrm{i} \rangle = g_r^{\mathrm{eff}}\,{}_0\langle \mathrm{f} \| \mathbf{t}ir\mathbf{Y}_1 \| \mathrm{i} \rangle_0 P + \Delta g_r^{\mathrm{eff}}\,{}_0\langle \mathrm{f} \| \mathbf{t}[\boldsymbol{\sigma} \times ir\mathbf{Y}_1]_1 \| \mathrm{i} \rangle_0 P, \tag{5.3.9}$$

$$g^{\mathrm{eff}} \equiv g\langle \mathrm{f} \| \mathbf{t}ir\mathbf{Y}_1 \| \mathrm{i} \rangle / ({}_0\langle \mathrm{f} \| \mathbf{t}ir\mathbf{Y}_1 \| \mathrm{i} \rangle_0 P) = g_r^{\mathrm{eff}} + \Delta g_r^{\mathrm{eff}}\,{}_0\langle \mathrm{f} \| \mathbf{t}[\boldsymbol{\sigma} \times ir\mathbf{Y}_1]_1 \| \mathrm{i} \rangle_0 / {}_0\langle \mathrm{f} \| \mathbf{t}ir\mathbf{Y}_1 \| \mathrm{i} \rangle_0. \tag{5.3.10}$$

Here $\Delta g_r^{\mathrm{eff}}$ stands for the effective term arising from the spin-dependent interaction. For a typical spin–flip transition $(n, l, +\frac{1}{2}) \to (n-l, l+1, -\frac{1}{2})$, where $(n, l, \pm\frac{1}{2})$ stands for a single-particle state with the number of radial nodes n, orbital angular momentum l and spin $J = l \pm \frac{1}{2}$, eqn (5.3.10) becomes

$$g_{\mathrm{s.f.}}^{\mathrm{eff}} = g_r^{\mathrm{eff}} + \Delta g_r^{\mathrm{eff}} \sqrt{\{2(l+1)\}}. \tag{5.3.11}$$

For a non-spin–flip transition $(n, l, +\frac{1}{2}) \to (n, l-1, +\frac{1}{2})$ it is

$$g_{\mathrm{n.s.f.}}^{\mathrm{eff}} = g_r^{\mathrm{eff}} + \Delta g_r^{\mathrm{eff}} / \sqrt{2}. \tag{5.3.12}$$

The value of $g_{\mathrm{s.f.}}^{\mathrm{eff}}$ is quite sensitive to the value $\Delta g_r^{\mathrm{eff}}$, while the value of $g_{\mathrm{n.s.f.}}^{\mathrm{eff}}$ is not. If experimental values for both the spin–flip ($|\mathrm{i}\uparrow\rangle \to |\mathrm{f}\downarrow\rangle$) and non-spin–flip ($|\mathrm{i}\uparrow\rangle \to |\mathrm{f}\uparrow\rangle$) transition matrix elements are available, then both g_r^{eff} and $\Delta g_r^{\mathrm{eff}}$ can be obtained. The (p, γ) reactions in heavy nuclei do not show any resonances for the spin–flip transitions, indicating upper limits of 0.3 to 0.2 on $\Delta g_r^{\mathrm{eff}}$. The quantitative arguments relating to $\Delta g_r^{\mathrm{eff}}$ remain open.

E1 γ transitions from non-IAS in medium–heavy nuclei are mostly j-forbidden. Some E1 transitions from collective octupole states were studied. A detailed analysis of E1 transitions from octuple–phonon coupled-states was made (Hertel, 1969) and effective E1 charges were deduced (Hamamoto, 1969/1970). Many single-particle E1 transitions are known in light nuclei as described in Section 2.3. As an example the single particle $2\mathrm{p}_{3/2} \to 2\mathrm{s}_{1/2}$ E1 γ transition in ^{29}P has $\langle \mathbf{T}_\gamma \rangle_{\mathrm{exp}} = 0.29 \pm 0.07$ efm (Ejiri *et al.*, 1964), which gives $g^{\mathrm{eff}}/(e/2) = 0.36 \pm 0.08$. This is consistent with the values $g^{\mathrm{eff}}/g = 0.25$ to 0.4 in the medium–heavy nuclei.

It is interesting to note that the phase of g^{eff} can be obtained from the interference between the E1 γ amplitudes for an IAR and the underlying E1 GR. The positive value of $0 < g^{\mathrm{eff}} < 1$ for ^{141}Pr is found to be consistent with the observed interference pattern (Ejiri and Bondorf, 1968b).

5.3.3 *Magnetic dipole transitions and magnetic moments*

Magnetic dipole (M1) transitions, M1 moments and analogous Gamow–Teller (GT) β transitions have been studied over a wide range of the mass number. The major component of their operators is the $\sigma\tau$ term. In-beam γ-ray spectroscopy has been used to determine M1 moments in many nuclei.

The observed M1 matrix elements are known to be reduced in comparison with shell model estimates, and have been analysed theoretically with various methods (Noya *et al.*, 1958; and see Section 5.2) The details of the experimental and theoretical aspects of M1 moments are described by Morinaga and Yamazaki (1976). In this section M1 matrix elements and the analogous GT ones are briefly discussed in view of the spin–isospin polarizations (susceptibilities) and the spin–isospin giant resonance.

5.3.3a *M1 moments and M1 transitions*

In the case of a simple shell-model nucleus the g-factor for a single particle with $j=l\pm\frac{1}{2}$ is written as

$$g=\pm\frac{1}{4l+2}(g_{s-}^{\text{eff}}\tau_3+g_{s+}^{\text{eff}})+\frac{2l+1\mp 1}{4l+2}(g_{l-}^{\text{eff}}\tau_3+g_{l+}^{\text{eff}}), \tag{5.3.13}$$

where $g_{s\pm}^{\text{eff}}=\{g_s(\text{n})\pm g_s(\text{p})\}^{\text{eff}}$ and $g_{l\pm}^{\text{eff}}=\{g_l(\text{n})\pm g_l(\text{p})\}^{\text{eff}}$ are the effective spin and orbital g-factors, respectively, and the subscripts + and − indicate the isoscalar and isovector components, respectively. As an typical example of M1 moments we consider the M1 moment of the $1\text{h}_{11/2}$ state in the semi-magic nuclei.

The g factor for the $1\text{h}_{11/2}$ quasi-proton state in ^{141}Pr ($N=82$) has been obtained as $g=1.30\pm0.08$ by measuring the perturbed angular distribution of the M2 γ rays from the $1\text{h}_{11/2}$ state populated by the (α, 2n) reaction (Ejiri *et al.*, 1974). The experimental method and the time spectrum are shown in Figs 3.70 and 3.71 (Chapter 3). The g-factors for the $1\text{h}_{11/2}$ quasi-neutron states in the Sn isotopes ($Z=50$) have been measured as $g\approx-0.23$ by using a similar method. A superconducting magnet was used since the small g-factors for neutron states required a high magnetic field (Brenn *et al.*, 1974; see Figs 3.73 and 3.74 for the experimental arrangement and the result). In eqn (5.3.13) g_l factors have been shown to be shifted slightly from the free nucleon values of $g_{l\pm}=\pm1$ by measuring the magnetic moments of high spin isomers (Yamazaki *et al.*, 1970; Nagamiya *et al.*, 1970). Using their $g_{l\pm}^{\text{eff}}$ values, the observed magnetic moments for the proton and neutron states lead to $g_{s-}^{\text{eff}}/g_{s-}\approx0.5$ (Ejiri *et al.*, 1974). Thus the M1 spin–isospin susceptibility for the M1 moment is given by

$$\frac{g_{s\pm}^{\text{eff}}}{g_{s\pm}}=\frac{1}{1+\kappa^-+\kappa^+}=\frac{1}{1+2\kappa^\pm}, \tag{5.3.14}$$

where $\kappa^-=\kappa^+=\kappa/2$ for the diagonal moment. The value $g_{s-}^{\text{eff}}/g_{s-}\approx0.5$

yields $\kappa = 2\kappa^{\pm}(\text{M1}) = 1$. This is much smaller than the values $\kappa(\text{M2}) = \kappa_{-}(\text{M2}) + \kappa_{+}(\text{M2}) \approx 3$ for the M2 moment of the same $1h_{11/2}$ state as described in Section 5.3.4.

The susceptibility κ for the M1 moment is evaluated from eqn (5.2.45) as (Bohr and Mottelson, 1969/1975),

$$\kappa = 2\kappa^{\pm}, \kappa^{\pm} = \kappa_p + \kappa_n = \Sigma N_j \alpha_j, \tag{5.3.15a}$$

$$\alpha_j = \frac{\chi_{\tau\sigma}}{\Delta E_j} \frac{|\langle j' = l - \frac{1}{2} | \boldsymbol{\tau\sigma} | j = l + \frac{1}{2} \rangle|^2}{3(2j+1)}. \tag{5.3.15b}$$

In semi-closed shell nuclei where both the $j_p = l_p + \frac{1}{2}$ proton shell and the $j_n = l_n + \frac{1}{2}$ neutron shell are closed, one gets $N_p = 2j_p + 1$ and $N_n = 2j_n + 1$. ^{141}Pr has the semi-closed core of ^{140}Ce with closed $(1g_{9/2})_p$ and $(1h_{11/2})_n$ shells. The spin–flip excitation energy is $\Delta E_j \approx 12(2l+1)\, A^{-2/3}$ MeV (Bohr and Mottelson, 1969/1975). Assuming that the Majorana-type interaction is dominant, the spin–isospin interaction strength $\chi_{\tau\sigma}$ is approximately given by the isospin one χ_τ, which is derived from the symmetry energy $V_\tau = 100\, \mathbf{tT}/A$ (MeV). Then one gets

$$\kappa = 2\kappa^{\pm} \approx \tfrac{16}{3} A^{-1/3}. \tag{5.3.16}$$

It is interesting to note that eqn (5.3.16) leads to $\kappa \approx 1$ and $g^{\text{eff}}/g \approx 0.5$ for medium–heavy nuclei ($A = 60$–200).

The arguments given above are indeed quite simplified, and many important effects associated with M1 diagonal and transition moments have to be considered. It is the M1 transitions and M1 moments that have been most extensively studied in the past two decades. They are analysed in terms of effects such as the spin–isospin core polarizations (see Section 5.2), tensor interactions (Shimizu *et al.*, 1974), M1 giant resonances (see Section 5.3.3b), meson exchange currents (Miyazawa, 1951; Chemtob, 1969) and even the Δ–isobar admixture (Bohr and Mottelson, 1981). Various terms contributing to the renormalization of the M1 operator are illustrated in Fig. 5.9. Taking

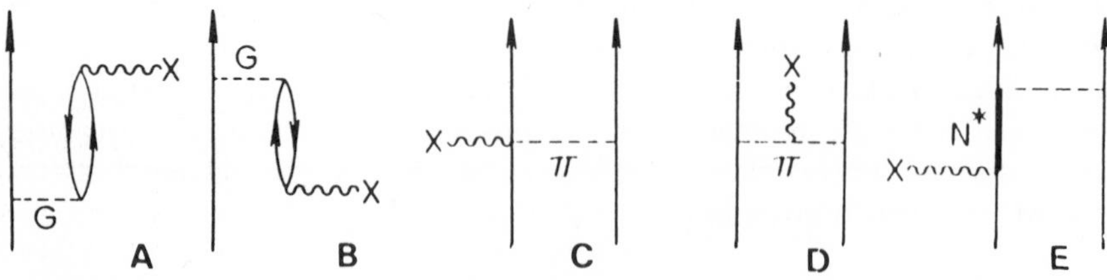

Fig. 5.9. The renormalization of the M1 operator. A and B: First-order corrections to the M1 transition process. C and D: Pion-exchange current processes. E: Nucleonic interaction current processes in which one nucleon is excited to the N* isobar. (Arima and Huang-Lin, 1972.)

into account these $\sigma\tau$ dependent effects the M1 operator is effectively described by (Bohr and Mottelson, 1969; Arima and Huang-Ling, 1972)

$$\mu^{\text{eff}} = (g_s + \delta g_s)\mathbf{s} + (g_l + \delta g_l)\mathbf{l} + \tfrac{1}{2}g_p[\mathbf{Y}^2 \times \mathbf{s}]_1, \qquad (5.3.17)$$

where $\mathbf{s} = \boldsymbol{\sigma}/2$ and δg_s, δg_l and g_p are the correction factors due to core-polarization, meson exchange currents, Δ-isobar effects and so on. The details of M1 transitions and diagonal moments, together with those of GT transitions, have been discussed in conference proceedings (Petrovich *et al.*, 1982).

M1 transitions from IAS to AIAS (anti-isobaric analogue state) are closely related to the GT β transitions. The GT matrix element is analogous to the spin–isospin matrix element of the M1 transition from the IAS. Thus, one may be known by measuring the other, as for the E1 transition from the IAS (see Section 5.3.2). The M1 transition from the IAS in 2s–1d shell nuclei has been discussed by Endt (1969) and that in medium nuclei by Sakai *et al.* (1970). M1 transitions from the IAR in ^{56}Fe and the corresponding GT transitions are shown in Fig. 5.10. The M1 operator contains the orbital term as well as the spin term, while the GT one contains only the spin term. Thus detailed comparison between them may give the orbital term in medium–heavy nuclei.

M1 transitions between single-particle states of $|j_i l_i\rangle$ and $|j_f l_f\rangle$ with $\Delta l = |l_i - l_f| = 2$ are l-forbidden because the M1 transition operators $\boldsymbol{\sigma}$ and $\mathbf{l}$ allow only transitions with $\Delta l = 0$. In fact l-forbidden M1 transitions have been observed in many nuclei, and they have been analysed in terms of the j–j coupling shell model with first order perturbation (Arima *et al.*,1957), and of the quasi-particle model coupled with the quadrupole phonon (Sorensen, 1963). The l-forbidden M1 transition is interesting because it is sensitive to the tensor term $g_p\ [\mathbf{Y}^2 \times \mathbf{S}]_1$ in eqn (5.3.17).

Since l-forbidden M1 transitions are retarded, the life-times have been measured directly by means of pulsed beam technique in in-beam γ-ray spectroscopy. An example of l-forbidden M1 transitions is shown in Fig. 5.11. Observed values for the polarization factor g_p are empirically expressed as $g_p \approx 10/A^{2/3}$, namely $g_p \approx 0.5$ for $2p_{3/2} \leftrightarrow 1f_{5/2}$ transitions (Nakayama *et al.*, 1978), and $g_p \approx 0.4$ for $2d_{5/2} \leftrightarrow 1g_{7/2}$ ones (Nagai *et al.*, 1977). The l-forbidden M1 matrix elements in the lead region have been analysed by Arima and Huang-Lin (1972).

A unique aspect of conversion electron spectroscopy is that the penetration term, which is due to the component of the electron wave function penetrating into the nucleus, explicitly includes the tensor term $[\mathbf{Y}^2 \times \boldsymbol{\sigma}]_1$. The M1 operator for the penetration term is given by (Church and Weneser, 1960)

$$\mathbf{M}_e = \sqrt{\frac{3}{4\pi}}\left(\frac{r}{R}\right)^2\left[\frac{5}{6}g_s\boldsymbol{\sigma} + g_l\mathbf{l} + \frac{\sqrt{2\pi}}{3}g_s[\mathbf{Y}^2 \times \boldsymbol{\sigma}]_1\right]. \qquad (5.3.18)$$

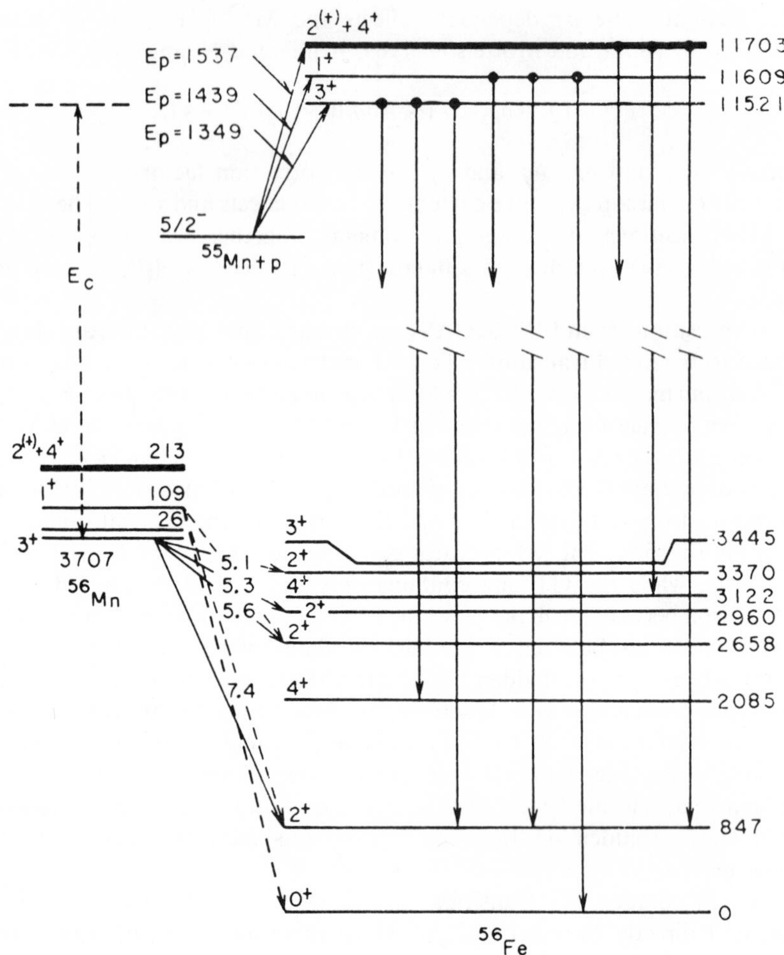

Fig. 5.10. The Gamow-Teller β decay of ^{56}Mn and the related M1 transitions from the isobaric analogue states in the ^{55}Mn(p, γ)^{56}Fe reaction. The excitation energies shown are in keV and E_c is the Coulomb displacement energy. (Sakai *et al.*, 1970.)

The third term thus gives a non-vanishing contribution to the *l*-forbidden M1 transition. The penetration matrix element $\langle \mathbf{M}_e \rangle$ is derived experimentally from the conversion coefficient α, given by

$$\alpha = \frac{\beta(0)(1 + b_1\lambda + b_2\lambda^2) + \delta^2\alpha(0)}{1+\delta^2}, \tag{5.3.19}$$

where $\lambda \equiv \langle M_e \rangle / \langle M_\gamma \rangle$ is the penetration factor, δ^2 is the E2/M1 mixing ratio, $\alpha(0)$ is the E2 conversion coefficient, $\beta(0)$ is the M1 conversion

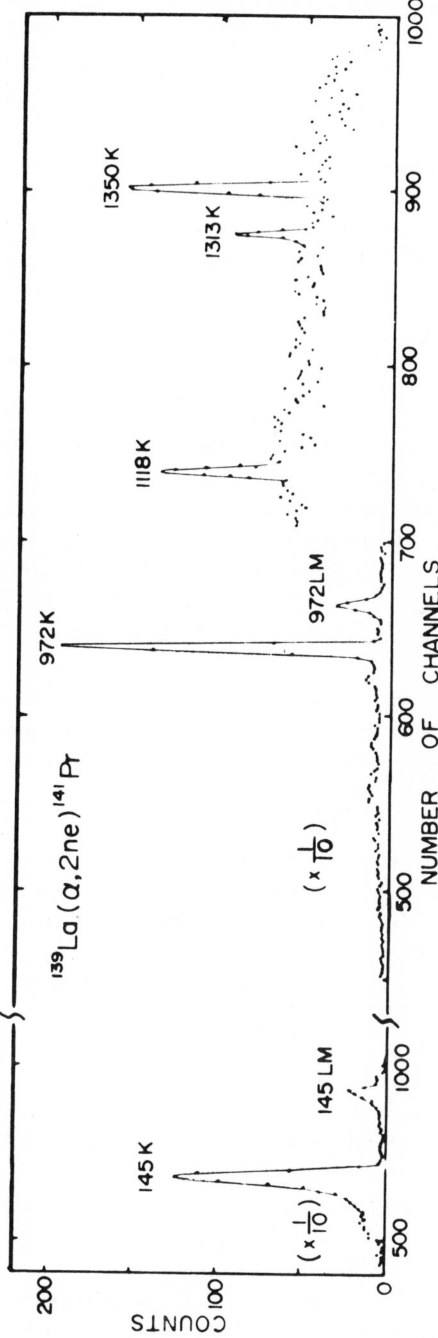

Fig. 5.11. The conversion electrons from the ^{139}La(α, 2n)^{141}Pr reaction. The transition energies in units of keV are indicated on the appropriate electron peaks. The 145 keV *l*-forbidden M1 transition from the 145 keV ($1g_{7/2}$) to the ground ($2d_{5/2}$) quasi-proton states, and the 972 keV M2 transition from the 1118 keV ($1h_{11/2}$) to the 145 keV ($1g_{7/2}$) quasi-proton states are clearly seen. (Nagai *et al.*, 1977.)

coefficient without including the penetration term, and b_1 and b_2 are the penetration coefficients. Since M1 is l-forbidden the E2 contribution becomes relatively important. The ratio δ is obtained from angular correlations and $|\langle \mathbf{M}_\gamma \rangle|^2$ is obtained from the partial γ-decay width. The penetration factor itself is of the order of 10^{-2}. Thus extraction of the penetration term requires high precision measurements of the conversion coefficient. The matrix element $\langle 2d_{5/2} \| \mathbf{M}_e \| 1g_{7/2} \rangle$ in ^{143}Pm with $N=82$ has been studied by using the triple-focussing electron separator TESS (Ejiri *et al.*, 1976). The observed value suggests that the $\langle \mathbf{Y}^2 \times \boldsymbol{\sigma} \rangle$ matrix element is renormalized by a factor $g^{\text{eff}}/g \approx 0.3$ (Nagai *et al.*, 1984).

5.3.3b *Gamow–Teller beta transitions*

The Gamow–Teller (GT) operator $t_\pm \boldsymbol{\sigma}$, which is analogous to the spin term of the isovector M1 operator $t_3\boldsymbol{\sigma}$ is related directly to the supermultiplet symmetry. Thus the GT matrix elements in heavy nuclei give a measure of the breaking of the supermultiplet symmetry (Morita *et al.*, 1971; Morita, 1973). The GT matrix elements in medium and heavy nuclei are reduced uniformly by factors $g^{\text{eff}}/g=0.2$ to 0.3 compared with the single quasi-particle (pairing model) values. The effective coupling constants for GT transitions in odd Sb–Sn nuclei are in accord with the values calculated by means of the commutator method based on the supermultiplets (Fujita and Ikeda, 1965b; Fujita *et al.*, 1967).

The GT state for the $t_+\sigma(\beta^+)$ in ^{148}Tb has been studied by means of β–γ spectroscopy combined with an on-line (in-beam) mass separator (Kleinheinz *et al.*, 1985). The neutron-deficient nucleus ^{148}Dy was excited by the (^{3}He, xn) reaction with a 240 MeV ^{3}He beam, and ^{148}Tb was separated by the ISOCELE II at Orsay. The 1^+ state with $[(1h_{11/2})_p(1h_{9/2})_n]$ is strongly fed by the EC-β^+ decay of ^{148}Dy with $\log ft=3.95$ as shown in Fig. 5.12. Since the 1^+ state is the only GT state in the $t_+\sigma$ channel, it corresponds to the pure GT β^+ giant state, but the strength found is only 15 per cent of the RPA calculation. The problem of the missing $t_+\sigma$ strength remains open.

Since the reduction of the GT matrix element is due to the absorption of the GT strength by the possible GT GR, the reduction factor depends on the excitation energy of the corresponding GT GR. In the commutator method the effective coupling constant for the β^+ transition has a smaller denominator $(\Delta_+ - Q_{\beta+} - mc^2)$ than that $(\Delta_- + Q_{\beta-} + mc^2)$ for the β^- transition, as shown in Fig. 5.13, resulting in larger g_A^{eff} value for the β^+ decay than that for the β^- decay. Actually the experimental ft values for the β^+ decay have been found to be uniformly smaller by a factor ~ 0.2 than the ft values for the β^- decay, namely $(ft)_-/(ft)_+ \sim 5$ (Ejiri, 1967). This fact is a nice indication of the GT core polarization effect (Mottelson, 1967).

The GT GR, with $J^\pi=1^+$, $l=0$, $S=1$ at $E_{\text{ex}} \approx 10$ MeV, has been found experimentally by means of (p, n) reactions (Doering *et al.*, 1975; Goodman *et*

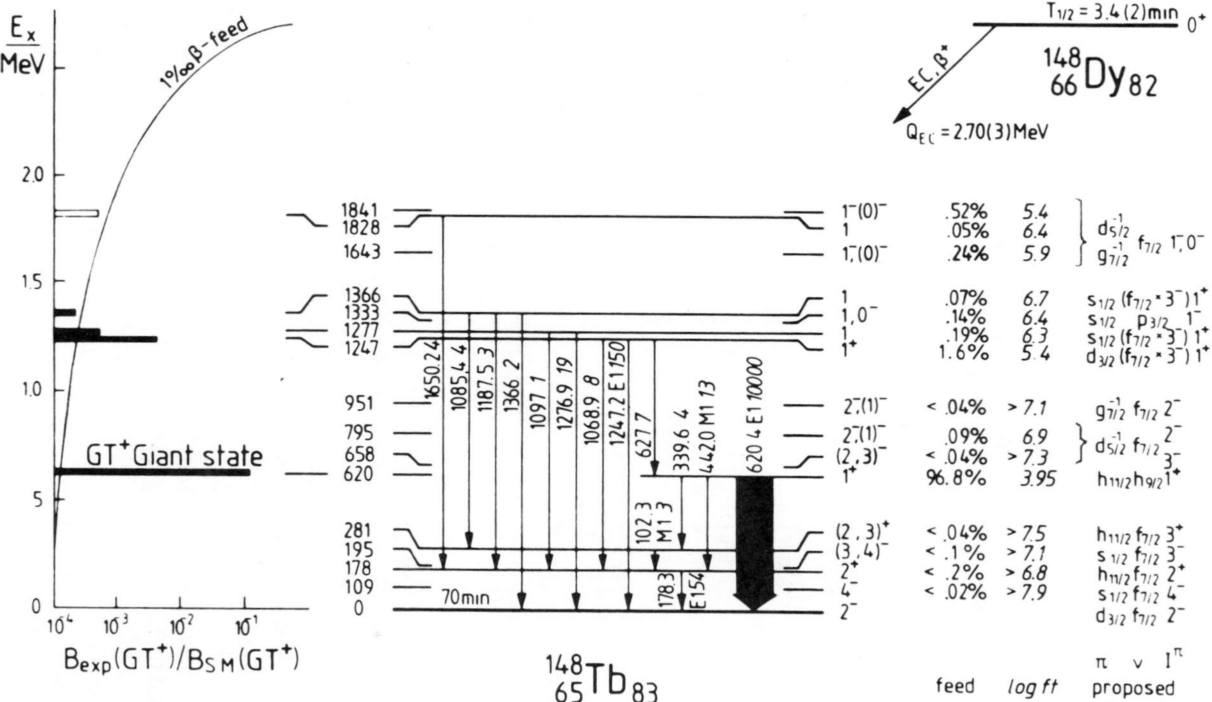

Fig. 5.12. Levels in ^{148}Tb observed in the decay of ^{148}Dy. Da-excitation γ rays are shown only for definite and possible 1^+ states. The transition multipolarities derived from the α_K-values are given together with the γ-ray energies and intensities. The measured GT^+ transition strengths to identified 1^+ levels are plotted to the left where the detection sensitivity limit for 1^+ strength is also included. (Kleinheinz *et al.*, 1985.)

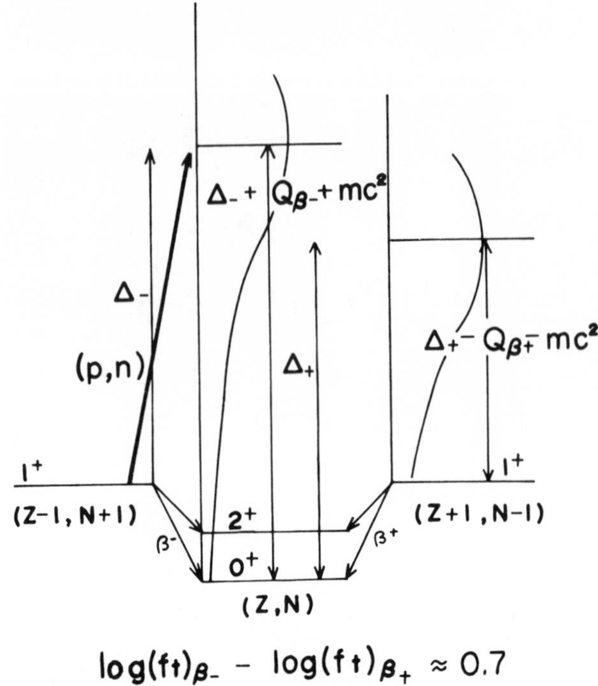

Fig. 5.13. The decay scheme of 1^+ odd–odd nuclei and the GT giant resonances. mc^2 is the electron mass, $Q_{\beta\pm}$ is the Q-value for the $\beta_\pm$ decay and $\Delta_\pm$ is the Δ in eqn (5.2.28) (see text) for the $\beta_\pm$ decay. (Ejiri and Fujita, 1978.)

al., 1980). The (p, n) reaction operator consists of $\boldsymbol{\tau}$ and $\boldsymbol{\tau\sigma}$ components. The t matrix for the $\tau\sigma$ channel with respect to that of the τ channel becomes very large at $E_p \approx 200$ MeV (Love and Franey, 1981). Thus the (p, n) reaction around $E_p = 200$ MeV excites the GT resonance in the $\boldsymbol{\tau\sigma}$ channel more strongly than it does the IAR in the τ channel, as shown in Fig. 5.14. Giant resonances with odd parity are seen in the higher excitation region. They correspond to the $\mathbf{t}[\mathbf{Y} \times \boldsymbol{\sigma}]_J$ mode GR's associated with the M2-γ and forbidden-β transitions (see Section 5.3.4). The GT strength in the GR peak region is found to be about 50–60 per cent of the GT sum rule limit (Gaarde, 1983). The origin of the missing strength has been discussed in terms of the possible GT strength remaining in the background region (Osterfeld, 1982), the GT strength being spread far beyond the GR region by a tensor-type $\sigma\tau$-dependent interactions (Bertsch and Hamamoto, 1982), and the non-nucleonic degree of excitation in the nucleon-hole Δ-isobar channel (Oset and Rho, 1979). The quenching of the $\sigma\tau$ strength has been extensively studied both experimentally and theoretically (see for example Petrovich *et al.*, 1982).

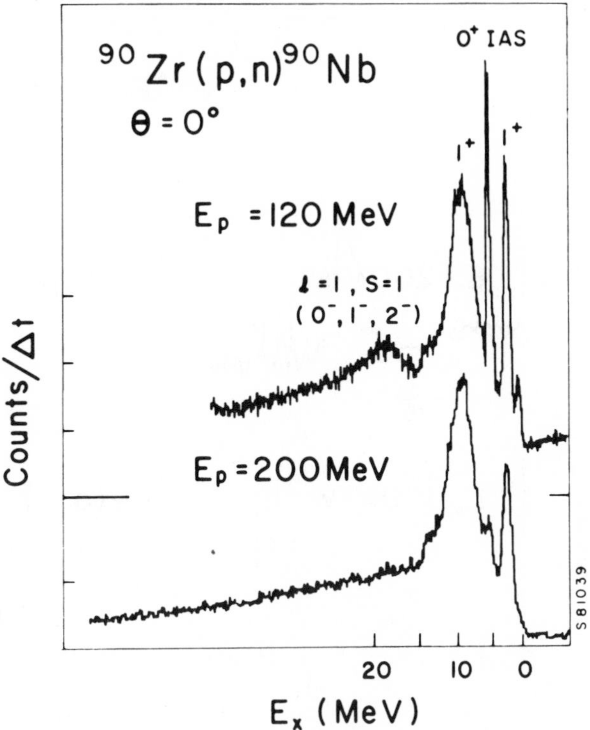

Fig. 5.14. Neutron spectra for ^{90}Zr at 120 and 200 MeV. The GT GR with $J^{\pi}=1^{+}$, $l=0$ at $E_{ex}\approx 10$ MeV is strongly excited. The $l=1$ resonance is only seen at 120 MeV because the angular distribution is narrower (in angle) at 200 MeV. (Gaarde, 1983.)

The M1 GR's have been studied by means of the (p, p′) reaction (Anantaraman *et al.*, 1981; Bertrand *et al.*, 1981). As in the (p, n) reaction for the GT mode, the (p, p′) reaction at around $E_p = 200$ MeV excites selectively the M1 mode in the $\tau_3\sigma$ channel. The M1 GR is found to appear systematically at around $E_x =$ 8 to 9 MeV excitation energy in medium–heavy nuclei, but the M1 strengths are much smaller than the predicted values. The M1 GR's observed by the (p, p′) and (e, e′) reactions show good correspondence to the GT resonance observed by the (p, n) reaction, as shown in Fig. 5.15.

The Δ isobar with $M = 1232$ MeV has quantum numbers $t = 3/2$ and $s = 3/2$, and is strongly excited in the spin–isospin channel by flipping the τ and σ of one of the three quarks in a nucleon. In this case any nucleon can participate in the excitation because there is no Pauli blocking effect on the Δ isobar. Then the nucleon–hole Δ isobar ($N^{-1}\Delta$) mode affects naturally the M1 ($t_3\sigma$) and the GT ($t_{\pm}\sigma$) modes as discussed extensively by Blin-Stoyle *et al.* (1978).

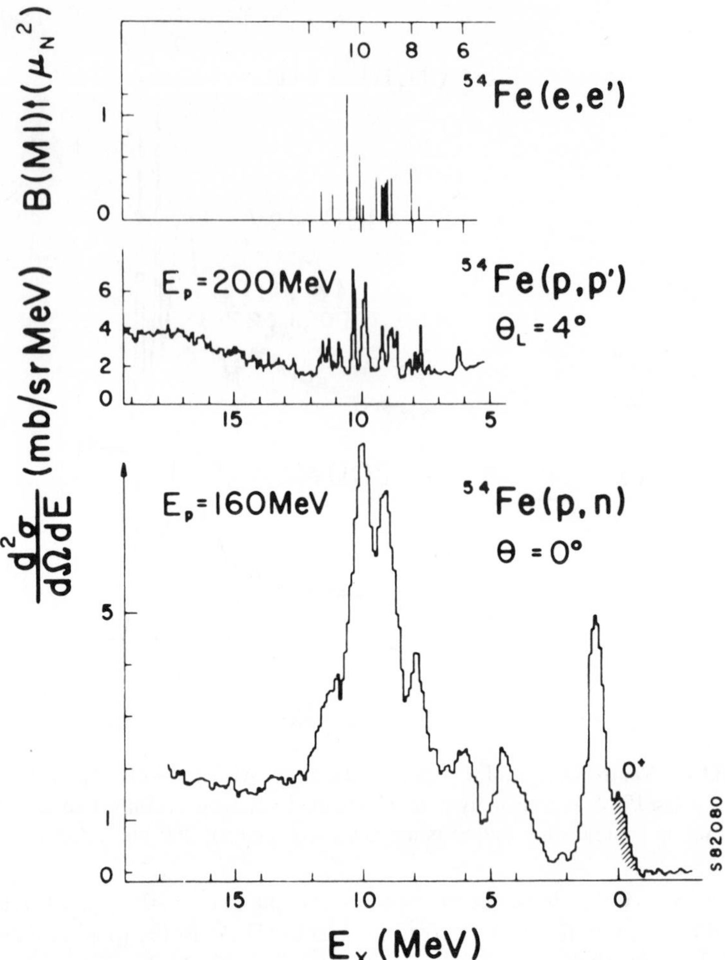

Fig. 5.15. Spectra from (p, n) and (p, p′) reactions on ^{54}Fe compared with the M1 strength function as determined from (e, e′) data. The resolution in the (p, n) data is ~600 keV, around 80 keV for the (p, p′) data and 30 keV in the (e, e′) experiments. All the structure in the (p, n) and (p, p′) spectra above the smooth continuum is observed to have an $l=0$ angular distribution. (Gaarde, 1983; Crawley *et al.*, 1982; Eulenberg *et al.*, 1982.)

The strength distribution of the nucleon–hole nucleon ($N^{-1}N$) and $N^{-1}\Delta$ transitions are evaluated on the basis of the quark model (Bohr and Mottelson, 1981). The result obtained shows a large strength for the $N^{-1}\Delta$, as illustrated in Fig. 5.16. The coupling to the Δ isobar mode explains consistently the observed quenching of the M1 and GT strengths.

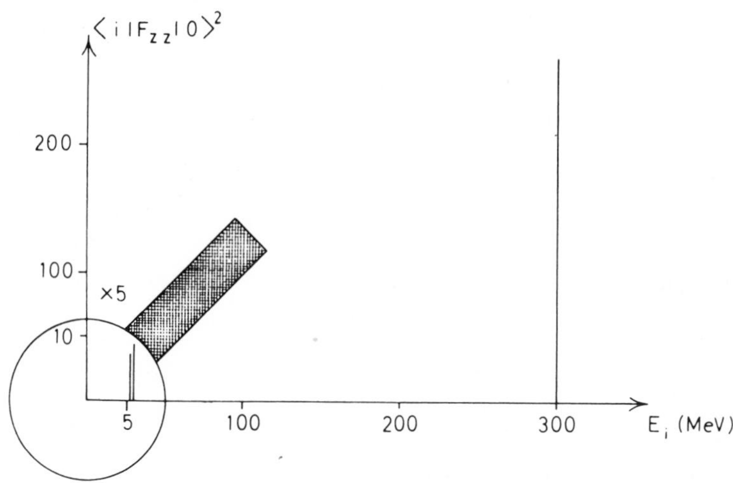

Fig 5.16. The strength distribution of $N^{-1}N$ and $N^{-1}\Delta$ transitions induced by the field $F_{zz} \equiv \sigma_z \tau_z$. (Bohr and Mottelson, 1981.)

5.3.4 *Magnetic quadrupole transitions*

5.3.4a *The M2 transition operators for spin-stretched transitions*

Magnetic quadrupole (M2) transitions have been investigated by studying spin-stretched M2 transitions between single (quasi)-particle states with $j = l + \frac{1}{2}$ and $j' = (l-1) - \frac{1}{2}$ (see Section 2.3). The stretched M2 transitions are quite adequate for investigating quantitatively the renormalization effect on the spin–isospin matrix elements $\boldsymbol{\tau}[\boldsymbol{\sigma} \times \mathbf{Y}_2]_2$ and $[\boldsymbol{\sigma} \times \mathbf{Y}_2]_2$. The reasons are as follows:

1. These are fairly pure single quasi-particle states with $j = l' + \frac{1}{2}$ and $j' = l' - 1 - \frac{1}{2}$. Here l' is the maximum orbital-angular momentum in the major shell, and the state with $j = l' + \frac{1}{2}$ is the intruder state (see Section 2.3). The stretched M2 transition between those states is a favoured single (quasi)-particle transition because of the large radial and angular overlap integrals.

2. The stretched M2 matrix element is dominated by the spin operator and is thus sensitive to the spin response in nuclei.

3. Since the pairing reduction factor for the quasi-particle magnetic transition is a constructive sum of the UU and VV factors—see eqn (2.3.6)—it is stable and rather insensitive to the uncertainty of the U and V factors.

4. Configuration mixings in one major shell caused by the quadrupole and other interactions with $\Delta\pi = \text{no}$ do not greatly disturb the M2 transition with $\Delta\pi = \text{yes}$.

The magnetic γ-transition operator with rank J is written as (see Section 2.1)

$$\mathbf{M}(\mathrm{M}J)=\frac{e\hbar}{2Mc}\{J(2J+1)\}^{1/2}r^{J-1}\cdot\mathrm{i}^{J-1}$$

$$\times\left\{\frac{1}{2}\left(g_s-\frac{2g_l}{J+1}\right)(\mathbf{Y}_{J-1}\times\boldsymbol{\sigma})_J+\frac{2g_l}{J+1}(\mathbf{Y}_{J-1}\times\mathbf{j})_J\right\}, \qquad (5.3.20)$$

where g_s and g_l are the g factors for the spin and orbital motions, respectively. For stretched single-particle transitions $J_\mathrm{i}\rightarrow J_\mathrm{f}=J_\mathrm{i}\pm J$ the second term in eqn (5.3.20) vanishes. Then eqn (5.3.20) is rewritten in terms of the coupling constants $g_{\gamma\mathrm{p}}(\mathrm{M}L)$ and $g_{\gamma\mathrm{n}}(\mathrm{M}L)$ for proton and neutron transitions, respectively, as

$$\langle\mathbf{M}(\mathrm{M}J)\rangle=\left\{g_{\gamma\mathrm{p}}(ML)\frac{1-\tau_3}{2}+g_{\gamma\mathrm{n}}(ML)\frac{1+\tau_3}{2}\right\}\langle\mathrm{i}^{J-1}[\mathbf{Y}_{J-1}\times\boldsymbol{\sigma}]_J\rangle, \qquad (5.3.21)$$

$$g_{\gamma\mathrm{p}}(\mathrm{M}L)=\frac{e\hbar}{2Mc}\{J(2J+1)\}^{1/2}\left(\mu_\mathrm{p}-\frac{1}{J+1}g_{l\mathrm{p}}\right), \qquad (5.3.22)$$

$$g_{\gamma\mathrm{n}}(\mathrm{M}L)=\frac{e\hbar}{2Mc}\{J(2J+1)\}^{1/2}\left(\mu_\mathrm{n}-\frac{1}{J+1}g_{l\mathrm{n}}\right), \qquad (5.3.23)$$

where $\mu_\mathrm{p}=2.79$, $\mu_\mathrm{n}=-1.91$. The transition matrix element is divided into the isovector and isoscalar parts (Warburton and Wenser, 1969),

$$\langle\mathbf{M}(\mathrm{M}J)\rangle=g_1(MJ)\langle\tau_3\mathbf{M}_{J-1\,1\,J}\rangle+g_0(MJ)\langle\mathbf{M}_{J-1\,1\,J}\rangle, \qquad (5.3.24)$$

where

$$\mathbf{M}_{J-1\,1\,J}=[\mathrm{i}^{J-1}r^{J-1}\mathbf{Y}_{J-1}\times\boldsymbol{\sigma}]_J, \qquad (5.3.25)$$

and $g_1(\mathrm{M}J)$ and $g_0(\mathrm{M}J)$ are the isovector and isoscalar coupling constants for the magnetic J-pole γ transition. They are expressed as

$$g_1(\mathrm{M}J)=\frac{e\hbar}{2Mc}\{J(J+1)\}^{1/2}\left(\mu_--\frac{1}{J+1}g_{l-}\right)\frac{1}{2}, \qquad (5.3.26a)$$

$$g_0(\mathrm{M}J)=\frac{e\hbar}{2Mc}\{J(J+1)\}^{1/2}\left(\mu_+-\frac{1}{J+1}g_{l+}\right)\frac{1}{2}, \qquad (5.3.26b)$$

where $\mu_-=\mu_\mathrm{n}-\mu_\mathrm{p}$, $\mu_+=\mu_\mathrm{n}+\mu_\mathrm{p}$, $g_{l-}=g_{l\mathrm{n}}-g_{l\mathrm{p}}$ and $g_{l+}=g_{l\mathrm{n}}+g_{l\mathrm{p}}$. The orbital part given by the second term of eqns (5.3.22), (5.3.23) and (5.3.26) is indeed small compared with the spin term given by the first term of these equations so that approximate values of $g_l=\frac{1}{2}-\tau_3/2$ and $g_{l\pm}=\pm 1$ are used. The renormalization factors $g_{\gamma\mathrm{p}}^{\mathrm{eff}}/g_{\gamma\mathrm{p}}$ and $g_{\gamma\mathrm{n}}^{\mathrm{eff}}/g_{\gamma\mathrm{n}}$ are expressed in terms of the isovector and isoscalar coupling constants,

$$\frac{g_{\gamma p}^{\rm eff}}{g_{\gamma p}}=\frac{g_{\gamma 1}^{\rm eff}}{g_{\gamma 1}}\cdot\frac{-1}{2}\left(\mu_{-}-\frac{1}{J+1}g_{l-}\right)\Big/\left(\mu_{\rm p}-\frac{1}{J+1}g_{l\rm p}\right)$$
$$+\frac{g_{\gamma 0}^{\rm eff}}{g_{\gamma 0}}\cdot\frac{1}{2}\left(\mu_{+}-\frac{1}{J+1}g_{l+}\right)\Big/\left(\mu_{\rm p}-\frac{1}{J+1}g_{l\rm p}\right), \tag{5.3.27}$$

$$\frac{g_{\gamma n}^{\rm eff}}{g_{\gamma n}}=\frac{g_{\gamma 1}^{\rm eff}}{g_{\gamma 1}}\cdot\frac{1}{2}\left(\mu_{-}-\frac{1}{J+1}g_{l-}\right)\Big/\left(\mu_{\rm n}-\frac{1}{J+1}g_{l\rm n}\right)$$
$$+\frac{g_{\gamma 0}^{\rm eff}}{g_{\gamma 0}}\cdot\frac{1}{2}\left(\mu_{+}-\frac{1}{J+1}g_{l+}\right)\Big/\left(\mu_{\rm n}-\frac{1}{J+1}g_{l\rm n}\right). \tag{5.3.28}$$

The renormalization factors $g_{\gamma 1}^{\rm eff}/g_{\gamma 1}$ and $g_{\gamma 0}^{\rm eff}/g_{\gamma}$ for the isovector and isoscalar modes are obtained from the $g_{\gamma p}^{\rm eff}/g_{\gamma p}$ and $g_{\gamma n}^{\rm eff}/g_{\gamma n}$ for the proton and neutron transitions in nuclei with similar mass numbers, provided that the renormalization factors are rather independent of the mass number (a correction for the small A-dependence is discussed later).

The unique $(J-1)$th forbidden β transition is analogous to the isovector component of the stretched MJ transition. The β-transition operator is written as

$$\mathbf{M}_{\beta}(\mathrm{A}J)=g_{\rm A}t_{\pm}\mathbf{M}_{L1J}, \tag{5.3.29}$$

where $g_{\rm A}$ is the axial–vector coupling constant and $L=J-1$. The effective coupling constant given by

$$\frac{g_{\beta \rm A}^{\rm eff}}{g_{\rm A}}=\frac{\langle t_{\pm}\mathbf{M}_{L1J}\rangle_{\rm exp}}{{}_0\langle t_{\pm}\mathbf{M}_{L1J}\rangle_0} \tag{5.3.30}$$

is analogous to that for the isovector MJ transition.

5.3.4b *The renormalization factors for M2 transitions and the analogous β transitions*

Stretched M2 transitions have been extensively investigated by means of in-beam e–γ spectroscopy. Their lifetimes are just in the region to permit direct timing experiments. The large conversion coefficients for M2 transitions make the identification of the M2 multipolarity easy by means of the conversion electron studies. The $1\mathrm{h}_{11/2}\rightarrow 1\mathrm{g}_{7/2}$ single-proton transitions have been extensively studied in nuclei with $N\approx 82$ (Ejiri and Shibata, 1975) as well as the $1\mathrm{h}_{11/2}\rightarrow 1\mathrm{g}_{7/2}$ single neutron transitions in nuclei with $Z=50$ (Brenn *et al.*, 1974). The transition schemes for proton transitions are shown in Fig. 2.15 in Section 2.3, and those for the neutron ones in Fig. 5.17. The time spectra for the M2 γ transition in ^{141}Pr are shown as an example in Fig. 5.18.

The unique first-forbidden β transition matrix elements are obtained from the $\log f^1 t$ values listed in the tables of Lederer and Shirley (1978). The

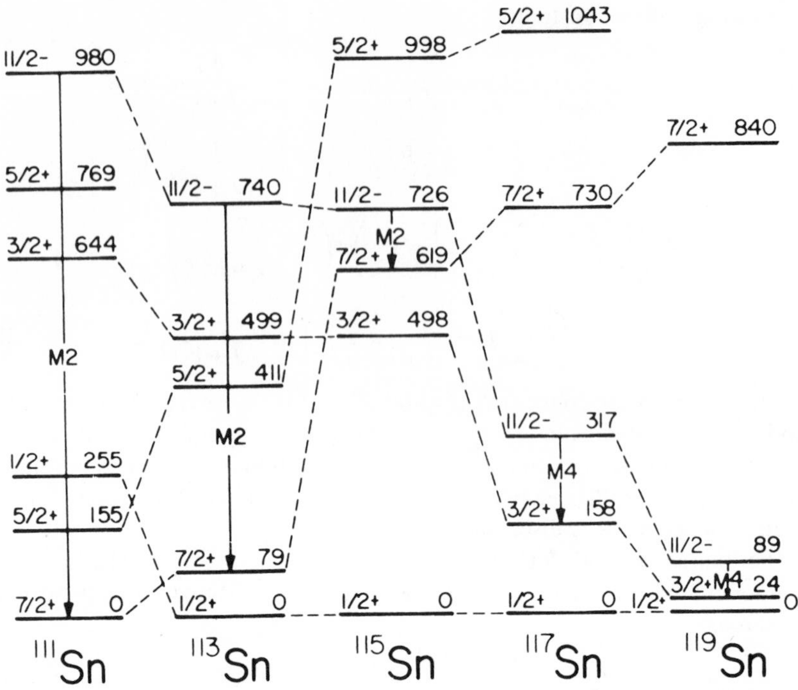

Fig. 5.17. The level schemes of the lighter odd tin isotopes, and $1h_{11/2} \leftrightarrow 1g_{7/2}$ single-neutron M2 transitions. (Brenn *et al.*, 1974.)

renormalization factors in nuclei with $N \approx 50$ have been investigated by Ejiri (1972). The values g^{eff} for single quasi-particle transitions are obtained as shown in Fig. 5.19. The values for the $1h_{11/2} \rightarrow 1g_{7/2}$ M2 γ transitions are plotted separately in Fig. 5.20. It is clear that the M2 transitions are uniformly reduced as given by the renormalization factors $g^{\text{eff}}/g = 0.25 \pm 0.05$ $\{\kappa(\text{M2}) \approx 3\}$. This is of the same order of magnitude as that for the isovector E1 mode, as discussed in Section 5.3.2, and is consistent with expectations based on the supermultiplet theory.

It is interesting to note that $\kappa(\text{M2})$ is larger by factors around 3 than $\kappa(\text{M1})$ as given in Section 5.3.3a. This is partly explained by the number of nucleons $n(\text{M}L)$ participating in the MJ polarization since $\kappa(\text{M}J)$ is proportional to $n(\text{M}J)$ (see eqn (5.2.45)). The M1 polarization consists mainly of spin–flip particle–hole excitations from $j = l + \frac{1}{2}$ to $j = l - \frac{1}{2}$ orbits, while the M2 one consists of $1\hbar\omega$-jump particle–hole excitations. Thus $n(\text{M1})$ is given by the number of nucleons in the $j = l + \frac{1}{2}$ shell and $n(\text{M2})$ is that of all nucleons in one major (harmonic oscillator) shell. Thus $n(\text{M2}) \approx (3 \text{ to } 4) \cdot n(\text{M1})$. It is interesting to note that the reduction rates for the proton transitions are about the same

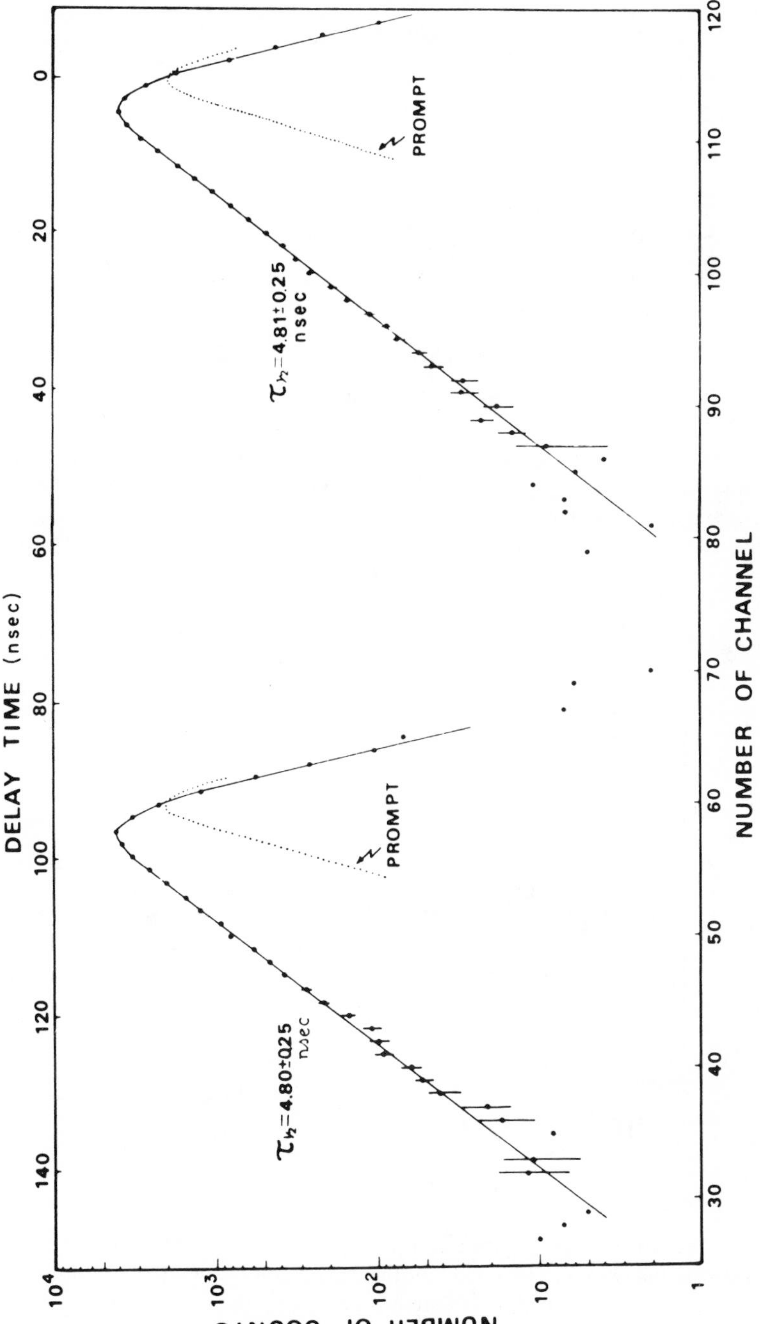

Fig. 5.18. Decay curves of the 972 keV M2 transition from the $h_{11/2}$ state in ^{141}Pr populated by the ^{139}La(α, 2nγ) reaction. Yields of the γ rays are plotted as a function of time between the starting pulse from the Ge detector for the γ ray and the stopping pulse triggered by every other RF signal of the cyclotron oscillator. (Ejiri *et al.*, 1973a.)

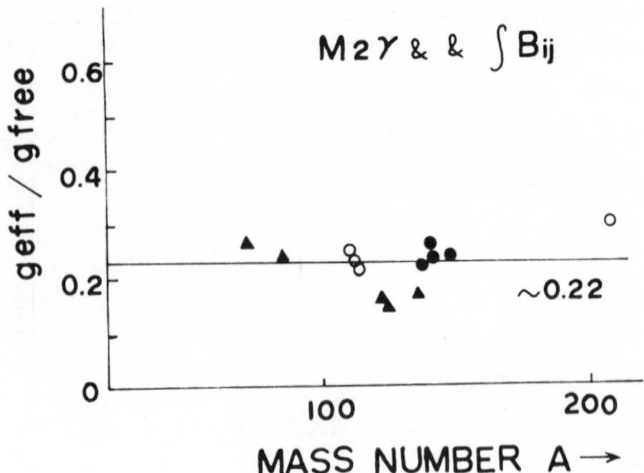

Fig. 5.19. The effective coupling constants for the stretched M2 γ transitions (circles) and the unique first-forbidden β transitions (triangles) between single quasi-particle states.The open and closed circles refer to the proton and neutron transitions, respectively. The solid line gives an approximate value ~ 0.22 as an eye-guide. (Ejiri and Fujita, 1978.)

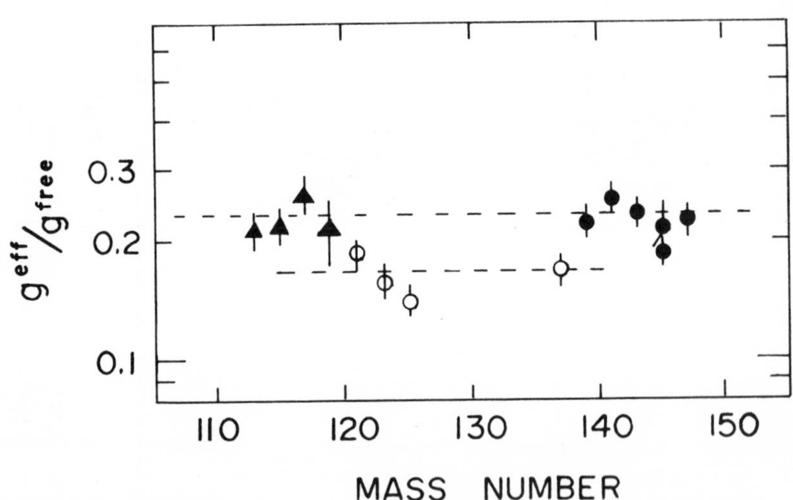

Fig. 5.20. The effective coupling constants for the $(1h_{11/2}) \rightarrow (1g_{7/2})$ M2 γ transitions. The closed circles and closed triangles are for the proton and the neutron transitions, respectively, and the open circles are for the corresponding β-decays.

as those for the neutron transitions. This fact suggests that the isoscalar and isovector components of the M2 γ transitions are reduced by about the same renormalization factors as expected on the basis of Majorana-type exchange interaction.

The $1h_{11/2} \leftrightarrow 1g_{7/2}$ M2 γ-matrix elements in semi-magic nuclei ($N=82$ or $Z=50$) are quite well studied for both the single quasi-proton states and the single quasi-neutron states. Furthermore, the precise values of the $1h_{11/2} \leftrightarrow 1g_{7/2}$ β decay matrix elements are available (Morita and Kume, 1974). Therefore it is interesting to obtain the effective coupling constants for the isovector and isoscalar γ modes, and compare them with the value for the β mode. The effective values $g_{\gamma 1}^{\text{eff}}$ and $g_{\gamma 0}^{\text{eff}}$ for the isovector and isoscalar M2 γ modes are obtained by inserting the measured values $g_{\gamma \text{p}}^{\text{eff}}$ and $g_{\gamma \text{n}}^{\text{eff}}$ into eqns (5.3.27) and (5.3.28). Here $g_{\gamma 1}^{\text{eff}}$ and $g_{\gamma 0}^{\text{eff}}$ at $A \approx 140$, where the values for $g_{\gamma \text{p}}^{\text{eff}}$ are obtained, may not be necessarily the same as those at $A \approx 120$, where the values $g_{\gamma \text{n}}^{\text{eff}}$ are obtained, if such effective constants may depend on the mass number (A) of the core. Although the A-dependence is small because of the uniform effect of the core polarization with $\Delta\pi = \text{yes}$, it has to be corrected for this effect in order to get reliable values of $g_{\gamma 1}^{\text{eff}}$ and $g_{\gamma 0}^{\text{eff}}$. Correcting for this A-dependence, the $g_{\gamma \text{p}}^{\text{eff}}/g_{\gamma \text{p}}$ and $g_{\gamma \text{n}}^{\text{eff}}/g_{\gamma \text{n}}$ are plotted against $g_{\gamma 1}^{\text{eff}}/g_{\gamma 1}$ and $g_{\gamma 0}^{\text{eff}}/g_{\gamma 0}$ in Fig. 5.21. The renormalization factors for the isovector modes and isoscalar modes are $g_{\gamma 1}^{\text{eff}}/g_{\gamma 1} \approx 0.24 \pm 0.01$ and $g_{\gamma 0}^{\text{eff}}/g_{\gamma 0} \approx 0.35$ to 0.4.

It should be noted here that the isovector M2 mode is reduced by about the same factor $g^{\text{eff}}/g \approx 0.25$ as the IAS E1 mode, and that the reduction factor for the isoscalar M2 γ mode is of the same order of magnitude as the isovector one. In other words the isospin–spin polarization effect reduces the isovector M2 γ mode as much as the isospin polarization effect does the isovector E1 γ mode, and the spin polarization effect on the isoscalar M2 γ mode is about the same as the isospin polarization effect. This is just what one expects from the major exchange interaction of the Majorana type.

The unique first-forbidden β-matrix elements for single quasi-particle $(1h_{11/2}) \leftrightarrow (1g_{7/2})$ transitions in nuclei with $N \approx 82$ and $Z \approx 50$ give the renormalization factors $g_{\beta \text{A}}^{\text{eff}}/g_{\beta \text{A}} \approx 0.16$ to 0.18. They are smaller than the values $g_{\gamma 1}^{\text{eff}}/g_{\gamma 1}$ for the analogous isovector M2 γ mode by 20 to 30 per cent (see Fig. 5.20). The susceptibilities and the effective coupling constants are calculated using the same polarization interaction strength χ'_{1112} as used for the isovector M2 γ mode. The calculated values are very close to the values for the M2 γ modes. This is due to the fact that the microscopic structure of the $t_{\pm}$ (β) mode core polarization is similar to that for the t_3 mode (γ) core polarization, namely the ratios $(g_{\beta \text{A}}^{\text{eff}}/g_{\beta \text{A}})_{\text{cal}}/(g_{\gamma 1}^{\text{eff}}/g_{\gamma 1})_{\text{cal}}$ are ≈ 0.9. The observed ratios (≈ 0.7), which are smaller than the calculated ones, are not explained by the simple polarization effect, and are thus suggestive of some effects such as the exchange current effect, the coupling between the exchange

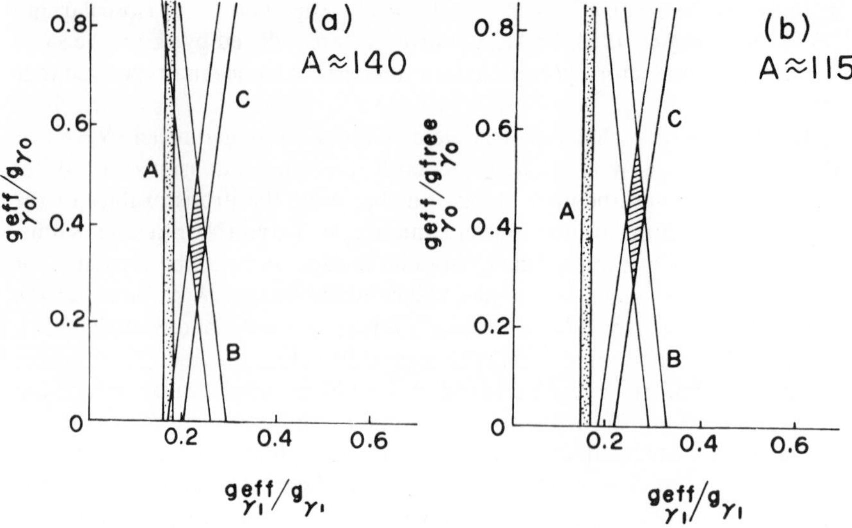

Fig. 5.21. The effective coupling constants for the isoscalar and isovector M2 γ transitions and the analogous axial β transitions between the $1h_{11/2}$ and $1g_{7/2}$ single quasi-particle states. *A* stands for β transitions, *B* for M2 transitions between the proton states, and *C* for γ transitions between the neutron states. The experimental values are corrected for the small effect of the *A*-dependence of the core polarization (see text). The hatched region meets the observed values for both the proton and the neutron transitions. (Ejiri and Fujita, 1978.)

current and the polarization effects, and some higher-order configuration mixing effects.

The first forbidden axial vector transitions $\langle t_{\pm}[ir\mathbf{Y}_1 \times \boldsymbol{\sigma}]_1\rangle$ with rank 1 are not well known because they are difficult to determine uniquely from the β decay data. It should be noted that the matrix element $\langle t_{-}[ir\mathbf{Y}_1 \times \boldsymbol{\sigma}]_1\rangle$ can be deduced from the in-beam IAS-γ matrix element $\langle t_{-}[ir\mathbf{Y}_1]\rangle$ by using a simple relation based on the conserved vector current theory (Fujita, 1962). The matrix element $\langle t_{-}[ir\mathbf{Y}_1 \times \boldsymbol{\sigma}]_1\rangle$ for the $2f_{7/2}\rightarrow 2d_{5/2}$ transition in ^{141}Ce was obtained by combining the β-decay data and the measured E1 matrix element for the IAS (Ejiri *et al.*, 1968/1969). The obtained renormalization factor $(g^{\mathrm{eff}}/g) = 0.204 \pm 0.05$ is close to the value 0.26 ± 0.05 for the $\langle t_{-} ir\mathbf{Y}_1\rangle$ in ^{141}Ce.

5.3.4c *Giant resonances and renormalization factors for spin–isospin operators with $\Delta l = 1$*

Spin–isospin operators with $\Delta l = 1$, such as M2 γ and unique forbidden β transitions are shown to be renormalized by factors of $g^{\mathrm{eff}}/g \approx 0.2$ to 0.3, as discussed in Section 5.3.4b. The renormalization is directly related to the relevant spin–isospin mode GR with $\Delta l = 1$, as discussed in Sections 5.1 and 5.2. Such a GR with $\Delta l = 1$ has been observed in (p, n) reactions (Horen *et al.*, 1980, 1981; Bainum *et al.*, 1980) as predicted by Ejiri *et al.*, 1968a). It is interesting to see how the renormalization (quenching) of the β and γ transitions and the GR's in the spin–isospin mode are explained consistently. The nucleon mode NN^{-1} spin–isospin strength pushes up the spin–isospin GR energy and quenches the single-particle $\tau\sigma$ transitions between low-lying states. On the other hand the high excitation spin–isospin strength such as the ΔN^{-1} pushes down the spin–isospin GR energy and quenches the single-particle $\tau\sigma$ transitions. Because of these different features of the NN^{-1} mode and the high excitation XN^{-1} mode (ΔN^{-1} etc), a proper analysis of these two kinds of data elucidates the effect of the XN^{-1}, if any, in the high excitation region (Ejiri, 1982, 1983). The transition scheme including the NN^{-1} and XN^{-1} modes is schematically illustrated in Fig. 5.22.

The spin–isospin modes with $\Delta l = 1$ are analysed in the framework of a simple schematic model. The $1\hbar\omega$ jump NN^{-1} has an excitation energy of

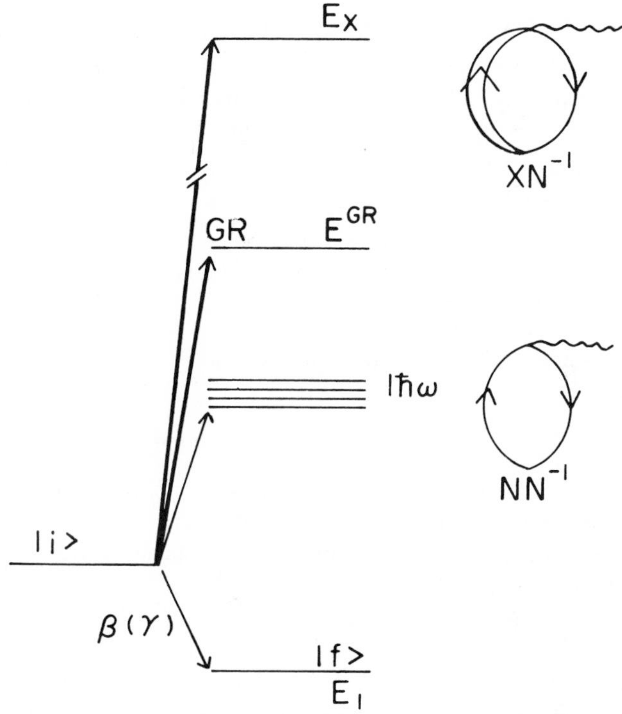

Fig. 5.22. Schematic picture for the first forbidden $\beta^-(\gamma)$ transition scheme, showing the single-particle state with E_1, the giant resonance with E^{GR} (mainly NN^{-1} strength) and the high-lying state with E^X (XN^{-1} strength where $X = \Delta$, etc).

about $1\hbar\omega$, which is much higher than the single-particle transition energy of about 1 MeV. Thus we use average values of $\bar{E}=\hbar\omega=41A^{-1/3}$ MeV and $\bar{G}^2$ (see Section 5.2.2d). Then we get for the isospin mode $\mathbf{T}_\alpha=\tau r\mathbf{Y}_1$ the dispersion equation for $\beta^\pm$,

$$f_\tau\left\{\frac{(1\mp k)}{(E^{\rm GR}/\hbar\omega)-1}-\frac{(1\pm k)}{(E^{\rm GR}/\hbar\omega)+1}\right\}=1, \qquad (5.3.31)$$

where k is the neutron excess $N-Z$ in units of the number n of the nucleons in one $\hbar\omega$ shell, $f_\tau=n\bar{G}^2/\hbar\omega$ is the $1\,\hbar\omega$ shell interaction strength, and $E^{\rm GR}_\pm$ is the $\beta^\pm$ GR excitation energy. The value $E^{\rm GR}_\pm$ is corrected for the constant term $\Delta_c-\Delta_s$, where Δ_c is the excitation energy of IAS (the Coulomb displacement energy) and Δ_s is the symmetry energy. The actual excitation energy is $E^{\rm GR}_{\rm ex}=E^{\rm GR}+\Delta_c-\Delta_s$ for the β^- GR. Then the GR energy $E^{\rm GR}$ and the susceptibility κ are given by

$$E^{\rm GR}_\pm/\hbar\omega=\mp kf_\tau+(k^2f_\tau^2+1+2f_\tau)^{1/2} \qquad (5.3.32)$$

$$\kappa_\tau=f_\tau\left[\frac{1+k}{1-(E_1/\hbar\omega)}+\frac{1-k}{1+(E_1/\hbar\omega)}\right]. \qquad (5.3.33)$$

In the case of the spin–isospin mode the stretched NN^{-1} $(l+\frac{1}{2}\rightleftarrows l-1-\frac{1}{2})$ state has a large transition matrix element and the energy is shifted from the average $1\,\hbar\omega$ value by the spin–orbit energy Δ_{sl}. Correcting for this shift the dispersion equation for β^- is

$$f_{\tau\sigma}\left(\frac{\hbar\omega(1+k-\eta)}{E^{\rm GR}_--\hbar\omega}-\frac{\hbar\omega(1-k)}{E^{\rm GR}_-+\hbar\omega}+\frac{\hbar\omega\eta}{E^{\rm GR}_--(\hbar\omega-\Delta_{sl})}\right)=1, \qquad (5.3.34)$$

where η is the ratio of the interaction strength of the stretched NN^{-1} to the total one. Approximate values are $\eta\approx0.41-0.14\,k$ and $\Delta_{sl}/\hbar\omega\sim0.6$.

$$\kappa_{\tau\sigma}=f_{\tau\sigma}\hbar\omega\left(\frac{(1+k)-\eta}{\hbar\omega-E_1}+\frac{1-k}{\hbar\omega+E_1}+\frac{\eta}{\hbar\omega-\Delta_{sl}-E_1}\right). \qquad (5.3.35)$$

The interaction strength $f_\tau=1.15$ for the isospin τrY_1 mode is obtained so as to reproduce the observed E1 giant resonance energy of $E^{\rm GR}_\gamma(\tau_3rY_1)\approx1.9\,\hbar\omega$. The same f_τ is used for the $\tau_\pm r\mathbf{Y}_1$ mode because of charge independence. The spin–isospin interaction strength $f_{\tau\sigma}\approx0.9$ is obtained so as to fit the observed $E^{\rm GR}$. Here the calculation was made for the centroid $E^{\rm GR}=\Sigma(2J+1)E^{\rm GR}(J^\pi)/9$ with $J^\pi=0^-$, 1^-, and 2^-.

It is remarkable to note that the $1\,\hbar\omega$ NN^{-1} polarization interaction $\mathbf{H}_\tau$ with $f_\tau=1.15$ reproduces both the renormalization (quenching) factor and the E1 γ giant resonance energy as shown in Fig. 5.23. For the spin–isospin modes, however, the observed renormalization factors are about 30 per cent smaller than the prediction based on the $1\,\hbar\omega$ jump NN^{-1} spin–isospin

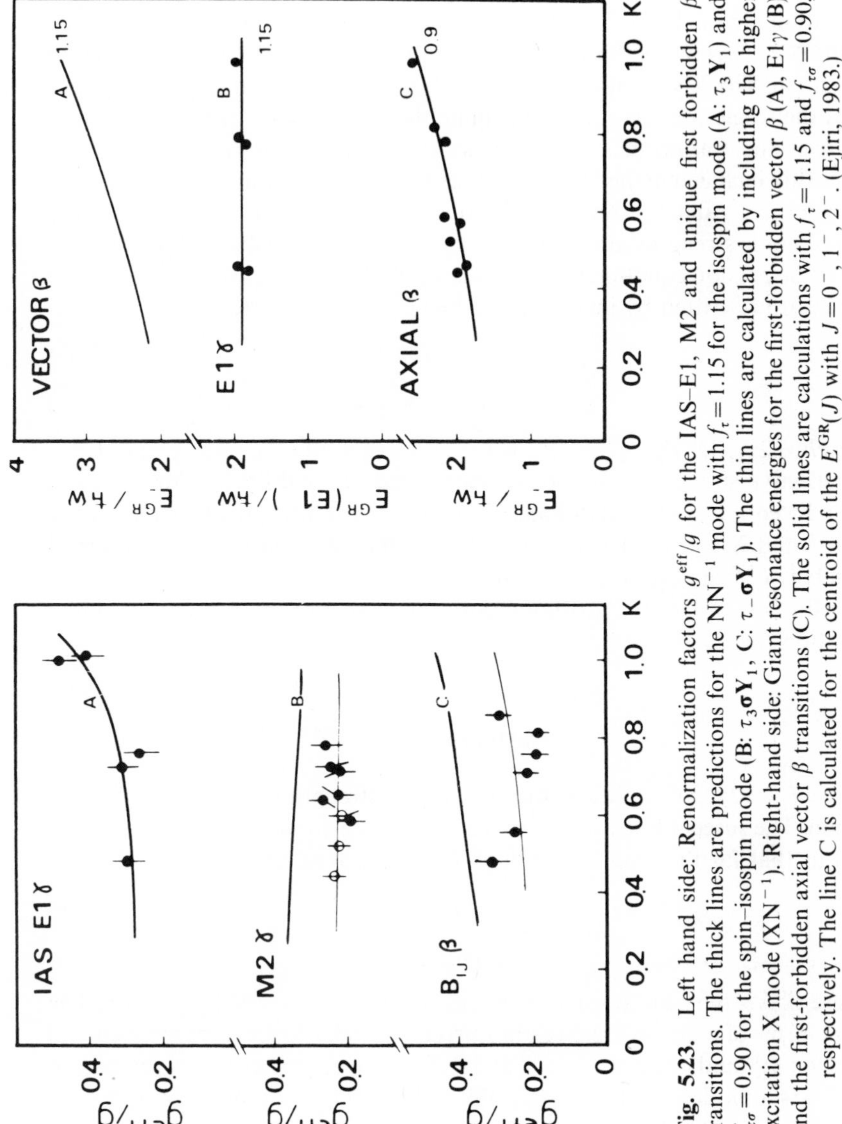

Fig. 5.23. Left hand side: Renormalization factors g^{eff}/g for the IAS–E1, M2 and unique first forbidden β transitions. The thick lines are predictions for the NN^{-1} mode with $f_\tau = 1.15$ for the isospin mode (A: $\tau_3 \mathbf{Y}_1$) and $f_{\tau\sigma} = 0.90$ for the spin–isospin mode (B: $\tau_3 \boldsymbol{\sigma} \mathbf{Y}_1$, C: $\tau_- \boldsymbol{\sigma} \mathbf{Y}_1$). The thin lines are calculated by including the higher excitation X mode (XN^{-1}). Right-hand side: Giant resonance energies for the first-forbidden vector β (A), E1γ (B) and the first-forbidden axial vector β transitions (C). The solid lines are calculations with $f_\tau = 1.15$ and $f_{\tau\sigma} = 0.90$, respectively. The line C is calculated for the centroid of the $E^{\text{GR}}(J)$ with $J = 0^-, 1^-, 2^-$. (Ejiri, 1983.)

polarization interaction $\mathbf{H}_{\tau\sigma}$ with $f_{\tau\sigma}=0.90$, while this $\mathbf{H}_{\tau\sigma}$ reproduces the giant resonance energy well. This indicates that there is an additional spin–isospin strength which is not effectively represented by the NN^{-1} mode.

Now let us introduce explicitly the spin–isospin strength $\chi_x G_x^2$ in the high excitation region. Here $\chi_x G_x^2$ stands for all $\tau\sigma$ strengths above the $\tau\sigma$ GR such as the $\Delta\mathrm{N}^{-1}$ strength, the strength shifted upwards by the tensor interaction and so on. The XN^{-1} mode reduces E^{GR} and removes the transition strength from the single-particle transition, while the NN^{-1} mode increases E^{GR} and removes the transition strength. Thus the high excitation X mode is not effectively included in the NN^{-1} mode as long as one requires reproduction of both the E^{GR} and the renormalization factors. In other words the X and NN^{-1} mode strengths are obtained from the two observables g^{eff}/g and E^{GR}.

The dispersion equations and the susceptibilities for the isospin mode with $\tau_- r\mathbf{Y}_1$ are rewritten by including the higher excitation strength as

$$f_\tau F_\tau(E_\tau^{\mathrm{GR}})-h_\tau=1, \tag{5.3.36}$$

$$\kappa_\tau=f_\tau K_\tau+h_\tau. \tag{5.3.37}$$

Here $h_\tau=2\chi_{x\tau}G_{x\tau}^2/E_{x\tau}$, where $E_{x\tau}$ is the effective excitation energy, stands for the effective strength in the isospin mode, and $f_\tau F_\tau$ and $f_\tau K_\tau$ are the terms as given by eqns (5.3.31) and (5.3.33), respectively. The strengths f_τ and h_τ for the NN^{-1} and XN^{-1} modes, respectively, are then derived from the observed variables E_τ^{GR} and κ_τ, as plotted in Fig. 5.24 (top).

In the case of the spin–isospin mode with $\tau_-[\boldsymbol{\sigma}\times r\mathbf{Y}_1]_2$ the corresponding equations are

$$f_{\tau\sigma}F_{\tau\sigma}(E_{\tau\sigma}^{\mathrm{GR}})-h_{\tau\sigma}=1 \tag{5.3.38}$$

$$\kappa_{\tau\sigma}=f_{\tau\sigma}K_{\tau\sigma}+h_{\tau\sigma} \tag{5.3.39}$$

where $h_{\tau\sigma}=2\chi_{x\tau\sigma}G_{x\tau\sigma}^2/E_{x\tau\sigma}$ is the effective $\tau\sigma$ strength in the high-excitation region. The susceptibility of the isovector M2 transition is written in the same way as that for the unique forbidden β transition as

$$f_{\tau\sigma}F_{\tau\sigma}^{\gamma}(E_{\gamma\sigma}^{\mathrm{GR}})-h_{\tau\sigma}=1, \tag{5.3.40}$$

$$\kappa_{\tau\sigma}^{\gamma}=f_{\tau\sigma}K_{\tau\sigma}^{\gamma}+h_{\tau\sigma}, \tag{5.3.41}$$

where $E_{\gamma\sigma}^{\mathrm{GR}}$ is the giant resonance energy for the isovector M2 transition. The exact forms of F and K are given in Ejiri (1982, 1983). Since there are plenty of data for $E_{\tau\sigma}^{\mathrm{GR}}$ and $\kappa_{\tau\sigma}^{\gamma}$, eqns (5.3.38) and (5.3.41) are plotted against $f_{\tau\sigma}$ and $h_{\tau\sigma}$ in Fig. 5.24 (middle), and eqns (5.3.38) and (5.3.39) are plotted in Fig. 5.24 (bottom).

The isospin mode is well explained by $f_\tau\approx 1.15$ with $h_\tau\approx 0$ as discussed before. The spin–isospin mode, however, indicates a finite strength of $h_{\tau\sigma}\approx 0.45$ in addition to the NN^{-1} mode strength with the new value

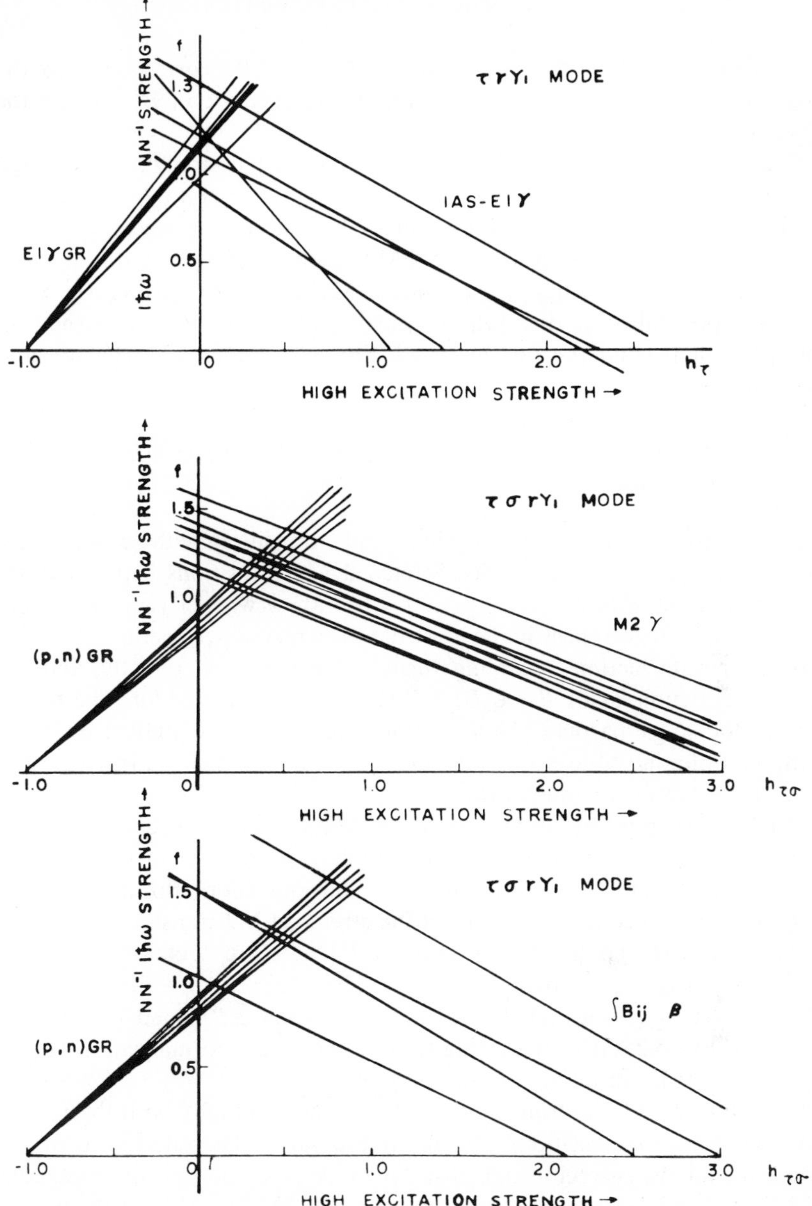

Fig. 5.24. The renormalization factors for isospin (top) and spin–isospin modes (middle, bottom), and relevant giant resonances. The vertical and horizontal axes are the nucleon mode $1\hbar\omega$ NN^{-1} strength and the high excitation XN^{-1} (such as ΔN^{-1}) strength. Each line shows experimental data. The crossing region of the E^{GR} data lines and the quenching factor lines indicates the $1\hbar\omega$ nucleon mode (NN^{-1}) strength f and the higher excitation mode strength h. (Ejiri, 1983.)

$f^{\mathrm{N}}_{\tau\sigma} \approx 1.35$. Then the previous value of the $f_{\tau\sigma} = 0.90$ corresponds to the effective value including the effect of the higher excitation (X) mode on the observed E^{GR}.

One possible origin for such a strong spin–isospin strength in the high excitation region is the isobar nucleon–hole strength. The ratio $h_{\tau\sigma}/f^{\mathrm{N}}_{\tau\sigma} \approx 0.3$ is of the same order of magnitude as the isobar effect suggested for the M1 moment and GT transitions (Bohr and Mottelson, 1981; Oset and Rho, 1979; Speth *et al.*, 1982). Using $E_x = E_\Delta \approx 300$ MeV, we get $\chi_\Delta G^2_\Delta/\chi_{\mathrm{N}} n \bar{G}^2_{\mathrm{N}} \approx 6.5$. Another possibility is due to the tensor interaction, which pushes up the strength beyond the one $\hbar\omega$ shell (Shimizu *et al.*, 1974; Bertsch and Hamamoto, 1982).

5.3.5 *Higher multipole magnetic γ transitions and higher forbidden β transitions*

Higher multipole single (quasi)-particle γ and β transitions with $\Delta\pi$ = yes have been investigated in some nuclei. Stretched M4 transitions between single quasi-particle states with $j = l' + \frac{1}{2}$ and $j' = (l' - 3) - \frac{1}{2}$, where l' is the maximum l in the major shell, appear in the low excitation region (see Section 2.3). They have moderate lifetimes enabling them to be measured directly, and the pairing reduction factor $P = U_i U_f + V_i V_f$ is easily corrected for. The reduction of the single-particle M4 transition rates have been studied by many authors (Bohr and Mottelson, 1969; Goldhaber and Sunyar, 1951; Horie and Oda, 1964; Kotajima and Morinaga, 1960).

The matrix element of the stretched M4 transition is given by eqns (5.3.25–5.3.28) with $J = 4$. The orbital coupling strength $(g_l/(J+1))$ is much smaller than that for the spin (μ^- and μ^+). Thus the renormalization effect on g_l may be neglected as in the case of the stretched M2 transitions.

The effective coupling constants g^{eff}_γ (M4) are obtained for the transitions $(1g_{9/2})_p \leftrightarrow (2p_{1/2})_p$, $(1g_{9/2})_n \leftrightarrow (2p_{1/2})_n$, $(2d_{3/2})_n \leftrightarrow (1h_{11/2})_n$ and $(2f_{5/2})_n \leftrightarrow (1i_{13/2})_n$. They all lie around the values $g^{\mathrm{eff}}_\alpha/g_\alpha = 0.32 \pm 0.06$ as shown in Fig. 5.25. It is interesting to note that the renormalization factors $g^{\mathrm{eff}}_\gamma(\mathrm{M4})/g_\gamma(\mathrm{M4})$ are about the same as the values $g^{\mathrm{eff}}_\gamma(\mathrm{M2})/g_\gamma(\mathrm{M2})$ for the lower multipole ($J = 2$) transitions, and that the values for both the proton and the neutron transitions are about the same. The effective coupling constants for the isovector and isoscalar modes are obtained by using eqns (5.3.27) and (5.3.28) from the observed values g^{eff}_γ (M4)/g_γ(M4) for the proton transitions of $(2p_{1/2})_p \leftrightarrow (1g_{9/2})_p$ and the neutron transitions of $(2p_{1/2})_n \leftrightarrow (1g_{9/2})_n$ in simple shell-model nuclei with $N \approx 50$ and $Z \approx 40$. Here the cores of nuclei with proton transitions are about the same as those for the neutron cases, and the core polarization effects in this case with $\Delta\pi$ = yes are not sensitive to the configuration of the valence nucleons. Therefore, the A-dependence of g^{eff}/g is neglected for simplicity and g^{eff}_1/g_1 and g^{eff}_0/g_0 are

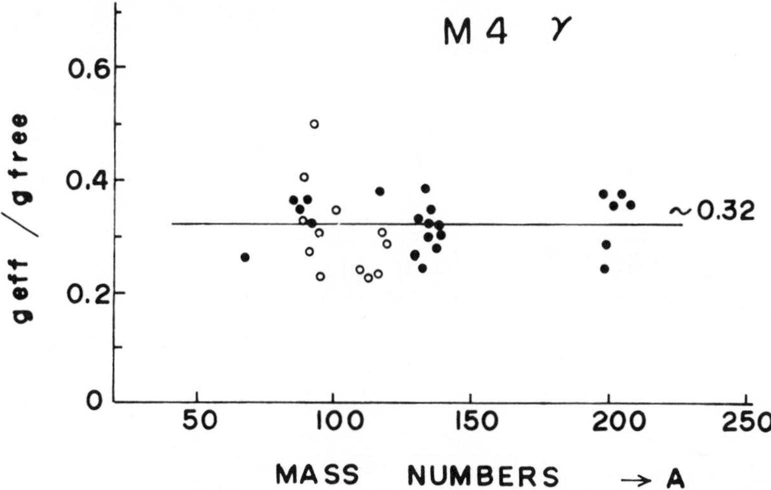

Fig. 5.25. The effective coupling constants for the M4 transitions between single quasi-particle states, $(1g_{9/2})_p \leftrightarrow (2p_{1/2})_p$, $(1g_{9/2})_n \leftrightarrow (2p_{1/2})_n$, $(2d_{3/2})_n \leftrightarrow (1h_{11/2})_n$ and $(2f_{5/2})_n \leftrightarrow (1i_{13/2})_n$. The open and closed circles are for the proton and the neutron transitions, respectively. (Ejiri and Shibata, 1975.)

obtained from the plot as shown in Fig. 5.26. They are $g_{\gamma 1}^{\text{eff}}/g_{\gamma 1} = 0.34 \pm 0.04$ and $g_{\gamma 0}^{\text{eff}}/g_{\gamma 0} = 0.3 \pm 0.2$ (Ejiri and Shibata, 1975). The proton and neutron transition moments in the ^{87}Sr (^{86}Sr core $+$p) and the ^{87}Y (^{86}Sr core + n), respectively, give $g_1^{\text{eff}}/g_1 = 0.34$ and $g_0^{\text{eff}}/g_0 = 0.27$. Therefore both the isoscalar and isovector M4 modes are reduced by approximately the same factor ≈ 0.3, as in case of the M2 γ transitions.

5.4 Spin–isospin polarization interactions and spin–isospin renormalization factors

5.4.1 *The renormalization factors for γ and β transitions in various spin–isospin modes*

It has been shown in previous sections that γ and β transitions in spin–isospin modes have been extensively investigated mainly by means of in-beam e–γ spectroscopy. The transition probabilities are uniformly quenched by factors of $g_\alpha^{\text{eff}}/g_\alpha \approx 0.2$ to 0.4. The average values of $g_\alpha^{\text{eff}}/g_\alpha$ for various spin–isospin modes of $\alpha = TLSJ$ are listed in Table 5.4.1. It is interesting to note that the renormalization factors are rather independent of the mode α, i.e. of the quantum numbers T, S, L, and J of the spin–isospin operator $g_\alpha \mathbf{t}^T[\mathrm{i}^L r^L \mathbf{Y}_L \times \boldsymbol{\sigma}]_J$.

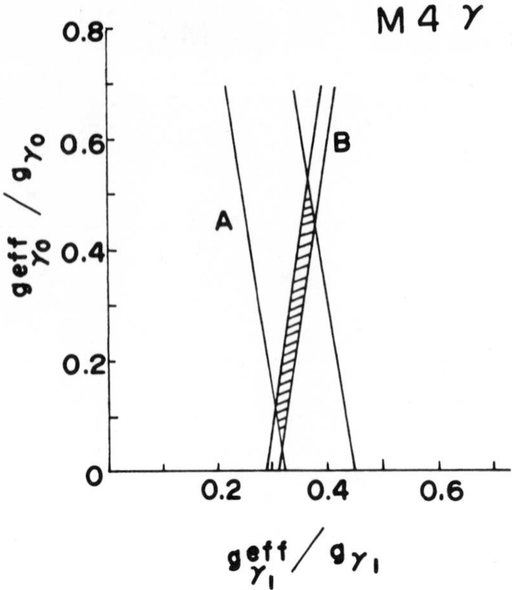

Fig. 5.26. The effective coupling constants for the isoscalar and isovector M4 γ transitions. The regions for the $(2p_{1/2})_p \leftrightarrow (1g_{9/2})_p$ and the $(2p_{1/2})_n \leftrightarrow (1g_{9/2})_n$ are indicated by A and B. The hatched region meets both proton and neutron transitions. The widths show the standard deviations. (Ejiri and Shibata, 1975.)

The renormalization factors $g_\alpha^{\text{eff}}/g_\alpha$ for the isovector electric ($T=1$, $S=0$) and isovector magnetic ($T=1$, $S=1$) transitions are about the same. The values for the $L=1$ and $L=3$ modes are nearly the same, so that they are rather insensitive to the multipolarities in the space coordinate. The isovector γ and the analogous β transitions have approximately the same renormalization factors. The isoscalar magnetic transitions with $T=0$ and $S=1$ are quenched as well as the isovector magnetic ones with $T=1$ and $S=1$, although the uncertainties for the isoscalar transitions are in general large.

Such uniform quenching factors for the spin–isospin responses indicate that spin–isospin core polarizations are induced mainly by the polarization interactions of the Wigner–Majorana type, and that they are thus closely related to the supermultiplet symmetry.

5.4.2 *Spin–isospin polarization interactions*

The uniform renormalization factor is analysed in terms of the spin–isospin core polarization with the separable form of the polarization interaction as given by eqn (5.2.33). In order to deal with various multipole moments, the

Table 5.4.1 Effective coupling constants g^{eff}/g and susceptibilities κ (polarizabilities) for the β and γ transition moments (Ejiri and Fujita, 1978).

Transition mode	Operator	g^{eff}/g	κ
Isovector spin†	$t_3\sigma$	~ 0.5	~ 1
Isovector E1 γ	$t_3 ir\mathbf{Y}_1$	$0.3\sim 0.4$	$1.5\sim 2$
First forbidden axial vector β	$t_{\pm}[ir\mathbf{Y}_1\times\boldsymbol{\sigma}]$	$0.2\sim 0.35$	$1.5\sim 4$
Isovector M2 γ	$t_3[ir\mathbf{Y}_1\times\boldsymbol{\sigma}]_2$	$0.2\sim 0.25$	$3\sim 4$
Isoscalar M2 γ	$[ir\mathbf{Y}_1\times\boldsymbol{\sigma}]_2$	$0.2\sim 0.5$	$1\sim 4$
Second forbidden axial vector†	$t_{\pm}[i^2r^2\mathbf{Y}_2\times\boldsymbol{\sigma}]_3$	~ 0.3	~ 2
Third forbidden axial vector	$t_{\pm}[i^3r^3\mathbf{Y}_3\times\boldsymbol{\sigma}]_4$	~ 0.4	~ 1.5
Isovector M4 γ	$t_3[i^3r^3\mathbf{Y}_3\times\boldsymbol{\sigma}]_4$	$0.3\sim 0.4$	$1.5\sim 2$
Isoscalar M4 γ	$[i^3r^3\mathbf{Y}_3\times\boldsymbol{\sigma}]_4$	$0.2\sim 0.6$	$1\sim 4$

†Isovector modes with $\Delta\pi=$ yes correspond to isovector M1 γ and GT β.

separable interaction for the $\alpha = TLSJ$ mode is rewritten in terms of the dimensionless operator as

$$\mathbf{H}_{\text{I}} = \tilde{\chi}_\alpha \tilde{\mathbf{T}}_\alpha \cdot \tilde{\mathbf{T}}_\alpha \tag{5.4.1}$$

with

$$\tilde{\mathbf{T}}_\alpha = \boldsymbol{\tau}^T[\boldsymbol{\sigma}^s \times (r/b_0)^L \mathbf{Y}_L]_J. \tag{5.4.2}$$

Here the oscillator parameter b_0 is given by $b_0 = \sqrt{(\hbar/M\omega)}$ where $\hbar\omega = 41A^{-1/3}$ MeV and M is the nucleon mass. The values for the isospin modes are derived as $\tilde{\chi}_\alpha \approx 320/A(2.1)^{2L}$ MeV for $L=0$ and $L=1$ so as to reproduce the observed quenching factors and the E1 giant resonance energies. This gives $\tilde{\chi}_{1000} = 320/A$ MeV for the $L=0$ channel, corresponding to the symmetry energy $100\mathbf{tT}/A$ MeV. The value $\tilde{\chi}_{111J} \approx 85/A$ MeV for $L=1$ channels is obtained so as to reproduce both the observed giant resonance energies and the quenching factors in terms of both the $1\hbar\omega$ p–h polarization and the higher excitation $\tau\sigma$ strength $h_{\tau\sigma}$ such as ΔN^{-1} and the second-order polarization (see Section 5.3.4).

The separable interaction is quite convenient to describe the essential features of collective motions such as spin–isospin polarizations and spin–isospin GR's. On the other hand, the Landau–Migdal interaction in the form of the delta-function is known to represent the short-range correlation as long as the physics is limited to small and medium-size momenta (Bäckman and Weise, 1979). Actually, GT GR energies and quenching phenomena of GT and M1 transition strengths have been systematically studied in terms of the Landau–Migdal interaction (Bertsch *et al.*, 1981).

The separable interaction and the Landau–Migdal interaction are quantitatively compared by Toki and Ejiri (1985). The spin–isospin part of the

Landau–Migdal interaction is given by

$$V(LM) = \frac{f_\pi^2}{m_\pi^2} g' \delta(\boldsymbol{r}_1 - \boldsymbol{r}_2) \boldsymbol{\sigma}_1 \cdot \boldsymbol{\sigma}_2 \, \boldsymbol{\tau}_1 \cdot \boldsymbol{\tau}_2, \tag{5.4.3}$$

where f_π is the πNN coupling constant and m_π the pion mass. The combined value is $f_\pi^2/m_\pi^2 = 396$ MeV·fm^3 in common units. The quantity g' is called the Landau parameter. Note that the Landau–Migdal interaction is constant in the momentum space.

The matrix elements of both the separable interaction and the Landau–Migdal interaction are evaluated by using a macroscopic model. Comparing the two matrix elements, the Landau parameter corresponding to the separable interaction strength of $\tilde{\chi}_{11 1J} = (85/A)$ MeV is obtained as $g' = 0.73$. On the other hand comparison of the p–h matrix elements of the Landau–Migdal and the separable interactions leads to $g' = 0.73$ to 0.80, corresponding to the same separable interaction strength. The Landau parameter $g' = 0.6$ has been deduced from the GT GR energy without explicitly taking into account the higher excitation strength $h_{\tau\sigma}$ (Bertsch *et al.*, 1981). Correcting for the contribution of $h_{\tau\sigma}$, the parameter g' is around 0.7, which is quite consistent with the values $g' = 0.73$ to 0.8 corresponding to the separable interaction.

5.5 References

Anantaraman, N., Crawley, G. M., Galonsky, A., Djalali, C., Marty, N., Morlet, M., Willis, A. and Jourdain, J.-C. (1981). *Phys. Rev. Lett.* **46,** 1318.

Anderson, J. D. and Wong, C. (1961). *Phys. Rev. Lett.* **7,** 250.

Arima, A. and Horie, H. (1954). *Progr. theor. Phys.* **11,** 509.

Arima, A., Horie, H. and Sano, M. (1957). *Progr. theor. Phys.* **17,** 567.

Arima, A. and Huang-Lin, L. J. (1972). *Phys. Lett.* **41B,** 429.

Arita, K., Ejiri, H. and Fujita, J. I. (1976). *Suppl. Progr. theor. Phys.* **60,** 47.

Arvieu, R. and Veneroni, M. (1960). *C.r. hebd. Séanc. Acad. Sci. Paris* **250,** 992.

Austin, S. M. (1972). *Proc. Symp. on the two-body force in nuclei* (eds. S. M. Austin and G. M. Crawley), p. 285. Plenum Press, New York.

Bäckman, S.-O. and Weise, W. (1979). *Mesons in nuclei* (eds. M. Rho and D. H. Wilkinson), p. 1097. North-Holland, Amsterdam.

Bainum, D. E., Rapaport, J., Goodman, C. D., Horen, D. J., Foster, C. C., Greenfield, M. B. and Goulding, C. A. (1980). *Phys. Rev. Lett.* **44,** 1751.

Baldwin, G. C. and Klaiber, G. S. (1947). *Phys. Rev.* **71,** 3.

Baranger, M. (1960). *Phys. Rev.* **120,** 331.

Bertrand, F. E., Gross, E. E., Horen, D. J., Wu, J. R., Tinsley, J., McDaniels, D. K., Swenson, L. W. and Liljestrand, R. (1981). *Phys. Lett.* **103B,** 326.

Bertsch, G. F. and Tsai, S. F. (1975). *Phys. Rev.* **18** No. 2, 125.

Bertsch, G. F., Cha, D. and Toki, H. (1981). *Phys. Rev.* **C24,** 533.

Bertsch, G. F. and Hamamoto, I. (1982). *Phys. Rev.* **C26,** 1323.

Blin-Stoyle, R. J. (1953). *Proc. Phys. Soc.* **A66,** 1158.

Blin-Stoyle, R. I., Arima, A., Ericson, M., Rho, M. and Green, A. M. (1978). *Progress in Particle and Nuclear Physics*, Vol. 1 (ed. D. H. Wilkinson), Pergamon Press.

Bohr, A. and Mottelson, B. (1953). *Dan. mat. Fys. Medd.* **27,** No. 16.

Bohr, A. and Mottelson, B. (1969/1975). *Nuclear structure*, Vols 1 and 2. Benjamin, New York.

Bohr, A. and Mottelson, B. (1981). *Phys. Lett.* **100B,** 10.

Brenn, R., Bhattacherjee, S. K., Sprouse, G. D. and Young, L. E. (1974). *Phys. Rev.* **C10,** 1414.

Brown, G. E. and Bolsterli, M. (1959). *Phys. Rev. Lett.* **3,** 472.

Brown, G. E., Castillejo, L. and Evans, J. A. (1966). *Nucl. Phys.* **22,** 1.

Chemtob, M. (1969). *Nucl. Phys.* **A123,** 449.

Church, E. L. and Weneser, J. (1960). *Ann. Rev. nucl. Sci.* **10,** 193.

Crawley, G. M., Anantaraman, N., Galonsky, A., Djalali, C., Marty, N., Morlet, M., Willis, A., Jourdain, J.-C. and Kitching, P. (1982). *Phys. Rev.* **C26,** 87.

Damgaad, I. and Winter, A. (1964). *Nucl. Phys.* **54,** 615.

Danos, M. and Steinwedel, H. (1951). *Z. Naturf.* **6a,** 217.

Delabaye, M. and Lipnik, P. (1966). *Nucl. Phys.* **86,** 668.

Doering, R. R., Galonsky, A., Patterson, D. M. and Bertch, G. F. (1975). *Phys. Rev. Lett.* **35,** 1691.

Ejiri, H., Ohmura, H., Nakajima, Y., Horie, K., Hashimoto, Y., Eto, K., Matsumoto, S. and Nogami, Y. (1964). *Nucl. Phys.* **54,** 561.

Ejiri, H. (1967). *J. Phys. Soc. Japan* **22,** 361.

Ejiri, H., Ikeda, K. and Fujita, J. I. (1968a). *Phys. Rev.* **176,** 1277.

Ejiri, H. and Bondorf, J. (1968b). *Phys. Lett.* **28B,** 304.

Ejiri, H., Richard, P., Ferguson, S., Heffner, R. and Perry, D. (1968/1969). *Phys. Rev. Lett.* **21,** 373; *Nucl. Phys.* **A128,** 388.

Ejiri, H., Ferguson, S. M., Halpern, I. and Heffner, R. (1969). *Phys. Lett.* **29B,** 111; *Proc. Int. Conf. on Nuclear Isospin*, (eds. J. D. Anderson, S. D. Bloom, J. Cerny and W. W. True), p. 65. Academic Press, New York.

Ejiri, H. (1971). *Nucl. Phys.* **A166,** 594.

Ejiri, H. (1972). *Nucl. Phys.* **A178,** 350.

Ejiri, H., Shibata, T. and Satoh, K. (1972). *Phys. Lett.* **38B,** 73.

Ejiri, H., Shibata, T. and Fujiwara, M. (1973a). *Phys. Rev.* **C8,** 1892.

Ejiri, H. (1973b). *Nucl. Phys.* **A211,** 232.

Ejiri, H., Shibata, T. and Takeda, M. (1974). *Nucl. Phys.* **A221,** 211.

Ejiri, H. and Shibata, T. (1975). *Phys. Rev. Lett.* **35,** 148.

Ejiri, H., Shibata, T., Nagai, Y. and Nakayama, S. (1976). *Nucl. Instr. Meth.* **134,** 107.

Ejiri, H. and Fujita, J. I. (1978). *Phys. Report* **38C,** 85.

Ejiri, H. (1982). *Phys. Rev.* **C26,** 2628.

Ejiri, H. (1983). *Nucl. Phys.* **A396,** 181C.

Endt, P. M. (1969). *Proc. Nuclear Isospin, Asilomar* (eds J. D. Anderson, S. D. Bloom, J. Cerny and W. W. True), p. 51. Academic Press, New York.

Ericson, M. and Ericson, T. E. O. (1966). *Ann. Phys.* **36,** 323.

Ericson, M., Figureau, A. and Thévbent, C. (1973). *Phys. Lett.* **45B,** 19.

Eulenberg, G., Sober, D. I., Steffen, W., Gräf, H.-D., Küchler, G., Richter, A., Spamer, E., Metsch, B. C. and Knüpfer, W. (1982). *Phys. Lett.* **116B,** 113.

Flynn, E. R. and Garrett, J. D. (1972). *Phys. Rev. Lett.* **29,** 1748.

Fox, J. D., Moore, C. F. and Robson, D. (1964). *Phys. Rev. Lett.* **12,** 198.

Freed, N. and Kisslinger, L. S. (1968). *Nucl. Phys.* **A116,** 401.

Fujita, J. I. (1962). *Phys. Rev.* **126,** 202.

Fujita, J. I. (1963/1967). Brookhaven National Lab. report 887 C39; *Phys. Lett.* **24B,** 123.

Fujita, J. I., Fujii, S. and Ikeda, K. (1964). *Phys. Rev.* **133B,** 549.

Fujita, J. I. and Ikeda, K. (1965). *Nuovo Cim.* **67,** 145.

Fujita, J. I. and Ikeda, K. (1965b). *Nucl. Phys.* **67,** 145.

Fujita, J. I., Futami, Y. and Ikeda, K. (1967). *Progr. theor. Phys.* **38,** 107.

Fujita, J. I. (1968). *Phys. Rev.* **172,** 1047.

Fujita, J. I., Hirata, M. and Shoda, K. (1970). *Phys. Lett.* **33B,** 550.

Gaarde, C., Kemp, K., Naumov, Y. V. and Amundsen, P. R. (1970). *Nucl. Phys.* **A143,** 497.

Gaarde, C. (1981). *Nucl. Phys.* **A369,** 258.

Gaarde, C. (1983). *Nucl. Phys.* **A396,** 127C.

Glassgold, A. F., Heckrotte, W. and Watson, K. M. (1959). *Ann. Phys.* **6,** 1.

Goldhaber, M. and Teller, E. (1948). *Phys. Rev.* **74,** 1046.

Goldhaber, M. and Sunyar, A. W. (1951). *Phys. Rev.* **83,** 906.

Goodman, C. D., Goulding, C. A., Greenfield, M. B., Rapaport, J., Bainum, D. E., Foster, C. C., Love, W. B. and Petrovich, F. (1980). *Phys. Rev. Lett.* **44,** 1755.

Grecksch, E., Knupeer, W. and Huber, M. G. (1975). *Lettere Nuovo Cim.* **14,** 505.

Halbleib Sr., J. A. and Sorensen, R. A. (1969). *Nucl. Phys.* **A98,** 542.

Hamamoto, I. (1965). *Nucl. Phys.* **62,** 49; *ibid.* **66,** 176.

Hamamoto, I. (1969/1970). *Nucl. Phys.* **A135,** 576; *ibid* **A148,** 465.

Hashinoff, M., Fisher, G. A., Kurjan, P. and Hanna, S. S. (1972). *Nucl. Phys.* **A195,** 78.

Hertel, J. W., Fleming, D. G., Schiffer, J. P. and Gove, H. E. (1969). *Phys. Rev. Lett.* **23,** 488.

Hirata, M. (1970). *Phys. Lett.* **32B,** 656.

Horen, D. J., Goodman, C. D., Foster, C. C., Goulding, C. A., Greenfield, M. G., Rapaport, J., Sugarbaker, E., Masterson, T. G., Petrovich, F. and Love, W. G. (1980, 1981). *Phys. Lett.* **95B,** 27; *ibid* **99B,** 383.

Horie, H. and Arima, A. (1955). *Phys. Rev.* **99,** 778.

Horie, H. and Oda, T. (1964). *Progr. theor. Phys.* **32,** 65.

Hsieh, S. T. and Ogata, H. (1970). *Phys. Rev.* **C2,** 2091.

Ikeda, K., Fujii, S. and Fujita, J. I. (1963). *Phys. Lett.* **3,** 271.

King, P. W. and Peaslee, D. C. (1954). *Phys. Rev.* **94,** 1284.

Kisslinger, L. S. and Sorensen, R. A. (1963). *Rev. mod. Phys.* **35,** 853.

Klapdor, H. V. (1971). *Phys. Lett.* **35B,** 405.

Kleinheinz, P., Zuber, K., Conic, C., Protop, C., Zuber, J., Liang, C. F., Paris, P. and Blomqvist, J. (1985). *Phys. Rev. Lett.* **55,** 2664.

Kotajima, K. and Morinaga, H. (1960). *Nucl. Phys.* **16,** 231.

Kurath, D. and Lawson, R. D. (1967). *Phys. Rev.* **161,** 915.

Lederer, C. M. and Shirley, V. S. (1978). *Table of Isotopes*, 7th Edition. John Wiley, New York.

Letessier, J. and Foucher, R. (1969). *Ann. Phys.* **4,** 55.

Levinger, J. S. (1960). *Nuclear Photo-disintegration*, Oxford University Press.

Lindgren, R. A., Bendel, W. L., Fagg, L. W. and Jones, Jr., E. C. (1975). *Phys. Rev. Lett.* **24,** 1423.

Lipnik, P. and Sunier, J. W. (1964). *Nucl. Phys.* **56,** 241.

Lipnik, P. (1966). *Nucl. Phys.* **86,** 668.

Liu, K. F. and Brown, G. E. (1976). *Nucl. Phys.* **A265,** 385.

Love, W. G. and Franey, M. A. (1981). *Phys. Rev.* **C24,** 1073.
Maripuu, S. (1970). *Phys. Lett.* **31B,** 181.
Marumori, T. (1960). *Progr. theor. Phys.* **24,** 331.
Migdal, A. B. (1944/1945). *J. Phys.* **8,** 331; *Zh. exp. theor. Fiz.* **15,** 81.
Migdal, A. B. (1964). *JETP* **19,** 1136.
Migdal, A. B. (1968). *Nuclear Theory, The Quasi-particle Method*, Benjamin.
Miyazawa, H. (1951). *Progr. theor. Phys.* **6,** 801.
Morinaga, A. and Hisatake, K. (1975). *J. Phys. Soc. Japan*, **38,** 322.
Morinaga, H. and Yamazaki, T. (1976). *In-Beam Gamma-Ray Spectroscopy*. North-Holland, Amsterdam.
Morita, M. (1967). *Phys. Rev.* **161,** 1028.
Morita, M., Yamada, M., Fujita, J. I., Fujii, A., Ohtsubo, H., Morita, R., Ikeda, K., Yokoo, Y., Hirooka, M. and Takahashi, K. (1971). *Progr. theor. Phys. Suppl.* No. 48, p. 41.
Morita, M. (1973). *Beta Decay and Muon Capture*. Benjamin, Reading, Mass.
Morita, M. and Kume, K. (1974). Private Communication.
Morita, M., Arita, K., and Ohtsubo, H. (1976). *Progr. theor. Phys. Suppl.* No. 60, p. 125.
Mottelson, B. R. (1967). *Proc. INS Summer School, Nikko*, p. 1. INS PT20.
Nagamiya, S., Kataou, T., Nomura, T. and Yamazaki, T. (1970). *Phys. Lett.* **33B,** 574.
Nagai, Y., Shibata, T., Nakayama, S. and Ejiri, H. (1977). *Nucl. Phys.* **A282,** 29.
Nagai, Y., Shibata, T., Okada, K., Ohsumi, H. and Ejiri, H. (1984). *Proc. Int. Symp. Nucl. Spectroscopy and Nucl. Interactions*, Osaka (eds. H. Ejiri and T. Fukuda), p. 72. World Scientific, Singapore.
Nakayama, S., Kishimoto, T., Itahashi, T., Shibata, T. and Ejiri, H. (1978). *J. Phys. Soc. Japan* **45,** 740.
Noya, H., Arima, A. and Horie, H. (1958). *Progr. theor. Phys. Suppl.* No. 8, p. 33.
Noziéres, P. and Pines, D. (1958). *Nuovo Cim.* **IX,** 470; *Phys. Rev.* **109,** 741.
Oset, E. and Rho, M. (1979). *Phys. Rev. Lett.* **42,** 47.
Osterfeld, F. (1982). *Phys. Rev.* **C26,** 762.
Paul, P. (1972). *Proc. Int. Conf. on Nuclear Structure Studies Using Electron Scattering and Photo Reactions*, Sendai (eds. K. Shoda and H. Ui), p. 343. Tohoku Univ.
Petrovich, F., Brown, G. E., Garvey, G. T., Goodman, C. D., Lindgren, R. A. and Love, W. G. (1982). *Proc. Conf. Spin Excitations in Nuclei, Telluride, Colorado*, Plenum Press, New York.
Richard, P., Gobbard, F., Porter, A. G. and Hopkins, F. F. (1969). *Phys. Lett.* **29B,** 649.
Ring, P., Bauer, R. and Speth, J. (1973). *Nucl. Phys.* **A206,** 97.
Rose, M. E. and Osborn, R. K. (1954). *Phys. Rev.* **93,** 1326.
Sakai, M. and Yoshida, S. (1964). *Nucl. Phys.* **50,** 497.
Sakai, M., Bertini, R. and Gehringer, C. (1970). *Nucl. Phys.* **A157,** 113.
Shibata, Y. (1967). *Phys. Lett.* **24B,** 557.
Shimizu, K., Ichimura, M. and Arima, A. (1974). *Nucl. Phys.* **A226,** 282.
Shoda, K., Suzuki, A., Sugawara, M., Saito, T., Miyase, H. and Oikawa, S. (1971). *Phys. Rev.* **C3,** 1999, 2006.
Snover, K. A., Armann, J. F., Hering, W. and Paul, P. (1971). *Phys. Lett.* **37B,** 29.
Sorensen, R. A. (1963). *Phys. Rev.* **132,** 2270.
Spector, R. M. (1960). *Nucl. Phys.* **40,** 338.
Speth, J., Krewald, S., Osterfeld, F. and Suzuki, T. (1982). *Proc. Int. Conf. Spin Excitations in Nuclei*, Telluride, Colorado (eds. F. Petrovich, *et al.*), p. 445. Plenum Press, New York.

Steinwedel, H., Jensen, J. H. D. and Jensen, P. (1950). *Phys. Rev.* **79,** 1019.
Takagi, S. (1959). *Prog. theor. Phys.* **21,** 174.
Toki, H. and Ejiri, H. (1985). *Nucl. Phys.* **A438,** 503.
Towner, I. S., Warburton, E. and Garvey, G. (1971). *Ann. Phys.* **66,** 674.
Towner, I. S. and Khanna, F. C. (1979). *Phys. Rev. Lett.* **42,** 51.
Überall, H. (1965). *Phys. Rev.* **B137,** 502; **B139,** 1239.
Überall, H. (1966). *Nuovo Cim.* **41B,** 25.
Warburton, E. K. and Weneser, J. (1969). *Isospin in Nuclear Physics* (ed. D. H. Wilkinson), p. 174. North-Holland, Amsterdam.
Wild, W. (1956). *Bayr, Akad, Wiss. Math. Nat. Kl.* **18,** 371.
Yamada, M. (1965). *Bull, Sci, Eng. Res. Lab. Waseda Univ.* No. 35, p. 146.
Yamazaki, T., Nomura, T., Katou, U., Inamura, T., Hashizume, A. and Tendou, Y. (1970). *Phys. Rev. Lett.* **24,** 317.
Yamazaki, T., Nomura, T., Nagamiya, S. and Katou, T. (1970). *Phys. Rev. Lett.* **25,** 547.
Yoshida, S. (1961). *Phys. Rev.* **123,** 2122.
Yoshida, S. (1962). *Nucl. Phys.* **38,** 380.

6

HIGH-SPIN STATES

6.1 General features of high-spin physics

The nuclear spectroscopy of high-spin states has undergone spectacular progress in the last decade due to the possibilities of accelerating heavy ions and the development of sophisticated detection techniques as well as nuclear theories. Much excitement was caused by the observations of nuclear states with very high angular momentum ($I\hbar \geqslant 40\hbar$) as discrete lines in the γ-ray spectra. They have been isolated in a domain of high excitation energy with generally high nuclear density. The excitement was not only due to the beautiful experimental selection techniques, but also to the possibility of studying new features of nuclear structure. The fast rotation of a nucleus indeed induces significant modifications of nucleonic motions and of nuclear stability, changes in the nuclear shape and in the deformation. It may also induce new phases of the strongly interacting nucleon system.

The atomic nucleus is the only strongly interacting many-body quantum system that can be given extremely high angular momentum and for which the influence of fast rotation can be studied. The angular momentum may amount even to $60 \sim 80\hbar$, while keeping the nucleons in a bound nuclear system. The nucleus with such a high angular momentum therefore attains a relatively long lifetime (10^{-15} to 10^{-13} s), being stable, at least in part, against nucleon emission and fission. Consequently, its nuclear structure can be well studied by means of electron and gamma spectroscopy. We realize, however, that highly excited nuclei with excitation energy above the nucleon-emission or fission threshold energy have even higher angular momentum. The dynamic properties of such high-spin unbound nuclei are discussed in Chapter 4. The large amounts of energy and angular momentum transferred in nuclear reactions are not only converted to external excitations (i.e. rotation of the nucleus as a whole) but also to internal degrees of freedom. The excitation of collective and single-particle degrees of freedom offer us the possibility of studying the interplay between classical and quantal phenomena. An illustrative example, discussed below, is the breaking of the pairing correlations between nucleons, due to the Coriolis interaction experienced by the nucleons in the fast rotating nucleus.

We can identify at least two different modes of rotation, both generating very high angular momentum. In a *collective rotation* many nucleons contribute coherently to the nuclear motion resulting in a rotation of a sizeable number of nucleons around an axis different from the nuclear symmetry axis. This type of excitation occurs in well-deformed nuclei since the rotation of the deformed nucleus, having a large moment of inertia, is so slow that the individual nucleonic motion remains undisturbed in the rotating deformed potential. A schematic example of a prolate nucleus rotating around an axis perpendicular to the nuclear symmetry axis is given in Fig. 6.1(a).

In spherical or weakly deformed nuclei, however, the alignment of the individual nucleonic orbitals along the nuclear symmetry axis appears to be the mechanism of building up high angular momentum. This single-particle type of nuclear motion is also called *non-collective rotation* and is illustrated schematically in Fig. 6.1(b) for a few valence nucleons moving in an oblate average potential in orbitals with positive angular momentum projections on the nuclear symmetry axis.

The two types of nuclear rotation lead to different lifetimes of the excited states because the γ-ray decay proceeds either via fast collective transitions, or in the case of non-collective rotation, by rearrangement of only a few particles (possibly with a large effective charge), via retarded transitions.

Two contrasting energy spectra of a prolate deformed nucleus ^{158}Er and of a spherical nucleus ^{147}Gd are shown in Fig. 6.2. The regular de-excitation pattern of ^{158}Er is characteristic of collective rotation. When the rotation is fast enough the Coriolis force may destroy the correlation between a pair of nucleons, causing the anomaly at $I^{\pi} = 14^{+}$ and (less pronounced) at 28^{+}. In ^{147}Gd, however, the individual nucleonic motion dominates and different excited states are formed by different occupations of nucleonic levels, causing the irregularity. The difference in decay probabilities of the two modes of

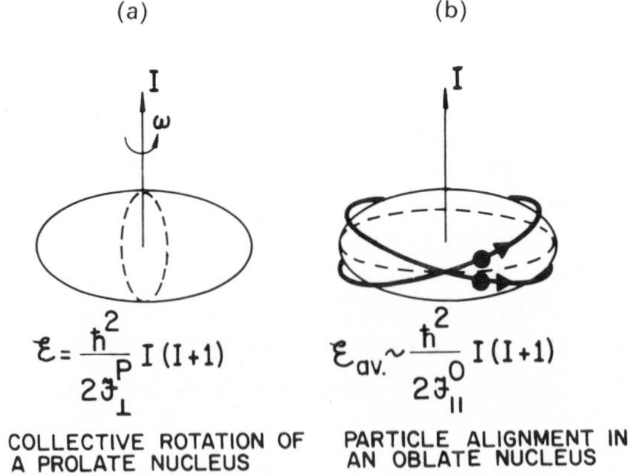

Fig. 6.1. Two different modes of rotation generating angular moentum. The quantities $\mathcal{J}_{\perp}^{P}$ and $\mathcal{J}_{\parallel}^{O}$ represent the moments of inertia of the two respective rigid bodies rotating around an axis perpendicular to (a) or parallel with (b), the symmetry axis.

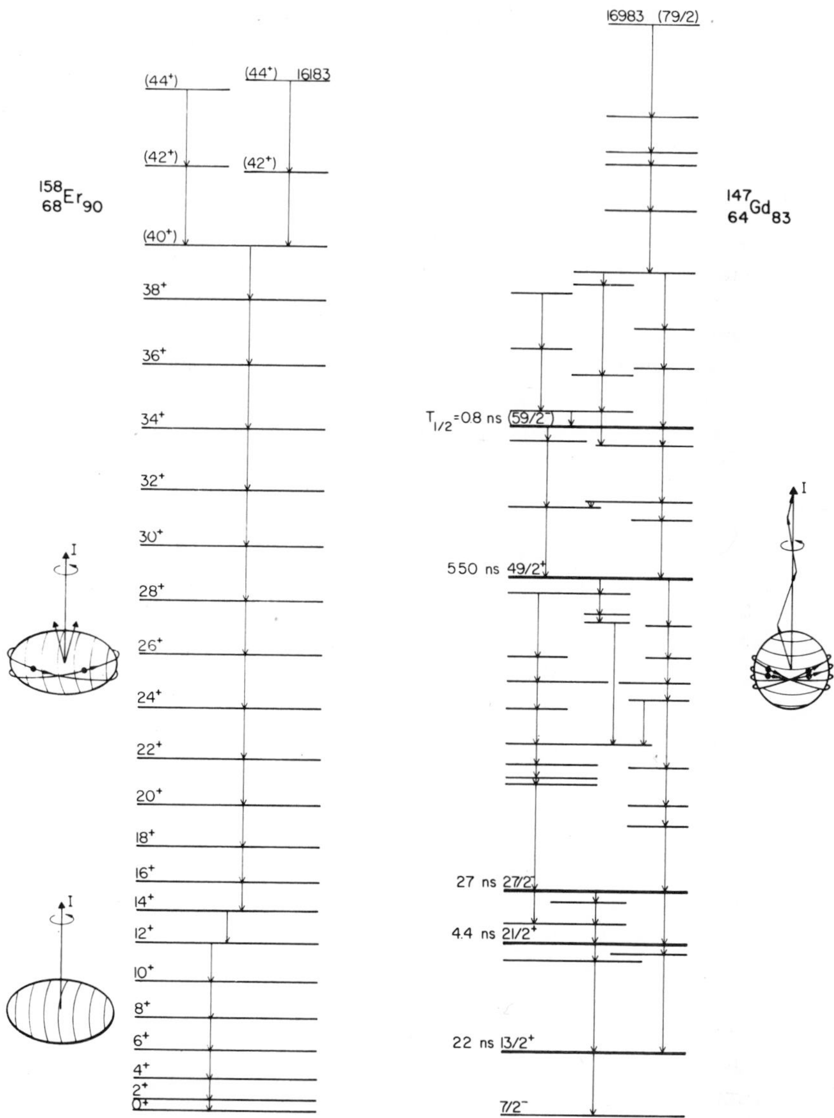

Fig. 6.2. The level schemes of ^{158}Er deduced from ^{40}Ar-induced fusion reactions (Burde *et al.*, 1982) and using ^{48}Ca beams (Simpson *et al.*, 1984) and of ^{147}Gd using α, ^{12}C and ^{30}Si beams (Broda *et al.*, 1982; Khoo *et al.*, 1983). The angular momentum I at low excitation energies in ^{158}Er is generated by collective rotation, as illustrated schematically on the side. At higher spins the alignment of quasi-particles also contribute to I. For ^{147}Gd the assumed shape is spherical and I can only be generated by individual nucleons. Isomeric states in ^{147}Gd are indicated by half-lives. (After Stephens, 1983.)

excitation is shown in Fig. 6.3 where electric quadrupole transition rates are collected for high-spin states in a number of rare-earth nuclei. Near the closed $N=82$ shell the transitions in the nuclei with spherical shape are strongly retarded with respect to those in deformed nuclei further away from the closed shell. This different behaviour of collective and non-collective electromagnetic transitions requires different experimental and analysis techniques which are discussed separately in Sections 6.3 and 6.4, respectively. It also appears that the balance and interplay between these two modes of rotation play an important role in the nuclear structure of high-spin states. Therefore the two Sections 6.3 and 6.4 cannot be seen as independent treatments of two aspects of high-spin physics, and so cross-references are made whenever relevant.

In fact, experimental results in recent years have provided evidence for transitions from one mode of rotation to the other, accompanied by drastic changes of the nuclear shape. Such features are expected to occur in the

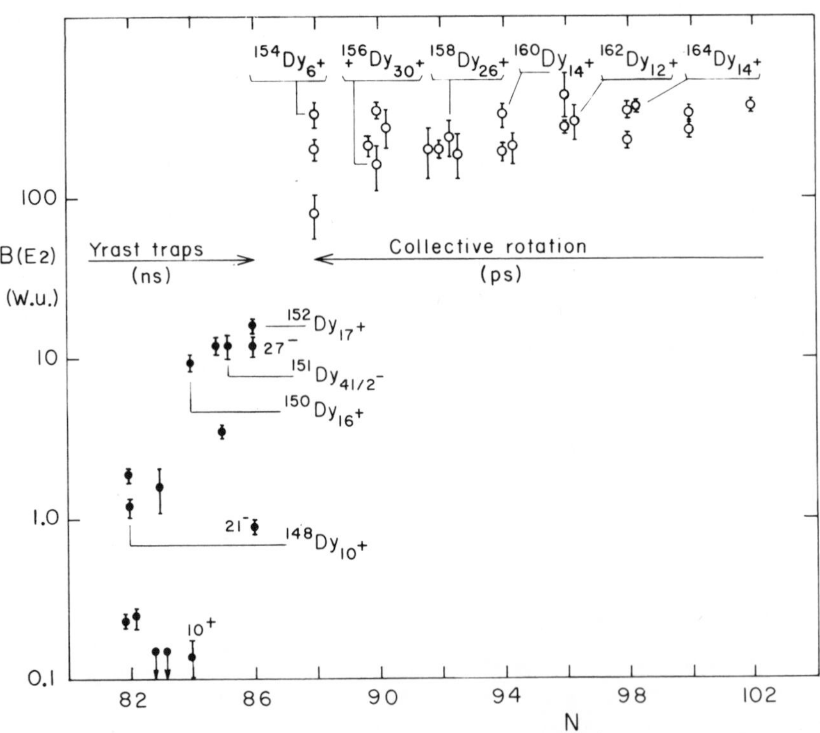

Fig. 6.3. Electric quadrupole transition probabilities (expressed in Weisskopf units) measured for a number of high-spin states in rare earth nuclei. For $^{148-164}$Dy isotopes the data points are marked with the spin and parity of the state. The black dots correspond to isomeric states, called yrast traps (see Section 6.4).

domain of very high angular momenta (30 to 40ħ) where the level density becomes so large that most of the decay transitions are unresolved, constituting a 'continuum' in the γ-ray spectrum. The study of the 'quasi-continuum' spectrum has established, in the past decade, a new field in nuclear spectroscopy requiring special techniques, such as crystal balls, multiplicity filters and sum spectrometers, as discussed in Chapter 3.

Modern experimental techniques have revealed rich spectra of various kinds of nuclear excitations, exhibiting rotational band structures up to high spin (~ 60). They enabled accurate measurements to be made of the energy systematics of quasiparticle excitations as a function of the rotational frequency. They also provided testing material for the simplified picture of the independent particle motion in the hierarchy of various nuclear potentials as illustrated in Fig. 6.4 for the 2s–1d shell states. The most simple assumption is the motion of nucleons in a harmonic oscillator potential with, for instance, 12 particles occupying the $N=2$ level with positive parity π (degeneracy $=(N+1)(N+2)=12$). Any distortion of this potential will lift the degeneracy with respect to the orbital angular momentum l, as is the case of a Woods–Saxon potential. Each level can be occupied by $2(2l+1)$ particles, thus there are two in the 2s ($l=0$) and ten in the 1d ($l=2$) level. A further

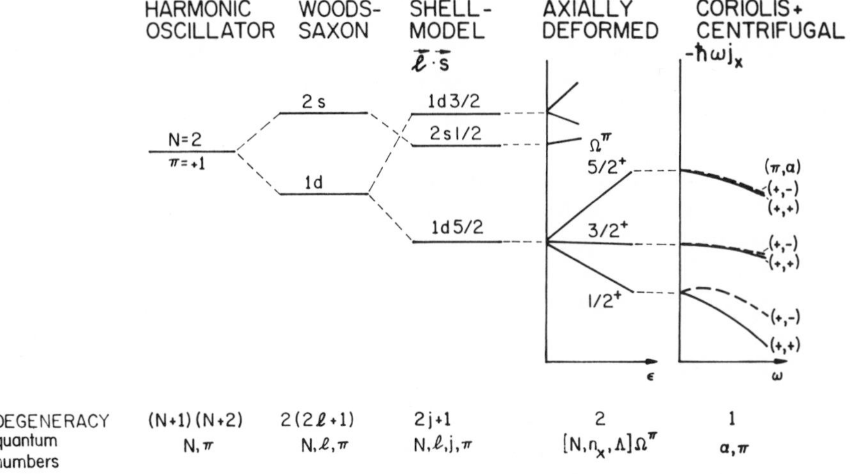

Fig. 6.4. A schematic energy spectrum of the s–d shell states generated by various nuclear potentials. For each potential the degeneracy and the characteristic quantum numbers are given. Here π is the parity, N the total oscillator quantum number, l the orbital and j the total angular momentum. In the case of deformed nuclei, n_3, Λ and Ω are the components along the symmetry axis of N, l and j, respectively. The signature $\alpha=\frac{1}{2}$ is related to the eigenvalue $r=\exp(i\pi\alpha)$ of the rotation operator $R_{x'}$ (see Section 6.3.2). (After Garrett *et al.*, 1983.)

modification is induced by the spin–orbit interaction as in the spherical shell model, leading to the correct shell structure and observed 'magic numbers' at the shell closures. The $2j+1$ fold degeneracy of the shell model states is further reduced for an axially deformed potential as in the Nilsson model. The six particles in the spherical $1d_{5/2}$ state, for instance, can occupy different states labelled with the asymptotic (Nilsson) quantum numbers for a certain deformation (see Bohr and Mottelson, 1975). Each state can be occupied by a maximum of 2 particles. Note that states with a projection of the particle spin j on the symmetry axis with $+\Omega$ and $-\Omega$ are indistinguishable.

When the deformed nucleus experiences a fast rotation, Coriolis and centrifugal forces act on the individual nucleons in the rotating frame and an extra term $-\hbar\omega j_x$ must be added to the intrinsic Hamiltonian for independent-particle motion. Here j_x represents the spin projection on the axis of rotation. The Coriolis interaction depends on the angular velocity and therefore violates time-reversal invariance and lifts the two-fold degeneracy of the Nilsson levels, each level being split into two components as shown in Fig. 6.4. Axially deformed nuclei are invariant under space inversion P and rotation R by an angle π around an axis perpendicular to the symmetry axis. These are the only symmetries left in fast rotating axially deformed nuclei and the resulting states are labelled by the eigenvalues of the two operators P and R, namely parity $\pi = \pm 1$ and signature $\alpha = \pm\frac{1}{2}$ (or $r = \mp i$).

Fast nuclear rotation thus offers the possibility of studying the single-(quasi) particle spectrum in its full complexity where all degeneracies are removed. Consequently, the twelve particles in the single $N=2$ harmonic oscillator level must each occupy an identifiable state in the rotating deformed potential.

The phenomena indicated above will be discussed extensively in the following sections after we have summarized in Section 6.2 some possibilities of populating high-spin states. For detailed nuclear structure theory the reader is referred to the standard works of Bohr and Mottelson (1969, 1975). An extensive theoretical treatment of fast nuclear rotation can be found in the book by Szymański (1983). An experimental and theoretical review of high-spin phenomena is given by de Voigt *et al.* (1983).

6.2 The population of high-spin states

6.2.1 *Nuclear reactions for populating high-spin states*

In studies of high-spin states one obviously searches for reactions which transfer to the nucleus the highest possible angular momentum l with the largest possible cross-section σ. It was found that fusion reactions provide a most efficient way to achieve this. In a complete-fusion (CF) reaction the incident particle is captured by the target nucleus to form a compound nucleus. The kinetic energy in the centre of mass (c.m.) system is converted into excitation energy of the compound nucleus, being shared among its constituent nucleons. One speaks about the formation of a compound nucleus only if the lifetime of the composite system (projectile + target nucleus) is long enough ($\geqslant 10^{-20}$ s) for thermodynamic equilibrium to be reached before it decays. During that time all memory of its formation process is lost, with the exception of the conservation of energy, angular momentum and parity.

The transferred angular momentum in the reaction is determined by the momentum of the projectile $m_p v_p$ relative to the target nucleus and by the impact parameter b, according the relation $l = m_p v_p b$. This simple relation shows that high-energetic heavy ions are the most suitable projectiles to excite high-spin states. There are, however, several restrictions on the CF process due to competing reaction mechanisms. It is shown in Fig. 6.5 that, for instance, complete fusion takes place only up to a certain limiting value of b and that other processes occur for larger values of b. The largest impact parameters and thus the shortest reaction times can be related to collisions in which the nuclear density distributions of projectile and target nucleus do not overlap. In these 'distant' collisions no mass is exchanged and the Coulomb force (which is long-range) determines exclusively the processes such as elastic (Rutherford) scattering and inelastic (Coulomb) excitation.

When projectile and target nucleus come into closer contact the nuclear interaction becomes important. In 'soft grazing' collisions energy and in some

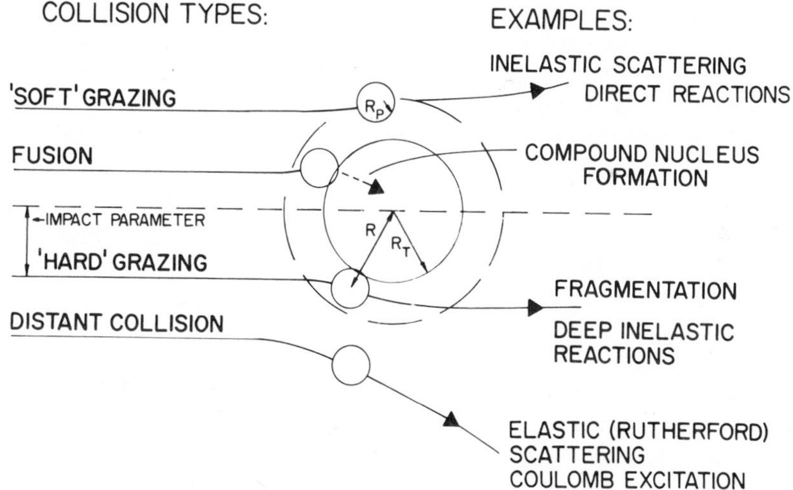

Fig. 6.5. A schematic illustration of heavy-ion collisions. The half-density radii of projectile and target nucleus are indicated by R_p and R_T, respectively. It is assumed that fusion takes place when the impact parameter is smaller than the distance of closest approach $R = R_p + R_T$ (see text). (Hageman, 1981.)

cases also mass will be exchanged but the system will keep its original asymmetry in kinetic energy and mass (inelastic scattering, direct reactions). The term 'hard grazing' collisions is used when the projectile penetrates deeper into the target nucleus. It may then be fragmented or a composite system formed for a short time (fragmentation, deep inelastic reactions). The reaction time will increase for this kind of collision and a substantial mass and/or energy transfer will occur but the mass and/or energy of the ejectiles will still correspond to that of the projectiles. If the projectile–target system is strongly asymmetric (e.g. $^{16}O + ^{146}Nd$) these collisions can be associated with processes in which a substantial mass transfer occurs from projectile to target nucleus, whereas the velocity of the ejectile remains close to the projectile velocity. These processes are called 'massive transfer' or 'incomplete fusion' (ICF). For more symmetric systems (e.g. $^{40}Ar + ^{122}Sn$) a large part of the kinetic energy of the projectile is converted into excitation energy of the target nucleus and/or the projectile, while the system keeps its original mass asymmetry ('strongly damped' or 'deep inelastic' reactions). These processes have essentially a non-equilibrium character. For even smaller impact parameters the projectile may fuse with the target nucleus to form a compound thermal equilibrium system if the nuclear attraction at a certain distance R is stronger than the sum of the Coulomb and the centrifugal repulsion. It is thus found that with increasing impact parameter the angular momentum increases, but that a limit may be reached where fusion is not any more the dominant reaction process. In fact the impact parameter b gives merely a measure of the dominant reaction process and accordingly the classification of the reaction process by the parameter b is not quite rigid. Even in the case of a central collision there is a finite probability that a number of nucleons (or a nuclear fragment) is ejected through the internucleon collision cascade before thermal equilibrium is attained. Therefore the angular momentum window for each reaction process is not sharp cut but rather overlaps with the next one. These features are unique for nuclear collisions associated with a finite number of nucleons and a finite range of the interaction (see Section 4.2; pre-equilibrium processes).

6.2.2 *Compound nucleus formation*

Once the compound nucleus is formed after a fusion process there is a finite possibility that fission takes place rather than particle and γ decay via high-spin states. The fission probability increases with the angular momentum and also depends on the mass of the compound nucleus, as pointed out by Cohen *et al.* (1974). Figure 6.6 shows the maximum angular momenta l_{max} that can be given to beta-stable nuclei. Above the line l_I nuclei are superdeformed and the line l_{II} corresponds to the fission limit. For nuclei with $A > 100$ the value of l_{II} decreases due to the increasing influence of the Coulomb force, favouring

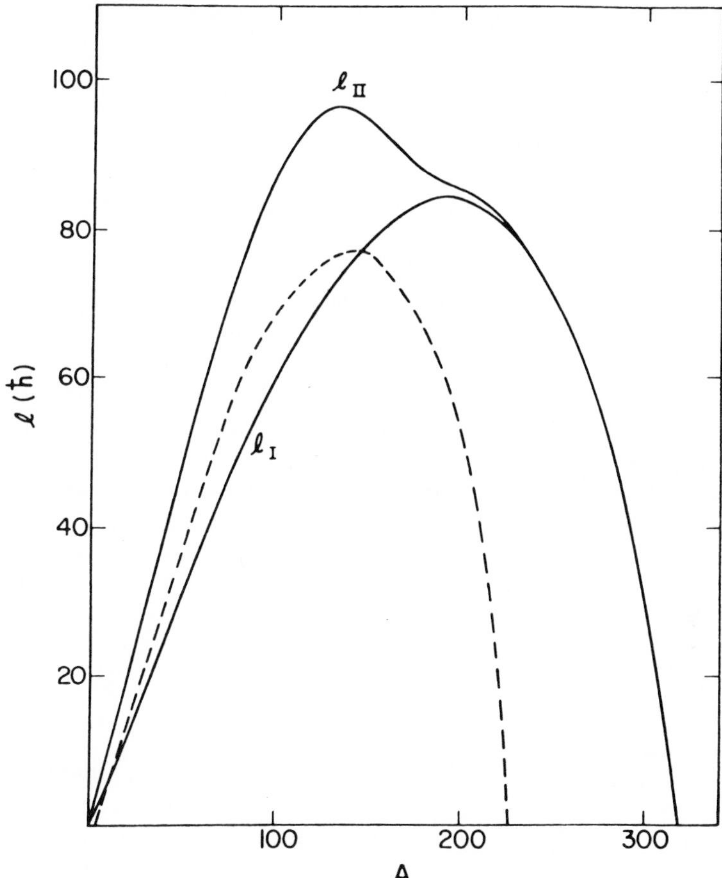

Fig. 6.6. The maximum angular momentum that can be accommodated in a β-stable rotating nucleus (classical estimates). Below the dashed curve the fission barrier is higher than 8 MeV (approximately equal to the nucleon binding energy). The region in between the l_{I} and l_{II} curves corresponds to superdeformed shapes; for further discussion see text. (Cohen *et al.*, 1974.)

fission. For rare earth nuclei $l_{\max}\hbar$ exceeds $90\hbar$ for a fission barrier $B_f = 0$ and $70\hbar$ for $B_f = 8$ MeV. From this we conclude that an upper limit for the angular momentum in compound nuclei is imposed by fission, which is largest in the rare earth region.

Let us return to the formation process of compound nuclei and try to determine quantitatively the maximum angular momentum $l_{\max}$ corresponding to a peripheral collision. We use for simplicity a sharp cut-off approximation and assume that no reaction will occur at $l > l_{\max}$; i.e. the transmission coefficient $T_l = 0$ for partial waves with $l > l_{max}$, and $T_l = 1$ for $l \leqslant l_{max}$. The total reaction cross-section σ_R can be written as a sum of the contributions from all partial waves up to l_{max} as:

$$\sigma_R = \pi \lambda\!\!\!^{-2} \sum_{l=0}^{l_{max}} (2l+1) T_l \simeq \pi \lambda\!\!\!^{-2} (l_{max}+1)^2, \tag{6.2.1}$$

where the reduced wavelength for the entrance channel is written

$$\lambda\!\!\!^{-} = \hbar / \sqrt{(2E_{cm}\mu)}. \tag{6.2.2}$$

Here E_{cm} denotes the centre-of-mass energy and $\mu = A_1 A_2/(A_1 + A_2)$ the reduced mass of the colliding nuclei with mass numbers A_1 and A_2.

From the requirement of energy conservation from infinity to the distance of closest approach R between the colliding nuclei we obtain the relationship:

$$l_{max}^2 = (2\mu R^2/\hbar^2)(E_{cm} - V_c), \tag{6.2.3}$$

where R is the maximum distance at which the collision leads to any reaction. Wilczyński (1973) derived an empirical expression for R as

$$R = 1.36(A_1^{1/3} + A_2^{1/3}) + 0.5 \text{ fm}. \tag{6.2.4}$$

The Coulomb energy V_c in eqn (6.2.3) is given by

$$V_c = 1.442 \frac{Z_1 Z_2}{R} \tag{6.2.5}$$

in units of MeV when R is in fm and Z_1 and Z_2 are the atomic numbers. In eqn (6.2.3) $2\mu R^2/\hbar^2$ is expressed in MeV^{-1} when $2/\hbar^2 = 0.04784$, and R is in fm.

The maximum angular momenta transferred in CF reactions using various projectiles, all leading to the formation of the compound nucleus ^{162}Er (or ^{164}Er), are shown in Fig. 6.7. The experimental data points were derived from measured total reaction cross-sections using eqn (6.2.1), or from measured γ-ray multiplicities according to procedures described in Section 6.5. The data are compared in Fig. 6.7 with l_{max} calculated according to eqn (6.2.3). The excitation energy E_{ex} is obtained from:

$$E_{ex} = E_{cm} + Q_{fus} \tag{6.2.6}$$

where Q_{fus} is the Q-value for CF.

At low bombarding energies, and thus also for small E_{ex}, the dominating reaction process is CF. At low energies the calculated l_{max} values are found to agree with the experimental data points. With increasing E_{ex} the transferred angular momentum increases and with it the repulsive centrifugal force. At higher energies CF may no longer be the dominant process because we assume that CF occurs in peripheral collisions only when the attractive nuclear force is larger than the repulsive Coulomb and centrifugal forces. At

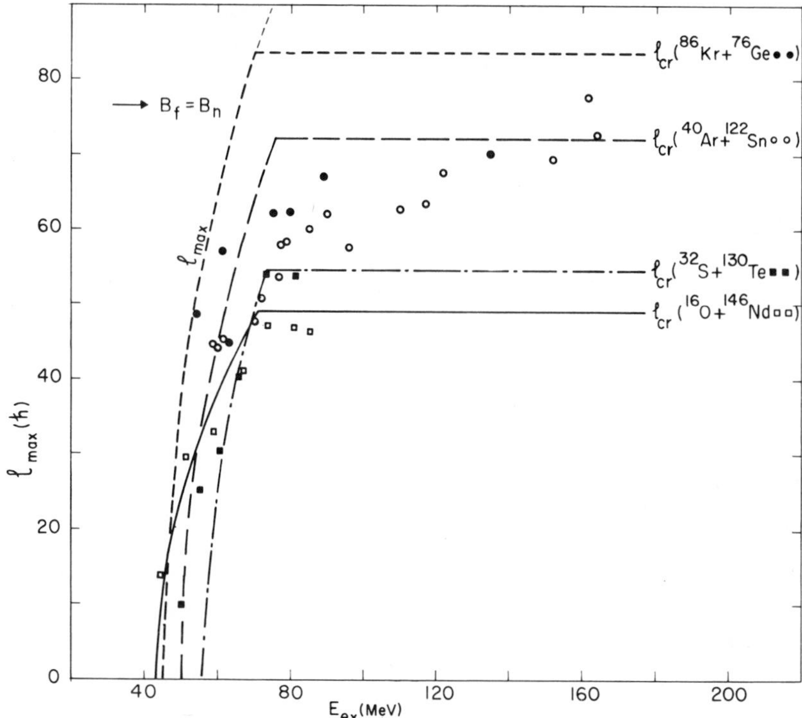

Fig. 6.7. Plot of the maximum angular momentum as a function of the excitation energy of the compound system ^{162}Er (^{164}Er) in heavy-ion collisions leading to complete fusion as derived from cross-section and γ-ray multiplicity measurements (Hillis *et al.*, 1979 and Garrett and Herskind, 1979). For comparison the l_{max} values are given as calculated according to eqn (6.2.3) and l_{cr} calculated with eqn (6.2.9). Also indicated is the angular momentum for which the fission barrier B_f of a rotating liquid drop with $A=162$ and $Z=68$ equals $B_n=9.2$ MeV, the neutron separation energy of ^{162}Er.

higher E_{ex} we see in Fig. 6.7 that the calculated l_{max} values are much larger than the experimental values. It is thus necessary to find another theoretical estimate in the case of high bombarding energies.

Following Wilczyński (1973) and Wilczyński *et al.* (1982) we may determine the critical angular momentum l_{cr}, the upper limit for CF to take place. At not too high bombarding energies l_{cr} may be calculated from eqn (6.2.3), replacing R by R_{cr} and V_c by V_{cr}, the potential value at R_{cr}. Here R_{cr} is the distance up to which the target and projectile can be considered as two separate ions. At high bombarding energies, however, CF can only occur up to a fixed angular momentum l_{cr}. The derivation of l_{cr} in this case is based on the liquid-drop model and the assumption that CF occurs when the projectile penetrates into the region of attraction in the total nucleus–projectile potential. Thus l_{cr} is determined in the case of a vanishing pocket in the total nuclear+Coulomb+centrifugal potential, assuming that in such a situation the colliding nuclei approach each other to $R_T + R_P$. These half-density radii can be approximated by

$$R_i = R_i'\{1-(1/R_i')^2 + \dots\} \tag{6.2.7}$$

with

$$R_i' = 1.28\ A_i^{1/3} + 0.8\ A_i^{-1/3} - 0.76\ \text{fm}. \tag{6.2.8}$$

One finds for l_{cr} (Wilczyński *et al.*, 1982)

$$(l_{cr}+\tfrac{1}{2})^2 = \frac{\mu(R_T+R_P)^3}{\hbar^2}\left\{4\pi\gamma\frac{R_T P_P}{R_T+R_P} - \frac{Z_1 Z_2 e^2}{(R_T+R_P)^2}\right\} \tag{6.2.9}$$

with $\gamma = 0.90-0.95$ MeV fm^{-2} being the surface tension of the compound nucleus.

It is seen in Fig. 6.7 that the experimental data points approach l_{cr} at high energies for the lighter projectiles, but for Kr ions l_{cr} is larger than the observed maximum angular momenta. The calculated l_{max} and l_{cr} values rely on rather crude models and can only be regarded as rough estimates, and this is probably the reason why the data do not follow the theoretical curves precisely but rather form a smooth transition from l_{max} to l_{cr}. Thus at high E_{ex} some data points show still an increase whereas l_{cr} remains constant. The data as well as the calculated l_{cr} (values near $E_{ex} \simeq 80$ MeV) show clearly an increase of the maximum angular momenta when the projectile mass increases for the same compound nucleus.

6.2.3 *Decay of a compound nucleus*

Let us now analyse the decay of a compound nucleus formed after fusion of a projectile and target nucleus as illustrated in Fig. 6.8 by the case of 147 MeV ^{40}Ar on a ^{124}Sn target, leading to the compound nucleus ^{164}Er at an excitation energy of 53.8 MeV. The evolution of compound-nucleus formation and its subsequent decay is schematically illustrated at the top and on the right-hand side of the figure. The time needed for the ^{40}Ar ion to pass the target nucleus is $\sim 10^{-22}$ s and a fast rotating dinuclear system can be formed that may lead to the formation of a compound nucleus if it survives fast fission. The rapid rotation with an angular velocity $\omega \sim 0.75$ MeV/$\hbar$ corresponds to 2×10^{20} revolutions/s of the compound nucleus ^{164}Er.

In the central part of the figure we show the results of a calculation based on a statistical model. The top part displays the cross-section as a function of the angular momentum l transferred to ^{164}Er. The cross-section is maximum at $l\hbar \simeq 25\hbar$ and the maximum angular momentum reaches $\sim 60\hbar$. The decay

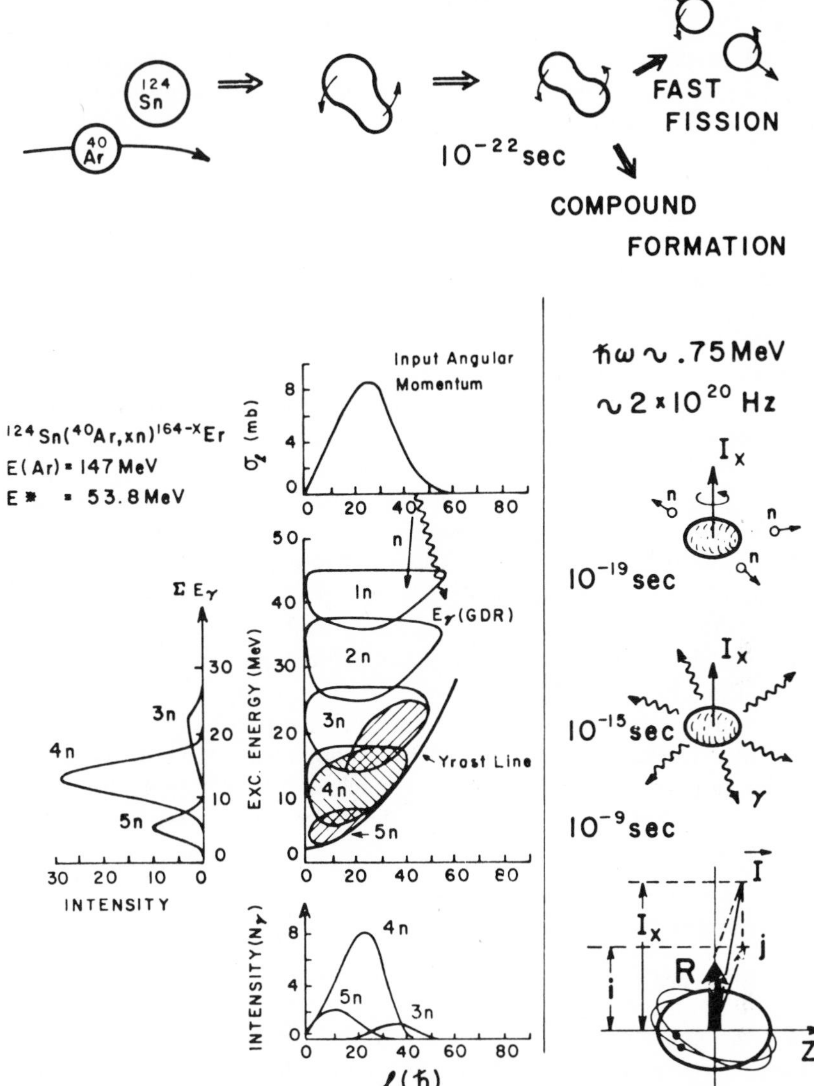

Fig. 6.8. Schematic illustration of the time evolution of the formation and decay of the ^{164}Er compound nucleus (at the top and on the right) and the results of a statistical model calculation of the decay (central part). The assumed population of the ^{164}Er compound nucleus is given as a function of the angular momentum in the top part of the figure. The calculated populations $\sigma(l, E_{ex})$ are indicated as functions of the excitation energy and angular momentum of the system after the emission of 1 to 5 neutrons. The shaded region of 3n–5n population (of entry states) shows the ranges in which γ-ray emission competes. The entry populations for the 3n–5n evaporation residues are indicated as functions of angular momentum and of excitation energy at the bottom and at the left-hand side of the figure. (Herskind, 1983.)

proceeds mainly via neutron emission, although high-energy γ-ray emission is not excluded, due to the enhancement of the giant dipole resonance, GDR, built on excited states, as discussed in Section 4.4. An emitted neutron lowers considerably the excitation energy, at least by its separation energy of 8–10 MeV, but to a much lesser degree the angular momentum of the system (on the average by about $1\hbar$). In general not only neutrons but also charged particles can be emitted from the highly excited nucleus.

After $\sim 10^{-15}$ s (or after $\sim 10^5$ collective revolutions) the excitation energy is lowered by neutron emission to a region of one neutron separation energy or less above the yrast line (shaded areas). The yrast line is a sequence of all states that have the highest angular momentum for a given energy (Swedish: *yr* 'dizzy', *yrast* 'the dizziest'). The projections of the shaded areas on the excitation energy and angular momentum axes show that evaporation of the least number of neutrons ($3n$) leaves the nucleus in states with the highest excitation energy and angular momentum. The decay of states in the shaded areas (called entry states) proceeds via γ-ray emission, reaching the ground state after $\sim 10^{-9}$ s or $\sim 10^{12}$ collective revolutions. If we realize that this is as many revolutions as our own planet has made since its formation or only ~ 10 times less than that of individual particles in orbital motion, then it becomes clear that classical centrifugal and Coriolis forces play an important role in this rotating quantum object.

The γ-ray spectroscopy probes the states in the region on and just above the yrast line in the E_{ex} versus I plane which are populated in the last 10^{-9} s of the de-excitation process of the compound nucleus. See Section 4.2 for a more detailed discussion of the γ and neutron de-excitation processes of compound and precompound nuclear systems.

6.3 Collective rotation

6.3.1 *Some general aspects of collective rotation*

It was recognized quite some time ago (Bohr, 1951) that atomic nuclei can absorb energy in one external degree of freedom, the collective rotation. If the nucleus, in the simplest approximation, is regarded as a rigid body then concepts like the moment on inertia $\mathscr{I}$, the rotational frequency ω and the angular momentum R may be applied. The classical relations between these quantities would yield a first estimate of the energy for a symmetric top as $E = \frac{1}{2}\mathscr{I}\omega^2$ or, using $R = \mathscr{I}\omega$, $E = (1/2\,\mathscr{I})R^2$. The real nucleus, however, is a quantum object and the variable R is not continuous. The states with energy E_{ex} are discrete with angular momentum I.

Let us consider an axially symmetric (prolate) deformed nucleus in the laboratory frame (x, y, z) with body-fixed coordinates (x', y', z') as schematically given in Fig. 6.9. The total angular momentum I results from the

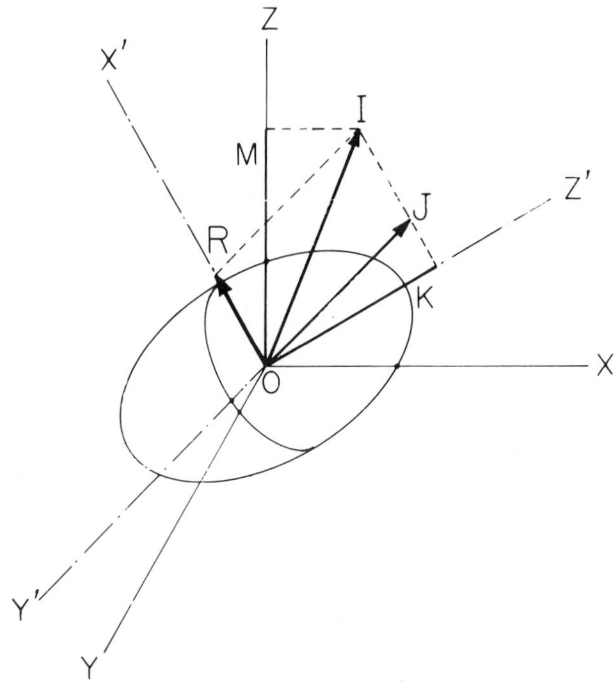

Fig. 6.9. An axially-symmetric deformed nucleus with body-fixed axes $O_{x'}$, $O_{y'}$ and $O_{z'}$ drawn in the space-fixed system (x, y, z). The collective rotation represented by the vector **R** can only take place around the axis $O_{x'}$, adding to the total angular momentum I with components $K\hbar$ along $O_{z'}$ and $M\hbar$ along O_z. The quantum number K is at the same time the component of the intrinsic angular momentum J along $O_{z'}$.

collective rotation R and intrinsic motion with an angular momentum J related to the symmetry axis $O_{z'}$, by its projection K. Since no collective rotation is possible about $O_{z'}$, R must always be perpendicular to the symmetry axis. This is a property of the quantum-mechanical description and expresses the impossibility of distinguishing different orientations when the nucleus rotates around the symmetry axis. The system is invariant under rotations about $O_{z'}$ and the component K of I is a constant of motion as is the component M along the space-fixed axis O_z. The state of motion is completely specified by I^2, M and K with

$$R = \hbar\sqrt{\{I(I+1) - K^2\}}. \tag{6.3.1}$$

This expression is valid for $J = K$ and for $K \neq 1/2$. When there is no intrinsic excitation ($K = 0$) we find a simple expression for the energies of a collectively rotating rigid nucleus, using the classical relation for the symmetric top and the expression for R from eqn (6.3.1):

$$E_{ex} = (\hbar^2/2\mathscr{I})\ I(I+1). \tag{6.3.2}$$

Rotational states were observed as early as 1953 by Huus and Zupancic (1953) by means of Coulomb excitation in which the electric field of an accelerated charged particle that is scattered by the target sets it into rotation. At that time rotational states had also been observed in alpha decay (Asaro and Perlman 1952, 1953). Systematic analysis of rotational motion, especially at relatively high spins, became possible with heavy ion-induced fusion reactions (see, for example, Stephens *et al.*, 1965). In-beam γ-ray spectroscopic measurements were initiated by Morinaga and Gugelot (1963) using (α, xnγ) reactions for the first time. They used NaI(Tl) detectors of which a γ-ray spectrum is displayed in Fig. 6.10(a). The bumps correspond to states belonging to the ground-state-rotational band in ^{160}Dy. Several Dy isotopes were investigated with ^{x}Gd (α, 4n) ^{x}Dy reactions and members of the ground-state-band (GSB) were traced up to 8^+, 10^+ or 12^+ as shown in Fig. 6.10(a). The rotational character of the states is reflected in the approximate $E \sim I^2$ dependence according to eqn (6.3.2).

The detection techniques have greately improved as can be seen in the spectrum for ^{160}Dy measured about ten years later in Stockholm. States were identified up to 18^+ and the anomaly in the increase of γ-ray energy with spin at 14^+ was the first observation of 'backbending' (Johnson *et al.*, 1972). Again about a decade later states have been observed up to spins of ~ 40 as is shown for ^{158}Er in Fig. 6.10(c). Besides the first backbending at 14^+ a second anomaly was seen at 28^+. From such a spectrum a level scheme can be deduced as shown in Fig. 6.2.

The levels in ^{158}Er also obey approximately the I^2 dependence expressed by eqn (6.3.2) except for levels near the anomalies at 14^+ and 28^+.

The E_{ex} vs. I dependence for yrast states in two light and two heavy Dy isotopes is shown in Fig. 6.11 (on a I^2 scale). For not too low spins the $E \sim I^2$ dependence is well obeyed for $^{156,\,158}$Dy and only approximately for $^{151,\,152}$Dy. The heavy isotopes are considered as prolate deformed nuclei and thus should exhibit the smooth rotational $E \sim I^2$ characteristics, according to eqn (6.3.2). The light isotopes, however, are not well deformed and deviations from the straight line are observed in the graph of Fig. 6.11. The approximate $E \sim I^2$ dependence cannot be explained on the basis of collective rotation but stems from a peculiar feature in particle excitations that will be explained in Section 6.4.

This figure shows that collective rotation is certainly not the only mode of excitation and even in deformed nuclei it may compete or coexist with other (intrinsic) degrees of freedom like quasi-particle excitations. The coexistence of such particle degrees of freedom with the collective degree of freedom is one of the unique features of the nucleus, the ensemble of a finite number of nucleons. The collective rotational states appear normally lowest in excitation energy for a given angular momentum provided that the individual particle motions remain unperturbed. There are, however, various types of particle excitations such as vibrations, particle–hole excitations, and other mechanisms which change the angular momentum and its direction. Some of them couple coherently with the collective rotation, and thus the excitation energies become low. Consequently such particle degrees of excitation, appearing in the low-excitation region also modify the collective rotation. The simple expression (6.3.2) thus has only a very limited value and has to be modified in many cases. The moment of inertia $\mathscr{I}$, for instance, may vary with I, even within the same collective band. We shall not discuss all possible modifications and refer the reader to various monographs on theoretical nuclear physics (i.e. Bohr and Mottelson, 1969, 1975).

As high-spin states are often connected with low-lying states via collective γ-ray transitions within rotational bands some knowledge of the elementary modes of collective excitation is necessary and will be provided in the next section.

6.3.2 *Collective bands*

6.3.2a *Rotational bands; symmetry properties*

The Hamiltonian for the rotation of an axially symmetric nucleus can be written by analogy to eqn (6.3.2) as

$$H = \frac{\hbar^2}{2\mathscr{I}}(\vec{I} - \vec{J})^2 + H_{\text{intr}}, \tag{6.3.2a}$$

where H_{intr} describes the nucleonic motion (with spin $\vec{J}$) within the body-fixed system of reference, which is valid only when the rotation is slow compared to the single-particle motion (the adiabatic condition).

First we assume for simplicity that there is no coupling of the rotational motion with the other degrees of freedom. The intrinsic structure, represented by H_{intr} (that is the particle degree of freedom) is in this case completely decoupled from the collective rotation.

The rotational state can be described approximately by a wave function Ψ_{IKM} which can be separated into two components:

$$\Psi_{IKM} = \Phi_K \phi_{IKM}(\phi, \theta, \psi), \tag{6.3.3}$$

where Φ_K represents the intrinsic motion of the nucleons in the coordinate system (x', y', z') and ϕ_{IKM} describes the rotation of the nucleus as a whole in the (x, y, z) frame. The orientation of the nucleus is specified by the Euler

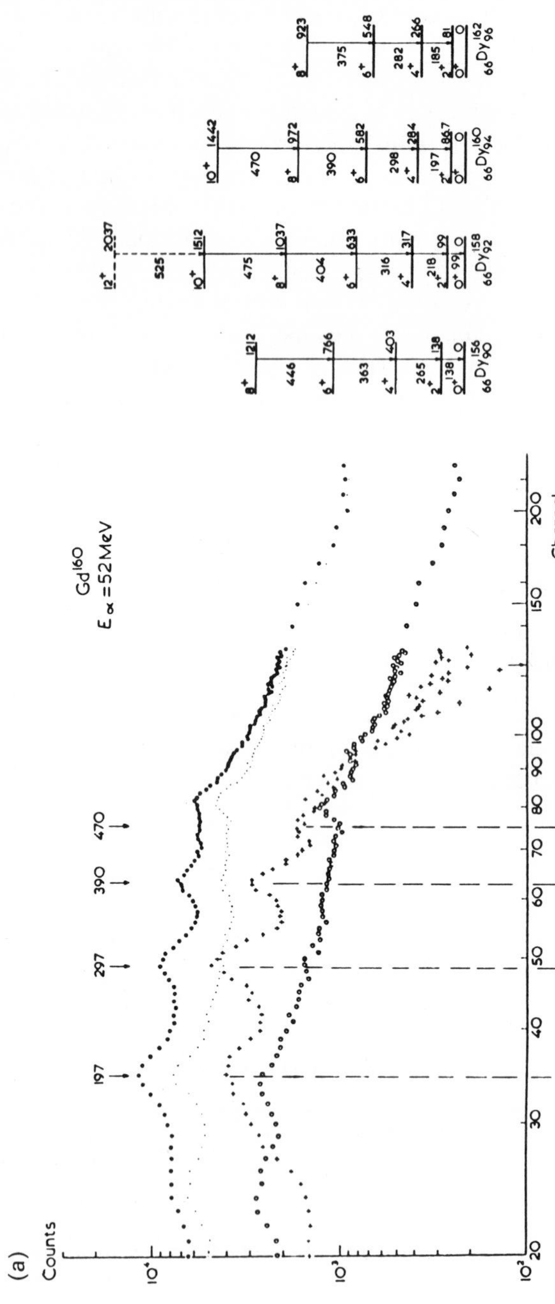
(a)
Counts
Channel
Gd160
Eα = 52 MeV
197
297
390
470
66Dy156 90
66Dy158 92
66Dy160 94
66Dy162 96

(b)

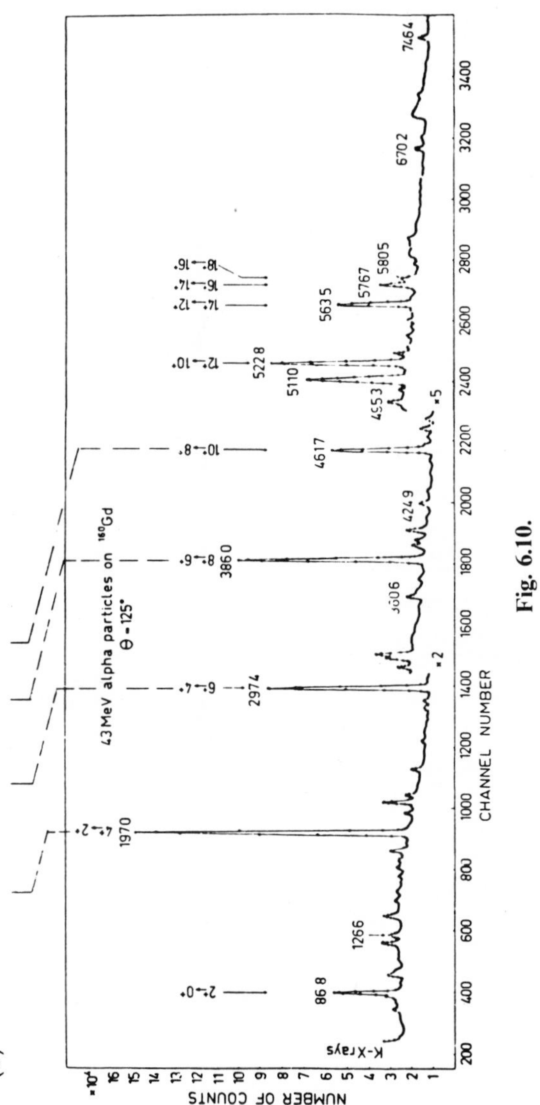

NUMBER OF COUNTS
×10⁴
CHANNEL NUMBER
43 MeV alpha particles on ¹⁶⁰Gd
θ = 125°
K-X rays
2⁺→0⁺
4⁺→2⁺
6⁺→4⁺
8⁺→6⁺
10⁺→8⁺
12⁺→10⁺
14⁺→12⁺
16⁺→14⁺
18⁺→16⁺
868
1266
1970
2974
3860
4249
4617
4553
5110
5228
5635
5767
5805
6702
7464
×2
×5

Fig. 6.10.

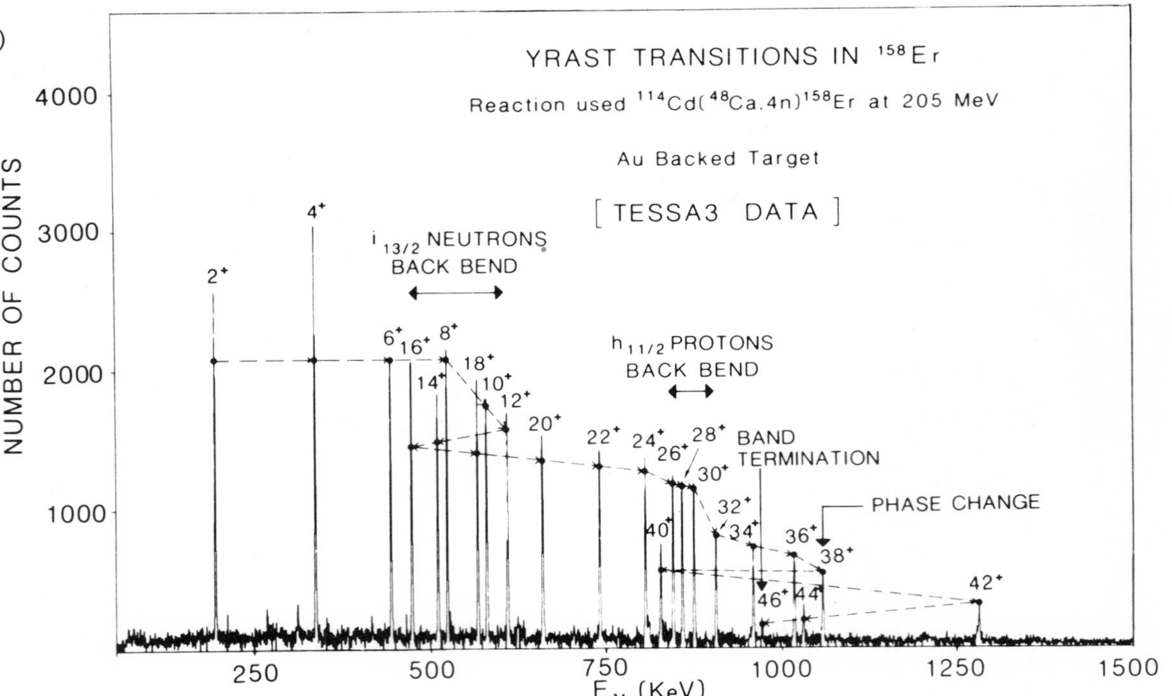

Fig. 6.10. Gamma-ray spectra measured in-beam in 1963 with ^{4}He projectiles and a NaI(Tl) detector (a) by Morinaga and Gugelot (1963) and in 1972 using a high-resolution Ge(Li) detector (b) by Johnson *et al.* (1972). The application of heavy ions as projectiles and multi-detector anti-Compton spectrometers lead to the observation of high-spin states as was done for ^{158}Er (c) with the TESSA III spectrometer at Daresbury (see Fig. 6.17b). The corresponding level scheme of ^{158}Er is given in Fig. 6.2. The level spectra for some Dy isotopes, deduced from the data in 1963 are displayed at the top right.

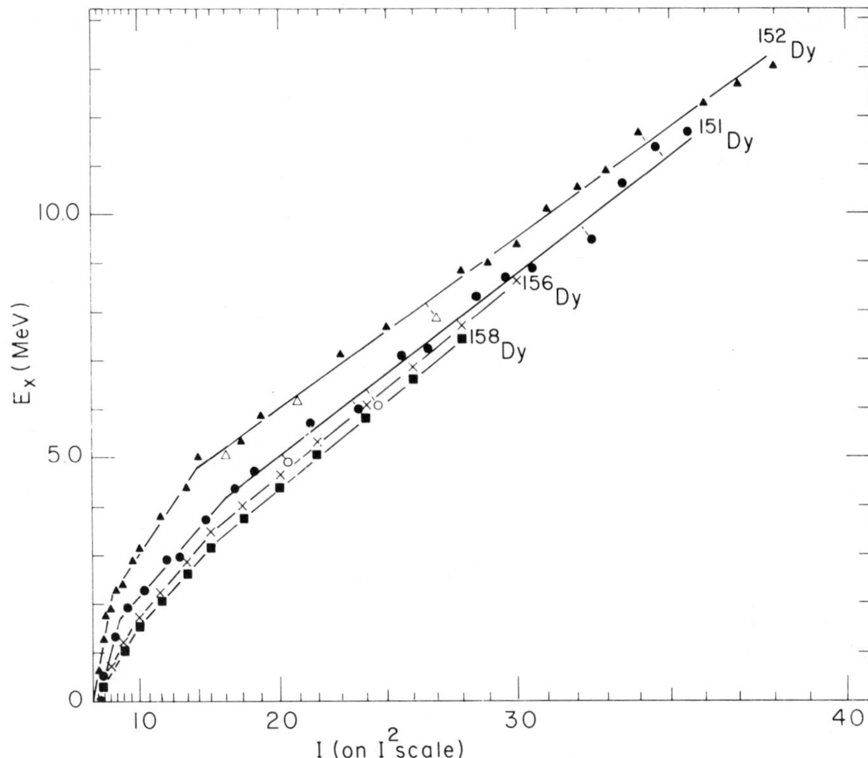

Fig. 6.11. The excitation energies of yrast states plotted as functions of I (on an I^2 scale) for various Dy isotopes. ^{156,158}Dy are considered as collectively rotating deformed nuclei while ^{151,152}Dy exhibit non-collective behaviour including restarted γ decay (yrast traps indicated by open symbols) as will be discussed in Section 6.4.

angles ($\omega = \phi, \theta, \psi$), and three angular momentum quantum numbers are needed to describe the state of motion. It is convenient to express the rotational part of the wave function by the well-known rotation matrices $D^I_{MK}(\phi, \theta, \psi)$. They are eigenfunctions of I, M, and K; M and K being good quantum numbers as was argued above. The rotational wave function ϕ_{IKM} in eqn (6.3.3) can thus be expressed in terms of the Euler angles $\omega = \phi, \theta, \psi$ and I, K, and M by

$$\phi_{IKM}(\phi, \theta, \psi) = \left(\frac{2I+1}{8\pi^2}\right)^{1/2} D^I_{MK}(\phi, \theta, \psi). \tag{6.3.4}$$

The projection K of I on the symmetry axis $O_{z'}$ (see Fig. 6.9) is not only a constant of motion; it is also the projection of J, the angular momentum of the intrinsic motion. The value of K is thus determined by the intrinsic motion and has a constant value within a rotational band. Thus the intrinsic structure specified in eqn (6.3.3) Φ_K should remain constant within a band, which in practice is approximately valid only in the low-spin region. At higher spins deviation from this picture occur because of the coupling of particle degrees of freedom with the collective motion.

The reflexion symmetry implies that the system is invariant to rotation by 180° about an axis perpendicular to the symmetry axis $O_{x'}$;

$$\mathrm{R}_{x'} = \mathrm{e}^{-i\pi I_{x'}}. \tag{6.3.5}$$

$(\mathrm{R}_{x'})^2$ is equivalent to a rotation of 2π and thus leaves the wave function unchanged for a nucleus with an even number of nucleons, but changes the sign for odd A nuclei;

$$(\mathrm{R}_{x'})^2 = (-1)^A. \tag{6.3.6}$$

It should be noted that the R invariance implies that the rotation R is part of the intrinsic degrees of freedom i.e.

$$\mathrm{R} = \mathrm{R}_i = \mathrm{R}_e, \tag{6.3.7}$$

where R_i and R_e refer to rotations R acting on intrinsic and external variables, respectively. Thus we can write

$$\mathrm{R}_i^{-1}\mathrm{R}_e\Psi_{IKM} = \Psi_{IKM}. \tag{6.3.8}$$

If we apply R_e on the rotational wave function we obtain:

$$\mathrm{R}_e D^I_{MK}(\omega) = \mathrm{e}^{-i\pi I_{x'}} D^I_{MK}(\omega) = (-1)^{I+K} D^I_{M-K}(\omega). \tag{6.3.9}$$

In order to satisfy reflection symmetry, and thus to satisfy eqn (6.3.7), and using eqns (6.3.4) and (6.3.9) we find for the total wave function (Bohr and Mottelson, 1975)

$$\Psi_{IKM} = 2^{-1/2}(1 + \mathrm{R}_i^{-1}\mathrm{R}_e)\left(\frac{2I+1}{8\pi^2}\right)^{1/2} \Phi_K D^I_{MK}(\omega)$$

$$= \left(\frac{2I+1}{16\pi^2}\right)^{1/2} \{\Phi_K D^I_{MK}(\omega) + (-1)^{I+K}\Phi_{\bar{K}} D^I_{M-K}(\omega)\}, \tag{6.3.10}$$

where $\Phi_{\bar{K}}$ corresponds to a projection of the angular momentum of $-K$ and is obtained from Φ_K by a rotation π around $O_{x'}$ for instance. It is thus a consequence of the R invariance that (for $K > 0$) the two degenerated intrinsic states Φ_K and $\Phi_{\bar{K}}$ with eigenvalues K and $-K$ of $J_{z'}$, respectively, constitute only one single series of rotational states with spins I:

$$I = K, K+1, K+2, \ldots, \tag{6.3.11}$$

(if we adopt $K > 0$ only).

The sequence of states Ψ_{IKM} based on a given intrinsic configuration Φ_K is called a rotational band with fixed K. The parity is determined by the intrinsic state as:

$$\mathrm{P}\Psi_{IKM} = \mathrm{P}\Phi_K = \pi\Phi_K, \text{ with } \pi = \pm 1. \tag{6.3.12}$$

The phase factor in (6.3.10);

$$\sigma = (-1)^{I+K} \tag{6.3.13}$$

is referred to as the signature (Bohr and Mottelson, 1975) and changes sign within a band for alternating spin values. The rotational band with states Ψ_{IKM} may thus be considered as composed of two sequences with $I = K$, $K+2, \ldots$ and $I = K+1$, $K+3, \ldots$ with opposite signature.

For the special case of $K = 0$ the D functions reduce to spherical harmonics $Y_{IM}(\phi, \theta)$ and eqn (6.3.4) reduces to

$$\Psi_{I, K=0, M}(\phi, \theta) = (2\pi)^{-1/2}\, Y_{IM}(\phi, \theta). \tag{6.3.14}$$

The rotation R_e will reverse the direction of the symmetry axis $O_{z'}$ $(\theta \rightarrow \pi - \theta$, $\phi \rightarrow \phi + \pi)$

$$\mathrm{R}_e Y_{IM}(\phi, \theta) = (-1)^I\, Y_{IM}(\phi, \theta). \tag{6.3.15}$$

Thus the eigenvalue of $R_e = R_i$ is:

$$r = (-1)^I \tag{6.3.16}$$

which in fact is identical to the signature quantum number σ defined by eqn (6.3.13) and is a consequence of the reflection invariance. The total wave function for the case of $K = 0$ can thus be written as:

$$\Psi_{r, I, K=0, M} = (2\pi)^{-1/2}\, \Phi_{r, K=0}\, Y_{IM}(\phi, \theta) \tag{6.3.17}$$

which leads to only two sequences of states:

$$I = 0, 2, 4, 6, \ldots \text{ for } K = 0 \text{ and } r = +1, \tag{6.3.18}$$

and

$$I = 1, 3, 5, 7, \ldots \text{ for } K = 0 \text{ and } r = -1. \tag{6.3.19}$$

Sometimes the signature quantum number α is used (see e.g. Bengtsson and Frauendorf, 1979), defined as

$$r = \mathrm{e}^{-i\pi\alpha}. \tag{6.3.20}$$

Thus $r = +1$ is equivalent to $\alpha = 0$ and $r = -1$ to $\alpha = 1$. Equations (6.3.18) and (6.3.19) are equivalent with

$$\alpha = I \bmod 2. \tag{6.3.21a}$$

It also follows from eqns (6.3.6) and (6.3.20) that the particle number A obeys the relation:

$$2\alpha = A \bmod 2. \tag{6.3.21b}$$

For odd-A nuclei the signature cannot be defined by eqn (6.3.16), but instead we apply eqn (6.3.20) and (6.3.21a) which again leads to two distinct level sequences:

$$I = 1/2, 5/2, 9/2, \ldots \quad \alpha = 1/2, \text{ thus } r = -i \tag{6.3.22}$$

and

$$I = 3/2, 7/2, 11/2, \ldots \quad \alpha = -1/2, \text{ thus } r = +i. \tag{6.3.23}$$

In further discussions we shall use as signature quantum number $\alpha = 0, 1$ for even nuclei and $\alpha = \pm 1/2$ for odd nuclei. The advantage of using α rather than r is that α is an additive quantity.

The usefulness of signature as a quantum number to classify rotational bands is illustrated in Fig. 6.12. The two bands in ^{159}Dy can be classified with $\alpha = \pm 1/2$ according to eqns (6.3.22) and (6.3.23). The insert in Fig. 6.12 shows that both bands obey the rotational relation (6.3.2), particularly well for the high-spin states. The signature seems to be not very meaningful in the low-spin region (slow rotation) but becomes a good quantum number at high spins. The Coriolis interaction modifies the energies so that signature splitting occurs, causing two rotational sequences of states with different moments of inertia, but constant within each band with definite signature.

So far we have assumed that the nuclear deformed average field is characterized by even-multipolarity (λ) components, i.e. quadrupole ($\lambda = 2$) and hexadecapole ($\lambda = 4$) deformations. Sometimes higher-order deformation of the octupole type ($\lambda = 3$) occurs, corresponding to a pear-shaped nucleus rather than an ellipsoid. Rotation of such an object leads to a $K^\pi = 0^-$ rotational octupole band with level sequences according to eqn (6.3.18) $I^\pi = 0^-, 2^-, 4^-, \ldots$ and the sequence according to eqn (6.3.19) of $I^\pi = 1^-, 3^-, 5^-, \ldots$. The former sequence may be suppressed with respect to the latter because of the unnatural parity.

6.3.2b *Coupling between rotation and other degrees of freedom*

We shall now discuss in more detail the interplay between a rotating deformed nuclear core and the individual nucleonic motion. The quantum numbers characterizing the intrinsic state may be obtained by considering the motion of individual nucleons in a deformed nuclear field (Nilsson, 1955). As discussed in the introduction to this chapter, states, can be labelled by the projection Ω of the particle angular momentum j onto the symmetry axis $O_{z'}$. The R invariance causes a two-fold degeneracy Ω and these states are thus filled pairwise. Consequently the ground-state bands in even–even nuclei have

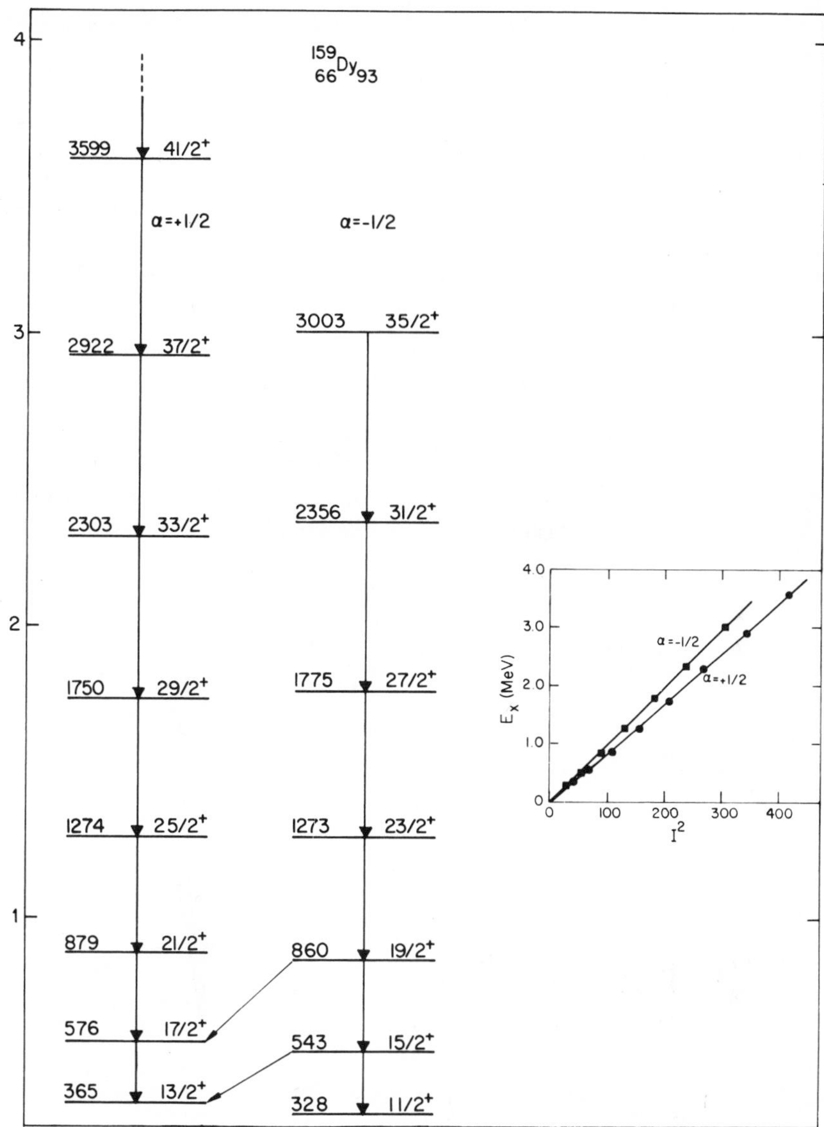

Fig. 6.12. The experimental level scheme of ^{159}Dy (Lee and Reich, 1979), illustrating the signature splitting in the spectrum of a rotating odd-A nucleus. The odd particle occupies either the lower or higher of the two orbitals with $\alpha = \pm\frac{1}{2}$ which are degenerate at $\omega = 0$ due to the Kramers degeneracy. Both signatures constitute a sequence of states with spin differences of 2 between the neighbouring states. The insert contains a plot of the excitation energy vs. I^2 for the two bands; it illustrates the rotational character of the bands and the fact that at high spins the signature is a good quantum number.

$K=0$ and parity $\pi = +1$ and in odd-A nuclei $K^\pi = |\Omega|^\pi$ corresponding to the odd nucleon.

For an odd valence nucleon moving in a rotating nuclear potential the two leading factors that determine the nucleonic motion are the deformed average nuclear field and the Coriolis force. In the *strong coupling* limit the motion is essentially determined by the momentary orientation of the deformed rotating field. This is the case when the deformation is large and simultaneously the rotation not too fast, which forces the particle to couple to the deformed core as shown in Fig. 6.13(a). Consequently the particle motion follows the rotating deformed potential, keeping the intrinsic structure unperturbed. In this case $K=\Omega$ and according to eqn (6.3.11):

$$I = K, K+1, K+2, \ldots \quad (K=\Omega). \tag{6.3.24}$$

Another extreme case arises when the odd nucleon moves in a weakly-deformed field and the rotation is so fast that the Coriolis forces largely determine the particle motion. This is especially the case for nucleonic orbitals with high-j and low-Ω because the Coriolis coupling is expressed by the

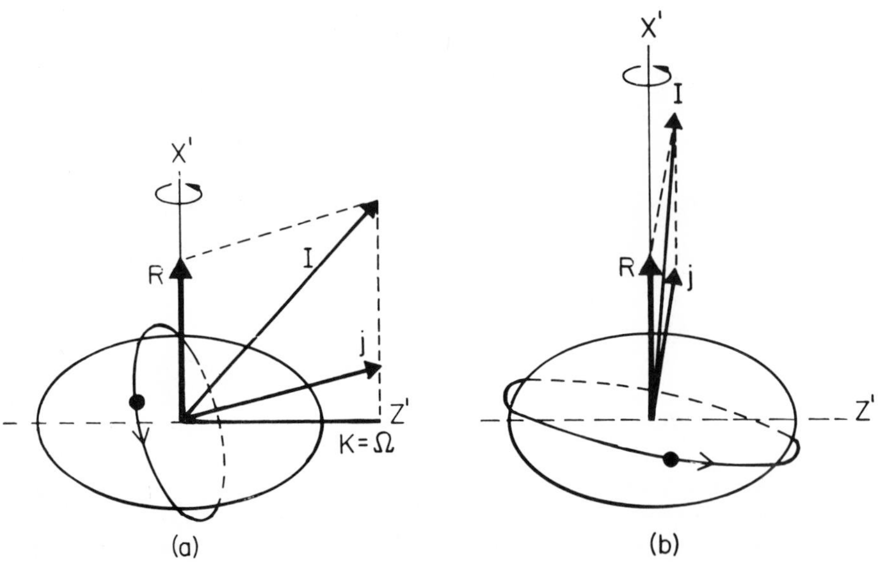

Fig. 6.13. Angular momentum coupling schemes for two extreme cases: (a) Strong-coupling or deformation aligned (DAL) scheme showing that the motion of the odd particle is determined by the (strongly) deformed core, and (b) Weak-coupling or rotation-aligned (RAL) scheme showing that the odd particle follows the rotation of the core. The RAL bands are sometimes called decoupled bands (decoupled from the deformation of the core).

matrix elements of the $j_{x'}$ operator in states $|j\Omega\rangle$ as:

$$\langle j,\Omega\pm 1|j_{x'}|j,\Omega\rangle = \tfrac{1}{2}\{(j\mp\Omega)(j\pm\Omega+1)\}^{1/2}. \qquad (6.3.25)$$

The Coriolis interaction favours the alignment of the nucleonic angular momentum j with that of the rotating core R. The nucleonic orbital is thus oriented in a plane approximately perpendicular to the rotation axis. This case is referred to as *rotation-aligned* and the nucleon becomes totally *decoupled* from the rotating core as shown in Fig. 6.13(b). In this description the band properties are solely determined by the wave function of the odd particle (with spin j) and the spin values of the band members are given by

$$I = j, j+2, j+4, \ldots. \qquad (6.3.26)$$

The energies of the rotational aligned states can be calculated from eqn. (6.3.2) taking into account the projection j_x of the odd particle on the rotation axis (Stephens, 1975):

$$E_{\text{rot.al.}} = (\hbar^2/2\mathscr{I})(I-j_x)(I-j_x+1). \qquad (6.3.27)$$

The lowest-lying rotational-aligned band is the one with complete alignment; i.e. $j_x = j$, called the favoured band. Bands with less alignment; i.e. $j_x = j-1$, have also been observed in (HI, xnγ) reactions, but with less intensity than the favoured bands. The spin sequence for the $j_x = j-1$ 'unfavoured' band is

$$I = j-1, j+1, j+3, \ldots. \qquad (6.3.28)$$

The two bands in ^{159}Dy displayed in Fig. 6.12, for example, can be regarded as the $j_x = j = 13/2^+$ favoured and $j_x = j-1 = 11/2^+$ unfavoured neutron bands.

Another explanation of rotational-aligned bands can be found in terms of the weak-coupling model assuming an interaction between the odd particle and even core. Therefore one also describes the bands as *weakly-coupled* bands in contrast to strongly-coupled bands.

Examples of the two different modes of excitation are shown in Fig. 6.14. It is known that the even–even nucleus ^{154}Dy is not very deformed as is indicated by its moment of inertia $\mathscr{I} = 9$ MeV$^{-1}\hbar^2$ and $B(\text{E2}) = 0.3e^2b^2$ for the $2^+ \to 0^+$ transition. The odd neutron occupies the [651]3/2 Nilsson orbital, and thus $j = 13/2$ and $\Omega = 3/2$. Therefore, it experiences a rather strong Coriolis force, which causes rotational alignment. The $i_{13/2}$ neutron band in ^{155}Dy indeed resembles quite closely the sequence in ^{154}Dy. The resemblance would even be more striking if one would take the average of ^{154}Dy and ^{156}Dy as core of ^{155}Dy. In fact the excitation energy $E_{\text{odd}}(I')$ in ^{155}Gd corresponds well to the excitation energy $E_{\text{even}}(I)$ of the level with $I = I'-j$ in ^{154}Dy (or ^{156}Dy). In the deformed ^{175}Hf nucleus ($\mathscr{I} = 33$ MeV$^{-1}\hbar^2$) the odd neutron occupies the [512]5/2$^-$ Nilsson orbital and is strongly coupled to the very deformed even core, ^{174}Hf. The band sequence obeys eqn (6.3.24) with $j = 9/2$, $K = \Omega = 5/2^-$. The relation of the

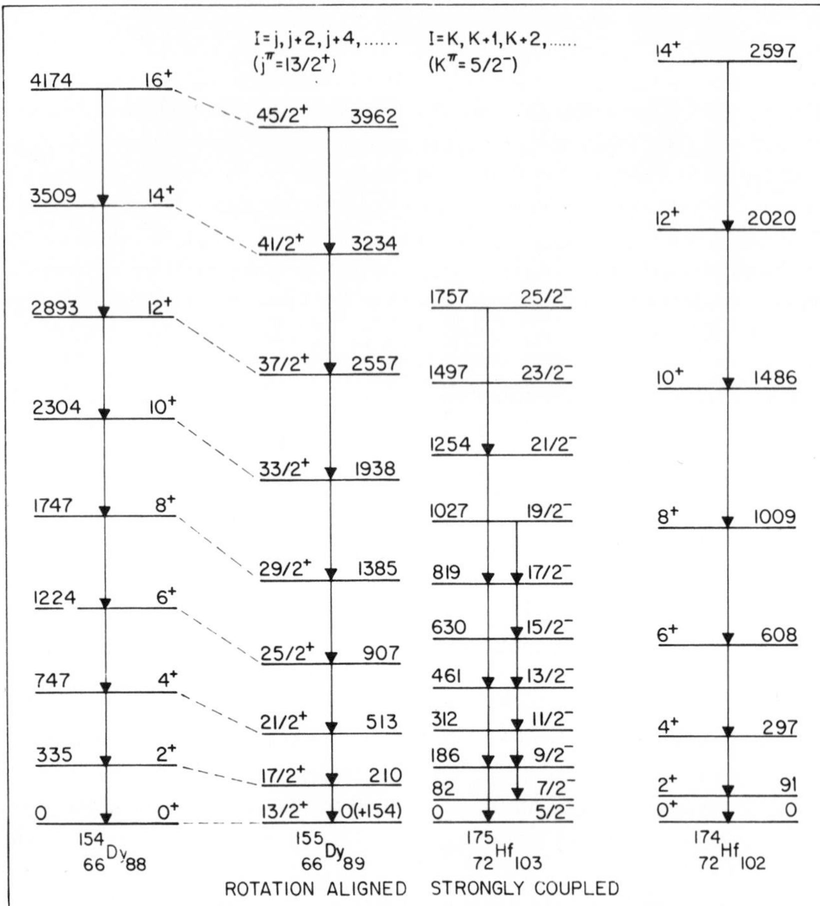

Fig. 6.14. An example of RAL (left) and DAL (right) bands in odd nuclei. For comparison the adjacent even nuclei in each case are also given. The $13/2^+$ state in ^{155}Dy is not the ground state and the energies of the other members are given relative to that of the $13/2^+$ state (add 154 keV to obtain the excitation energies). The data on ^{155}Dy are from Beuscher *et al.* (1975) and on ^{175}Hf from Hultberg *et al.* (1973).

odd nucleus with the even core, in this case, is reflected in the energy difference between band members with $\Delta I = 2$, which should be the same for the two bands, particularly at high spin. For example $E(25/2^-) - E(21/2^-) = 503$ keV can be compared with $E(10^+) - E(8^+) = 477$ keV.

It should be emphasized that these are only two limiting cases with a broad region of intermediate situations in between. Actually there are many particle (hole) states with different j and (or) Ω. Rotational levels based on these particle (hole) states are mixed when the Coriolis interaction is switched on. Thus the energy levels are modified more or less depending on the j, Ω, and the excitation energy. One extreme case of weak modification is the band with $K^{\pi}=5/2^{-}$ in ^{175}Hf, and another one of strong modification is the band with $j^{\pi}=13/2^{+}$ in ^{155}Dy discussed above. A smooth transition from one case to the other is possible depending on the deformation and the particular orbital of the single nucleon; this is discussed, for instance, by Lieder and Ryde (1978).

Another mode of collective excitation, different from rotation, is the *vibration* of the nuclear shape. Assuming a drop of incompressible liquid, the shape of the nuclear surface (or potential) can be expressed in terms of spherical harmonics and a few shape parameters $a_{\lambda\mu}$ depending on the multipolarity λ of the vibrations:

$$R=R_o\left\{1+\sum_{\lambda\mu}a_{\lambda\mu}Y_{\lambda\mu}(\theta,\phi)\right\}. \tag{6.3.29}$$

Here R_o represents the radius at spherical equilibrium (taking into account the condition of volume conservation).

The most common deformation is of the quadrupole ($\lambda=2$) type. The five parameters $a_{2\mu}$ can be reduced to two shape parameters β and γ since $a_{21}=a_{2-1}=0$ and $a_{22}=a_{2-2}$ are required by reflection symmetry of the deformation. They are defined as:

$$a_{20}=\beta\cos\gamma;\quad a_{22}=(1/\sqrt{2})\,\beta\sin\gamma. \tag{6.3.30}$$

The parameter β (sometimes denoted by β_2) represents the magnitude of the nuclear deformation (elongation or flattening) and γ the degree of nonaxiality. For a spheroidal shape the axially symmetrical deformation is specified by:

$$\delta=\Delta R/R_o, \tag{6.3.31}$$

where ΔR is the difference between the major and minor semi-axes of the ellipse. In this case $\beta\simeq1.06\,\delta$. Possible nuclear shapes can be specified in the β–γ plane as shown in Fig. 6.15. All possible shapes can be specified within the range $\gamma=-2\pi/3,\pi/3$. Axial symmetry exists for $\gamma=0,\pm\pi/3,\pm2\pi/3,\pi$. For a certain mass and deformation β the liquid-drop model yields the largest moment of inertia for $\gamma=60°$, a smaller value at $\gamma=0°$, a still smaller value for $\gamma=-60°$ and the smallest value in the case of $\gamma=-120°$ (Stephens, 1979). The largest moments of inertia correspond to rotations with the lowest energies, according to eqn (6.3.2), and it is therefore not surprising that most nuclei attain shapes corresponding to the sector between $\gamma=0°$ and $60°$. The adjacent sector corresponds to collective rotation with prolate nuclei ($\gamma=0°$) favoured energetically over those with an oblate deformation ($\gamma=-60°$). We concentrate our discussion on prolate nuclei in this section,

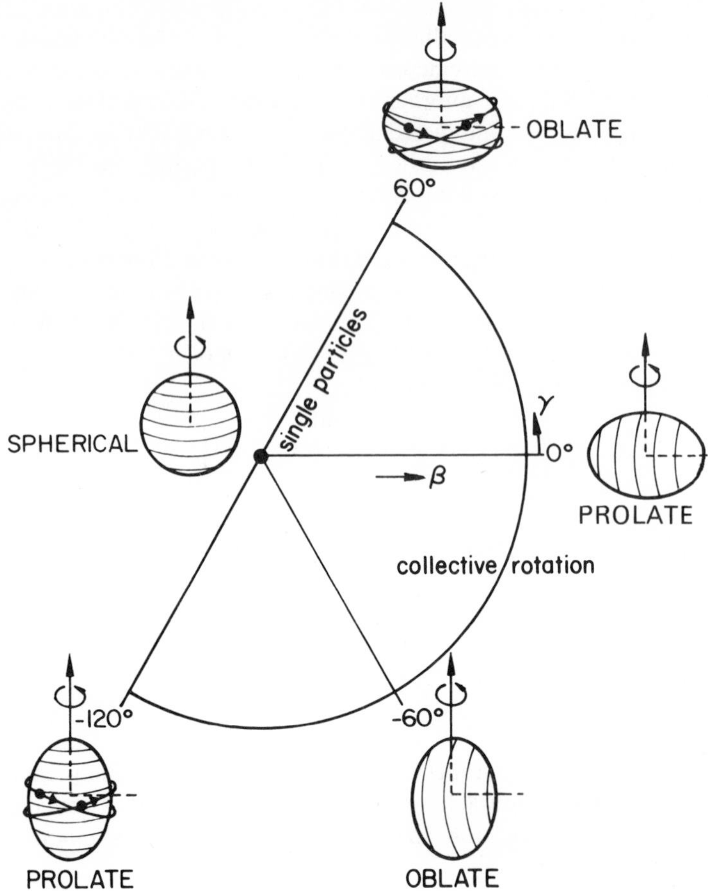

Fig. 6.15. Various nuclear shapes expressed by the deformation parameter β (plotted radially) and the non-axiality parameter γ measured as an angle from 0–2π. The maximum collectivity is found for $\gamma = -60° - 0°$ whereas individual nucleon excitations occur at shapes with $\gamma = 60°$ and $\gamma = -120°$ (also for hole excitations).

because they occur more frequently than oblate nuclei in the case of collective rotation. The conclusion here that rotation around the principal axis of inertia is most favoured (lower energy) is based on the classical picture of motion. In the case of quantal effects, such as shell corrections, other situations may occur. We shall discuss in Section 6.4 the particular situation of non-collective rotation, corresponding to the two extremes in Fig. 6.15; $\gamma = 60°$ and $-120°$.

Besides vibrations around a spherical equilibrium shape, oscillations of a permanently deformed shape may also be possible. In a non-spherical field the phonon (vibration) quantum number λ is not specified, but instead one uses the projection ν of the phonon angular momentum along the symmetry axis. In the case of $\nu = 0$ (β-vibrations) axial symmetry is conserved ($\gamma = 0$). A rotational band built on the β vibration can be characterized, using eqn (6.3.18), by $K = 0$;

$$I^\pi = 0^+, 2^+, 4,^+, \ldots \quad (K = 0, \nu = 0). \tag{6.3.32}$$

Oscillations with $\nu = \pm 2$ (γ-vibrations) break the axial symmetry and a rotational band built on such vibration is characterized, using eqn (6.3.11), by $K = 2$;

$$I^\pi = 2^+, 3^+, 4^+, 5^+, \ldots \quad (K = 2, \nu = \pm 2). \tag{6.3.33}$$

Oscillations with higher multipolarity, e.g. octupole ($\lambda = 3$) and hexadecapole ($\lambda = 4$) vibrations, etc., also occur, but we refer the reader for that subject and for many other details on collective excitation to Bohr and Mottelson (1969, 1975).

6.3.3 *Bandcrossing and independent quasi-particle motion*

In this section we discuss rotational bands based on some intrinsic excitation and the phenomena accompanying the crossing of two bands. A crossing at a certain angular momentum I_{cross} is defined when below I_{cross} one band is the lowest in energy and above I_{cross} the other band is lowest, with respect to states with the same spins of the two bands.

6.3.3a *The quasi-particle spectrum*

Let us consider two quasi-particle states as intrinsic excitations, and then discuss the spectrum of single-particle states in a rotating deformed potential with axial symmetry. A typical single-neutron spectrum, shown on the left-hand side of Fig. 6.16, can be generated by the Nilsson Hamiltonian $\mathbf{H}_{\text{sp}}$ (see also Fig. 6.4 and corresponding text). The total one-particle Nilsson potential is written as:

$$\mathbf{H}_{\text{sp}} = \frac{\vec{p}^{\,2}}{2m} + \frac{1}{2} m(\omega_x^2 x^2 + \omega_y^2 y^2 + \omega_z^2 z^2) + v_{ls}\hbar\omega_0 \vec{l}\cdot\vec{s} + v_{ll}\hbar\omega_0(\vec{l}^{\,2} - \langle \vec{l}^{\,2} \rangle). \tag{6.3.34a}$$

The quantities $\mathbf{p}$ and x, y, z denote the momentum and space vectors, respectively, of the particle with mass m, spin $\vec{s}$ and with orbital angular momentum $\vec{l}$. The quantity $\langle \vec{l}^{\,2} \rangle$ is averaged over all states in one major oscillator shell in the case of spherical symmetry. The frequency for a spherical oscillator potential is written as ω_0. In the case of axially symmetrical deformation the harmonic oscillator frequency ω_0 has two components $\omega_\perp$

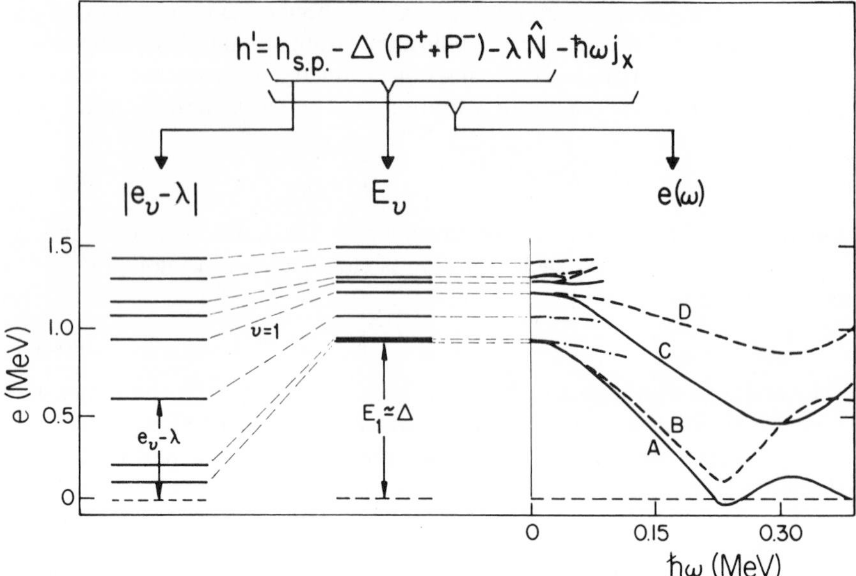

Fig. 6.16. The spectrum of single-particle states in a deformed (Nilsson) potential described by the single-particle Hamiltonian h_{sp} (left). The effect of including the pairing term is shown in the central part and that of rotation on the right. The quasi-particle diagram on the right-hand side is calculated in terms of the cranked shell model (CSM) for ^{168}Hf (Arciszewski *et al.*, 1983) with parameters given at Fig. 6.23.

and ω_z, expressed as

$$\omega_\perp = \omega_0(\varepsilon, \varepsilon_4)\,(1+\tfrac{1}{3}\varepsilon), \tag{6.3.34b}$$

$$\omega_z = \omega_0(\varepsilon, \varepsilon_4)(1-\tfrac{2}{3}\varepsilon). \tag{6.3.34c}$$

It should be noted that the particular choice of single-particle potential is rather important for a proper description of high-spin states. It appears, however, that the presence of the $\vec{l}^2$ term in the potential introduces some systematic deviations from observed quantities. Therefore one may prefer the Woods–Saxon potential, which produces better results. An analytical approach, like that used with the Nilsson potential, is not possible (Szymański, 1983).The average spacing of the single-neutron states is smaller than the neutron pairing correlation energy $\Delta \simeq 1$ MeV. This energy is required to break a pair of neutrons before they can be excited, each into some quasi-particle level. The effect of pairing correlations on the single-particle spectrum is explicitly taken into account in the Hamiltonian by adding the term $\Delta(\mathbf{P}^+ + \mathbf{P}^-)$ where the operators $\mathbf{P}^+$ and $\mathbf{P}^-$ represent combinations of quasi-particle creation and annihilation operators. Conservation of the number of neutrons is approximately achieved by adding the term $\lambda\mathbf{N}$ with an appropriate choice of the chemical potential λ, $\mathbf{N}$ being the particle number operator. The quasi-neutron energies are then given by:

$$E_\nu = \{\Delta^2 + (e_\nu - \lambda)^2\}^{1/2} \tag{6.3.35}$$

where e_ν is the single-particle energy.

The effect of pairing on the single neutron spectrum is shown in the central part of Fig. 6.16. The single neutron states with seniority 1 are shifted upwards with respect to the 'vacuum' seniority 0 state, i.e. the ground state of the adjacent even nucleus. The ground state of the odd nucleus has an energy $E_1 \simeq \Delta$ with respect to the vacuum state. A second effect is the compression in excitation energy of levels near the Fermi surface.

It was shown in Fig. 6.4 (see also the corresponding text) that rotation of the deformed nucleus modifies the single-particle motion which can be accounted for by adding the 'cranking' term $-\hbar\omega j_x$ (here j_x is the quasi-particle angular momentum projection on the rotation axis x'). The effect of rotation is demonstrated in the right-hand side of Fig. 6.16 which shows the quasi-particle diagram, calculated in terms of the Cranked-Shell-Model (CSM); i.e. the quasi-particle energies $e(\hbar\omega)$ as a function of the rotational frequency. The negative term $-\hbar\omega j_x$ in the Hamiltonian causes in general the level energies to decrease with increasing ω. The signature splitting of the two-fold degenerated Nilsson levels, as discussed before, is also illustrated by, for example, level A ($\alpha = +\frac{1}{2}$) and level B ($\alpha = -\frac{1}{2}$).

6.3.3b *Bandcrossing*

The 'vacuum' state of an even nucleus corresponds to the situation in which all levels below the Fermi surface $\{e(\omega)=0\}$ are occupied and all those above it are empty. For small values of ω the ground-state rotational band constitutes the vacuum state (with no intrinsic excitation). The first excited two-quasi-particle state is formed when levels A and B are occupied, with total signature $\alpha=0$. A rotational band based on that state consists of levels with spin values according to eqn (6.3.18) $I=0, 2, 4, \ldots$. With increasing ω the excitation energy of the two-quasi-particle state $E_{AB}=E_A+E_B$ is lowered with respect to the vacuum state. In other words the Coriolis interaction, $\mathbf{H}_c = -(\hbar^2/2\mathscr{I})\vec{I}\cdot\vec{j}$, lowers the energy of quasi-particle levels with angular momentum component j_x along the axis of collective rotation. At a critical frequency $\omega^* \simeq 0.26\,\mathrm{MeV}\hbar^{-1}$ (see Fig. 6.16) the energy of that state is minimal ($E_{AB} \simeq 0$) and excitation of two quasi-particles into levels A and B is energetically favourable. In fact the pairing gap for those two particles has disappeared and so the pair can be broken up. The two particles occupying levels A and B experience the strong Coriolis interaction and will consequently align their spins along the rotation axis. In this way a rotational aligned two-quasi-particle state is created with the maximum angular momentum and the least energy. This intrinsic state can be the basis of a rotational band. The discontinuity of the levels A and B at $\omega^* \simeq 0.26$ MeV$\hbar^{-1}$ represents the *crossing* of the two-quasi-particle excited band with the GSB. Beyond ω^* the AB configuration has even less energy than the GSB and constitutes the yrast (vacuum) state whereas the GSB becomes an excited configuration. The crossing can thus be defined as a point of instability in the vacuum with respect to the two-quasi-particle excitation. The yrast sequence below ω^* is thus formed by the GSB and above ω^* by the AB configuration. The crossing of the two bands has been observed experimentally for the first time in ^{160}Dy at Stockholm in 1972 (see Fig. 6.10(b)). Therefore, the AB-excited 2-quasi-particle band is called the Stockholm band or 'S-band'.

Let us now turn to experimental data and notice that the development of high-resolution–low-background multi-detector spectrometers has recently pushed up the maximum spin-value of an observed discrete state beyond 40. The simplified level scheme of ^{158}Er, given in Fig. 6.2, shows a regular structure in the excitation energy up to $I^\pi = 40^+$, with a strong anomaly at 14^+ and a less pronounced irregularity at 28^+. The corresponding γ-ray spectrum, displayed in Fig. 6.10(c), shows that the anomalies at 14^+ and 28^+ are even more pronounced in the γ-ray energies, which correspond to the first derivative of the excitation with respect to the angular momentum.

The spectrum of Fig. 6.10(c) was obtained with the TESSA-III spectrometer at Daresbury. The TESSA spectrometers are schematically shown in Fig. 6.17. TESSA-II consists of six high-resolution Ge detectors surrounded by NaI(Tl) Compton suppression shields. In the centre 50 bismuth-germanate (BGO) scintillation detectors are closely packed to serve as a sum-energy/multiplicity array (see Section 6.5 for more details and Chapter 3 for details on detectors). Those BGO detectors facilitate the selection of the region of high-spin states populated in heavy-ion fusion reaction. TESSA-III contains twelve Ge(BGO) assemblies and an inner BGO ball like TESSA-II.

With the multi-detector Compton-suppression spectrometers it became possible to measure simultaneously the correlation between pairs of γ-rays with high resolution and efficiency and low background. For $n = 6$ detectors one high-efficient spectrometer not only makes it possible to observe states with very high spin, but also many (sometimes weakly-excited) side bands. An example is presented in Fig. 6.18 for ^{168}Hf measured at Daresbury with the TESSA-II spectrometer (Chapman *et al.*, 1983). In addition to the GSB up to 22^+ and the yrast band up to 34^+ two negative-parity sidebands are observed. The interesting aspect of the observation of sidebands is the possibility of studying band interactions and thus revealing the nature of intrinsic (e.g. quasi-particle) excitations.

The interactions between bands of different nature is best studied in the vicinity of the bandcrossing, as illustrated in Fig. 6.19. The upper curves in

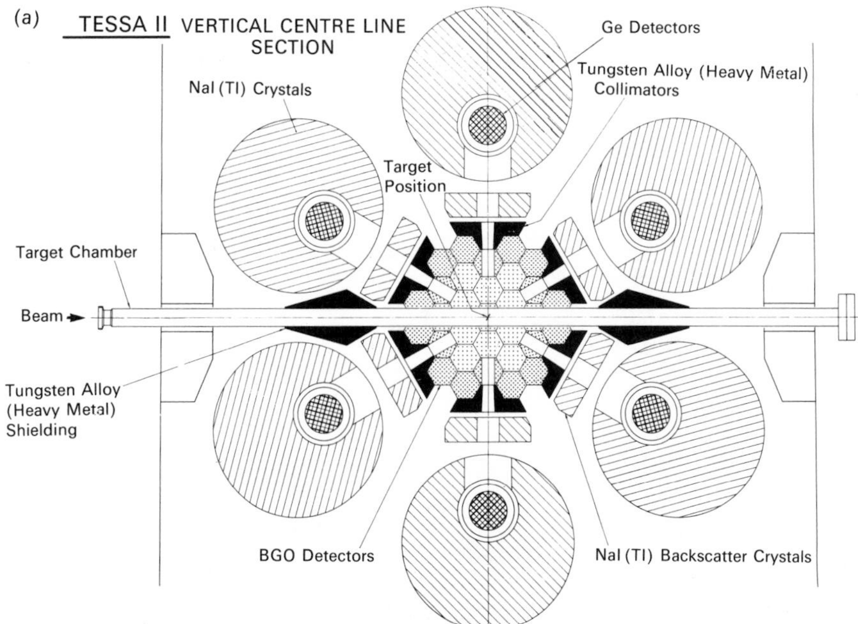

Fig. 6.17(a). The multi-detector γ-spectrometer TESSA-II used at Daresbury in studies of high-spin states (see Twin, 1983). In the centre 50 BGO detectors are installed as a sum-energy multiplicity array, surrounded by six Ge–NaI(Tl) Compton suppression spectrometers.

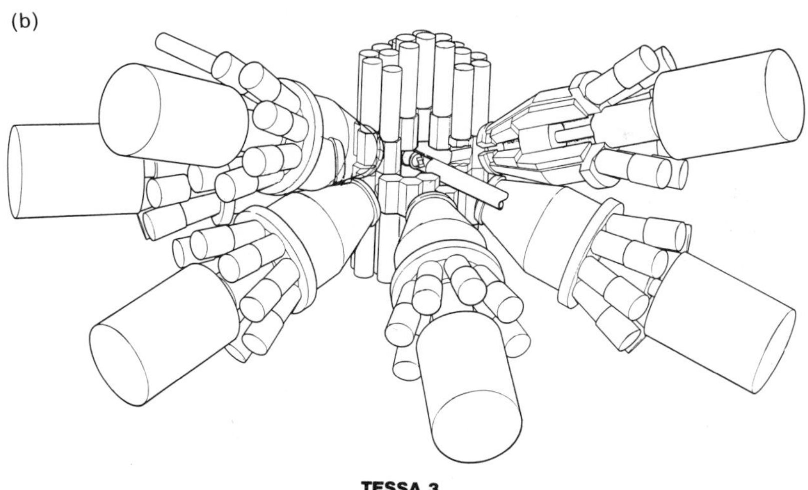

Fig. 6.17(b). The TESSA-III spectrometer with a similar inner BGO ball to TESSA-II (see above), but with twelve Ge detectors with symmetric BGO Compton suppression shields. (Figures obtained from J. Sharpey-Schafer, Liverpool.)

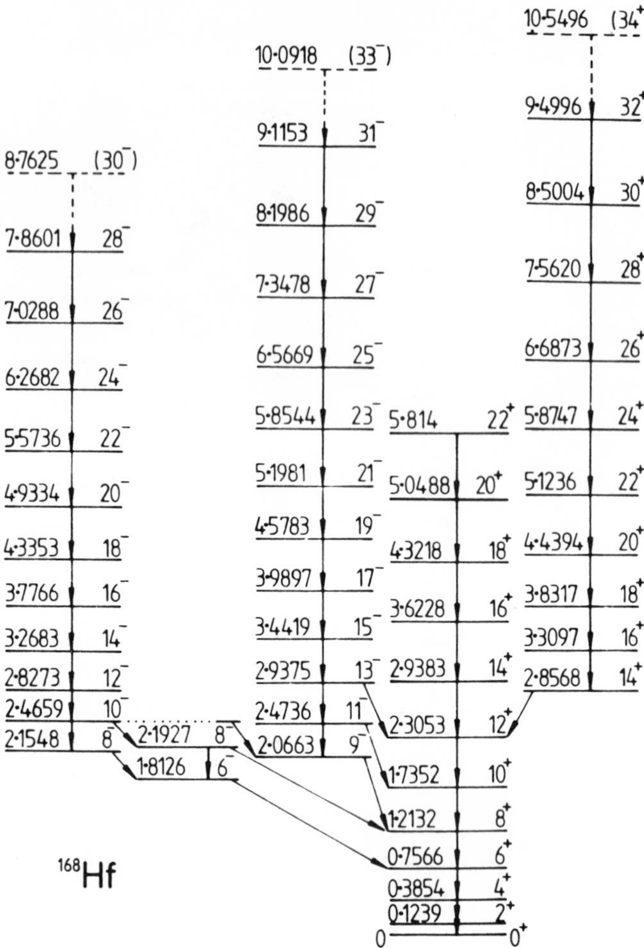

Fig. 6.18. The level scheme of ^{168}Hf obtained from experiments using the reaction ^{124}Sn(^{48}Ti, 4n)^{168}Hf at a bombarding energy of 216 MeV and the TESSA-II spectrometer (see Fig. 6.17a). (Chapman *et al.*, 1983.)

Fig. 6.19 represent the GSB and the two-quasi-particle excited band or *S*-band, as discussed above. At a certain angular momentum I_{cross} the two curves cross each other in the case of weak interaction or are repelled by the strong interaction. In the vicinity of the crossing some nuclear properties change drastically, such as the moment of inertia $\mathscr{I}$, which is best seen when plotted versus the rotational frequency ω. By analogy with classical mechanics the rotational frequency ω can be defined as the derivative of the energy E with respect to the angular momentum I. Quantum-mechanically one should use

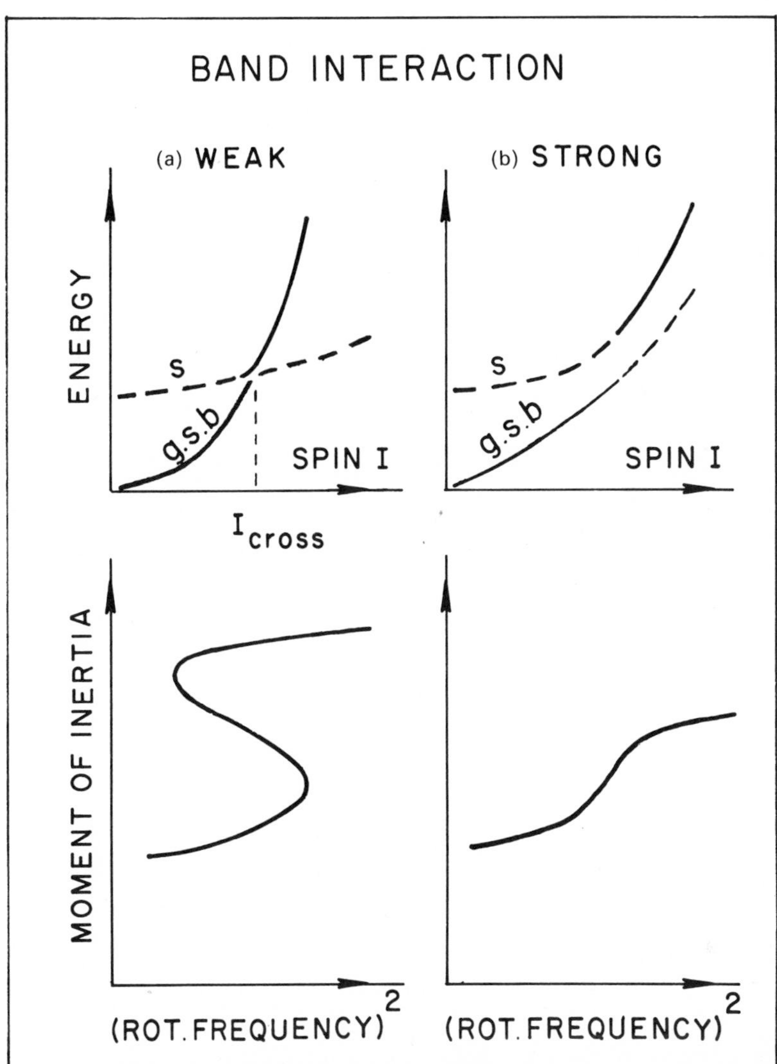

Fig. 6.19. Schematic illustration of the weak (a) and strong (b) band interaction leading to two different shapes of the curve of the moment of inertia plotted vs. the rotational frequency ω^2. For weak interaction a 'backbending' of the curve is seen and for strong interaction an 'upbending'. The distance of closest approach of the two upper curves is related to the interaction strength $|V|$ discussed in the text (to a good approximation, this distance equals $2|V|$).

the expectation value of the angular momentum $\hbar\{I(I+1)\}^{1/2}$

$$\omega = \hbar^{-1}\, dE/d\{I(I+1)\}^{1/2}. \tag{6.3.36a}$$

For $I \gg 1$ $\quad \omega \simeq \hbar^{-1}\, dE/dI$, and for $I \gg K$, $\omega \simeq \hbar^{-1} dE/dI_x$. (6.3.36b)

The latter equality follows from eqn (6.3.45) for $K=0$ (see below). The rotational frequency can be derived from the experimental energy spacings $E_\gamma = E(I) - E(I-2)$ between rotational states of angular momenta I and $I-2$. Differentiating eqn (6.3.2) and taking into account the discrete character of E and I one obtains for a $K=0$ band

$$2\mathscr{I}/\hbar^2 = \left(\frac{dE}{dI(I+1)}\right)^{-1} = \left(\frac{E_\gamma}{4I-2}\right)^{-1}. \tag{6.3.37}$$

From eqn (6.3.36a) we get

$$\hbar\omega = 2\{I(I+1)\}^{1/2}\frac{dE}{dI(I+1)}. \tag{6.3.38}$$

If we determine the derivative at an angular momentum I' lying in the middle of I and $I-2$ and substitute in eqn (6.3.38) the average value of

$$I'(I'+1) = I^2 - I + 1 \tag{6.3.39}$$

we then obtain for the rotational frequency:

$$\hbar^2\omega^2 = (I^2 - I + 1)\left(\frac{E_\gamma}{2I-1}\right)^2. \tag{6.3.40}$$

For spin values larger than $I\hbar \simeq 6\hbar$ we find the simple approximation:

$$\hbar\omega \simeq \tfrac{1}{2}E_\gamma. \tag{6.3.41}$$

The rotational frequency is thus directly given by the γ-ray transition energy in the case of a $K=0$ rotational band.

Figure 6.19 demonstrates that for $I < I_{\text{cross}}$ the yrast states correspond to members of the GSB, but beyond I_{cross} the S-band becomes yrast, as explained above in the independent quasi-particle picture (see Fig. 6.16).

In the GSB configuration all nucleons are assumed to be paired to spin 0 under influence of the pairing-correlation interaction. The nucleus can be regarded as a superfluid (with an intrinsic energy well below the pairing gap), and thus has low rigidity and correspondingly a small moment of inertia. With increasing rotational velocity, however, the Coriolis force tends to counteract the pairing correlations so as to increase the rigidity of the nucleus. The observed smooth increase of the moment of inertia with increasing ω can indeed be attributed to the Coriolis anti-pairing (CAP) effect.

A more dramatic effect is observed at the crossing ($I \simeq I_{\text{cross}}$) of the GSB and the S-band: namely a sudden increase of the moment of inertia and a slowing down of the collective rotation (see Fig. 6.19). In the case of strong interaction the crossing of the S-band with the GSB is less sharp and the accompanying effects less dramatic.

We will now relate the distance of closest approach of the two bands, represented in the E vs I plane, to the interaction H_r. Let E_g represent the energies of states belonging to the GSB and E_s to those of S-band. After switching on the interaction H_r, states with the same spin and parity belonging to the GSB and the S-band are repelled and shifted to energies E^l and E^u, respectively. The values E^l and E^u are obtained by solving the determinental equation

$$\begin{vmatrix} E_s - E & V \\ V & E_g - E \end{vmatrix} = 0, \tag{6.3.42}$$

where $V = \langle I_s | H_r | I_g \rangle$ is the interaction matrix between the GSB $|I_g\rangle$ and the S-band $|I_s\rangle$.

Assuming $E_s \simeq E_g$ at $I = I_{\text{cross}}$ one finds

$$E^u - E^l = 2|V|. \tag{6.3.43}$$

The interaction thus gives rise to mixing of the two bands and the mixing coefficient is proportional to the interaction strength.

6.3.3c *Experimental Routhians*

We shall now outline the procedure used to analyse collective rotational bands in terms of independent quasi-particles. The dynamical coupling between the quasi-particle degrees of freedom with those characterizing nuclear rotation can be described by the cranking Hamiltonian:

$$\mathbf{H}^{\omega} = \mathbf{H}' - \hbar\omega I_x, \tag{6.3.44a}$$

where I_x represents the projection of the total angular momentum I on the axis of collective rotation O_x and H' is the Hamiltonian in the intrinsic (rotating) system. For convenience we let the body-fixed axis $O_{x'}$ coincide with the space-fixed axis O_x (see Fig. 6.9).

The second term in H^{ω} is a consequence of the transformation of the time-dependent Schrödinger equation

$$i\hbar \frac{\partial \Psi}{dt} = \mathbf{H}\Psi, \tag{6.3.44b}$$

from the laboratory to the intrinsic rotating frame of reference. The wave function Ψ and Hamiltonian $\mathbf{H}$ in the laboratory system are transformed by means of the rotation operator

$$\mathrm{R} = \mathrm{e}^{-iI_x\omega t}, \tag{6.3.44c}$$

where $I_x = I_{x'}$, denotes the component of the total angular momentum, giving

$$\Psi = R\Psi' \tag{6.3.44d}$$

$$\mathbf{H} = \mathrm{R}\mathbf{H}'\mathrm{R}^{-1}. \tag{6.3.44e}$$

The quantities Ψ' and $\mathbf{H}'$ refer to the body-fixed (rotating) coordinate frame. Within this frame one obtains, after substituting eqns (6.3.44d, e) into eqn (6.3.44b)

$$i\hbar\frac{\partial\Psi'}{\partial t} = (\mathbf{H}' - \hbar\omega I_{x'})\Psi'. \tag{6.3.44f}$$

From this equation we obtain the expression of eqn (6.3.44a) for $\mathbf{H}^{\omega}$. The term $\hbar\omega I_x$ in eqns (6.3.44a) and (6.3.44f) is like the Coriolis and centrifugal forces in classical mechanics—see e.g. de Voigt *et al.* (1983).

By analogy to eqn (6.3.1) we obtain for I_x the value

$$I_x = \{I(I+1) - K^2\}^{1/2} \simeq \{(I+\tfrac{1}{2})^2 - K^2\}^{1/2}. \tag{6.3.45}$$

The cranking Hamiltonian $\mathbf{H}^{\omega}$ can be obtained by a simple addition of the single quasi-particle Hamiltonians h' given in Fig. 6.16. The nuclear quantities can be calculated for fixed angular momentum using solutions of the cranking Hamiltonian of eqn (6.3.44a) in the theoretical framework of the Hartree–Fock–Bogolyubov–Cranking approximation (HFBC) or of the Cranked-Shell-Model (CSM). (See Szymański (1983), also for references to both approaches.) If the energy E^{ω} is an eigenvalue of $\mathbf{H}^{\omega}$ then I_x can be found as the derivative of eqn 6.3.44a:

$$\mathrm{d}E^{\omega}/\mathrm{d}\omega = \langle \mathrm{i}|\partial\mathbf{H}^{\omega}/\partial\omega|\mathrm{i}\rangle = -\hbar I_x. \tag{6.3.46}$$

However, the single quasi-particle solutions do not specify the energies in the laboratory frame, but rather in the intrinsic (rotating) frame of reference (for a full theoretical treatment of this problem see Szymański, 1983). Instead of calculating the true energies and comparing them with experimental data a different procedure, which we will follow below, has been suggested by Bengtsson and Frauendorf (1977, 1979).

The experimental data are first transformed from the laboratory to the intrinsic frame of reference and then compared with the theoretical results. In this way calculated independent quasi-particle diagrams can directly be tested. For the transformation of experimental data from the laboratory to the intrinsic frame it is convenient to plot them as functions of the rotational frequency $\hbar\omega$ (in MeV), as illustrated for the yrast sequence in ^{168}Hf in Fig. 6.20. The lowest curve represents the excitation energies E_x plotted in the usual way versus the angular momentum I, and showing the crossing of the GSB with the S-band.

The central curve represents a plot of E_x vs. $\hbar\omega$, obtained from eqn (6.3.40), displaying the backbending more prominently than the first curve. The

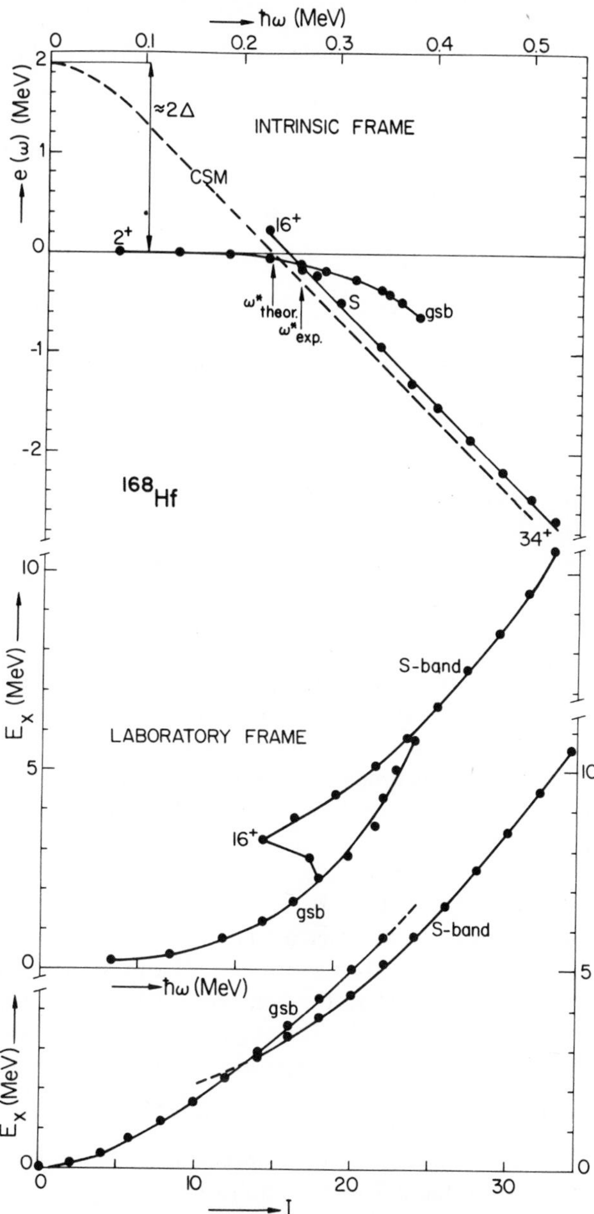

Fig. 6.20. Experimental excitation energies (Janssens *et al.*, 1981) of the GSB and S-band in ^{168}Hf plotted vs. I (bottom curve) and ω (central part). The transformation of these total energies into quasi-particle Routhians $e(\omega)$ is achieved by using eqn (6.3.47) as explained in the text. The top part of the figure displays the experimental values of $e(\omega)$ along with a theoretical curve obtained in terms of the cranked shell model.

transformation of the laboratory energies $E_x(I)$ to those in the intrinsic (rotating) frame $E_x^\omega(I)$ is accomplished using (see e.g. Szymański, 1983)

$$E_x^\omega(I) = E_x(I) - \hbar\omega(I) I_x(I). \tag{6.3.47}$$

We notice that $\hbar\omega(I)$ according to eqn (6.3.40) is defined at an angular momentum I lying in the middle of the interval $I+1$ and $I-1$. Thus we rewrite eqn (6.3.47) accordingly and obtain

$$E_x^\omega(I) = (1/2)\{E_x(I+1) + E_x(I-1)\} - \hbar\omega(I) I_x(I). \tag{6.3.48}$$

The energy of a rotational state in the intrinsic frame of the reference is called the 'Routhian'. This name is chosen by analogy with the 'Routh' functions in classical mechanics and emphasizes that for a rotating system the Hamiltonian is not the same as the energy (see e.g. Landau and Lifshitz, 1965).

The total Routhian $E_x^\omega(I)$ contains a contribution from the collective rotation and the intrinsic excitation of the quasi-particles. In order to obtain the quasi-particle Routhian $e(\omega)$, as shown in Fig. 6.16, one has to subtract a reference energy $E_{\text{ref}}^\omega(I)$. {Remember that ω is defined at each I via eqn (6.3.40)}:

$$e(\omega) = E_x^\omega\{I(\omega)\} - E_{\text{ref}}^\omega\{I(\omega)\}. \tag{6.3.49}$$

As a reference rotor one may select the GSB. However, the levels of the GSB near the backbending are already disturbed by the interaction with the S-band. Therefore one may also take into account the high-lying levels of the S-band when fitting the experimental data (see below).

For the reference rotational band (i.e. the GSB) it is convenient to introduce a kinematical moment of inertia $\mathscr{I}_{\text{ref}}^{(1)}$ (see Section 6.3.4) that is linear in ω^2:

$$\mathscr{I}_{\text{ref}}^{(1)} = \mathscr{I}_0^{(1)} + \mathscr{I}_2^{(1)}\omega^2. \tag{6.3.50}$$

This expression is equivalent to the one used in the variable moment of inertia model—i.e. the two-parameter Harris formula; see for instance Harris (1965) and Mariscotti *et al.* (1969). Here the constants $\mathscr{I}_0^{(1)}$ and $\mathscr{I}_2^{(1)}$ have to be adjusted for each nucleus separately. The angular momentum of the reference rotor along the axis of rotation $I_{x,\text{ref}}(\omega)$, denoted by $I_{\text{ref}}(\omega)$, is given by:

$$I_{\text{ref}}(\omega) = \hbar^{-1}\mathscr{I}_{\text{ref}}^{(1)}\omega = \hbar^{-1}(\mathscr{I}_0^{(1)} + \mathscr{I}_2^{(1)}\omega^2)\omega. \tag{6.3.51}$$

The reference energy in eqn (6.3.49) can be obtained by integrating eqn (6.3.51) as can be seen e.g. from the differential eqn (6.3.46):

$$E_{\text{ref}}^\omega\{I(\omega)\} = -\hbar\int I_{\text{ref}}(\omega)\,\mathrm{d}\omega = -\tfrac{1}{2}\mathscr{I}_0^{(1)}\omega^2 - \tfrac{1}{4}\mathscr{I}_2^{(1)}\omega^4 + \hbar^2/(8\mathscr{I}_0^{(1)}). \tag{6.3.52}$$

where the integration constant is added to assure that the ground-state energy $E_{\text{ref}}^\omega(I=0)$ is equal to zero (Bengtsson and Frauendorf, 1979b). This is approximately correct for small ω, when the ω^4 term in eqn (6.3.52) and the ω^2 term in eqn (6.3.50) can be neglected. The constant is then found when one obtains $I_{\mathrm{ref}} = \frac{1}{2}$ for $I = 0$ (and $K = 0$) from eqn (6.3.45).

A fit to the lowest GSB and highest S-band members of ^{168}Hf yield: $\mathscr{I}_0^{(1)} = 23\ \mathrm{MeV}^{-1}\,\hbar^2$ and $\mathscr{I}_2^{(1)} = 90\ \mathrm{MeV}^{-3}\,\hbar^4$ for the example in Fig. 6.20. Using these values the quasi-particle Routhian $e(\omega)$ has been calculated with eqn (6.3.49) and (6.3.52). The result is shown in the top part of Fig. 6.20. In this plot the critical frequency at the crossing point can be determined quite precisely as $\omega^*(\mathrm{exp}) = 0.26\ \mathrm{MeV}\,\hbar^{-1}$. A direct comparison with the two-quasi-particle Routhian obtained from the cranked shell model is possible. The theoretical crossing occurs at $\omega^*(\mathrm{theor}) = 0.23\ \mathrm{MeV}\,\hbar^{-1}$. This slight deviation from experiment (Janssens *et al.*, 1981) as well as the slight shift of the theoretical curve with respect to that of the S-band data depends strongly on the chosen Δ parameter in the calculation (see Fig. 6.20).

6.3.3d *Aligned angular momentum*

Any contribution to the total angular momentum $I(\omega)$ of the rotating nucleus originating from quasi-particles will show itself by a deviation from the smooth increase described by eqn (6.3.51). The quasi-particle aligned angular momentum $i(\omega)$ is thus obtained as:

$$i(\omega) = I_x(\omega) - I_{\mathrm{ref}}(\omega). \tag{6.3.53}$$

For the same example used in Fig. 6.20 of ^{168}Hf the angular momenta I_x, from eqn (6.3.45), are plotted versus $\hbar\omega$, from eqn (6.3.40), in Fig. 6.21. The above values for $\mathscr{I}_0^{(1)}$ and $\mathscr{I}_2^{(1)}$ were used which were fitted to assure an approximate constant aligned angular momentum $i(\omega)$ for the S-band as displayed in the central part of the figure. This curve illustrates clearly a sudden increase of $i(\omega)$ by $\Delta i \cong 10$ at the crossing frequency $\omega^* \simeq 0.26\ \mathrm{MeV}\,\hbar^{-1}$. This jump in $i(\omega)$ must be attributed to the excitation of two quasi particles with their angular momenta approximately aligned along the axis of rotation. It will be explained below that the $\mathrm{i}_{13/2}$ quasi-neutron orbitals are the only candidates to generate such high $i(\omega)$. The backbending curves of the moments of inertia plotted versus $\hbar\omega$ (top part of Fig. 6.21) will be discussed in the next Section 6.3.4.

To get a better understanding of the band-crossing phenomena on the basis of the independent quasi-particle picture one should also consider odd nuclei. The quasi-particle aligned angular momenta $i(\omega)$ for the yrast bands in^{167,168,169}Hf as functions of $\hbar\omega$ are therefore given in Fig. 6.22.

The backbending in ^{168}Hf occurs at $\omega^* = 0.26\ \mathrm{MeV}\,\hbar^{-1}$ as discussed above. This is also seen in Fig. 6.22, showing at the crossing a gain in aligned angular momentum of $\Delta i = 9.4$. The upbeding in the odd isotopes occurs at higher rotational frequencies; i.e. at $\omega = 0.33\ \mathrm{MeV}\,\hbar^{-1}$ in ^{167}Hf and at

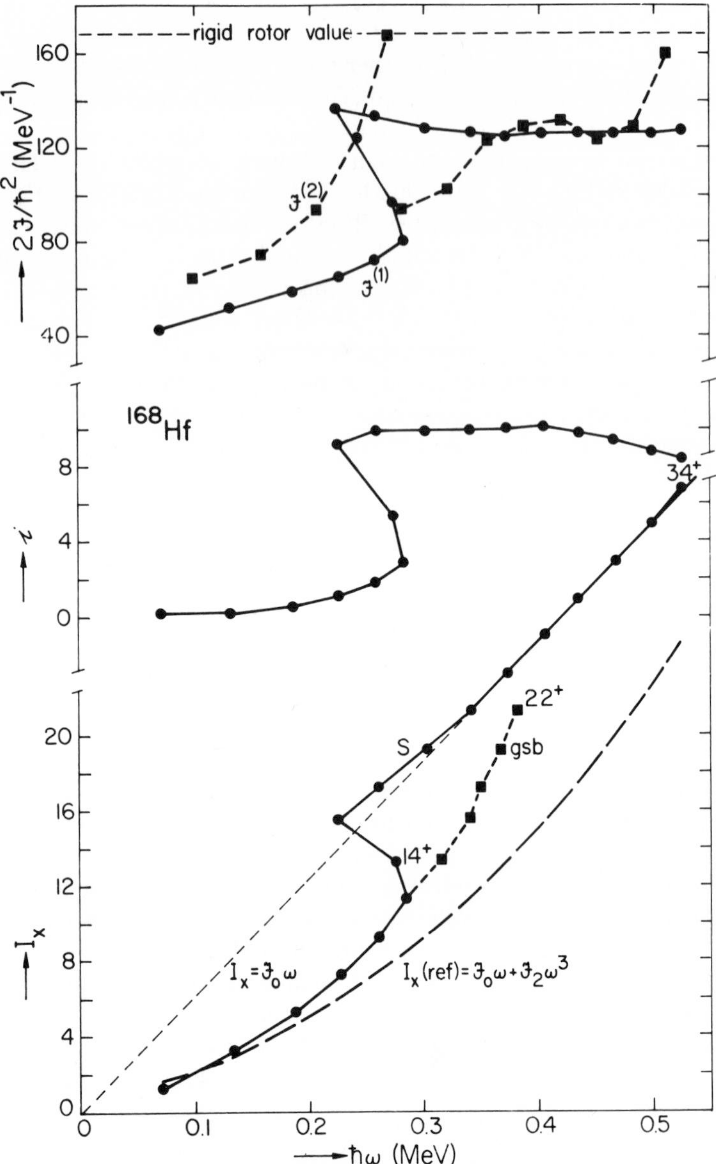

Fig. 6.21. The component of the angular momentum along the rotation axis I_x (bottom), the quasi-particle aligned angular momentum i (central part) and the kinematical $\mathscr{I}^{(1)}$ and dynamical $\mathscr{I}^{(2)}$ moments of inertia (multiplied by $2/\hbar^2$, top) plotted vs. the rotational frequency. The curve representing a reference rotor (bottom) is obtained from a fit to the GSB and assuming that $i(\omega)$ is approximately $i(\omega)$ in the S-band with $\mathscr{I}_0^{(1)} = 23\ \text{MeV}^{-1}\ \hbar^2$ and $\mathscr{I}_2^{(1)} = 90\ \text{MeV}^{-3}\ \hbar^4$ (see also Fig. 6.20 and text). The straight dashed line in the bottom plot represents the classical relation $I_x = \mathscr{I}_0\omega$ with a constant $\mathscr{I}_0$ for the S-band.

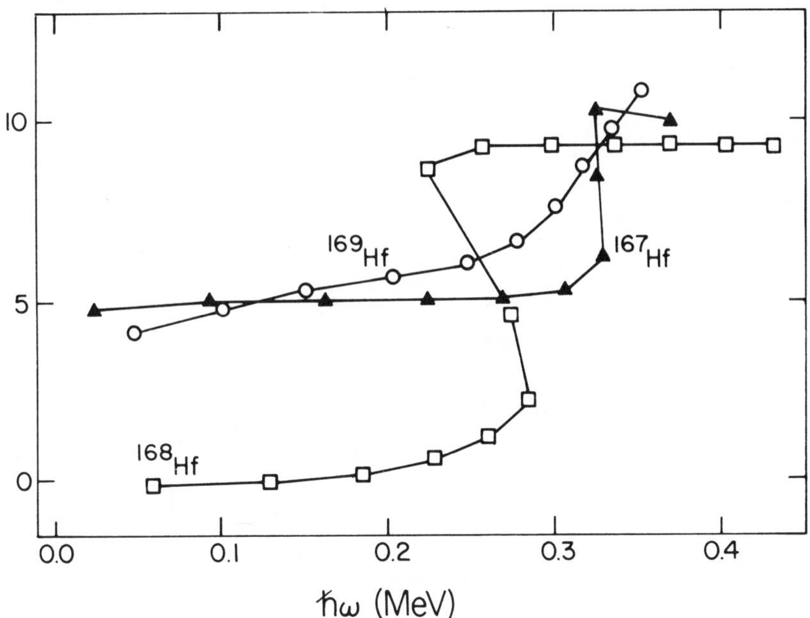

Fig. 6.22. Plot of the aligned angular momentum $i(\omega)$ versus $\hbar\omega$ for the yrast bands in ^{167,168,169}Hf. The data for the odd-A nuclei concern the favoured ν $i_{13/2}$ bands. (Arciszewski *et al.*, 1983.)

$\omega > 0.30$ MeV$\hbar^{-1}$ in ^{169}Hf. The initial alignment at low rotational frequencies in both odd isotopes is $i(\omega) \sim 5.0$, which implies a nearly complete alignment of the odd neutron in an $i_{13/2}$ orbital with the axis of rotation. The gain in angular momentum at the crossing in the odd isotopes is about half the gain in ^{168}Hf, namely $\Delta i = 5.1$ in ^{167}Hf and > 5.0 in ^{169}Hf. All those observations should be explicable in terms of the independent quasi-particle picture.

6.3.3e *The quasi-particle diagram*

The neutron quasi-particle diagram resulting from cranked shell model calculations is shown in Fig. 6.23. In this context, the 'favoured' and 'unfavoured' bands in the odd-N isotopes are based on the intrinsic states formed by a neutron occupying the lowest-lying quasi-particle trajectories with positive energy and parity, and positive and negative signature, respectively. In Fig. 6.23 these trajectories are indicated by A(positive signature) and B(negative signature) with single quasi-particle energies e_A and e_B. In ^{168}Hf a strong backbending is predicted to occur in the yrast band with $\omega_1 = 0.23$ MeV$\hbar^{-1}$ due to the alignment of a pair of quasi-particles into the

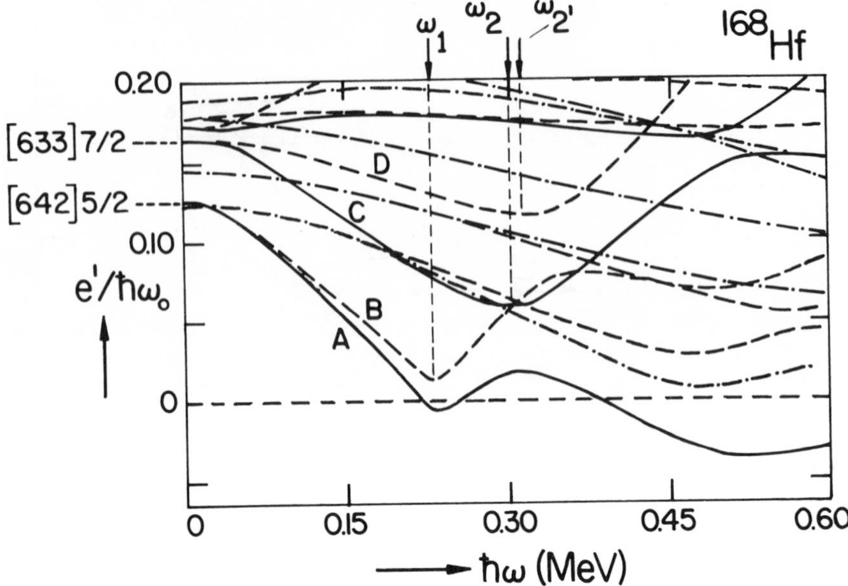

Fig. 6.23. Quasi-particle diagrams for neutrons in ^{168}Hf calculated in terms of the cranked shell model according to the procedure of Bengtsson and Frauendorf (1979). The four lowest positive-parity trajectories with lowest energy at $\hbar\omega = 0$ originating from the ν $i_{13/2}$ orbit are denoted A, B, C, and D. The trajectories with positive signature ($\alpha = +\frac{1}{2}$) are indicated by solid lines and those with negative signature ($\alpha = -\frac{1}{2}$) by dashed lines. The indicated frequencies refer to the irregularities in the yrast band for the even–even nucleus (ω_1) and for the odd-N nuclei (ω_2). The quasi-particle energies are relative to $\hbar\omega_0 = 41A^{-1/3}$ MeV. The parameters used in the calculation are: $\varepsilon_2 = 0.235$, $\varepsilon_4 = 0.005$, $\Delta = 0.914$ MeV, $\lambda = 6.494$, $\omega_0 = 7.43$ MeV$\hbar^{-1}$, $V(\omega_1) = 0.036$ MeV and $V(\omega_2) = 0.153$ MeV. (Arciszewski *et al.*, 1983.)

trajectories A and B. A two-quasi-particle state is formed with energies E_{AB} and collective rotation based on this state constitutes the *S*-band. At the crossing frequency the two-quasi-particle energy $E_A + E_B \sim 0$ with respect to the reference configuration, implying a crossing of the *S*-band (*AB*) and the ground-state (reference) band.

A similar comparison between theory and experiment is also possible for the odd isotopes ^{167}Hf and ^{169}Hf. The absence of backbending at the frequency ω_1 where it occurs in ^{168}Hf can be explained in terms of the *blocking effect*. At low frequencies the trajectory A is occupied in the case of the 'favoured' band ($\alpha = +\frac{1}{2}$) and the trajectory B in the case of the 'unfavoured' band ($\alpha = -\frac{1}{2}$) as already discussed. At the frequency $\omega_1 = 0.23$ MeV$\hbar^{-1}$ the alignment of the AB quasi-neutron pair in these cases is thus impossible ('blocked'). Therefore in the odd isotopes the trajectories A and B do not experience a discontinuity at ω_1, but continue to slope down (not drawn here). At $\omega_2 = 0.30$ MeV$\hbar^{-1}$, however, the two-quasi-particle energy $E_B + E_C \simeq 0$ and the alignment of a BC neutron pair is favourable in the odd isotopes (with a neutron in the orbital A as spectator). The blocking effect explains in this way why the crossing frequencies in the odd isotopes are larger than in the adjacent ^{168}Hf nucleus.

This analysis reveals the nature of the backbending on a microscopic basis; i.e. why the neutron $i_{13/2}$[642]5/2 orbitals are responsible for the first bandcrossing in ^{168}Hf and the [642]5/2, $\alpha = -\frac{1}{2}$ and [633]7/2, $\alpha = +\frac{1}{2}$ orbitals for those in ^{167}Hf and ^{169}Hf.

The next observable to be analysed is the aligned angular momentum i. From eqn (6.3.46) we learn that the slope of the trajectories in the quasi-particle diagram yield directly the aligned angular momentum (with a negative sign). It was explained before that at the crossing the yrast configuration in ^{168}Hf is rearranged; i.e AB forms the new yrast state after the crossing. This new yrast configuration can be considered as a new vacuum (see e.g. Szymański, 1983). The gain in angular momentum can thus be calculated from the slopes of trajectories A and B before the crossing as:

$$\Delta i = -\hbar^{-1}\left(\frac{de_A}{d\omega} + \frac{de_B}{d\omega}\right)_{\omega \leqslant \omega^*}, \qquad (6.3.54)$$

where ω^* denotes the crossing frequency. As a result of the theoretical analysis it appears that after the crossing the slopes of trajectories with opposite signatures are interchanged with reversed signs—see Szymański, (1983) for a full discussion of this problem. Therefore Δi can also be obtained as the change in slope of one trajectory (A or B) at ω^*.

The slopes of the trajectories A, B, C and D in Fig. 6.23 correspond to: $i = 5.3$, 4.3, 3.5 and 1.6, respectively. This yields an initial aligned angular momentum for the favoured bands in ^{167,169}Hf of $i = 5.3$, a gain at ω_1 in ^{168}Hf of $\Delta i = 5.3 + 4.3 = 9.6$ and at ω_2 in the odd isotopes of $\Delta i = 4.3 + 3.5 = 7.8$. These calculated values are in very good agreement with the experimental data, except for the somewhat too large Δi value calculated for the odd isotopes.

Finally, it may be noticed that upbending in the unfavoured bands of the odd isotopes (B being occupied and alignment of an AD neutron pair) is predicted at $\omega^* \simeq 0.33$ MeV$\hbar^{-1}$ with a gain $\Delta i = 5.3 + 1.6 = 6.9$. Although those bands have been observed the high-spin members are still lacking, which prevents this prediction being checked.

A second backbending in ^{168}Hf has not yet been observed up to 34^+, although the next anomalies in the quasi-particle diagrams are calculated at $\omega \simeq 0.45$ MeV$\hbar^{-1}$ due to alignment of an $h_{11/2}$ proton pair and at still higher frequencies for next neutron trajectories. Very similar data as that discussed here for the Hf isotopes have been observed for nuclei with 2 protons more, namely ^{169,170,171}W (Arciszewski *et al.*, 1983). Instructions for the use or interpretation of quasi-particle diagrams are given by de Voigt *et al.* (1983).

6.3.3f *Residual interactions and pairing correlations*

So far we have only compared excitations in odd and even nuclei due to quasi-particles in the $\nu i_{13/2}$ Nilsson orbitals mainly with $\Omega \ll j$, which have a large spatial overlap in the prolate deformed potential. It is instructive to consider also quasi-particle excitations to other orbitals with, for instance, negative parity, as they contain interesting information on the residual interactions between the valence nucleons and on pairing correlations. Here the residual interaction is considered without including the pairing correlations.

In the independent quasi-particle picture it is assumed that two-body interactions are small with respect to the one-body nucleon interactions with the average main nuclear field. It is possible to estimate from experimental data (expressed in the intrinsic frame of reference) the two-body interactions $V_{\mu\nu}$ as the magnitude of the residual interactions between valence nucleons

$$e(\omega) = \sum_{\mu} e_{\mu}(\omega) + \sum_{\mu<\nu} V_{\mu\nu}(\omega). \tag{6.3.55}$$

Experimental Routhians, deduced for odd- and even-N neighbouring nuclei can be used as inputs in a series of equations of the type:

$$e_{\mathrm{AB}} = e_{\mathrm{A}} + e_{\mathrm{B}} + V_{\mathrm{AB}} + 2\Delta. \tag{6.3.56}$$

The quantity V_{AB} can thus be obtained as a function of ω, and in the same way V_{AE} and V_{BE} as well (for negative parity bands with one of the quasi neutrons occupying a negative-parity level E). Here a Δ is assumed constant, which is certainly not valid at higher frequencies ($\omega > 0.3$ MeV) as discussed below. Such an analysis has been carried out for different configurations in ^{160,161}Yb involving A[651,3/2] $\alpha = +\frac{1}{2}$, B[651,3/2] $\alpha = -\frac{1}{2}$, E[532,3/2] $\alpha = +\frac{1}{2}$ and F[532, 3/2] $\alpha = -\frac{1}{2}$ neutron orbitals. Negative $V_{\mu\nu}$ values were found, corresponding to attractive residual interaction strengths of about 250 keV between neutrons in those orbitals (see Garrett *et al.*, 1983).

A change of the quasi-particle configuration (i.e. at a bandcrossing) will change the energy of the system. Since the quasi-particle energies are very sensitive to the pairing correlation energy (Δ) the effects of the alignment of a quasi-particle pair in an odd nucleus may be different from that in the neighbouring even nucleus. The 'spectator particle' in the odd nucleus is excluded from the pairing correlations. Moreover, since the pairing correlation is a collective effect the blocking of that occupied orbital affects the basis of the other orbitals, depending on the spatial overlap between them; an effect called 'pairing blocking'. Notice that this blocking is of a different kind than the one discussed above for $\nu i_{13/2}$ trajectories.

Let us consider an odd nucleus with an occupied negative-parity orbital E. At the first bandcrossing, for instance, two neutrons are aligned in the $\nu i_{13/2}$ orbitals and the one-quasi-particle state (E) continues as a 3-quasi-particle state (EAB). In this example the E quasi-neutron is blocked and therefore a slower nuclear rotation will compensate for the smaller pairing correlation energy as compared to the even nucleus. Indeed, the crossing in the odd nucleus ^{163}Yb (EAB configuration) occurs at a lower frequency than in ^{162}Yb (AB configuration).

This pair-blocking effect has been observed consistently in the rare earth region as a shift in the crossing frequencies $\delta\hbar\omega^*$ between neighbouring even and odd-N nuclei. The crossing concerns the alignment of a pair of $i_{13/2}$ neutrons, thus an $0\rightarrow$AB excitation as compared with an E$\rightarrow$EAB excitation. A few exceptions concern nuclei where the odd particle occupies an orbital which has only a small overlap with those involved in the two-quasi-particle alignment—for example the [505]11/2$^-$ orbital in ^{161}Er. The data are shown in Fig. 6.24. The shifts in frequency $\delta\hbar\omega^*$ are related to the residual interactions $V_{\mu\nu}$ discussed above. In the expression for the total quasi-particle energy the odd nuclei have two terms in addition to those of the even nuclei, namely $V_{\rm AE}$ and $V_{\rm BE}$ (with negative values). A shift in ω can be translated into a shift in $e(\omega)$ in the quasi-particle diagram by multiplying by $\mathrm{d}e(\omega)/\mathrm{d}\omega$. Using eqn (6.3.46) one finds:

$$V_{\rm AE}+V_{\rm BE}=\delta\hbar\omega^*\overline{\Delta i}. \qquad (6.3.57)$$

Here $\overline{\Delta i}$ is the average gain in angular momentum in the EAB and AB alignments. The average residual interaction for the highly aligned orbitals is found to be $\langle V_{\mu\nu}\rangle=\frac{1}{2}(V_{\rm AE}+V_{\rm BE})\simeq 230$ keV (from $\delta\hbar\omega^*\simeq 40$ keV and $\overline{\Delta i}=8$ to 10). This value is in agreement with the above-mentioned value for the ^{160,161}Yb system deduced from the quasi-particle Routhians (Garrett *et al.*, 1983).

The analysis discussed above is based on the transformation of observables from the laboratory system of reference to the rotating intrinsic frame, with a reference rotor with a variable moment of inertia. At higher rotational frequencies ($\omega\geqslant 0.3$ MeV$\hbar^{-1}$) the second term in eqn (6.3.50) yields an unrealistically large $\mathscr{I}^{(1)}_{\rm ref}(\omega)$ and does not account for the observed moments of inertia (see next Section 6.3.4). Therefore the quasi-particle Routhians and the aligned angular momenta have also been analysed in the intrinsic frame of reference with a constant moment of inertia (cf. Herskind, 1983). In such an analysis indications were found that in rare earth nuclei the pairing correlation energy decreases with increasing ω. Depending on the choice of the moment of inertia, the pairing-gap parameter Δ vanishes in, for instance, ^{167,168}Hf near $\omega\simeq 0.4$ MeV$\hbar^{-1}$. This is in line with the HFBC calculation including quadrupole pairing (Diebel, 1984), but is at variance with that of

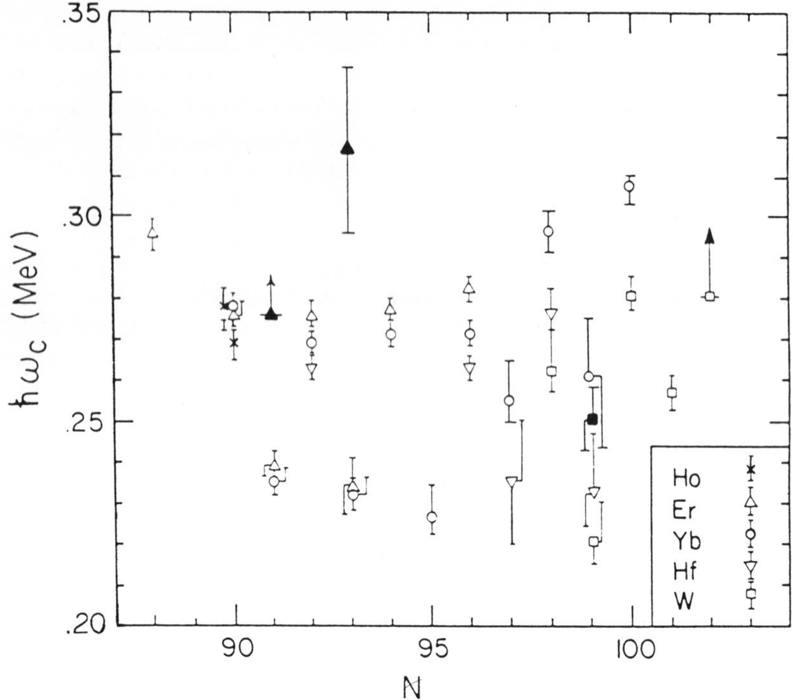

Fig. 6.24. The experimentally-determined crossing frequencies for the ($v=0\rightarrow v=2$, $0\rightarrow$AB) and negative parity ($v=1\rightarrow v=3$, $E\rightarrow$EAB) $i_{13/2}$ neutron alignments in a series of Ho, Er, Yb, Hf, and W nuclei. All the low points at $\hbar\omega\simeq 0.23$ MeV are for the odd isotopes, whereas the high points at $\hbar\omega\simeq 0.27$ MeV are for the even isotopes. (Garrett *et al.*, 1983.)

Mütz and Ring (1984). Reduced pairing has also been reported by Dudek *et al.* (1983) for actinide nuclei. They used the HFBC approach with a Woods–Saxon potential and found vanishing 'self-consistent' neutron pairing Δ above a certain critical frequency in attempts to explain anomalies in the observed moments of inertia (see next section). Future experiments will hopefully reveal detailed information on a microscopic basis about these interesting phenomena which are possibly related to pairing collapse.

6.3.4 *The moments of inertia and γ–γ-energy correlations*

The moment of inertia of a rotating deformed nucleus is expected to be strongly affected by changes in the pairing correlations. It has been explained in the previous section that the Coriolis anti-pairing (CAP) effect tends to break a pair and to cause an increase of the moment of inertia with increasing rotational frequency. There are also other dynamical variables which influence the moment of inertia such as deformation changes, shell effects and in particular in the bandcrossing region the spin alignment of quasi-particles. It is often very difficult to associate a certain observed change in the moment of inertia with one specific aspect of nuclear dynamics, and therefore it is convenient to introduce two possible definitions for the moment of inertia reflecting two different classes of nuclear dynamical properties (Bohr and Mottelson, 1981).

6.3.4a *The kinematical and dynamical moments of inertia*

The first definition of moment of inertia is already contained in eqn (6.3.37) as a quantity proportional to the inverse of the slope (first derivative) of the E vs. $I(I+1)$ curve. This quantity $\mathscr{I}^{(1)}$ is called the kinematical moment of inertia to emphasize its relation with the kinetics (motion) of the nucleus expressed in the ratio I/ω (see below). The second definition $\mathscr{I}^{(2)}$ as the dynamical moment of inertia expresses the response of the nucleus to a force and is related to the curvature (second derivative) of the E vs. $I(I+1)$ curve. Such definitions are analogous to the two possible definitions for the effective mass of an electron in a crystal lattice. In this case the first derivative of energy with respect to momentum describes properties like level density, specific heat etc. and the second derivative yields an effective mass that expresses dynamical properites.

From eqn (6.3.37) and from the definition of the rotational frequency ω expressed by eqn (6.3.36a) and using the identity $I_x = \{I(I+1)\}^{1/2}$ from eqn (6.3.45) for $K=0$, one obtains:

$$\text{Kinematical:}\quad \mathscr{I}^{(1)} = \frac{\hbar^2}{2}\left\{\frac{\mathrm{d}E}{\mathrm{d}(I_x)^2}\right\}^{-1} = \hbar^2 I_x\left(\frac{\mathrm{d}E}{\mathrm{d}I_x}\right)^{-1} = \hbar\frac{I_x}{\omega}. \qquad (6.3.58)$$

The second definition as $\mathrm{d}I_x/\mathrm{d}\omega$, using again the expression for ω in eqn (6.3.36a) yields

$$\text{Dynamical:}\qquad \mathscr{I}^{(2)} = \hbar^2\left(\frac{\mathrm{d}^2E}{\mathrm{d}I_x^2}\right)^{-1} = \hbar\frac{\mathrm{d}I_x}{\mathrm{d}\omega}. \qquad (6.3.59)$$

One may take $I = I_x$ for $I \gg 1$. Then $\mathscr{I}^{(1)}$ in eqn (6.3.58) agrees with the classical expression for the moment of inertia and $I\hbar$ with the classical angular momentum.

It can be seen in Fig. 6.21 that a variable moment of inertia for the reference rotor that increases quadratically with ω, according to eqn (6.3.50), does not account very well for the observations. At low ω the fit to the GSB is rather poor and at high ω the value of $\mathscr{I}^{(1)}_{\text{ref}}$ becomes unrealistically high. A peculiar aspect of the S-band in ^{168}Hf (and also in ^{170}W) is the fact that the data points are well described by the classical relation $I_x = \mathscr{I}_0\omega$ with constant $\mathscr{I}_0$. This behaviour suggests that after the first alignment of a neutron pair the pairing correlations are significantly diminished, driving the nucleus towards rigid (classical) rotation. Indeed, the deduced moments of inertia (see below) are surprisingly constant after the backbending as can be seen in Fig. 6.21. It is noticeable that the value of $\mathscr{I}^{(1)}$ is $\sim$ 25 per cent below the one of a rigid rotor, suggesting the existence of some paired nucleons after the backbending.

The dynamical moment of inertia shows an anomalous behaviour in the bandcrossing region as expected from its definition as second derivative of the yrast line—eqn (6.3.59)—and this is shown in Fig. 6.21. It is thus impossible to deduce from the experimental data the values of $\mathscr{I}^{(2)}$ at the bandcrossing.

Let us examine the relation between $\mathscr{I}^{(1)}$ and $\mathscr{I}^{(2)}$. From eqn (6.3.58) and (6.3.59) we get

$$\mathscr{I}^{(2)}(\omega) = \mathscr{I}^{(1)}(\omega) + \omega \frac{\mathrm{d}\mathscr{I}^{(1)}(\omega)}{\mathrm{d}\omega}. \tag{6.3.60}$$

From this equation it is clear that in the case of rigid rotation, i.e. when $\mathscr{I}^{(1)}$ is constant; $\mathscr{I}^{(2)} = \mathscr{I}^{(1)} = \mathscr{I}_{\mathrm{rig}}$. This seems to be the final result when the pairing correlations have collapsed. In the case of ^{168}Hf we indeed find $\mathscr{I}^{(2)} \simeq \mathscr{I}^{(1)}$ beyond the bandcrossing, but this is still well below the $\mathscr{I}_{\mathrm{rig}}$ value. In general it can be stated that the equality of $\mathscr{I}^{(1)}$ and $\mathscr{I}^{(2)}$ holds if the Hamiltonian contains only a kinetic term and the total energy is given by an expression like eqn (6.3.2) with a constant $\mathscr{I}$ for a rigid rotor, as follows from eqns (6.3.58) and (6.3.59). Any additional I-dependent term in the Hamiltonian will cause a difference between $\mathscr{I}^{(1)}$ and $\mathscr{I}^{(2)}$. This is the case, for example, for the Coriolis force ($\sim \vec{\mathrm{I}} \cdot \vec{\mathrm{j}}$ term) causing the CAP effect. This also explains why $\mathscr{I}^{(1)} = \mathscr{I}^{(2)}$ when the pairing correlations have collapsed.

6.3.4b *The band and envelope moments of inertia*

In defining moments of inertia it should also be recognized that the nucleus may possibly increase its angular momentum in two very different ways. One possibility is to increase the rotational frequency for a frozen intrinsic excitation, thus forming one particular rotational band. In this case the appropriate inertia constants are $\mathscr{I}^{(1)}_{\mathrm{band}}$ and $\mathscr{I}^{(2)}_{\mathrm{band}}$ to emphasize the (collective) property of the band (see Fig. 6.25a). In most cases, however, the nucleus also changes its intrinsic structure with increasing angular velocity and gains angular momentum at bandcrossings due to particle alignments. The overall variation of the angular momentum I with frequency ω is characterized by the envelope of the various bands in the E vs. I plot (Fig. 6.25b). The appropriate moments of inertia are in this case indicated by $\mathscr{I}^{(1)}_{\mathrm{env}}$ and $\mathscr{I}^{(2)}_{\mathrm{env}}$, which are sometimes called the effective moments of inertia.

The relation between the band and envelope moments of inertia is found in the following way. Consider the total spin increase ΔI over an interval $\Delta\omega$ including a bandcrossing; i.e. $\Delta I/\Delta\omega = \hbar^{-1}\mathscr{I}^{(2)}_{\mathrm{env}}$ according to eqn (6.3.59). To this increase the band contributes $\hbar^{-1}\mathscr{I}^{(2)}_{\mathrm{band}}$ and the particle alignment

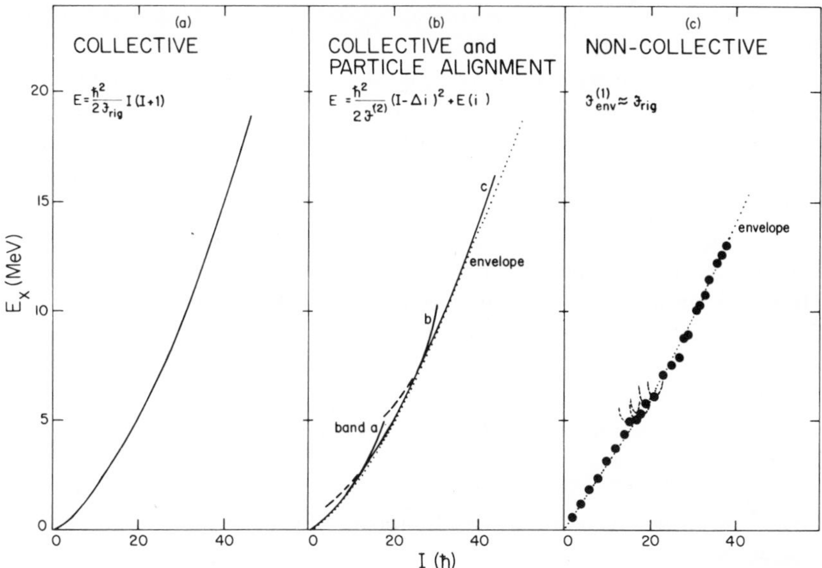

Fig. 6.25. The yrast states represented by an E vs. I plot for collective rotation (a) particle alignment (b) and non-collective excitation (c). In case (b) three different bands a, b, and c are drawn with the envelope curve. At each bandcrossing an angular momentum Δi is generated by the particle alignment. The expression given for the energy in the case of collective (rigid) rotation (a) is modified for the particle alignment in (b). The other extreme of collective rotation is the individual nucleon excitation (c). Each state represents a different intrinsic excitation ('degenerated bands', schematically indicated by 'infinitesimal small' parabolas in the centre of the curve). The 'envelope' curve through the data points is characterized by an effective $\mathscr{I}^{(1)}_{env} \simeq \mathscr{I}_{rig}$. (Bohr and Mottelson, 1969.) (See Section 6.4 for further discussion of the non-collective rotation.)

contributes $\Delta i/\Delta\omega$. From this we find the relation

$$\mathscr{I}^{(2)}_{env} = \mathscr{I}^{(2)}_{band} + \hbar\frac{\Delta i}{\Delta\omega} \quad \text{or} \quad \frac{\Delta i}{\Delta I} = 1 - \frac{\mathscr{I}^{(2)}_{band}}{\mathscr{I}^{(2)}_{env}}, \tag{6.3.61}$$

leading to the simple results:

$$\mathscr{I}^{(2)}_{band} = \mathscr{I}^{(2)}_{env} \quad \text{for } \Delta i = 0, \tag{6.3.62a}$$

$$\mathscr{I}^{(2)}_{band} < \mathscr{I}^{(2)}_{env} \quad \text{for } \Delta i > 0. \tag{6.3.62b}$$

The first equality expresses the trivial situation that without crossing with another band the nuclear excitation follows the one band coinciding with the envelope (namely the GSB). The inequality (6.3.62b) also follows from the definitions: the inverse curvature of the envelope is always larger than that of the bands.

The two extreme ways of generating high angular momentum, collective rotation and individual nucleon excitation, are illustrated in Fig. 6.25 with the particle alignment as a sort of intermediate situation. Several rotational bands constitute an envelope curve. In the case of non-collective rotation (see Section 6.4) each state can be regarded as a degenerated band with only one state, or in other words each state represents a bandhead. The rotational levels based on those states may exist, but presumably they have high energy. The envelope curve connects the infinitesimally small parabolas representing these degenerated bands. The only appropriate inertia parameter in this case is the envelope moment of inertia $\mathscr{I}^{(1)}_{\text{env}}$ and this appears to be of the order of the rigid-body value, $\mathscr{I}_{\text{rig}}$ (Bohr and Mottelson, 1969). The parameter $\mathscr{I}^{(2)}_{\text{env}}$ in this case is not meaningful because of the strong local fluctuations of the yrast line (see Section 6.4).

Let us now present here some experimental methods to determine the various moments of inertia discussed above. In general one may follow two different ways;

(1) the integral method, using two related quantities, E_γ and I, and

(2) using the second derivative $\mathrm{d}^2E/\mathrm{d}I^2$ in a differential method.

In the first method $\mathscr{I}^{(1)}_{\text{env}}$ is obtained from the transition energy E_γ (as the first derivative of eqn (6.3.2), corresponding to the decay of a state with spin I, and using eqn (6.3.37). In this way it is assumed that the state is part of a decay path that follows an envelope along several rotational bands and crossings. In other words the γ-ray decay path is assumed to follow the yrast states which constitute the envelope. If only one particular rotational band is selected by the observation of discrete γ rays then $\mathscr{I}^{(1)}_{\text{band}}$ is obtained and the inequality $\mathscr{I}^{(1)}_{\text{band}} < \mathscr{I}^{(1)}_{\text{env}}$, which may depend on spin, can be tested. It is easy to see that eqn (6.3.37) is a special approximation to the more general expression eqn (6.3.58), in the case of $\Delta I = 2$ transitions:

$$E_\gamma = \Delta E = \frac{\mathrm{d}E}{\mathrm{d}I}\Delta I = \frac{\hbar^2}{2\mathscr{I}^{(1)}_{\text{env}}}4I. \tag{6.3.63}$$

The differential method leads in a similar way, using eqn (6.3.59), to:

$$\Delta E_\gamma = \frac{\mathrm{d}E_\gamma}{\mathrm{d}I}\Delta I = \frac{\mathrm{d}^2E}{\mathrm{d}I^2}(\Delta I)^2 = 8\,\frac{\hbar^2}{2\mathscr{I}^{(2)}_{\text{env}}}. \tag{6.3.64}$$

If the decay follows an envelope (yrast) line then as discussed above one obtains $\mathscr{I}^{(2)}_{\text{env}}$ using eqn (6.3.64) from the differences ΔE_γ of consecutive γ-rays in the cascade. Notice that one obtains $\mathscr{I}^{(1)}_{\text{band}}$ or $\mathscr{I}^{(2)}_{\text{band}}$ from eqn (6.3.63) and (6.3.64), respectively, when the envelope coincides with the band. It is an interesting fact that the only observable required to deduce $\mathscr{I}^{(2)}_{\text{env}}$ is ΔE_γ and not the angular momentum! This is the basis of the γ–γ-energy correlation method discussed below.

In the case of quasi-continuum γ-rays (see Section 6.5) one cannot deduce ΔE_γ from discrete γ rays, but rather from measurements of the average number of γ rays ΔN in a given interval ΔE_γ. If one assumes $\Delta I \simeq 2\Delta N$ (stretched quadrupole transitions) and $E_\gamma \simeq 2\hbar\omega$ {eqn (6.3.41)} then $\mathscr{I}^{(2)}_{\text{env}}$ can be deduced from the γ-ray spectrum using eqn (6.3.59) (Stephens, 1983) in the form

$$\frac{\Delta N}{\Delta E_\gamma} \simeq \frac{\Delta I}{4\hbar\Delta\omega} \simeq \frac{\mathrm{d}I}{4\hbar \mathrm{d}\omega} = \frac{1}{8}\left(\frac{2\mathscr{I}^{(2)}_{\text{env}}}{\hbar^2}\right). \tag{6.3.65}$$

This method makes it possible to estimate $\mathscr{I}^{(2)}_{\text{env}}$ up to very high rotational frequencies (in the quasi-continuum).

6.3.4c *The γ–γ-energy correlations*

The method to be discussed finally is that of γ–γ-energy correlations (Herskind, 1980) to obtain the $\mathscr{I}^{(2)}_{\text{band}}$. This technique becomes particularly useful when the decay of a highly excited nucleus can proceed via many bands, each of them populated not frequently enough to observe the individual transitions as discrete γ-ray lines (see also Section 6.5). This is the case at very high spins (> 40) at high angular velocities, or for decay pathes well above the yrast line. In those cases one would still like to obtain information about the average behaviour of the collective and individual nucleonic motion. As the decay is believed to proceed mainly via various bands, measured correlations between two γ rays will most likely correspond to transitions within the same band. Therefore the analysis of γ-ray energy differences ΔE_γ will yield an average $\mathscr{I}^{(2)}_{\text{band}}$ for all the bands contributing to the correlations.

For the experiments at least two γ-ray detectors are needed to detect the energies of two of the many γ-rays emitted simultaneously by the excited nucleus. The first experiments have been carried out with low resolution but highly efficient NaI(Tl) detectors which are sufficient to observe gross structures in the correlation patterns (Herskind, 1980). For more detailed information, however, high-resolution germanium detectors are needed. The smaller efficiency can be compensated for by using a large number of detectors, which at the same time facilitates the measurement of higher-order correlations (e.g. triple coincidences) as discussed in more detail in Chapter 3. Contemporary experimental arrangements consist of a number of Compton-suppressed (BGO) Ge detectors combined with many-detector multiplicity/sum energy arrays. These arrays serve the purpose of selecting a certain (high) spin region and eliminating some of the unwanted background radiation (see e.g. Fig. 6.17).

Let us analyse in detail the correlation matrix obtained from each pair of detectors in the experiment. Figure 6.26(a) displays a correlation matrix for the hypothetical situation that all the γ rays belong to only four bands with constant $\mathscr{I}^{(2)}_{\text{band}}$, that is for a perfect rotor. The two coordinates of the matrix represent the γ-ray energies in the two detectors. Each point in the matrix corresponds to a measured coincidence between the two detectors, one registering $E_{\gamma 1}$ of a particular cascade and the other $E_{\gamma 2}$ of the same cascade ($E_{\gamma 1} \neq E_{\gamma 2}$ for a perfect rotor). The matrix exhibits a very regular structure of intensity points (ridges), which can be easily understood. The diagonal ($E_{\gamma 1} = E_{\gamma 2}$) is free of counts (valley). The first ridges on each side of the diagonal correspond to adjacent transitions in all participating cascades, the next ridges to transitions further apart and so on. The distances between adjacent ridges thus correspond to the difference of adjacent γ rays in the

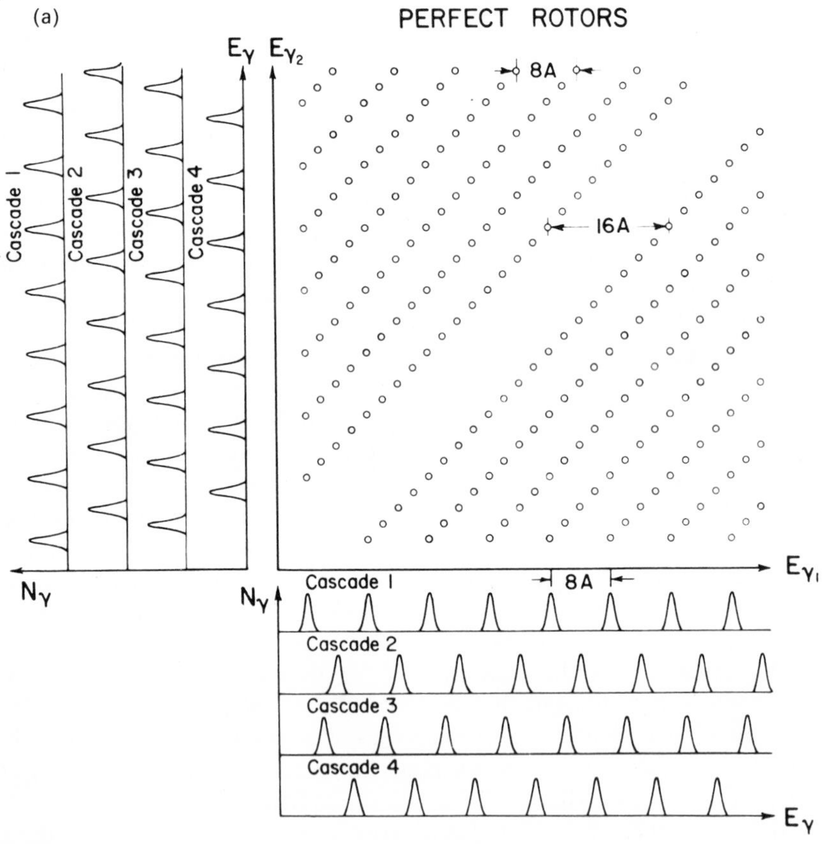

Fig. 6.26(a).

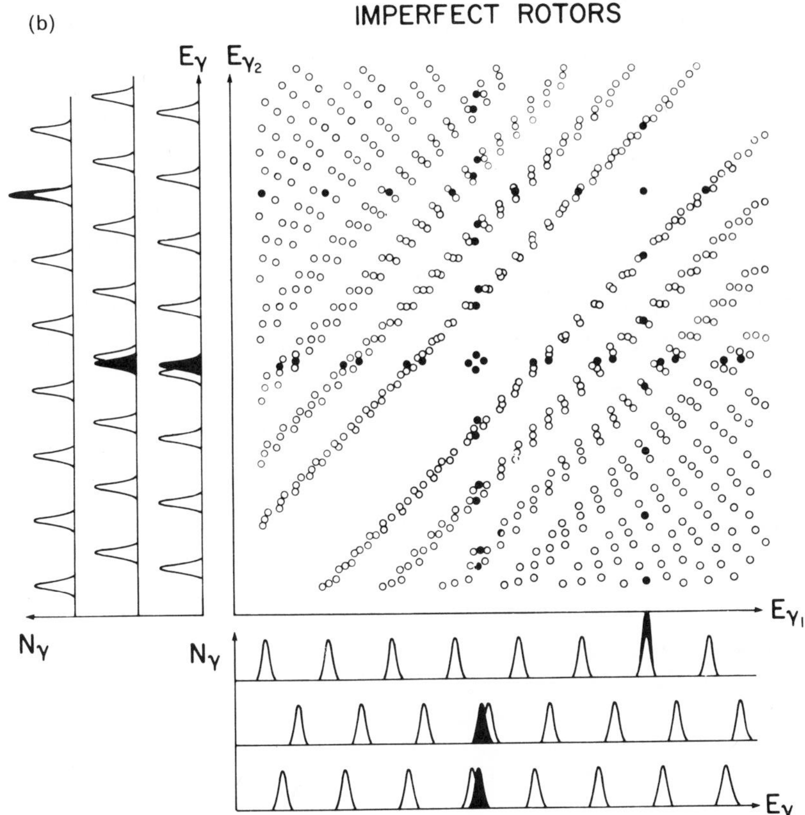

Fig. 6.26. A model for an $E_{\gamma 1}$–$E_{\gamma 2}$ correlation matrix as a coincidence pattern for the four cascades indicated with constant moment of inertia (a). The distance between the inner ridges along the 45° valley is $16A = 16 \cdot \hbar^2/(2\mathscr{I}^{(2)}_{\text{band}})$. On the right-hand side is shown (b) a coincidence pattern for the four bands shown in (a) plus 8 additional bands with 10 per cent different moment of inertia from those shown above. Non-rotational features (upbend or backbend) are also shown by shaded transitions in the cascades and solid dots in the matrix. (Herskind, 1980.)

cascades; i.e. ΔE_γ, the width of the valley W being twice as large. The measured width W can thus be related with $\mathscr{I}^{(2)}_{\text{band}}$ using eqn (6.3.66):

$$W = 2\Delta E_\gamma = 16\frac{\hbar^2}{2\mathscr{I}^{(2)}_{\text{band}}} = 16A. \tag{6.3.66}$$

The constant $A = \hbar^2/2\mathscr{I}^{(2)}_{\text{band}}$ can thus directly be obtained from the correlation matrix, shown in Fig. 6.26(a) (Herskind, 1980).

A situation reflecting reality somewhat more closely is displayed schematically in Fig. 6.26(b). Here eight bands with ± 10 per cent difference in the moments of inertia are added, causing a broadening of the ridges. Also band crossings are included causing at the backbending two different γ rays to coincide closely in energy. Such a crossing shows up in the matrix as an intensity (bump) on or near the diagonal. These important observations are thus possible even for unresolved γ rays.

In practice, however, the situation is much worse if we consider the detector response function. We are in fact only interested in correlated events with a deposition of the full γ-ray energies in the detectors. A typical germanium γ-ray detector has a photopeak efficiency of ~ 10 per cent; ~ 90 per cent being distributed in a wide region below the Compton edge. The full energy correlated events thus constitute only $\sim (10 \text{ per cent})^2 = 1$ per cent of all the intensity in the correlation matrix. A subtraction technique to eliminate uncorrelated events has been proposed by Andersen *et al.* (1979), as discussed in Section 3.2.6.

This is why the application of the recently-developed small size (BGO) Compton suppression shields surrounding the Ge detectors, are vital in these experiments (see Chapter 3). With an average suppression factor of ~ 6 the full-energy contribution amounts to ~ 40 per cent of all counts in a single spectrometer. In the coincidence matrix we will find $(40 \text{ per cent})^2 \sim 16$ per cent of the intensity in the full-energy correlations, 2×24 per cent in photopeak–Compton coincidence ridges (rows and columns) and 36 per cent due to Compton–Compton coincidences (smeared out over the entire region of the matrix below the photopeak energies). An early example of a measured correlation matrix with two conventional Ge-[NaI(Tl)] Compton suppression spectrometers is given in Fig. 6.27. The central valley along the diagonal is observed clearly up to $E_\gamma \simeq 800$ keV ($\hbar\omega \simeq 400$ keV). An additional advantage of the high-resolution Ge detectors is the possibility of observing the highly intense discrete γ-rays due to the decay of the low-spin yrast states. Besides useful information on the discrete states one may also use discrete γ-ray lines to check the analysis procedures before applying them to the unresolved part of the matrix.

Enhanced intensities in the valley of the matrix in Fig. 6.27 are observed at $E_{\gamma 1} = E_{\gamma 2} \simeq 460$, 520 and 680 keV, and possibly also at 840 and 1040 keV, corresponding to $\hbar\omega \simeq 0.23$, 0.26, 0.34 and to 0.42 and 0.52 MeV $\hbar^{-1}$, respectively. The first two frequencies correspond to the first backbending in ^{168}Hf and the third to the upbending in ^{167}Hf as discussed in Section 6.3.3. The possible upper two intensities may be attributed to bandcrossings involving $\pi h_{1/2}$ or higher orbitals (see also Lee *et al.*, 1977; Beck *et al.*, 1979; and Deleplanque *et al.*, 1980).

The width of the central valley of the Hf matrix decreases smoothly with increasing angular velocity, which implies a steady increasing $\mathscr{I}^{(2)}_{\text{band}}$ approaching the value $\mathscr{I}^{(1)}_{\text{env}}$ as shown in Fig. 6.21 at the highest frequencies of ~ 0.4 MeV $\hbar^{-1}$, being well below the value of the rigid rotor: de Voigt *et al.* (1981, 1983).

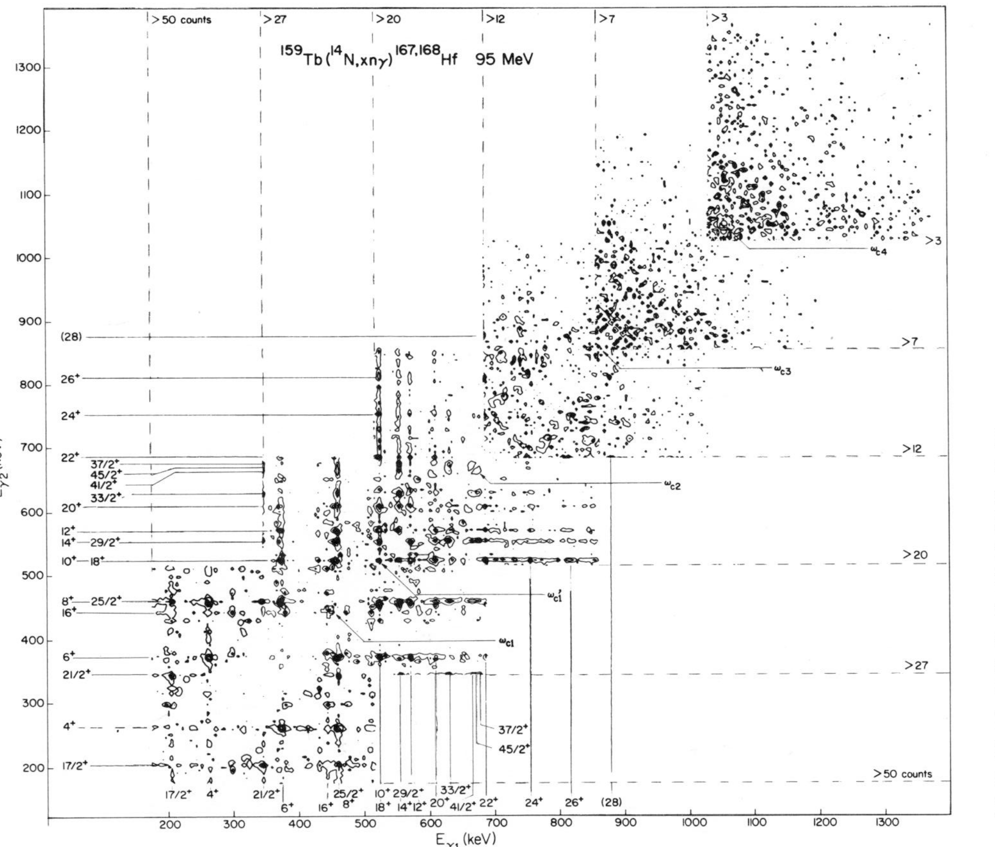

Fig. 6.27. Contour plot of the symmetrized γ–γ correlation matrix, symmetric with respect to the 45° diagonal. Six different intensity thresholds are indicated by the numbers at the dashed lines. Enhanced intensities in the valley are indicated by $\omega_{c1}, \ldots, \omega_{c4}$. The lowest frequencies ($\omega_{c1}, \omega_{c1'}$) and ω_{c2} are due to known backbending effects in ^{168}Hf and ^{167}Hf, respectively. The known discrete yrast-band transitions in these nuclei are indicated on the left and in the bottom part by the spin of the initial levels—de Voigt *et al.* (1981, 1983).

It was discussed in Section 6.3.3 that the approximate condition $\mathscr{I}^{(1)} \simeq \mathscr{I}^{(2)}$ is indicative of reduced pairing, which has also been proposed for very similar data on Yb isotopes by Herskind (1983). The matrix for ^{166,167,168}Yb is shown in Fig. 6.28 as deduced from experiments using the ^{124}Sn (^{48}Ca, xn)$^{172-x}$Yb reaction at 201 MeV beam energy and the Tessa II spectrometer (see Section 6.3.3). Note the clear valley structure observed up to high γ-ray energies of $\sim$1.2 MeV, corresponding to an angular velocity of $\sim$ 0.6 MeV $\hbar^{-1}$. The left-hand side of the figure contains 45° cuts through the matrix (perpendicular to the diagonal) over 36 keV intervals. The decrease in valley width from $E_\gamma = 796$ to 1120 keV corresponds to an increase of $\mathscr{I}^{(2)}_{\text{band}}$ from 62.5 to 70 MeV$^{-1}\,\hbar^2$.

6.3.5 *Electromagnetic moments and transition probabilities; multiple Coulomb excitation*

In the previous sections it has been shown that an impressive amount of nuclear-structure information can be deduced from the transition energies within the yrast band and side bands. We have discussed the example of the strongly-deformed nucleus ^{158}Er and the spherical nucleus ^{147}Gd (Fig. 6.2) on the basis of level systematics in the beginning of this chapter. In addition it was shown with Fig. 6.3 that B(E2) transition probabilities are very different for collective and non-collective nuclear motion.

In this section we will present several methods of obtaining static and dynamic (transition) electromagnetic multipole moments. We will also show with a few examples that they contribute to a more detailed understanding of the structure of high-spin states.

Collectively rotating nuclei are characterized by strong electric quadrupole transitions (E2) within the rotational bands. The reduced transition probabilities B(E2) are generally deduced from the measured lifetimes (see below). The lifetimes of strongly-deformed states are relatively short (in the picosecond region or less) especially at high spins when the transition energy becomes relatively large because of the $E_\gamma^{2\lambda+1}$ proportionality of the transition probabilities, λ being the multipolarity. Modifications of the commonly used recoil-distance method bring the measurable lifetimes down to a few tenths of a picosecond (see Section 6.3.5b below and also Section 6.5). For the experiments one needs to have a large velocity for the recoil ions ($v/c \simeq 10$ per cent) which can be achieved in compound nuclear (fusion) reactions with heavy-ion beams bombarding light target nuclei (inverse reaction). The changes of the B(E2) values in a particular band with increasing spin are indicative of the change in the internal structure, for instance when band-crossing occurs.

A sensitive parameter characterizing the nuclear shape is the static electric quadrupole moment $Q(I)$. Measurements of this quantity for high-spin states

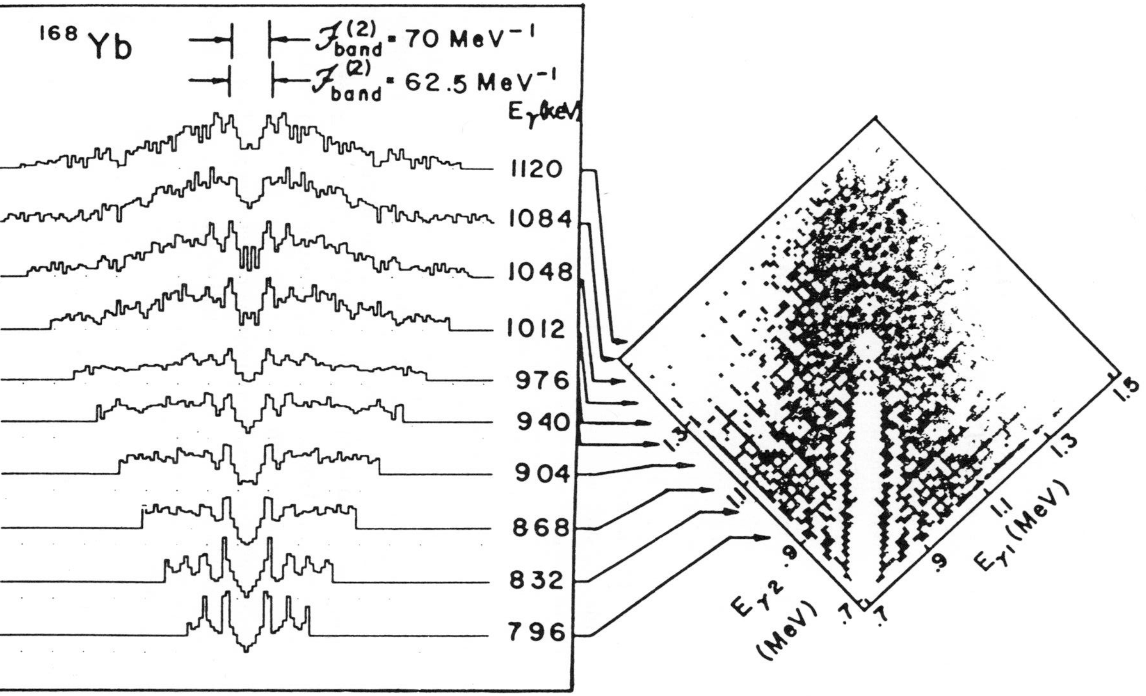

Fig. 6.28. The right side of the figure shows $E_{\gamma 1}$–$E_{\gamma 2}$ correlation data for ^{168}Yb. The data have been unfolded and background subtracted for uncorrelated events, according to the method of Andersen *et al.* (1979). The left side of the figure shows 45° cuts across the valley with 36 keV intervals from $E_{\gamma 1} = E_{\gamma 2} = 796$ keV until $E_{\gamma 1} = E_{\gamma 2} = 1120$ keV. The valley-ridge structure is clearly seen to go through all the cuts (discussed by Herskind, 1983).

are not so easy. We will discuss the method used in Multiple Coulomb Excitation (MCE) which has contributed significantly to the detailed understanding of high-spin phenomena, particularly in the actinide and rare earth regions.

We have learned from the previous sections that the alignment of quasi particles becomes increasingly important when the nucleus rotates faster and faster. The phenomena are accompanied by changing static and dynamic electromagnetic multipole moments. In contrast to E2 moments, the magnetic dipole transition probabilities $B(\mathrm{M1})$ are sensitive to the single-particle motion. They are also deduced from lifetime measurements, often combined with branching-ratio measurements. The M1 transition strengths are indicative of the interplay of collective and single-particle excitations as they appear, for instance, in transitional nuclei. They are apparently also sensitive to the presence of collective rotation based on a quasi-particle excitation with a different signature. This effect is discussed below for odd nuclei.

We will also devote some attention to the measurements of the static magnetic dipole moment $\vec{\mu}$ or the related gyromagnetic factor g. They are particularly sensitive to the detailed single-particle structure of excited states. The conventional techniques are based on the measurement of the Larmor precession ω_L under the influence of an external magnetic field B. The technique involves the measurement of the spatial γ-ray intensity distribution, which rotates with an angular velocity

$$\omega_L = \hbar^{-1} g B \mu_N, \tag{6.3.67}$$

where $\mu_N = e\hbar/(2mc)$ is the nuclear magneton and $\mu = gI\mu_N$ (see also eqn 3.5.11). The initial orientation of the excited nuclei needed to create an asymmetric γ-ray angular distribution can be produced in nuclear reactions. For the observation of a sizable precession the nulcear lifetimes must be at least in the nanosecond region, due to the limited strengths of external magnetic fields. This situation occurs frequently in non-collectively rotating nuclei, the subject of the next section.

In the present case (collective rotation), however, lifetimes less than a few picosecond require a different technique. We will discuss below a method utilizing the transient magnetic field interaction for ionized nuclear recoils passing a thin foil of magnetized material. The internal magnetic fields generated are extremely high, producing measurable precessions for very short-lived states. The excitation process can be initiated both in heavy-ion induced (inverse) fusion reactions and in MCE.

6.3.5a *Electromagnetic multipole moments*

Before discussing the experimental methods and the related physics we have to be familiar with some basic properties of gamma radiation that follow from electromagnetic theory. We present here only the relevant results applicable to rotational bands. For the underlying theory and derivation of formulae the reader is referred to other relevant works like those of Pelte and Schwalm (1982), Alexander and Forster (1978) or Bohr and Mottelson (1975), see also Section 2.1.

The transition probability $T(\lambda)$ for radiating a photon of multipolarity λ with energy E_γ is related to the nuclear (mean) lifetime $\tau(\lambda)$ or the half-life $\tau_{1/2}(\lambda)$ as shown in Section 2.1.

We will restrict ourselves to transitions within the same band, thus $K_i = K_f$ and to M1 and E2 transitions, which are most frequently occurring. The multipole operators $\mathcal{M}(\lambda)$ yield both the transition probabilities and the static properties of intrinsic nuclear states Φ_K. They are related to the intrinsic quadrupole moment $Q_0(K)$ due to the electric charge distribution and to the gyromagnetic factor g_K, due to the electric currents generated by the nuclear motion:

$$eQ_0(K) = \left(\frac{16\pi}{5}\right)^{1/2} \langle \Phi_K | \mathcal{M}(\text{E2}) | \Phi_K \rangle \tag{6.3.68}$$

$$Kg_K = \left(\frac{4\pi}{3}\right)^{1/2} \langle \Phi_K | \mathcal{M}(\text{M1}) | \Phi_K \rangle \mu_N^{-1}. \tag{6.3.69}$$

(For the explicit terms of the multipole operators see references mentioned above). The quantities $Q_0(K)$ and g_K may be calculated using some model wave function Φ_K or when possible they should be determined experimentally.

In the case of a rotational band, characterized by the quantum number K ($K \neq \frac{1}{2}$) simple relations can be derived for the static electric quadrupole moment $Q(I)$ and the magnetic dipole moment $\mu(I)$ for states with spin I and for the reduced transition probabilities $B(\lambda)$:

$$B(\text{E2};\, I_\text{i} K \to I_\text{f} K) = \frac{5}{16\pi} \langle I_\text{i} K 20 | I_\text{f} K \rangle^2 e^2 Q_0^2(K) \tag{6.3.70}$$

$$Q(I) = \frac{3K^2 - I(I+1)}{(I+1)(2I+3)} eQ_0(K) \tag{6.3.71}$$

$$B(\text{M1};\, I_\text{i} K \to I_\text{f} K) = \frac{3}{4\pi} \langle I_\text{i} K 10 | I_\text{f} K \rangle^2 (g_K - g_R)^2 K^2 \mu_N^2 \tag{6.3.72}$$

$$\mu(I) = \left[g_R I + (g_K - g_R) \frac{K^2}{I+1} \right] \mu_N, \tag{6.3.73}$$

where $\langle | \rangle$ denotes the vector addition or Clebsch–Gordan coefficient. The unit for $Q(I)$ is eb ($1\ eb = 10^2\ e\text{fm}^2$). The unit for $\mu(I)$ is the nuclear magneton $\mu_N = e\hbar/(2m_p) = 5.0508 \cdot 10^{-27}\ \text{JT}^{-1}$. It should be noted that the observable quadrupole moment is $Q(I)$ rather than $Q_0(K)$ and similarly $\mu(I)$ rather than

g_K and g_R. In fact $Q(I)$ is the quadrupole moment in the space-fixed frame of reference (laboratory frame) and Q_0 in the body-fixed frame. For small K (i.e. $K = 0, 1$) or for $I \gg K$ bands $Q(I)$ and Q_0 have opposite signs (see eqn 6.3.71). For $K = \frac{1}{2}$ bands the last two expressions have an extra term (see e.g. Pelte and Schwalm, 1982).

The gyromagnetic factor g_R characterizes the collective rotation and can be approximated by

$$g_R \simeq k\frac{Z}{A}, \quad \text{with } k = 0.8 \sim 1.0. \tag{6.3.74}$$

In the rare-earth region $g_R \simeq 0.35$. In many cases rotational bands are based on quasi-particle excitations. In odd nuclei, for example, with $K = |\Omega|$ the intrinsic magnetic dipole moment can be ascribed to the motion of the unpaired nucleon and eqn (6.3.69) simplifies to:

$$Kg_K = \langle \Phi_K | g_1 \hat{\mathrm{l}} + g_S \hat{S} | \Phi_K \rangle. \tag{6.3.75}$$

Here $\hat{S}$ and $\hat{\mathrm{l}} = \hbar^{-1} m(\vec{r} \times \vec{v})$ are the spin and orbital angular momentum operators of the particle. The corresponding gyromagnetic factors g_S and g_1 have to be found for the nucleon moving inside the nucleus. Comparison of the measured g_K with calculated g_1 and g_S within a certain model yields direct information on the intrinsic wave function Φ_K of the single-particle state. The values for a free proton (neutron) are $g_1 = 1$ (0) and $g_S = 5.58$ (-3.82). Although the values for bound nucleons can be significantly different, one observes consistently positive g-factors for proton states and negative ones in the case of neutrons.

In the case of high-spin states ($I \gg K$) the vector addition coefficients in eqns (6.3.70–6.3.72) can be approximated by

$$\langle I_i K 2 0 | I_f K \rangle = \begin{cases} (3/8)^{1/2} & \text{for } I_f = I_i \pm 2 \\ \pm(3/2)^{1/2}(K/I_i) & \text{for } I_f = I_i \pm 1 \\ -\frac{1}{2} & \text{for } I_f = I_i \end{cases} \tag{6.3.76a}$$

$$\langle I_i K 1 0 | I_f K \rangle = (\tfrac{1}{2})^{1/2} \qquad \text{for } I_f = I_i \pm 1. \tag{6.3.6b}$$

Inserting these values in eqns (6.3.70–6.3.72) we obtain very simple expressions for the high-spin limit:

$$B(\mathrm{E2};\, I_i \to I_f = I_i \pm 2) \simeq \frac{5}{16\pi}\frac{3}{8} e^2 Q_0^2(K), \tag{6.3.77}$$

$$B(\mathrm{E2};\, I_i \to I_f = I_i \pm 1) \simeq \frac{5}{16\pi}\frac{3}{2}\left(\frac{K}{I_i}\right)^2 e^2 Q_0^2(K), \tag{6.3.78}$$

$$Q(I) \simeq -\tfrac{1}{2} e Q_0(K), \tag{6.3.79}$$

$$B(\mathrm{M1}; I_\mathrm{i} \to I_\mathrm{f} = I_\mathrm{i} \pm 1) \simeq \frac{3}{4\pi} \frac{K^2}{2} (g_K - g_R)^2 \mu_N^2, \tag{6.3.80}$$

with $\mu(I) \simeq g_R I \mu_N$.

From these expressions it becomes evident that for high-spin members of rotational bands the B(M1), $B(\mathrm{E2}; \Delta I = 2)$, and $Q(I)$ values are approximately constant (independent of I), and $\Delta I = 2$ transitions are favoured over $\Delta I = 1$ transitions. In fact the $B(\mathrm{E2}; \Delta I = 1)$ values should decrease with increasing I according to eqn (6.3.78).

In analysing properties of the GSB and S-bands commonly occurring in even–even deformed nuclei (i.e. $K = 0$, $I_\mathrm{f} = I_\mathrm{i} - 2$), it is convenient to calculate the B(E2) values with respect to that for the $2^+ \to 0^+$ transition, using eqn (6.3.70) in the form:

$$B(\mathrm{E2}; I \to I_\mathrm{f} = I - 2, K = 0) = \frac{5}{16\pi} \langle I\, 020 | (I-2) 0 \rangle^2 e^2 Q_0^2$$

$$= 5 \langle I\, 020 | (I-2) 0 \rangle^2 B(\mathrm{E2}; 2^+ \to 0^+), \tag{6.3.81}$$

with

$$\langle I\, 020 | (I-2) 0 \rangle^2 = \frac{3I(I-1)}{2(2I-1)(2I+1)}. \tag{6.3.82}$$

It should be noted that all the formulae (6.3.70)–(6.3.82) apply in the rotational limit only. With increasing angular velocity the Coriolis and centrifugal forces may lead to major modifications in the expressions for the electromagnetic multipole moments. Besides possible mixing of states with different K, new coupling schemes may develop such as rotational alignment leading, for instance, to decoupled bands in odd-A nuclei (see Section 6.3.2).

It is not always possible to measure lifetimes for all nuclear states of interest and to deduce absolute values for the reduced transition probabilities. Measured γ-ray intensity ratios for states decaying by more than one γ ray (branching ratios) may already supply interesting information. For example, the intensity ratio (branching) of two γ rays with multipolarity λ originating from the same state $I_\mathrm{i} K_\mathrm{i}$ and decaying to different states of the same band $I_\mathrm{f} K_\mathrm{f}$ and $I'_\mathrm{f} K_\mathrm{f}$ is given by the ratio of the vector addition coefficients squared according to eqns (6.3.70) and (6.3.72):

$$\frac{B(\lambda; I_\mathrm{i} K_\mathrm{i} \to I_\mathrm{f} K_\mathrm{f})}{B(\lambda; I_\mathrm{i} K_\mathrm{i} \to I'_\mathrm{f} K_\mathrm{f})} = \frac{\langle I_\mathrm{i} K_\mathrm{i} \lambda (K_\mathrm{f} - K_\mathrm{i}) | I_\mathrm{f} K_\mathrm{f} \rangle^2}{\langle I_\mathrm{i} K_\mathrm{i} \lambda (K_\mathrm{f} - K_\mathrm{i}) | I'_\mathrm{f} K_\mathrm{f} \rangle^2}. \tag{6.3.83}$$

The underlying assumption is that the intrinsic structure, specified by $Q_0(K)$, g_K and g_R, for the states I_f and I'_f is approximately the same. This and similar intensity rules are very helpful in identifying rotational bands.

We shall now present some expressions needed to deduce B(E2) and B(M1) values from observed nuclear properties. When an excited state with mean life

τ decays via an E2 γ branch with strength $B_\gamma = Y_\gamma/\Sigma_{\gamma'} Y_{\gamma'}$ (Y representing the γ-ray intensity) and energy E_γ, then we find for the experimental reduced transition probability:

$$B(\mathrm{E2}) = \frac{0.08156\, B_\gamma}{E_\gamma^5\, \tau(1+\alpha_{\mathrm{tot}})}\, e^2 \mathrm{b}^2. \tag{6.3.84}$$

The result is obtained in units of $e^2 b^2 = 10^4 e^2 \mathrm{fm}^4$ when E_γ is given in MeV and τ in ps. The probability that the transition proceeds via internal conversion electrons rather than via γ rays is accounted for by the total conversion coefficient α_{tot}. Similarly one finds for an M1 transition:

$$B(\mathrm{M1}) = \frac{0.05697\, B_\gamma}{E_\gamma^3\, \tau(1+\alpha_{\mathrm{tot}})}\, \mu_N^2. \tag{6.3.85}$$

It is sometimes convenient to express the transition probabilities in terms of the reduced transition quadrupole moment Q_t defined as

$$Q_t(I \to I-2) = 0.907(\tau E_\gamma^5)^{-1/2}/\langle I020|(I-2)0\rangle, \tag{6.3.86}$$

where Q_t is in eb when the mean lifetime τ is given in ps and E_γ in MeV. The relation between Q_t and B(E2) follows directly from eqns (6.3.84) and (6.3.86). For a $K=0$ band we find from eqns (6.3.86), (6.3.84), and (6.3.81) the relation $Q_t = eQ_0$. The information would even be more complete when an independent measurement of the static quadrupole moment $Q(I)$, as defined by eqn (6.3.71), can be made, which is outside the present possibilities for high-spin states. The reduced static quadrupole moment Q_d is related to $Q(I)$ by

$$Q_d(I) = -Q(I)(2I+3)/I. \tag{6.3.87}$$

Formulae for other types of radiation as well as expressions in terms of single-particle estimates (Weisskopf units, Wu) can be found for example in Alexander and Forster (1978).

6.3.5b *Lifetime measurements*

Equations (6.3.84) and (6.3.85), given above show that the reduced transition probability can be determined if the lifetime τ is measured, provided that E_γ, B_γ and α_{tot} are known. The experimental method used to measure nuclear lifetimes in the picosecond region by means of the recoil-distance (plunger) technique is explained in Section 3.5.1. Here we will present some results obtained for high-spin states.

Let us illustrate the Doppler shifts measured by the reaction ^{25,26}Mg (^{136}Xe, xn)^{156,157,158}Dy with beam energies of 557 and 642 MeV obtained with the UNILAC accelerator at GSI, Darmstadt (Emling *et al.*, 1984). The velocity of the ^{158}Dy recoil ions, for instance, was $v_0/c = 0.078$ which corresponds to a flight time of 0.43 ps/10 μm. We also notice that a γ-ray of 0.5 MeV will experience a maximum shift (at $\theta = 0°$) of 39 keV. This shift is far more than the energy resolution of the commonly used Ge detectors ($\leqslant 2.0$ keV) and is therefore easily observed. An example of γ-ray spectra taken at three different distances d between target and stopper is given in Fig. 6.29. The figure shows clearly the changing relative intensity of the γ-ray lines corresponding to stopped (s) and moving (m) recoils when the distance d varies.

Some decay curves for the high-spin states in ^{156}Dy plotted as the fraction $R(t)$ versus the flight time are given in Fig. 6.30. The method is explained in Section 3.5.1b. It shows the accuracy of the plunger method applied to heavy-ion inverse reactions and that the decay of such very short-lived nuclear states can be followed over a time region of only a few ps. The solid curves represent the fits to the data employing a simple rotational feeding model as explained in Section 6.5.5e. The first results of these experiments are summarized in Fig. 6.31 (Grosse, 1979; Emling *et al.*, 1981), which shows lifetimes down to 0.21 ps for the 26^+ state. The deduced B(E2) values contain nuclear structure information that will be discussed in Section 6.3.5e. With some assumptions the feeding times of yrast states can also be extracted and hence information about the quasi-continuum states (see Section 6.5).

Even shorter lifetimes than 1.0 ps can be measured with a modified technique, using a retardation foil between the target and stopper as discussed in Section 3.5.1. The lifetimes and feeding times of levels in the nuclei ^{156,157,158}Dy have been measured down to ~ 0.1 ps using the inverse reaction of the type Mg(^{136}Xe, xn) as in the experiment mentioned above (Emling *et al.*, 1984). The state analysed with the highest spin, $I^\pi = 30^+$, in ^{156}Dy has a lifetime of $\tau = 0.16 \pm 0.08$ ps, while a feeding time for that state was deduced as 0.39 ± 0.19 ps. The deduced lifetimes show retarded E2 transition probabilities for high-spin states above the backbending region and fast E2 components of the pre-yrast decay, as discussed below in Section 6.3.5e.

Let us conclude this section with an example of the DSAM using the first results obtained with the Berkeley 21 Compton-shielded Ge assembly (Bacelar *et al.*, 1986). Due to the many detectors a very high doubles coincidence count rate is achieved and also a high triples rate. Consequently one is able to gate on many discrete peaks and observe very cleanly in coincidence the high-spin states as discrete lines. This allows, for the first time, detailed lineshape analyses of the highest observed discrete states in the DSAM technique as discussed in Section 3.5.1c. Comparing the decay times of subsequent levels one may then obtain the feeding- and lifetimes of each high-spin state.

The nucleus ^{166}Yb has been studied with a 180 MeV ^{40}Ar beam of the Berkeley 88″ cyclotron. Two experiments have been carries out; one with thin targets of 200–360 μg/cm^2 deposited on very thin gold foils and one with a 1 mg/cm^2 thick target on a 12 mg/cm^2 gold backing. The 21 detectors subtend in eight different angles and the spectra obtained with the thin targets

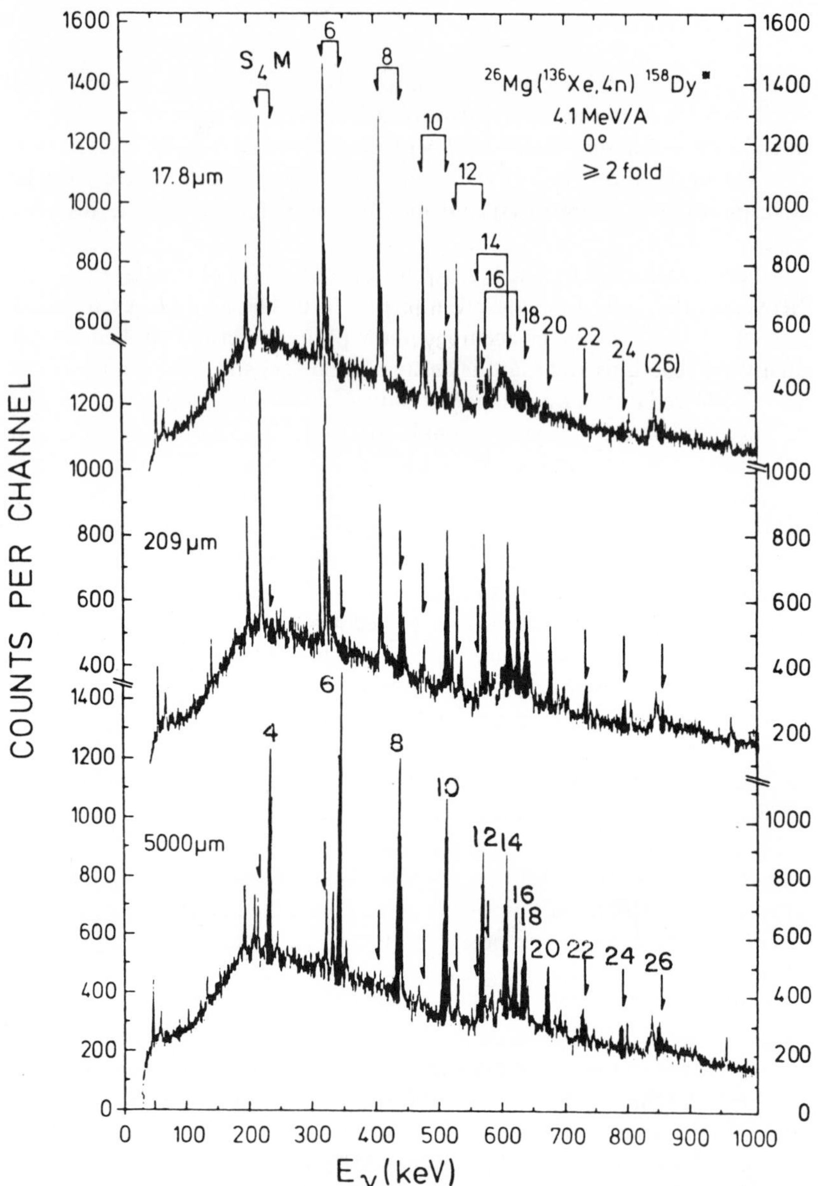

Fig. 6.29. Gamma-ray spectra observed with a Ge(Li) detector placed at $\Theta = 0°$ with respect to the beam direction for three different target-stopper distances d (target thickness: 400 μg/cm^2). The spectra were recorded in coincidence with at least two out of four NaI(Tl) detectors. At the short distance we observe mainly the stopped (unshifted) component and at the large distance the moving (shifted) component (black peaks in the spectra, and indicated by the spin value of the upper state in the transition at the bottom spectrum). The s-peaks are indicated with an left-arrow and the m-peaks with a right-arrow. (Grosse, 1979.)

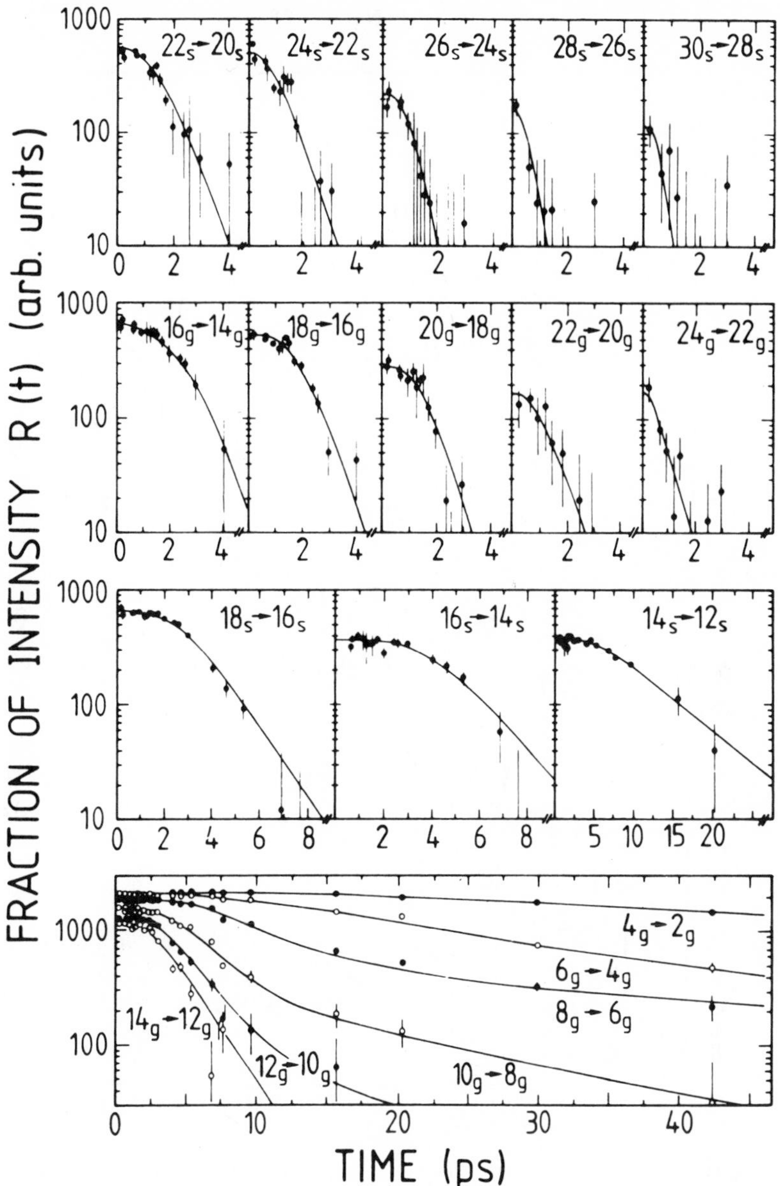

Fig. 6.30. Experimental decay curves for transitions in ^{156}Dy. The quantity $R(t)$ is equal to the intensity of the moving component or equal to the total intensity (taken from the fit) minus the intensity of the retarded component. When both components could be analysed a weighted average has been taken for clarity of presentation. The data obtained from the two Ge detectors have been averaged. The solid curves represent the fitted time distribution obtained by employing feeding models. The same averaging procedure was applied as for the experimental data. (Emling *et al.*, 1984.)

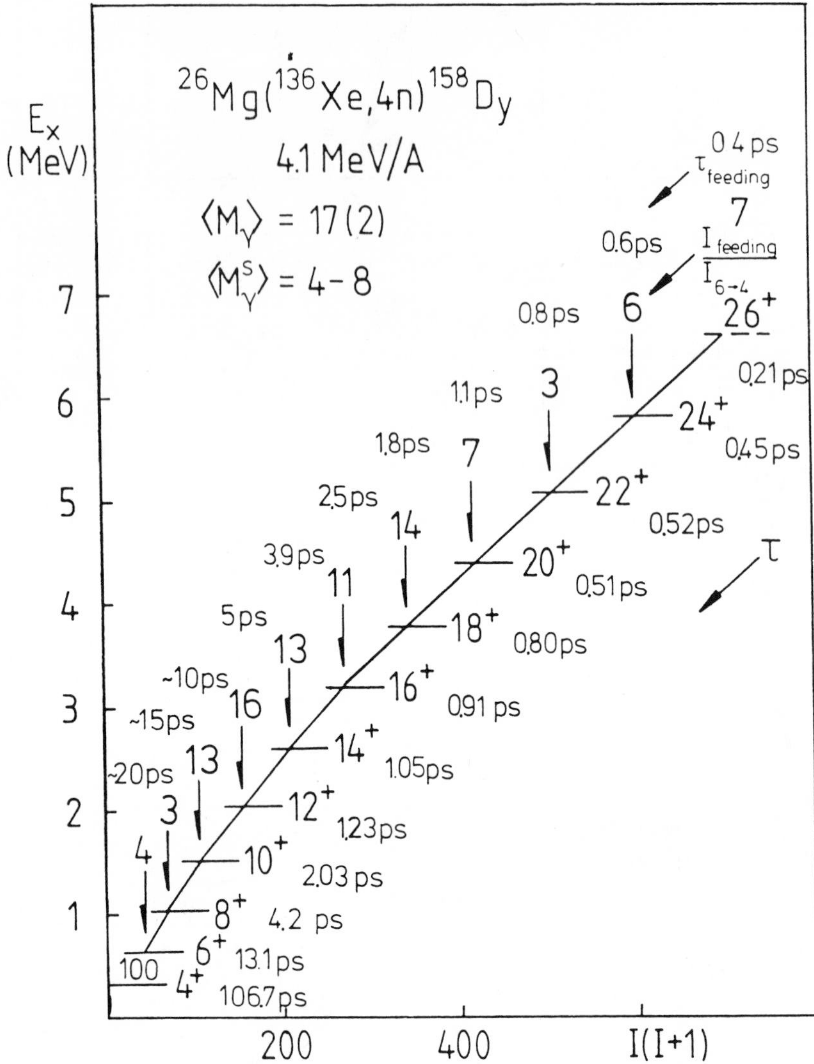

Fig. 6.31. Lifetimes of nuclear states in ^{158}Dy and the yrast-states feeding times of quasi-continuum γ rays deduced from recoil distance measurements. The intensities relative to the one of the $6^+ \to 4^+$ transition of the feeding γ rays are indicated. All times are given in picoseconds. (Grosse, 1979 and Emling *et al.*, 1981.)

have to be corrected for the Doppler shifts. Three spectra, corresponding to the high-spin states, are shown in Fig. 6.32. They have been measured by requiring coincidences with γ rays emitted from the 16^+, 18^+, and 20^+ states. Figure 6.32a shows a spectrum for the thin target, summed over all 21

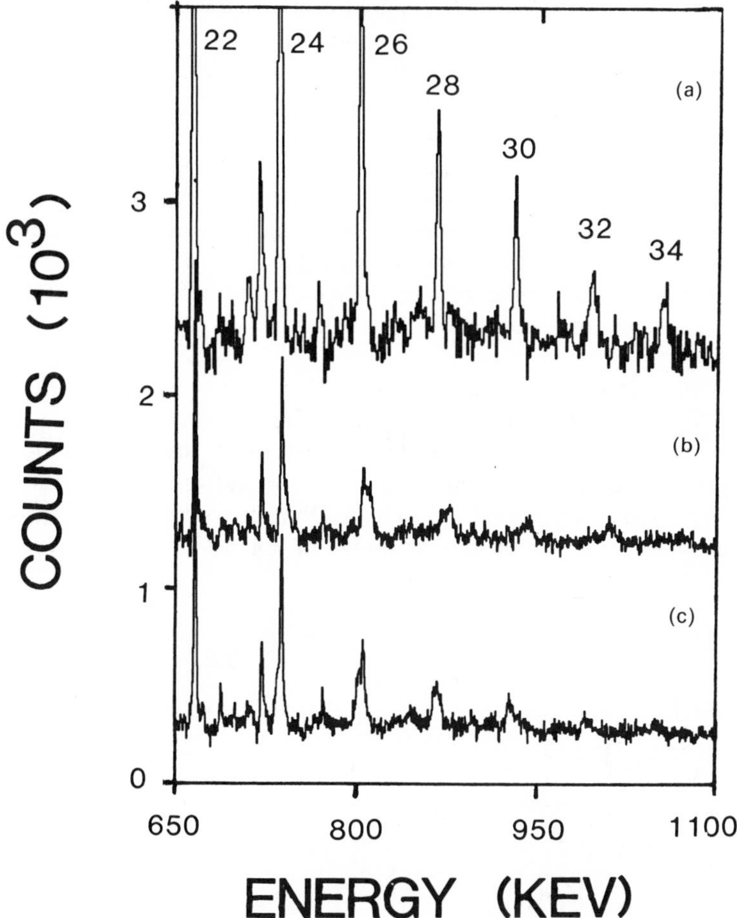

Fig. 6.32. Spectra gated on the yrast cascade of ^{166}Yb. The upper spectrum (a) was taken with a thin target and the two lower ones with a thick target. Spectrum (b) was taken with forward detectors and (c) with backward ones. (Bacelar *et al.*, 1986.)

detectors after a Doppler-shift correction. All the peaks corresponding to the 22^+ to 34^+ states are sharp, indicating rather short lifetimes for all states. Indeed, the analysis gave lifetimes ranging from 0.32 ± 0.04 to 0.05 ± 0.01 ps for the 22^+ to 34^+ states, respectively.

Some spectra obtained with the thick target by adding the data of the most forward ($\sim 40°$) and of the most backward ($\sim 140°$) detectors are shown in Fig. 6.32(b) and (c). The line-shapes of the peaks shown in these two figures consist of two components, and from their positions it is inferred that the side-feeding corresponds to the slow time component (non-shifted) in contrast to the fast (shifted) component due to in-band transitions.

In this analysis the sidefeeding to each observed discrete state has been assumed to proceed via a rotational band with constant quadrupole moment Q_0. In the slowing-down process only the electronic stopping power was used with a mean stopping time of 0.5 ps of ^{166}Yb in gold. A fit to all lineshapes observed with 21 detectors yielded both the lifetimes of the discrete states (mentioned above) as well as the sidefeeding times. The latter were significantly less than the lifetimes, by about a factor of three. The consequences for nuclear structure are discussed in Section 6.5.

6.3.5c *Multiple Coulomb Excitation* (*MCE*)

The method and technical details of MCE are discussed in Section 3.5.1d. Here we will present some results of experiments carried out at GSI, Darmstadt on some Dy isotopes using the set-up shown in Fig. 3.69. We will start with the data obtained for ^{164}Dy. A Doppler corrected γ-ray spectrum from the coincidence experiment is shown in Fig. 6.33.

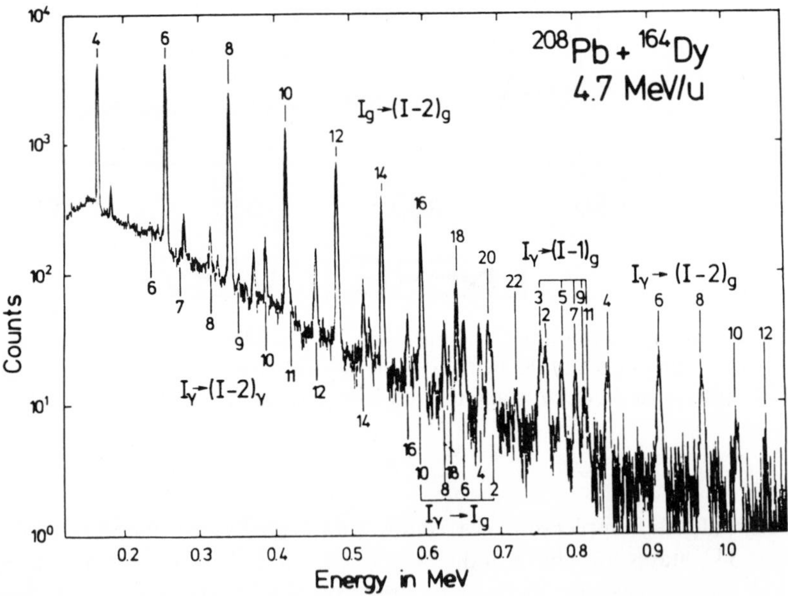

Fig. 6.33. A Doppler-corrected γ-ray spectrum measured in coincidence with recoiling ions and scattered projectiles in the MCE process. The ^{208}Pb projectiles were measured at c.m. angles $64° \leqslant \theta_1 \leqslant 146°$. The marked γ lines were attributed to decays of states belonging to the GSB or the γ-band in ^{164}Dy. (Schwalm, 1981.)

The analysis of the data starts with the determination of the γ-ray intensities as a function of the projectile c.m. scattering angle θ_1. The result for ^{164}Dy is shown in Fig. 6.34, compared with the result of the rotational model. They are already in good agreement. Usually one wants to extract the E2 matrix elements in a model-independent way. This is done by a complicated χ^2-procedure by varying all the matrix elements between the states involved. In this case the system has to be overdetermined which is achieved by using all observed γ-ray yields at as many scattering angles as possible. The sensitivity may be improved by extending the range of scattering angles, and this can be done by measuring double coincidences between the γ rays and the recoils only detected in either detector I or II (Fig. 3.69), in addition to the triple coincidences. In the case of double coincidences the observed Doppler shifts are used to distinguish the detection of a projectile from a recoil. In the calculational procedure, e.g. using the de Boer–Winther code (Alder and Winther, 1975) many effects have to be accounted for. There are effects, for instance, due to other bands and other matrix elements such as the giant dipole resonances as intermediate states and the deorientation effect. In the analysis it appears that at particular scattering angles certain matrix elements become very sensitive and they can thus be determined accurately. In this way one may also be able to deduce diagonal E2 matrix elements corresponding to the quadrupole moments $Q(I)$ as was done in the work of Stachel *et al.* (1982).

So far we have discussed the determination of E2 matrix elements from the observed Coulomb excitation probabilities. A different approach can be followed by measuring the nuclear lifetimes and γ-ray branching ratios and subsequently deduce $B(\lambda)$ values already discussed using the recoil-distance technique or the DSAM as discussed in Section 3.5.1.

Before discussing some of the implications for nuclear structure at high spin as deduced from measured B(E2) values, we want to present some methods of measuring magnetic moments.

6.3.5d *Determination of magnetic moments*

The sensitivity of magnetic moments or g-factors to changes in the pairing correlations of high-spin states arises from the fact that they deviate from the collective value $g_R \sim 0.35$ when proton or neutron pairs become rationally aligned. It has been shown by Frauendorf (1981) that $g \sim 0.2$ in the case of a neutron-aligned pair and $\sim +1.2$ when a proton pair is aligned.

The method of the transient-field technique to measure magnetic moments is outlined in Section 3.5.2. Here we will present the results for ^{158}Dy using MCE with 4.7 MeV/A ^{208}Pb projectiles and the experimental arrangement shown in Fig. 3.75. The results are included in Fig. 6.35 and will be discussed after the presentation of a second experimental possibility.

For high-spin states, including those in the quasi-continuum (qc), not accessible by MCE one has to use (HI, xn) reactions. The states are mostly of

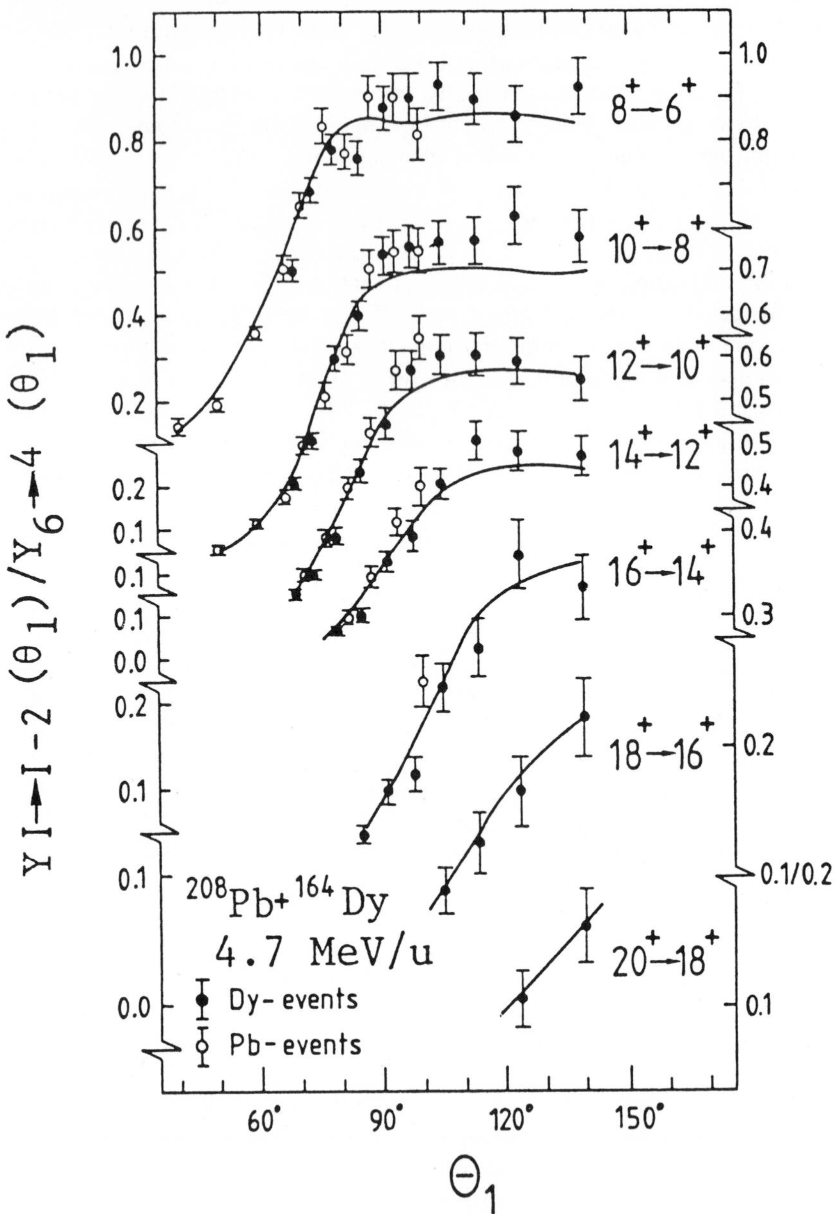

Fig. 6.34. The observed intensities of γ rays relative to the $6^+ \to 4^+$ transition in ^{164}Dy as a function of the projectile c.m. scattering angle θ_1 in the MCE process. The full lines correspond to calculations in the rotational model. (Schwalm, 1981.)

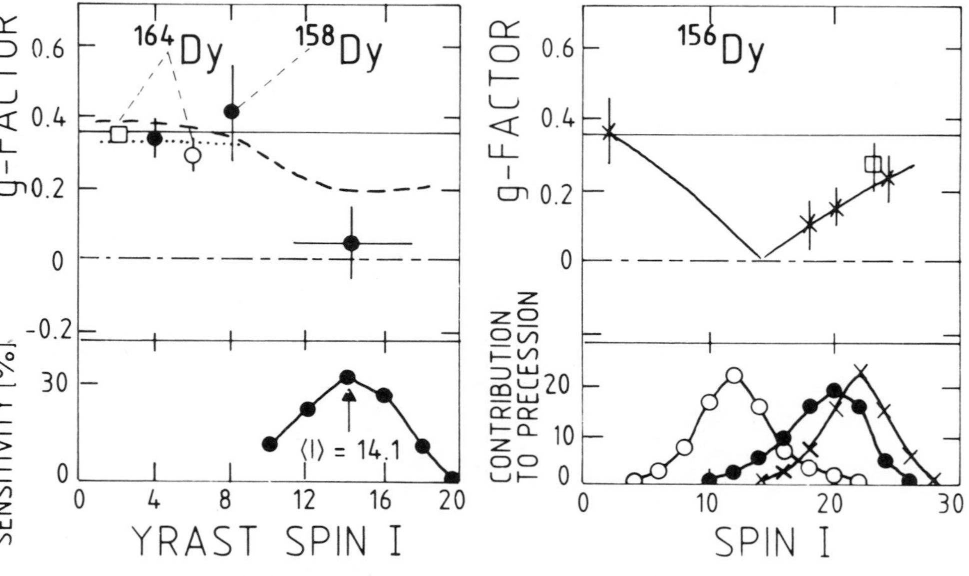

Fig. 6.35. The g-factors deduced from MCE experiments with a 4.7 MeV/A ^{208}Pb beam at GSI (left). The g-factors are given for discrete states except for the high-spin data point which is averaged over the spin region indicated below (see text and Seiler-Clark *et al.*, 1983). The g-factors for ^{156}Dy are deduced in the spin regions indicated below obtained from multiplicity measurements (see Section 6.5). The reaction used was ^{24}Mg(^{136}Xe, 4n)^{156}Dy at ~600 MeV. The open squares are from Häusser *et al.* (1983) and the crosses from Taras *et al.* (1985). The dashed line (left) represents a theoretical result in terms of a Cranked Shell Model (Diebel *et al.*, 1980) and of HFB calculations on the right.

very short life (< 1 ps) and decay as they pass through the magnetized Fe foil, Andrews *et al.* (1982). In an experiment carried out at Chalk River (Rud *et al.*, 1981) using the reaction ^{24}Mg(^{136}Xe, 4n)^{156}Dy at 544, 570 and 650 MeV beam energy a four-layer target assembly was used. After the ^{24}Mg target the fast ^{156}Dy recoils with $v/c \simeq 8$ per cent were slowed down in a layer of lead to ~ 5 per cent in order to be in the range of calibrated transient fields. The third layer was the Fe foil with a thick copper backing. In the Fe medium the recoils obtained their precession when decaying through qc states. Before the decay reached the discrete yrast states the recoils entered the Cu backing. Here no magnetic field influenced the precession accumulated in the Fe foil and the decaying discrete states thus carried the precession of the qc states.

The g-factors for the high-spin qc states could thus be determined via the precession carried by the discrete probe states. The results are presented in Fig. 6.35. Three different spin regions were populated at the three different bombarding energies. The initial spin populations were derived from measured γ-ray multiplicities with a NaI(Tl) detector.

We notice in Fig. 6.35 a g-factor of ~ 0.35 for low-spin state which is in line with $g_R \simeq 0.35$ for deformed collective rotating nuclei. In the region of $I \simeq 14$ the significant drop of the g-factor to a value of $g \simeq 0.04$ must be attributed to the excitation of neutrons. Indeed, we learned in the previous sections that at these spins backbending in rare-earth nuclei is due to the alignment of an $i_{13/2}$ neutron pair.

Two additional experiments have been carried out to find the g-factors of qc states at high spin. In the work of Häusser *et al.* (1983) a similar technique has been applied as in the first investigation discussed above, but using Gd as ferromagnetic medium instead of Fe. This raised the transient field by a factor of ~ 1.4 with respect to the one in iron at the higher recoil velocities, while keeping the density of polarized electrons about the same. The result was $\langle g \rangle = 0.21 \pm 0.03$ at $\langle I \rangle = 22.8$, as given in Fig. 6.35. A third experiment has been carried using the same technique and a Gd ferromagnetic medium in the ^{24}Mg(^{136}Xe, 4n) ^{156}Dy reaction at 600 and 620 MeV. The main difference from the second experiment was the application of a sum spectrometer (Taras *et al.*, 1985), to select certain spin ranges. The results are included in Fig. 6.35 and fall in the range of $\langle g \rangle = 0.1$ to 0.2 for $\langle I \rangle = 19$ to 23. This established the lower g-values at high-spin, in accordance with HFB calculations.

6.3.5e *Nuclear structure information deduced from transition probabilities*

The examples discussed above of measurements of B(E2) values and g-factors concern mainly Dy isotopes (^{156,158,164}Dy). From energy systematics on these isotopes (including also ^{154}Dy) we know that ^{164}Dy shows a very regular behaviour, consistent with a strongly deformed stiff rotor. The nucleus ^{154}Dy, on the contrary, shows a backbending at $\omega \simeq 0.28$ MeV $\hbar^{-1}$ and a second anomaly at $\omega \simeq 0.4$ MeV$\hbar^{-1}$ (Pakkanen *et al.*, 1982). The nucleus ^{156}Dy also shows a backbending at $\omega \simeq 0.28$ MeV$\hbar^{-1}$ where ^{158}Dy shows only a smooth anomaly. There seems to exist a transition from the soft nucleus ^{154}Dy via ^{156,158}Dy to the stiff rotor ^{164}Dy. The backbending at $\omega = 0.28$ MeV$\hbar^{-1}$ has been explained in terms of the alignment of a quasi-neutron pair. This explanation agrees with the measured reduced g-factor for ^{158}Dy in the bandcrossing region, as discussed above.

Measured B(E2) values can be helpful to answer the question of shape changes. This is important because with the alignment of quasi-particles the nuclear shapes may also be modifed. To study possible shape-changes in deformed nuclei it is convenient to express the B(E2) values relatively to those for a rigid rotor using eqn (6.3.81). The results derived from the lifetime measurements, discussed above, are summarized in Fig. 6.36 for the yrast states in the three isotopes ^{154,156,158}Dy. The values fluctuate around 1.0 up to $I^\pi \simeq 20^+$, but they all drop below 1.0 for larger spin values. A reduced E2 strength is also measured for the $16^+ \rightarrow 14^+$ transition, particularly pronounced in ^{156}Dy. This drop appears to coincide with the alignment of a pair of quasi-neutrons at the first bandcrossing ($I^\pi \simeq 16^+$), and is apparently followed by a reduced collectivity at higher spins. The effect seems to be most dramatic in ^{154}Dy where one has also observed in the level scheme a second backbending at $I^\pi = 28^+$. It is speculated that this is due to the alignment of a second $i_{13/2}$ neutron pair or an $h_{11/2}$ proton pair. (Khoo *et al.*, 1983). It is not excluded that at $\omega = 0.4$ MeV$\hbar^{-1}$ even all four $i_{13/2}$ neutrons and two $h_{11/2}$ proton orbitals take part in the alignment. This in turn might well induce an oblate deformation resulting in very much reduced B(E2) values. The B(E2) values given in Fig. 6.36 for ^{154}Dy indeed almost reach the low single-particle estimates. These data show that the measured lifetimes and deduced B(E2) values are very helpful in identifying possible shape changes, which will be discussed in more detail in Section 6.5.6.

We will conclude this section by showing the relation of $B(M1)$ values to particle alignment in the presence of collective rotations and their sensitivity to nuclear shapes.

In the case of rotational alignment of protons with spin $\vec{j}_p$ and neutrons with spin $\vec{j}_n$ the magnetic dipole moment has three components

$$\vec{\mu}(I) = g_R \vec{R} + g_p \vec{j}_p + g_n \vec{j}_n. \tag{6.3.88}$$

In this schematic representation the expression for B(M1) as given by eqn (6.3.72) has to be modified accordingly, taking into account the aligned angular momenta. See Dönau and Frauendorf (1983).

B(M1) values have been measured for $I\,(\alpha = \frac{1}{2}) \rightarrow I - 1\,(\alpha = -\frac{1}{2})$ transitions and vice versa between two bands in ^{165}Lu with opposite signature (Jónsson *et al.*, 1984). The qualitative agreement between the measured B(M1) values and those calculated within the schematic representation of Dönau and

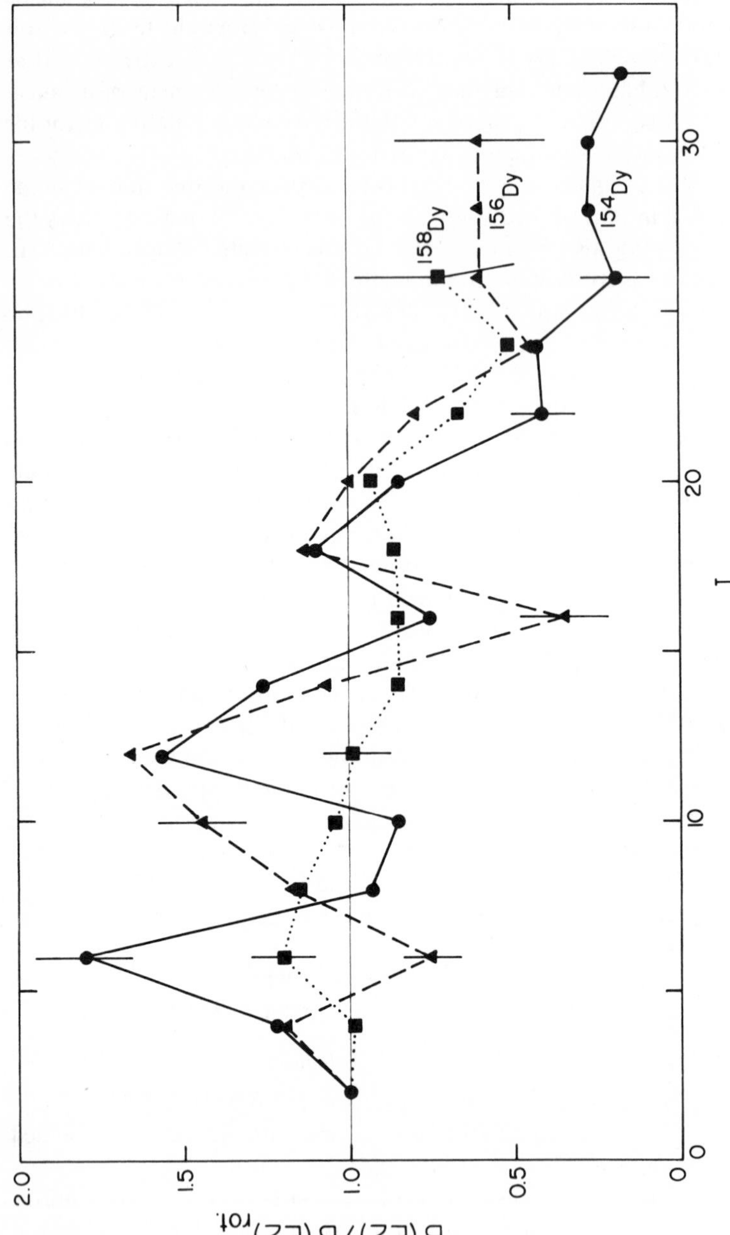

Fig. 6.36. Reduced E2 transition probabilities relative to the rotational value according to eqn (6.3.81) plotted as a function of spin of the upper yrast level in the transitions in ^{154,156,158}Dy. The data for ^{154}Dy are from Khoo *et al.*, 1983 and for ^{156,158}Dy from Emling *et al.* (1981, 1984).

Frauendorf (1983) proves the influence of quasi-particle alignment on the magnetic properties of nuclei.

A strong dependence of the M1 transition probability on signature has been observed for rotational bands based on the $\pi 7/2^-[523]$ configuration in ^{155,157}Ho (Hagemann *et al.*, 1984). The signature-splitting observed in ^{155}Ho is much larger than in ^{157}Ho bringing the $13/2^-$ state in that nucleus even above the $15/2^-$ state in energy. Assuming a very similar proton configuration for both nuclei one must hold a shape change responsible for the observed phenomenon.

The observed disappearance of the signature splitting at rotational frequencies where a pair of $i_{13/2}$ neutrons align has been analysed in terms of changing triaxiality by Frauendorf and May (1983). In Figure 6.37 the $B(\mathrm{M1}; I \to I-1)$ values relative to the $B(\mathrm{E2}; I \to I-2)$ values are shown as functions of spin for ^{155,157}Ho. A strong signature-dependence is observed below the bandcrossing in ^{155}Ho and a very weak dependence in ^{157}Ho. After the $i_{13/2}$ neutron alignment the splitting has disappered. From the measured lifetimes in ^{157}Ho very little dependence of the E2 transition probabilities on signature or spin was found, and therefore a strong dependence of the $B(\mathrm{M1})$ values on signature has been inferred. In the explanation of Frauendorf and May (1983) triaxiality with $\gamma = -7°$ below the backbending was assumed and $\gamma = +4°$ thereafter. This phenomenon in the γ-soft nuclei is suggestive for the driving force of $i_{13/2}$ neutrons towards $\gamma > 0$, counterbalancing the equilibrium deformation of $\gamma < 0$ induced by the $h_{11/2}$ proton (large $\Omega^\pi = 7/2^-$). From those data in combination with measured lifetimes in ^{157}Ho one is led to the conclusion that large signature effects are present in the $\Delta I = 1$ transitions of both M1 and E2 character. A detailed understanding, however, is still lacking, and more experimental and theoretical work is needed.

Although we have presented here only a very few specific examples we hope that the importance of electromagnetic properties for the understanding of nuclear structure has been demonstrated.

6.3.5f *Interband γ-transition from S-band levels*

Interband γ-transitions reflect microscopic configurations of levels associated with the transitions. They are thus quite sensitive to the changes of internal structure. The coupling between the ground-state rotational band (g) and the spin-aligned s-band (s) is rather weak in the sharp back-bending region. The levels are then represented by a superposition of the unperturbed s-band level $|I_s\rangle_0$ and the g-band level $|I_g\rangle_0$,

$$|I_s\rangle = a_I|I_s\rangle_0 - \sqrt{(1-a_I^2)}|I_g\rangle_0, \tag{6.3.89}$$

$$|I_g\rangle = a_I|I_g\rangle_0 + \sqrt{(1-a_I^2)}|I_s\rangle_0, \tag{6.3.90}$$

where a_I is the mixing coefficient of the levels with spin $I = I_s = I_g$. The

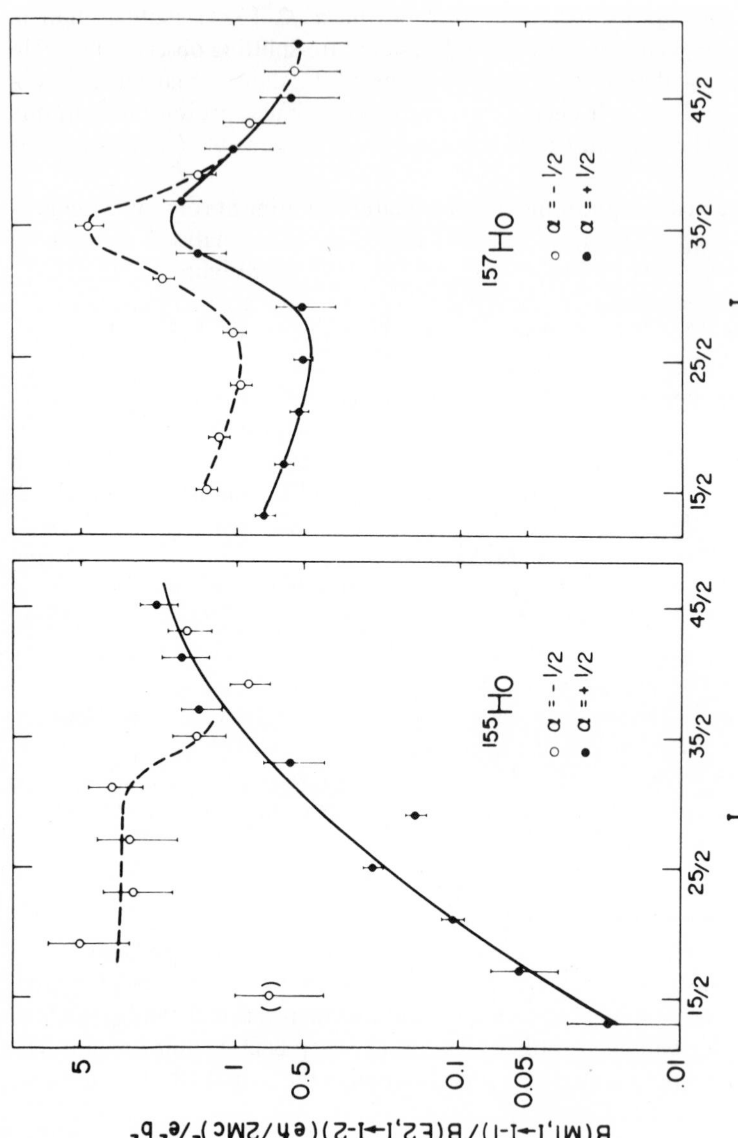

Fig. 6.37. Experimental $B(\mathrm{M1})/B(\mathrm{E2})$ ratios for states with different signature belonging to rotational bands based on the $\pi 7/2^{-}[523]$ configuration plotted as functions of spin. The curves are drawn to guide the eye. (Hagemann *et al.*, 1984.)

reduced transition probabilities for the interband transitions are

$$B(\mathrm{M1}, I_s \rightarrow I_g) = \frac{3}{4\pi} a_I^2 (1 - a_I^2) I_s^2 (g^s - g^g)^2, \tag{6.3.91}$$

$$B(\mathrm{E2}, I_s \rightarrow I_g) = \frac{5}{16\pi} a_I^2 (1 - a_I^2) (Q_s - Q_g)^2, \tag{6.3.92}$$

$$B(\mathrm{E2}, I_s \rightarrow (I-2)_g) = \frac{5}{16\pi} \{a_I \sqrt{(1 - a_{I-2}^2)} Q_s' - a_{I-2} \sqrt{(1 - a_I^2)} Q_g'\}^2, \tag{6.3.93}$$

where g^α, Q_α and Q_α' (with $\alpha \equiv s, g$) are the g factor, the intrinsic quadrupole moment and the E2 transition moment, respectively. It is interesting to note that the interband transition between the same spin states is very sensitive to the mixing coefficient and to differences in the electromagnetic moments Q and g between the two bands. In fact the Q moments for the s- and g-bands are similar as shown by the similar values for their moments of inertia, while the spin alignment of the two quasi-particles in the s-band changes the g factor substantially. The spin alignment of the two neutrons at the first backbending leads to $g^s \simeq 0$, while that of the two proton at the second backbending leads to $g^s \simeq 1$. Since $g^g \simeq Z/A \sim 0.4$ in medium-heavy nuclei, the difference $g^s - g^g$ is of the same order of magnitude as g^g. Consequently M1 radiation dominates the interband transition between levels of the same spin. The branching ratio $B(\mathrm{M1}, I_s \rightarrow I_g)/B(\mathrm{E2}, I_s \rightarrow (I-2)_g)$ gives the g-factor of the s-band level at the backbending, provided that the Q moments and the mixing coefficients are known.

The interband γ-ray yields are only a few per cent of the major yrast γ-ray yields in the (a, xnγ)-type compound reactions, and so studies of the electromagnetic properties of such weak γ rays require high precision e–γ spectroscopy. The γ-transitions from the s-band levels to the g-band levels in ^{164}Er and ^{166}Yb have been studied by measuring very precisely the γ-ray angular distributions and internal conversion electrons (Nagai *et al.*, 1983). The interband transitions of $12_s \rightarrow 12_g$ are predominantly of M1 type with E2 being less than 10 per cent. The g-factor for the 12^+ s-band level is extracted from the observed branching ratio $B(\mathrm{M1}, 12_s \rightarrow 12_g)/B(\mathrm{E2}, 12_s \rightarrow 10_g)$ and the mixing coefficient. It is noted that the mixing is related to the energy difference ΔE_I and the interaction matrix $\langle V_{sg} \rangle$ by

$$a_I \sqrt{(1 - a_I^2)} = \langle V_{sg} \rangle / \Delta E_I. \tag{6.3.94}$$

The g-factors of $g(12_s) = 0.07 \pm 0.03$ for ^{164}Er and -0.14 ± 0.04 for ^{166}Yb, are deduced using $Q_g \simeq Q_s = 7$ b and $g^g = 0.35$ (Nagai *et al.*, 1983). The g-factor for the s-band is obtained from

$$g^s = g^g + (g^i - g^g) I_a / I, \tag{6.3.95}$$

where g^i and I_a are the g-factor and the angular momentum for the aligned two quasi-particles. The present values are in accord with the calculated value of $g^s(12_s) \simeq 0$ for the aligned $i_{13/2}$ two-neutron state with $g_s^{\text{eff}} = 0.7 g_s(\text{n})$, where $g_s(\text{n})$ is the free neutron g_s.

6.4 Independent particle motion generating high spin

6.4.1 *Non-collective nucleonic motion*

The observations of irregular decay patterns and of retarded transitions in nuclei located mainly near closed shells have been understood in terms of single-particle motion, sometimes called non-collective rotation for high-spin states. This type of nuclear motion is characterized by the alignment of the total angular momentum with the symmetry axis of axially symmetric nuclei, as shown in Fig. 6.1(b).

As an example, the level scheme of the nucleus ^{147}Gd in Fig. 6.2 exhibits the irregularity characteristic of non-collective rotation. Many nuclei near closed shells show similar features, notably isomeric states. The corresponding transition probabilities are strongly retarded with respect to collective transitions, as can be seen in Fig. 6.3 for several rare-earth nuclei.

The notion of rotation around the nuclear symmetry axis may however be somewhat misleading. It was argued in Section 6.3 that no collective rotation is possible around the symmetry axis as a result of the quantum-mechanical description. The only possibility of building up high angular momentum in the case of non-collective rotation is by aligning the spins of individual nucleons.

As a result of the impossibility of having collective rotation around the symmetry axis the concepts of rotational frequency ω and of moment of inertia $\mathcal{I}$ do not seem to be very meaningful. Nevertheless it appears that an overall characterisitc of yrast states is provided by the average slope of the curve representing their energies as a fucntion of I^2. It can be seen in Fig. 6.11 that in the case of non-collective rotation ($^{151,\,152}$Dy) the curve is a straight line over a relatively large range of angular momentum. This behaviour can be ascribed to a peculiar property of the individual nucleonic motion as follows. The angular momentum I increases roughly linearly with the number of quasi-particles, which constitute the yrast states. The average quasi-particle energy is also approximately proportional to the number of quasi-particles, because of the exclusion principle and considering the single-particle levels in a rotating potential, with generally strongly decreased energies for large ω. Therefore the total excitation energy can be approximated by $E \sim I^2$ (Bohr and Mottelson, 1981). The parameter which governs this quadratic proportionality may be called a moment of inertia to restore the analogy with collective rotation.

In Section 6.3.4 the 'envelope' moment of inertia was introduced to describe the ability of the nucleus to increase its energy and angular momentum by means of configuration rearrangements. In the case of non-collective rotation the moment of inertia is determined as an average property of many states, each of them representing a different configuration. We shall denote this moment of inertia by $\mathscr{I}^{(1)}_{\text{env}}$ to indicate that it is determined by the envelope curve connecting the different states when represented in an E vs. I plot (see Fig. 6.25c).

A particular consequence of non-collective rotation is the occurrence of high-spin isomeric yrast states, often referred to as 'yrast traps'. They are of much interest for nuclear structure studies in the domain of high angular momentum, and can be located relatively easily using various timing techniques because of their delayed decay. The retarded transitions also make possible the measurement of electromagnetic multipole moments and these provide direct information about the single-particle structure associated with high-energy excitations. These topics will be discussed below in Sections 6.4.2 and 6.4.3. A brief outline of the theoretical methods used to describe single-particle configurations is presented in Section 6.4.4.

The delayed transitions may also serve as a time selection for certain decay paths of interest. In that case, for example, the yrast traps may act as stepping stones to states with the highest possible spins. Experiments have revealed shape-changes in the high-spin region ($I > 40$) of some rare earth nuclei. We will discuss a few examples of this use of yrast traps in Section 6.5.

6.4.2 *The nature of yrast traps*

Isomeric states are characterized by their retarded decay. The retardation can be caused by several different mechanisms. One mechanism is the nature of the electromagnetic radiation, with an emission rate proportional to $(kR)^{2l+1}$. The wave number $k = E_\gamma/(\hbar c)$ shows the strong dependence on the γ-ray energy. On the other hand when $kR \ll 1$, for radiation in the region of a few MeV and $R = 10^{-14}$ m for a medium-size nucleus, the emission rate decreases strongly with increasing multipolarity l. Thus in the case of a large spin change or of a low-energy transition the emission rate will be retarded and the decaying level becomes long-lived, and is called isomeric.

Another mechanism that may also lead to the formation of isomeric states is due to significant differences in the intrinsic structure of the initial and possible final states involved in the γ-ray transitions. In some cases the transitions require rearrangements of one particle with a large spin change or of several particles simultaneously. As the electromagnetic transition operators are single-particle operators these transitions will be hindered.

The irregular dependence of energy on angular momentum also causes local minima in the yrast line. This means that states in those minima cannot decay to a lower-lying state via a small multipolarity transition or only with small energy (sometimes the energy difference can even be negative). The γ-decay, proceeding preferentially via the yrast line, is thus trapped in those yrast states, leading to the name of 'yrast traps'. They are sometimes also called 'superdizzy' states, owing to the underlying irregular individual particle motion.

Owing to the two mechanisms of retardation described above one may distinguish energy traps (or spin traps because of the large spin change) from structure traps. In our convention we will call any isomeric yrast state with a lifetime $T_{1/2} \geqslant 1$ ns an yrast trap, and will consider in this chapter only states with spin $I \geqslant 10$. We realize that this convention has no physical basis, but it is introduced for convenience so as to reflect some common ideas in the literature.

Isomeric states were first found several decades ago. They occur in nuclei throughout the periodic table, but mainly in nuclei near the various closed shells. In the high-spin region ($I > 10$) isomers were found a long time ago near the $Z = 82$, $N = 126$ closed shells, for instance in ^{211,212}Po. Early shell-model calculations related those isomeric states to the single-particle level spectrum and the nucleonic residual interaction (Glendenning, 1962; Auerbach and Talmi, 1964). Another type of isomer that has been known for a long time is the K-isomer that will be discussed below in Section 6.4.4.

A first attempt to construct high-spin states out of many shell-model orbitals, all aligned with their angular momenta parallel to the nuclear symmetry axis, was made by Grover (1967) (see also Grover and Gilat, 1967a–c). This was related to the nuclear shape by Cohen *et al.* (1974) suggesting the existence of oblate shapes in many nuclei. The importance of individual particle motion with spins in the direction of the axis of axial symmetry for the formation of non-collective high spin states was emphasized by Bohr and Mottelson (1974). This triggered a new interest in the possible existence of yrast traps resulting from a nuclear coupling scheme of pure individual-particle nature.

The relation between the nuclear motion and the nuclear shape is schematically represented in Fig. 6.15. The rotation axis coincides with the axis of axial symmetry of a deformed nucleus with oblate shape at $\gamma = 60^\circ$ and with prolate shape at $\gamma = -120^\circ$. Although both possibilities represent single-particle motion, oblate shapes are energetically favourable, as was argued in Section 6.3.2. In prolate nuclei one normally finds that the collective modes of excitation corresponding to rotation perpendicular to the symmetry axis ($\gamma = 0^\circ$) are lower in energy than the single-particle states. In some cases, however, shell effects bring the single-particle states below the collective bands, thus favouring $\gamma = -120^\circ$, and then isomeric single-particle states may be observed. By shell effects we mean here the changes in the shell structure caused by certain deformations in some nuclei.

The normal situation in oblate nuclei is that the aligned single-particle states are lower in energy than the collective mode of excitation. In this case the single-particle motion has a larger overlap with the oblate (attractive) nuclear potential than with the collective motion (see Fig. 6.15), and therefore the single-particle excitation occurs lowest in energy. The yrast line then consists of those single-particle states and some of them are the most likely candidates to form yrast traps. This may also happen in nuclei that are spherical or transitional at low spin, because they tend to attain oblate shapes at higher spins. In that case the spin-aligned nucleons move predominantly in orbitals near the equatorial plane and thus may induce quadrupole polarization, leading to oblate deformation.

It is thus important to know the locations of deformed nuclei with axial symmetry on the nuclear chart, particularly of those with oblate shapes, as well as to have some idea of the variation of shape with angular momentum. The most direct experimental information on the nuclear shape can be derived from the electric quadrupole moments. They are measured for the first excited 2^+ states in many even–even nuclei and for several other cases such as for some high-spin states and for states in several odd nuclei. These measurements helped us to establish the locations of the closed shells and to find the regions with prolate and oblate shapes at low spin throughout the periodic table. Unfortunately, no clear-cut experimental determination of nuclear shapes at high spin exists, except in a very few cases such as ^{147}Gd, discussed below in Section 6.4.3.

Therefore we have to resort to the results of theoretical calculations as a guideline for the location of nuclear shapes and their evolution with angular momentum. A rough idea about those features can be obtained, for instance, by applying the Strutinsky renormalization procedure to a reasonable nuclear potential like the Nilsson or Woods–Saxon potential (see Szymański, 1983 for details).

Calculations have been carried out by Andersson *et al.* (1978) to find the energy minima for axially-symmetrical nuclei with the rotation axis coinciding with the symmetry axis, both for $\gamma = 60^\circ$ and -120°. Their results are given in Fig. 6.38 for $I = 20$ and in Fig. 6.39 for $I = 40$. The open areas correspond to collective rotation, discussed in Section 6.3. A number of interesting features can be seen in those figures. Obviously one finds the spherical nuclei (with $\varepsilon \leqslant 0.05$) near the closed shells $N = 50$, 82 and 126 at low spin ($I = 20$). At spin $I = 40$, however, many nuclei near the $N = 50$ and 126 closed shells change from spherical to oblate shapes. We may also notice that oblate shapes dominate strongly in the regions where non-collective rotation is known to occur, i.e. near the closed shells. This arises from the fact that just beyond the closed shells there are available high-spin orbitals with large spin projection on the symmetry axis for oblate shapes ($\gamma = 60^\circ$) and particles can be promoted to these orbitals at relatively low energy cost. In

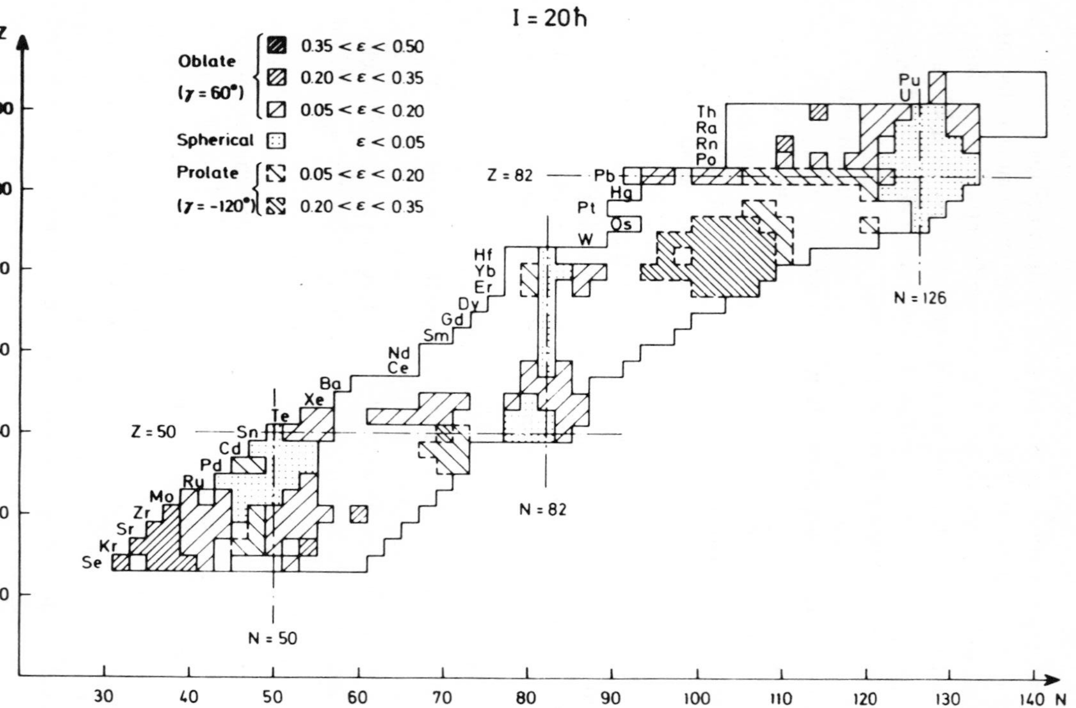

Fig. 6.38. Various types of nuclear axially-symmetric shapes in the (Z, N) plane for angular momentum $I\hbar = 20\hbar$ aligned with the symmetry axis, $\gamma = 60^\circ$ (oblate) and $\gamma = -120^\circ$ (prolate). Calculations have been made using the Nilsson potential. The shadowed regions correspond to those areas where yrast traps might a priori be expected. The regions which are left blank are most favourable for the collective rotation and yrast traps are there highly unlikely. (Anderson *et al.*, 1978.)

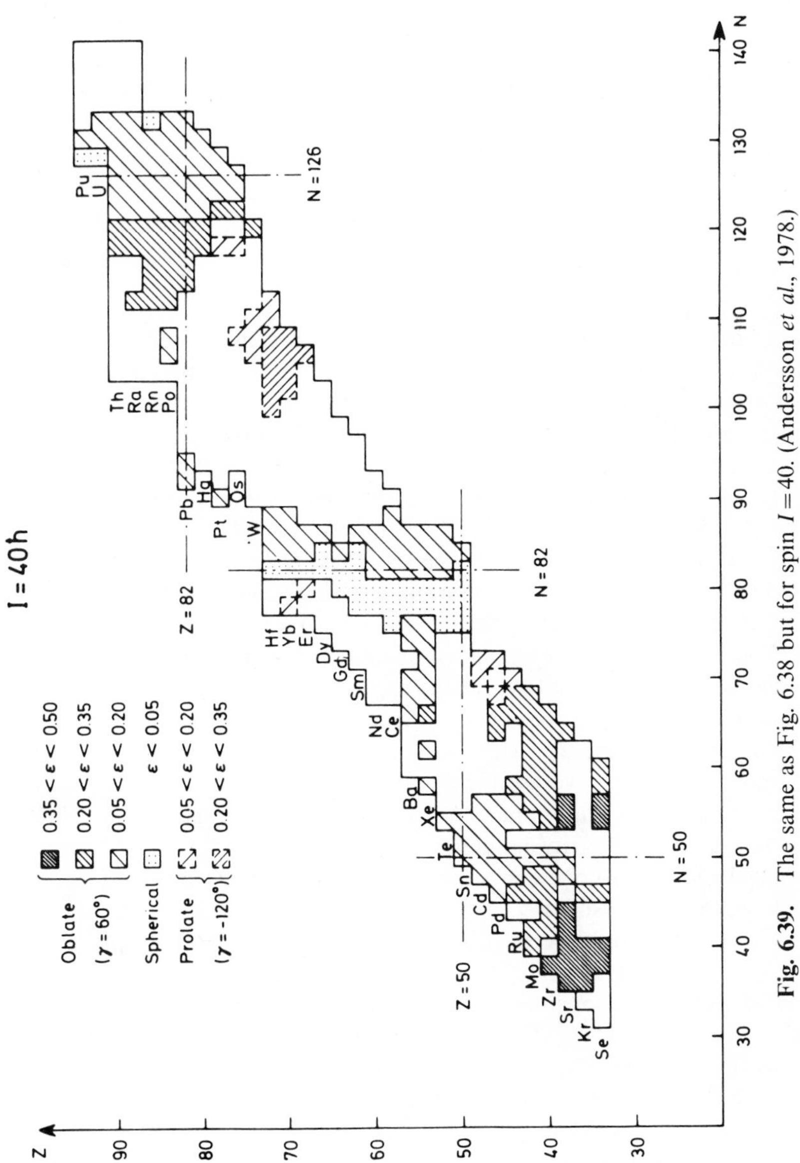

Fig. 6.39. The same as Fig. 6.38 but for spin $I = 40$. (Andersson *et al.*, 1978.)

other words the shell correction energy (see Section 6.5.4) decreases rapidly with increasing rotational frequency, causing minima in the total energy surface for oblate shapes beyond the shell closures at high spin, notably for $N \geqslant 82$. The only exception according to this calculation is found in the Hf region for $I = 20$ with prolate deformation and somewhat extended at $I = 40$. We shall discuss the non-collective features of some nuclei with oblate shapes in the next experimental section and present the example of ^{176}Hf with prolate shape at the end of Section 6.4.4.

6.4.3 *Experimental investigation of yrast traps*

6.4.3a *Properties of high-spin isomers*

The most powerful method of populating high-spin isomeric states is the fusion reaction with heavy-ion projectiles. For the location of isomeric states and for the study of their properties conventional γ–γ-coincidence techniques can be applied. The use of high resolution (Ge) detectors emphasizes the important role of characteristic de-excitation γ-rays in the process of identifying the states and in the determination of their properties. Other techniques have also been developed to select high-spin states and to determine the gross properties of the delayed γ-ray cascades. These techniques include large sum spectrometers, many-detector multiplicity filters and small size Compton-suppression spectrometers (see Section 3.2). We will illustrate the various methods, using actual experiments on some rare-earth nuclei as examples.

The difficulties encountered in the experimental investigations are mainly connected with the fact that in fusion reactions several isotopes are produced which makes it hard to assign isomeric transition to the proper isotopes. Moreover, it is often very difficult to deduce unambiguous decay schemes from the data. This causes uncertainties in the excitation energy, in the lifetime and in the spin and parity assignments to isomeric states. Nevertheless, many groups have tried to overcome the problems in various ways because of the great interest in yrast traps.

A first systematic experimental search for nuclei with high-spin isomeric states over a broad region of the nuclear chart between Ba and Pb has been carried out by a Copenhagen–Darmstadt collaboration (Pedersen *et al.*, 1977). They used ^{40}Ar, ^{50}Ti, and ^{65}Cu from the UNILAC accelerator at GSI in Darmstadt to produce compound nuclei in fusion reactions with angular momenta as high as $60\hbar$. The rapidly recoiling nuclei were stopped in a catcher foil positioned 15 cm behind the target (see Fig. 6.40a) in order to separate the delayed γ-rays from those which are emitted promptly in the reactions. This technique eliminated the strong prompt γ radiation from all kinds of reaction products. On the other hand it required a minimum flight time of $\sim$ 10 ns for the recoils to reach the catcher, thus missing all the rapidly

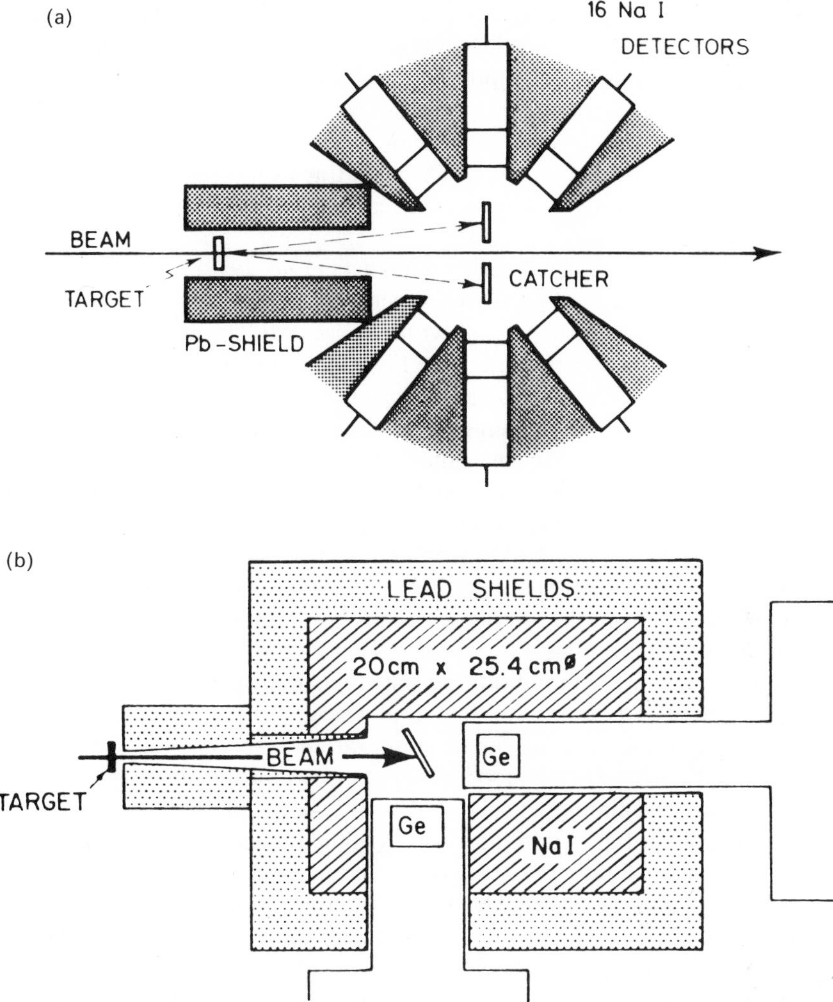

Fig. 6.40. A schematic view of the recoil catcher and part of the sixteen NaI detectors surrounded by lead shielding (a), used for the discovery of the island of high-spin isomers near $N=82$ (Pedersen *et al.*, 1977). In a follow up to that experiment the multiplicity filter was replaced by a sum spectrometer with two Ge-detectors (b) to identify the isomeric states more precisely. (Borggreen *et al.*, 1980.)

decaying ($<$ 10 ns) isomers. In that experiment a multiplicity filter was used, consisting of sixteen $5.1\phi \times 5.1\ \text{cm}^3$ NaI(Tl) detectors, to record the (delayed) γ rays emitted from the catcher foil. Cascades with many γ rays (and thus with high multiplicity) were selected by requiring that at least a few detectors fired simultaneously. In this way, as each γ ray carried off on the average 1 to 2 units of angular momentum, a selection of high-spin isomeric states was achieved. The reactions did produce compound nuclei leading to evaporation residues (final reaction products) with γ-ray transitions of at least a 10 ns delay. This experiment produced a very interesting result although the evaporation residues were not located precisely. The yrast traps appeared to occur only in a narrow region of neutron-deficient rare earth nuclei between $N = 82$ and 86 for isotopes ranging from Gd to Lu. This region overlapped almost entirely the theoretical one predicted earlier (Cerkaski *et al.*, 1977).

In an extension of the first search experiment, described above, the same group surrounded the catcher foil with a large NaI(Tl) crystal acting as a total-energy or sum spectrometer (see Fig. 6.40b). Discrete γ-ray lines were observed using Ge-detectors placed at angles of 0° and at 90° with respect to the beam direction in the openings of the sum spectrometer. The background from prompt reactions or from Coulomb excitation in the target was thus eliminated by recording only the γ rays coming from the catcher foil. This lead foil also stopped the beam without generating disturbing γ radiation. The isomeric nuclei in this case had to survive a flight of $\sim$ 30 ns before they were caught on the foil. Both the energy and time signals from the sum spectrometer have been used in the analysis. Gamma-ray transitions from high-lying and thus high-spin states have been selected with the Ge detectors by imposing a high-energy condition on the sum-spectrometer signal.

The different conditions on the total energy measured with the sum spectrometer did lead to different evaporation residues, because a high neutron multiplicity (xn) reaction leaves less energy for the γ decay than a low (xn) reaction. The timing signals of the sum spectrometer and of the Ge-detectors were used to obtain the lifetimes of the isomeric states and to aid in the identification of more than one isomer in one isotope. That work resulted in the identification of final nuclei which contained isomeric states in several cases. The excitation energies of the isomers have been obtained from the measured sum energy of the delayed γ-ray cascades, and the spins have been estimated on the basis of the rotational relation between energy and angular momentum—eqn (6.3.2)—assuming a reasonable moment of inertia. This last procedure can be questioned, particularly here in the case of non-collective rotation. More reliable spin values should be deduced from discrete-line spectroscopy.

Information on the spin of the isomeric states may also be obtained from the measured average multiplicity M_γ of the delayed γ-ray cascade. The main difficulty is to transfer a M_γ value into spin I. For rotational bands in even–even nuclei the simple relation $I = 2M_\gamma$ is valid because the γ rays have predominantly the stretched E2 character. In the case of non-collective rotation severe deviations from that expression may occur. In the decay scheme of ^{152}Dy, for example, several $\Delta I = 1$ transitions are present (see Fig.

6.43), so that one has to modify that relation on the basis of experimental data. In the work of Jastrzebski *et al.* (1980) an experimental relation was derived between the spin I of an isomer and the number of delayed transitions M_γ between that isomer and the ground state with spin I_g. The relation is $I = I_g + M_\gamma \Delta L$, where ΔL is the average angular momentum removed by one transition. The value of ΔL varies between 1.6 and 1.9 as obtained from the known level schemes of ^{154}Er, ^{151}Ho, ^{150,151,152}Dy. This empirical relation is consistent with the data set for $^{146-150}$Gd and $^{148-152}$Dy isotopes as shown in Fig. 6.41. The figure also indicates upper and lower limits for the spins I_{up} and I_{low} for certain measured multiplicities. Note that the line indicated by I_{up} corresponds roughly to $\Delta L \cong 2.0$ and the one with I_{low} to $\Delta L \cong 1.4$.

The systematic searches for yrast traps discussed above have also been carried out by several other groups employing high-resolution Ge-detectors

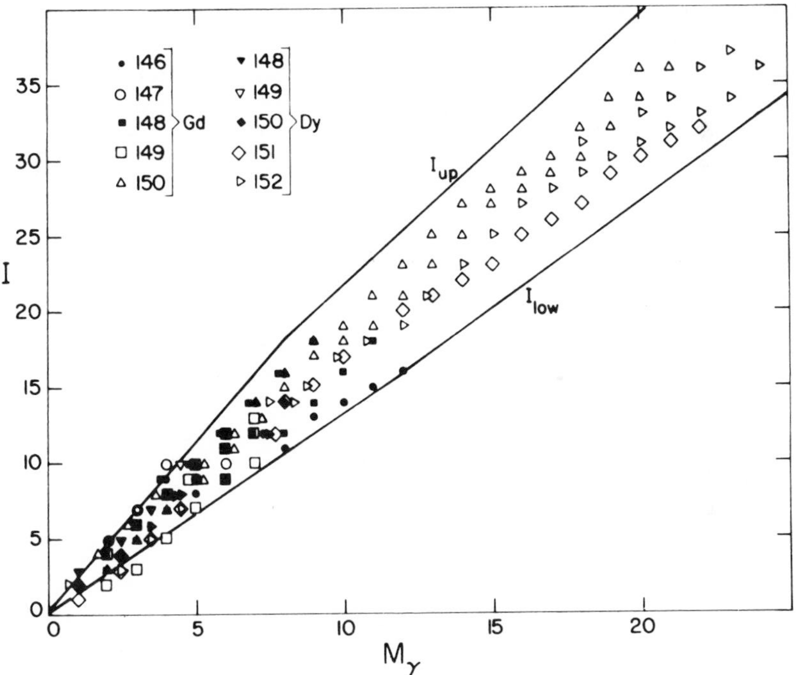

Fig. 6.41. The empirical relation between the spins of isomeric states and the γ-ray multiplicity of the de-exciting cascades extracted from cross experiments in various Gd and Dy isotopes. The spins and multiplicities were determined independently. The solid lines indicate the upper and lower estimates for spin assignments, reflecting uncertainties in the decay schemes used. These systematics may be used to assign spin limits to isomeric states from measured multiplicities. (Hageman, 1981.)

Z ↑								
Yb 70							156 3.03 11^- 6.0	
Tm 69			151 (2.66 $\frac{27}{2}^+$) 470					
Er 68			150 2.80 10^+ 2.6 μs	151 >2.5 ≥ $\frac{27}{2}$ 620 ms	152 4.52 16^+ 1.5 4.68 (14) 6.0 8.97 (24) 32 >12 ≥33 4	153 2.75 $\frac{27}{2}^+$ 373 5.18 ($\frac{43}{2}$)270 (5.20)	154 3.02 ($\frac{10^-}{11^-}$) 50	
Ho 67			149 (2.74 $\frac{27}{2}^-$) 59	150 2.63 17^+ 92 (7.2) >10 50	151 >6.1 ($\frac{67}{2}$) 20	152 3.4 ≥ 19 47 6.0 ≥ 16 70 (3.7) (>500)	2.8 ($\frac{33}{2}$) 231 ($\frac{35}{2}$) 3.7 ≥ $\frac{27}{2}$ 56	
Dy 66	146 2.94 10^+ 150 ms		148 2.92 10^+ 470 ms	149 2.66 $\frac{27}{2}^-$ 500 ms 7.41 ($\frac{47}{2}$) 29	150 3.03 10^+ 1.1 4.57 16^+ 1.7	151 2.96 $\frac{27}{2}^-$ 1.3 4.90 $\frac{41}{2}^{(-)}$ 6.1 6.03 $\frac{49}{2}^+$ 13	152 3.16 11^- 4.3 5.09 17^+ 60 6.13 21^- 10 7.88 27^- 1.6	

	80	81	82	83	84	85	86	87
Tb 65			147 2.99 $(\frac{23}{2})$ 3.5	148 7.55 (30) 1.3 μs				
Gd 64	144 3.45 10^+ 135		146 8.92 ≤ 20 4.0	147 2.76 $\frac{21}{2}^+$ 4.8 3.58 $\frac{27}{2}^-$ 27 8.59 $\frac{49}{2}^+$ 550 11.0 $(\frac{59}{2}^-)$ 0.8		149 3.39 $\frac{27}{2}^+$ 6.0		
Eu 63				146 2.5 14 8				

N →

Fig. 6.42. Yrast traps located in 25 different nuclei near the closed $N = 82$ shell and the $Z = 64$ subshell. The indicated properties of the isomers are (from left to right) the excitation energy (MeV), the spin and parity, and the half-life in nanoseconds, unless indicated otherwise. Isomeric states are only included if $I \geqslant 10$ and the half-life $\geqslant 1$ ns (as an arbitrary criterium).

in addition to multiplicity filters in order to identify the evaporation residues (e.g. Hageman *et al.*, 1979; and Jastrzebski *et al.*, 1980). Most of the yrast traps discovered so far in the neutron-deficient rare earth nuclei are summarized in Fig. 6.42, which includes data from many experiments, referenced in de Voigt *et al.*, 1982). An arbitrary choice was made to include yrast traps in Fig. 6.42 only if the measured half-life was greater than 1.0 ns and the spin $I \geqslant 10$. A remarkable feature is the abrupt ending of the observation of yrast traps at $N = 87$ nuclei, only five neutrons away from the closed neutron shell. This indicates that a prolate deformed region, favouring collective rotation, sets in at $N = 87$, which also agrees with the theoretical results given in Figs 6.38 and 6.39.

A nucleus that has been found in several searches to contain high-spin isomeric states and that has been subject of extensive experimental as well as theoretical studies is ^{152}Dy. As illustrative examples we shall discuss in detail in this section the experiments performed to determine the properties of this nucleus below $I = 40$, leaving the region of very high spins ($I > 40$) to Section 6.5. A level scheme of ^{152}Dy is presented in Fig. 6.43, resulting from many experimental investigations . The first high-spin isomeric state found in the neutron-deficient light rare-earth mass region has indeed been reported for this nucleus ^{152}Dy in 1974 by the Groningen–Warsaw collaboration (Jansen *et al.*, 1974; see also Jansen *et al.*, 1976, 1979). The finding of the $T_{1/2} = 60$ ns isomer at $E_x = 5$ MeV with spin $I = 15$–18 was in line with a theoretical calcualtion of the Warsaw–Lund collaboration (Cerkaski *et al.*, 1977) showing a local minimum at $I = 15, 16$ in the yrast line of ^{152}Dy as discussed in Section 6.4.4 below.

In this experiment the conventional $E_{\gamma 1}$–$E_{\gamma 2}$-time coincidence technique with registration of the time between the high-resolution Ge(Li) detectors has been applied using the ^{154}Gd(α, 6nγ)^{152}Dy reaction at $E_\alpha = 80$ MeV. The spins and parities have been determined from the angular distributions and internal conversion coefficients. The electron spectra have been measured in-beam with a 3 mm thick, 300 mm^2 Si(Li) detector placed inside a mini-orange electron spectrometer (see Section 3.4). This spectrometer contained permanent SmCo magnets with an adjustable configuration to select a certain energy transmission region.

The electron spectrometer is a useful tool for high-spin studies because it can be included in multi-detector assemblies owing to its very small size (10 cm in diameter). A conversion electron spectrum obtained for the ^{154}Gd(α, 6n)^{152}Dy reaction is shown in Fig. 6.44. All the lines indicated in the figure could be assigned to the decay of the 60 ns isomer in ^{152}Dy or to other Dy isotopes. The work has been extended by employing ^{12}C beams and a sixteen-NaI(Tl)-detector multiplicity filter in coincidence with a Ge(Li) detector. The multiplicity filter used is shown in Fig. 3.15 (and see Section 3.2.2).

Beams of ^{12}C (or heavier ions), populate states with higher spins than do alpha-particle beams. The use of a multiplicity filter will also enhance the high-spin region with respect to γ rays from low-spin states. Figure 6.45 demonstrates the possibility of observing only those transitions which feed the 60 ns isomer in ^{152}Dy via prompt coincidences between the Ge(Li) detector and the filter or those decaying from that isomer via delayed coincidences. In that work also a second isomeric state has been observed at $E_x = 6$ MeV with a half-life of 9.5 ns.

The main problem in the identification of isomers as yrast traps is the possible existence of unobserved isomeric transitions (presumably with low energy), for instance in the decay of the 60 ns isomer in ^{152}Dy. This is a more general difficulty encountered in establishing decay schemes of several nuclei. The problem of the unobserved isomeric transition in ^{152}Dy was solved by Nagai *et al.* (1980) who carried out a multi-Ge-detector coincidence experiment, including a low-energy photon detector. They used the (α, 4n) reaction at 60 MeV to excite ^{152}Dy and utilized the catcher-foil technique (7 cm distance between target and catcher) to eliminate the strong prompt radiation from the target. A peak at 53.4 keV was observed in the γ-ray spectrum which also appeared to be in coincidence with the other delayed transitions in ^{152}Dy. The multipolarities of the transitions have been deduced from intensity ratios with respect to Dy KX- and LX-rays. The results are E2 for the 53.4 keV line and M1 for the 604 keV line feeding the 14^+ state. That established the energy, spin, and parity of the 60 ns isomeric state in ^{152}Dy as $E_x = 5089$ keV, $I^\pi = 17^+$ and thereby the character of the isomer as a yrast trap as defined in Section 6.4.2.

The existence of the 53.4 keV isomeric transition in ^{152}Dy has been confirmed by the Argonne–Chalk River–Strasbourg collaboration who carried out extensive experiments on light Gd and Dy isotopes using ^{12}C, ^{28}Si and $^{32,34}S$ beams (Khoo *et al.*, 1978; Haas *et al.*, 1979, 1981). They applied a four-NaI(Tl)-detector multiplicity filter as well as a sum spectrometer in coincidence with Ge-detectors for timing and selection purposes. The measurements used $E_{\gamma 1}$–$E_{\gamma 2}$-time and E_γ-time coincidences to establish decay schemes using the $^{122}Sn(^{34}S, 4n)^{152}Dy$ reaction at 144 MeV. A recoil-distance (plunger) technique was used to measure the lifetimes, and the angular distributions and linear polarizations of γ rays to establish the spins and parities of levels. The observation of the 53.4 keV line is shown in Fig. 6.46. The prompt Dy KX-rays have been subtracted from the delayed spectrum which allows the 53.4 keV line stand out clearly. The time spectrum is shown on the right-hand side of Fig. 6.46. The measured decay time for the 53.4 keV transition as well as the associated total γ-ray energy measured with the sum spectrometer supported the assignment as being the isomeric transition in ^{152}Dy.

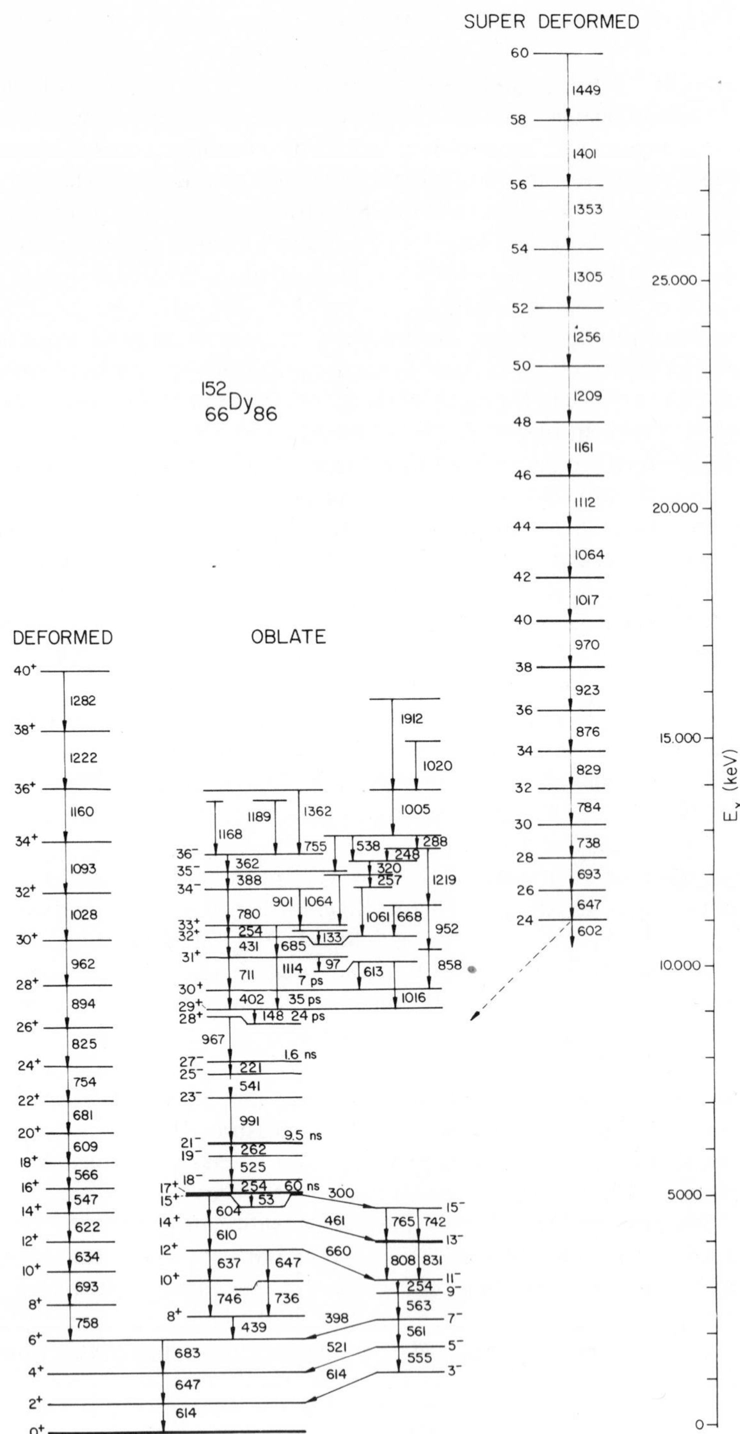
SUPER DEFORMED
$^{152}_{66}Dy_{86}$
DEFORMED
OBLATE
E_x (keV)
25.000
20.000
15.000
10.000
5000
0

Important determinations of the multipolarities of transitions up to the $I = 37$ state have been obtained from the γ-ray linear polarization measurements carried out by Haas *et al.* (1979, 1981) as well as by Merdinger *et al.* (1979). The latter work also included the measurement of the g-factor of the 10 ns yrast trap in ^{152}Dy as discussed below.

The level scheme of ^{152}Dy shown in Fig. 6.43 is one of the nicest examples of the possibilities of the present spectroscopic techniques employed by various groups using light-ion and heavy-ion accelerators. These results enable us to make a detailed comparison with theory as will be presented in Section 6.4.4 below.

6.4.3b *Lifetime measurements in the nano- and picoseconds regions*

A brief description of the possible electronics involved in the measurements of lifetimes is given in Section 3.5.1a. As examples for the determination of lifetimes in the nano- and picoseconds regions we shall give here the results for some Dy isotopes, in particular for ^{152}Dy from the work of Haas *et al.* (1981). The time spectra recorded after fusion reactions with a 145 MeV ^{32}S beam on ^{120,124}Sn targets, with the beam burst as reference, are shown in Fig. 6.47 (Haas *et al.*, 1981). The lifetimes of the yrast traps in ^{148,149,151,152}Dy are determined from the slopes of the curves.

In the picosecond region one most often uses the recoil-distance technique as discussed in Sections 3.5.1b and 6.3.5b. In the case of ^{152}Dy the velocity of the recoiling nuclei was $v/c = 1.7$ per cent. The decay curves for several transitions following the decay of various isomeric states are given in Fig. 6.48 (see Fig. 6.43 for the decay scheme). Short-lived states produce curves with a steep slope as compared to longer-living states, as can be seen by comparing e.g. the curves for the 431 keV line (6.2 ps) and the 254 keV line (10 ns). The best fits to the data are represented by the solid lines, and these obtained by taking into account the observed high-lying isomers and a fixed side-feeding time of 0.5 ps to each discrete state included in the fit. The fitting procedure yielded lifetimes which are given in Fig. 6.43.

This experiment established the level scheme of ^{152}Dy up to $I = 38$ and in addition to the two known nanoseconds isomers another yrast trap was found with $T_{1/2} = 1.6$ ns as well as six other isomers in the picosecond region. The highest-lying yrast trap has spin 37, which is certainly a unique situation. It

---

Fig. 6.43. The level scheme of ^{152}Dy as deduced from the combined evidence of several experiments leading to the oblate structure in the centre (Jansen *et al.*, 1976, 1979; Khoo *et al.*, 1978; Haas *et al.*, 1979, 1981; Merdinger *et al.*, 1979; Nyako *et al.*, 1986). The band-like structure on the left has been observed with an (α, 4n) reaction at 60 MeV bombarding energy up to 18^+ (Styczen *et al.*, 1983), and up to 40^+ with a 205 MeV ^{48}Ca beam (Nyako *et al.*, 1986) as well as the superdeformed band on the right (Twin *et al.*, 1986).

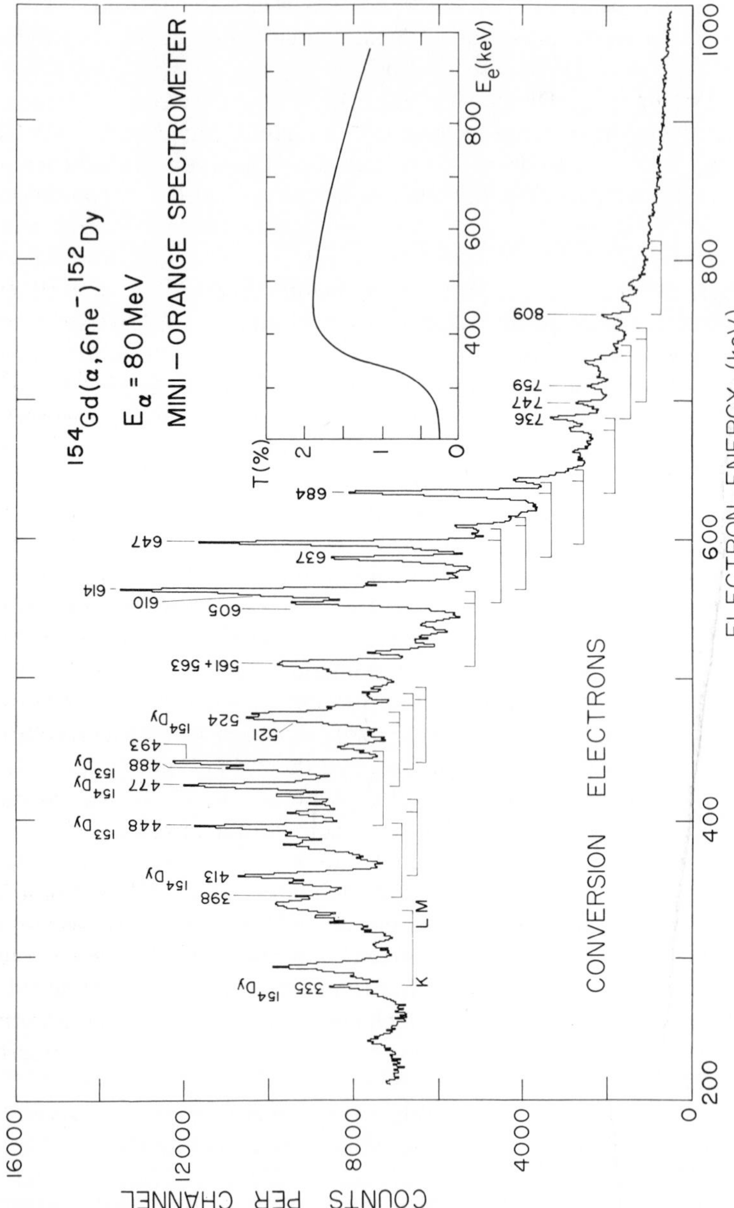

Fig. 6.44. The conversion electron singles spectrum of ^{152}Dy observed with a mini-orange spectrometer. The insert shows the transmission curve measured for the configuration of the small SmCo permanent magnets. (Jansen *et al.*, 1979.)

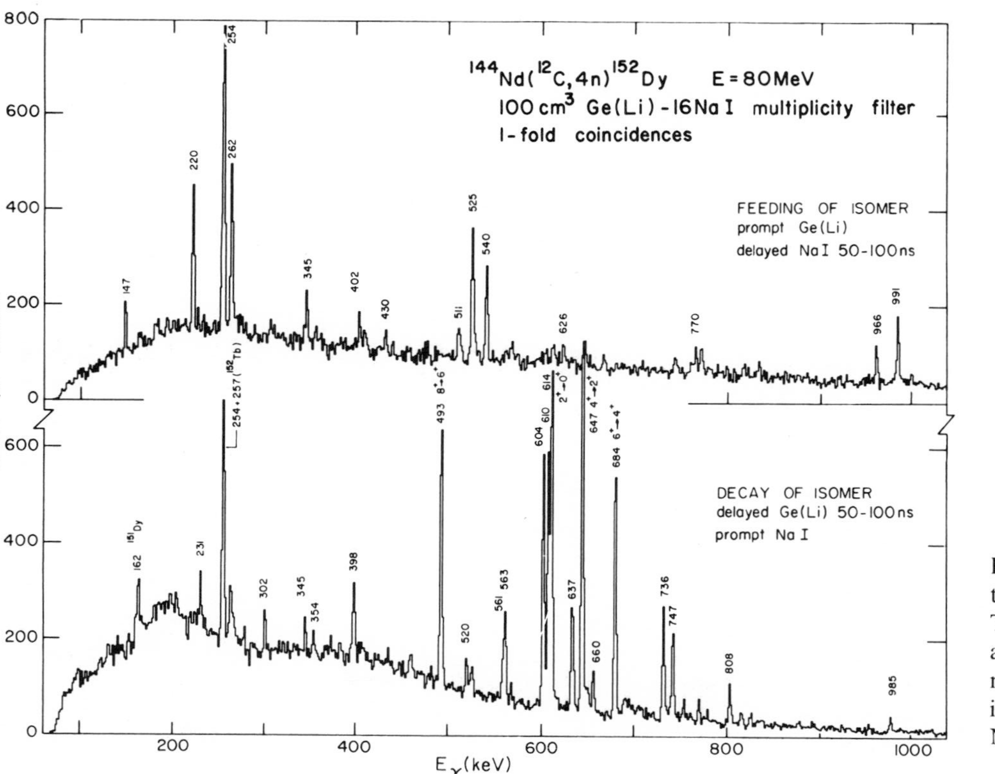

Fig. 6.45. An illustration of time selection of γ rays with a multiplicity filter. The gamma rays of ^{152}Dy feeding (top) and de-exciting (bottom) the 60 ns isomer, as observed with a Ge(Li) detector in coincidence with one out of sixteen NaI(Tl) detectors using a 100 MeV ^{12}C beam. (Jansen *et al.*, 1979.)

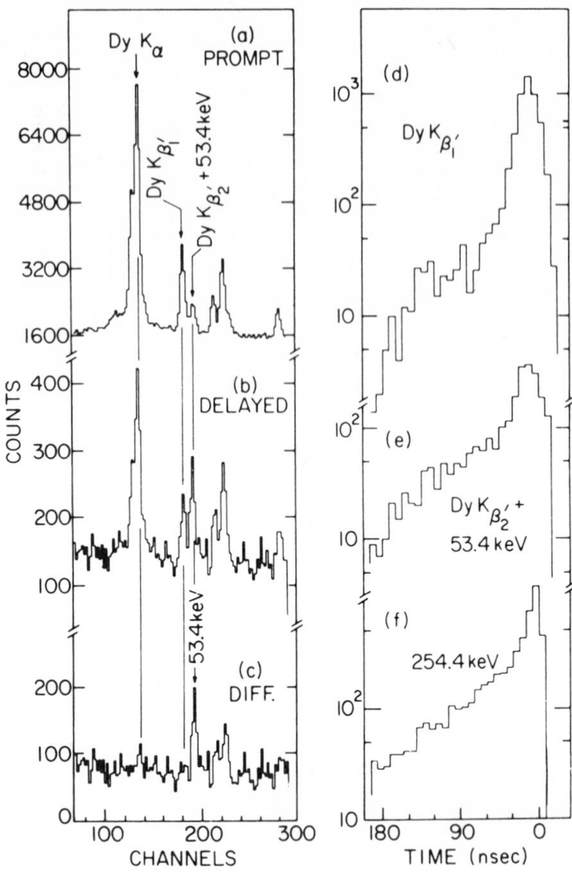

Fig. 6.46. Observation of the 53.4 keV isomeric transition in ^{152}Dy. The spectra of prompt (a) and delayed (b) transitions in ^{152}Dy; the difference, (c), is obtained by subtracting from (b) the fraction of (a) necessary to eliminate K_α X-rays. The time spectra for the Dy $K_{\beta 1'}$, $K_{\beta 2'}$ + 53.4 keV transition, and a 254 keV γ-ray in ^{152}Dy are shown in (d)–(f). The time between an intrinsic Ge and NaI(Tl) sum spectrometer was measured. (Haas *et al.*, 1981.)

indicates that ^{152}Dy has its angular momentum aligned with the nuclear symmetry axis ($\gamma = 60°$), which is characteristic for non-collective rotation at least up to $I = 40$. This is consistent with the E2 transition probabilities deduced from the measured lifetimes (see Section 6.3.5 for the formalism). On the average the E2 strengths for yrast traps amount to $\sim$ 1 Wu, which is about two orders of magnitude less than one would expect for a collective transition (see also Fig. 6.3). The transitions must therefore be due to just a single particle. For even smaller transition strengths, $\ll$ 1 Wu, several single

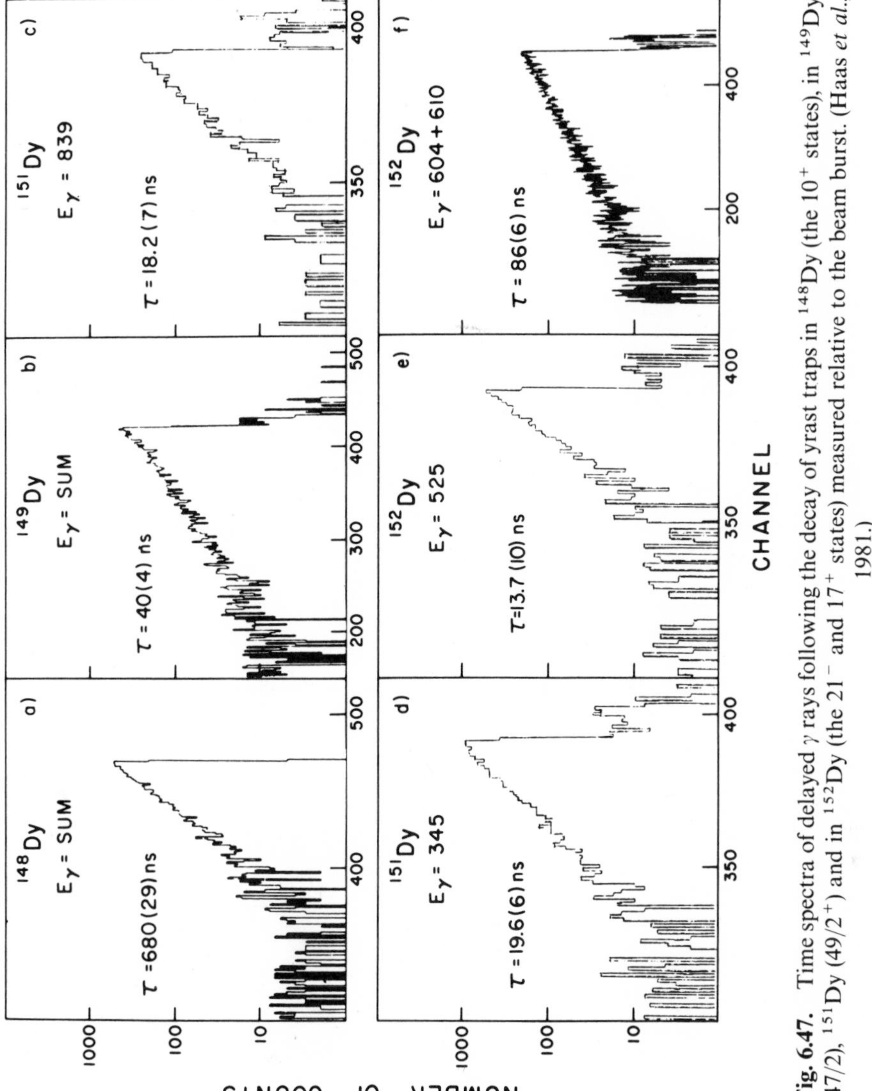

Fig. 6.47. Time spectra of delayed γ rays following the decay of yrast traps in ^{148}Dy (the 10^+ states), in ^{149}Dy (47/2), ^{151}Dy ($49/2^+$) and in ^{152}Dy (the 21^- and 17^+ states) measured relative to the beam burst. (Haas *et al.*, 1981.)

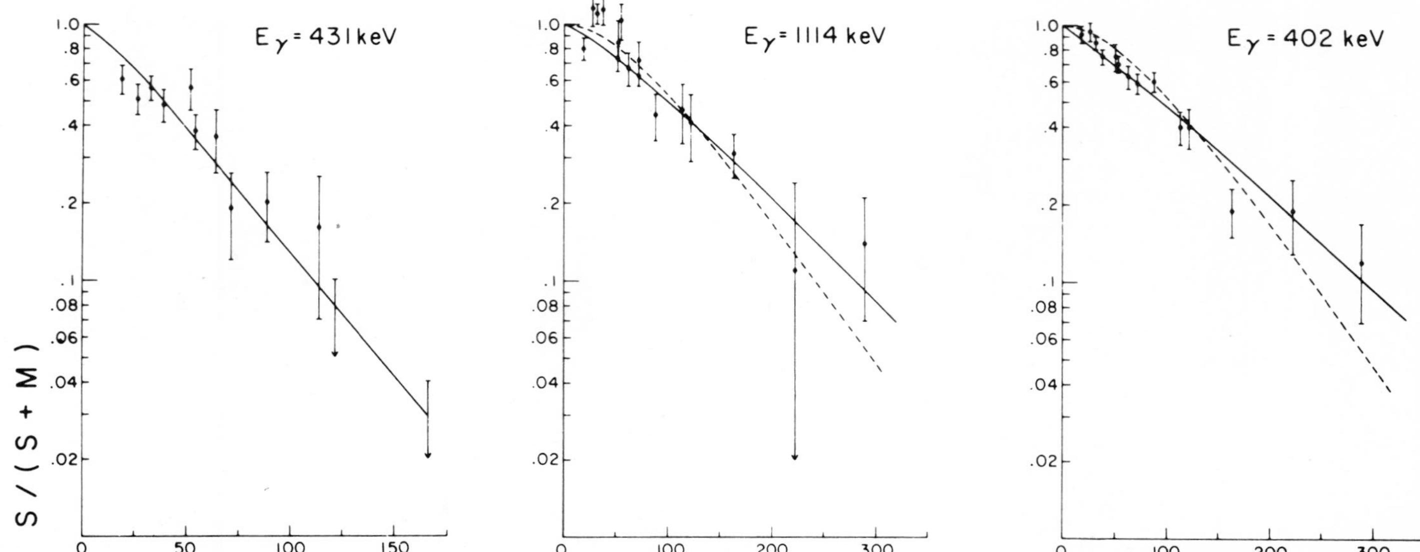

S / (S + M)
Eγ = 431 keV
Eγ = 1114 keV
Eγ = 402 keV
1.0
.8
.6
.4
.2
.1
.08
.06
.04
.02
0
50
100
150
200
300

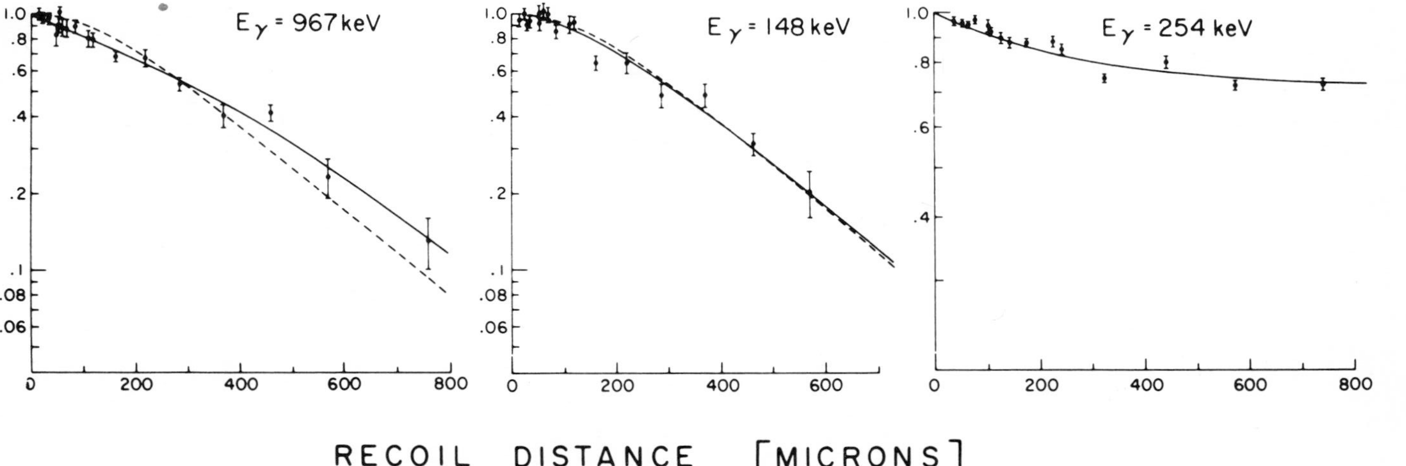

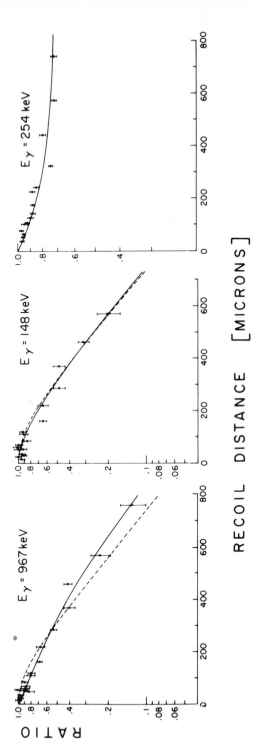

Fig. 6.48. Examples of recoil distance spectra obtained in ^{152}Dy with the ^{122}Sn(^{34}S, 4n) reaction at 150 MeV bombarding energy. S and M refer to the yield in the stopped and moving peaks, respectively. The solid curves correspond to the best fits assuming that the side-feeding time was 0.5 ps and the dashed lines assume that the side-feeding time structure was the same as that entering the states from its observed precursor. The curve for $E_\gamma = 245$ keV resulted from the triplet origin of that γ ray. The decay curve is seen to be constant after ~ 300 μm distance. This is due to the decay of the 253.5 and 254.4 keV components which lie below the 13.7 ns isomer and hence a 13.7 ns feeding line. The initial decay corresponds to the 3rd component of γ 254 (254.2 keV) which depopulates the level at 10.365 MeV (cf. Fig. 6.43). Hass *et al.*, 1981).

particles must be involved or the hindrance must be due to a large differences in spin of the initial and final states (energy trap).

It is thus shown here that measurements of nuclear lifetimes yield important information distinguishing collective rotation from other degrees of freedom. We will show in Section 6.5 that at the highest possible spin values $I\hbar = 40\text{–}60\,\hbar$ evidence has been obtained of a drastic shape change.

6.4.3c *Measurements of magnetic dipole moments*

The method of TDPAD is discussed in Section 3.5.2a1, where it is also argued that knowledge of the magnetic hyperfine fields in the slowing-down material is required. The magnetic hyperfine fields for Gd in three metallic slowing-down materials have been reported by Häusser *et al.* (1979). They also measured g-factors in ^{144,146,147,148}Gd in addition to the relaxation times τ_2, discussed in Section 3.5.2. The high-spin aligned isomeric states have been populated in ^{4}He and ^{28}Si-induced fusion reactions. The recoiling nuclei were stopped in thick ^{144}Sm, ^{120,122,124}Sn targets or they were implanted in Sm or in Pb backings behind a thin target in the case of the ^{28}Si beam. The applied external magnetic fields varied from 3.0 tesla at room temperature to 1.6 Tesla at target temperatures between 293° and 968°K.

The time distributions of the delayed γ rays were recorded by two Ge(Li) detectors placed at $\theta = \pm 135°$, using the RF (beam burst) signal as a time reference. If both the nuclear lifetime and the relaxation time are long enough then one may even observe more than one Larmor precession period. This is shown in Fig. 6.49 for states in ^{144,146}Gd. The ratio function $R(t)$ derived from the γ-ray yields in the two detectors is plotted as a function of time, with $t = 0$ representing the time of population of the isomeric state studied:

$$R(t) = \{N(\theta_1, t) - N(\theta_2, t)\}/\{N(\theta_1, t) + N(\theta_2, t)\}. \qquad (6.4.1)$$

The rotation of the angular distribution pattern can be followed in both cases for more than six turns, which accurately determines the Larmor frequency. The shorter lifetime and relaxation time in the case of ^{144}Gd with respect to ^{147}Gd is clearly shown by the steeper exponential decrease of the intensity with time.

The measured g-factors lead directly to very reasonable suggestions for the particle-hole configurations of the yrast traps. The results for Gd isotopes are summarized in Table 6.4.1. The theoretical g-factors used in the comparison with experiment are based on effective g-factors for the single-particle orbitals involved, as given in the footnote to Table 6.4.1. The four yrast traps in ^{147}Gd also shown in Fig. 6.2 up to the $49/2^+$ state are thus explained in terms of a few single-particle orbitals. It may be noticed in Table 6.4.1 that the yrast traps are based mainly on the alignment of a few (high-spin) single-particle orbitals with approximately equal contributions from protons and neutrons.

Table 6.4.1 The g-factors and quadrupole moments of isomeric states in Gd isotopes[†]

Nucleus	$T_{1/2}$ (ns)	J^{π}	E_{γ} (keV)	$a_2(t=0)$	Suggested[‡] main configuration	g_{main}[§]	g_{exp}[¶]	$Q(I)$[‖]	Theory[‡‡] g	$Q(I)$	β_2
^{144}Gd	130 ± 10	10^+	328	-0.60	$(\pi d_{5/2}^{-2} h_{11/2}^{2})_{10}(\nu h_{11/2}^{-2})_0$		1.276 ± 0.014				
^{146}Gd	6.7 ± 0.1	7^-	324	$+0.24$	$(\pi d_{5/2}^{-1} h_{11/2})_7$	1.39	1.283 ± 0.027				
		7^-			$(\pi g_{7/2}^{-1} h_{11/2})_7$	1.14					
	4.1 ± 0.2	(19^+)	865	-0.19	$(\pi d_{5/2}^{-2} h_{11/2}^{2})_{10}(\nu h_{11/2}^{-1} f_{7/2})_9$	0.58	0.63 ± 0.09				
^{147}Gd	22.2 ± 1.5	$13/2^+$	997	0.45	$(\nu i_{13/2}) + (3^- \times \nu f_{7/2})_{13/2}$		-0.037 ± 0.011	-0.73 ± 0.07	-0.18	-0.44	-0.03
	5 ± 1	$21/2^+$	1491	0.24	$(\pi d_{5/2}^{-1} h_{11/2})_7(\nu f_{7/2})$	0.75	0.72 ± 0.11				
	26.8 ± 0.7	$27/2^-$	183	-0.10	$(\pi d_{5/2}^{-2} h_{11/2}^{2})_{10}(\nu f_{7/2})$	0.88	0.840 ± 0.017	-1.26 ± 0.08	0.81	-1.25	-0.06
	550 ± 30	$49/2^+$	254	0.17	$(\pi d_{5/2}^{-2} h_{11/2}^{2})_{10}(\nu h_{9/2} i_{13/2} f_{7/2})_{29/2}$	0.47	0.446 ± 0.008	-3.14 ± 0.17	0.47	-2.99	-0.17
			339	-0.15							
^{148}Gd	16.5 ± 0.3	9^-	785	0.26	$(\nu i_{13/2} f_{7/2})_9$	-0.12	-0.028 ± 0.009				

[†] The g-factors and $|Q(I)|$-values are from Häusser *et al.* (1979, 1980).
[‡] Relative to a closed ^{146}Gd core ($N = 82$, $Z = 64$). The configurations are from Häusser *et al.* (1979) except that for the $49/2^+$ state which is from Dudek (1980), see Fig. 6.59.
[§] The following basic g-factors were used: for protons $d_{5/2}^{-1}$, 1.66; $h_{11/2}$, 1.30; for neutrons $h_{11/2}^{-1}$, -0.18; $f_{7/2}$, -0.30; $i_{13/2}$, -0.04.
[¶] Paramagnetic correction factors have been applied as explained in the text.
[‖] Absolute values from Häusser *et al.* (1979), signs from Dafni *et al.* (1984, 1985).
[††] Microscopic calculation using deformed orbitals by Dössing *et al.* (1980, 1981).

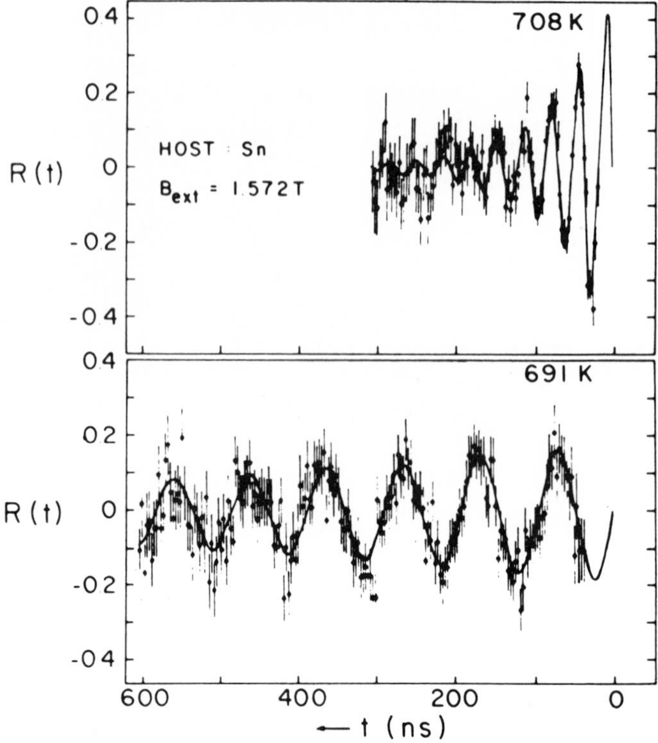

Fig. 6.49. Larmor precession patterns $R(t)$ for M1–E2 transitions ($E_\gamma = 231, 328$ and 545 keV) de-exciting the 130 ns, $J^\pi = 10^+$ isomer in ^{144}Gd (upper panel). The lower panel shows $R(t)$ for the 254 keV γ ray, one of about fifteen de-excitation γ rays exhibiting the effect of the half-life and g-factor of the 550 ns yrast trap in ^{147}Gd. (Häusser *et al.*, 1979.)

Let us now return to ^{152}Dy and discuss the measurement of the g-factor for the 6.13 MeV, 21^-, $T_{1/2} = 9.5$ ns yrast trap carried out by Merdinger *et al.* (1979). They used the reaction ^{140}Ce(^{16}O, 4n)^{152}Dy at 88 MeV and applied the same technique as that described above. The excited recoil ions were stopped in a lead foil behind the 1 mg/cm^2 thick target. Lead at a temperature of 260° C was chosen as an appropriate slowing-down material with a cubic structure and a known paramagnetic correction factor $\beta = 0.27$ (see Section 3.5.2). The applied external field of 0.9 Tesla was perpendicular to the detection plane and was reversed every minute. The measured $R(t)$ function is plotted in Fig. 6.50 and shows a clear oscillation, which determines the Larmor frequency as $\omega_L = (8.88 \pm 0.53)\ 10^{-7}$ rad s^{-1}. The deduced g-factor $g = +0.55 \pm 0.06$ indicates a mixed proton (π)-neutron (ν) character, presumably $\pi h^2_{11/2}\nu(i_{13/2}h_{9/2})$ as discussed in Section 6.4.4 below.

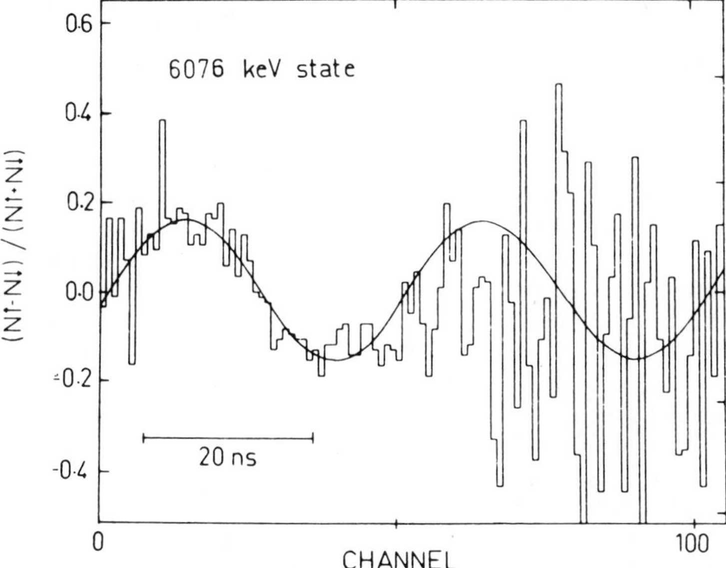

Fig. 6.50. Experimental TDPAD data combined for the 262 and 525 keV γ rays in the decay of the 21^- yrast trap of ^{152}Dy. The opposite signs of the α_2 angular distribution coefficients of the two γ rays were taken into account. The solid lines represent a least-squares fit yielding the Larmor frequency. (Merdinger *et al.*, 1979.)

We have shown here that g-factor measurements contribute greatly to our knowledge of the single-particle configurations. The question of possible increased deformation at high spin has to be answered by measuring the quadrupole moments of the excited states.

6.4.3d *Measurements of electric quadrupole moments*

It was mentioned above in Section 6.4.2 that the locations of the nuclear regions with spherical, oblate and prolate shapes at low spin have been established by many experiments, most notably by measurements of quadrupole moments.

Quadrupole moments for yrast traps in ^{147}Gd up to the $49/2^+$ isomer have been measured using the TDPAD in the ^{124}Sn(^{28}Si, 5n)^{147}Gd reaction at 144 MeV (Häusser *et al.*, 1980). The ^{147}Gd nuclei produced in a 2 mg/cm^2 target recoiled into the backing that consisted of a 1 mm thick single Gd crystal. The target and backing assembly was heated to temperatures between 324° and 469°K, well above the Curie temperature for Gd (293°K) to ensure paramagnetic conditions. The quadrupole modulation pattern expressed by the $R(t)$ function—see eqn (6.4.1) was measured with two Ge(Li) detectors placed at $\theta_1 = 0°$ and $\theta_2 = 90°$ with respect to the beam direction.

We shall now give the results of a few experimental investigations of the nucleus ^{147}Gd. A detailed level scheme up to the $49/2^+$ isomeric state is given in Fig. 6.51. The measured $R(t)$ functions are displayed in Fig. 6.52 for two transitions de-exciting the 550 ns, $49/2^+$ isomer in ^{147}Gd. The difference between the two upper curves reflects the opposite signs of the corresponding a_2 angular distribution coefficients (see Table 6.4.1) for the two transitions. Note that the 254 keV line corresponds to a stretched E2 transition and the 339 keV line to a dipole one (see Fig. 6.51). The pronounced interference pattern in the first time interval of $t = 0$–200 ns is due to the specific orientation of the single Gd crystal with its symmetry axis ($\hat{c}$) at 45° to the beam axis, which appears to produce a more distinct modulation pattern than a polycrystalline host material.

The single crystal also makes it possible to switch off the quadrupole interaction QI by turning the $\hat{c}$-axis along the beam axis. This was done in the case of the $R(t)$ function, given in the bottom part of Fig. 6.52, showing that the interference pattern is indeed washed out. The quadrupole modulation for the 27 ns, $27/2^-$ isomer in ^{147}Gd is shown in Fig. 6.53. It shows that the relative short lifetime for this isomer restricts the observation of the interference pattern to a shorter period than is the case for the transitions in Fig. 6.52.

The solid lines through the data points in Figs 6.52 and 6.53 show theoretical fits using the product of the electric-field gradient (EFG) and $Q(I)$ as a free parameter. The value of $Q(I)$ can thus be obtained if one knows the magnitude and temperature dependence of the EFG.

In the case of ^{144,147}Gd both the sign and magnitude of quadrupole moments have been measured by Dafni *et al.* (1984, 1985) employing the tilted multifoil technique as discussed in Section 3.5.2. In the experimental arrangement (see Fig. 3.76) four Ge(Li) detectors were used and the tilt (polarization) direction was altered periodically between $+90°$ and $-90°$ with respect to the beam direction. From the measured yields in the four detectors and the two tilt directions two times four combinations of yields can be added to construct a $R(t)$ function similar to eqn (6.4.1). The result is shown in Fig. 6.54 for the same transitions as in Fig. 6.52 corresponding to the decay of the $49/2^+$ yrast trap in ^{147}Gd. The solid line represents a theoretical fit which yields the magnitude and the sign of $Q(I)$ using the known (positive) EFG for Gd in a Gd matrix. That measurement confirmed the values of $Q(I)$ quoted in Table 6.4.1 and established the negative signs for the quadrupole moments of the ^{147}Gd yrast traps. This in turn fixes the sign of the β_2 deformation parameter as negative and thereby the shape as oblate for all three yrast traps in ^{147}Gd.

The shell-model values for $Q(I)$ in Table 6.4.1 have been calculated with typical effective charges $e_p(\text{eff}) = 0.5$ for the protons and $e_n(\text{eff}) = 1.0$ for the neutrons, yielding total charges of $e_p = 1.5$ and $e_n = 1.0$. Even with these

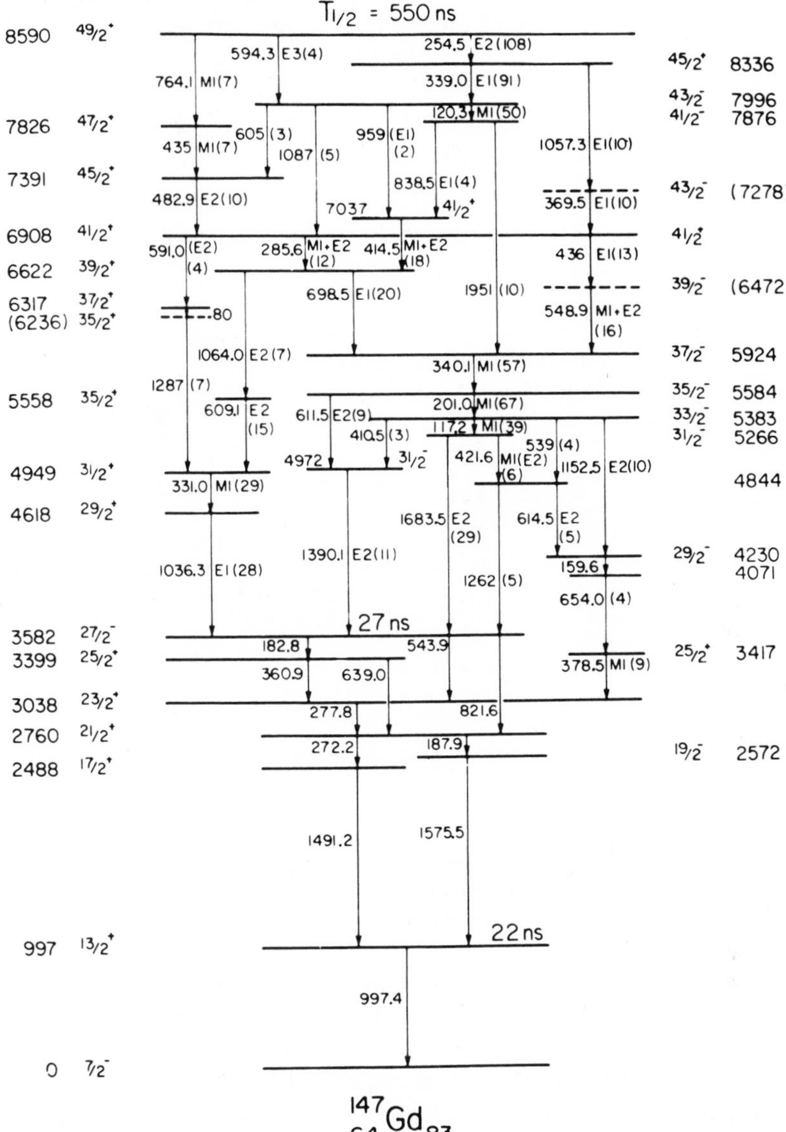

Fig. 6.51. The level scheme of ^{147}Gd as deduced from pulsed beam γ-ray and electron spectroscopy, employing the reaction ^{135}Ba(^{16}O, 4nγ) at bombarding energies of 77–84 MeV and the reaction ^{124}Sn(^{28}Si, 5nγ) at 128–140 MeV. The transitions are indicated by their energies, and above the 27 ns isomer also by the multipolarities and intensities in brackets, including conversion. (Baklander *et al.*, 1982.)

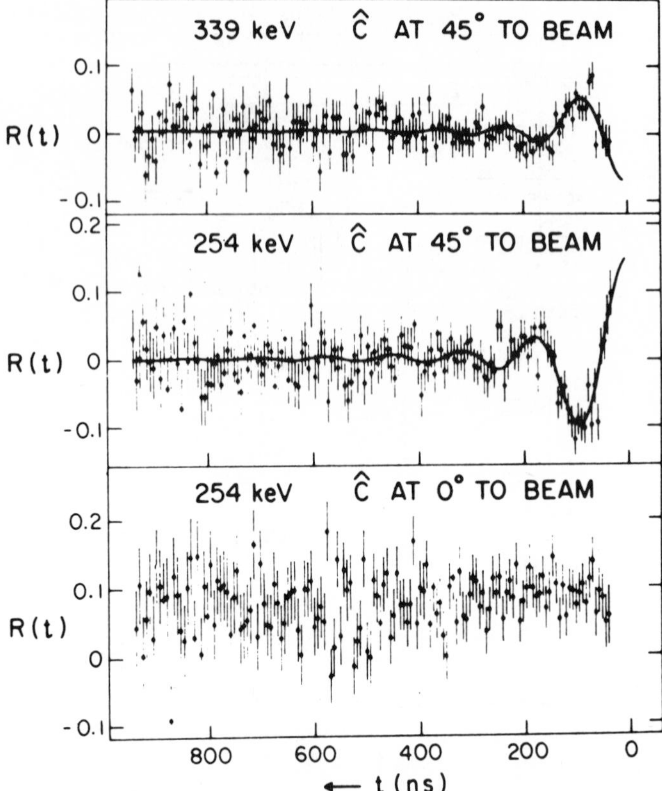

Fig. 6.52. The modulation patterns resulting from quadrupole interaction of the 550 ns isomer in ^{147}Gd in a single crystal of Gd at 413° K. The orientation of the single hexagonal Gd host crystal is indicated by the orientation of the $\hat{c}$-axis with respect to the beam. (Häusser *et al.*, 1980.)

charges the estimated quadrupole moments were much smaller than the experimental values for the $13/2^+$ and $27/2^-$ states. Unrealistic effective charges of 2.0 and 2.9, respectively, would be required to fit these data and values > 3.1 to fit the quadrupole moment of the $49/2^+$ state.

The β_2 values in Table 6.4.1 have been inferred from the experimental $Q(I)$ values. Theoretical calculations also produce negative β_2 values, very close to the experimental data (Døssing *et al.*, 1980, 1981). Those calculations indicate that the rather sudden increase in (oblate) deformation at the $49/2^+$ yrast trap in ^{147}Gd is due to the promotion of a neutron pair across the $N = 82$ gap. The quadrupole moments have also been explained with bare-nucleon charges by assuming that the quasi-particles producing the high spin move outside a deformed core (Häusser *et al.*, 1980).

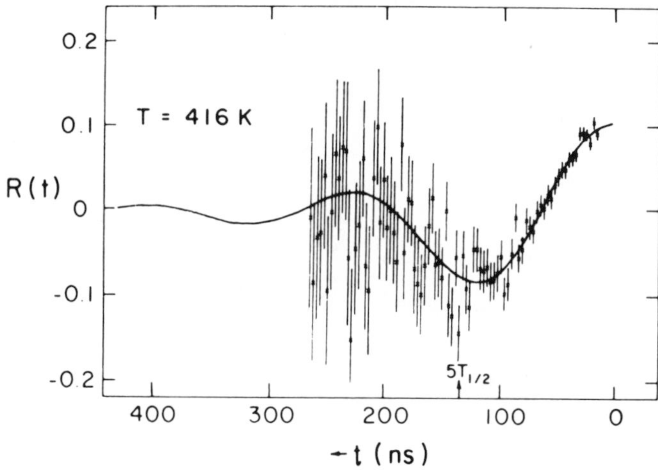

Fig. 6.53. The quadrupole modulation of the 27 ns, $J^\pi = 27/2^-$ isomer in ^{147}Gd observed via the $21/2^+ \to 17/2^+$, 272 keV E2 transition. (Häusser *et al.*, 1980.)

It is thus shown here that refined experimental techniques reveal the detailed properties of high-spin yrast traps. Those properties serve as a crucial testing ground for nuclear models and for our understanding of the underlying nuclear structure of high-spin excitations. We shall therefore conclude this section with a brief account of the current theoretical descriptions of yrast traps and of some of their properties.

6.4.4 *Identification of the single-particle configurations of yrast traps*

In the previous sections we have presented an extensive set of experimental data on the properties of yrast traps in ^{147}Gd and ^{152}Dy, as examples of contemporary techniques. In this section we will explore simple theoretical considerations to reveal the single-particle structure of those yrast traps. Again our aim will be to provide examples which may be helpful in quite different investigations and also to illustrate the current understanding of noncollective high-spin phenomena. For detailed theoretical treatments of specific nuclear properties the reader is referred to textbooks on nuclear theory, for example, Bohr and Mottelson (1969, 1975) and Szymański (1983) and to references quoted therein.

Two basically different schemes have been employed, namely the independent single-particle motion in a deformed potential and in a spherical one. They are commonly referred to as a deformed (orbital) and a spherical shell model. It will become clear below that the two different descriptions lead to similar results. That agreement is not so surprising because the deformation

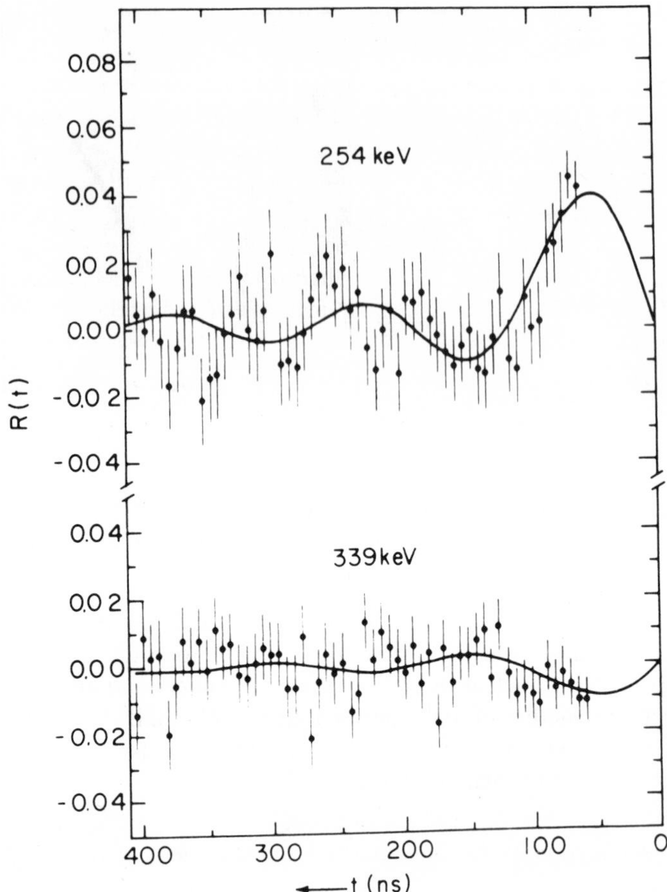

Fig. 6.54. Experimental ratio functions and fitted curves for the 254 and 339 keV transitions from the decay of the ^{147}Gd ($\frac{49}{2}^+$) isomer. The opposite sign and different magnitudes of the effects for the two γ rays are due to different α_2 coefficients. (Dafni *et al.*, 1984, 1985.)

arises from the residual interactions between the particles, and therefore the deformed field represents the effect of those interactions used in the spherical shell model. In other words the quadrupole type (Y_2) of polarization induced by valence particles in the spherical potential can be effectively represented by a simple deformation. For the description of states with very high spin, however, deformed potentials may be preferred because of the presence of (oblate) deformation as has been inferred from quadrupole moments, as discussed above.

In the work of Døssing *et al.* (1980, 1981) the region around ^{146}Gd has been described in terms of the deformed shell-model. Empirical single-particle energies have been used, including the pairing force in the BCS approximation with variation of the pairing gap parameters after particle-number projection and Strutinsky renormalization (see e.g. Szymański, 1983, for details).

A crucial procedure in all calculations involves the proper choice of the active single-particle levels and the corresponding energies relative to a certain core. The relevant levels for light rare-earth nuclei are shown in Fig. 6.55, which is mainly based on experimental data. The relative energies of the five proton levels resulted from a model fit to the experimental values for ^{146}Gd ($Z=64$, $N=82$). The centroid of the five levels was fixed at the calculated position given by a Woods–Saxon potential. The relative energies of the lowest three neutron levels have been constructed on the basis of data

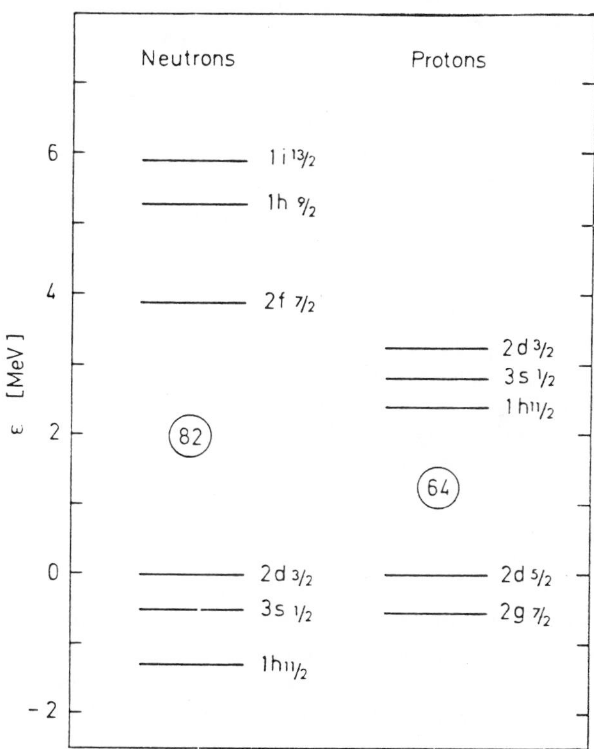

Fig. 6.55. The spherical single-particle spectrum around the $N = 82$ and $Z = 64$ closed shells, obtained in a phenomenological way. The relative single particle energy ε is taken with arbitrary zero point. (Døssing *et al.*, 1981.)

for neutron pick-up reactions from even $N = 82$ targets. The upper three neutron levels follow from the stripping data with some small adjustments. The distance between the two sets of neutron levels, forming the $N = 82$ gap, was determined as 3.7 MeV from the difference in neutron separation energies in ^{147}Gd and ^{146}Gd. This procedure of Døssing *et al.* (1981) shows how the single-particle level scheme can be constructed on the basis of experimental data and reasonable theoretical notions.

Next we have to consider the $2j+1$ substate of a single j shell in an oblate nuclear potential. The projection of j on the symmetry axis has the values $m = -j, -j+1, \ldots +j$. The total angular momentum generated by N particles in the direction of the symmetry axis is given by

$$I = \left|\sum_{\nu}^{N} m_{\nu}\right|. \tag{6.4.2}$$

The single-particle energies e_{ν} depend on m_{ν}, and for a deformed quadrupole field with strength parameter k they are given by:

$$e_{\nu}(m_{\nu}) = -k\{3m_{\nu}^{2} - j(j+1)\}/\{j(j+1)\}^{1/2}. \tag{6.4.3}$$

The single-particle energies are represented by points on inverted parabolas in the $e_{\nu}(m_{\nu})$ plane (see Fig. 6.56). In the case of oblate deformation (k positive) the levels with highest m come lowest in energy, as can be seen from eqn (6.4.3). This is due to the large overlap of the oblate potential with the single-particle orbitals, with large spin projection (m_{ν}) along the symmetery axis. The attractive nuclear interaction makes the corresponding energies low.

One has to search for the minimum energy in occupying the different m_{ν} states in order to generate states with a certain angular momentum I for a particular nucleus. A useful procedure is provided by the tilted Fermi surface method. We will briefly outline the method here, again using ^{147}Gd and ^{152}Dy as examples.

The total single-particle energy in the rotating frame with respect to the Fermi energy λN (see Section 6.3.3 and Fig. 6.16) is given (when pairing is neglected) by

$$E_{\nu} - \lambda N = \sum_{\nu}(e_{\nu}^{\omega} - \lambda) = \sum_{\nu}(e_{\lambda} - \lambda - \hbar\omega m_{\nu}). \tag{6.4.4}$$

The last equality follows directly from the transformation of the single-particle energies from the laboratory to the rotating frame of reference—see eqn (6.3.47). In the search for yrast states with spin I one has to minimize the energy given by eqn (6.4.4), while satisfying the constraint expressed by eqn (6.4.2) and at the same time preserving the particle number. In this minimization procedure the quantities λ and $\hbar\omega$ play the role of Lagrange multipliers. The minimization can be accomplished graphically as illustrated in Fig. 6.56 for ^{152}Dy.

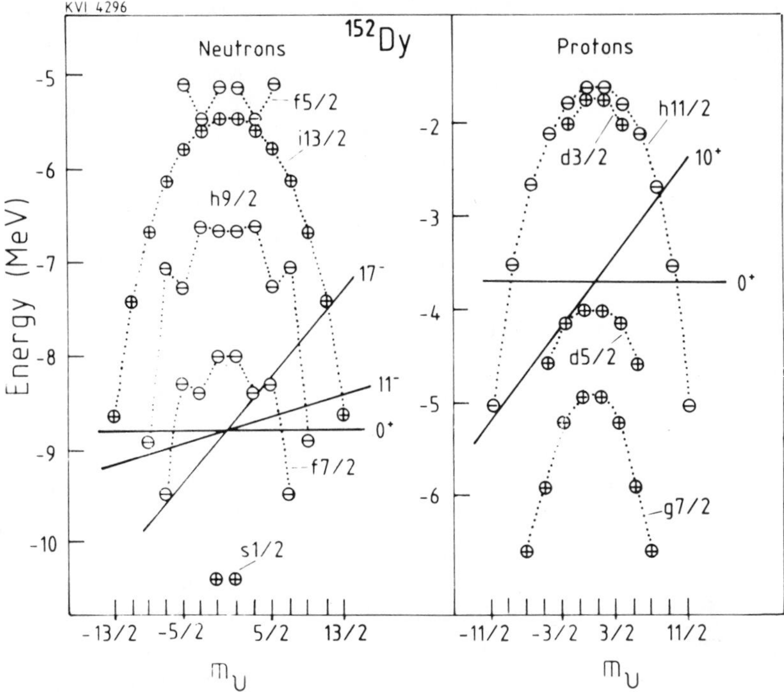

Fig. 6.56. Single-particle orbital energies versus angular momentum component m_ν plotted for $A = 152$ with deformation parameters $\beta_2 = -0.16$ and $\beta_4 = 0.03$. The tilted Fermi surfaces correspond to different optimal configurations. To demonstrate the neutron spin of 17^- the $\nu(f_{7/2})_{-5/2\ \ldots\ +5/2}$ states have been shifted downward slightly and for the 10^+ proton spin also the $\nu(d_{5/2})$ states with respect to the calculation of Cerkaski *et al.* (1979.)

The yrast states with spin $I = I(\omega)$ which satisfy the constraint expressed by eqn (6.4.2) are optimal states and are fitted in the $e_\nu(m_\nu)$ plane by a straight line

$$e_\nu = \lambda + \hbar\omega m_\nu. \tag{6.4.5}$$

This relation thus represents the least possible energy for that given spin I according to eqn (6.4.4). The straight line represents the tilted Fermi surface with the value of λ determined at $m = 0$ and a (positive) slope $\hbar\omega$ determined by the (spin) rotation. The angular momentum generated along the rotation (and symmetry) axis is given by eqn (6.4.2) by adding all m_ν values *below* the tilted Fermi surface, also taking into account the sign of m_ν. Note that orbitals located entirely below the Fermi surface contribute zero angular momentum.

Let us try to construct the single-particle configuration of some yrast traps in ^{152}Dy. With $Z = 66$ and $N = 86$ at least two protons must occupy the levels above the $Z = 64$ gap and four neutrons those above the $N = 82$ gap (see Fig. 6.55). The (horizontal) line in Fig. 6.56 for the 0^+ ground state determines the value of λ. A favourable configuration for the two protons is $(h_{11/2})^2$ coupled to 0^+ (anti-stretched) or 10^+ (stretched coupling of spins). In the case of 10^+ the substates $m_\nu = 11/2$ and $9/2$ are occupied and other $h_{11/2}$ levels are empty. In the case of 0^+ only the $m_\nu = \pm 11/2$ levels are occupied.

For the neutrons three lines are drawn to illustrate how the four neutrons can occupy different levels to couple to 0^+, 11^- or 17^-. For instance, the 17^- state is formed by two neutrons in the $f_{7/2}$ orbital with $m_\nu = 5/2, 7/2$, one in the $h_{9/2}$, $m_\nu = 9/2$, and one in the $i_{13/2}$, $m_\nu = 13/2$ level. This simple coupling scheme thus results in configurations for the ^{152}Dy yrast traps as summarized in Table 6.4.2.

In the theoretical works of Cerkaski *et al.* (1979) and of Døssing *et al.* (1981) both the 21^- and 27^- yrast traps have been predicted correctly. They also predicted the occurrence of yrast traps with spins of 31^+ and 33^- with specified configurations. On the other hand the spin and/or parity of the 17^+ trap has not been reproduced by any of the calculations so far (see Table 6.4.2).

The configurations represented by a straight line as tilted Fermi surface have minimum energy according to eqn (6.4.4) and are called 'optimal states'. Only one optimal state with a certain spin and parity can be found in this way, which then forms an yrast trap.

It should be realized that many states cannot be bordered by a straight line in the e_ν vs. m_ν plot and these are particle–hole states with respect to the nearest optimal configurations, and are called 'non-optimal' states. It appears that the tilted Fermi surface method does not always provide a solution for a certain spin, because of its discrete character.

Table 6.4.2 Properties of some yrast traps in ^{152}Dy

E_x (MeV)	I	$T_{1/2}$ (ns)	g^{exp}	g^{theor}	Configuration
3.16	11^-	4.3			$(h_{11/2})^2_{0^+}(h_{9/2})_{9/2^-}(i_{13/2})_{13/2^+}$
5.09	17^+	60		0.44†	$(h_{11/2})^2_{0^+}[(f_{7/2})^2{}_{5/2^-7/2^-=6^+}$ $(h_{9/2})_{9/2^-}(i_{13/2})_{13/2^+}]_{17^-}$
6.13	21^-	9.5	0.55‡	0.66†	$(h_{11/2})^2_{10^+}(h_{9/2})_{9/2^-}(i_{13/2})_{13/2^+}$
7.88	27^-	1.6		0.39†	$(h_{11/2})^2_{10^+}(f_{7/2})^2_{5/2^-7/2^-=6^+}$ $(h_{9/2})_{9/2^-}(i_{13/2})_{13/2^+}$

† Calculations from Cerkaski *et al.* (1979). The experimentally-determined spin and parity 17^+ of the 60 ns isomer was calculated as 16^+ and the given g-factor thus corresponds to a different configuration than displayed above, which is taken from Døssing *et al.* (1981) leading to a 17^- state.

‡ Measurement of Merdinger *et al.* (1979).

The energies of yrast states in ^{152}Dy calculated by Cerkaski *et al.* (1979) are shown in Fig. 6.57. They exhibit the characteristic irregular dependence on angular momentum, forming yrast traps at spin values where they have been found experimentally, or very near to it. On the other hand experiment indicates their character as structure traps rather than their behaviour as energy traps (or spin traps) indicated by the theory (see Section 6.4.2 for the definition of those traps). The theory also deviates significantly from experiment in giving a less steep slope of the yrast line, shown in a E_x vs. I plot. This may be attributed to the neglect of the pairing interaction, because pairing tends to reduce the moment of inertia and thereby to increase the slope of the yrast line. Much improvement has been achieved by including the pairing interaction, but probably more importantly also by improving the parametrization of the single-particle energies (Døssing *et al.*, 1981). The two different results obtained with and without pairing and with different single-particle energies are compared in Fig. 6.58.

The comparison of theoretical calculations with experimental data thus gives a detailed insight into the structure of the high-spin states in ^{152}Dy. The

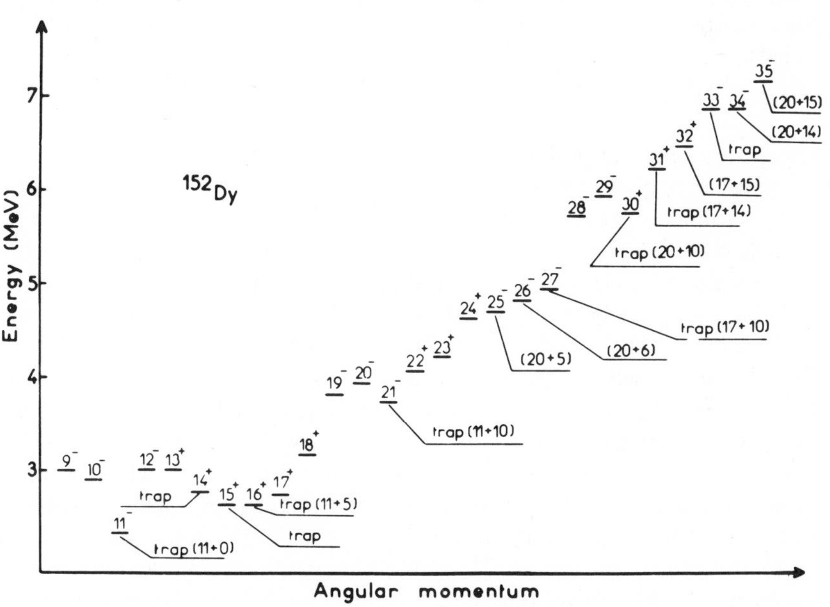

Fig. 6.57. Yrast energy versus angular momentum plot for the nucleus ^{152}Dy. Each level is labelled by angular momentum and parity I^{π}. All the double optimal states are marked by $I_n + I_p$, where I_n and I_p are neutron and proton contributions to the total spin, respectively. The predicted yrast traps are indicated. Location of the origin of the energy scale is irrelevant. (Cerkaski *et al.*, 1977.)

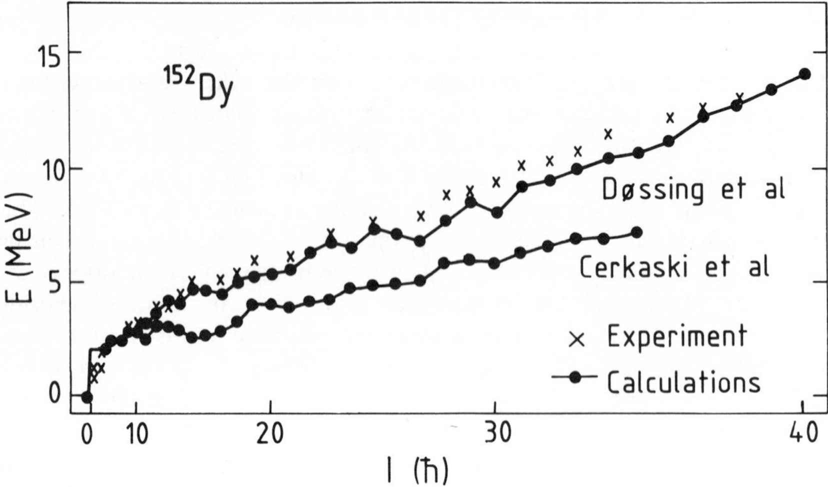

Fig. 6.58. Comparison of two calculated yrast lines with experimental data. The importance of a particular choice of single-particle energies and of the inclusion of pairing correlations (Døssing *et al.*, 1981) is seen with respect to results which did not include pairing (Cerkaski *et al.*, 1979). Both calculations, however, do produce yrast traps of approximately the correct spin values (see also Fig. 6.57.)

yrast traps are particularly helpful in the sense that they facilitate the measurements of electromagnetic moments. The method of the tilted Fermi surface seems to provide a transparent way of finding the important single-particle configurations without performing too complicated calculations.

The single-particle configurations for yrast traps in Gd isotopes presented in Table 6.4.1 have been suggested by Häusser *et al.* (1979, 1980) on the basis of the spherical shell-model in an attempt to explain the measured g-factors. Calculations based on the deformed orbitals are also available for some Gd isotopes (Døssing *et al.* (1980) and Dudek (1980)). The $21/2^+$, $27/2^-$ and $49/2^+$ yrast traps in ^{147}Gd have been analysed with the tilted-Fermi-surface method and the results are shown in Fig. 6.59. The occupied proton and neutron substates m_ν below the straight lines appear to coincide with the shell-model configurations given in Table 6.4.1, except for a small difference for the $49/2^+$ state. The configuration assignments to states in ^{147}Gd above the $49/2^+$ isomer have been suggested by an experimental investigation of Sletten *et al.* (1984). Their assignments are based on deformed orbital model calculations of Døssing *et al.* (1981) and of Neergaard *et al.* (1981). It appears possible to generate fully-aligned spins in terms of particles and holes up to the highest observed level of $I^\pi = (79/2^-)$ at $E_x = 16.983$ MeV (see Fig. 6.2). A suggested configuration for that state,

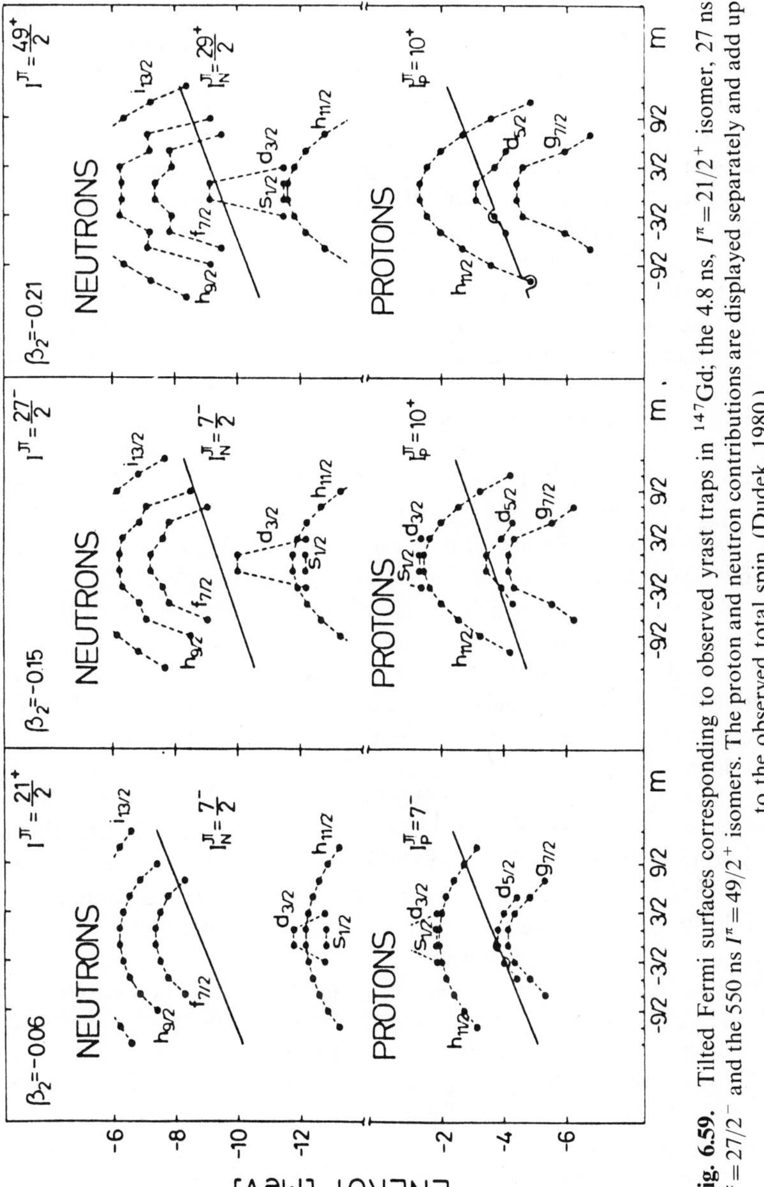

Fig. 6.59. Tilted Fermi surfaces corresponding to observed yrast traps in ^{147}Gd; the 4.8 ns, $I^\pi = 21/2^+$ isomer, 27 ns $I^\pi = 27/2^-$ and the 550 ns $I^\pi = 49/2^+$ isomers. The proton and neutron contributions are displayed separately and add up to the observed total spin. (Dudek, 1980.)

$(d_{5/2})^{-3}(h_{11/2})^3(d_{3/2})^{-1}(f_{7/2})(h_{9/2})(i_{13/2})^2$, illustrates the possibility of generating very high angular momenta by the alignment of single-particle orbitals.

A systematic search for spherical shell-model states in the light Dy isotopes has been carried out by Horn *et al.* (1983). They have compared high-spin states in $^{148-152}$Dy assuming the same proton core for each of the isotopes and adding valence neutrons to the closed $N=82$ shell. The results are given in Fig. 6.60. The $\pi(h_{11/2})^2 10^+$ configuration is taken as a basis, with subsequent addition of 0, 1, 2, 3 and 4 neutrons for the different Dy isotopes. The aligned neutron configurations thus become $(f_{7/2})_{7/2^-}$, $(f_{7/2}h_{9/2})_{8^+}$, $(f_{7/2}h_{9/2}i_{13/2})_{29/2^+}$ and $[(f_{7/2})^2h_{9/2}i_{13/2}]_{17^-}$, and these configurations lead to the 10^+, $27/2^-$, 18^+, $49/2^+$ and 27^- states in ^{148,149,150,151,152}Dy, respectively. They have all been found experimentally to be yrast traps, except that in ^{150}Dy which is not yet observed. The fully-aligned particle configurations for a certain spin are lowest in energy for that spin, thus being the optimal states or yrast traps.

With further increase of spin the aligned neutron configurations will presumably stay the same, but the $Z=64$ proton core will experience a particle–hole excitation. The extra $\pi(d_{5/2}^{-1}h_{11/2})$ configuration will contribute to the spin by an amount ranging from 1^- to 7^-. Note that all states above the basis state (which is the lowest in energy in Fig. 6.60) have the same parity, which is opposite to that of the basis state because negative parity is added to the basis state. The levels which do not follow this systematic behaviour are in this figure indicated by short lines not connected with other levels.

Shell-model calculations have been carried out by the same group. They limited the configuration space to a certain maximum occupation of the same single-particle levels as shown in Fig. 6.55. The nucleus $^{146}Gd_{82}$ has again been taken as a closed core and empirical single-particle energies have been used following Kleinheinz (1979). An effective residual interaction was used with adjustment of some particle–hole interactions. The results are shown in Fig. 6.60 (bottom part). The quantitative agreement between experimental and theoretical level energies is on the average within 150 keV. In this way we obtain a simple and systematic explanation of many high-spin states in the neutron-deficient rare-earth nuclei.

The possibility of having a less obvious non-collective excitation in a nucleus like ^{152}Dy is found in the work of Styczen *et al.* (1983). They studied the ^{152}Gd(α, 4n)^{152}Dy reaction at 60 MeV by means of four-parameter $E\gamma$–$E\gamma$-coincidence techniques and other conventional methods. A band-like structure was found on top of the known 8^+ state as indicated in the level scheme, given in Fig. 6.43. This structure has not been observed before with HI-induced reactions or with higher-energy alpha-particle beams. Although a band is suggested ranging from the ground state up to 18^+, the deviations from a smooth increase of energy with spin are still too large for this to be associated with pure collective motion. On the other hand the regularity of the

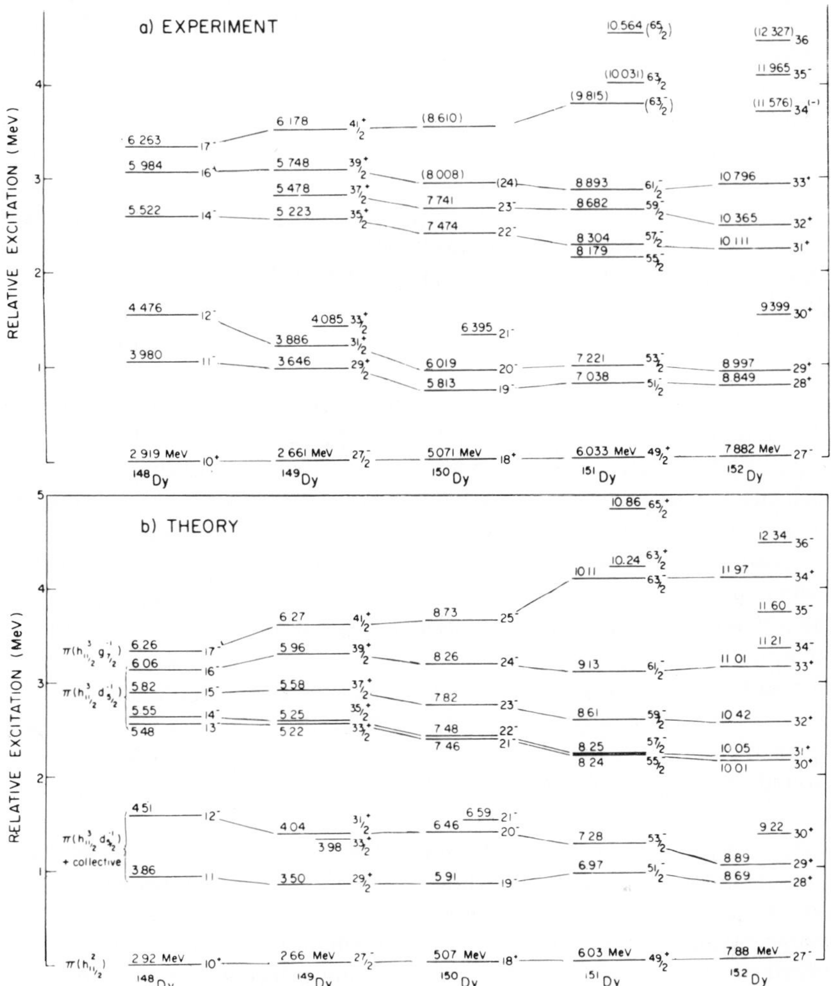

Fig. 6.60. (a) Experimental energy levels of light Dy isotopes in the high-spin region. The excitation energies in MeV relative to the aligned valence nucleon configurations (see text) are indicated by the scale at the left. For each isotope, the absolute level energies are also noted, and the uncertain values are in parentheses. The full-length lines represent the levels of ^{148}Dy whose structure is manifested in the other isotopes as valence neutrons are added. Additional energy levels are shown by half-length lines. (b) Shell-model calculations of energy levels of light Dy isotopes. Additional energy levels (half-length lines) are those produced by the substitution of an $i_{13/2}$ neutron for the $f_{7/2}$ neutron. These levels are shown only where one would expect them to be fed strongly by the main gamma-ray cascade. (Horn *et al.*, 1983.)

band is certainly much larger than of the remaining decay scheme, so that one may consider this as a first indication of some collective excitation in ^{152}Dy, coexisting with the predominant non-collective degree of freedom. Of course, the transition strengths have to be measured before such a picture can be established. In Section 6.5 we shall discuss the shape change from oblate to prolate above spin 40, thus changing the shape of the nucleus ^{152}Dy from spherical at low spin to prolately deformed at high spin.

We realize that we have greatly restricted the discussion, taking a few rare earth nuclei as examples. Nuclei in the lead region, for example, also show similar single-particle features, as has been studied, for instance, by Blomqvist (1979). The reader is referred to the review article of de Voigt *et al.* (1983) for details of nuclear properties in the high-spin region in this and other mass regions.

We should like to conclude this section by briefly discussing the second possibility of the non-collective rotation of a prolate nucleus with $\gamma = -120°$, as already mentioned in Section 6.4.2. It is shown in Fig. 6.61 that very strong shell effects may lower in energy the non-collective states ($\gamma = -120°$) in the prolate nucleus so much that they become the yrast states, thus constituting the yrast line. This line in prolate nuclei is normally composed of collective states ($\gamma = 0°$). It is explained in Section 6.5.6a that the shell energy E_{shell} lowers the total energy. The effect is apparently larger in the case of non-collective rotation than in the collective situation (see the discussion in Section 6.4.2). The bottom part of Fig. 6.61 shows that the shell effects in the case of ^{176}Hf ($\gamma = 120°$) are exceptionally large. This brings the single-particle excitation below the collective one, leading to the preference of the system to rotate around the symmetry axis ($\gamma = -120°$) rather than the axis perpendicular to it ($\gamma = 0°$).

Isomeric states have been indeed located in several Hf isotopes, notably the long-living 14^-, 19^+ and 22^- states in ^{176}Hf, shown in the level scheme in Fig. 6.62 (see Khoo *et al.*, 1976). They seem to have a four or six quasi-particle character, all including the $(g_{7/2}h_{11/2})_{8^-}$ configuration. The states are well described by the Nilsson model with high-spin projection on the symmetry axis of the prolately deformed nucleus (high Ω). By summing Ω over all the quasi-particles one reaches relatively high K-values (up to $K = 22$). Those high-K states form the basis for rotational bands. Because of the lowering in energy described above the band heads appear as a peculiar kind of yrast trap. They can only decay by K-forbidden transitions. Thus they become isomeric. The Hf isotopes are thus illustrative of the coexistence of collective motion (rotational bands) as well as the non-collective features of the K-isomeric band heads. Another unusually high-K isomeric state has been found with $I = (20)$ in ^{182}Os (Pedersen *et al.*, 1985). Instead of a K-forbidden transition a decay has been found proceeding via many collective levels. This was interpreted as evidence for a triaxial shape of ^{182}Os.

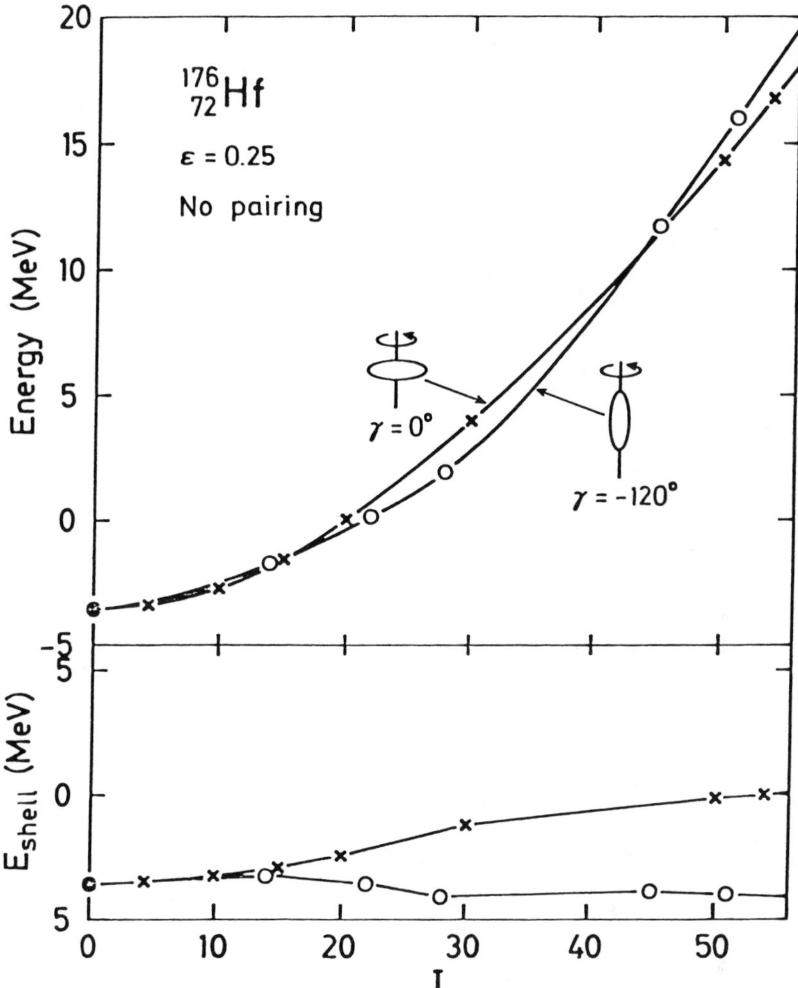

Fig. 6.61. The total energy (upper part) and the shell energy (lower part) plotted vs. the total angular momentum I for ^{176}Hf at a fixed prolate deformation. The rotation takes place around the axis with the largest moment of inertia ($\gamma = 0°$, collective rotation, crosses) or, alternatively, around the axis with the smallest moment of inertia ($\gamma = -120°$, non-collective rotation, open circles). The calculation was performed with the Nilsson potential. (Åberg, 1978.)

We have tried to illustrate the richness of the nuclear phenomena accompanying high-spin single-particle excitation (non-collective rotation). Although we have gained a detailed insight in the microscopic nuclear motion, much remains still to be learned. In this section we have restricted the

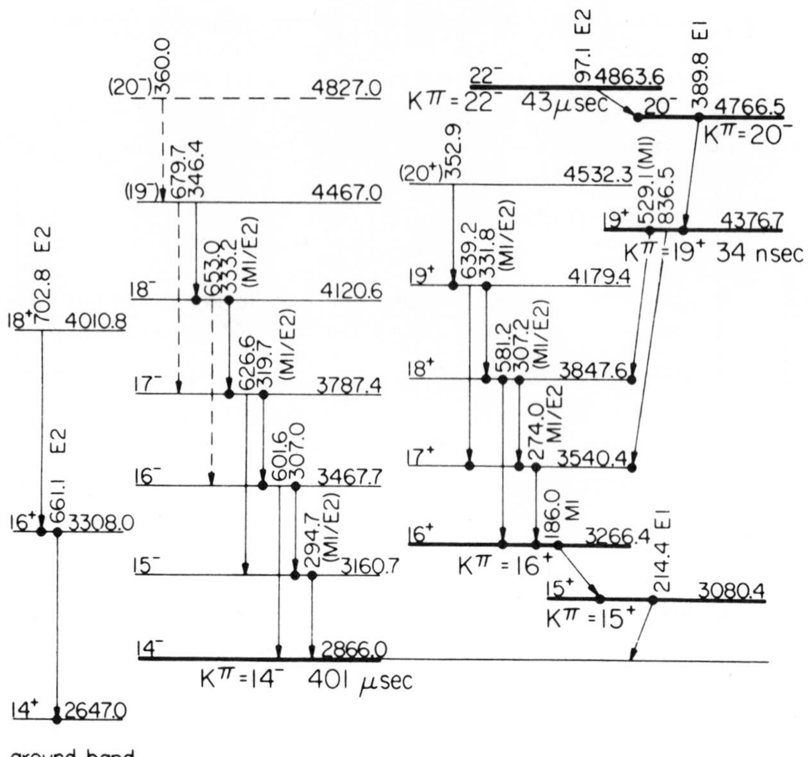

Fig. 6.62. Partial level scheme for ^{176}Hf showing four- and six-quasi-particle excitations and the upper portion of ground band. The assignments in parentheses are tentative. The filled circles indicate γ rays entering and leaving a level in prompt coincidence. (Khoo *et al.*, 1976.)

discussions to states with spin 40 which could be observed as discrete lines in the spectra. As drastic changes in the nuclear structure are expected at higher spins we will continue the discussion by considering quasi-continuum states in the next section.

6.5 Quasi-continuum γ-ray spectroscopy

6.5.1 *General features*

Quasi-continuum γ-ray spectroscopy probes those states in the regime of rather high excitation energy with high spins (fast rotation), where the level density is too high to observe the γ decay of individual states separately.

We have seen in the previous sections that significant modifications of the nucleonic motion have been observed with increasing rotational frequencies. The related phenomena are discussed in terms of the interplay between collective and single-particle degrees of freedom (see Sections 6.3 and 6.4).

With further increase of the rotational frequency significant changes in the behaviour of nuclear matter are expected to occur such as:

(1) shape changes with increasing angular momentum up to the maximum value imposed by fission (~ 70 in rare-earth nuclei);

(2) an increasing competition between single-particle alignments and excitations and collective rotation, possibly influenced by shape oscillations;

(3) the possible disappearance of nuclear superfluidity at fast rotations; i.e. the disappearance of pairing correlations reflected in the moment of inertia, or more precisely when $\mathscr{I}^{(1)}_{\text{env}} = \mathscr{I}^{(2)}_{\text{env}} = \mathscr{I}_{\text{rigid}}$ (see Section 6.3.4).

The first indications of shape changes (as discussed in Section 6.5.6 below) as well as of the breakdown of pairing correlations have been observed in the spectra of discrete γ rays, corresponding to states with spins of ~ 40 or less (see Section 6.3).

The interest in very fast nuclear rotation had already been triggered in 1974 by the analysis of Cohen *et al.* They treated the rotating nucleus as a liquid drop having a spherical shape when at rest and acquiring deformed shapes under rotation, according to the laws of classical mechanics. At slow rotation (frequency $< 10^{20}$ Hz of the nucleus) the most likely shape of a liquid drop is close to that of an oblate ellipsoid, because that gives the largest moment of inertia when rotating around the symmetry axis (see Fig. 6.15). With increasing rotational frequency, however, the oblate object may lose its stability in favour of a rather elongated prolate shape, rotating about an axis perpendicular to the symmetry axis. The transition between the two different shapes may proceed via a series of triaxial shapes. This picture may to some extent apply to nuclei but several other features like shell effects, for instance, cause differences from the above picture such as low-lying excitations with a prolate shape.

The regions of nuclear stability in terms of the limiting angular momenta are indicated in Fig. 6.6 (see Section 6.2). The studies of Cohen *et al.* (1974) also indicated the possibility of superdeformed shapes (a ratio of long to short nuclear axis of 2:1) before the nucleus becomes unstable with respect to fission. Indeed, the possible existence of superdeformation now poses one of the most intriguing questions about atomic nuclei in states of high angular momentum.

Instabilities due to deformation occur in astronomical objects (Jacobi instabilities) and, rotating stars may change their initially favourable oblate shapes (McLaurin spheroids) via a series of triaxial forms into more elongated shapes (Jacobi ellipsoids), giving birth to double stars. Instabilities may result from a competition between the centrifugal-stretching and long-range gravitational forces.

In atomic nuclei, however, at least two essentially different aspects can be indicated, apart from the Coulomb repulsion between the protons. Firstly, the nuclear interaction has a short-range character; and secondly, the quantal shell structure also plays an important role.

The quantal character of the nuclear motion has been taken into account by Bohr and Mottelson (1974) in predicting a picture of possible nuclear phases as a function of angular momentum and excitation energy. Most of the excitation energy of states on or near the yrast line is concentrated in only a single (rotational) degree of freedom in the case of prolate ($\gamma = 0°$) deformation (see Section 6.3). In the case of oblate deformation the yrast line represents only a few single-particle degrees of freedom (see Section 6.4). This implies that in yrast states, even at high excitation energy, the nucleus remains rather cold, and quantal effects are expected to dominate as is the case for states near the ground state.

At low excitation energy the superfluid pair correlations are important in most nuclei and (prolate) shape deformations can be induced by shell effects. Rotation of a prolately deformed nucleus about an axis perpendicular to the symmetry axis, seems energetically favourable in most cases as argued in Section 6.3. This gives rise to regular structures of rotational bands as schematically indicated for region A in Fig. 6.63. With increasing angular momentum, crossing of bands may become a dominant feature and the nucleus may acquire triaxial shapes (region B in Fig. 6.63). This is due to alignment of particles with their spins along the axis of rotation, causing more nucleons to move in the plane perpendicular to that axis than in other planes. At faster rotation the centrifugal and Coriolis forces may become strong enough to destroy the pairing between nucleons completely. Subsequently the spins of the individual nucleonic orbitals can be aligned with their equatorial planes perpendicular to the rotation axis as a predominant excitation mode, causing oblate shapes (region C in Fig. 6.63). At still higher angular momenta (beyond region C) superdeformed states may occur unless the stability is lost, leading to the fission process. It should be realized that different sequences of shape changes may also occur like the one from oblate at low energy to prolate (superdeformed) at high energy.

Quasi-continuum γ-ray spectroscopy aims at detecting signals from those rather drastic changes in structure, shape and deformation in the very high-spin region. The γ rays, possibly carrying those signals, are schematically indicated in Fig. 6.63. They constitute the final steps of the decay process of the compound nucleus towards the yrast line, after the particle evaporation as discussed in Section 6.2.3. The contour lines in Fig. 6.63 indicate population of levels in the entry regions in the E_x vs. I plane, where the probability for γ

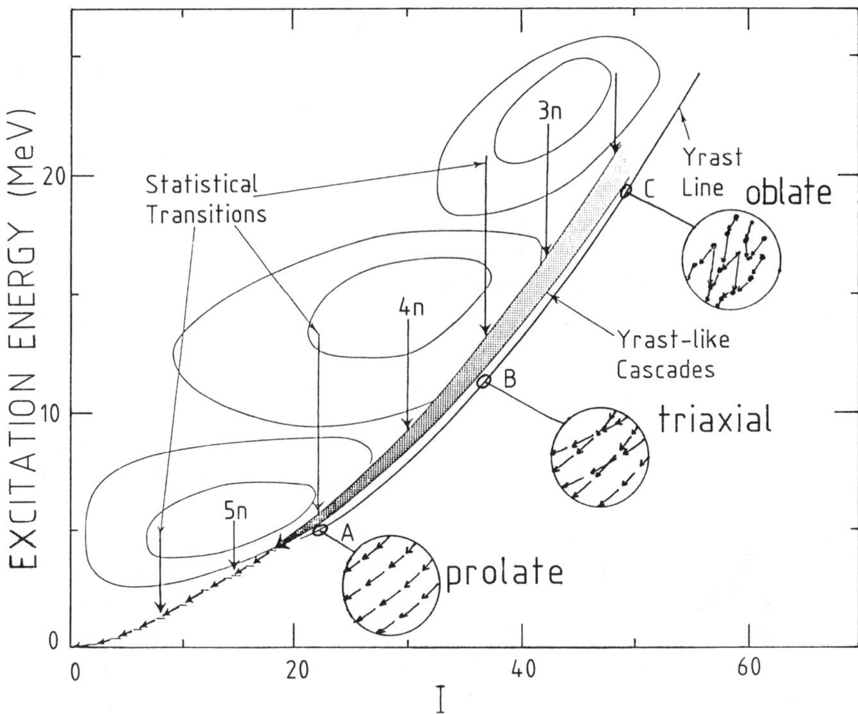

Fig. 6.63. Illustration of the de-excitation process of the ^{164}Er compound nucleus according to a statistical-model calculation for 147 MeV ^{40}Ar ions bombarding ^{124}Sn. The contours enclose regions corresponding to γ rays emitted from the (3–5)n evaporation residues (Hillis *et al.*, 1979). A possible change of structure along the yrast line is schematically indicated by the magnified areas A, B, and C (according to Bohr and Mottelson, 1974.)

decay becomes comparable with that for neutron emission. The determination of the entry regions is important for the understanding of the nature of decaying γ-ray cascades and of the population mechanism of high-spin states, as discussed below in Section 6.5.2. States in the entry region (entry states) are, strictly speaking, in most cases located at energies above the neutron threshold and are thus energetically unbound. They are, however, effectively bound because neutron decay with a large angular-momentum change is hindered by the centrifugal barrier. Therefore γ-ray emission from entry states can compete with their neutron decay.

The high level density above the yrast line creates many possible pathways for the γ decay of the states in the entry region, so that the number of decaying γ rays is so large that none of them acquires enough intensity for observation as a discrete line in the γ-ray spectrum. Estimates of the level densities do not however imply significant overlap of the levels within their natural widths, which are of the order of 10^{-3} to 10^{-1} eV. This leads to the notion of 'quasi-continuum spectrum' (qcs) in which discrete γ rays are not resolved experimentally. In a simple picture one may think of the qcs as being composed of a mixture of so-called statistical γ-ray cascades, governed mainly by the level densities, and yrast-like cascades. The latter cascades follow approximately the yrast line and are typically of a collective nature, as is indicated in Fig. 6.63. The special features of the qcs are discussed in Section 6.5.3 below. We should realize that even a small part of the qcs contains γ decays from numerous states and the deduced quantities are thus averages for those many states. Therefore one may only expect to obtain the bulk properties of highly excited nuclei, although the rapid progress in experimental techniques is expected to help in revealing more and more details.

The development of experimental techniques in recent years has been so diverse that various properties of the γ rays in the qcs can be measured. They are the average number (multiplicity) of γ rays in the cascades, the electromagnetic character (multipolarity) of the radiation and the average lifetimes, all discussed in Sections 6.5.4 and 6.5.5 below. In those sections the physics is emphasized, leaving many instrumental details to Chapter 3. Finally, this chapter concludes with an account on the most exciting findings of recent qcs γ-ray spectroscopy, namely the evolution of nuclear shapes with increasing angular momentum (Section 6.5.6) and evidence for superdeformation (Section 6.5.7).

6.5.2 *Entry regions for gamma-ray cascades*

The entry regions are defined in the E_x vs. I plane as the ensemble of states populated after compound nuclear reactions and after termination of particle emission (mostly neutrons in medium and heavy nuclei), from where the γ-ray cascades start.

Assuming the statistical character of the particle decay one may expect that the centroids of the entry regions are located near a line in the E_x vs. I plane where the particle and gamma decay widths become equal. This line is called the entry line. One would expect that for all reaction channels the entry line is located approximately one particle (neutron) separation energy above the yrast line. Deviations from this expectation may occur when non-statistical particle emission or other processes remove more energy and angular momentum than the statistical model predicts. This effect can be studied by varying the bombarding energy and then comparing the measured and calculated entry lines, using the statistical model.

Several methods can be applied to measure the average energy and angular momentum of the entry region. In many cases one would also require the selection of a single exit channel (evaporation residue) out of several possibilities after the nuclear reactions. The instruments used are in general γ-ray detector assemblies such as multiplicity filters, crystal balls and sum spectrometers, i.e. large single crystals acting as total-energy detectors. These instruments are most commonly used in coincidence with one or more high-resolution (germanium) detectors. (See Section 3.2 for instrumental details).

We shall now discuss in some detail a few examples to illustrate the various techniques and problems involved in extracting information from the qcs. The average excitation energy of entry states can be measured directly by means of a sum spectrometer (see e.g. Folkmann *et al.*, 1981) or indirectly with a multiplicity filter (see e.g. de Voigt *et al.*, 1979) or a crystal ball (see e.g. Sarantites *et al.*, 1982).

Entry lines have been deduced for ^{72}Se and ^{71}As populated in the (^{16}O, 2pγ) and the (^{16}O, 3pγ) reactions using a ^{58}Ni target and bombarding energies in the range 45 to 77 MeV. In the work of Simpson *et al.* (1977) use was made of nine 5.1 cm × 5.1 cm ϕ NaI(Tl) detectors in coincidence with a Ge(Li) detector, whereas Tjøm *et al.* (1978) used a sum spectrometer of 20 cm × 25 cm ϕ NaI(Tl). The two methods could be compared in the case of ^{72}Se and in both works proton energy spectra have also been measured to obtain the excitation energies independent of the γ-ray methods. The average excitation energy obtained from the multiplicity filter is given by the product of the measured average γ-ray energy $\langle E_\gamma \rangle$ and the average multiplicity $\langle M \rangle$;

$$\langle E_x \rangle = \langle E_\gamma \rangle \times \langle M \rangle \tag{6.5.1}$$

The results deduced from the data obtained with the multiplicity filter agreed, within the experimental errors, with those obtained more directly with the sum spectrometer. Both sets of data were also in agreement with the energies deduced from the proton spectra.

We shall now discuss a particular example of deducing the energies and angular moments of the entry region for the compound nucleus ^{160}Er, produced in heavy-ion compound-nuclear reactions, and employing a sum spectrometer—Folkmann *et al.* (1981). The sum spectrometer is a large NaI(Tl) crystal designed to detect the full energy of the total γ-ray emission from a target inside the detector. The total energy is derived from the measured pulse height after correction for solid angle, efficiency, X-ray emission and neutrons. It was shown in Section 3.2.3 that for a finite (<100 per cent) detection efficiency the pulse height is still approximately proportional to the total energy, but that the resolution is very much affected by the efficiency.

The advantage of additional (small) detectors is the possibility of measuring the average energies of the γ-ray transitions and the average multiplicities of the cascades. For the latter quantity one has also to record the coincidences between the detectors. The average multiplicity can be deduced from the coincidence to singles ratio as will be shown below in Section 6.5.4.

In the work of Folkmann *et al.* (1981) two small NaI(Tl) detectors were placed in spaces in the sum spectrometer at 0° and 90° with respect to the beam direction. The beam was stopped in the backing of the target. From that combination of various detectors it was possible to deduce the multiplicities as a function of energy deposited in the sum crystal. The results for $^{160,\,164}$Er are shown in Fig. 6.64 including data taken with a fourteen-detector multiplicity filter (Hillis *et al.*, 1979). The angular momenta were derived from the measured multiplicities as a function of total γ-ray energy deposited in the sum crystal. The relation between M and I is discussed below in Section 6.5.4. The solid and dashed lines in the figure represent the results of the statistical model. The discrepancy between the model calculations and the data may indicate either inadequate parameters used in the model or the importance of non-statistical processes. One should realize, however, that no exit channel

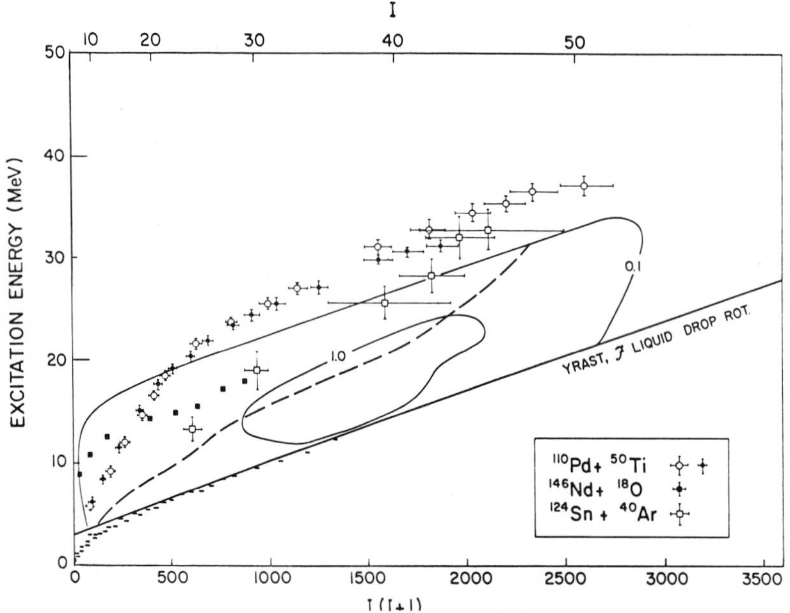

Fig. 6.64. The entry lines for the evaporation residues of ^{160}Er* and ^{164}Er* measured with ^{50}Ti (dots for 223 MeV, open circles for 204 MeV) and ^{18}O beams, respectively (Folmann *et al.*, 1981), are shown together with the centroids of exit-channel selected entry regions from the ^{40}Ar + ^{124}Sn reaction at 161 and 236 MeV (Hillis *et al.*, 1979). The contours marked 0.1 and 1.0 (mb/MeVħ) correspond to the calculated population of the evaporation residues. The entry line calculated with a statistical model (Folkmann *et al.*, 1981) is shown as a broken line.

selection has been applied in the sum-spectrometer experiment, thus yielding entry-state properties averaged over all evaporation residues.

The interesting features of the results in Fig. 6.64 are:

(1) the slopes of the entry lines, being nearly parallel to the yrast line above $I \sim 20$ and

(2) a maximum angular momentum of entry states of ~ 50.

The first observation implies a moment of inertia for the highly-excited (entry) states close to that of the rotating liquid drop. Note, however that the location of the entry lines of ~ 15 MeV above the yrast line seems rather difficult to understand in view of the particle binding energy (~ 10 MeV). The maximum angular momentum observed for discrete yrast states is at least 20 units lower than the above value for populated entry states. This implies that the quasi-continuum cascades must carry off a significant amount of angular momentum, thus revealing some information on the structure of the states above the yrast line (see Section 6.5.3).

An extensive study of entry lines after (α, xn) compound nucleus reactions and with exit-channel selection has been carried out at the KVI, Groningen, using a 16 NaI(Tl) detector multiplicity filter as displayed in Fig 3.15 (Ockels *et al.*, 1978; Hageman, 1981). The total energy and angular momentum released in the decay process of the compound nucleus could be compared with the calculated values, using the reaction kinematics, only after the properties of the emitted neutrons were measured. Neutron time-of-flight measurements after light-ion induced reactions have been carried out extensively at the RNCP, Osaka (see e.g. Ejiri *et al.*, 1978).

In the analysis of γ decay it is convenient to distinguish between cascades following essentially the yrast line, passing through observable discrete states, and cascades feeding the yrast states from quasi-continuum states, called side-feeding (sf) cascades. The entry states which feed a particular yrast state can be located by measuring the properties of the cascades in coincidence with a discrete transition de-exciting that yrast state. For that purpose the multiplicity filter mentioned above has always been operated in coincidence with at least one high-resolution Ge-detector acting as a gating detector. For each reaction event (i.e. a selected γ ray in the gating detector) the number of activated NaI(Tl) detectors, hit by one or more γ rays, was recorded as well as the energy disposed in each detector. This made it possible to determine the multiplicity as a function of γ-ray energy.

The results for the ^{160}Gd(α, 4n)^{160}Dy reaction are presented in Fig. 6.65, showing the entry lines deduced from the properties of the side-feeding cascades for different alpha-particle bombarding energies in the range 40 to 70 MeV. Points on the entry line have been retrieved by gating on the discrete transitions deexciting the 4^+ to 14^+ yrast states. For those states the excitation energies E_x(yr) and spins I(yr) were precisely known. The total energy released by the sf-cascades has been deduced from the average γ-ray energy $\langle E_\gamma(i)\rangle$ and multiplicity $\langle M_\gamma(i)\rangle$ in four energy bins i, adding up with E_γ(yr) to the excitation energy

$$E_x = E_\gamma(\mathrm{yr}) + \sum_i \langle E_\gamma(i)\rangle\langle M_\gamma(i)\rangle. \tag{6.5.2}$$

In this method the possible problem of strongly varying E_γ and M_γ is significantly reduced with respect to the situation assumed for the relation of eqn (6.5.1) discussed earlier, because of the very limited energy range of the bins. For each energy bin an angular momentum release per transition was estimated on the basis of earlier measured conversion electron coefficients (see Section 6.5.5b). Those values ranged typically from $l=2$ at low γ-ray energy to $l=1$ or 0.5 at high energy. The result thus reads:

$$I = I(\mathrm{yr}) + \sum_i \langle M_\gamma(i)\rangle l(i), \tag{6.5.3}$$

where $l(i)$ represents the average angular momentum release per γ ray in bin i.

Two striking features can be seen in Fig. 6.65;

(1) the drastic increase in energy of the entry line with the increase of bombarding energy up to $\sim$50 MeV and

(2) the relatively small change thereafter.

The first effect is due apparently to the statistical decay properties of the excited nuclei. More energy is available for γ-emission in this case when the bombarding energy E_α increases, assuming a constant energy release by neutrons for a particular exit channel. As a consequence one measures an increase in the multiplicity M^s of the high-energy statistical transitions as indicated in the figure (Ockels *et al.*, 1978). This feature has also been observed consistently for heavy-ion beams such as ^{12}C and ^{20}Ne (Sarantites *et al.*, 1978) and five years later for ^{28}Si and ^{32}S beams (Ward *et al.*, 1983).

The second feature seems to be attributed to competing reaction processes with particle emission that takes away part of the angular momentum. It appears that the angular momentum, and thus also the multiplicity M_γ, does not increase as much as $(E_\alpha)^{1/2}$. In fact a saturation of M_γ is observed with increasing E_α, or more precisely a decrease in multiplicity of the collective transitions, which carry off most of the angular momentum, and a slight increase in M^s (Ockels *et al.*, 1978). Indeed, measurements of particle spectra do indicate the importance of pre-equilibrium effects at the higher bombarding energies, as discussed in detail in Chapter 4. The neutron time-of-flight spectra measured at Osaka for the ^{158}Gd(α, xn)$^{162-x}$Dy reaction at 70 MeV

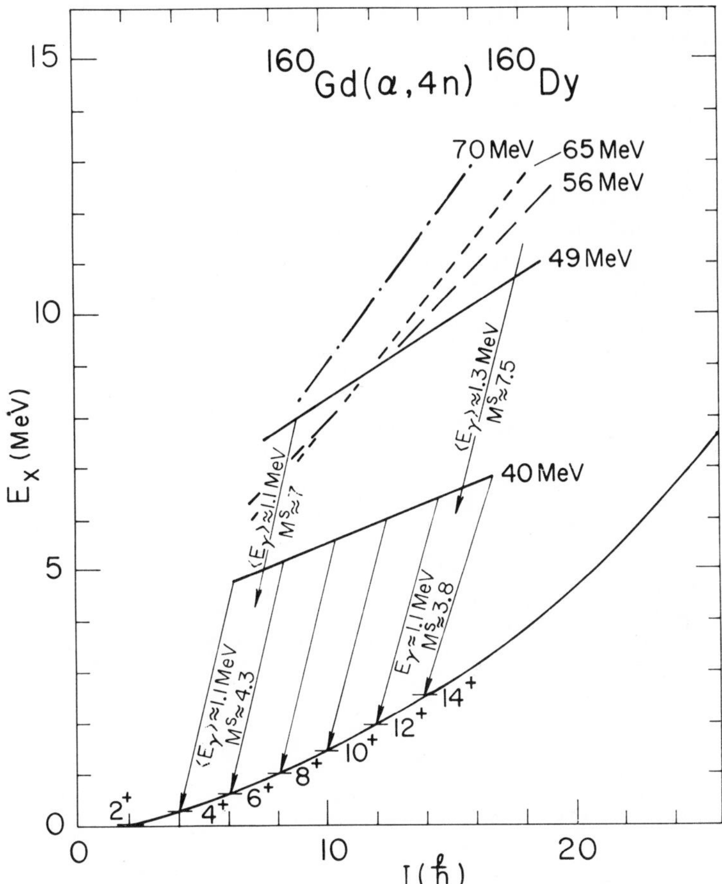

Fig. 6.65. Energy-spin 'entry lines'. The energy values were obtained from average multiplicities and γ-ray energies, and the corresponding spin values were obtained assuming a certain spin release per γ-ray transition (see text). M^s and E_γ represent average quantities of the γ-ray cascades connecting the entry lines and the ground-state band. (Ockels *et al.*, 1978.)

did yield energies and angular momenta consistent with the Groningen γ-ray data for exactly the same reaction (de Voigt *et al.*, 1982).

One may notice that the entry lines at the higher bombarding energies are located at about 8–10 MeV above the yrast line. This establishes roughly the neutron drip line as a function of angular momentum. On the drip line the neutrons have just enough energy (their binding energy) to leave the nucleus and thus to drip out. It is not surprising that we find the drip line just one neutron binding energy above the yrast line. The interesting feature is that this is one of the few cases that a drip line has been observed as a function of angular momentum.

The same method as described above has been applied to ^{12}C-induced reactions at Groningen, with the modification that the multiplicity and energy have been deduced averaged over all the side-feeding cascades (Hageman *et al.*, 1981). The method used to deduce the properties of these continuum-feeding cascades is discussed in Section 6.5.4. The results culminate in entry centroids rather than regions or lines, as is illustrated in Fig. 6.66. Features similar to those observed in the (α, xn) reactions also show up in this case, such as the initial increase of the energy of the entry centroids, followed by a decrease with increasing bombarding energy. For comparison, the average value of the (α, 6n) reaction at $E_\alpha = 70$ MeV is included in the figure, which is

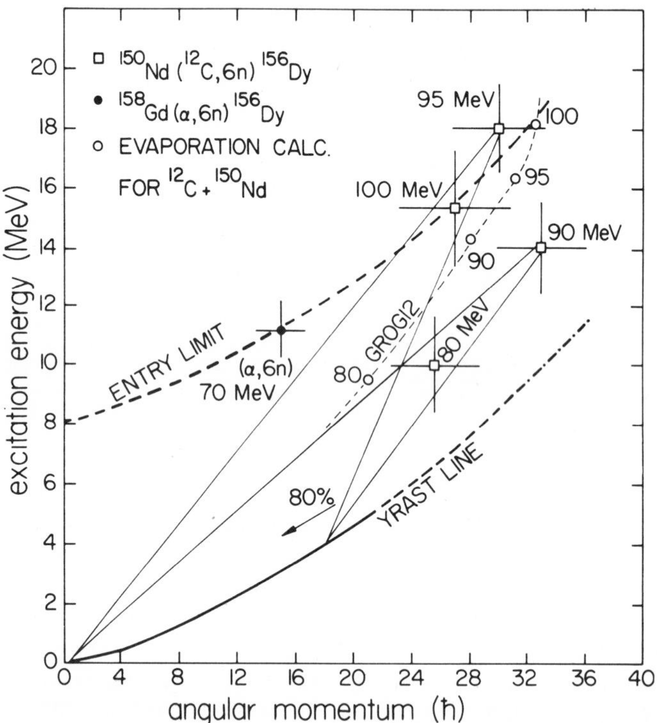

Fig. 6.66. Centroids for the entry regions populated in ^{156}Dy by the ^{150}Nd + ^{12}C reaction at the indicated beam energies. The yrast line consists of states populated in these reactions (connected by the full line), of states previously known (dashed line) and an extrapolation (dashed-dotted line). The entry limit is a line parallel to the yrast line, 8 MeV above it, just to indicate where the values for the mean entry energies roughly saturate. (Hageman *et al.*, 1981.)

consistent with the concept of drip line discussed above (entry limit in the figure).

The decrease in angular momentum at the higher bombarding energies has been attributed to competing incomplete-fusion reactions (Hageman *et al.*, 1981). This could be taken as a sign that high bombarding energies do not necessarily lead to the excitation of high-spin states (see Chapter 4 for further discussion).

The location of entry regions has become much more reliable in recent years with the development of crystal balls with near-4π geometry, as discussed in detail in Chapter 3. From the measured total pulse-height one deduces the total energy, and from the recorded number of detectors hit by γ rays, the total multiplicity. The peculiar feature here is that these quantities can be deduced for each reaction event separately, while the other methods yield only averages over many events, deduced after complicated data manipulation. The results for ^{20}Ne-induced reactions obtained with the Oak Ridge spin spectrometer are shown in Fig. 6.67 (Sarantites *et al.*, 1982). The fascinating aspect is the determination of the population of entry states for the 5, 6 and $7n$ exit channels in the two dimensions, E_x and I, in a straightforward way. The agreement of the results of the statistical model calculations with the data proves that at the bombarding energy of 136 MeV compound (statistical) reactions are predominant.

The many-detector assembly (70 detectors in this case) also allowed the deduction of the multiplicity distribution, that is the cross-section as a function of multiplicity. Similarly also the cross section was deduced as a function of excitation energy. Those results are presented in Fig 6.68, and are compared with the statistical-model results. We shall not discuss the comparison in too much detail (see the original paper; Sarantites *et al.*, 1982), instead we shall point out a few interesting aspects. First, the angular momentum distribution can be retrieved from the data, and it shows a pronounced dependence on the exit channel, which is selected by discrete transitions in a Ge detector. The total spin distribution after the reaction appears asymmetric with a kind of sharp cut-off edge at I–50 (to be compared with the entrance value of $\sim$60).

We conclude this section by stating that it now seems possible to locate rather precisely the entry states in energy and angular momentum, for selected evaporation residues and even for certain populated yrast states. This is very helpful in understanding the population of high-spin states and in deducing nuclear-structure information from the qcs.

6.5.3 *The quasi-continuum gamma-ray spectrum*

It has already been mentioned that one observes two very different types of γ-ray cascades in the qcs. The probabilities of so-called statistical transitions are

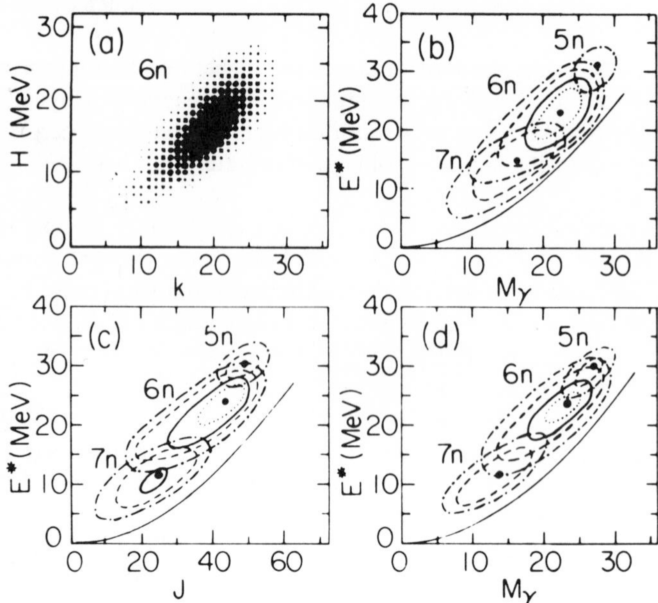

Fig. 6.67. The entry-state distributions for xn products from 136 MeV ^{20}Ne on ^{146}Nd. (a) Experimental density map in (H, k) space for the 6n channel. Here H denotes pulse height and k the number of active detectors. (b) Contour maps of experimental results in (E^*, M_γ) space for the strongest xn channels. (c) Contour maps of calculated results in (E^*, J) space from the statistical model. (d) Contour map of results in (E^*, M_γ) space from the same calculations as in (c). The cross-section contours in (b), (c) and (d) decrease going outward by the factors of 1.4, 2.0, 4.0 and 8.0 relative to the peak value of the 6n channel and are represented by the dotted, full, dashed and dashed-dotted curves, respectively. The heavy dots locate the maximum intensity for each channel. The ^{160}Yb yrast line used in the calculations is shown by the curve below the contours. (Sarantites *et al.*, 1982.)

mainly determined by the densities of final states available for γ-decay. Far above the yrast line, statistical transitions may cause changes of the total nuclear spin by ΔI which can be either positive or negative, although $\Delta I = \sim 1$ always dominates because of the nature of the statistical decay. This can be seen from the general expression of the level density $\rho(E)$ for a Fermi gas

$$\rho(E) = \exp\,(E/kT) = \exp\,(aE)^{1/2}, \tag{6.5.4}$$

where the relation is used between energy E and temperature T

$$E = a(kT)^2. \tag{6.5.5}$$

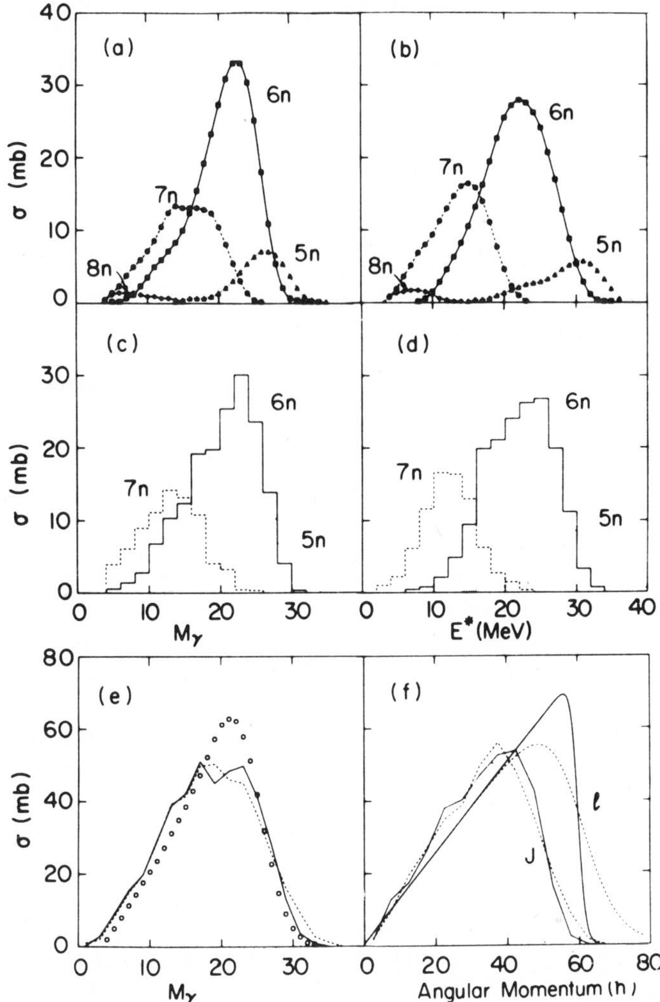

Fig. 6.68. Projections of entry-state distributions. (a) Experimental cross-sections of the xn channels as a function of M_γ. (b) Experimental cross-sections of the xn channels as a function of E^*. (c) Calculated cross-sections of the xn channels as a function of M_γ. (d) Calculated cross-sections of the xn channels as a function of E^*. The calculations are based on an entrance l-distribution with $l_{fus} = 59.5$ and a diffuseness $d = 1$. (e) Experimental cross-sections (points) and calculated J distributions (lines) for the sum of the xn and αxn channels as a function of M_γ. (f) Calculated entrance-channel l distributions and the resulting J distributions summed over all xn and αxn channels. The calculated results in (e) and (f) are for $d = 1$ (solid lines) and $d = 5$ (dashed lines). (Sarantites *et al.*, 1982.)

The parameter a is related to the one-particle level density at the Fermi surface and can be approximated by $a = A/10\ \text{MeV}^{-1}$. The energy E represents the internal excitation energy above the yrast line and one should subtract the rotational energy from the total excitation energy to obtain

$$E = E_x - I(I+1)\hbar^2/(2\mathscr{I}). \tag{6.5.6}$$

The transition probability $P(E;\ I \to I')$, in the case of statistical decay, is proportional to the level density of the final state. Thus the probability ratio of the $\Delta I = -1$ to $+1$ transitions can be calculated from eqn (6.5.4) and using eqn (6.5.6) gives

$$P(I \to I-1)/P(I \to I+1) = \exp\left[2\hbar^2 I/(\mathscr{I}kT)\right]. \tag{6.5.7}$$

For a reasonable moment of inertia parameter, $\hbar^2/(2\mathscr{I}) = 0.0067$ MeV it follows from eqn (6.5.7) that at high spin, $I > 20$, the $\Delta I = -1$ transitions dominate even at a high temperature of $kT = 1$ MeV.

Close to the yrast line $\Delta I < 0$ transitions will reach regions with a higher-level density than $\Delta I > 0$ transitions and they will thus become even more probable. Consequently, the statistical cascades will thus release some amount of angular momentum. The picture sketched above leads to a mixture of stretched ($\Delta I = +1$), anti-stretched ($\Delta I = -1$) and non-stretched ($\Delta I = 0$) dipole transitions, although quadrupole admixture should not be excluded. Another effect close to the yrast line, due to the low temperature, is the increasing probability of higher multipole ($l = 2$) transitions which follows from a similar expression to eqn (6.5.7). Any collectivity of the nucleus will show up in bands, located along or parallel with the yrast line. In that case the dominant character of the transitions will be electric quadrupole with relatively small energy as compared to the statistical transitions. Such a chain of E2 transitions is commonly referred to as 'collective cascade' or 'yrast-like' cascade, as is illustrated in Fig. 6.63.

The collective transitions span in general only a limited energy region of the qcs (extending up to 1 or at most 1.5 MeV), while the statistical ones may be spread over several MeV. Therefore the resulting γ-ray spectrum can in most cases be approximated by a statistical distribution (see below) with a broad bump on the low-energy side. The bump is so pronounced in collectively rotating nuclei because a larger number of collective transitions (in the bands) occur in a smaller energy interval than is the case for the statistical transitions. On top of that structure one may observe discrete transitions due to de-excitation of the yrast states. Such spectral shapes have been observed consistently for all kinds of compound nuclear reactions. Examples are given in Fig. 6.69 for ^{40}Ar and ^{16}O-induced reactions carried out at Berkeley (Simon *et al.*, 1977).

On the logarithmic scale the high-energy part of the spectrum exhibits an approximately linear dependence on the transition energy—see eqn (6.5.8)

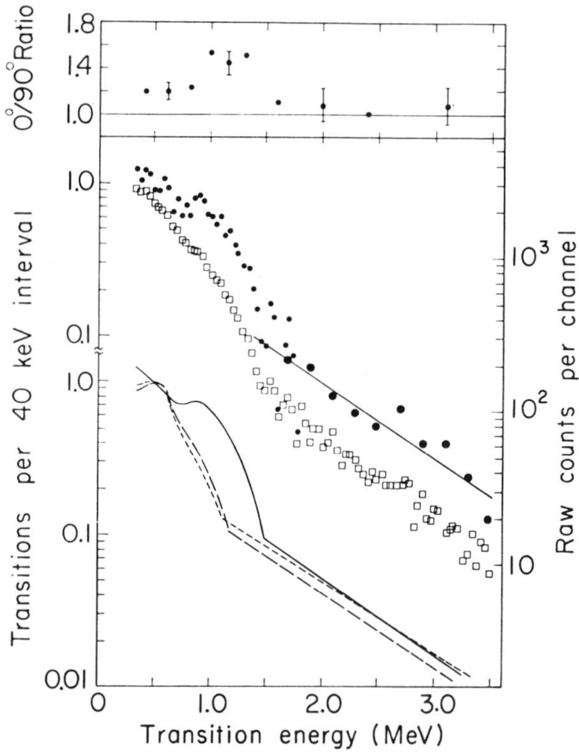

Fig. 6.69. The γ-rays qcs from the reaction $^{126}Te(^{40}Ar, 4n)^{162}Yb$ at 181 MeV is given in the form of raw (squares) and also normalized unfolded (small black dots) data. The larger dots are five-channel averages. At the top the 0°/90° ratio for unfolded spectra is given; at the bottom the experimental unfolded spectra for the same case (solid line) and for the reactions $^{126}Te(^{40}Ar, 4n)^{162}Yb$ at 157 MeV (long-dashed line) and $^{150}Sm(^{16}O, 4n)^{162}Yb$ at 87 MeV (short-dashed line) (Simon *et al.*, 1977) are schematically illustrated.

below; the intensity in Fig. 6.69 indeed decreases exponentially at high E_γ. The corresponding angular distribution (or the 0° to 90° ratio), as indicated at the top of the figure, is isotropic in the high-energy region and is thus consistent with the nature of statistical transitions, as pointed out above. The observed asymmetry for the bump, however, indicates a stretched E2 character for the collective transitions.

Obviously, the two types of γ cascades mentioned above do not always have to occur in a definite order in time. It may also happen that after the emission of a few statistical γ rays, one or more collective E2 transitions take place, followed by other statistical γ rays, and vice-versa.

In order to analyse the γ-ray spectrum in terms of the two different types of cascades one may try to fit a theoretical statistical γ-ray distribution to the high-energy part of the spectrum. Before doing so it is advisable to first unfold the spectrum, using a known response function for the detector, in order to retrieve the full energies of the γ rays which have been Compton scattered. Sometimes it is also possible to subtract the discrete lines if their energies and intensities are known from measurements with additional Ge detectors.

In Figure 6.70 a γ-ray spectrum is presented, measured for a ^{12}C-induced reaction (Łukasiak *et al.*, 1982). The spectrum shows similar features to the one in Fig. 6.69. The solid line represents a fit to the data at high energy, where the bump is assumed to have vanished. The fitted curve obeys the usual statistical relation:

$$N^{\text{stat}} = E^{2L+1} \exp\{-(E/T)\}. \tag{6.5.8}$$

The only parameter in the fitting procedure is the nuclear temperature T (or more precisely kT), which turned out to be 0.31 MeV in this case. The multipolarity L could seemingly be chosen as 1 for dipole transitions. However, the possible influence of the giant dipole transition may lead to a value $2L+1=5$ rather than 3 (Koeling, 1979). One can find in the literature $(2L+1)$ values ranging from 0 to 5, depending on the type of nuclei under investigation. The fit in Fig. 6.70 has been obtained with $2L+1=5$. After the subtraction of the (calculated) statistical transitions in the low-energy region of the spectrum one obtains the amount of collective transitions.

The methods used to determine special properties of the qc γ-rays are discussed below in separate subsections for multiplicities and multipolarities.

The properties of the γ-ray cascades in the quasi-continuum may also to a large extent determine the population of the low-lying yrast states, observed via discrete lines in the spectra. Obviously, the reaction mechanism is another important factor, which largely determines the location of the entry regions. Therefore it is illustrative to compare the intensities for two very different nuclei which are populated in similar reactions. Figure 6.71 shows the intensities of observed (discrete) yrast states in ^{152,156}Dy populated in several reactions. The ^{156}Dy nucleus is known to be well-deformed in the low-spin region. We can see in Fig. 6.71(a) that the intensities do not depend very much on the reaction or beam energy. Thus they are rather independent of the location of the entry regions. We can understand this behaviour in the following way. The deformation of ^{156}Dy implies that there are definite rotational bands along the yrast line in the qcs region, presumably coming down to $I \sim 20$ to 30. Most of the intensity originating in different entry regions is therefore guided along those bands towards the low-spin region. This feature has been pointed out by Ferguson *et al.* (1972) who proposed a K-band de-excitation model for the qc decay in rotational nuclei. The model assumes a number of rotational bands above the yrast line with high K

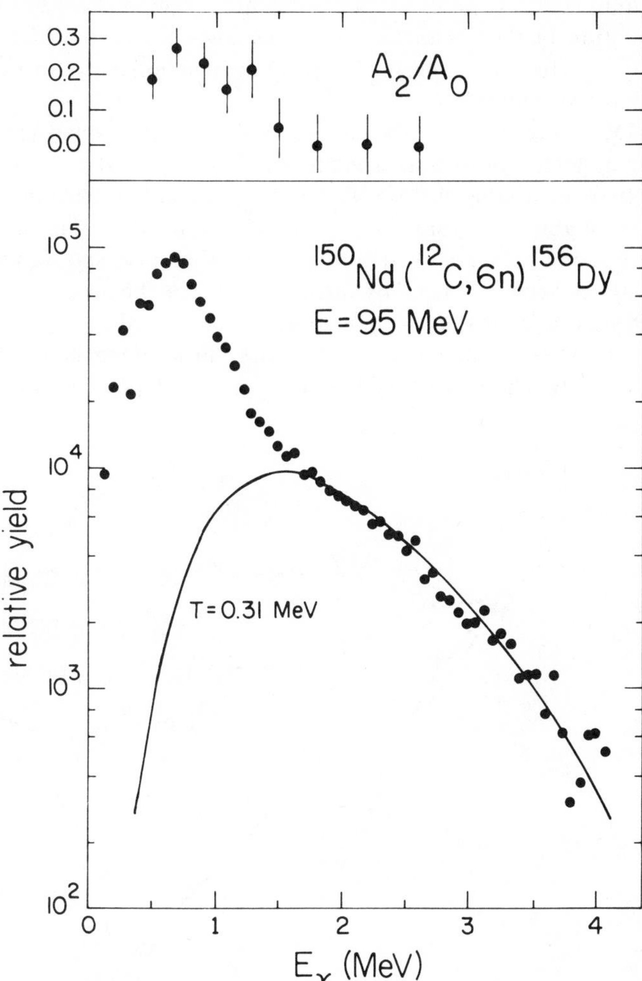

Fig. 6.70. The unfolded energy spectra of gammay-rays observed in the NaI(Tl) detectors. A theoretical statistical spectrum discussed in the text is fitted to the high energy part of the spectrum. The indicated temperature T is related to the mean energy of the statistical gamma-rays. At the top the anisotropy coefficients A_2/A_0 obtained from the angular distributions are displayed. The measured gamma-ray energy spectrum and the anisotropy coefficients deviate from the concept of a statistical gamma-decay process for gamma-ray energies below 1500 keV. (Łukasiak *et al.*, 1982.)

quantum number, built on intrinsic states of quasi-particle nature. The transitions within the bands are strongly favoured over the statistical transitions towards the yrast states because the latter ones are K-forbidden. When the γ-flow hits those rotational bands then much angular momentum is released until the yrast states are populated in the region of $I = 20$ to 30. Thus the population of the various levels in the low-spin part of the spectrum ($I = 2, 4, 6, \ldots 20$) is little affected either by the reaction mechanism or by the ion bombarding energy.

In ^{152}Dy, however, yrast states are observed with spins close to the maximum angular momentum brought into the compound nucleus. We see in Fig. 6.71(b) that with the increase of projectile mass or bombarding energy the population of yrast states shifts towards higher spins. This nucleus is known to be spherical or only weakly deformed in a rather large spin range up to at least $I \sim 40$ (see Section 6.4). Consequently, there may be no rotational bands along the yrast line, or they are not so well pronounced up to $I \sim 40$. Hence there is no such channeling mechanism of the γ-flow above the yrast line as is the case in ^{156}Dy. Therefore the side-feeding pattern depends more directly on

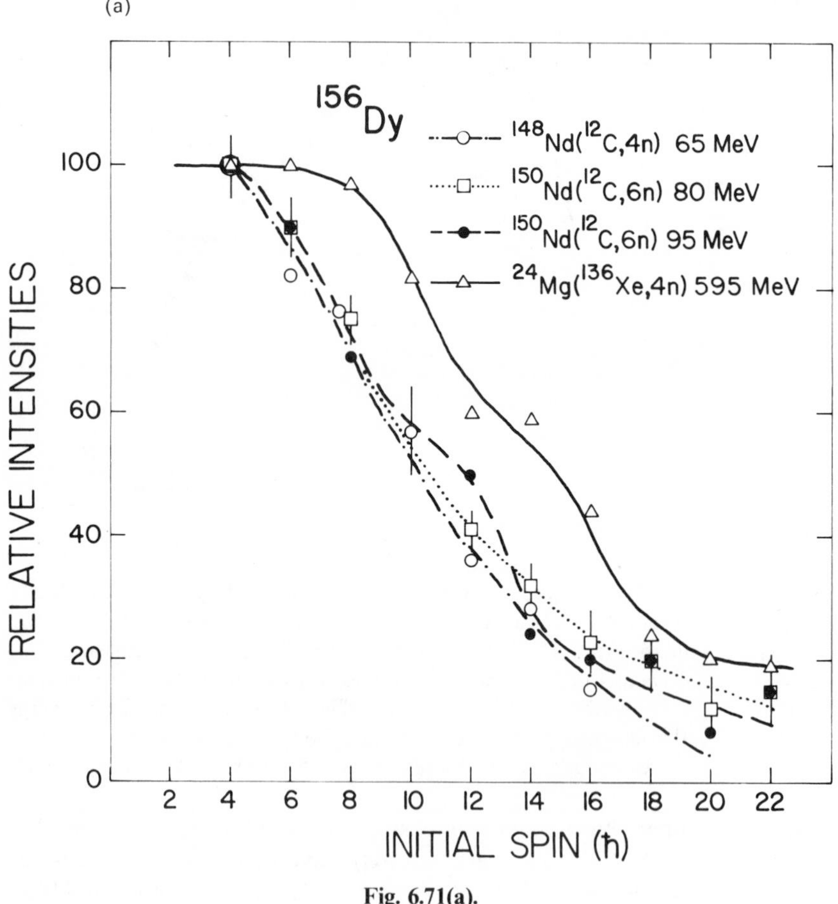

Fig. 6.71(a).

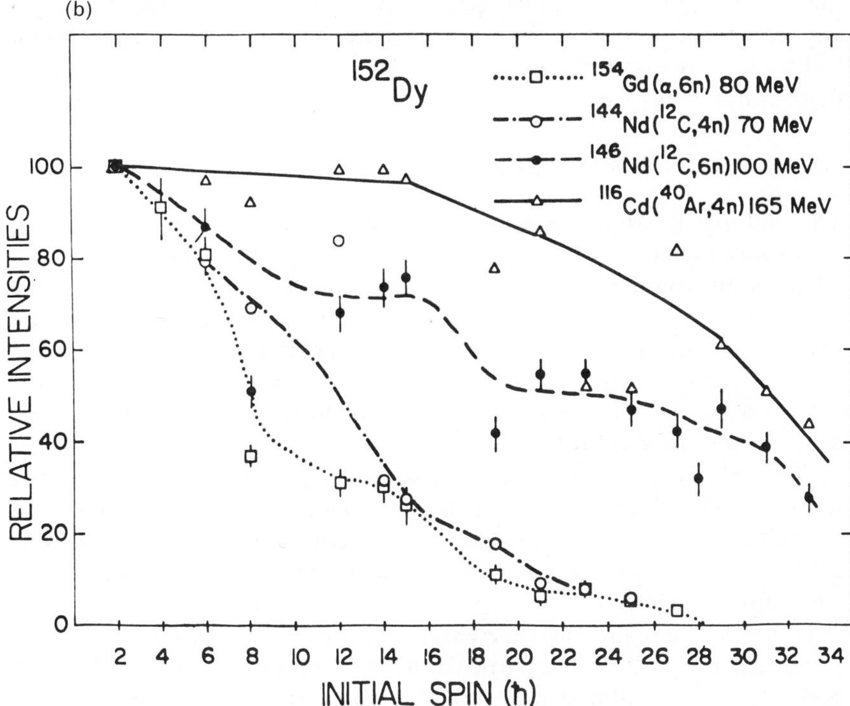

Fig. 6.71. Total γ-ray intensities in ^{156}Dy (a) excited in the (^{12}C, 4n) reaction (Sujkowski *et al.*, 1979), in (^{12}C, 6n) reactions (Hageman *et al.*, 1981), and in the (^{136}Xe, 4n) reaction (Ward *et al.*, 1979), and in ^{152}Dy (b) excited in (α, 6n) and (^{12}C, xn) reactions (Hageman *et al.*, 1981) and in the (^{40}Ar, 4n) reaction (Cornelis *et al.*, 1980.)

the reaction mechanism and on the projectile mass and energy. The yrast population thus reflects the entry population directly because no large angular momentum difference between the two regions has been observed (small E2 component) at least up to spin 40. Above that spin the situation is different, as will be discussed in Section 6.5.7.

Besides the reaction mechanism and the quasi-continuum decay, the level density is also of importance in the population of yrast states. The level density increases at high spin when the number of active nucleonic orbitals increases, thereby creating a favourable situation for statistical decay. As the high spins in ^{152}Dy have to be generated by individual particles, the level densities may also be higher than in collectively-rotating nuclei at high spin on the yrast line.

We have seen that several interesting features are connected with the qcs and thus there are many open questions. There is a growing interest from a theoretical point of view in the qcs. For a summary of theoretical concepts applied to the qcs we refer to Szymańsky (1983). Many experimental techniques have been developed recently to reveal detailed properties of quasi-continuum γ rays as will now be discussed in the following sections.

6.5.4 *Gamma-ray multiplicities*

The multiplicity M is the number of γ rays in a cascade. In the quasi-continuum spectrum (qcs) we may determine only the average multiplicity $\langle M \rangle$, as an average over many cascades. The most useful feature of the multiplicity is its direct relation to the angular momentum. The precise relationship between M and I, however, depends to some extent on the nuclear structure and on the γ-de-excitation process as well as on the region in the E–I plane. Therefore I is not easily derived from M and some procedures to do so are proposed below.

The multiplicity, with other observables such as the γ-ray energy, the multipolarity and the lifetimes together yield information on the nature of the de-exciting γ-ray cascades. It may become even more useful when correlations are measured between the multiplicity and γ-ray energy as will be shown in a few examples below.

To a first approximation one can determine $\langle M \rangle$ from the count rates measured, for instance, for a particular discrete transition with a Ge detector, both singles (N^{s}) and in coincidence (N^{c}) with a NaI(Tl) detector

$$\langle M \rangle - 1 = N^{\mathrm{c}}/(N^{\mathrm{s}}\Omega). \tag{6.5.9}$$

Here Ω represents the efficiency of the NaI(Tl) detector, namely the average detection probability for one of the remaining $\langle M \rangle - 1$ transitions of the cascades (see also Section 3.2.2 and eqn 3.2.7).

In general one may also be interested in the multiplicity distribution $\rho(M)$, that is the probability of γ-ray emission as a function of multiplicity. For that purpose more than two detectors are required, the number depending on the required accuracy of the mean value $\langle M \rangle$ and the central moments μ of the multiplicity distribution. The second central moment determines the width or standard deviation σ and the third central moment gives the skewness s. Since the early work of Hagemann *et al.* (1975) several papers have been devoted to the description of multi-detector systems. The many-detector arrays are called multiplicity filters with reference to their common use as a selection device or 'filter' for multiplicity, and thus also for angular momentum. In most cases six detectors are sufficient, although the statistical accuracy obviously increases with the number of detectors. There are several filters in use, consisting of a large number (10 to 16) small-size detectors, e.g. 5.1 cm $\phi \times 7.6$ cm NaI(Tl) crystals (see e.g. Fig. 3.15). With increasing multiplicity one also measures a higher multifold coincidence probability. Mathematical formalisms have been developed to calculate the multiplicity distribution from the measured coincidence probabilities as discussed separately in Section 3.2.2.

To obtain detailed information about the γ-cascades in the qcs one aims at determining the multiplicity parameters $\langle M \rangle$, σ, and s for the multiplicity distribution of cascades terminating at the various yrast states, separately. This is done by requiring coincidences between a high-resolution Ge(Li) detector and several detectors of the multiplicity filter. With the Ge(Li) detector discrete yrast transitions are observed and $\rho(M)$ can be determined for selected cascades which include those transitions.

An example of a data set obtained with the 16-detector multiplicity filter (shown in Fig. 3.15) is presented in Fig. 6.72. A rather smooth increase of $\langle M \rangle$ is seen with increasing spin of the discrete yrast transition, observed with a Ge(Li) detector, depending on the bombarding energy and on the reaction. The variation of $\langle M \rangle$ with the bombarding energy is discussed in terms of the reaction mechanism in Section 6.5.2 above. The multiplicity of cascades

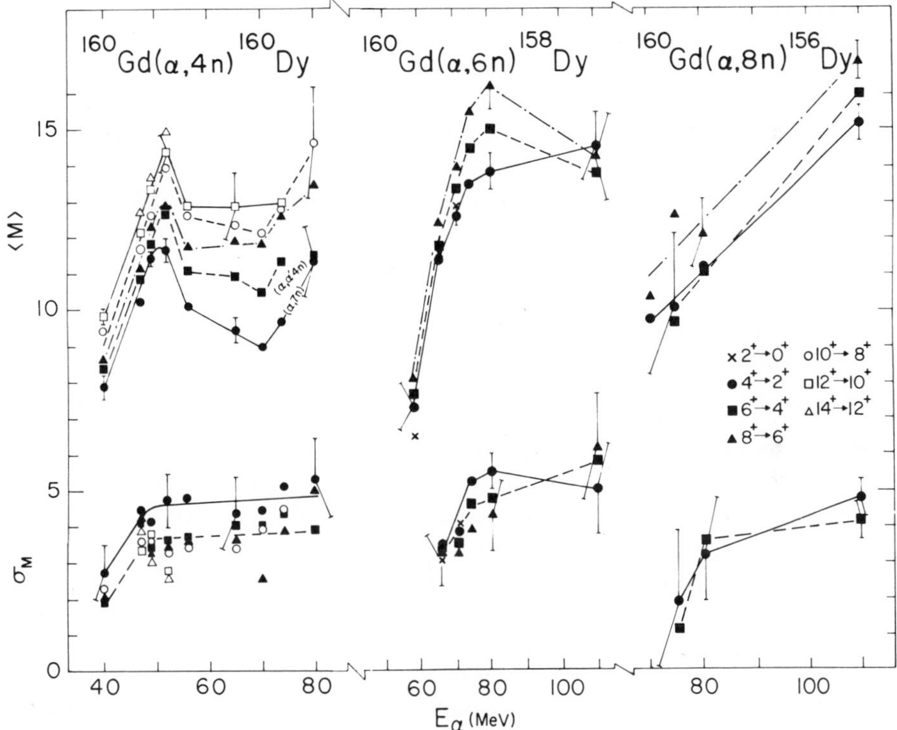

Fig. 6.72. The average total γ-ray multiplicity $\langle M \rangle$ and standard deviation σ_M for the various GSB members in dysprosium isotopes. The experimental errors are only given for some data points to illustrate the significance of the data. (de Voigt *et al.*, 1979.)

passing through high-spin yrast states is thus larger than of those passing through low-spin states. This appears as a general feature in many different reactions. It can be understood from the fact that high-spin states must be fed from the upper part of the entry region, and that the entry lines are more or less parallel to the yrast line (see Section 6.5.2). This point is clearly illustrated in Fig. 6.65 for an (α, 4n) reaction. The sidefeeding multiplicity M^{sf} is rather independent of spin and is added to the multiplicity of the yrast cascade. The values of $\langle M \rangle$ are thus to a large extent determined by the lengths of the yrast cascades themselves, which have known multiplicities of $I/2$ in even–even rotational nuclei. The total multiplicities in HI-induced incomplete fusion reactions are less spin-dependent, because the entry regions are generally narrow in angular momentum and are located at high spin in that case.

From the above discussion it is clear that the total multiplicities $\langle M \rangle$ are not very sensitive to specific properties of states in the qcs but rather reflect the gross angular momentum. The sidefeeding multiplicities (M^{sf}), on the contrary, are more sensitive to the de-excitation mechanism (Westerberg *et al.*, 1977). Therefore we shall present some procedures to deduce from the data the M^{sf} values associated with γ-rays in the qcs.

Let $M(I)$ be the total multiplicity associated with an yrast state of spin I, obtained by gating on its decaying transition. Note that correction for conversion electrons is important in the case of small γ-ray energies such as for $2^+ \to 0^+$ transitions in deformed nuclei. The total multiplicity (M^f) feeding that state is easily obtained by subtracting the number of transitions below the level I in the case of a single cascade with $\Delta I = 2$ transitions:

$$M^f(I) = M(I) - I/2. \qquad (6.5.10)$$

If $N(I)$ represents the intensity of the de-exciting transition of the level I, then one obtains the side-feeding multiplicity in the form (Hagemann *et al.*, 1975):

$$M^{sf}(I) = [M^f(I)N(I) - \{M^f(I+2)+1\}N(I+2)]/\{N(I) - N(I+2)\}. \qquad (6.5.11)$$

This method may lead to large uncertainties because the side-feeding intensities $\alpha = N(I) - N(I+2)$ are not always accurately known, particularly at high spin. When the relative intensities drop below 10 per cent then the errors in α may amount to 50 per cent or more. Therefore one may want to obtain a quantity, less specific, but also less uncertain namely the continuum-feeding multiplicity M^{cf} which represents the average of all side-feeding transitions above a specified level with spin I. One simply substracts from the feeding multiplicity all the known yrast transitions above level with spin I, weighted by their intensities. If the decaying transition of level I has an intensity $N(I)$ and the levels above with spin J have intensities $N(J)$ then one calculates the continuum-feeding multiplicity of cascades above level I from (Łukasiak *et al.*, 1982):

$$M^{cf}(I) = M^f(I) - \sum_{J>I} N(J)/N(I). \qquad (6.5.12)$$

Let us show the usefulness of the continuum-feeding multiplicities for the two nuclei ^{152,156}Dy which are discussed in the above example. The continuum-feeding multiplicities $\langle M^{cf} \rangle$, averaged over all discrete states above the 60 ns isomer in ^{152}Dy, have been deduced as a function of γ-ray energy (Hageman *et al.*, 1981; Łukasiak *et al.*, 1982). As discussed above, the low-energy region is associated with collective transitions and the high-energy region (>1 MeV) with statistical transitions. It appears that $\langle M^{cf} \rangle = 5$ to 8 for the statistical transitions, both in ^{152}Dy and ^{156}Dy, excited in ^{12}C-induced reactions at similar energies. The number of collective transitions, however, were very different: ~ 10 in ^{156}Dy and ~ 0 to 3 in ^{152}Dy. Thus the collectivity in the qcs of ^{156}Dy is much larger than that in ^{152}Dy, at least below spin 40, as discussed above.

More detailed information can be extracted from energy-correlated multiplicity spectra. The first measurements have been reported by Newton *et al.* (1977). Information on the collective behaviour of nuclei in states of high angular momentum can be deduced from the shape of such spectra (see e.g. Deleplanque *et al.*, 1978; Hillis *et al.*, 1979 and Stwertka *et al.*, 1985). A spectrum showing a bump at low energy is given in Fig. 6.69. The absence of a bump, however, cannot unambiguously be interpreted as the absence of collectivity, because it may well be that the entry region does not populate or only weakly populates the region with collective bands.

In the work of Ward *et al.* (1983) several light rare earth nuclei have been studied, which are known to be undeformed or only weakly deformed at low spin. The use of ^{28}Si and ^{32}S beams enabled them to observe the qcs corresponding to states up to $I \sim 55$. The γ-ray anisotropies, multiplicities and energy spectra, measured at the highest bombarding energies (151 MeV for ^{152}Dy) are given in Fig. 6.73. The multiplicities have been obtained by direct division, channel by channel, of the coincidence and singles spectra, according to eqn (6.5.9).

The multiplicity spectrum of ^{152}Dy shows clearly a bump around ~ 1 MeV, but only for the highest bombarding energy. The upper edge of the bump in the energy spectrum at 1.4 MeV corresponds to an angular momentum $I \sim 39$ (calculated with eqn 6.3.37, and an assumed moment of inertia $2\mathscr{J}/\hbar^2 = 113\ \mathrm{MeV}^{-1}$). Thus the collectivity in ^{152}Dy was suggested to occur at spins above $I \sim 39$. This finding supplements nicely the conclusions of several years ago, discussed above, that no collectivity would be present below $I \sim 40$. These results illustrate to what extent detailed nuclear structure information can be extracted from the quasi-continuum data.

So far we have assumed a direct relation between the measured multiplicity and the angular momentum of the nucleus when emitting γ rays. When dealing with the qcs, however, the relation between M and I is not straightforward. Several methods of deriving such a relation can be found in the literature (Banaschik *et al.*, 1975; Simpson *et al.*, 1977; Andersen *et al.*, 1978; Körner *et al.*, 1979; de Voigt *et al.*, 1979; Folkmann *et al.*, 1981).

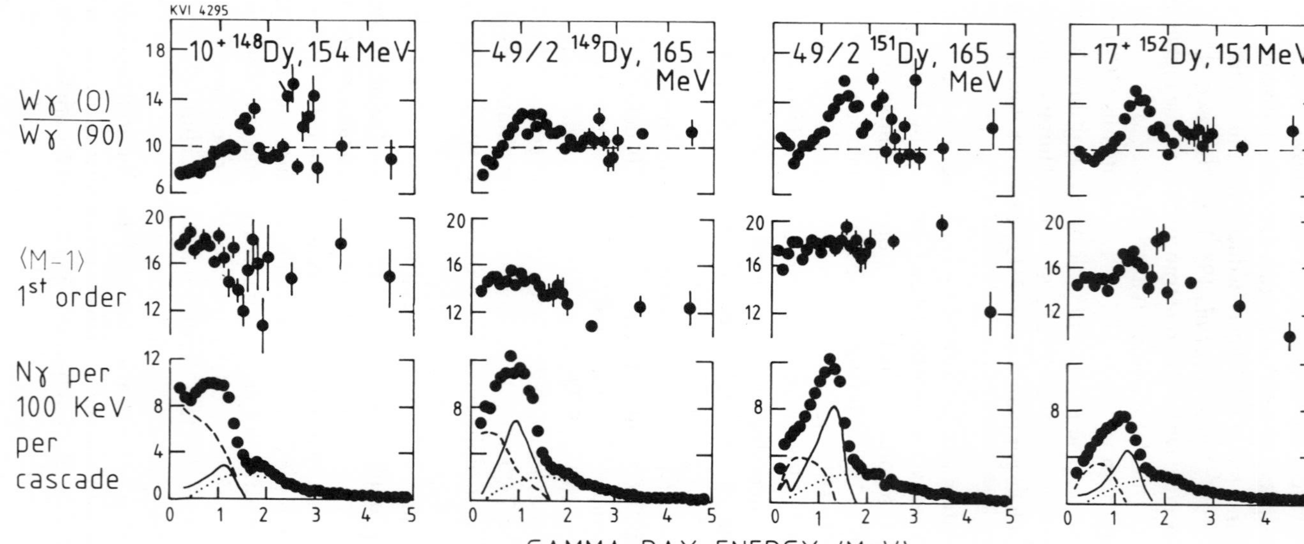

Fig. 6.73. The angular asymmetries (top), multiplicities (middle) and intensities (bottom) of prompt γ rays feeding high-spin isomers in Dy isotopes. The asymmetries have been obtained by channel division of the spectra measured with detectors at 0° and 90° and the multiplicities by coincidence and singles spectra measured with the same detector. The intensity distributions are deconvoluted into three components: statistical γ rays (dotted line), stretched quadrupole ones (solid line) and a low-energy dipole component (dashed line). (Ward *et al.*, 1983.)

For collectively rotating even–even nuclei the relation may be generalized by the simple formula: $I=2(M-M^s)$ (Simon *et al.*, 1976; Sarantites *et al.*, 1976). With this relation it is assumed that the statistical transitions, represented by M^s, do not carry off angular momentum, and usually $M^s=4$ was taken. In reality, however, some angular momentum can be released and M^s may vary with the total multiplicity M and with the reaction mechanism (Sie *et al.*, 1981; Łukasiak *et al.*, 1982). Another assumption in the above relation, namely that the γ rays release either two or zero units of angular momentum seems to be rather crude. The multipolarity of γ rays in the qcs has now been measured for several different cases, and varies with the γ-ray energy (see Section 6.5.5 below). The following procedure takes into account the energy dependence of the multipolarity and of the side-feeding multiplicity (de Voigt *et al.*, 1979; Łukasiak *et al.*, 1982).

When no detailed information on the multipolarity of the qcs γ rays is available the relation between multiplicity and average angular momentum removed by γ rays, may be approximated by

$$\langle l_\gamma \rangle = 2(M-M^s)+0.5M^s. \tag{6.5.13}$$

Here M represents the measured total γ-ray multiplicity and M^s the number of statistical transitions. The estimated release of 0.5 units of angular momentum is based on the average result of many investigations. The quantity M^s should be estimated from the measured multiplicities of the high-energy ($E_\gamma > 1.5$ MeV) γ rays, and by fitting a statistical intensity distribution to the high-energy part of the measured γ-ray spectrum as discussed above.

When the multipolarity of the qcs γ rays is known then one can apply a more detailed method (see Section 6.5.5). Let us consider an yrast state with known spin I_j, for which the side-feeding multiplicity M^{sf} has been measured in various (n) γ-ray energy regions i. The angular momentum released by cascades passing through level j is then given by:

$$l_j = \sum_{i=1}^{n} M^{sf}(i)\Delta I(i) + I_j, \tag{6.5.14}$$

where the summation is performed over the total number n of energy regions with known average angular-momentum release of $\Delta I(i)$ in each region i. Note that ΔI is not necessarily the same as the γ-ray multipolarity unless the γ-decay proceeds along collective bands via stretched transitions. The values of ΔI range from ~ 0.5 at the high energy region (several MeV), where statistical transitions dominate, to 2.0 at low energy (a few hundreds of keV), where rotational transitions are predominant. The precise values obviously depend on the reaction and on the final nucleus. To obtain the average angular momentum of the entry region, one has to average the values calculated with eqn (6.5.14) over all observed yrast states. That quantity can also be obtained in a more direct way by using the continuum-feeding multiplicities $M^{cf}(i)$ instead of $M^{sf}(i)$ in eqn (6.5.14).

6.5.5 *Gamma-ray multipolarities*

Many experiments have been carried out to determine the multipolarity of γ rays in the qcs, employing various methods. This information is crucial in order to reveal the properties of the qcs and also to establish the relation between measured multiplicities and the corresponding angular momentum carried off by the γ rays (see Section 6.5.4).

A first indication, for instance, of the collective or non-collective behaviour of the qcs can be derived from the shape of the γ-ray energy and multiplicity spectra as discussed above. To be more conclusive, however, one needs to know at least the multipolarity of the radiation as a function of the γ-ray energy. We shall describe the various methods of determining the multipolarity and demonstrate that unambiguous results can only be expected from the combined evidence of several methods.

6.5.5a *Angular correlations*

The angular asymmetry relative to the beam axis in the γ-ray intensity is directly related to the change in spin by the corresponding transition and thus in most cases to the multipolarity of the radiation. In the commonly used experimental methods the angular distribution can be expanded in terms of Legendre polynomials P_k as:

$$W(\theta)=\sum_k \alpha_k P_k(\cos\theta). \tag{6.5.15}$$

The coefficients α_k depend on the spins of the states involved in the transition and on the angular momentum carried off by the γ rays. They can be determined from a least-squares fit of eqn (6.5.15) to the data. The result of such a procedure is given in the top part of Fig. 6.70, obtained with a 16-detector filter (see Fig. 3.15) in coincidence with discrete γ rays measured with a Ge-detector to identify the reaction channel (Hageman *et al.*, 1981). In coincidence experiments one measures a correlation rather than a distribution of γ rays as represented by eqn (6.5.15). In practice, however, the two patterns are very similar because the detection of one discrete γ ray in the Ge detector does not change the alignment in the high-spin region (Bauer, 1979; Deleplanque *et al.*, 1978a), as was the case in the above example.

In a simpler approach one compares the intensities measured with detectors, which are positioned at $\theta=0°$ and $90°$ with respect to the beam axis. For stretched quardupole radiation, for instance, one obtains a ratio $W(0°/90°)>1$ (usually ~ 1.5) and for stretched dipoles $W(0°/90°)<1$ (~ 0.7). For the examples discussed in the previous Section 6.5.4 asymmetries are given in the top parts of Figs 6.69 (Simon *et al.*, 1977) and 6.73 (Ward *et al.*, 1983). All these results are consistent with a pronounced stretched quadrupole component in the qcs around ~ 1 MeV and statistical transitions $\{W(0°/90°) \cong 1\}$ at high energy (>1.5 MeV), as already indicated in earlier reports (see e.g. Banaschik *et al.*, 1975; Newton *et al.*, 1977). A significant dipole component at low energy (<0.5 MeV) has also been reported by several groups (see e.g. Newton *et al.*, 1977; Ward *et al.*, 1983).

Let us illustrate the decomposition of the qcs into components with different multipolarities for a few light rare-earth nuclei as discussed in the example in Section 6.5.4 (Ward *et al.*, 1983). The energy spectra (bottom part of Fig. 6.73) have been measured with a 7.6 cm $\phi \times 7.6$ cm NaI(Tl) detector placed inside an anti-Compton shield. Discrete lines have been subtracted and a deconvolution procedure corrected for Compton-scattered γ rays. First a fit to the high-energy region of 2.2 to 3.5 MeV has been made using the statistical approach of eqn (6.5.8) with $2L+1=3$ yielding a distribution $\omega(E)$ represented by the dotted curve in Fig. 6.73. Next, the measured anisotropy (top part of the figure) was approximated by:

$$W(0°/90°) = \alpha(E)W_D + \beta(E)W_Q + \omega(E)W_S, \tag{6.5.16}$$

where $W_D(0°/90°) = 0.68$ for stretched dipoles, $W_Q = 1.48$ for stretched quadrupoles and $W_S = 1$ for statistical transitions. The resulting dipole, $\alpha(E)$, quadrupole, $\beta(E)$, and statistical, $\omega(E)$, distributions are drawn in the bottom part of Fig. 6.73. These numbers are valid for assumed prompt (<1 ps) transitions and for full alignment. De-alignment brings the values for W_D and W_Q closer to 1. In some of the cases the stretched-dipole component at low energy, as discussed above, is clearly signified by the dip in the anisotropy as opposed to the bump for stretched quadrupoles. This dip can also partly be caused by dipole yrast transitions.

A detailed angular correlation study has been reported for rotational Er and Hf nuclei and for non-collective Te and Gd nuclei excited by 131–219 MeV ^{48}Ca and ^{40}Ar beams (Deleplanque *et al.*, 1978a). The experimental arrangement consisted of a multiplicity array with six 7.6 cm $\phi \times 7.6$ cm NaI(Tl) detectors and three additional detectors in the reaction plane and one out of that plane, as illustrated in Fig. 6.74. The figure also contains the result of a calculation of angular correlation functions for the various possible multipolarities and angular momentum changes. The multiplicity and corresponding γ-ray spectra for multipolarity $L=1$ and $L=2$ are shown in Fig. 6.75. The data confirm again the presence of the low-energy E2 stretched component, which is particularly pronounced in the evaporation products of the rotational compound nuclei ^{164}Er and ^{170}Hf. A strong stretched dipole component below $E_\gamma = 1$ MeV is present in the spectra of the evaporation products of the compound nuclei ^{122}Te and ^{150}Gd, which have nearly closed shells. The stretched quadrupole component amounts to ~ 50 per cent of the

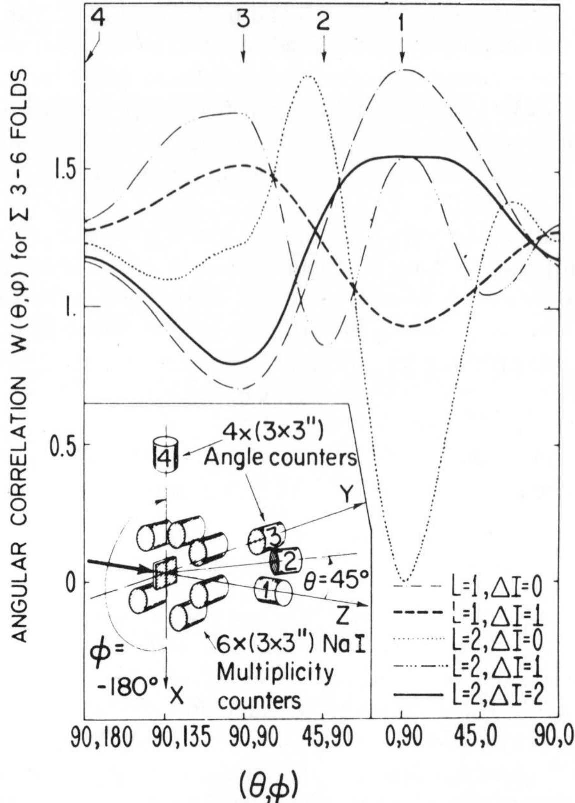

Fig. 6.74. The angular correlation functions are shown for the five transition types considered, with the assumption that three or more of the multiplicity counters fired. The numbers 1–4 indicate the actual counter position and the inset shows the experimental arrangement. The angle θ is the polar angle from the z-axis (beam direction), and ψ is the azimuthal angle measured from the x-axis. The values of $W(\theta, \psi)$ are traced in the xy plane from the 'up' position (counter 4) to the y-axis (counter 3), then in the yz plane to the z-axis (counter 1), and from there, in the xz plane to the x-axis. (Deleplanque *et al.*, 1978a.)

total intensity in the case of the near-spherical nuclei and to ~80 per cent for the rotational nuclei.

An example of the use of several detectors with a sum spectrometer is provided by the work of Folkmann *et al.* (1981). High-spin states in evaporation residues of ^{100}Ru and ^{160}Er compound nuclei were selected by gating on the high energy part of the total energy measured with the sum spectrometer. The anisotropies were measured by two detectors placed at 0° and 90° with respect to the beam axis. Multipole ($L=1$ and 2) spectra at

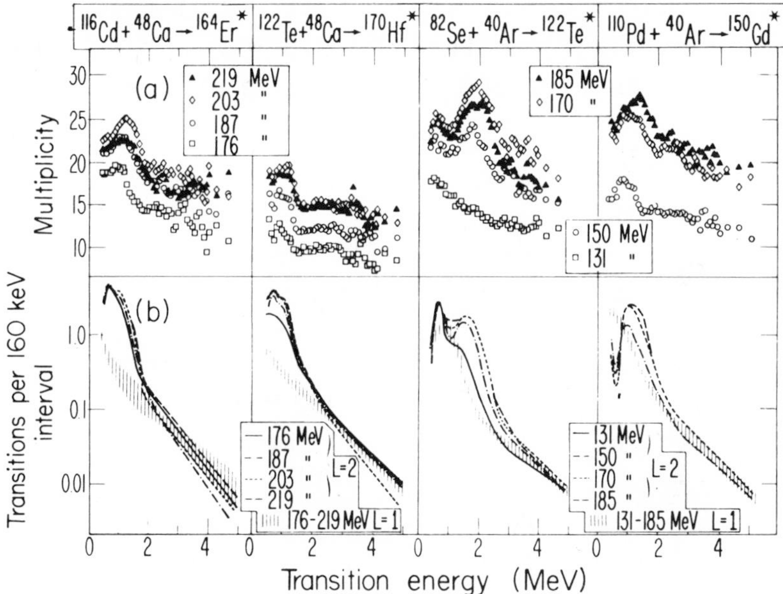

Fig. 6.75. The multiplicity spectra for the four indicated systems at several bombarding energies. (b) The corresponding multipole spectra are shown for stretched quadrupole and dipole components. The lines show the averaged quadrupole component for each bombarding energy (the individual points would have been too dense), while the hatched area covers the full range of the dipole spectra for all bombarding energies listed. (Deleplanque *et al.*, 1978a.)

low energy are presented in Fig. 6.76. They exhibit two bumps of mixed quadrupole and dipole character for the evaporation residues of both compound nuclei.

The quadrupole components below 1 MeV could be attributed to the discrete states; i.e. due to their vibrational character in the residues of ^{100}Ru and due to the known rotational character in those of ^{160}Er. Also in that investigation a strong stretched-dipole component at low γ-ray energies was found. The intensity of that component increased with the sum energy, indicating that the dipole transitions predominantly occur in the highest spin and energy region. The enhanced stretched quadrupole intensity above $E_\gamma = 1$ MeV in both the ^{100}Ru and ^{160}Er compound systems has also been attributed to that high-spin region. The high-energy transitions appeared to have nearly isotropic angular distributions, which is consistent with a mixture of stretched and non-stretched dipole transitions in the ratio of about 2:1.

It should be pointed out that the angular distribution results are generally far from being unambiguous. One of the typical ambiguities is due to the fact

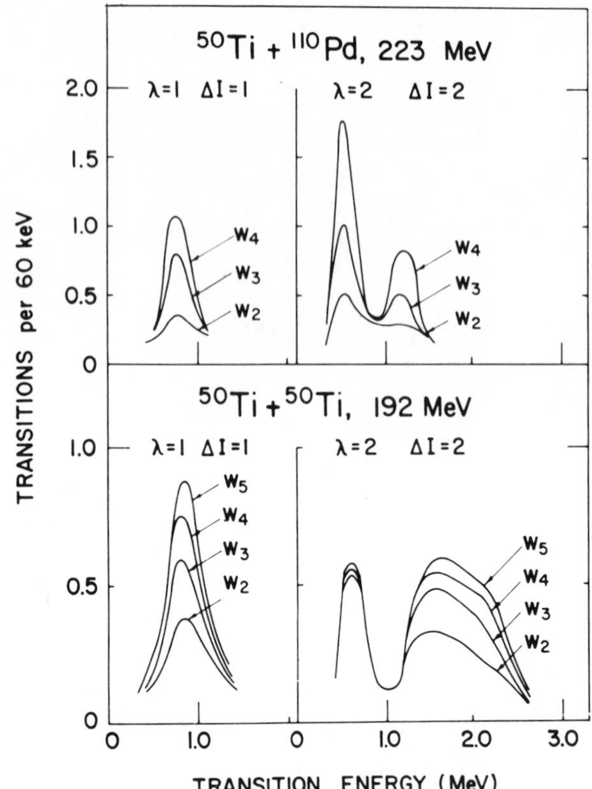

Fig. 6.76. Comparison of the $\lambda=1$, $\Delta I=1$ and $\lambda=2$, $\Delta I=2$ components in the multiplicity-normalized transition energy spectra for various windows W_i in the sum spectrometer energy signal. Here $i=1, 2, 3, 4, 5$ stands for increasing energy from 1 to 5. The upper part corresponds to the ^{50}Ti + ^{110}Pd reaction at 223 MeV and lower part to the ^{50}Ti + ^{50}Ti reaction at 192 MeV . (Folkmann *et al.*, 1981.)

that stretched quadrupole and non-stretched dipole transitions have about the same correlation function (see Fig. 6.74), which makes it difficult to distinguish between them. The same holds for non-stretched quadrupole and stretched dipole transitions, both of which have negative anisotropies.

Other problems may arise when exit channel selection is not applied, which leaves it unclear in what final nuclei certain effects occur. Also the influence of the discrete lines in the spectra is hard to eliminate. Unfolding and subtraction procedures have been applied, but they also introduce uncertainties. Generally speaking, angular correlation results concern averages over many states in the qcs, perhaps including some discrete states, and thus should be interpreted with care.

6.5.5b *Conversion electrons*

Additional information on the electromagnetic character of γ radiation can be obtained from the internal conversion electron coefficients (ICC). The first experiment to observe conversion electrons from the decay of states in the qcs has been carried out with a mini-orange spectrometer at Groningen for the reaction ^{160}Gd(α, 4n)^{160}Dy at 47 MeV bombarding energy (Feenstra *et al.*, 1977). The conversion coefficients have been determined from simultaneously measured electron and γ-ray energy spectra, both in coincidence with ground-state-band transitions measured with a Ge(Li) detector.

In that experiment the mini-orange spectrometer consisted of a magnetic filter with a central absorber, made of gold and lead, and of a cooled Si(Li) detector, as discussed in Section 3.4.3. The spectrometer is particularly suited for use in an experiment like this, because of the large opening angle ($\sim 65°$), giving a large solid angle, and the compact construction which leaves much space free for other detectors. The reduction of background is essential in view of the small conversion coefficients ($\sim 10^{-3} - 10^{-4}$). Nevertheless, the background has to be measured in the actual experiment, and this was done in the above experiment by withdrawing the mini-orange spectrometer from the target, which effectively shifted the transmission to higher energies. Consequently, one has to correct for the changed solid angle by measuring the spectra of calibration sources in both cases. Four different magnet configurations have been used to cover the desired range of energies.

The measured ICC are given in Fig. 6.77. The figure includes the results of ^{160}Dy and also those of a subsequent investigation of the rotational nucleus ^{162}Yb excited by a 105 MeV ^{20}Ne beam from the Louvain–La-Neuve cyclotron (Feenstra *et al.*, 1979). The figure also shows that gating on discrete transitions is essential, not only to select the proper residual nucleus, but more so to reduce the background and enhance the reliability. Here the gates were opened widely to show the difference to the deduced coefficients when no channel selection is made. These experiments confirm the dipole character of the statistical transitions at high energy, identifying them as E1, and the E2 character at low energy for the collective transitions. The same method has been applied in an investigation of high-spin superdeformed states as discussed in Section 6.5.7.

The use of a magnetic solenoid spectrometer has been reported in an investigation of the multipolarity of qc transitions in any evaporation residue of ^{170}Yb produced with 115 and 130 MeV ^{20}Ne beams from the Oak Ridge facility (Westerberg *et al.*, 1978). Their results are similar to the ones discussed above, but for the high-energy region where they found $\sim$55 per cent E2 admixture. It should be noted, however, that no channel selection was applied in that work. The latter work also indicated some possible M1 strength at low energy, a finding that has also been reported in very different investigations and mainly for less deformed nuclei (see discussion above in Section 6.5.5a, and below in the next subsection).

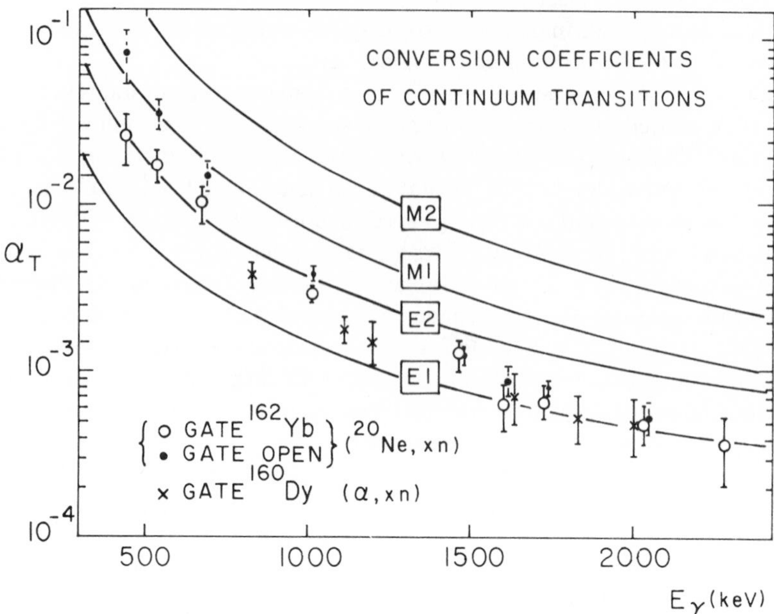

Fig. 6.77. Measured conversion coefficients α_T for the ^{146}Nd (^{20}Ne, 4n)^{162}Yb reaction (circles) compared with theoretical E1, E2 and M1 values (curves) and with 'pseudo'-coefficients obtained with a broad gate setting (dots). The mean-energy values correspond to various mini-orange settings, each covering a typical energy width of 30 per cent. The crosses represent the results obtained with the ^{160}Gd (α, 4n)^{160}Dy reaction at 47 MeV. (Feenstra *et al.*, 1977, 1979.)

6.5.5c *Linear polarizations*

The linear polarization of γ rays in the qcs has been measured for several nuclei employing the usual method with three-detector Compton polarimeters. Most experiments carried out so far yield zero or small polarization of the high-energy γ rays up to 2.5 MeV, consistent with the statistical character as a mixture of stretched and non-stretched E1 transitions or as a mixture of quadrupole and dipole radiation. In most experiments a small E2 admixture (<27 per cent) in the high-energy cascades could not be excluded. In the low-energy region stretched E2 transitions were found to constitute the bump around 1 MeV in all polarization measurements. So far those experiments confirmed the nature of the two different cascades in the qcs as discussed in the previous subsections.

The nucleus ^{152}Dy has been studied together with the well-deformed nucleus ^{160}Er with 85 and 89 MeV ^{16}O beams (Vivien *et al.*, 1979). The ^{151,152}Dy and ^{156,158}Er nuclei have been investigated with a 150 MeV ^{32}S beam (Trautmann *et al.*, 1979). The well-deformed ^{164}Er compound nucleus and the near closed-shell compound nucleus ^{150}Gd have been studied using a 170 MeV ^{40}Ar beam (Hübel *et al.*, 1980).

These experiments provided some additional information on various types of nuclei. In the near-spherical nucleus ^{152}Dy, for instance, the low-energy bump included up to 40 per cent E1/M1 admixtures, possibly due to known discrete transitions above the $I^{\pi}=17^{+}$, 60 ns isomeric state (Vivien *et al.*, 1979). For that nucleus a negative polarization was reported for γ rays below $E_{\gamma}=300$ keV, which indicates an M1 character. This finding seems consistent with the observation of low-energy dipole components for non-deformed nuclei, as discussed in the previous subsections. In the same experiment a much weaker polarization was found for the low-energy γ rays in the deformed nucleus ^{160}Er. The results obtained by Hübel *et al.* (1980) are summarized in Fig. 6.78. The yrast ($L=2$) bump around 1 MeV is particularly pronounced in the multipole spectrum of the ^{164}Er system and has a stronger $L=1$ admixture for the ^{150}Gd system. These multipole spectra have been deduced from the angular correlation measurements. The measured linear polarizations have values in between the stretched E2 and M1. A fit to the data shows a larger stretched M1 contribution in the energy region of 500 to 1600 keV of ^{150}Gd than of ^{164}Er.

A rather peculiar finding has been reported for ^{152}Dy, namely increased E2 strengths in the energy region $E_{\gamma}=1.0$ to 1.5 MeV (Trautmann *et al.*, 1979). This component is believed to be due to increased collectivity in the spin region above 40. Confirmation of this finding has more recently been derived from multiplicities, as discussed above in Section 6.5.4. We shall see in the final part, Section 6.5.6 of this chapter that more evidence for shape changes in ^{152}Dy has been added very recently.

We have illustrated here with a few examples that linear polarization measurements can reveal the electromagnetic character of γ rays, even in the qcs. Nevertheless, several ambiguities in the results are very hard to remove because of the lack of sharp channel selection and the influence of discrete lines, particularly at low energies. Another uncertainty arises from the common assumption of the absence of non-stretched dipole transitions of which even a weak admixture would affect the linear polarization significantly.

6.5.5d *K-shell ionization*

The average multipolarity of γ radiation from the qcs can be inferred from the observation of the characteristic K X-rays from the reaction residues. Each conversion electron creates a vacancy in the atomic (K) shells. The conversion electron process depends very strongly on the multipolarity of the transition, as was shown above in Section 6.5.5b. The inner-shell vacancy is filled in a time much shorter than the nuclear lifetime and the effect will thus be additive

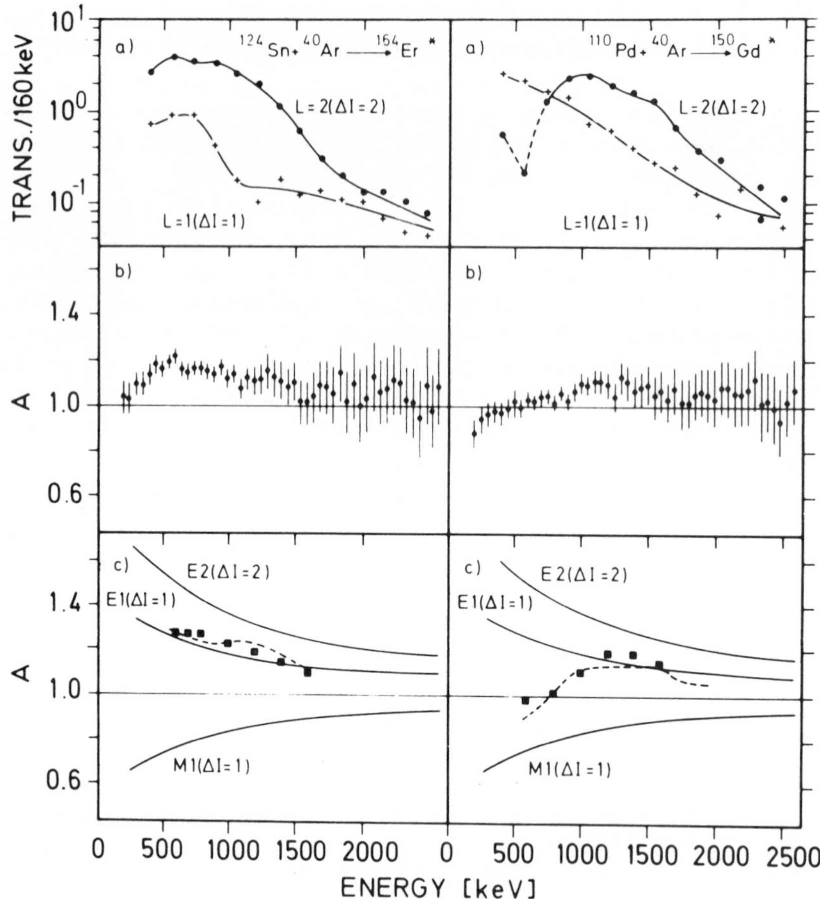

Fig. 6.78. (a) Multipole spectra for stretched quadrupole and dipole components from angular correlation measurements. (b) Linear polarization asymmetry double ratios $A = \dfrac{N_\perp(90^\circ)}{N_\parallel(90^\circ)} \Big/ \dfrac{N_\perp(0^\circ)}{N_\parallel(0^\circ)}$ from measurements at $\theta = 90^\circ$ and 0° to the beam direction (direct experimental points). (c) Linear polarization asymmetry double ratios A after unfolding (closed squares), and calculations for pure stretched M1, E1, and E2 (solid curves) and for a mixture of the E2 and M1 components of the decay (dashed curve) with the $L = 2$ and $L = 1$ amounts taken from the angular correlation experiment (upper part of the figure). (Hübel *et al.*, 1980.)

for multiple conversions. Therefore each conversion is approximately accompanied by a K X-ray emission or in the case of low-energy transitions by L X-rays. This method can only be applied when the level schemes of the reaction products are well known. The conversion electron probability due to discrete transitions then can be calculated and the resulting K X-ray yield can be subtracted from the number of observed K X-rays, attributing the remaining yield to the qcs. This method is particularly sensitive for low-energy M1 radiation because of its high conversion probability. The qc contribution, particularly in well-deformed nuclei, can be much smaller than that of the discrete transitions, and this makes the experiments difficult.

The method was first applied in an investigation of well-deformed Dy isotopes excited in alpha-particle and ^{12}C-induced reactions (Sujkowski *et al.*, 1979; Chmielewska *et al.*, 1981). The observed K-shell ionization yields were completely accounted for by the known electron conversions due to discrete lines for a wide range of bombarding energies. From that observation it was concluded that transitions in the qcs with high conversion probability, such as low-energy M1 transitions, must be weak or absent.

At about the same time as the above investigation a study has been carried out on transitional nuclei in the mass region around 200, using 55 to 124 MeV ^{6}Li beams (Karwowski *et al.*, 1979, 1982). Their results were expressed in X-ray multiplicities obtained from coincidences between two or three X-ray detectors or between a γ-ray and an X-ray detector. They found relatively large values, which could only be accounted for when the qcs would contribute e.g. by means of low-energy M1 radiation. A drop in the K X-ray multiplicity was observed for nuclei near the deformed or spherical symmetric regions.

A specific study has been devoted to the non-deformed nucleus ^{153}Ho excited by 150 to 175 MeV ^{37}Cl beams from the Argonne superconducting linac accelerator (Janssens *et al.*, 1985). They used a 16-detector multiplicity filter, two Ge detectors, one for γ rays and one for X-rays, and a 7.6 cm ϕ × 7.6 cm NaI(Tl) detector for continuum γ rays. High-spin states were selected by the requirement that at least three detectors of the filter fired in prompt coincidence with one or more of the additional detectors and one fired in delayed coincidence. The latter condition selects the feeding of the known 229 ns isomer with spin $31/2^{+}$. Such a selection is very effective and is nowadays applied whenever possible. The measured K X-ray yield increased with bombarding energy, corresponding to an angular momentum of $\sim 50\hbar$ at the highest energy. That yield exceeded the one calculated from the well-known level scheme, particularly at the highest bombarding energies. The conclusion thus pointed to the occurrence of low-energy (<350 keV) M1 transitions in the qcs.

A quite different method has been applied in a study of $^{154-158}$Er nuclei by measuring the atomic charge distribution after fusion reactions with ^{32,34}S beams (Cormier *et al.*, 1984). The reaction products were identified with a recoil spectrometer, after they have recoiled into vacuum with approximately equilibrated atomic charge distributions. It was estimated that a single conversion electron, occurring during the flight of the ion, changed the charge state of the ion by ~ 5 to 6 units (including ~ 4 to 5 Auger electrons). From the shift in the mean charge distribution one may determine the number of conversions. The method is thus severely limited to cases with significant conversion of at least a few units.

In conclusion one observes rather consistently low-energy M1 radiation from the decay of weakly-deformed states and no or only weak dipole radiation from deformed states. Possible interpretations are presented in Section 6.5.6. It should be clear from the above discussion that the K X-ray method offers only the possibility of implicitly observing the M1 component in the qcs. For unambiguous conclusions combined evidence with other methods is required, as already pointed out earlier in this chapter.

6.5.5e *Lifetime measurements*

The multipolarity of γ radiation determines to a large extent the transition probability (and thereby the lifetime) of a nuclear level (see Section 6.3.5). The lifetime, being the inverse of the transition probability, thus also reveals in many cases the multipolarity.

The lifetimes of nuclear levels in the qcs have already been measured some time ago by means of the recoil distance Doppler shift technique. Most of those early experiments indicated upper limits of only a few tens of a picosecond (Diamond *et al.*, 1969; Kutschera *et al.*, 1972; Newton *et al.*, 1973; Ward *et al.*, 1973; Rud *et al.*, 1973; Bochev *et al.*, 1975). This led to the conclusion that transitions in the qcs with an energy of ~ 1 MeV could have only dipole or strongly enhanced electric quadrupole character. More recently, some more refined techniques could be applied and, for instance, the feeding times into specific high-spin states could be determined.

The recoil distance technique has been applied to a study of $^{156-158}$Dy isotopes, using the inverse reactions ^{24,25}Mg(^{136}Xe, xnγ) at ~ 4.1 MeV/A bombarding energy (Emling *et al.*, 1981, 1984). The technique used in the early investigation of discrete states is described in Section 6.3.5. In that work, feeding times to high-spin states shorter than 1 ps have already been reported (see Fig. 6.31). The slowing-down time of Dy recoils in solid material, however, amounts to a few picoseconds. Consequently, the high-spin γ transitions taking place during the slowing-down process result in Doppler-broadened lines. In later work they have modified the technique in order to measure shorter lifetimes with better precision as described in Section 3.5.

To deduce the side-feeding times from the data one has to assume a certain model for the feeding cascades. Usually one assumes the feeding to proceed via rotational cascades with $\Delta I = 2$ transitions, parameterized by e.g. the moment of inertia $\mathscr{J}$ and an average transition quadrupole moment Q_t. The relation between the lifetime τ and the quadrupole moment Q_t is given by eqn (6.3.86). Here the γ-ray energy E_γ and angular momentum I, as deduced from the measured multiplicities, are related within the rotational model by eqn

(6.3.37). In the data fitting procedure the initial population of the feeding cascades and their ratio $Q_t^2/\mathscr{J}^5$ were treated as free parameters. The number of transitions in the feeding cascades were given by the side-feeding multiplicity M^{sf}. Feeding times as short as 0.3 ps to the highest-spin states (28^+, 30^+ in ^{156}Dy) have been deduced. The average quadrupole moments Q_t of the side-feeding cascades in the spin region 16 to 42 amounted to 5 to 6 eb, about the same as those of the collective yrast bands (Emling *et al.*, 1984). These findings imply the existence of rotational bands in the qcs roughly parallel to the yrast line with average collective properties similar to the yrast sequences. This is a confirmation of earlier indications, as discussed in the previous subsections.

Techniques similar to those described here have also been applied to the less-deformed nuclei ^{152,154}Dy. In ^{152}Dy the feeding times to high-spin states were found to be significantly longer, $\sim$10 ps, than for the heavier Dy isotopes, whereas in ^{154}Dy both slow ($\sim$10 ps) and fast (<1 ps) components have been found (Azgui *et al.*, 1985). The implications for the structure of the qcs are discussed in Section 6.5.6.

A very different method has to be applied to measure lifetimes much shorter than 1 ps. This can be done with the Doppler shift attenuation method as already discussed in Section 3.5. With this technique the γ ray is attenuated during the slowing-down process in a thick target or backing and the result is a shift in energy and a broadening of the line in the spectrum. Both the shift and line shape can be calculated if the stopping powers are known, with the decay time as a free parameter. We will illustrate this method with two examples; one measuring directly the γ rays from the qcs and another measuring discrete γ rays from high-spin states and subsequently deducing the feeding times.

The quasi-continuum γ spectrum has been measured for deformed residues of the ^{163}Ho and ^{164}Er compound nuclei excited by a 700 MeV ^{136}Xe beam from the LBL superHILAC at Berkeley (Hübel *et al.*, 1978). The experiment has been repeated with 720 MeV ^{129}Xe and 740 MeV ^{136}Xe beams on a 2.3 mg/cm^2 thick ^{28}Si target with and without a 25 mg/cm^2 thick gold backing, producing ^{157}Er and ^{164}Er compound nuclei (Hübel *et al.*, 1982). The evaporation residues of ^{164}Er are known to be deformed and those of ^{157}Er to be non-deformed.

The basic idea was to compare spectra measured at 0° with and without the backing. In the case without the backing the γ rays are emitted from the recoil ion at full velocity ($v/c \sim 0.085$) and are thus fully shifted in energy. In the case with backing, retardation causes less energy shift for longer lifetimes.

The experimental arrangement consisted of a six 7.6 cm $\phi \times$ 7.6 cm NaI(Tl) detector multiplicity filter, a 12.7 cm $\phi \times$ 15.2 cm NaI(Tl) detector at 0° and one at 90°, for normalization purposes, and a Ge(Li) detector to determine the reaction residues (see Fig. 6.79). Data were recorded only when two detectors of the filter were in coincidence with one of the additional ones. To show the

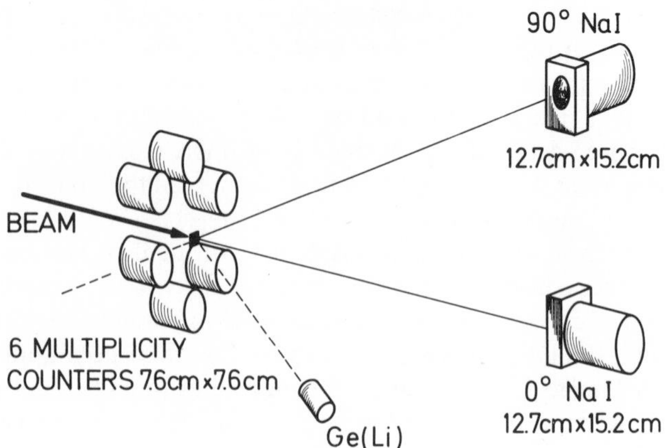

Fig. 6.79. The multiplicity filter and NaI(Tl) detectors to measure lifetimes of states in the qcs of evaporation residues of the $^{157,\,164}$Er compound nuclei. Using the Ge(Li) detector those residues could be identified. (Hübel *et al.*, 1982a.)

effect, the ratios of the normalized spectra (unbacked/backed) are given in Fig. 6.80 for the two systems. Qualitatively, one can already conclude that the decay times in the collective bump around 1 MeV must be longer (positive ratio) than at high energy (no difference). It is also clear that the times involved in the high-energy decay of the light system are longer than in the heavy one. The solid lines in the figures represent the results of calculations based on known stopping powers (from the tables of Northcliffe and Schilling, 1970; Blaugrund, 1966) and a decay model. The model basically included rotational feeding cascades with a certain moment of inertia and an average intrinsic quadrupole moment, fitted to the data. In the light (less deformed) system, however, dipole radiation had also to be taken into account.

The reported lifetimes in the collective yrast bumps are 0.055 ± 0.020 and 0.07 ± 0.05 ps in the heavy and light Er systems, respectively. These values can be related using eqn (6.3.86) to the reduced transition quadrupole moments $Q_t = 6.1$ and 5.4 eb, or 170 and 150 Wu., respectively (Hübel *et al.*, 1982). The high-spin states in the qcs decaying by yrast-like cascades thus have collectivities similar to those of the ground-state bands of the known deformed rare-earth nuclei. This agrees with the observations of Emling *et al.* (1984), discussed above. The peculiar feature of the enhanced E2 transitions here is that this seems also be the case in the light Er isotopes, which exhibit non-collective behaviour at low spin, implying a shape change at high spin (>40). On the other hand lifetimes associated with the statistical cascades appear much longer in the lighter Er system (~ 0.6 ps). In the heavy system they cannot be measured beyond 1.6 MeV. It may be that those somewhat retarded

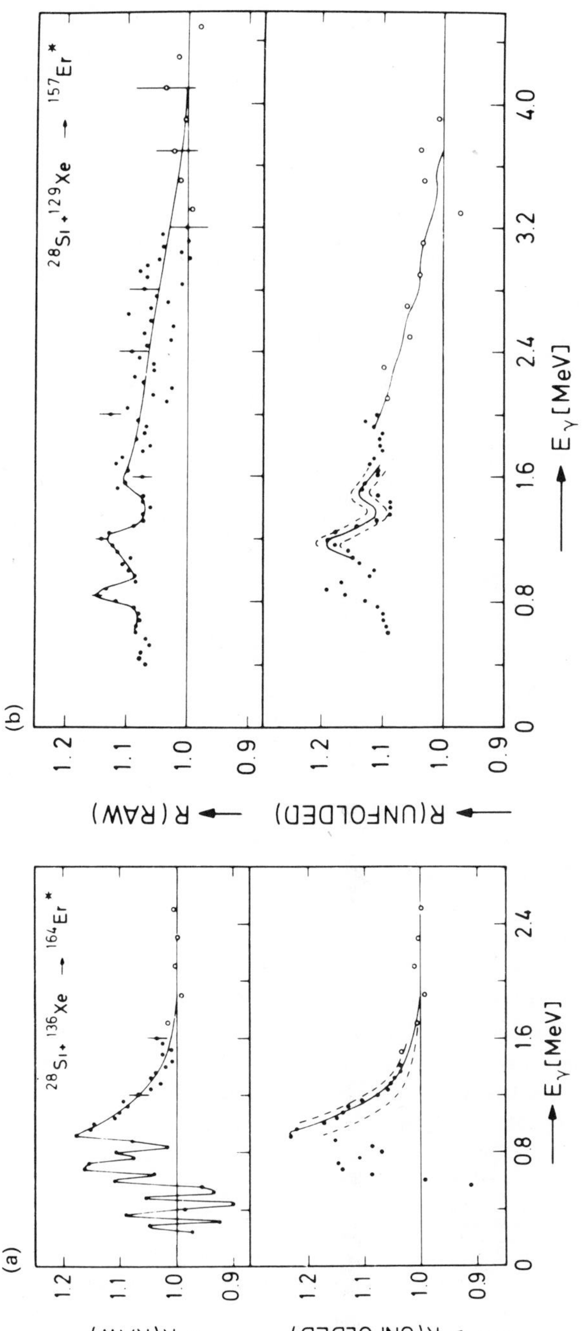

Fig. 6.80. (a) Upper part: Ratio of the raw (not unfolded) spectra measured at 0° with the ^{136}Xe beam on a self-supporting ^{28}Si target and a ^{28}Si target on a Au backing. The spectra were normalized using the integrated counts of the 90° detector. The curve is drawn to guide the eye. Lower part: Ratio of the unfolded experimental (dots) and calculated spectra (solid curves). The dashed curves were calculated for the extreme values of the intrinsic quadrupole moment Q_0, possible in the fit. (b) The same for the lighter ^{157}Er* system excited with a ^{129}Xe beam on the same target as for (a) (Hübel *et al.*, 1982a.)

statistical transitions connect the collective states at the highest spins in the light Er isotopes with the less deformed yrast states.

Let us conclude this section with the first results obtained with the Berkeley High-Energy-Resolution-Assembly (HERA) consisting of 21 BGO Compton-shielded Ge detectors, as discussed in Section 3.2. The many detectors give a very high doubles coincidence count rate and even an adequate triples rate. Consequently one is able to gate on many discrete peaks and observe the high spin states very cleanly in coincidence as discrete lines. This allows, for the first time, a detailed lineshape analysis of discrete states with spins up to 34 with the DSA technique.

The nucleus ^{166}Yb has been studied with HERA using the 180 MeV ^{40}Ar beam of the Berkeley 88″ cyclotron (Bacelar *et al.*, 1986). Thin ^{130}Te targets (of 360 and 200 μg/cm^2 stacked together) were used to observe sharp lines corresponding to very fast transitions at high spin, as shown in the top part of Fig. 6.32. The topmost transition in that figure corresponding to the decay of the 32^+ state exhibits a Doppler shift of only 11 keV, to be compared with a full shift of 18 keV in the case of negligible lifetime. From this number a feeding time 0.4 ps was derived for the cascades feeding the high-spin levels.

In addition, experiments have been carried out with a thick target (1 mg/cm^2) that was deposited on a gold backing to stop the recoils. The selected clean gating transitions were slow enough to be stopped in the backing and consequently had sharp lines. The spectra measured in coincidence and added together for the most forward ($\sim 40°$) and the most backward ($\sim 140°$) detectors are also presented in Fig. 6.32. The low-energy peaks are sharp and are not shifted in the two spectra, indicating that the corresponding lifetimes are longer than the slowing-down time. The high-energy peaks, however, are shifted and broadened. In fact we can observe that some of them consist of two components and this implies that the side feeding into these states is different from the in-band feeeding. From the intensities of the two components it has been inferred that the slow component (left-side in forward and right-side in backward direction) is associated with the side feeding. The actual lineshapes can be analysed with the Doppler Shift Attenuation Method (DSAM) as discussed in Section 3.5. In the analysis the same model has been used as discussed above; i.e. rotational side-feeding cascades with the same energies as the yrast sequence. Each cascade feeding an yrast state has been fitted with the intrinsic quadrupole moment as a single parameter. The deduced lifetimes varied from 0.19 ps for the 24^+ state to 0.05 ps for the 34^+ state, corresponding to 190 and 120 Wu, respectively. The lifetimes for the side-feeding cascades tuned out to be longer, corresponding to B(E2) values of 80 and 40 Wu, respectively. The remarkable decrease in collectivity towards high spin has been interpreted in terms of a possible shape change from prolate to triaxial.

We may have noticed in the discussions in this section that more and more evidence of shape changes has been obtained. They seem to occur at the highest observed discrete states or just above them. More details about these shape changes will be discussed below.

6.5.6 *Very fast nuclear rotation; changes in nuclear shape*

At very high spin, nuclear (pairing) superfluid correlations tend to disappear and consequently the kinematical moment of inertia is expected to approach the corresponding value for a rigid body (see Section 6.3.4). The quantum character of fast nuclear rotational motion then manifests itself by the alignment of the nucleonic angular momenta with the rotation axis. We have seen in Section 6.3.2 that in the case of prolate deformations such an alignment may result from the crossings of single-particle Routhians. It is well known by now from studies of rotating single-particle prolate potentials that particularly high-j low-Ω orbital energies decrease very much with increasing rotational frequency, due to the Coriolis interaction (see Section 6.3.3). Consequently, particles in those orbitals overcome most easily the pairing gap and one can expect an increased number of these high-spin levels near the Fermi surface (yrast state) in the case of fast rotation.

The single-particle alignments may induce shape polarization of individual nucleonic orbitals, which provides a link between the individual nucleonic alignments and the shape evolution of the nucleus as a whole. The nuclear shape evolution as a function of angular momentum may then be very sensitive to the particular N and Z because the positions of the single-particle orbitals are N and Z-dependent. Thus the experimental investigation of the nuclear shape as a function of N and Z is important in order to reveal the quantal effects (shell structure). It is another manifestation of the fast rotating nucleus that quantal effects are so crucial in determining the collective (shape) properties.

Transitional nuclei provide favourable conditions to study shape changes. We will refer to the rare-earth nuclei in between the double-closed shell $N=82$, $Z=64$ and the deformed nuclei, $N\sim 90$ and $64<Z<70$ to illustrate the various methods, because those nuclei have been studied extensively. At very high spin (>50) almost all these nuclei exhibit appreciable collectivity, culminating in possible super deformation in ^{152}Dy (Nyako *et al.*, 1984). At lower spins various shapes seem to compete, as indicated by discrete γ-ray spectroscopy. Above spin 40 no resolved lines have yet been observed, except in a few cases, and therefore most of the information has to be deduced from the quasi-continuum spectra. In the discussion of these spectra we will try to link the information as much as possible to the discrete line spectroscopy and therefore restrict ourselves to only a few examples. We will first present a very brief account on some relevant theoretical notions and predictions of shape evolution with increasing angular momentum. Thereafter the experimental evidence obtained so far will be discussed with the exciting possibility of super deformed high-spin states at the end (Section 6.5.7).

6.5.6a *Theoretical calculations of shape evolution*

A most convenient way of representing theoretical results is the graphical picture of potential-energy surfaces in the (β, γ) plane as a function of angular momentum for each nucleus. The connections between the equilibrium minima for each spin value form the spin trajectories and show the shape evolution as a function of angular momentum. Here we restrict ourselves to some basic notions and the most recent data. For details we refer to Szymański (1983).

One of the most successful approaches in high-spin nuclear physics today is the self-consistent cranked Hartree–Fock–Bogoliubov method, HFBC (see for an application to shape evolution Tanabe and Sugawara-Tanabe, 1984). Alternatively, the Cranking approximation is commonly applied using the Strutinsky shell-correction method and adequate single-particle potentials (see e.g. Dudek and Nazarewicz, 1985).

At high angular momenta the sum of the single-particle energy eigenvalues as obtained from the Cranking approach (see Section 6.3.3a) does not yield the total energy of the nucleus. Another problem is the poor description of the average nuclear properties. A better approach is the shell-correction method where a separation is accomplished between the average nuclear effects and those sensitive to single-particle configurations (see Myers and Swiatecki, 1966, 1967; Strutinsky, 1967, 1968). The method consists in replacing the density of discrete single-particle levels by a smoothed one, and this is accomplished by introducing a smooth function like a Gaussian distribution for each level. Two density functions are needed, namely ρ_1 for the description of the level densities and ρ_2 for the spin densities. In addition the single-particle energies should be replaced by the Routhians e^{ω} (see Section 6.3.3).

The smoothing has to be carried out so that the total number of particles A and the total angular momentum I are preserved. They are calculated in the non-smoothed system as:

$$A = \int_{-\infty}^{\lambda} \rho_1(e^{\omega})\, \mathrm{d}e^{\omega} \tag{6.5.17}$$

$$I = \int_{-\infty}^{\lambda} \rho_2(e^{\omega})\, \mathrm{d}e^{\omega}. \tag{6.5.18}$$

The energy is then given by:

$$E = \int_{-\infty}^{\lambda} e^{\omega} \rho_1(e^{\omega})\, de^{\omega} + \hbar\omega \int_{-\infty}^{\lambda} \rho_2(e^{\omega})\, de^{\omega}$$

$$= \sum_{\nu=1}^{A} e^{\omega} + \hbar\omega I. \tag{6.5.19}$$

In the smooth system one uses the smooth densities $\tilde{\rho}_1$ and $\tilde{\rho}_2$ and finds different values for the Fermi energy $\tilde{\lambda}$ and rotational frequency $\tilde{\omega}$. The smooth energy $\tilde{E}$ is found by substituting $\tilde{\rho}_1$, $\tilde{\rho}_2$, $\tilde{\lambda}$ and $\tilde{\omega}$ into these equations.

The total energy of the nucleus is obtained by superimposing the shell correction $\delta E = E - \tilde{E}$ on the (classical) energy for the macroscopic system E_{macr}:

$$E_{\mathrm{tot}} = \delta E + E_{\mathrm{macr}}. \tag{6.5.20}$$

The macroscopic energy can be written as the sum of a rotational energy (see eqn 6.3.2) and the energy as estimated in terms of the liquid drop model:

$$E_{\mathrm{macr}} = E(Z, N, \beta, \gamma) + (\hbar I)^2/\{2\mathscr{I}(\beta, \gamma)\}. \tag{6.5.21}$$

In this expression both terms depend on a set of deformation parameters, for instance, β and γ. In terms of the liquid drop model one may use the hydrodynamical moment of inertia:

$$\mathscr{I} = \mathscr{I}_{\mathrm{hydr}} = (2/5)AM(\Delta R)^2, \tag{6.5.22}$$

where ΔR represents the difference between the major and minor semiaxes of the spheroid. This is the moment of inertia for the irrotational flow of a liquid drop with deformation $\beta = 1.06\Delta R/R$. In the Cranking approach, however, the rigid-body value seems a more realistic one to use. The general programme consists in finding a minimum energy from eqn (6.5.20) for each fixed angular momentum I, thus yielding the deformation parameters, β and γ.

The energy surfaces for many rare-earth nuclei have been calculated as a function of angular momentum using the method described above by several groups (see e.g. Andersson *et al.*, 1976). The shape evolution of transitional Gd, Dy, Er, and Yb isotopes have been calculated in the spin range 30 to 80 with special emphasis on the description of the single-particle levels using a Woods–Saxon potential (Dudek and Nazarewicz, 1985). They used a generalized Strutinsky method in a Cranking approach which allowed them to keep track of separate rotational bands with definite parity and signature. The pairing effect was ignored as it is expected to be negligible at high spin (> 30). The results for the Dy isotopes are shown in the form of an E vs. I plot in Fig. 6.81. The characteristic features are the dominant oblate shape ($\gamma = 60°$) at low spins in the light isotopes, changing to prolate in ^{154,156}Dy. With

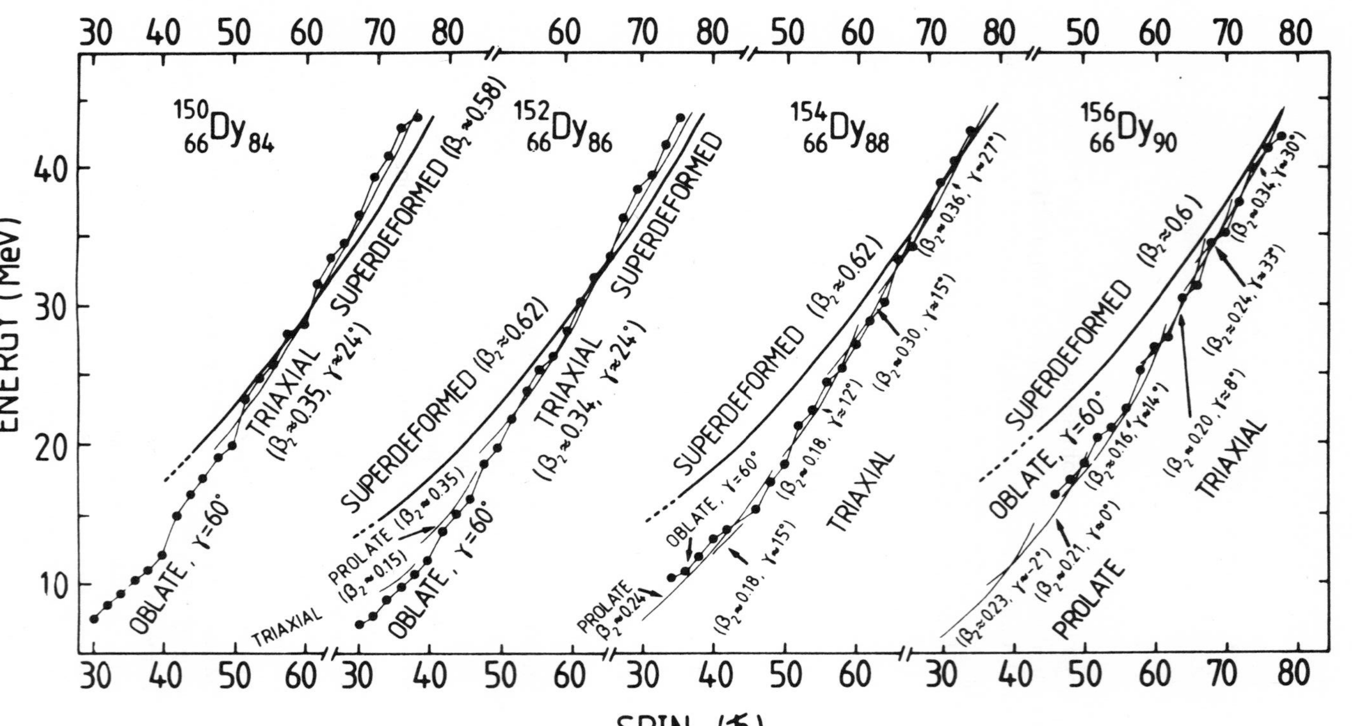

Fig. 6.81. Results of cranking model calculations on the high-spin behaviour of $^{150-156}$Dy. The low-lying superdeformed band in ^{150}Dy should be particularly noted. Pairing is ignored. Here, only the states with total parity $\pi = \pi$ (protons)π(neutrons) $= +1$ and the total signature $r = r$(protons) r(neutrons) $= 1$ are displayed. It has been checked, however, that these results are representative of the shape evolution in most of the structures with the remaining parity-signature combinations, i.e. $r = -1$, $\pi = -1$. (Dudek and Nazarewicz, 1985.)

increasing spin, the energies of the triaxial and superdeformed shapes are close to each other and become near-yrast in ^{150,152}Dy at rather high spin ($\sim$ 50 to 60). The lowest states in the figure are considered as yrast states. Lower spin values ($\sim$ 45 to 55) for this feature have been reported, for instance by Åberg (1982). The superdeformed shapes appear at much higher energies above the yrast line in the heavier Dy isotopes. These calculations thus indicate that the light Dy isotopes are good candidates for the observation of superdeformed states, as discussed below in Section 6.5.7. The nuclei ^{154,156}Dy, on the other hand, exhibit a change from prolate to oblate as I increases beyond 40 to 50, a feature that has already been observed (at lower spin, $\sim$30) in ^{154}Dy by means of discrete-line spectroscopy (Pakkanen *et al.*, 1982).

The lowest-lying superdeformed bands are calculated for the light Gd isotopes, particularly in ^{144}Gd (Dudek and Nazarewicz, 1985). In microscopic terms this may be attributed to the large proton energy gap at $Z = 64$ at very large deformation because shell effects here cause secondary minima in the total energy surfaces. A shape change from prolate to oblate was calculated to occur at $I \sim 30$ in ^{156}Er, which is in line with observations of discrete γ rays (see Section 6.5.6b below).

Let us now present some graphs of the energy surfaces using the Er isotopes as an example (Cwiok *et al.*, 1983). In the procedure sketched above they include, besides the usual quadrupole deformation β_2, also hexadecapole deformations β_4. The results for ^{152,154,158}Er are shown in Fig. 6.82. Note that in the full β–γ plane only one sector is given; from $\gamma = 0^\circ$ (prolate) to -60° (oblate, rotating around the symmetry axis, which corresponds to $\gamma = +60^\circ$ in our convention of Fig. 6.15). Both ^{152}Er and ^{154}Er exhibit a distinct minimum corresponding to oblate shapes, but they develop triaxial and prolate minima for $I > 50$. The oblate to prolate shape transition occurs in ^{146}Er at spin $\sim$ 40 and thus at increasing spin values with increasing neutron number. In ^{158}Er a distinct prolate-to-oblate shape transition is predicted to occur at $I \sim 46$, again in agreement with the observations discussed below.

The nucleus ^{158}Er has also been studied by Tanabe and Sugawara-Tanabe (1984) together with ^{160}Yb in the HFBC approximation. They added the quadrupole-pairing interaction (QPI) to the monopole-pairing-plus-quadrupole interaction. The QPI seemed to cause a sharp backbend and to shift the critical frequency ω_{cr} (see Section 6.3.3) to a higher value.

The QPI between neutrons and protons was found to play an essential role for $I < 30$ in the decrease in energy of single-particle levels near the Fermi surface, particularly of the $\pi h_{11/2}$ under certain conditions. Consequently, the alignment of these nucleonic orbitals becomes very important for the change towards oblate nuclear shapes. The shape evolution of the two nuclei as the angular momentum change is presented in the form of spin trajectories in Fig. 6.83. Both nuclei have prolate deformations with $\beta \sim 0.2$ up to spin $\sim$ 40; they

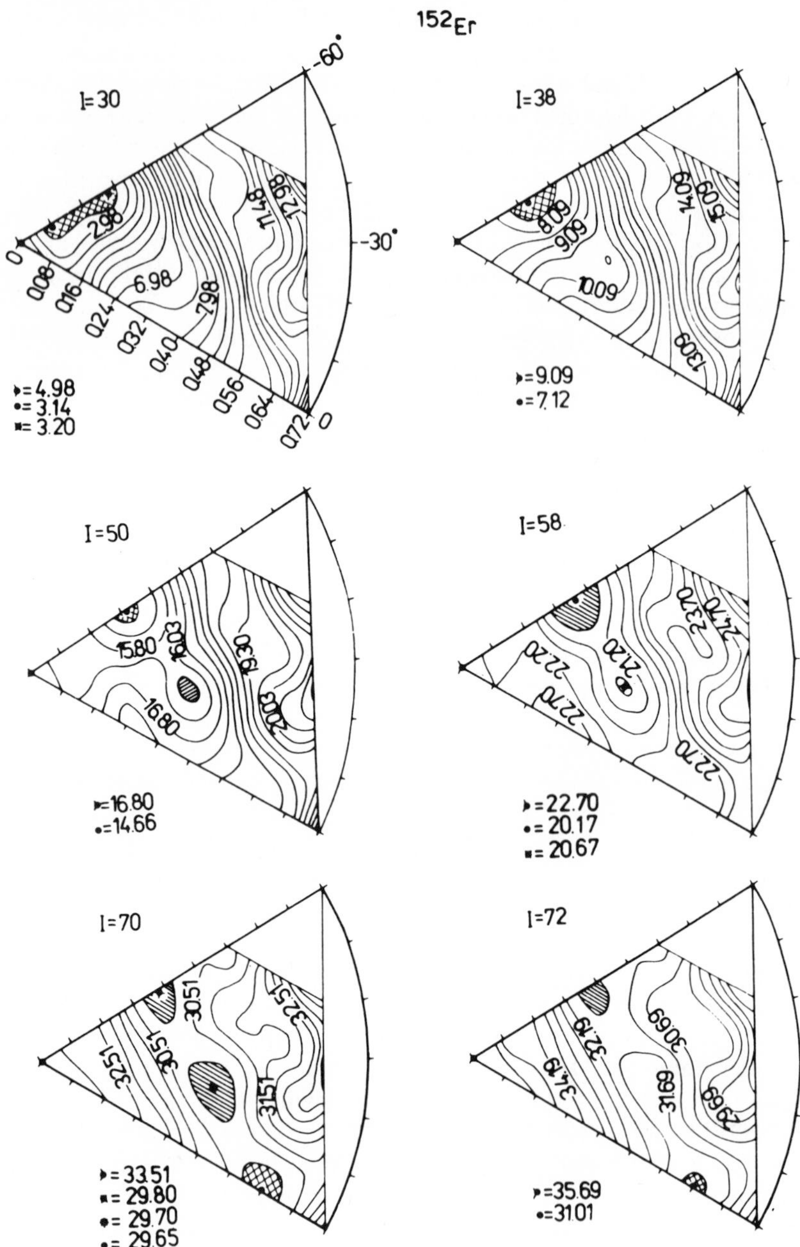
152Er
I=30
I=38
I=50
I=58
I=70
I=72
-60°
-30°
▸=4.98
•=3.14
▪=3.20
▸=9.09
•=7.12
▸=16.80
•=14.66
▸=22.70
•=20.17
▪=20.67
▸=33.51
▪=29.80
●=29.70
•=29.65
▸=35.69
•=31.01

Fig. 6.82.

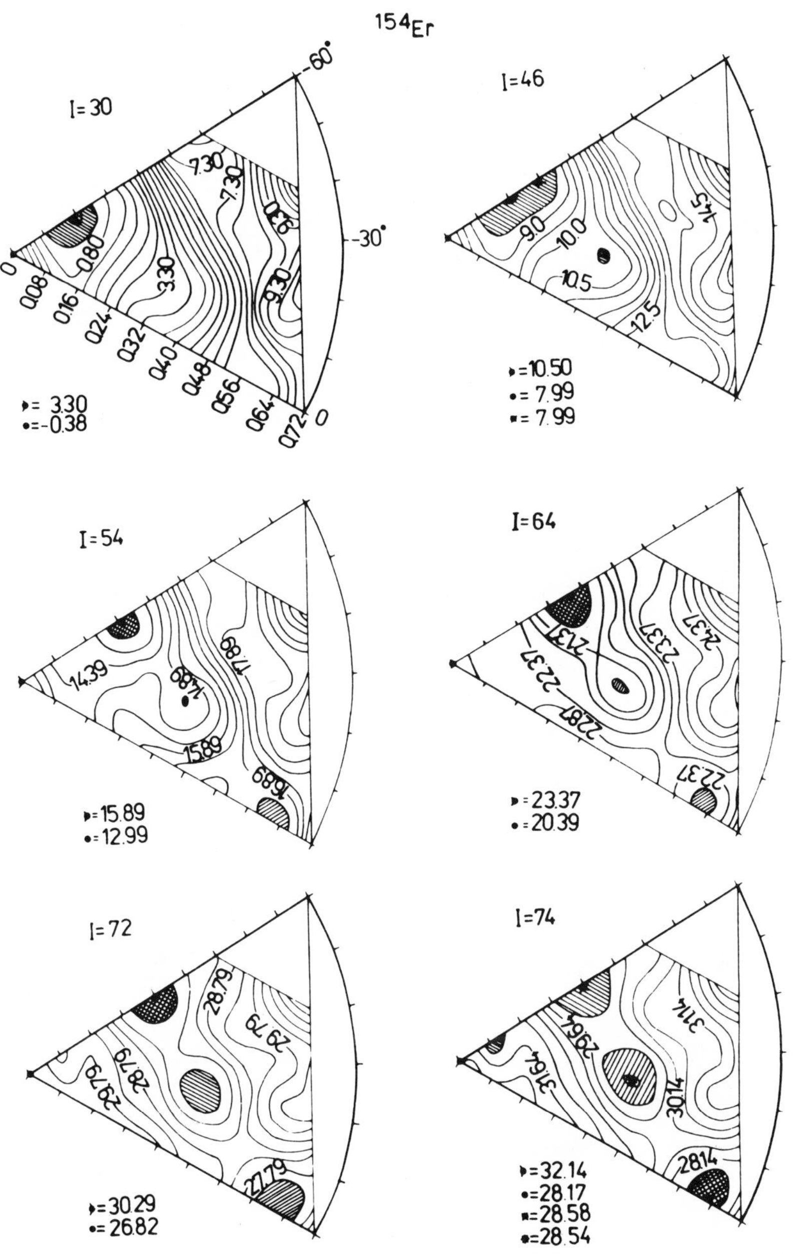

^{154}Er
I=30
I=46
I=54
I=64
I=72
I=74
-60°
-30°
0
0.08
0.16
0.24
0.32
0.40
0.48
0.56
0.64
0.72
▸= 3.30
•=-0.38
▸=10.50
•= 7.99
▪= 7.99
▸=15.89
•=12.99
▸=23.37
•=20.39
▸=30.29
•=26.82
▸=32.14
•=28.17
▪=28.58
•=28.54

Fig. 6.82.

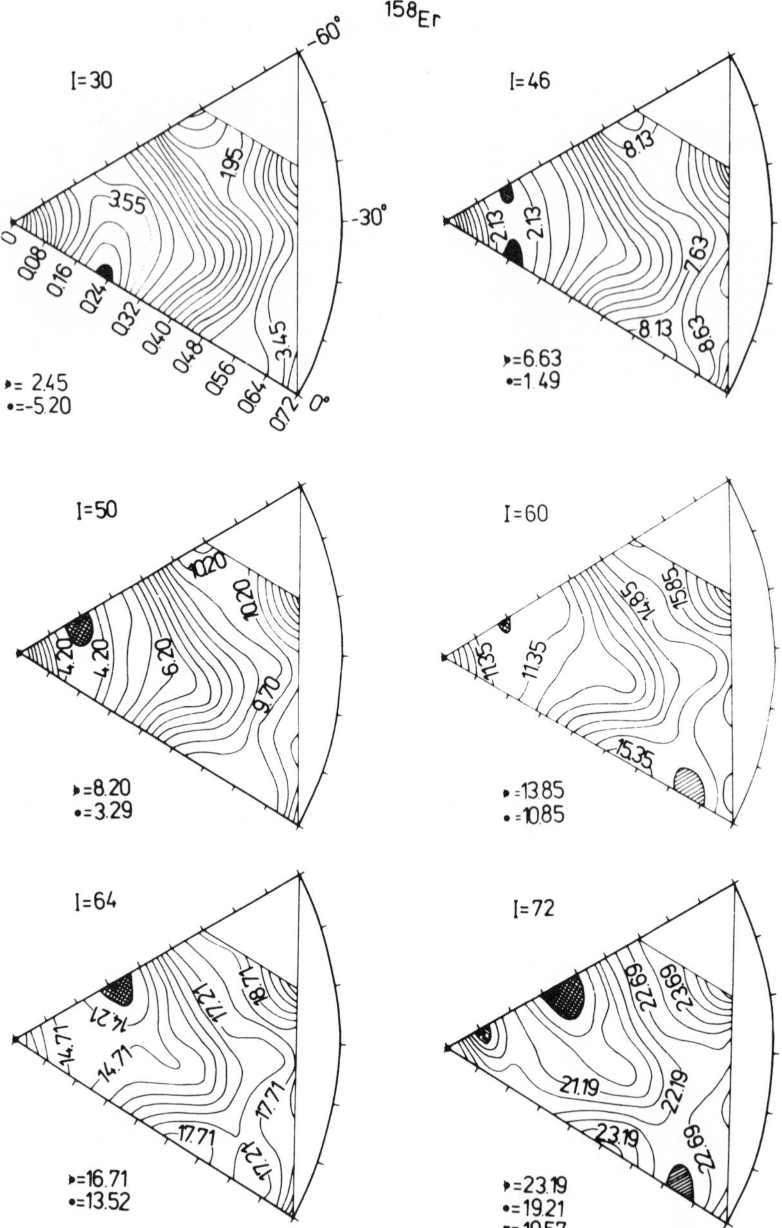

Fig. 6.82. The free-energy surfaces for $^{152,\,154,\,158}$Er in the β_2–γ plane. The left vertex corresponds to the spherical shape ($\beta = 0$). The prolate configurations rotating perpendicular to the symmetry axis are represented by the lower edge of the triangular

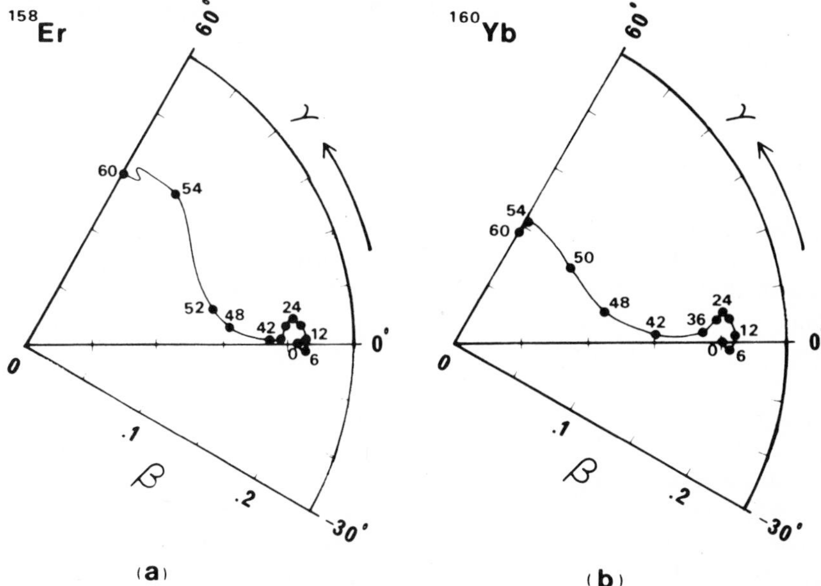

Fig. 6.83. The behaviour of yrast states in the β–γ plane calculated with the stronger proton pairing forces for ^{158}Er(a) and ^{160}Yb(b). (Tanabe and Sugawara–Tanaba 1984.)

attain triaxial shapes around $I \sim 50$ and thereafter they rotate with an oblate shape around the symmetry axis. Experimental evidence for those changes will be presented in the next section. It is interesting that the results of the above calculation indicate the existence of two oblate bands around spin 40, one of a collective nature and the other of a non-collective nature. For the particular case of ^{152}Dy the generalized Strutinsky shell correction method has been applied, and use was made of a deformed Woods–Saxon potential with optimized parameters (Schutz *et al.*, 1982). For another approach, but with similar results, a rotating Nilsson potential has been used (Ragnarsson *et al.*, 1980, Åberg, 1982).

Secondary minima often occur in the total energy surfaces, and these are due to the shell structure. They are particularly pronounced for nuclei with Z

Fig. 6.82 (*Contd.*)

charts ($\gamma = 0°$), and the oblate configurations rotating around the symmetry axis are represented by the upper edge ($\gamma = -60°$). The distance between the lines corresponds to 0.5 MeV. The valleys in the landscape are indicated by shaded areas, the darker ones corresponding to deeper minima. Below the plots the values of the free-energy of spherical configuration (>) and at the minimum (●) are given. Here the temperature is taken as $t = 0.2$ MeV if not stated otherwise. (Cwiok *et al.*, 1983.)

and N close to the magic nucleon numbers at large deformation (see the examples discussed above). The secondary minima may become the lowest in energy above some critical spin value I^*. Consequently one can expect that above I^* most of the γ rays in the qcs come from states in the secondary minimum rather than in the primary one. Thus for $I > I^*$ a drastic shape change may occur from small equilibrium deformation at low spin to a very large one at high spin.

The shape evolution of ^{152}Dy as a function of angular momentum is shown in Fig. 6.84. At spin 38 a distinct minimum corresponds to an oblate shape with small deformation. With increasing spin the nucleus acquires a triaxial shape up to high spin. But from $I \sim 52$, a secondary minimum develops corresponding to a prolate shape ($\gamma \sim 0^\circ$) and large deformation ($\beta \sim 0.6$). The primary ($\beta \sim 0.2$), the triaxial ($\beta \sim 0.3$) and the prolate ($\beta \sim 0.6$) minima are compared as a function of spin in Fig. 6.85. It was estimated that already at spin 50 to 60 some fraction (~ 0.1) of the γ-ray intensity may go via the secondary minimum relative to that for the triaxial minimum (see insert to Fig. 6.85). The calculation indicates that only at $I \sim 88$ does the secondary minimum become the lowest of all. The experimental search for those superdeformed structures is discussed in Section 6.5.7.

6.5.6b *The experimental evidence for shape changes*

In the previous sections it was shown that prolate nuclear shapes most commonly display collective motion. The resulting level schemes are regular and the transitions fast. In the case of oblate deformations one may expect dominant single-particle motion, resulting in irregular level schemes and retarded transitions. A shape change can thus be identified when the level schemes are known up to high spin and the lifetimes are measured.

The available experimental evidence shows that the light Gd, Dy, Er, and Yb nuclei ($64 < Z < 70$) with $82 < N < 88$ have non-collective behaviour (oblate or spherical shapes) along the decay pathways. The non-collective transitions (of the order of only a few single-paricle units or less) are indicated by their relatively slow feeding times (> 1 ps). This is in contrast with the very fast feeding times ($\ll 1$ ps) of the collective transitions in the deformed rare earth nuclei ($90 < N < 110$).

Let us present here the data for nuclei in the intermediate mass region, i.e. ^{156,158}Er, to continue the discussion of the examples used in the previous subsection. The first indication of a change from prolate to oblate above $I \sim 40$ in ^{158}Er is based on the observation of a breakdown of the regular level scheme (Simpson *et al.*, 1984). In a subsequent experiment at Berkeley a fast and a slow branches have been identified that feed the regular yrast sequence at spin 40, by using the ^{122}Sn (^{40}Ar, 4n) reaction at 175 MeV and employing a 14 Ge detector (Compton-suppressed) array (Tjøm *et al.*, 1985). The upper part of the level scheme is shown in Fig. 6.86. The fast 1203, 1210 keV cascade

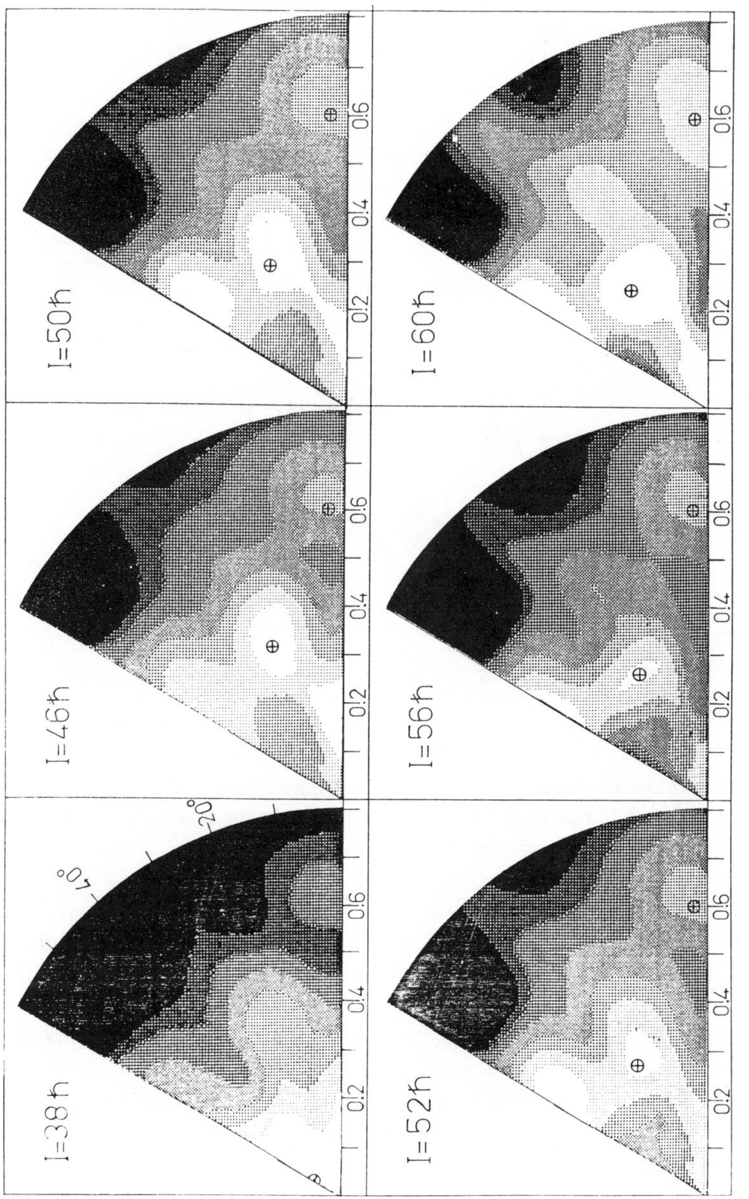

Fig. 6.84. Selected total-energy surfaces for ^{152}Dy. The separation between the stripes is 2 MeV and the crosses indicate the minima. (Schutz *et al.*, 1982.)

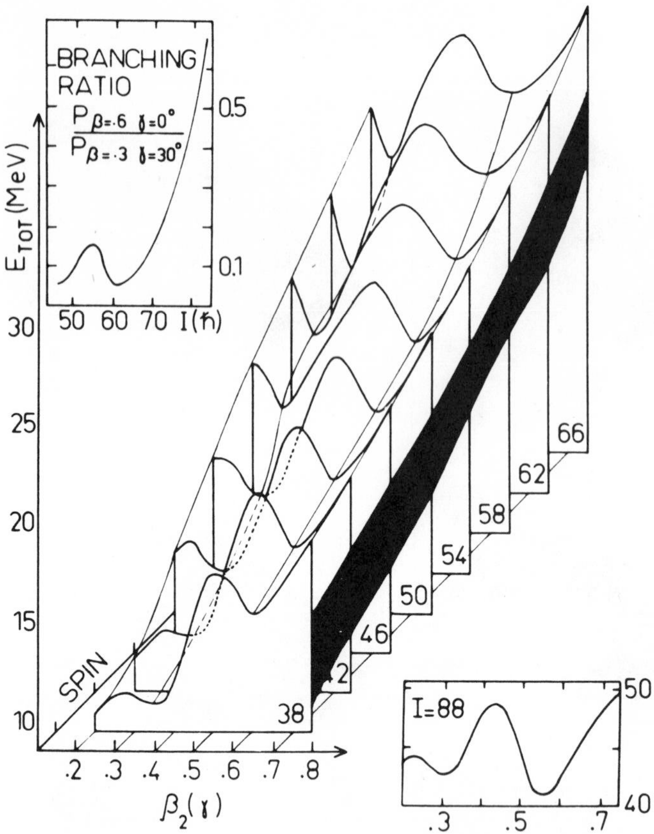

Fig. 6.85. The total energy of ^{152}Dy along the minimum valley. The upper insert gives an estimate of the relative feeding probability for the two collective minima. (Schutz *et al.*, 1982.)

and the slow 827, 1280, 1031, 971 keV cascade have been identified by comparing the γ-ray spectra measured with a self-supporting thin target and a gold or lead-backed target, as shown in Fig. 6.87. It is seen that all the lines in the lower spectrum (backed target) are sharper than in the upper one (unbacked target). This indicates that the corresponding γ rays must have been emitted after the recoils have stopped in the backing (lifetimes > 1 ps). The 1203 and 1210 keV lines, however, are missing from that lower spectrum, indicating that they are emitted from levels with lifetimes plus feeding times that are of the order of the mean slowing-down time (~ 0.5 ps). Therefore the γ-lines are Doppler-broadened and smeared out in the background. It may be inferred from those observations that the fast branch represents a continuation of the collective motion of the yrast sequence, whereas the slow branch is

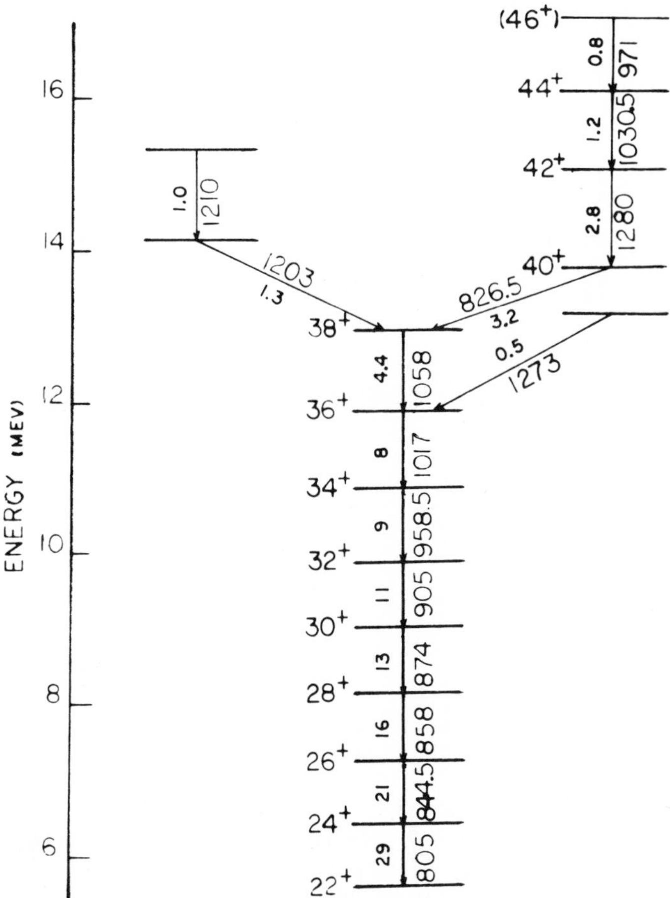

Fig. 6.86. The top-part of the level scheme for the yrast band in ^{158}Er. (Tjøm *et al.*, 1985.)

a strong evidence for a shape change accompanied by noncollective motion, which would confirm the theoretical results discussed above.

It is interesting to note that for this nucleus ^{158}Er the first oblate shape on the yrast line may be the fully aligned 46^+ state with a configuration (based on the ^{146}Gd core):

$$\pi[(h_{11/2})^4]_{16}\,\nu[(f_{7/2})^3\,(h_{9/2})^3\,(i_{13/2})^2]_{30}.$$

The appearance of oblate shapes (fully-aligned states) is accompanied by noncollective motion, thus exhibiting irregular level schemes which imply the termination of the collective bands.

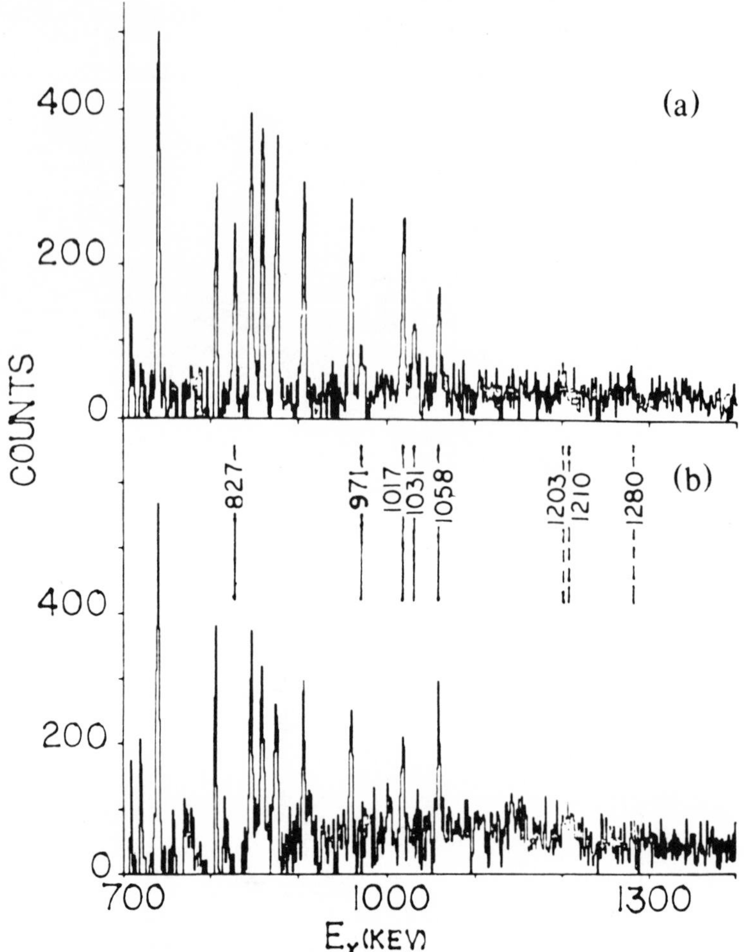

Fig. 6.87. Coincident gamma-ray spectra from the reaction ^{122}Sn(^{40}Ar, 4n)^{158}Er: (a) Transitions in coincidence with a gate on the 1058 keV γ-ray depopulating the spin 38^+ level, with self-supporting target. (b) Transitions in coincidence with the same gate as (a), but gold-backed target. (Tjøm *et al.*, 1985.)

The termination of a band at spin 38 has possibly been observed in ^{156}Er. The level scheme presented in Fig. 6.88 shows a band structure at high spin on top of the 30^+ state and terminating at 38^+ (Stephens *et al.*, 1985). Another noteworthy feature of the spectrum given in Fig. 6.88 is the abundant cross-feeding between bands, even those with different signatures. As the regularity of the bands above the region of cross-feeding is less than that below, it has

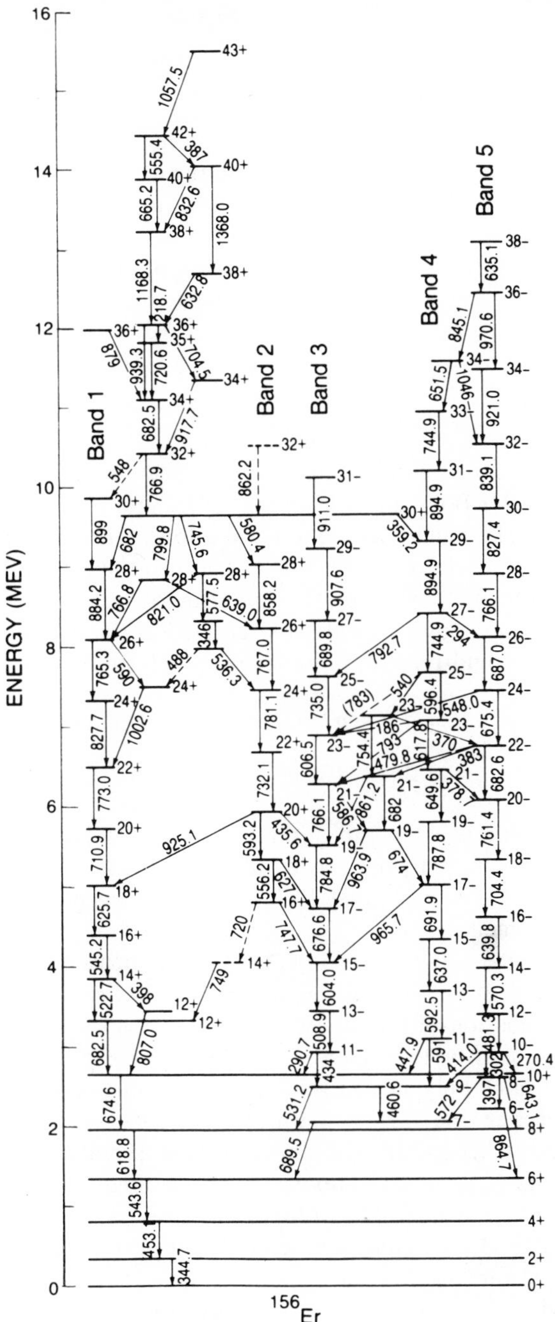

Fig. 6.88. The level scheme of ^{156}Er as deduced from the reaction ^{120}Sn(^{40}Ar, 4n) at 170 MeV. (Stephens *et al.*, 1985.)

been suggested that structural changes occur, possibly due to reduction of neutron pairing.

Next, we want to continue the discussion of the $^{156-158}$Dy nuclei, using the information on the lifetimes measured with the recoil-distance technique (see Section 6.5.5e). From the lifetimes of yrast states up to $I \sim 30$ the B(E2) values have been deduced, using eqn (6.3.84).

The reduced transition and static quadrupole moments Q_t and Q_d, as defined by eqns (6.3.86 and 6.3.87), can be related to the nuclear shape, defined by the shape parameters β and γ (see Section 6.3)

$$Q_t = 2(3/5\pi)^{1/2}\, ZeR_0^2\beta\cos(30^\circ+\gamma) \tag{6.5.23}$$

$$Q_d = 6/(5\pi)^{1/2}\, ZeR_0^2\beta\sin(30^\circ+\gamma). \tag{6.5.24}$$

The convention for γ is shown in Fig. 6.15. (See for the derivation of eqns (6.5.23 and 6.5.24) Bohr and Mottelson, 1975, and note the different convention for γ.) The equations are also valid for a triaxial nucleus with very high spin and it has been shown that this is also the case at low rotational frequencies with small axial asymmetries (Ring *et al.*, 1981).

A deformation change may also be reflected in the moment of inertia as derived with the rigid-rotor or liquid-drop models (see e.g. Szymański, 1983).

$$\mathscr{I}_{\mathrm{RR}} = (2/5)AMR_0^2\{1+(5/4\pi)^{1/2}\beta\sin(30^\circ+\gamma)\} \tag{6.5.25}$$

$$\mathscr{I}_{\mathrm{hydr}} = (3/2\pi)AMR_0^2\beta^2\cos^2(30^\circ+\gamma). \tag{6.5.26}$$

The latter formula is a generalization of eqn (6.5.22). Note that the nuclear shape parameters β and γ are related to the moment of inertia in a model-dependent fashion, whereas they have a model-independent relation to the quadrupole moment.

The dependence of the reduced quadrupole moments and of the moments of inertia on γ for fixed β is shown in Fig. 6.89 (Emling *et al.*, 1984). Although a unique determination of β and γ would require the simultaneous measurement of both quadrupole moments, a shape-change can be inferred from the measurement of Q_t only, as can be seen in this figure. The Q_t values for the yrast sequences in the various Dy isotopes as deduced from the lifetime measurements discussed in Section 6.5.5e, are presented in Fig. 6.90. A marked drop in Q_t is observed above the critical frequency of the first band crossing (ω_{c1}) in the light Dy isotopes. The values below that frequency (ground-state bands) are consistent with enhanced collectivity, similar to the more deformed rare-earth nuclei. The behaviour after the band crossings is consistent with a change to triaxiality because of the increased γ-value, provided that β remains constant. The triaxial shapes are presumably due to a polarizing effect on the core induced by the successive alignments of quasi-particles.

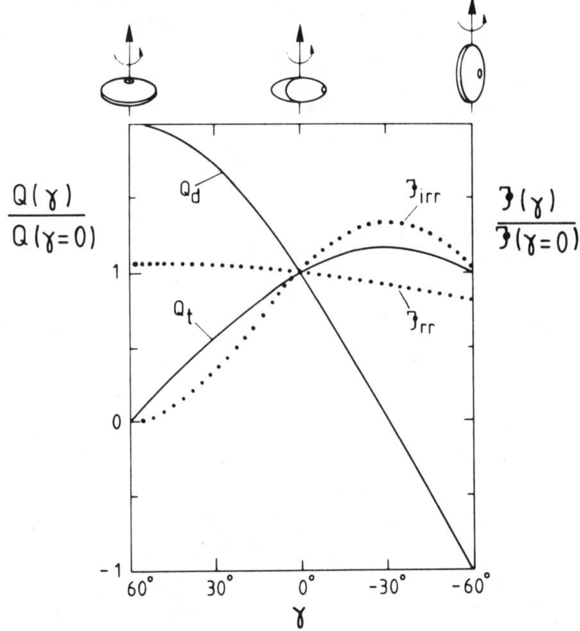

Fig. 6.89. The dependence of the quadrupole moments Q_d and Q_t and of the irrotational-flow and rigid-rotor moments of inertia $\mathscr{J}_{\text{irr}}$ and $\mathscr{J}_{\text{RR}}$, respectively, on the deformation parameter γ in the sector $\gamma \leqslant 60°$. All quantities are normalized to the corresponding values at $\gamma = 0°$; a value of $\beta = 0.35$ was used to calculate $\mathscr{J}_{\text{RR}}$; the other moments being independent of β owing to the normalization. (Emling *et al.*, 1984.)

In addition to the lifetimes of the yrast states feeding times have also been deduced from the data of the experiment discussed above. The average reduced transition quadrupole moments for qc cascades feeding the GSB in ^{156}Dy amount to $Q_t = 6.4$ eb, to 5.3 eb for those feeding the S-band and to 5.8 eb for cascades feeding the GSB in ^{158}Dy. From the measured side-feeding multiplicities it was estimated that these cascades are located in the region of $I = 20$ to 40. Their enhanced collectivity, corresponding to B(E2) values up to 0.7 times the rigid-rotor values, lead to a picture of rotational bands in the qcs parallel to the yrast line and terminating at spin ~ 20. This explains why the observation via discrete lines of yrast states above spin 20 becomes increasingly difficult towards the high-spin states.

It is now interesting and instructive to compare the above results with those obtained for the lighter isotopes ^{152,154}Dy. The nucleus ^{154}Dy shows rotational band structures at moderate spin values which breaks down around spin 30 (Pakkanen *et al.*, 1982), whereas ^{152}Dy shows non-collective

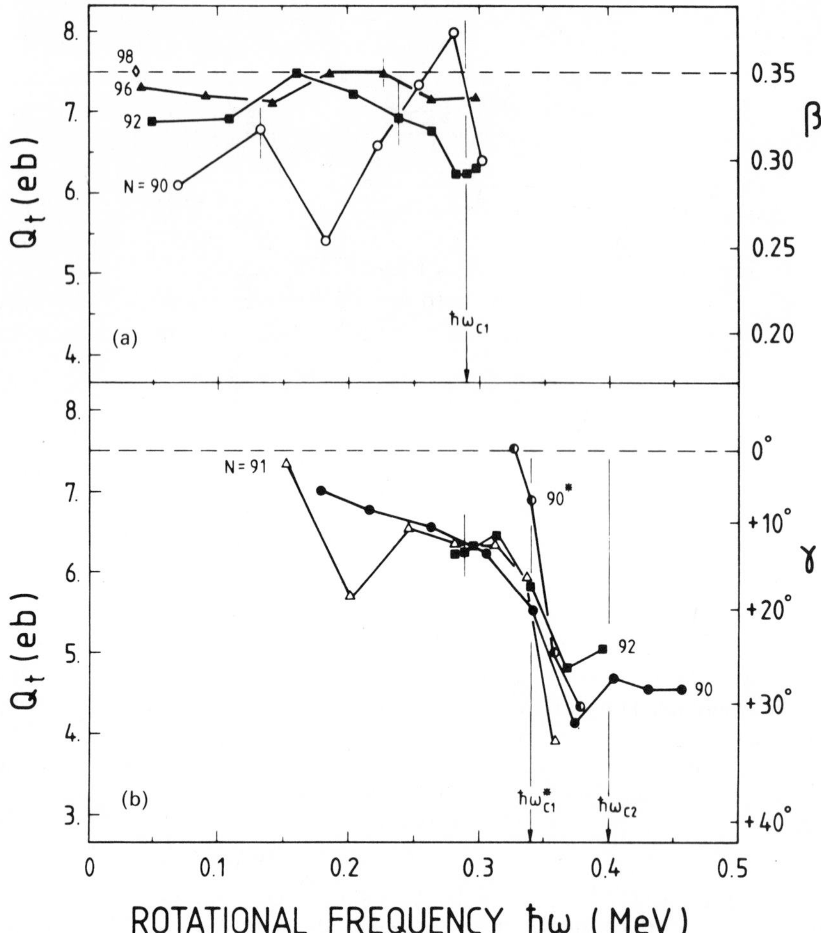

Fig. 6.90. (a) Quadrupole transition moments Q_t as function of the rotational frequency for bands below the first critical frequency $\hbar\omega_{c1}$. The deformation parameter β (right-hand scale) corresponds to the axial-symmetric case ($\gamma = 0°$). (b) Quadrupole transition moments Q_t versus rotational frequency $\hbar\omega$ for rotational states in ^{156,157,158}Dy obtained from the experiment. The deformation parameter γ (right-hand scale) was deduced assuming $\beta = 0.35$. (Emling *et al.*, 1984.)

behaviour up to the highest observed yrast state ($I = 37$). The lifetimes and feeding times have been deduced using similar reactions and techniques as discussed above (Azgui *et al.*, 1985). For ^{152}Dy a modification of the experimental arrangement made possible the detection of delayed transitions emitted after the isomeric decay of recoils in a catcher 70 cm behind the retardation foil.

The decay curve for ^{154}Dy given in Fig. 6.91 shows two components; a fast decay of < 1 ps and a slower component of about 10 ps. A curve for ^{156}Dy which is associated with fast (< 1 ps) decay is included for comparison. The fast component in ^{154}Dy is due to side-feeding. The slow component supports the earlier indication of a change from a prolate shape at low spin to an oblate shape at high spin.

The results for ^{152}Dy, however, show only long (> 10 ps) feeding times for qc cascades terminating above the $I^{\pi} = 17^{+}$ and 21^{-} isomeric states. This again confirmed earlier indications (see e.g. Hageman *et al.*, 1981) of the absence of collectivity in ^{152}Dy up to spin 40. The effective decay times T_{eff}, i.e. the times needed to depopulate the levels by 50 per cent, and the half-lives

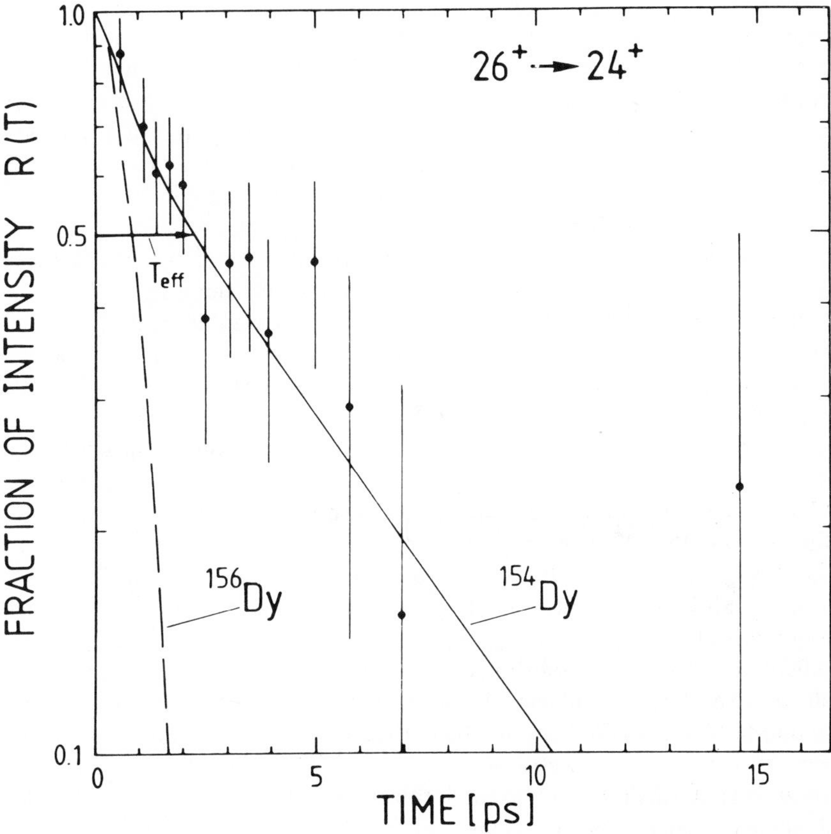

Fig. 6.91. The decay curve for the $26^{+} \rightarrow 24^{+}$ transition in ^{154}Dy obtained from the ^{25}Mg(^{134}Xe, 5n) reaction. The solid line represents the best fit and the dashed line the decay curve of the corresponding transition in ^{156}Dy. The arrow indicates the deduced value of the effective decay time T_{eff}. (Azgui *et al.*, 1985.)

for the lightest Dy isotopes discussed above are shown in Fig. 6.92. This illustrates nicely the change from the more prolate nucleus ^{156}Dy to the spherical one ^{152}Dy, with ^{154}Dy as an intermediate case.

Another piece of information on nuclear behaviour at high spin is contained in the moment of inertia as discussed in Section 6.3.4. In general, moments of inertia alone do not provide conclusive evidence for a definite nuclear shape. This is due to the fact that they are not only sensitive for the shape, but also to other phenomena such as the nuclear superfluidity (pairing). An increase of the moment of inertia, for instance, can be caused either by the increase of the nuclear deformation or by the decrease or disappearance of the pairing correlations.

The moments of inertia $\mathscr{J}^{(1)}$ over a very broad spin region (2–65) have been deduced for ^{162,164}Er compound nuclei, produced in ^{40}Ar and ^{86}Kr-induced reactions (Hillis *et al.*, 1979). The values have been deduced from the γ-ray energy corresponding to the upper edge of the collective bump in the multiplicity spectra (multiplicity plotted versus E_γ). In this procedure the rotational relation between E_γ and I (derived from measured multiplicities) as expressed by eqn (6.3.37) was used. A maximum angular momentum of $\sim 63\hbar$ was deduced from the measured multiplicities in the reaction ^{124}Sn(^{40}Ar, xn)$^{164-x}$Er at 236 MeV. The results are shown in Fig. 6.93, which also contains the moments of inertia for yrast states in ^{158}Er deduced from discrete-line spectroscopy. It appears that the moments of inertia in the high-spin region (40 to 60) follow within the experimental errors the calculated values for a rotating liquid drop with β varying smoothly from 0.2 at low rotational frequency (ω) to 0.5 at high ω. This indicates that no drastic shape changes occur in the evaporation products of those Er compound nuclei.

The less-deformed evaporation residues of the compound nucleus ^{100}Ru have been compared with those of ^{160}Er, as already discussed in the examples of Section 6.5.5e (Folkmann *et al.*, 1981). The moments of inertia deduced for the first system vary from $2\mathscr{J}^{(1)}_{\text{env}}/\hbar^2 = 50\ \text{MeV}^{-1}$ at the lowest spin to about 75 MeV^{-1} near $I = 42$. The latter value exceeds the value obtained from the constant slope of the entry line (see Fig. 6.64) by about 30 per cent, which amounts only to $\sim$ 58 MeV^{-1}, being close to the rigid-body value. From that difference it was suggested that some of the evaporation residues in excited states with spins $I \sim 40$ may be superdeformed (ratio of major to minor semiaxes of $\sim$ 2:1). On the other hand the spread of the second E2 bump (see Fig. 6.76) in the Ru spectra was interpreted as due to the decay of both collective and single-particle excited states. All those features may be regarded as some evidence for shape changes in the ^{100}Ru evaporation residues when the angular momentum increases up to $I \sim 40$. It is also clear that the information is only implicitly contained in the data and that the ambiguities in the interpretations are rather large.

Some different methods of obtaining more detailed information on the possible existence of superdeformed high-spin states will now be discussed.

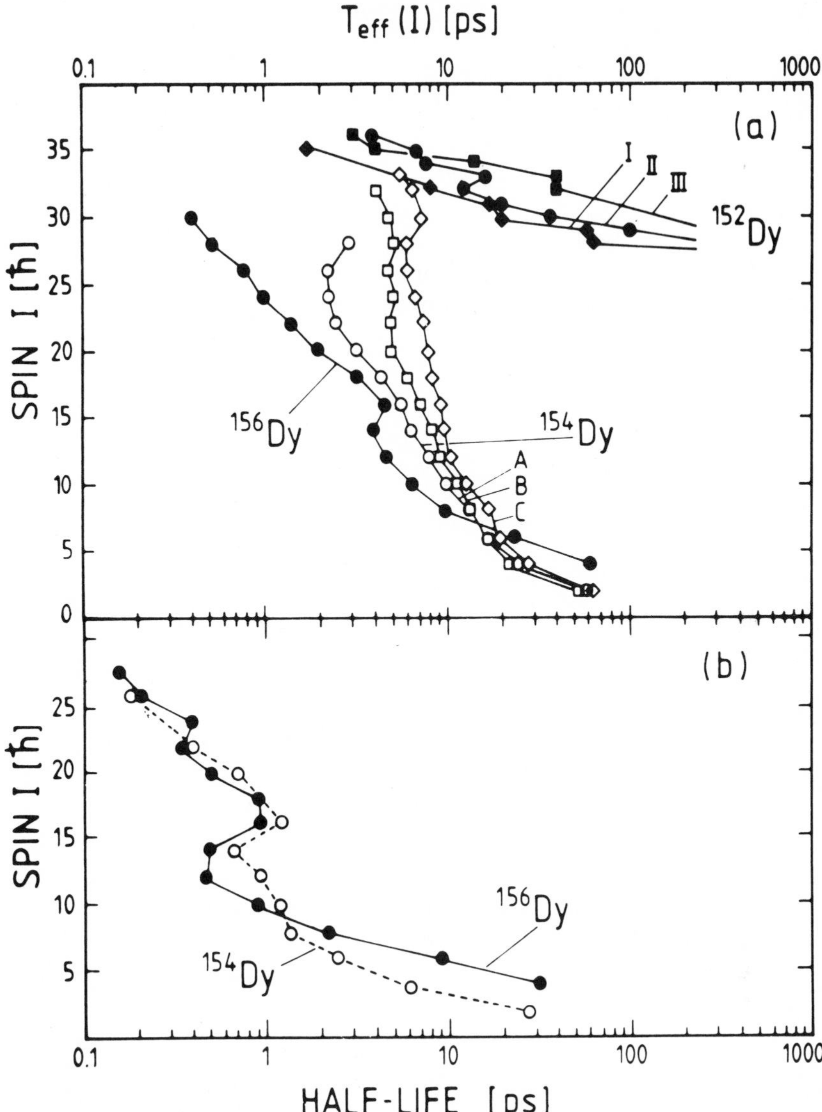

Fig. 6.92. (a) Effective decay times in ^{152}Dy ($N = 86$), ^{154}Dy ($N = 88$) and ^{156}Dy ($N = 90$) as a function of spin. The reactions ^{122}Sn(^{34}S, 4n), ^{25}Mg(^{132}Xe, 5n) and ^{122}Sn(^{34}S, 4n) leading to ^{152}Dy are denoted by I, II and III, respectively. The reactions ^{25}Mg(^{134}Xe, 5n) and ^{124}Sn(^{34}S, 4n) and ^{124}Sn(^{34}S, 4n) leading to ^{154}Dy are denoted by A, B and C, respectively. The data for ^{156}Dy were obtained in a ^{25}Mg(^{136}Xe, 5n) reaction. (b) Half-lives of positive-parity yrast states in ^{154,156}Dy as function of spin. The value shown at $I = 26$ for ^{154}Dy is an average value for the spins $I = 25$–28. (Azgui *et al.*, 1985.)

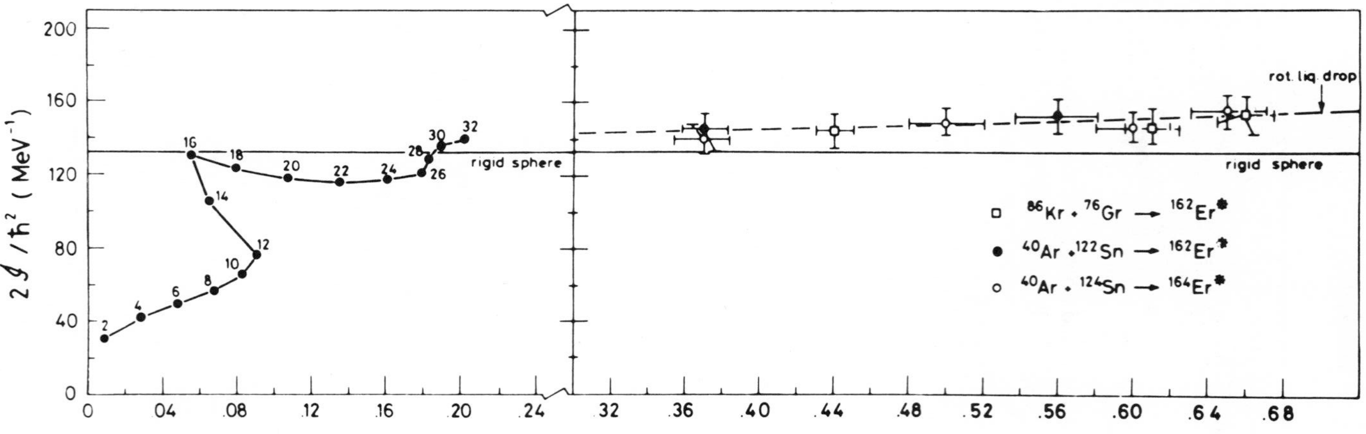

Fig. 6.93. Backbending plot for the Er systems produced with 161–236 MeV ^{40}Ar and 314–376 MeV ^{86}Kr beams (Hillis *et al.*, 1979). The data on the left-hand side of the figure are obtained from known transitions between discrete yrast states in ^{158}Er (Lee *et al.*, 1977.) The data on the right-hand side of the figure represents the extension of rotational bands into the quasicontinuum derived from the 'collective' bump in the multiplicity spectra. The moments of inertia calculated for a rigid sphere and a rotating liquid drop are also indicated.

6.5.7 *Correlations in the quasi-continuum; superdeformation*

A very useful method of studying properties of the qcs is the γ–γ energy correlation technique as discussed in Section 6.3.4c. We have discussed several data sets for ^{152}Dy in Sections 6.5.3–5, showing consistently the absence of collectivity at least up to spin 40. Above spin 40, however, we have given experimental evidence for collectivity. Several correlations studies have now been made in attempts to reveal the nature of the qc cascades at high spin in ^{152}Dy.

Recent investigations have focussed on the search for superdeformed high-spin states. Very large deformations have been found in the past for fission isomers with a ratio of 2:1 for the long to the short nuclear symmetry axes. At a deformation $\varepsilon = 0.6$ corresponding to the integer ratio 2:1 it follows from the properties of the harmonic oscillator potential that a degeneracy of single-particle levels occurs similar to that for spherical nuclei (Bohr and Mottelson, 1975). This can be understood by considering the spherical case which exhibits a strong degeneracy and thus a closed shell structure because the radial and angular frequencies of the oscillator quanta have the integer ratio of 2:1. This makes the oscillator orbitals closed, so that all levels within a major shell (characterized by N) have the same energy. With increasing deformation the integral ratio is no longer valid and the degeneracy is lifted and the shell structure is smeared out. In the case of a specific deformation of a spheroid with an integral ratio of the long to short symmetry axis (3:2, 2:1, 3:1, . . .) the oscillator frequencies along those axes will also have the same integral ratios. Therefore the oscillator orbitals will be closed again and the shell structure will reappear (Bohr and Mottelson, 1975). Therefore one may hope to study shell effects for high-spin states which are superdeformed.

A first attempt to observe superdeformation was made by Schutz *et al.* (1982) using the ^{124}Sn (^{32}S, 4n) reaction at 160 MeV. The use of three Ge(Li) detectors without Compton suppression made it rather difficult to recognize correlated events in the matrix. The theoretical part of that work is discussed in Section 6.5.6a. The next experiment was carried out with the TESSA2 detector at Daresbury, using the reaction ^{108}Pd(^{48}Ca, 4n)^{152}Dy at a bombarding energy of 205 MeV (Nyako *et al.*, 1984). It was estimated that ^{152}Dy recoils were produced with angular momenta up to $I \sim 63$. Owing to the effective Compton suppression of the six detectors used, 32 per cent of the raw data corresponded to real correlated events, compared with $\sim$ 2 per cent in the case of bare Ge(Li) detectors. In addition to the high-resolution Ge detectors, a multiplicity-filter-sum-spectrometer was used, consisting of 50 BGO detectors (see Section 3.2 for details of the detectors). The geometry was chosen so that delayed transitions following the decay of isomeric states could not be recorded by the Ge detectors. The BGO crystal ball discriminated the events corresponding to the 5n reaction (producing ^{151}Dy) down to a contribution of ~ 10 per cent. A clear valley-ridge structure observed from ~ 800 to ~ 1350 keV was a convincing indication of collective rotation and the rather constant width of the valley indicated a constant moment of inertia of $\mathscr{I}^{(2)}_{\text{band}} = 85 \pm 2$ MeV$^{-1}\hbar^2$. This moment of inertia exceeds the rigid-body value by a factor of 1.4, indicating a large deformation of $\varepsilon = 0.51$. The valley-ridge structure covers a range in angular momentum of about 40 to 58, as estimated from the range in γ-ray energies (800 to 1350 keV) and assuming $I\hbar = \mathscr{I}\omega$ (rigid rotation).

More details about the collective structure in ^{152}Dy have been obtained with TESSA-III at Daresbury (Twin *et al.*, 1986) and with the HERA spectrometer at Berkeley (de Voigt *et al.*, 1986, 1987). The High Energy Resolution Array (HERA) consists of 21 BGO Compton-suppressed Ge detectors to measure $E_\gamma(1)$–$E_\gamma(2)$ correlations. Thin targets of enriched ^{114,116}Cd and ^{118,120}Sn were bombarded with 180 MeV ^{40}Ar beams. The recoils were stopped in a lead catcher, positioned at a distance of 20 cm behind the target which corresponds to a flight time of ~ 28 ns in the case of ^{151,152}Dy and at a distance of 13 cm in the other cases. The cylindrical BGO detector at 0°, surrounding the beam pipe with catcher foil, provided a signal on delayed γ-ray emission from the recoils. All recorded energy signals of the Ge detectors were corrected for the full Doppler shifts. Only triple and higher-fold coincidences were written on magnetic tape. Typical event rates amounted to 2000 s^{-1}, with a beam current of 15 nA. Two days of running time yielded ~ 200 million triple and higher-fold events or about 600 million double coincidences in the case of ^{152}Dy and about one third of that in the other searches.

The valley-ridge structure is most clearly observed in the case of ^{152}Dy in the γ-ray energy region 800 to 1350 keV. An $E_\gamma(1)$–$E_\gamma(2)$ correlation matrix which covers the energy region 500 to 1500 keV, symmetrized and corrected for uncorrelated events, is shown in Fig. 6.94a. The results for the other nuclei are compared with those for ^{152}Dy in Fig. 6.94. The average distance between the central ridges in ^{152}Dy over the whole region is 94 ± 4 keV, yielding an average moment of inertia of $\mathscr{I}^{(2)}_{\text{band}} = 85 \pm 4$ MeV$^{-1}\hbar^2$, using eqn (6.3.64) as explained in Section 6.3.4c. The projections in the energy region 800 to 1340 keV perpendicular to the diagonal in the full matrix (raw data) as well as in the matrices gated on discrete lines above and below the 17^+ isomer in ^{152}Dy are shown in Fig. 6.95(a), (b) and (c), respectively. In the spectrum of Fig. 6.95(c) the ridges are seen with the correct separation of 94 keV which indicates that a significant fraction of the γ-ray flow from superdeformed (SD) states bypasses the isomeric state. This was also found by a comparison of intensities using matrices updated for prompt as well as for delayed events.

The result from gating on discrete lines in ^{151}Dy is shown in Fig. 6.95(d). The valley seems to be present as well as the ridges, 94 keV apart. Although this is consistent with the existence of SD bands in ^{151}Dy, the poor statistics in

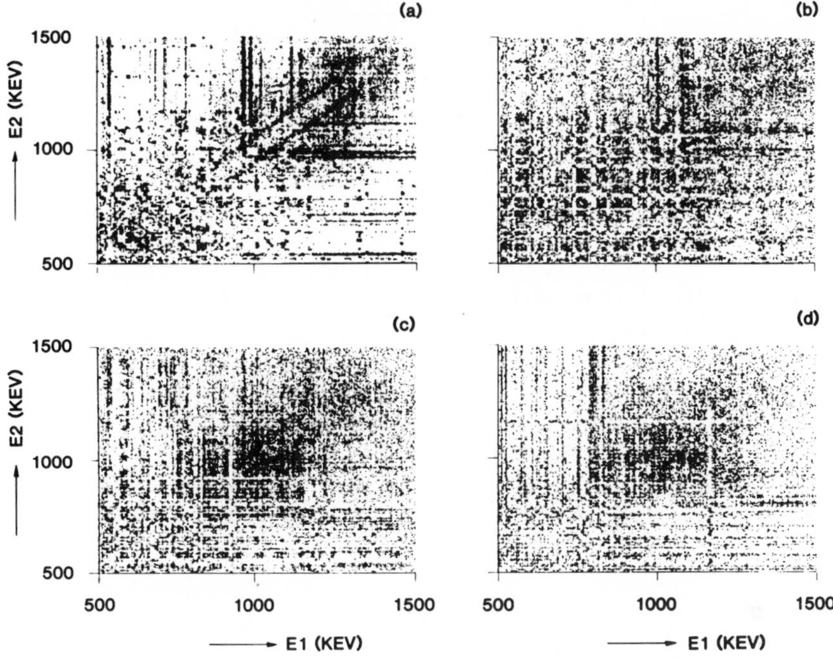

Fig. 6.94. Symmetrized $E_\gamma(1)$–E_γ (2) correlation matrices for the energy region of $E_\gamma = 500$–1500 keV (2keV/ch), obtained from (^{40}Ar, xn) reactions at 180 MeV. The matrices are corrected for uncorrelated events and the intensity threshold is 93 counts/ch for ^{152}Dy (a) and ~ 25 c/ch for the other cases. The matrix (a) for ^{152}Dy contains 192 million events generated from 96 million double coincidences in the displayed energy region. The two strong lines in the centre are due to the known 968 and 991 discrete transitions in ^{152}Dy. Matrix (b) for ^{150}Dy contains 73 million events, (c) for ^{154}Er contains 64 million events and (d) for ^{156}Er has 76 million events. (de Voigt *et al.*, 1986, 1987.)

the matrix gated on discrete lines does not allow definite assignments. The valley-ridge structure in ^{150}Dy and in ^{154}Er is present, and also in the case of ^{156}Er, but only below $E_\gamma = 1200$ keV—see Fig. 6.94(b), (c) and (d). It seems that in ^{150}Dy the valley is present, but with almost no ridges. In ^{154}Er, however, the valley is shallow and the ridges somewhat smeared out. This can be seen in more detail in Fig. 6.95(e), (f), and (g), which shows projected spectra in the energy region $E_\gamma = 1000$ to 1350 keV for ^{150}Dy, ^{154}Er and ^{156}Er, respectively. It is tempting to interprete a sharp ridge as due to collective rotation corresponding to a distinct minimum in the potential energy at large

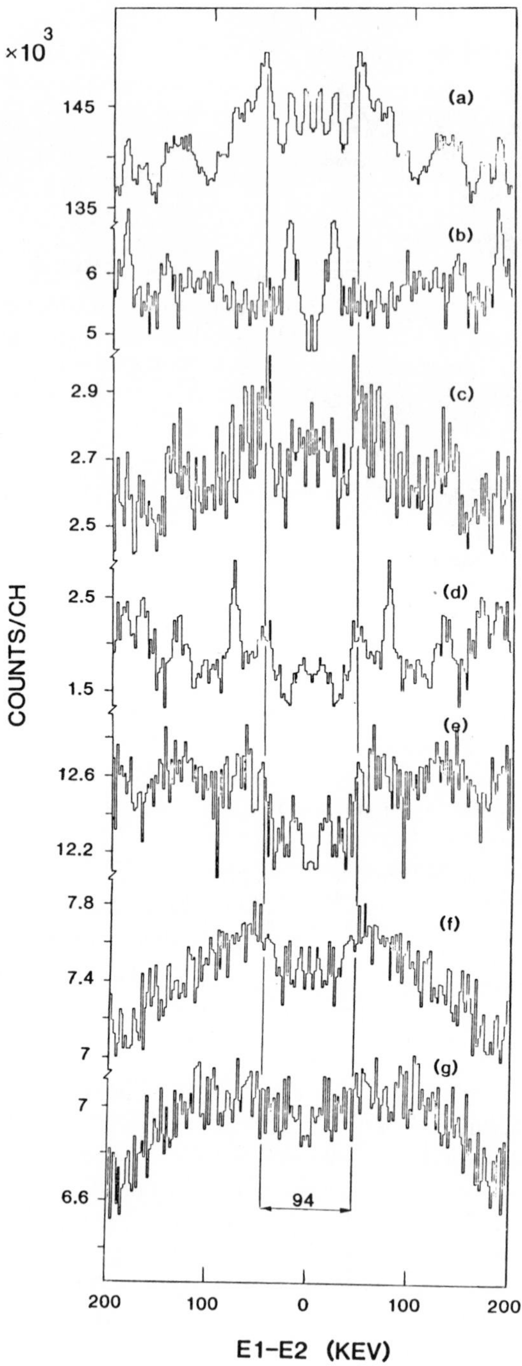
×10³
145
135
6
5
2.9
2.7
2.5
2.5
1.5
12.6
12.2
7.8
7.4
7
7
6.6
COUNTS/CH
(a)
(b)
(c)
(d)
(e)
(f)
(g)
94
200
100
0
100
200
E1−E2 (KEV)

deformation. The smearing-out effect may be due to a fluctuation of the moments of inertia associated with a flatter superdeformed potential pocket, for instance a superdeformed region at high temperature. A shallower valley may be caused by a stronger damping of the rotational motion (Bacelar *et al.*, 1986; Draper *et al.*, 1986). This would mean that the four nuclei in Fig. 6.94 represent limiting cases in some sense, because various combinations of valley-ridge structures occur. The measured moment of inertia $\mathscr{I}^{(2)}_{\text{band}} = 85 \pm 4\ \text{MeV}^{-1}\hbar^2$ corresponds to a deformation with a ratio of the long to the short nuclear semiaxes of 1.65, which is indeed not too far from the ratio 2 for the previously-observed SD fission isomers.

The weaker ridges for the other nuclei studied here is in contrast with the theoretical results, which showed SD bands in ^{150}Dy at lower energies than in ^{152}Dy and at similar energies in ^{154}Er (Dudek and Nazarewicz, 1985). The population of SD bands in the other nuclei was thus expected to be as strong as in ^{152}Dy, which apparently is not the case.

A striking feature in the matrix of ^{152}Dy (in Fig. 6.94a) is the absence of second and higher ridges, which should be present if the SD bands show a strong intraband decay (see Fig. 6.26). The first ridge must contain the discrete SD band which was observed (Twin *et al.*, 1986) from spin ~ 22 to ~ 60 and shows little side-feeding or out-branching. This band should thus show all the higher ridges. However the intensity of $\leqslant 1$ per cent is weak compared with the intensity of the first ridge of 5 ± 2 per cent (de Voigt *et al.*, 1986, 1987), and therefore the higher ridges of the discrete band do not show up in the (ungated) matrix of Fig. 6.94(a). The absence of higher ridges implies that the remaining intensity of the first ridge ($\sim 4\%$) has to be ascribed to SD bands which exhibit a strong out-branching.

An explanation of the observed different behaviours of the unresolved SD bands and the discrete SD band has been proposed (de Voigt *et al.*, 1986, 1987) in terms of accidental overlaps between SD states and less-deformed states at high level density, causing mixing between the different states. It is assumed that the discrete SD band is lowered in energy with respect to the unresolved SD structures, as expected from shell effects (Ragnarsson and Åberg, 1987) or from some remaining proton pairing (Riezebos *et al.*, 1987). The larger

---

Fig. 6.95. Projections (in 2 keV/ch) perpendicular to the diagonal in the raw correlation matrices imposing the following conditions: (a) $E_\gamma = 800$–1340 keV in ^{151,152}Dy; all data, no selections. $E_\gamma = 800$–1360 keV gated on discrete transitions in ^{152}Dy above (b) and below (c) the 17^+, 60 ns isomeric state. The strong peaks in the centre ($\Delta E_\gamma = 23$ keV) of (b) are due to the 968–991 keV coincidences (see Fig. 6.94). $E_\gamma = 900$–1360 keV, gated on discrete lines in ^{151}Dy (d). $E_\gamma = 1000$–1340 keV, all data for the three nuclei ^{150}Dy (e), ^{154}Er (f) and ^{156}Er (g). Note that in these cases no special selection has been applied and the nuclei indicated are the most strongly excited ones in the reactions. (de Voigt *et al.*, 1986, 1987.)

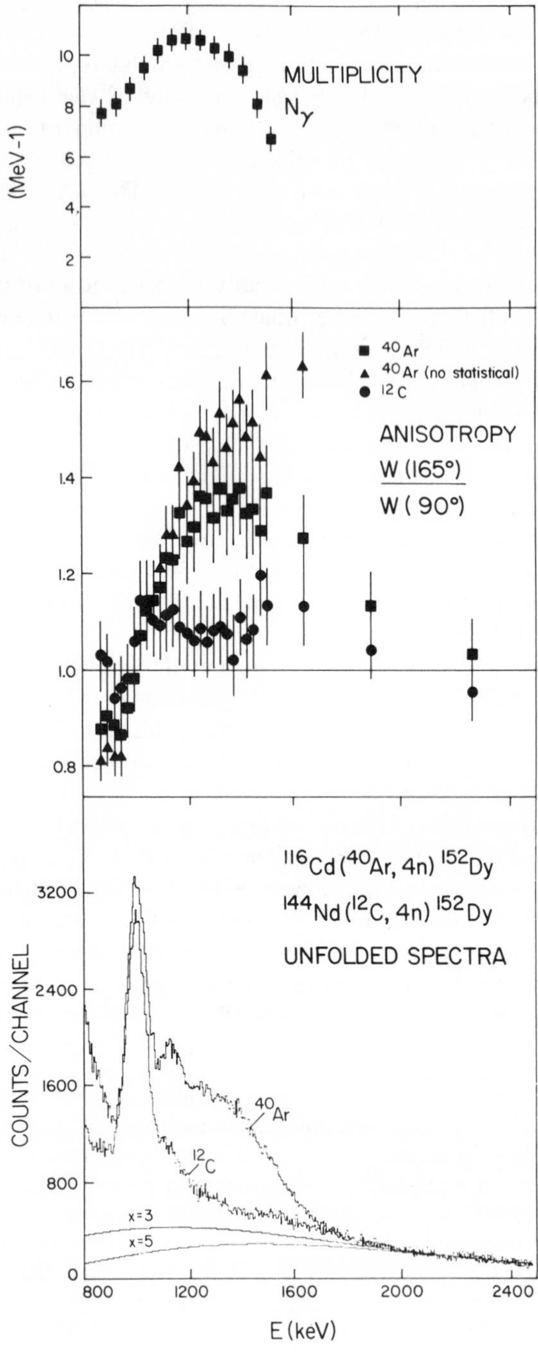

MULTIPLICITY
N_γ
(MeV-1)
■ ^{40}Ar
▲ ^{40}Ar (no statistical)
● ^{12}C
ANISOTROPY
W (165°) / W (90°)
^{116}Cd (^{40}Ar, 4n) ^{152}Dy
^{144}Nd (^{12}C, 4n) ^{152}Dy
UNFOLDED SPECTRA
COUNTS / CHANNEL
^{40}Ar
^{12}C
x=3
x=5
E (keV)

moment of inertia of the discrete SD band with respect to the yrast sequence causes the lower-spin states of that band to be higher in energy above the yrast level than the high-spin members. Thus the low-spin members lie in a region of high-level density with an increased probability of mixing. This explains why the outband decay towards the yrast sequence takes place only at the lowest-spin members. Similarly, the unresolved SD bands are located at still higher level density and thus have a larger probability of an accidental overlap. All those states are thus presumably mixed, which explain the strong outband decay.

The multipolarity is an important parameter for the identification of γ rays associated with the decay of superdeformed (SD) states, which should exhibit stretched E2 character. The angular anisotropy of the γ rays as well as the conversion electron coefficients, measured with a Mini-Orange spectrometer, have been reported by Riezebos *et al.* (1987) for ^{152}Dy. Conversion electron and γ-ray spectra have been measured for ^{152}Dy excited by 82 MeV ^{12}C and 180 MeV ^{40}Ar beams from the KVI cyclotron. The ^{40}Ar ion beams transfer enough angular momentum (upto 70$\hbar$) to excite the superdeformed states in the same way as in the Berkeley investigation described above, while the ^{12}C ion beams cannot reach the high-spin region of 40 to 60$\hbar$. A sum spectrometer has been used to enhance the high-spin region of ^{152}Dy in the spectra along with the catcher foil technique. The γ-ray spectra, angular γ-ray anisotropy and quasi-continuum multiplicities N_γ measured in ^{40}Ar and ^{12}C-induced reactions, covering the energy region where the γ rays from SD states have been found, is shown in Fig. 6.96. The anisotropy in the case of ^{40}Ar-induced reactions reaches the value of stretched E2 transitions between fully aligned states when the statistical γ rays are subtracted (dahsed line). The ^{12}C data show considerably less anisotropy. Comparison between the spectra and anisotropies obtained with the ^{40}Ar and ^{12}C beams reveals strong indications of rotational motion at spins above $\sim$40 in ^{152}Dy. The multiplicities are maximum around $E_\gamma = 1200$ keV, consistent with increased collectivity.

The conversion electron yield has been determined from the spectra with different transmission settings of the MO spectrometer. The background was determined from the yield with a transmission outside the region of interest, after normalization with a ^{56}Co source. The results are shown in Fig. 6.97 along with the theoretical values for E1, E2, and M1 transitions. In the case of

---

Fig. 6.96. The unfolded gamma-ray spectra of the ^{144}Nd(^{12}C, 4n) reaction and the ^{116}Cd(^{40}Ar, 4n) reaction to ^{152}Dy (bottom). The corresponding gamma-ray anisotropies defined as the gamma-ray yield at 165° divided by the one at 90° (centre). The triangle points show the anistotropy of the quasi-continuum for the ^{40}Ar reaction in the case that the statisticals are subtracted from the total yield (assumed isotropically distributed). Quasi-continuum multiplicities N_γ per MeV γ-ray energy in the (^{40}Ar, 4n) reaction (top). (Riezebos *et al.*, 1986.)

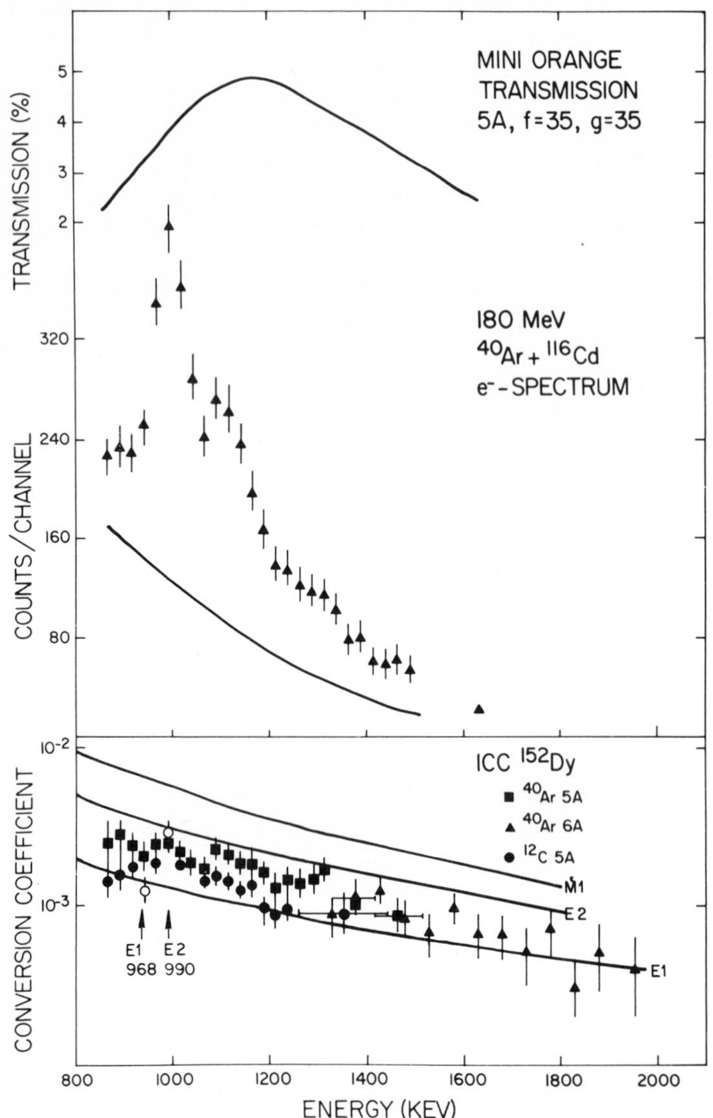

Fig. 6.97. The measured internal conversion coefficients (bottom) for the $^{116}Cd(^{40}Ar, 4n)$ reaction (squares for the 5A MO low-energy setting and triangles for the 6A high-energy setting) and for the $^{144}Nd(^{12}C, 4n)$ reaction (circles). The positions of the discrete lines of 968 keV (E1) and 990 keV (E2) are shown and open circles show the measured ICC of these lines in the ^{12}C-induced reaction. The measured conversion-electron spectrum (5A setting) is given in the centre and the 5A transmission curve at the top. (Riezebos *et al.*, 1986.)

^{12}C (small Doppler shift) it was possible to deduce the internal conversion coefficients (ICC) for two discrete lines, the 968 keV E1 transition and the 991 keV E2 transition (open-circles in the figure). Their agreement with the theoretical values proves that the method is sound, both for the ^{12}C and the ^{40}Ar data which have been obtained in the same experiment. The ^{40}Ar data have been processed with a gate on the high-sum energy as to enhance the highest spin region above 40 and to select predominantly ^{152}Dy. In the case of the ^{12}C beam ^{152}Dy has been populated as the predominant evaporation product and no sum-energy selection was necessary. The high-energy points have been measured with a different MO transmission setting.

The results show strong E2 admixtures in the γ-ray energy region where the decay from SD states has been observed, particularly around 1 MeV. Note that also M1 transitions will produce high ICC, but the observed anisotropy (Fig. 6.96) allows only non-stretched M1 transitions, which seems unlikely for high-energy γ rays (> 1 MeV) in the high-spin region. The ^{12}C data are consistent with nearly pure E1 character, as expected for non-collective motion. Therefore the high ICC, exclusively obtained with the ^{40}Ar beam, are attributed to the deformed quasi-continuum in ^{152}Dy. The measured intensities and multiplicities of the E2 bump in the quasi-continuum spectrum of ^{152}Dy imply that the bump contains a maximum of ~ 10 per cent intensity from SD states (Riezebos *et al.*, 1987). The remaining intensity is presumably associated with mainly triaxial rotational bands (Dudek and Nazarewicz, 1985).

A discrete superdeformed band in ^{152}Dy has been observed up to spin 60 (Twin *et al.*, 1986). The experiment was carried out with the TESSA-III spectrometer at Daresbury, using the ^{108}Pd (^{48}Ca, 4n)^{152}Dy reaction at 205 MeV. The spectrometer consisted of 12 BGO–Ge anti-Compton spectrometers outside a crystal ball of 50 BGO detectors (see Section 3.2.5). The catcher-foil technique was employed; that is a registration of delayed γ rays emitted from recoils caught by a foil positioned 5 cm downstream from the two thin targets.

Two matrices of Ge–Ge coincidences were constructed; i.e. a prompt matrix with the requirement that no delayed γ rays from the catcher were recorded by the inner BGO ball and an isomer matrix with the condition of the presence of delayed γ rays within 150 ns with respect to the prompt Ge–Ge event. Two spectra associated with the prompt and isomer matrices are shown in Fig. 6.98. These were obtained by summing gates on most of the transitions within the band. Several interesting conclusions were drawn from the two spectra such as the correspondence of the regular differences of the transition energies of about 47 keV with a large moment of inertia of $\mathscr{J}^{(2)}_{\text{band}} = 85\ \text{MeV}^{-1}\,\hbar^2$. This corresponds to a very large deformation of $\beta \simeq 0.6$.

The moments of inertia and the spins of the superdeformed band and of a low-deformation band ($\beta = 0.2$) in ^{152}Dy (see Fig. 6.43) are plotted versus the

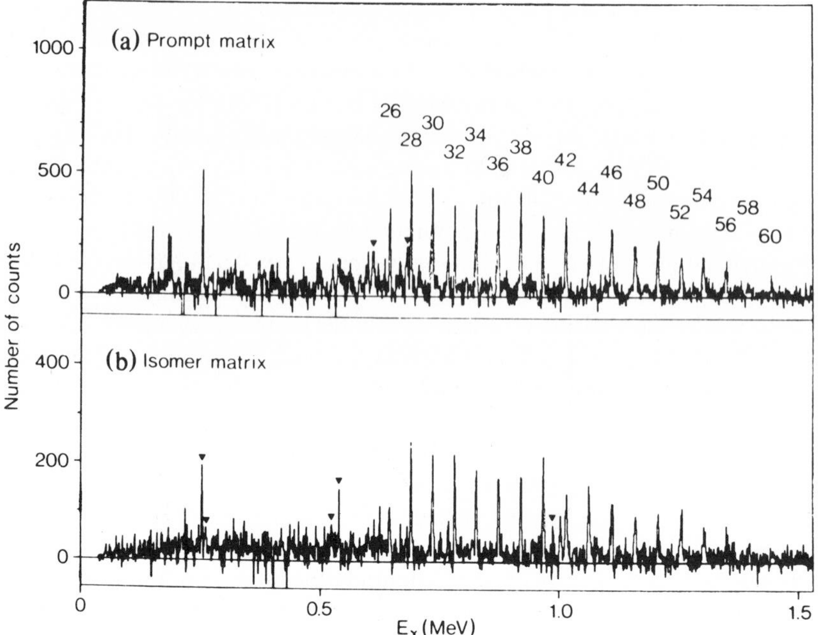

Fig. 6.98. The gamma-ray spectra in ^{152}Dy obtained by summation of gates set on most members of the superdeformed band. (a) The spectrum generated from the 'prompt' matrix, with gates set on all band transitions between states with spins 26 and 52; (▼) indicates 614 keV ($2^+ \to 0^+$) and 685 ($6^+ \to 4^+$) γ-rays. (b) The spectrum generated from the 'isomer' matrix with gates set on the 647 keV, 693 keV, 829 keV, 876 keV, 923 keV, 1017 keV, 1161 keV and 1209 keV transitions; (▼) indicates 254 keV, 262 keV, 525 keV, 541 keV and 991 keV transitions between the yrast oblate states. (Twin *et al.*, 1986.)

rotational frequency in Fig. 6.99. An interesting feature is the linear increase of I with ω for the upper parts of the bands, which projects back quite close to the origin. This linear behaviour is also discussed in Section 6.3 for ^{168}Hf (see Fig. 6.21) and finds a simple explanation by considering the nucleus as a classical rotor. The value of the moment of inertia is consistent with that of a (rigid) nucleus when pairing has diminished significantly or even collapsed.

So far the nuclear shape of ^{152}Dy has been derived from the measured moment of inertia. We have shown in Section 6.3 that particle alignment, coupled to the rotating core, also causes a large increase of the moment of inertia. Thus the moment of inertia is an ambiguous parameter to characterize the nuclear shape.

Clear-cut experimental evidence for the superdeformed shape of ^{152}Dy has been obtained from the measurement of lifetimes of the high-spin SD states

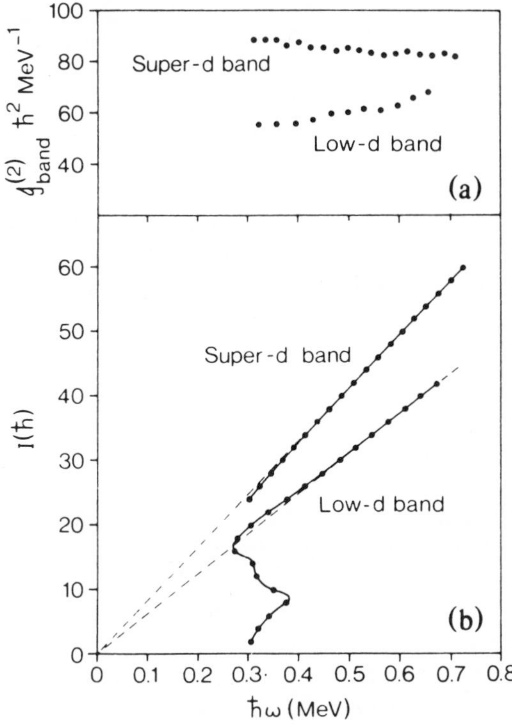

Fig. 6.99. (a) Plots of moment of inertia $\mathscr{J}^{(2)}$ versus frequency for the superdeformed band and the low-deformation band in ^{152}Dy. (b) Plots of spin against frequency for the superdeformed band and the low-deformation band in ^{152}Dy. The linear part of each graph is extrapolated back towards the origin. (Twin *et al.*, 1986.)

(Twin *et al.*, 1985 and Bentley *et al.*, 1987). In the first experiment the lifetimes were found to be less than 100 fs for γ-ray transitions correspoonding to the first ridge in the γ–γ-energy correlation matrix above 900 keV. The B(E2) values amount to over 2000 times the single-particle value, which is an order of magnitude larger than for bands with normal ($\varepsilon = 0.2$) deformation. The corresponding deformation amounts to $\beta > 0.4$ (Twin *et al.*, 1985). In a later experiment the same group used the TESSA-III spectrometer and the DSAM to study discrete transitions in the SD band in ^{152}Dy (Bentley *et al.*, 1987). The data indicate a full Doppler shift at the highest spins (50 to 60), which implies feeding times of less than a few fs. Towards the lower spin states a gradual decrease of the fractional shift was observed. The full data set could be fitted by assuming a constant quadrupole moment of $Q_0 = (19 \pm 3)$ eb, which is much larger than the value of 5 eb, commonly found for rotational bands in the rare-earth region with normal deformations. The large quadrupole moment thus establishes unambiguously the superdeformed shape.

It has been noticed by Swiatecki (1987) that the surprisingly constant and large moment of inertia of the superdeformed discrete band in ^{152}Dy can be associated with a very stiff nuclear potential. The stiffness is proportional to the negative inverse of the second derivative of the energy with respect to the angular-momentum factor $I(I+1)$. Note that for a rigid rotor this derivative is zero. The stiffness of the discrete SD band changes from the low-spin members (at $I = 24$) over two orders of magnitude to a large value close to that of a rigid rotor at the high-spin members ($I = 50$ to 60). The change in the stiffness of the band can be associated with a liquid-to-solid phase transition in nuclear matter (Swiatecki, 1987). In the liquid phase the individual nucleons are allowed to redistribute themselves into the lowest states at each deformation. In contrast, the response of the nucleus to deformation is solid-like if the stretching takes place with each particle frozen in its original state.

Both discrete and unresolved superdeformed bands have now been observed in several light rare-earth nuclei, particularly in some Ce, Nd, Gd and Dy isotopes.

We have concluded this part of the discussion of quasi-continuum physics with an example that illustrates how discrete γ-ray spectroscopy now reveals properties of states which were accessible a few years ago only with quasi-continuum spectroscopy, and leads to more precise and also to very new and surprising conclusions.

A problem with the γ–γ energy correlation technique, noticed by several authors, is the relatively weak signal corresponding to the ridges in the matrices. Indeed in most of the published data the ridges seem to contain only a small fraction of the total γ-ray intensity, and the second ridge expected for coincidences between γ rays separated by one transition is almost always very weak or absent. One experiment has been devoted explicitly to measure the intensity of the ridge structure in ^{130}Ce using the ^{36}S + ^{98}Mo reaction at 155 MeV with the TESSA-II spectrometer (Love *et al.*, 1985). The average intensity of the coincidences contributing to the ridge was found to vary from ~ 20 per cent at $E_\gamma = \sim 1$ MeV to ~ 5 per cent at ~ 1.5 MeV, relative to the $4^+/2^+$ coincidence yield in ^{130}Ce. It was concluded that the average probability for in-band collective E2 transitions in rotational sequences above spin 30 must be smaller than 20 per cent.

It has been pointed out by Døssing (1985) that the broadening of the ridges and the filling up of the valleys may well be caused by a damping of rotational states at high level density. With increasing rotational frequency rotation-aligned states will be produced due to the Coriolis interaction. We may thus expect the presence of many rotational bands in the quasi-continuum region, each with a (different) aligned-particle configuration. The moments of inertia of all those bands may differ owing to differences in the alignment, pairing, shape or something else. Therefore they may all have different slopes when their energies are plotted against the angular momentum, thus crossing each other to some extent. States of different bands, but with the same spin, may have common (single-particle) components in their wave functions and therefore mix. When the interaction strength is strong enough interband transitions may become significant and consequently the $B(E2)$ strength will be spread over several or many transitions. The width of this spread (in energy) is called the damping width Γ_{rot}. In the undamped (unmixed) case $\Gamma_{rot} = 0$, and this may be mostly the case at zero to low temperatures. It has been pointed out by Stephens *et al.* (1986, 1987) that Γ_{rot} rises sharply near some critical temperature (corresponding to an excitation energy in the region of 0.8–1.3 MeV) where the damping sets in and Γ_{rot} may attain values of ~ 100 keV. At higher temperatures Γ_{rot} may become significantly larger. Determination of Γ_{rot} may be possible by using gated spectra. The transition in the gate will be missing in the spectrum and will cause a dip at the gate position. When $\Gamma_{rot} = 0$ the dip will be sharp and deep (a cool effect). With increasing Γ_{rot} the dip will be washed out and its width and depth are to some extent a measure of Γ_{rot}. It is believed that a large width (with almost no dip observable) is due to damping (Stephens *et al.*, 1987). It may thus be possible to obtain information on the rotational damping in the quasi-continuum by measuring the width and the depth of the valleys. The technique of γ–γ energy correlation, using many high-resolution detectors with Compton suppression, may thus reveal interesting details of those features of the qcs.

An experiment has been carried out at Berkeley with eight large NaI detectors and a Ge(Li) detector (to identify the reaction products) using the reaction ^{124}Sn(^{40}Ar, 4n)^{160}Er (Draper *et al.*, 1986). Coincidence spectra for adjacent gates were compared with variable gate width. From the observed dips in the spectra (at the gate energy) information has been obtained about the γ rays depopulating rotational states at high spin. The width of the distribution of γ-ray energies, on the average emitted by each rotational state, varied from < 40 keV at $E_\gamma = 720$ keV to ~ 125 keV at $E_\gamma = 1200$ keV, with corresponding fractions of uncorrelated γ rays of 91 per cent to 83 per cent, respectively.

The smearing-out effects of the valley-ridge structure have been explained in terms of a spreading width of the rotational (quadrupole) decay strength by Bacelar *et al.*, 1985. The data have been obtained with the TESSA-II set-up at Darebury, using the reaction ^{124}Sn(^{48}Ca, xn)$^{172-x}$Yb at 200 MeV bombarding energy. Selections of total energy and multiplicity enhanced the ^{168}Yb contribution to the correlations to 66 per cent. The weak ridges and shallow valley at γ-ray energies above 780 MeV were explained by a model that included the spread of rotational decay strength over $\Gamma_{rot} = 75$–110 keV at excitation energies in the range of $U_0 = 1.25$ to 1.0 MeV above the yrast line. This explanation is interesting because earlier attempts to explain the uncorrelated intensity by introducing a spread in the moments of inertia required an unreasonably large range of values to wash out the valley-ridge structure as observed in the experiments.

We have shown that the qcs exhibits a wide variety of nuclear phenomena, including collective and single-particle degrees of freedom and with shape changes as well as possible phase transitions. Revealing details about those phenomena may provide unique information about nuclear interactions and challenge nuclear theory to provide adequate descriptions. It has also become clear that at present only the gross features have been measured and that more detailed information can be expected in the near future from the refined high-resolution, low-background many-detector assemblies. This possibility has been demonstrated by the example of the superdeformed band in ^{152}Dy, discussed above.

6.6 References

Åberg, S. (1982). *Physica Scripta* **25,** 23.

Alder, K. and Winther, A. (1975). *Electromagnetic Excitations.* North-Holland, Amsterdam.

Alexander, T. K. and Forster, J. S. (1978). *Adv. Nucl. Phys.* **10,** 197.

Andersen, O., Bauer, R., Hagemann, G. B., Halbert, M. L., Herskind, B., Neiman, M. and Oeschler, H. (1978). *Nucl. Phys.* **A295,** 163.

Andersen, O., Garrett, J. D., Hagemann, G. B., Herskind, B., Hillis, D. L. and Riedinger, L. L. (1979). *Phys. Rev. Lett.* **43,** 687.

Andersson, C. G., Larsson, S. E., Leander, G., Möller, P., Nilsson, S. G., Ragnarsson, I., Åberg, S., Bengtsson, R., Dudek, J., Nerlo-Pomorska, B., Pomorski, K. and Szymański, Z. (1976). *Nucl. Phys.* **A268,** 205.

Andersson, C. G., Hellström, G., Leander, G., Ragnarsson, I., Åberg, S., Krumlinde, J., Nilsson, S. G. and Szymański, Z. (1978). *Nucl. Phys.* **A309,** 141.

Andrews, H. R., Häusser, O., Ward, D., Taras, P., Nicole, R., Keinonen, J., Skensved, P. and Haas, B. (1982). *Nucl. Phys.* **A383,** 509.

Arciszewski, H. F. R., Aarts, H. J. M., Kamermans, R., Van der Poel, C. J., Holzmann, R., Van Hove, M. A., Vervier, J., Huyse, M., Lhersonneau, G., Janssens, R. V. F. and de Voigt, M. J. A. (1983). *Nucl. Phys.* **A401,** 531.

Arciszewski, H. F. R. (1984). Ph.D. thesis, University of Utrecht, The Netherlands.

Asaro, F. and Perlman, I. (1952, 1953). *Phys. Rev.* **87,** 393; **91,** 763.

Auerbach, N. and Talmi, I. (1964). *Phys. Lett.* **10,** 297.

Azgui, F., Emling, H., Grosse, E., Michel, C., Simon, R. S., Spreng, W., Wollersheim, H. J., Khoo, T. L., Chowdhury, P., Frekers, D., Janssens, R. V. F., Pakkanen, A., Daly, P. J., Kortelahti, M., Schwalm, D. and Seiler-Clark, G. (1985). *Nucl. Phys.* **A439,** 573.

Bacelar, J. C., Hagemann, G. B., Herskind, B., Lauritzen, B., Holm, A., Lisle, J. C. and Tjøm, P. O. (1985). *Phys. Rev. Lett.* **55,** 1858.

Bacelar, J. C., Holm, A., Herskind, B., Beck, E. M., Deleplanque, M. A., Diamond, R. M., Stephens, F. S. and Draper, J. (1986). *Phys. Rev. Lett.* **57,** 3019.

Baklander, O., Baktash, C., Borggreen, J., Jensen, J. B., Kownacki, K., Pedersen, J., Sletten, G., Ward, D., Andrews, H. R., Häusser, O., Skensved, P. and Taras, P. (1982). *Nucl. Phys.* **A389,** 93.

Banaschik, M. V., Simon, R. S., Colombani, P., Soroka, D. P., Stephens, F. S. and Diamond, R. M. (1975). *Phys. Rev. Lett.* **34,** 892.

Bauer, R. (1979). *Phys. Rev.* **C19,** 399.

Beck, F. A., Bożek, E., Byrski, T., Gehringer, C., Merdinger, J. C., Schutz, Y., Styczen, J. and Vivien, J. P. (1979). *Phys. Rev. Lett.* **42,** 493.

Bengtsson, R. and Frauendorf, S. (1977). *Proc. Int. Symp. on High-Spin States and Nuclear Structure*, Dresden, ZFK Rossendorf, p. 74.

Bengtsson, R. and Frauendorf, S. (1979a). *Nucl. Phys.* **A314,** 27.

Bengtsson, R. and Frauendorf, S. (1979b). *Nucl. Phys.* **A327,** 139.

Bentley, M. A., Ball, G. C., Cranmer-Gordon, H. W., Forsyth, P. D., Howe, D., Mokhtar, A. R., Morrison, J. D., Sharpey-Schafer, J. F., Twin, P. J., Fant, B., Kalfas, C. A., Nelson, A. H., Simpson, J. and Sletten, G. (1987). *Phys. Rev. Lett.* **59,** 2141.

Beuscher, H., Davidson, W. F., Lieder, R. M., Neskasis, A. and Mayer-Böricke, C. (1975). *Nucl. Phys.* **A249,** 379.

Blaugrund, A. E. (1966). *Nucl. Phys.* **88,** 591.

Blomqvist, J. (1979). *Symp. on High-Spin Phenomena in Nuclei*, Argonne, (ed. T. L. Khoo), p. 155. print ANL/PHY-79-4, Argonne.

Bochev, B., Karamyan, S. A., Kutsarova, T. and Subbotin, V. G. (1975). *Yad. Fiz.* **22,** 665—*Sov. J. Nucl. Phys.* **22,** 343 (1976).

Bohr, A. (1951). *Phys. Rev.* **81,** 134.

Bohr, A. and Mottelson, B. R. (1969 and 1975). *Nuclear Structure*, Vols 1 and 2. Benjamin, New York.

Bohr, A. and Mottelson, B. R. (1974). *Physica Scripta* **A10,** 13.

Bohr, A. and Mottelson, B. R. (1981). *Physica Scripta* **24,** 71 (Nuclei at Very High Spin-Sven Gösta Nilsson in Memoriam, Proc. Nobel Symp., Orenäs, 1980; ed. Georg Leander and Hans Ryde).

Borggreen, J., Bjørnholm, S., Christensen, O., Del Zoppo, A., Herskind, B., Pedersen, J., Sletten, G., Folkmann, F. and Simon, R. S. (1980). *Z. Phys.* **A294,** 113.

Broda, R., Kleinheinz, P., Lunardi, S., Styczen, I. and Blomqvist, J. (1982). *Z. Phys.* **A305,** 281.

Cerkaski, M., Dudek, J., Szymański, Z., Andersson, C. G., Leander, G., Åberg, S., Nilsson, S. G. and Ragnarsson, I. (1977). *Phys. Lett.* **B70,** 9.

Cerkaski, M., Dudek, J., Szymański, Z. and Nilsson, S. G. (1979). *Nucl. Phys.* **A315,** 269.

Chapman, R., Lisle, J. C., Mo, J. N., Paul, E., Simcock, A., Wilmott, J. C., Leslie, J. R., Price, H. G., Walker, P. M., Barcelar, J. C., Garrett, J. D., Hagemann, G. B., Herskind, B., Holm, A. and Nolan, P. J. (1983). *Phys. Rev. Lett.* **25,** 2265.

Chmielewska, D., Sujkowski, Z., Janssens, R. V. F. and de Voigt, M. J. A. (1981). *Nucl. Phys.* **A366,** 142.

Cohen, S., Plasil, F. and Swiatecki, W. J. (1974). *Ann. Phys.* **82,** 557.

Cornelis, K., Holzman, R., Huyse, M., Janssens, R. V. F., Lhersonneau, G., Łukasiak, J., Michel, C., Sujkowski, Z., Van Hove, M. A., Verplancke, J. and Vervier, J. (1980). *Proc. Int. Conf. on Nuclear Behaviour at High Angular Momenta*, Strasbourg, (1980). p. 75. print CRN, Strasbourg.

Cormier, T. M., Stwertka, P. M., Herman, M. and Nicolis, N. G. (1984). *Phys. Rev.* **C30,** 1953.

Cwiok, S., Mikhailov, I. N. and Briançon, Ch. (1983). *Z. Phys.* **A314,** 337.

Dafni, E., Bendahan, J., Broude, C., Goldring, G., Hass, M., Naim, E., Rafailovich, M. H., Chasman, C., Kistner, O. C. and Vajda, S. (1984). *Phys. Rev. Lett.* **26,** 2473; and (1985) *Nucl. Phys.* **A443,** 135.

Deleplanque, M. A., Lee, I. Y., Stephens, F. S., Diamond, R. M. and Aleonard, M. M. (1978). *Phys. Rev. Lett.* **40,** 629.

Deleplanque, M. A., Byrski, Th., Diamond, R. M., Hübel, H., Stephens, F. S., Herskind, B. and Bauer, R. (1978a). *Phys. Rev. Lett.* **41,** 1105.

Deleplanque, M. A., Stephens, F. S., Andersen, O., Garrett, J. D., Herskind, B., Diamond, R. M., Ellegaard, C., Fossan, D. B., Hillis, D. L., Kluge, H., Neiman, M., Roulet, C. P., Shih, S. and Simon, R. S. (1980). *Phys. Rev. Lett.* **45,** 172.

Diamond, R. M., Stephens, F. S., Kelly, W. H. and Ward, D. (1969). *Phys. Rev. Lett.* **22,** 546.

Diebel, M., Mantri, A. N. and Mosel, U. (1980). *Nucl. Phys.* **A345,** 72.

Diebel, M. (1984). *Nucl. Phys.* **A419,** 221.

Dönau, F. and Frauendorf, S. (1983). *Proc. Int. Conf. on High Angular Momentum Properties of Nuclei*, Oak Ridge, p. 143.

Døssing, T., Neergaard, K. and Sagawa, H. (1980). *J. Phys. (Paris) Colloq.* 41–10, 79 (*Int. Conf. on Nuclear Behaviour at High Angular Momentum*, Strasbourg).

Døssing, T., Neergaard, K. and Sagawa, H. (1981). *Physica Scripta* **24,** 258 (*Nuclei at Very High Spin*—Sven Gösta Nilsson in Memoriam, Orenäs, 1980. (ed. Georg Leander and Hans Ryde).

Døssing, T. (1985). *Proc. Niels Bohr Centennial Conf. on Nuclear Structure* Copenhagen, Denmark, May 20–25, p. 379.

Draper, J. E., Dines, E. L., Deleplanque, M. A., Diamond, R. M. and Stephens, F. S. (1986). *Phys. Rev. Lett.* **56,** 309.

Dudek, J. (1980). *J. Phys. (Paris) Colloq.* 41–10, 18 (*Int. Conf. on Nuclear Behaviour at High Angular Momentum*, Strasbourg).

Dudek, J., Nazarewicz, W. and Szymański, Z. (1983). *Physica Scripta* **T5,** 171. (*Proc. 4th Nordic meeting on Nuclear Physics, Fuglsø*, 1982).

Dudek, J. and Nazarewicz, W. (1985). *Phys. Rev.* **C31,** 298.

Ejiri, H., Shibata, T., Itahashi, T., Nagai, T., Sakai, H., Nakayama, S., Kishimoto, T., Maeda, K. and Hoshi, M. (1978). *Nucl. Phys.* **A305,** 167.

Emling, H. (1981). *Proc. XIVth Masurian Summer School on Nuclear Physics*, Mikolajki, Poland, and GSI preprint 81-40.

Emling, H., Grosse, E., Schwalm, D., Simon, R. S., Wollersheim, H. J., Husar, D. and Pelte, D. (1981). *Phys. Lett.* **B98,** 169.

Emling, H., Grosse, E., Kulessa, R., Schwalm, D. and Wollersheim, H. J. (1984). *Nucl. Phys.* **A419,** 187.

Feenstra, S. J., Ockels, W. J., Van Klinken, J., de Voigt, M. J. A. and Sujkowski, Z. (1977). *Phys. Lett.* **B69,** 403.

Feenstra, S. J., Van Klinken, J., Pijn, J. P., Janssen, R. V. F., Michel, C., Steyaart, J., Vervier, J., Cornelis, K., Huyse, M. and Lhersonneau, G. (1979). *Phys. Lett.* **B80,** 183.

Ferguson, S. M., Ejiri, H. and Halpern, I. (1972). *Nucl. Phys.* **A188,** 1.

Folkmann, F., Garrett, J. D., Hagemann, G. B., Harakeh, M. N., Herskind, B., Hillis, D. L., Ogaza, S., Emling, H., Grosse, E., Schwalm, D., Simon, R. S. and Tjøm, P. O. (1981). *Nucl. Phys.* **A361,** 242.

Frauendorf, S. (1981). *Phys. Lett.* **100B,** 219.

Frauendorf, S. and May, F. R. (1983). *Phys. Lett.* **125B,** 245.

Garrett, J. D. and Herskind, B. (1979). *Proc. Daresbury Study Weekend*, Daresbury Laboratory report (1980) p. 30.

Garrett, J. D. (1983). *Physica Scripta* **T5,** 21.

Garrett, J. D., Hagemann, G. B. and Herskind, B. (1983). *Nucl. Phys.* **A400,** 113c. (*Proc. Int. Conf. on Nucleus–Nucleus Collisions*, Michigan State University, 1982).

Glendenning, N. K. (1962). *Phys. Rev.* **127,** 923.

Grosse, E. (1979). *Proc. Symp. on High-Spin Phenomena in Nuclei*, Argonne ANL/PHY-79-4, p. 223.

Grover, J. R. (1967). *Phys. Rev.* **157,** 832.

Grover, J. R. and Gilat, J. (1967a). *Phys. Rev.* **157,** 802.

Grover, J. R. and Gilat, J. (1967b). *Phys. Rev.* **157,** 814.

Grover, J. R. and Gilat, J. (1967c). *Phys. Rev.* **157,** 823.

Haas, B., Andrews, H. R., Häusser, O., Horn, D., Sharpey-Schafer, J. F., Taras, P., Trautmann, W., Ward, D., Khoo, T. L. and Smither, R. K. (1979). *Phys. Lett.* **B84,** 178.

Haas, B., Ward, D., Andrews, H. R., Häusser, O., Ferguson, A. J., Sharpey-Schafer, J. F., Alexander, T. K., Trautmann, W., Horn, D., Taras, P., Skensved, P., Khoo, T. L., Smither, R. K., Ahmed, I., Davids, C. N., Kutschera, W., Levenson, S. and Dors, C. L. (1981). *Nucl. Phys.* **A362,** 254.

Hageman, D. C. J. M., de Voigt, M. J. A. and Jansen, J. F. W. (1979). *Phys. Lett.* **B84,** 301.

Hageman, D. C. J. M. (1981). Ph.D. thesis, University of Groningen, The Netherlands.

Hageman, D. C. J. M., Janssens, R. V. F., Łukasiak, J., Ockels, W. J., Sujkowski, Z. and de Voigt, M. J. A. (1981). *Physica Scripta* **24,** 145—*Nuclei at Very High Spin*—Sven Gösta Nilsson in Memoriam, Proc. Nobel Symposium, Orenäs, 1980 (ed. Georg Leander and Hans Ryde).

Hagemann, G. B., Broda, R., Herskind, B., Ischihara, M., Ogaza, S. and Ryde, H. (1975). *Nucl. Phys.* **A245,** 166.

Hagemann, G. B., Garrett, J. D., Herskind, B., Sletten, G., Tjøm, P. O., Henriques, A., Ingebretsen, F., Rekstad, J., Løvhøiden, G. and Thorsteinsen, T. F. (1982). *Phys. Rev.* **C25,** 3224.

Hagemann, G. B., Garrett, J. D., Herskind, B., Kownacki, J., Nyako, B. M., Nolan, P. L., Sharpey-Schafer, J. F. and Tjøm, P. O. (1984). *Nucl. Phys.* **A424,** 365.

Harris, S. M. (1965). *Phys. Rev.* **138,** B509.

Häusser, O., Taras, P., Trautmann, W., Ward, D., Alexander, T. K., Andrews, H. R., Haas, B. and Horn, D. (1979). *Phys. Rev. Lett.* **42,** 1451.

Häusser, O., Andrews, H. R., Mahnke, H. E., Sharpey-Schafer, J. F., Swanson, M. L., Taras, P., Ward, D. and Alexander, T. K. (1980). *Phys. Rev. Lett.* **44,** 132.

Häusser, O., Andrews, H. R., Ward, D., Rud, N., Taras, P., Nicole, R., Keinonen, J., Skensved, P. and Stager, C. V. (1983). *Nucl. Phys.* **A406,** 339.

Herskind, B. (1980). *J. Phys. (Paris) Colloq.* **41–10,** 106 (*Int. Conf. on Nuclear Behaviour at High Angular Momentum*, Strasbourg).

Herskind, B. (1983). *Proc. Int. Conf. on Nuclear Physics*, Florence, Italy, p. 117.

Herskind, B. (1984). *Ann. Israel Phys. Soc.* **7,** 3; *Proc. Workshop on Electromagnetic Properties of High Spin Levels*, Rehovoth (ed. G. Goldring and M. Hass).

Hillis, D. L., Garrett, J. D., Christensen, O., Fernandez, B., Hagemann, G. B., Herskind, B., Back, B. B. and Folkmann, F. (1979). *Nucl. Phys.* **A325,** 216.

Horn, D., Häusser, O., Towner, I. S., Andrews, H. R., Lone, M. A. and Taras, P. (1983). *Phys. Rev. Lett.* **50,** 1447.

Hübel, H., Smilansky, U., Diamond, R. M., Stephens, F. S. and Herskind, B. (1978). *Phys. Rev. Lett.* **41,** 791.

Hübel, H., Diamond, R. M., Stephens, F. S., Herskind, B. and Bauer, R. (1980). *Z. Phys.* **A297,** 237.

Hübel, H., Deeman, D. J., Grawe, H., Haas, H. and Maier, K. H. (1982). *Nucl. Phys.* **A382,** 56.

Hübel, H., Diamond, R. M., Aguer, P., Ellegaard, C., Fossan, D. B., Kluge, H., Schück, C., Shih, S., Stephens, F. S. and Smilanski, U. (1982a). *Z. Phys.* **A304,** 225.

Hultberg, S., Rezanka, I. and Ryde, H. (1973). *Nucl. Phys.* **A205,** 321.

Huus, T. and Zupancic, C. (1953). *K. Dan. Vidensk. Selsk. Mat. Phys. Medd* **28,** No. 1, 1.

Jaäskeläinen, M., Sarantites, D. G., Dilmanian, F. A., Woodward, R., Puchta, H., Beene, J. R., Hattula, J., Halbert, M. L. and Hensley, D. C. (1982). *Phys. Lett.* **119B,** 65.

Jansen, J. F. W., de Meijer, R. J., Sujkowski, Z., Chmielewska, D. and Grabowski, J. (1974). *Proc. Int. Conf. Nucl. Structure Spectrosc.*, Scholars Press, Amsterdam.

Jansen, J. F. W., Sujkowski, Z., Chmielewska, D. and de Meijer, R. J. (1976). *Proc. Third Int. Conf. on Nuclei Far from Stability*, Cargèse (CERN 76-13), p. 415.

Jansen, J. F. W., de Voigt, M. J. A., Sujkowski, Z. and Chmielewska, D. (1979). *Nucl. Phys.* **A321,** 365.

Janssens, R. V. F., de Voigt, M. J. A., Sakai, H., Aarts, H., Van der Poel, C. J., Arciszewski, H. F. R., Scherpenseel, D. E. C. and Vervier, J. (1981). *Phys. Lett.* **B106,** 475.

Janssens, R. V. F., Radford, D. C., Ahmed, I. and van den Berg, A. M. (1985). *Phys. Lett.* **152B,** 167.

Jastrzeþski, J., Kaczarowski, R., Łukasiak, J., Moszyński, M., Preibisz, Z., André, S., Genevey, J., Gizon, A. and Gizon, J. (1980). *Phys. Lett.* **B97,** 50.

Johnson, A., Ryde, H. and Hjorth, S. (1972). *Nucl. Phys.* **A179,** 753.

Jónsson, S., Lyttkens, J., Carlén, L., Roy, N., Ryde, H., Waluś, W., Kownacki, J., Hagemann, G. B., Herskind, B., Garrett, J. D. and Tjøm, P. O. (1984). *Nucl. Phys.* **A422,** 397.

Karwowski, H. J., Vigdor, S. E., Jacobs, W. W., Kailas, S., Singh, P. P., Soga, F. and Ploughe, W. D. (1979). *Phys. Rev. Lett.* **42,** 1732.

Karwowski, H. J., Vigdor, S. E., Jacobs, W. W., Kailas, S., Singh, P. P., Soga, F., Throwe, T. G., Ward, T. E., Wark, D. L. and Wiggins, J. (1982). *Phys. Rev.* **C25,** 1355.

Khoo, T. L., Benthal, F. M., Robertson, R. G. H. and Warner, R. A. (1976). *Phys. Rev. Lett.* **37,** 823.

Khoo, T. L., Smither, R. K., Haas, B., Häusser, O., Andrews, H. R., Horn, D. and Ward, D. (1978). *Phys. Rev. Lett.* **41,** 1027.

Khoo, T. L., Chowdhury, P., Emling, H., Frekers, D., Janssens, R. V. F., Kühn, W., Pakkanen, A., Chung, Y. H., Daly, P. J., Grabowski, Z. W., Helppi, H., Kortelahti, M., Bjørnholm, S., Borggreen, S., Pedersen, J. and Sletten, G. (1983). *Physica Scripta*, **T5,** 16.

Kleinheinz, P. (1979). *Proc. Conf. on High-Spin Phenomena in Nuclei*, Argonne (ed. T. L. Khoo), print ANL/PHY-79-4, Argonne, p. 125.

Klinken, J. van, Feenstra, S. J. and Dumont, G. (1978). *Nucl. Instr. Meth.* **151,** 483.

Koeling, T. (1979). *Nucl. Phys.* **A315,** 169.

Körner, H. J., Hillis, D. L., Roulet, C. P., Aguer, P., Ellegaard, C., Fossan, D. B., Habs, D., Neiman, M., Stephens, F. S. and Diamond, R. M. (1979). *Phys. Rev. Lett.* **43,** 490.

Kutschera, W., Dehnhardt, D., Kistner, O. C., Kump, P., Povh, B. and Sann, H. J. (1972). *Phys. Rev.* **C5,** 1658.

Landau, L. D. and Lifschitz, E. M. (1965). *Mechanika*, Nauka, Moscow.

Love, D. J. G., Nelson, A. H., Nolan, P. J. and Twin, P. J. (1985). *Phys. Rev. Lett.* **54,** 1361.

Lee, I. Y., Aleonard, M. M., Deleplanque, M. A., El-Masri, Y., Newton, J. O., Simon, R. S., Diamond, R. M. and Stephens, F. S. (1977). *Phys. Rev. Lett.* **38,** 1454.

Lee, M. A. and Reich, C. W. (1979). *Nucl. Data Sheets* **27,** 155.
Lieder, R. M. and Ryde, H. (1978). *Adv. Nucl. Phys.* **10,** 1.
Łukasiak, J., Hageman, D. C. J. M., Janssens, R. V. F., Sujkowski, Z. and de Voigt, M. J. A. (1982). *Nucl. Phys.* **A379,** 125.
Mariscotti, M. A., Scharff-Goldhaber, G. and Buck, B. (1969). *Phys. Rev.* **178,** 1864.
Merdinger, J. C., Beck, F. A., Byrski, T., Gehringer, C., Vivien, J. P., Bożek, E. and Styczen, J. (1979). *Phys. Rev. Lett.* **42,** 23.
Morinaga, H. and Gugelot, P. C. (1963). *Nucl. Phys.* **46,** 210.
Mütz, U. and Ring, P. (1984). *J. Phys. G* **10,** L39.
Myers, W. D. and Swiatecki, W. J. (1966). *Nucl. Phys.* **81,** 1.
Myers, W. D. and Swiatecki, W. J. (1967). *Ark. Fys.* **36,** 343.
Nagai, Y., Styczen, J., Piiparinen, M. and Kleinheinz, P. (1980). *Z. Phys.* **A296,** 91.
Nagai, Y., Shibata, T., Ohsumi, H. and Ejiri, H. (1983). *Phys. Rev. Lett.* **51,** 1247.
Neergaard, K., Døssing, T. and Sagawa, H. (1981). *Phys. Lett.* **B99,** 191.
Newton, J. O., Lee, I. Y., Simon, R. S., Aleonard, M. M., El Masri, Y., Stephens, F. S. and Diamond, R. M. (1977). *Phys. Rev. Lett.* **38,** 810.
Newton, J. O., Stephens, F. S. and Diamond, R. M. (1973). *Nucl. Phys*, **A210,** 19.
Nilsson, S. G. (1955). *Mat. Fys. Medd. Dan. Vid. Selsk.* **29,** No. 16.
Northcliff, L. C. and Schilling, R. F. (1970). *Nucl. Data Tables* **A7,** 256.
Nyako, B. M., Cresswell, J. R., Forsyth, P. D., Howe, D., Nolan, P. J., Riley, M. A., Sharpey-Schafer, J. F., Simpson, J., Ward, N. J. and Twin, P. J. (1984). *Phys. Rev. Lett.* **52,** 507.
Nyako, B. M., Simpson, J., Twin, P. J., Howe, D., Forsyth, P. D. and Sharpey-Schafer, J. F. (1986). *Phys. Rev. Lett.* **56,** 2680.
Ockels, W. J., de Voigt, M. J. A. and Sujkowski, Z. (1978). *Phys. Lett.* **B78,** 401.
Pakkanen, A., Chung, Y. H., Daly, P. J., Faber, S. R., Helppi, H., Wilson, J., Chowdhurry, P., Khoo, T. L., Ahmed, I., Borggreen, J., Grabowski, Z. W. and Radford, D. C. (1982). *Phys. Rev. Lett.* **48,** 1530.
Pedersen, J., Back, B. B., Bernthal, F. M., Bjørnolm, S., Borggreen, J., Christensen, O., Folkmann, F., Herskind, B., Khoo, T. L., Neiman, M., Pühlhofer, F. and Sletten, G. (1977). *Phys. Rev. Lett.* **39,** 990.
Pedersen, J., Back, B. B., Bjørnholm, S., Borggreen, J., Diebel, M., Sletten, G., Azgui, F., Emling, H., Grein, H., Seiler-Clark, G., Spreng, W., Wollersheim, H. J., Walker, P. J. and Åberg, S. (1985). *Phys. Rev. Lett.* **54,** 306.
Pelte, D. and Schwalm, D. (1982). *Heavy-Ion Collisions*, vol. 3 (ed. R. Bock, North-Holland, Amsterdam.
Ragnarsson, I., Bengtsson, T., Leander, G. and Åberg, S. (1980). *Nucl. Phys.* **A347,** 287.
Ragnarsson, I. and Åberg, S. (1986). Lund preprint MPH-86/11.
Riezebos, H. J., Balanda, A., Dudek, J., van Klinken, J., Nazarewicz, W., Sujkowski, Z. and de Voigt, M. J. A. (1987). *Phys. Lett.* **B183,** 277.
Ring, P., Hagaski, A., Harn, K., Emling, H. and Grosse, E. (1981). *Phys. Lett.* **110B,** 423.
Rud, N., Ward, D., Andrews, H. R., Graham, R. L. and Geiger, J. S. (1973). *Phys. Rev. Lett.* **31,** 1421.
Rud, N., Ward, D., Andrews, H. R., Häusser, O., Taras, P., Keinonen, P., Neiman, M., Diamond, R. M. and Stephens, F. S. (1981). *Phys. Lett.* **B101,** 35.
Sarantites, D. G., Barker, J. H., Halbert, M. L., Hensley, D. C., Dayras, R. A., Eichler, E., Johnson, N. R. and Gronemeyer, S. A. (1978). *Phys. Rev.* **C14,** 2138.

Sarantites, D. G., Westerberg, L., Dayras, R. A., Halbert, M. L., Hensley, D. C. and Barker, J. H. (1978). *Phys. Rev.* **C17,** 601.

Sarantites, D. G., Westerberg, L., Halbert, M. L., Dayras, R. A., Hensley, D. C. and Barker, J. H. (1978). *Phys. Rev.* **C18,** 774.

Sarantites, D. G., Jaäskeläinen, M., Woodward, R., Dilmanian, F. A., Hensley, D. C., Barker, J. H., Beene, J. R., Halbert, M. L. and Milner, W. T. (1982). *Phys. Lett.* **B115,** 441.

Schutz, Y., Vivien, J. P., Beck, F. A., Byrski, T., Gehringer, C., Merdinger, J. C., Dudek, J., Nazarewicz, W. and Szymański, Z. (1982). *Phys. Rev. Lett.* **48,** 1534.

Schwalm, D. (1981). *Nuclear Structure* (eds. K. Abrahams, K. Allaart and A. E. C. Dieperink). Plenum Press, New York.

Sie, S., Ward, H. D., Geiger, J. S., Graham, R. L. and Andrews, H. R. (1977). *Nucl. Phys.* **A291,** 443.

Seiler-Clark, G., Pelte, D., Emling, H., Bałanda, A., Grein, H., Grosse, E., Kulessa, R., Schwalm, D., Wollersheim, H. J., Hass, M., Kumbartzki, G. J. and Speidel, K. H. (1983). *Nucl. Phys.* **A399,** 211.

Simon, R. S., Banaschik, M. V., Colombani, P., Soroka, D. P., Stephens, F. S. and Diamond, R. M. (1976). *Phys. Rev. Lett.* **36,** 359.

Simon, R. S., Banaschik, M. V., Diamond, R. M., Newton, J. O. and Stephens, F. S. (1977). *Nucl. Phys.* **A290,** 253.

Simpson, J. J., Tjøm, P. O., Espe, I., Hagemann, G. B., Herskind, B. and Neiman, M. (1977). *Nucl. Phys.* **A287,** 362.

Simpson, J. J., Riley, J., Cresswell, J. R., Forsyth, P. D., Howe, D., Nyako, B. M., Sharpey-Schafer, J. F., Bacelar, J., Garrett, J. D., Hagemann, G. B., Herskind, B. and Holm, A. (1984). *Phys. Rev. Lett.* **53,** 648.

Sletten, G., Bjørnholm, S., Borggreen, J., Pedersen, J., Chowdhurry, P., Emling, H., Frekers, D., Janssens, R. V. F., Khoo, T. L., Chung, Y. H. and Kortelahti, M. (1984). *Phys. Lett.* **135,** 33.

Stachel, J., Kaffrell, N., Grosse, E., Emling, H., Folger, H., Kulessa, R. and Schwalm, D. (1982). *Nucl. Phys.* **A383,** 429.

Stephens, F. S., Lark, N. L. and Diamond, R. M. (1965). *Nucl. Phys.* **63,** 82.

Stephens, F. S. (1975). *Rev. mod. Phys.* **47,** 43.

Stephens, F. S. (1979). *Proc. Symp. on High-Spin Phenomena in Nuclei*, Argonne ANL/PHY-79-4, p. 335.

Stephens, F. S. (1983). *Physica Scripta* **T5,** 5. *Proc. 4th Nordic meeting on Nuclear Physics*, Fuglsø, 1982.

Stephens, F. S. (1985). *Proc. Niels Bohr Centennial Conf. on Nuclear Structure*, Copenhagen, Denmark, May 20–25, p. 363.

Stephens, F. S., Draper, J. E., Bacelar, J. C., Beck, E. M., Deleplanque, M. A. and Diamond, R. M. (1987). *Phys. Rev. Lett.* **58,** 2186.

Stephens, F. S., Deleplanque, M. A., Diamond, R. M., Macchiavelli, A. O. and Draper, J. E. (1985). *Phys. Rev. Lett.* **54,** 2584.

Stephens, F. S., Draper, J. E., Egido, J. L., Bacelar, J. C., Beck, E. M., Deleplanque, M. A. and Diamond, R. M. (1986). *Phys. Rev. Lett.* **57,** 2912.

Stephens, F. S., Draper, J. E., Bacelar, J. C., Beck, E. M., Deleplanque, M. A. and Diamond, R. M. (1987). *Phys. Rev. Lett.* **58,** 2186.

Strutinsky, V. M. (1967). *Nucl. Phys.* **A95,** 420.

Strutinsky, V. M. (1968). *Nucl. Phys.* **A122,** 1.

Stwertka, P. M., Cormier, T. M., Herman, M. G. and Nicolis, N. G. (1985). *Phys. Rev. Lett.* **54,** 1635.

Styczen, J., Nagai, Y., Piiparinen, M., Ercan, A. and Kleinheinz, P. (1983). *Phys. Rev. Lett.* **50,** 1752.

Sujkowski, Z., Chmielewska, D., Janssens, R. V. F. and de Voigt, M. J. A. (1979). *Phys. Rev. Lett.* **43,** 998.

Swiatecki, W. J. (1987). *Phys. Rev. Lett.* **58,** 1184.

Szymański, Z. (1983). *Fast Nuclear Rotation.* Oxford University Press, Oxford.

Tanabe, K. and Sugawara-Tanabe, K. (1984). *Phys. Lett.* **135B,** 353.

Taras, P., Häusser, O., Andrews, H. R., Ward, D., Deleplanque, M. A., Diamond, R. M., Macchiavelli, A. O. and Stephens, F. S. (1985). *Nucl. Phys.* **A435,** 294.

Tjøm, P. O., Espe, I., Hagemann, G. B., Herskind, B. and Hillis, D. L. (1978). *Phys. Lett.* **72B,** 439.

Tjøm, P. O., Diamond, R. M., Bacelar, J. C., Beck, E. M., Deleplanque, M. A., Draper, J. E. and Stephens, F. S. (1985). *Phy. Rev. Lett.* **55,** 2405.

Trautmann, W., Sharpey-Schafer, J. F., Andrews, H. R., Haas, B., Häusser, O., Taras, P. and Ward, D. (1979). *Phys. Rev. Lett.* **43,** 991.

Twin, P. J. (1983). *Proc. Int. Conf. on Nuclear Physics,* Florence, Italy. p. 527.

Twin, P. J., Nelson, A. H., Nyako, B. M., Howe, D., Cranmer-Gordon, H. W., Elenkov, D., Forsyth, P. D., Jabber, J. K., Sharpey-Schafer, J. F., Simpson, J. and Sletten, G. (1985). *Phys. Rev. Lett.* **55,** 1380.

Twin, P. J., Nyako, B. M., Nelson, A. H., Simpson, J., Bentley, M. A., Cranmer-Gordon, H. W., Forsyth, P. D., Howe, D., Mokhtar, A. R., Morrison, J. D., Sharpey-Schafer, J. F. and Sletten, G. (1986). *Phys. Rev. Lett.* **57,** 88.

Vivien, J. P., Schutz, Y., Beck, F. A., Bożek, E., Byrski, T., Gehringer, C. and Merdinger, J. C. (1979). *Phys. Lett.* **B85,** 325.

Voigt, M. J. A. de, Ockels, W. J., Sujkowski, Z., Zglinski, A. and Mooibroek, J. (1979). *Nucl. Phys.* **A323,** 317.

Voigt, M. J. A. de, Janssens, R. V. F., Sakai, H., Aarts, H. J. Van de Poel, C. J., Scherpenzeel, D. E. C. and Arciszewski, H. F. R. (1981). *Phys. Lett.* **B106,** 480.

Voigt, M. J. A. de, Maeda, K., Ejiri, H., Shibata, T., Okada, K., Motobayashi, T., Sasao, M., Kishimoto, T., Suzuki, H., Sakai, H. and Shimizu, A. (1982). *Nucl. Phys.* **A379,** 160.

Voigt, M. J. A. de, Dudek, J. and Szymański, Z. (1983). *Revs. Mod. Phys.* **55,** 949.

Voigt, M. J. A. de, Riezebos, H., van Klinken, J., Bałanda, A., Sujkowski, Z., Bacelar, J. C., Beck, E. M., Deleplanque, M. A., Diamond, R. M. and Stephens, F. S. (1986, 1987). *Int. Symp. on Weak and Electromagnetic Interactions in Nuclei,* Heidelberg, July 1986; *Phys. Rev. Lett.* **59,** 270.

Ward, D., Andrews, H. R., Geiger, J. S. and Graham, R. L. (1973). *Phys. Rev. Lett.* **30,** 493.

Ward, D., Andrews, H. R., Häusser, O., El-Masri, Y., Aleonard, M. M., Yang-Lee, I., Diamond, R. M., Stephens, F. S. and Butler, P. A. (1979). *Nucl. Phys.* **A332,** 433.

Ward, D., Andrews, H. R., Haas, B., Taras, P. and Rud, N. (1983). *Nucl. Phys.* **A397,** 161.

Westerberg, L., Sarantites, D. G., Lovett, R., Hood, J. T., Barker, J. H., Currie, C. M. and Mullani, N. (1977). *Nucl. Instr. Meth.* **145,** 295.

Westerberg, L., Sarantites, D. G., Geoffroy Young, K., Dayras, R. A., Beene, J. R., Halbert, M. L., Hensley, D. C. and Barker, J. H. (1978). *Phys. Rev. Lett.* **41,** 96.

Wilczyński, J. (1973). *Nucl. Phys.* **A216,** 386.

Wilczyński, J., Siwek-Wilczyńska, K., Van Driel, J., Gongrijp, S., Hageman, D. C. J. M., Janssens, R. V. F., Łukasiak, J., Siemssen, R. H. and van der Werf, S. Y. (1982). *Nucl. Phys.* **A373,** 109.

7

GAMMA-RAY SPECTROSCOPY FOR WEAK INTERACTION AND HYPERNUCLEAR SPECTROSCOPY

7.1 Fundamental interactions studied by in-beam and off-beam e–γ spectroscopy

7.1.1 *Left–right symmetry in nuclear beta decay*

Since the success of the standard electro-weak theory there have been many experimental and theoretical studies of further unified models beyond the standard $SU(2)_L \times U(1)$ model. In the standard theory the weak interaction consists of the pure left-handed current of the V-A type (vector-axial vector currents) and the parity violation is maximal. Thus the magnitude of the longitudinal polarization of the β rays is 100 per cent, i.e. $P_F = P_{GT} = 1$, for both the Fermi and the Gamow–Teller decays. As extensions of the standard theory several models with global symmetries have been proposed, some incorporating left–right symmetry.

In the left–right-symmetry model of $SU(2)_L \times SU(2)_R \times U(1)$ a right-handed (V + A) current is included, and thus a non-maximal degree of the longitudinal polarization (Senjanović, 1979). The degrees of polarizations are then given as

$$P_F = \frac{2\varepsilon}{1+\varepsilon^2} \cdot \frac{(1-\delta)(1+\varepsilon^2\delta)}{1+\varepsilon^2\delta^2}, \tag{7.1.1}$$

$$P_{GT} = \frac{2\varepsilon}{1+\varepsilon^2} \cdot \frac{(1-\delta)(\varepsilon^2+\delta)}{(\varepsilon^2+\delta^2)}, \tag{7.1.2}$$

where δ is the square of the mass ratio for the left- and right-handed weak bosons and ε is written in terms of the mixing angle ζ as $\varepsilon = (1+\tan\zeta)/(1-\tan\zeta)$.

The contributions of the right-handed current can be studied by measuring precisely the deviations of P_F or P_{GT} from unity. Because of the different effects of the δ and ε on P_F and P_{GT}, the ratio $R = P_F/P_{GT}$ deviates from unity. The quantities δ and ε, which give evidence for the break-down of the standard theory, are very small if they are non-zero. Thus the experimental studies of them require high-precision measurements of electrons (β rays) with high-intensity short-lived β-sources. Since such sources are produced by nuclear reactions, high-precision in-beam e spectroscopy is quite useful for studying the longitudinal polarization.

van Klinken *et al.* (1983) have studied the ratio P_F/P_{GT} by the in-beam study of the β-ray polarization since a comparative measurement of the P_F and P_{GT} is quite advantageous from the experimental point of view. For studying the Fermi and Gamow–Teller decays, ^{26}Al (6.3 s) and ^{30}P (1.5 m) sources were produced by the (p, n) reactions on enriched ^{26}Mg and ^{30}Si targets, respectively. In-beam measurements of the longitudinal polarizations were carried out by means of the mini-orange polarimeter (see Section 3.4). The positrons from the β-ray source, produced by the reaction, were analysed in energy by the mini-orange spectrometer, and were guided to the Bhabha scattering polarimeter, where the scattered positron–electron pairs were detected by plastic scintillators. The spins of the scattered electrons were inverted several times per second by changing the direction of the magnetizing current of the iron scatterer.

A fourfold Bhabha polarimeter designed to improve statistics and precision, as shown in Fig. 7.1 (Wichers *et al.*, 1985, 1987), was used in the measurements, and the result shows no finite deviation of the ratio $R = P_F/P_{GT}$ from unity beyond the statistical error. The observed value is $R = 1.003 \pm 0.004$. The constraints in the δ–ζ plane imposed by the measured ratio and other data are illustrated in Fig. 7.2.

7.1.2 *Double beta-gamma spectroscopy and lepton number conservation*

7.1.2a *Double beta–gamma spectroscopy and the neutrino*

Precise spectroscopic studies of neutrino-less double beta decays ($0\nu\beta\beta$) have recently attracted much attention in view of grand unified theories and astrophysical constraints on the cosmological mass. The normal double beta ($\beta\beta$) decay is accompanied by two neutrinos, and conserves the lepton number. On the other hand the $0\nu\beta\beta$ decay violates the lepton number conservation law. The two decay modes are written

$$2\nu\beta\beta;\ \mathrm{A} \to \mathrm{B} + \beta_1 + \beta_2 + \bar{\nu}_1 + \bar{\nu}_2, \tag{7.1.3}$$

$$0\nu\beta\beta;\ \mathrm{A} \to \mathrm{B} + \beta_1 + \beta_2. \tag{7.1.4}$$

The $0\nu\beta\beta$ process is a second-order weak process, where the neutrino emission and re-absorption take place between two nucleons (quarks) in a nucleus. Thus helicity mixing of such a neutrino in the $0\nu\beta\beta$ even leads to a finite Majorana mass (m_ν) of the electron neutrino and/or a finite right-handed current (V + A) in the nuclear beta-decay (Primakoff and Rosen, 1969/1981; Doi *et al.*, 1981, 1983; Kotani *et al.*, 1986). The $0\nu\beta\beta$ process has also been discussed in terms of the Majoron emission mechanism (Georgi *et al.*, 1981, Gelmini and Roncadelli, 1981), and the Higgs boson exchange

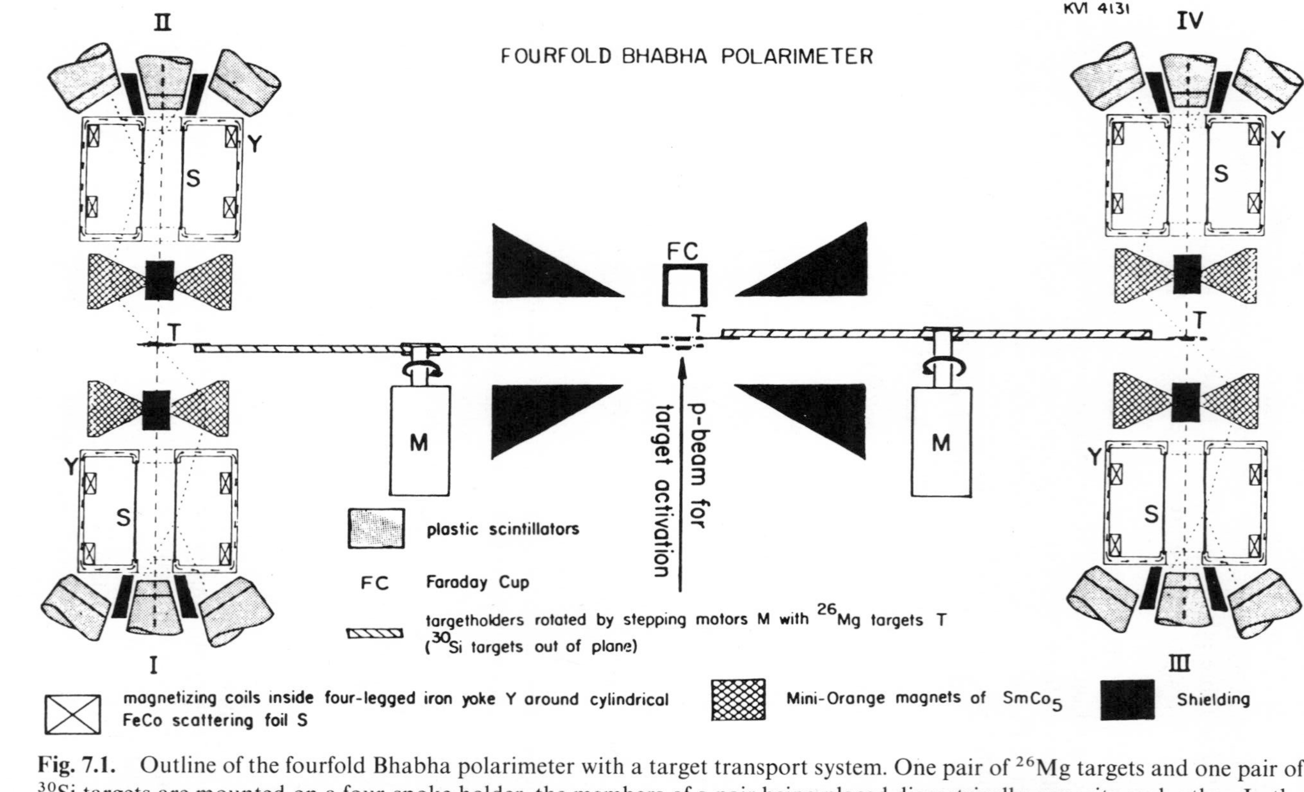

Fig. 7.1. Outline of the fourfold Bhabha polarimeter with a target transport system. One pair of ^{26}Mg targets and one pair of ^{30}Si targets are mounted on a four-spoke holder, the members of a pair being placed diametrically opposite each other. In the run for the Fermi decay the ^{26}Mg targets are activated alternately by the (p, n) reaction, and measured at the two measuring sites, and in the run for the Gamow–Teller decay the ^{30}Si targets are used by rotating the holder by 90°. (Wichers *et al.*, 1985.)

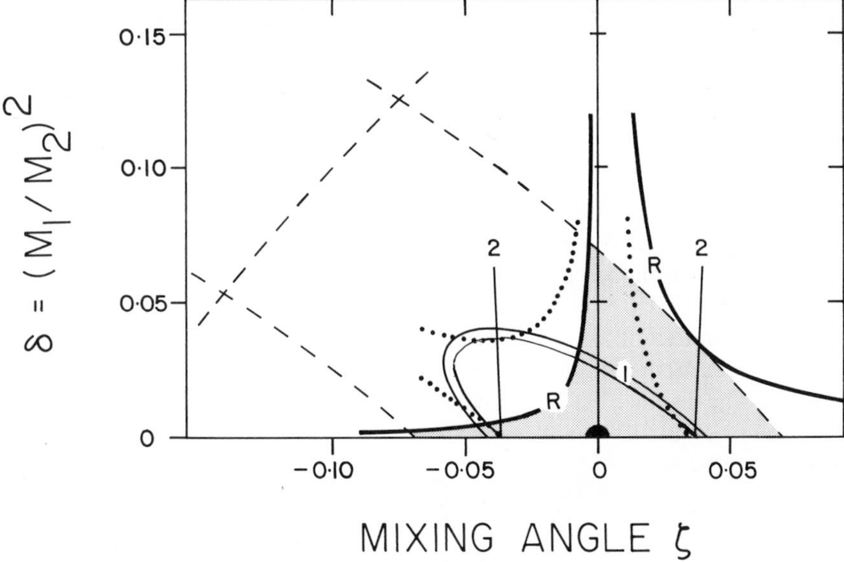

Fig. 7.2. The constraints at 90 per cent confidence on the $W_{1,2}$ (weak gauge boson) mass-squared ratio δ and the mixing angle ζ by R-values from the present work (bold curve), and from previous β-polarization measurements (dashed curves). Also indicated are the bounds from measurements of the e^+ endpoint in muon decay (outer and inner curves 1), and from the Michel parameter (curve 2). The dotted curve comes from an analysis of the decay of polarized ^{19}Ne. (Wichers *et al.*, 1987; and references therein.)

mechanism (Mohapatra and Vergados 1981; Wolfenstein 1982). Here the Majoron is the Goldston boson arising from the spontaneous symmetry-breaking of global baryon number-lepton number symmetry. Hereafter we discuss the Majorana neutrino for the $0\nu\beta\beta$ process given in eqn (7.1.4).

The finite neutrino mass and the right-handed weak interaction are beyond the minimal standard theory of $SU(2)_L \times U(1)$, and are discussed in terms of grand unified theories with the lepton–hadron universality and of the left–right symmetry models. Neutrinos with $m_\nu \approx 10 \sim 20$ eV would constitute the major fraction of the mass of the cosmos which is filled with neutrinos, and accordingly would lead to the stability of the cosmos. The $0\nu\beta\beta$ process is very sensitive to lepton number non-conservation, the neutrino mass and the right-handed weak current.

The $\beta\beta$ decay is a very rare process. The $\beta\beta$–γ spectroscopy is a kind of nuclear spectroscopy, where a nucleus is used as a quite good quantum system for studying such a rare and fundamental process. The unique and essential features of the $\beta\beta$ spectroscopy are as follows.

1. One may select an even nucleus with (N, Z) such that a single β decay to an odd nucleus with $(N \pm 1, Z \mp 1)$ is energetically forbidden and the $\beta\beta$ decay to an even nucleus with $(N \pm 2, Z \mp 2)$ is allowed, because of the pairing interaction in the even–even nuclei. Then the $\beta\beta$-events are not obscured by the single β-events, which would otherwise be much stronger than the $\beta\beta$-events by an order of 10^{25-28}. Typical $\beta\beta$-decay nuclei are shown in Fig. 7.3.

2. The $\beta\beta$-process changes two units of isospin ($\Delta T_z = \pm 2$), and thus it involves two nucleons (2n) with isospin 1/2 in a nucleus or two quarks in a Δ isobar with isospin 3/2 (Primakoff and Rosen, 1969/1981; Doi *et al.*, 1981, 1983). Thus the $0\nu\beta\beta$ process is the two-step (second order) process where a neutrino emitted from one nucleon (quark) is reabsorbed into another nucleon (quark) in a single nucleus (baryon). The decay schemes are shown in Fig. 7.4. The distance between the two nucleons (quarks) is much shorter than the wave length of the neutrino emitted in a normal (lepton number conserving) $2\nu\beta\beta$ process. Thus the $0\nu\beta\beta$ process is enhanced by the kinematical factor of the order of 10^{6-8} over the $2\nu\beta\beta$ process. This corresponds to an amplification factor of the nuclear microscope used to search for the very rare $0\nu\beta\beta$ event.

3. The initial and final states associated with the $\beta\beta$ decay are good eigenstates with definite spins, isospins and energies. Thus spectroscopic studies of energy and angular correlations of the two beta-rays give details of the interaction terms contributing to the $0\nu\beta\beta$.

4. The $0\nu\beta\beta$ process gives a discrete line at $Q_{\beta\beta}$ in the summed energy spectrum of two β rays. Thus high-resolution measurement of the energy spectrum can single out the rare $0\nu\beta\beta$ event without disturbance of the continuum component of the $2\nu\beta\beta$ process below the $Q_{\beta\beta}$.

5. The $0\nu\beta\beta$ process, if it exists, is a very rare process with a half-life $T_{1/2} \gtrsim 10^{22-24}$ year and the energy release is as small as a few MeV. Thus very elaborate spectroscopic techniques are needed to search for such low-energy rare events.

7.1.2b *Double beta–gamma spectrometers for rare ββ decays*

Even though the $\beta\beta$ decay has the unique advantage of using a nuclear system to study the fundamental properties of neutrinos, specially designed spectrometers have to be used to search for very rare and low-energy $\beta\beta$ events among a huge number of background events. The lower limit of the detectable true event rate is given by the statistical fluctuation of the background rate. Then the observable limit of the decay probability $\tilde{T}_{0\nu}$ is expressed in terms of the detector sensitivity S_D as follows.

$$\tilde{T}_{0\nu} = 1/(S_D \cdot \sqrt{t}), \tag{7.1.5}$$

$$S_D = 0.76 N_0 k / \sqrt{(N_{BG} \cdot \Delta E)}, \tag{7.1.6}$$

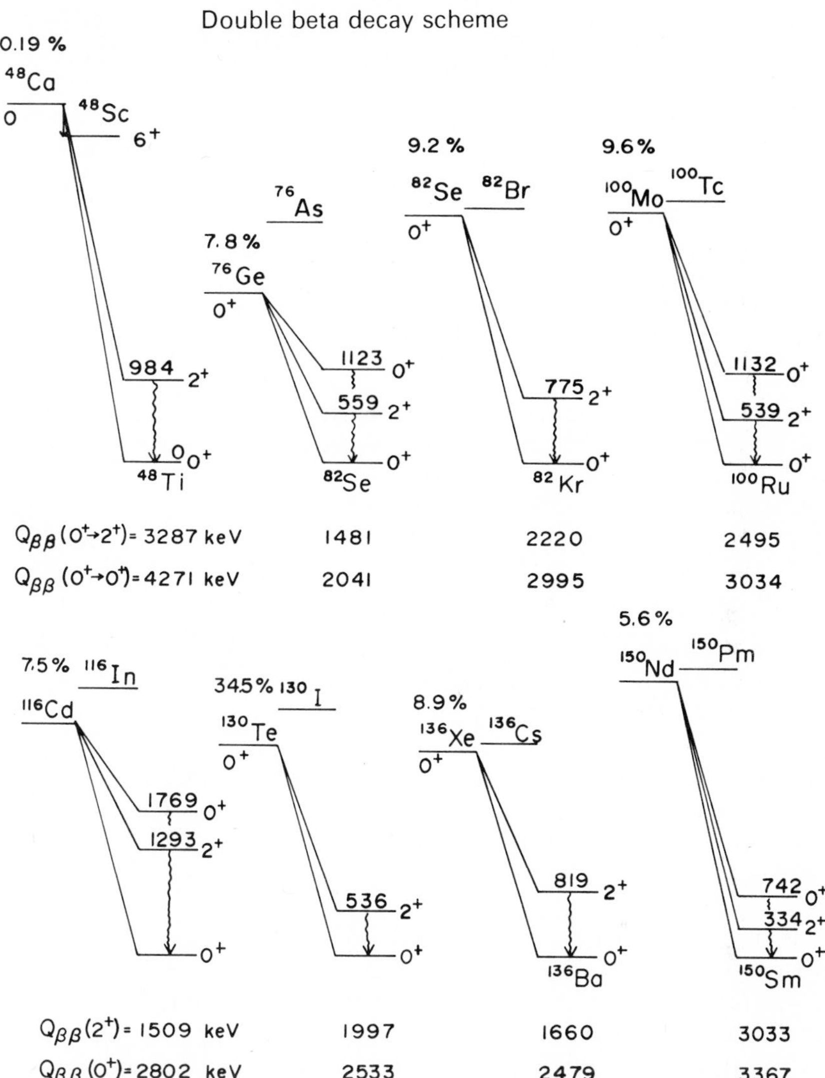

Fig. 7.3. Typical level schemes of double beta decays, where single beta decays are forbidden because of negative Q_β values or the large angular momentum change. (Ejiri, 1986.)

where N_0 is the number of the $\beta\beta$-source nuclei, k is the detection efficiency, N_{BG} is the background count rate per keV, ΔE is the energy resolution, in units of keV, and t is the running time. The coefficient 0.76 is the probability of the true event falling in the energy window ΔE. For given N_{BG} in units of $(\text{keV}\cdot\text{year})^{-1}$ and ΔE in units of keV, S_D is given in units of $(\text{year})^{1/2}$ and $\tilde{T}_{0\nu}$ in

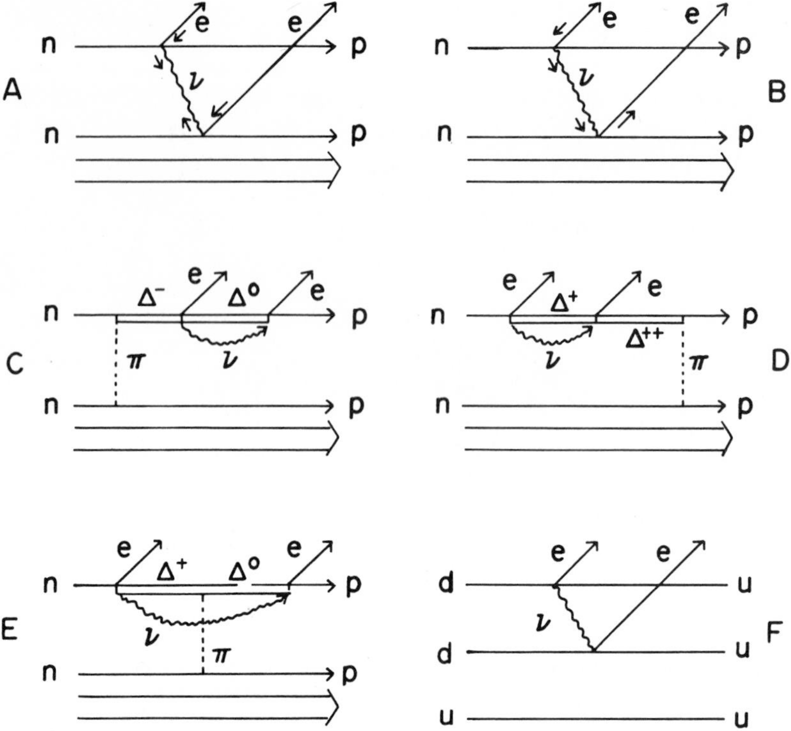

Fig. 7.4. Schematic diagrams of neutrino-less double beta decays. A: 2n with m_ν, B: 2n with V + A, C: Δ^- in the initial state, D: Δ^{++} in the final state, E: $\Delta^+\Delta^0$ process, F: 2 quark process in one Δ (C, D, E). the arrows in A and B give the helicity. (Ejiri, 1986; Doi *et al.*, 1985.)

units of $(\text{year})^{-1}$. Then the maximum value for the half-life to be measured is given as $\tilde{T}_{1/2} = (0.693/\tilde{T}_{0\nu})$ in units of year.

Much effort has to be made to enlarge t, N_0 and k and to reduce N_{BG} and ΔE. In a geochemical method one counts the $\beta\beta$ decay product nuclei for a geological period of $t \sim 10^{7-8}$ years. Although this method measures the sum of the $0\nu\beta\beta$ and $2\nu\beta\beta$ products, the ratio of two isotopes with different $Q_{\beta\beta}$ values is used to evaluate the $0\nu\beta\beta$ rate (Takaoka and Ogata, 1966; Hennecke *et al.*, 1975, 1978; Kirsten *et al.*, 1983). Counter experiments measure directly the $0\nu\beta\beta$ events although the running time t is in the order of one year. Large volume track detectors have a large N_0 and reduce the N_{BG} by the track identification (Bardin *et al.*, 1970; Cleveland *et al.*, 1975; Moe and Lowenthal, 1980; Elliott *et al.*, 1987).

Small volume semiconductor detectors have a small ΔE and a small N_{BG}. Natural germanium contains 7.76 per cent of the $\beta\beta$ decay ^{76}Ge isotope.

Therefore a Ge semiconductor detector is used for the high resolution β-ray detector as well as for the $\beta\beta$ source. The Milan group has used low-background Ge spectrometers made of specially-selected materials in the Mont Blanc tunnel, where cosmic-ray background counts are negligible (Fiorini *et al.*, 1973; Bellotti *et al.*, 1984, 1986). The detector system is shown in Fig. 7.5. Several groups have used low-background Ge detectors to search for $0\nu\beta\beta$ events of ^{76}Ge (Forster *et al.*, 1984; Simpson *et al.*, 1984). The sensitivity achieved by the Milan group by using the 143 cm^3 Ge detector is around $1.6 \cdot 10^{23} y^{1/2}$ for the ^{76}Ge $0^+ \to 0^+$ $0\nu\beta\beta$ process. Here the background rate at the $Q_{\beta\beta} = 2.041$ MeV is around 3.2/keV$\cdot y$, which corresponds to the observable limit of $\tilde{T}_{1/2} \sim 0.85 \cdot 10^{23} y$ for about one years observation.

A beta–gamma coincidence method (Avignone *et al.*, 1983/1985; Leccia *et al.*, 1983/1985; Ejiri *et al.*, 1984a, 1984b, 1986a, 1986b, 1987; Caldwell *et al.*, 1985/1986) is quite useful to select the true $0\nu\beta\beta$ events among other background events. Most of the natural backgrounds are due to radioactive decays of the Th, U, and Ac chain isotopes, which emit several γ rays characteristic of the decaying isotopes. The Compton electrons are always accompanied by the Compton-scattered γ rays. The $\beta\beta$ decays to the 2^+ first excited state and the 0^+ second excited state are followed by the $2^+ \to 0^+$ γ ray and the $0^+ \to 2^+ \to 0^+$ cascade γ rays, respectively, while the $\beta\beta$ decay to the 0^+ ground state is not accompanied by any γ ray. Consequently the efficient measurement of these γ-ray energies and of the multiplicities, in addition to the electron (β ray) measurement, is a very powerful way to identify the true $\beta\beta$ signal and to reject background γ signals. Ejiri *et al.* (1984a, 1986a) have used the intrinsic Ge detector surrounded by the 4π–γ counter which consists of six-segment NaI crystals. The detector system ELEGANTS (electron gamma-ray neutrino spectrometer) is shown in Fig. 7.6. All energy and time signals from the Ge and NaI detectors are recorded on a magnetic tape for event selection by means of a computer. Then the candidates for the $0\nu\beta\beta$ decay to the ground state are such signals from the Geβ counter that have a discrete energy of $Q_{\beta\beta}(0^+ \to 0^+)$ and are accompanied by no γ signals from any NaI segments. Those to the 2^+ excited state are the ones that have a discrete energy of $Q_{\beta\beta}(0^+ \to 2^+)$ and are accompaneid by the ^{76}Se $2^+ \to 0^+$ γ ray from one of the NaI segments. Such event selection reduces the background count rate due to radioactive isotopes around the Ge detector (internal background sources) by factors of 10^2 to 10^3. This β–γ event selection method is also effective for rejecting backgrounds due to nuclear capture of neutrons and μ-mesons since they are followed by many γ rays.

The β–γ spectroscopic method used for identifying true event signals may be called an intelligent (active) filter for background reduction, while various kinds of shields (including thick rocks covering underground laboratories) are called non-intelligent (passive) filters. Passive filters are useful for rejecting external backgrounds due to radioactivities in the room and to cosmic rays.

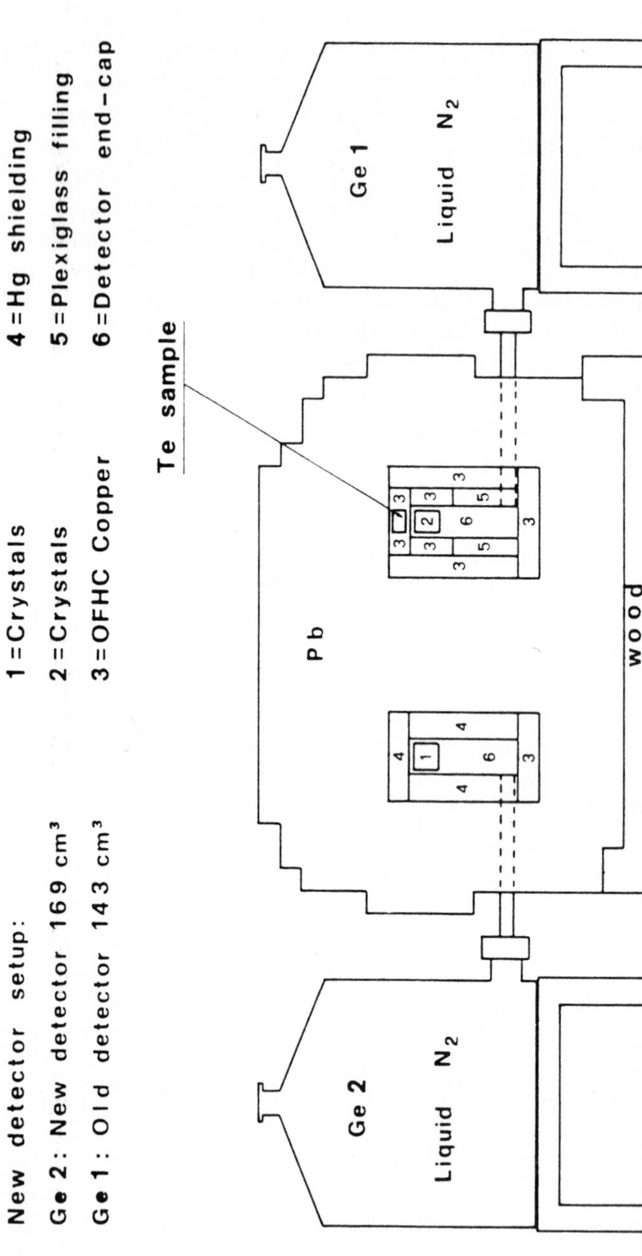

Fig. 7.5. Ultra-low background Ge detectors to search for low-energy rare events of the ^{76}Ge $0\nu\beta\beta$ process used in the Mont Blanc tunnel. OFHC stands for oxygen-free high-conductive copper, which is quite free from radioactive isotopes. (Bellotti *et al.*, 1984.)

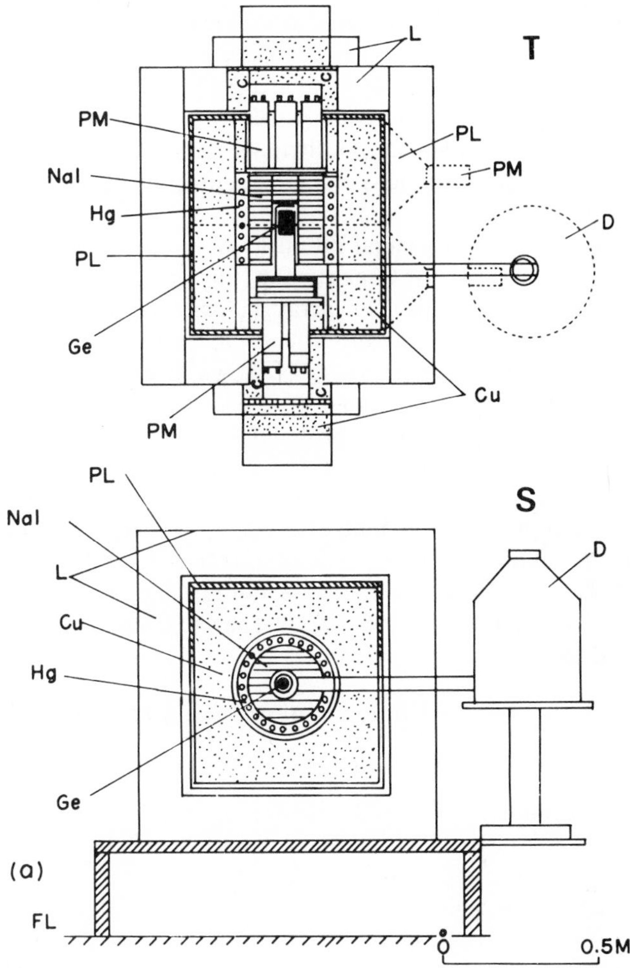

Fig. 7.6. Schematic views of the ELEGANTS (Electron gamma-ray neutrino spectrometer). (a): side view, and (b): top view, Ge: 171 cm³ Ge detector, D: dewar, NaI: 4π–NaI detectors of 10 in. ϕ × 12 in. and 8 in. ϕ × 3 in. PL: plastic scintillators, PM: photomultiplier, Hg: purified mercury, Cu: oxygen-free high conductive copper, L: lead. ELEGANT III (Ejiri *et al.*, 1984a, 1986a; Kamikubota *et al.*, 1986.)

Active filters are essential for rejecting the internal backgrounds due to radioactivities in the materials used for the detector (counter) and in the source material itself.

The sensitivities achieved by means of β–γ spectroscopic methods are $S_D \approx 1.3$ to $1.7 \cdot 10^{23} y^{1/2}$ for the ^{76}Ge $0^+ \rightarrow 0^+$ $0\nu\beta\beta$ process and

$S_D \approx 10^{23} y^{1/2}$ for the ^{76}Ge $0^+ \to 2^+$ $0\nu\beta\beta$ process (Ejiri *et al.*, 1984a, 1986a, 1986b; Caldwell *et al.*, 1985/1986).

Another detector which contains some fraction of the $\beta\beta$-source is a xenon detector. Natural xenon contains 8.9 per cent of ^{136}Xe. It may be a good candidate because of the larger Q-value of 2.48 MeV. Several groups (Caltech, Irvine, Milan, etc.; see refs in Kotani *et al.*, 1986) are constructing large TPC's (time projection chambers) or multiwire chambers with xenon. The merits of these detectors are the larger number of source nuclei and the two beta-ray track identification.

Detector systems for general $\beta\beta$ nuclei have been developed. The system ELEGANT IV used by the Osaka group (Ejiri *et al.*, 1984a, 1986c; Watanabe *et al.*, 1987) has a multisilicon array in place of the Ge detector of the ELEGANTS III. Thin foils of enriched $\beta\beta$-source nuclei are sandwiched between the multi-silicon detectors as shown in Fig. 7.7. Then one can use such $\beta\beta$-source nuclei as ^{100}Mo, ^{150}Nd etc. that have large $Q_{\beta\beta}$ values and accordingly large $\beta\beta$-transition probabilities.

7.1.2c *Double β–γ spectroscopy and lepton number conservation*

Theories of the neutrinoless double beta decay have been developed by several groups (Primakoff and Rosen, 1969/1981; Doi *et al.*, 1981, 1983, 1985; Haxton *et al.*, 1981/1982, 1984). Hereafter we use formulae and notations used by Kotani's group (Doi *et al.*, 1983/1985). The weak interaction Hamiltonian is expressed as

$$\mathbf{H}_w = G(\cos\theta/\sqrt{2})\left(j_L^u J_L^{u+} + \kappa j_L^u J_R^{u+} + \eta j_R^u J_L^{u+} + \lambda j_R^u J_R^{u+} \right) + hc, \tag{7.1.7}$$

where j_R (J_R) and j_L (J_L) are right-handed and left-handed leptonic (hadronic) currents, respectively. The first term is the left-handed V−A interaction and the others are the right-handed V+A interaction terms. The λ term is due to the gauge boson W_R^u, and λ is given by the square of the ratio of the two boson masses, $\lambda = (M_W^L/M_W^R)^2$. η and κ are the left-right coupling coefficients. The leptonic current is expressed as

$$j_L^u = \bar{e}\gamma^u(1-\gamma_5)\nu_{eL}, \quad j_R^u = \bar{e}\gamma^u(1+\gamma_5)\nu'_{eR}, \tag{7.1.8}$$

where γ^u and γ_5 are the Dirac matrices, and the neutrinos are given by the superposition of the mass eigenstates N_j,

$$\nu_{eL} = \Sigma U_{ej} N_{jL} \quad \text{and} \quad \nu'_{eR} = \Sigma V_{ej} N_{jR}. \tag{7.1.9}$$

The $0\nu\beta\beta$ process, which violates the lepton number conservation law, requires the Majorana neutrino with helicity mixing. The helicity-mixing originates from the finite neutrino mass m_ν in the first term (V−A term) and/or from the second, third and fourth terms (V+A terms) in eqn (7.1.7).

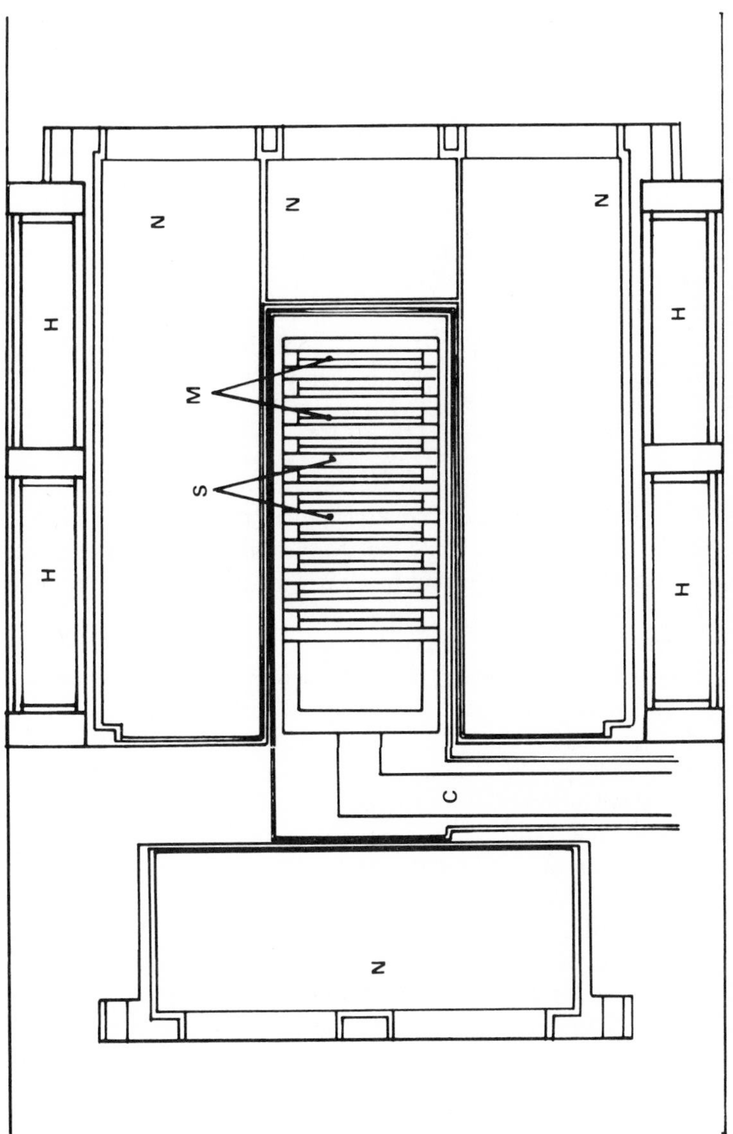

Fig. 7.7. Si-detector set inside the 4π–NaI detector. The arrangement is the same as in Fig. 7.6, except for the Ge detector, which is replaced by the Si detector array. N: NaI detectors, C: cold finger, S: Si detectors, and M: ^{100}Mo foils. ELEGANT IV (Ejiri *et al.*, 1984a, 1986c; Watanabe *et al.*, 1987.)

The $0\nu\beta\beta$ decay probability for the $0^+ \to 0^+$ transition is written as

$$T_{0\nu}(0^+ \to 0^+) = G_0(x, y, z) \cdot |M_{GT}^{0\nu}|^2, \tag{7.1.10}$$

$$G_0(x, y, z) = c_1' x^2 + c_2' xy + c_3' xz + c_4' y^2 + c_5' z^2 + c_6' yz \tag{7.1.11}$$

$$x = |\Sigma m_j U_{ej}^2| / m_e \equiv \langle m_\nu \rangle / m_e, \tag{7.1.12}$$

$$y = \lambda |\Sigma U_{ej} V_{ej}| \, |g_v'/g_v| \equiv \langle \lambda \rangle \tag{7.1.13}$$

$$z = \eta |\Sigma U_{ej} V_{ej}| \equiv \langle \eta \rangle, \tag{7.1.14}$$

where $M_{GT}^{0\nu}$ is the GT-type nuclear matrix element. The coefficients c_i' in eqn (7.1.11) include the nuclear and kinematical terms.

The $0\nu\beta\beta$ process for the $0^+ \to 2^+$ transition is due to the right-handed terms. In this case the Δ isobar with $J^\pi = 3/2^+$ contributes in addition to the two nucleons. The transition probability is then written

$$T_{0\nu}(0^+ \to 2^+) = (D_1 y^2 + D_2 yz + D_3 z^2), \tag{7.1.15}$$

$$D_1 = D_1' + D^*, \quad D_2 = D_2' + d^* D^*, \quad D_3 = D_3' + D^*, \tag{7.1.16}$$

where the D_i' includes the nuclear and kinematical terms and the nuclear matrix element for the 2n mode, while the D^* includes those for the isobar mode and d^* is a coefficient. The isobar process is the two-quark (2q) process in the isobar (see Fig. 7.4).

The kinematical factor G_{2N} for the 2n-mode is inversely proportional to the square of the two-nucleon distance r_{nn} in the nucleus, while that for the Δ(2q) mode is proportional to the square of the two-quark distance r_{qq} in the isobar. The factor (r_{nn}^2/r_{qq}^2) amounts to ~ 20. Furthermore the overlap integral for the 2q-mode is much larger than that for the 2n-mode. These two factors may compensate the small probability of the isobar in the initial or final state of the nucleus. Then the Δ-mode may contribute to the $0\nu\beta\beta$ $0^+ \to 2^+$ transition as much as the 2n-mode to the $0\nu\beta\beta$ $0^+ \to 0^+$ and $0^+ \to 2^+$ transitions (Primakoff and Rosen, 1969/1981; Doi *et al.*, 1983, 1985).

The kinematical and nuclear terms and the nuclear matrix elements have been discussed in detail by several groups (Doi *et al.*, 1983, 1985; Haxton *et al.*, 1981/1982, 1984; Klapdor and Grotz, 1984; Grotz and Klapdor, 1985).

It should be noted that the energy and angular distributions of the two β-rays depend on the terms contributing to the $0\nu\beta\beta$ process. Therefore beta–gamma spectroscopic studies of the $0\nu\beta\beta$ $0^+ \to 0^+$ and $0^+ \to 2^+$ transitions elucidate relative contributions of the neutrino-mass term and the right-handed interaction terms (Doi *et al.*, 1985).

Several groups are continuing precise measurements of the $0\nu\beta\beta$ process, mostly at underground laboratories. So far no appreciable peaks corresponding to the $0\nu\beta\beta$ transitions have been found beyond backgrounds. Typical examples of energy spectra of $E(\beta_1) + E(\beta_2)$ for ^{76}Ge $\beta\beta$ decays are shown in Fig. 7.8.

Lower limits on the half-lives for the ^{76}Ge $0\nu\beta\beta$ transition have been obtained as $1.2 \cdot 10^{23}\,y$ (Bellotti *et al.*, 1984), $2.5 \cdot 10^{23}\,y$ (Caldwell *et al.*, 1985/1986), $0.73 \cdot 10^{23}\,y$ (Ejiri *et al.*, 1987) and $1.16 \cdot 10^{23}\,y$ (Avignone *et al.*, 1983/1985) for the $0^+ \rightarrow 0^+$ transitions. Those for the $0^+ \rightarrow 2^+$ transition are $2.2 \cdot 10^{22}\,y$ (Bellotti *et al.*, 1984), $5.7 \cdot 10^{22}\,y$ (Ejiri *et al.*, 1987) and $5 \cdot 10^{22}\,y$ (Caldwell *et al.*, 1985/1986).

Lower limits on the half-lives give upper limits on the neutrino mass and the right-handed interaction terms. The values, however, depend on the nuclear matrix elements used and on the way of calculating the kinematical and nuclear factors. The upper limit on the neutrino mass obtained so far from the ^{76}Ge experiments discussed above are $\langle m_\nu \rangle / m_e \lesssim 2 \sim 10 \cdot 10^{-6} (\langle m_\nu \rangle \lesssim 1 \sim 5\,\text{eV})$ and $\langle \lambda \rangle$ or $\langle \eta \rangle \lesssim 2 \sim 10 \cdot 10^{-6}$ (see refs in Doi *et al.*, 1983 and those in Ejiri, 1986b, and Fig. 7.9).

Recently a lower limit of $T_{1/2} > 5 \cdot 10^{23}$ years has been reported for the ^{76}Ge $0^+ \rightarrow 0^+$ $0\nu\beta\beta$ (Caldwell *et al.*, 1987). This corresponds to the upper limits of $\langle m_\nu \rangle < 0.7 \sim 1.8\,\text{eV}$, depending on the $\beta\beta$ matrix elements.

If the $0\nu\beta\beta$ process is due to the mass term $\langle m_\nu \rangle / m_e$, then the transition probability $T_{0\nu}$ is written in terms of the nuclear sensitivity S_N as

$$T_{0\nu} = S_\text{N} |\langle m_\nu \rangle|^2 / m_e^2 \tag{7.1.17}$$

$$S_\text{N} = c_1' |M_\text{GT}^{0\nu}|^2 \tag{7.1.18}$$

Using S_D as defined by eqn (7.1.6), the lower limit on $|\langle m_\nu \rangle|^2 / m_e^2$ to be measured is written as

$$|\langle \tilde{m}_\nu \rangle / m_e|^2 = \frac{1}{S_\text{N} S_\text{D} \sqrt{t}} \tag{7.1.19}$$

The nuclear sensitivity depends on the matrix element and the kinematical factor c_1', and c_1' is roughly proportional to $Q_{\beta\beta}^5$. Thus it is important to use nuclei with large S_N, namely with large $M_\text{GT}^{0\nu}$ and $Q_{\beta\beta}$, as well as to use detectors with large S_D. Noting that nuclei with nucleons (quarks) in the quantum system are regarded as a good micro-laboratory for studying fundamental interactions, selection of good $\beta\beta$ nuclei with large S_N correspond to selection of nuclear laboratory with high sensitivity (large amplification factor).

As given in eqn (7.1.19), the overall sensitivity is expressed as $S = S_\text{N} \cdot S_\text{D}$. ^{76}Ge emphasizes the large S_D because of the high resolution (small ΔE) and the high efficiency (see refs. in Kotani *et al.*, 1986). Recently several groups are developing new detector systems with large overall S. They may cast light in near future on the unexplored regions of $\langle m_\nu \rangle / m_e$ and V+A coupling coefficients of the order of 10^{-6}.

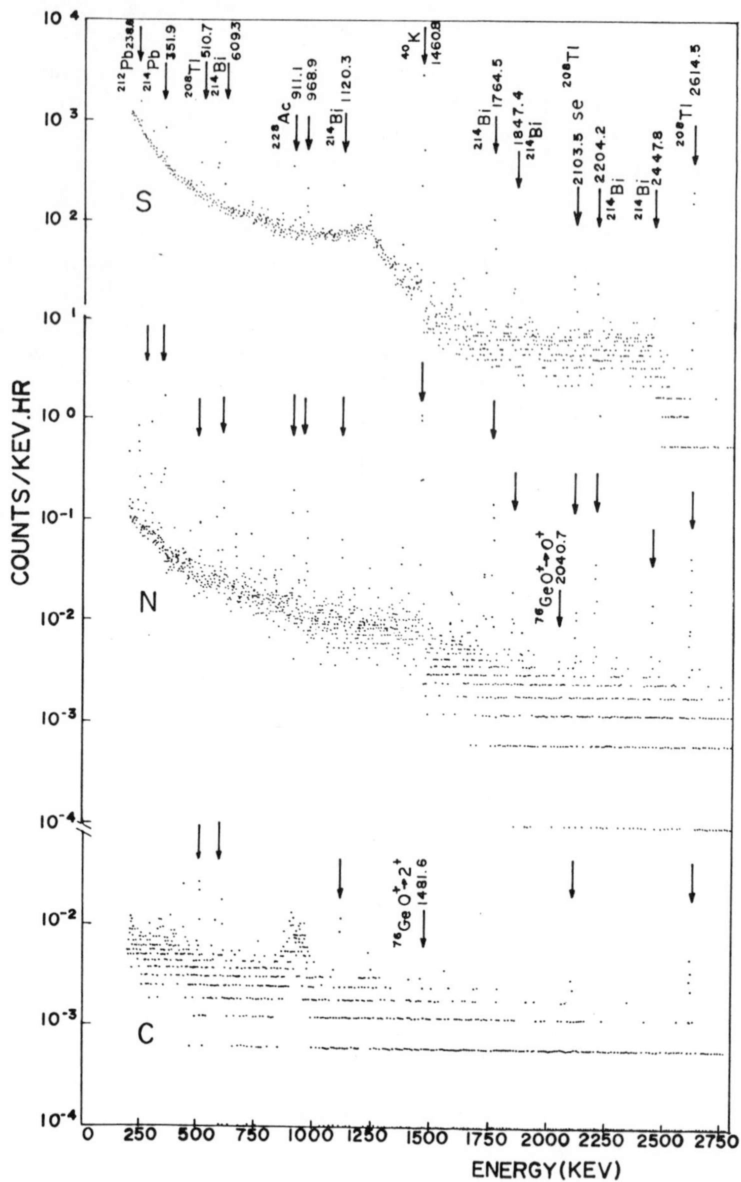

Fig. 7.8. Energy spectra measured by the Ge detector at the Kamioka underground laboratory. S; for singles spectrum without any shields, N; for the $0^+ \rightarrow 0^+$ followed by no signals from any NaI segments, C; for the $0^+ \rightarrow 2^+$ followed by the 559.1 keV γ signals from one of the NaI segments. N and C are with passive and active shields. (Ejiri *et al.*, 1986a.)

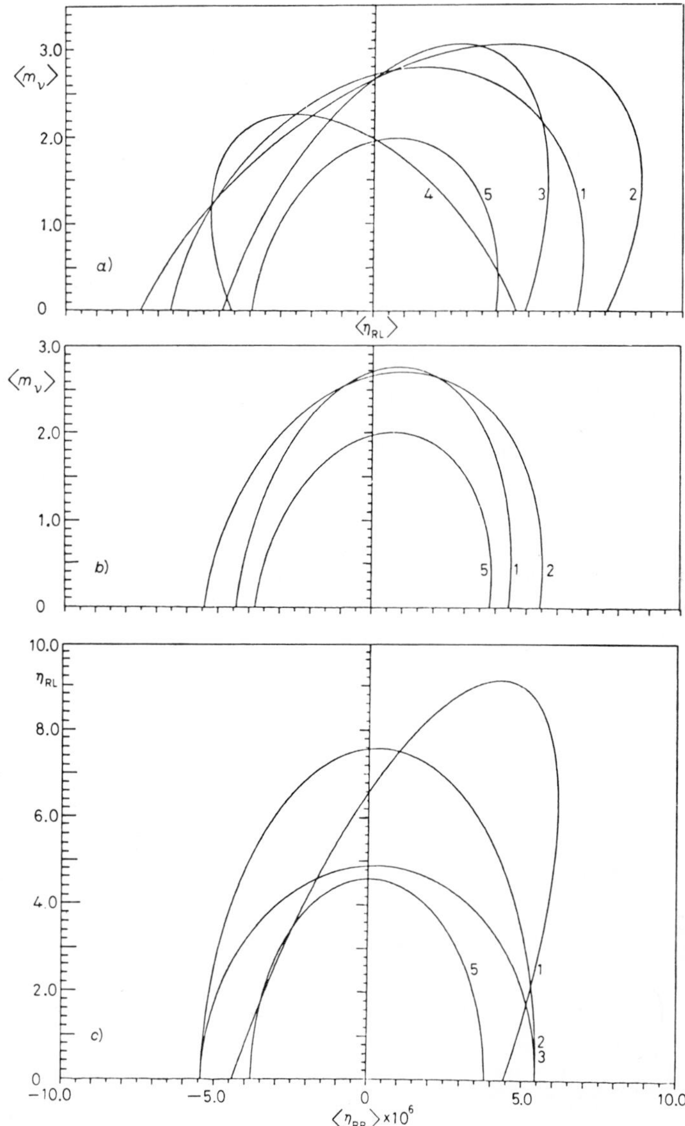

Fig. 7.9. Limits on the lepton-nonconserving parameters extracted from the ^{76}Ge data of $T_{1/2} > 3.3 \cdot 10^{23}$ y. 1) Los Alamos group ($\times 10^6$), 4) Bogdan *et al.* ($\times 10^6$) (only for $\langle m_\nu \rangle$ and $\langle \eta_{LR} \rangle$), Osaka group, 2) p-effect. corr., 3) p-effect. corr + nucleon recoil ($\times 10^7$) and 5) Tübinger-Jülich group ($\times 10^8$). Nuclear matrix elements from Haxton *et al.* (1984) only were used. m_ν is given in unit of eV. (Bellotti *et al.*, 1986 and ref. therein.)

7.1.3 *Parity non-conservation in nuclear interactions studied by in-beam γ spectroscopy*

7.1.3a *Gamma decays of parity-mixed nuclear states*

The parity non-conserving (PNC) nuclear interaction is very important in view of the neutral-current weak interaction between quarks and of the weak interaction of hadrons. An extensive review of parity violation in the nucleon–nucleon interaction has been made by Adelberger and Haxton (1985). The PNC force itself, which is due to the weak interaction, is of the order of 10^{-7} with respect to the normal (parity-conserving) nuclear force. Thus studies of the PNC require high-sensitive detection methods. Spectroscopic studies of nuclear γ rays have played important roles in investigation of such an important PNC interaction. In this subsection we give some examples of the in-beam γ-ray studies used for investigating the PNC nuclear interaction.

A nuclear state with the PNC interaction $\mathbf{H}_{\mathrm{PNC}}$ is written

$$\psi = \psi_0 + \varepsilon_1 \psi_1 \tag{7.1.20}$$

where ψ_0 is the major-parity component of the wave function, and $\varepsilon_1 \psi_1$ is an admixture of the opposite-parity component with the same J (spin). The mixing amplitude ε_1 is written

$$\varepsilon_1 = \frac{\langle \psi_1 | \mathbf{H}_{\mathrm{PNC}} | \psi_0 \rangle}{\Delta E}, \tag{7.1.21}$$

where $\Delta E = E_0 - E_1$ is the energy difference between the two states ψ_0 and ψ_1. The mixing amplitude can then be determined by measuring the circular polarization $P_\gamma = \langle \boldsymbol{\sigma}_\gamma \cdot \mathbf{P}_\gamma \rangle$ of the γ ray from the unpolarized state ψ. Here the pseudoscalar term is due to the interference between the PNC and PC transition amplitudes, and thus it is written

$$P_\gamma = 2\mathrm{Re}\left(\frac{\varepsilon_1 \langle \psi' | \mathbf{T}_\gamma(1) | \psi_1 \rangle}{\langle \psi' | \mathbf{T}_\gamma(0) | \psi_0 \rangle} \right), \tag{7.1.22}$$

where $\mathbf{T}_\gamma(0)$ and $\mathbf{T}_\gamma(1)$ are the γ-transition operators for ψ_0 and ψ_1, respectively. If the state ψ is spin-polarized, one can measure the PNC contribution by measuring the asymmetry A_γ of the γ ray with respect to the polarization axis. The asymmetry is essentially given by the same expression as in eqn (7.1.22) for P_γ.

The in-beam γ-spectroscopic study has several unique features characteristic of the nuclear micro-laboratory for testing the PNC interaction:

1. one can select doublet states with the same J and the small energy spacing ΔE. Then ε_1, as given by eqn (7.1.21) becomes large, and thus contributions from other states may be neglected.

2. By selecting γ transitions with a large ratio $R = \langle \psi' | \mathbf{T}_\gamma(1) | \psi_1 \rangle / \langle \psi' | \mathbf{T}_\gamma(0) | \psi_0 \rangle$, the observed quantity (P_γ or A_γ) is greatly enhanced by the factor R. There are several examples of γ transitions in light nuclei where E1 decays are strongly hindered but M1 decays are not hindered. Then the PNC observable is enhanced by the factor $R \approx \langle \mathrm{M1} \rangle / \langle \mathrm{E1} \rangle \gg 1$. The γ-matrix elements are obtained experimentally from the lifetimes (γ-decay widths) of the states ψ_0 and ψ_1.

3. The gamma transition operators involved are simple and unique. Mostly E1 and M1 matrix elements are involved. By using the nuclear doublet states with simple configurations and good isospins, one can investigate the PNC interaction and its isospin component.

4. The polarized state ψ is obtained by using nuclear reactions with a polarized beam or a polarized target.

Some nuclear doublet states and enhancement factors R are shown in Fig. 7.10. The doublet states in these nuclei have small energy spacings and large enhancement (amplification) factors, and thus provide very sensitive nuclear systems for detecting the small PNC effect. The observable P_γ is rewritten by introducing the nuclear sensitivity S_N as follows.

$$P_\gamma = \mathrm{Re}(S_\mathrm{N} \langle \psi_0 | \mathbf{H}_\mathrm{PNC} | \psi_1 \rangle) \tag{7.1.23a}$$

$$S_\mathrm{N} = 2 \frac{\langle \psi' | \mathbf{T}_\gamma(1) | \psi_1 \rangle}{\langle \psi' | \mathbf{T}_\gamma(0) | \psi_0 \rangle} \cdot \frac{1}{\Delta E}. \tag{7.1.23b}$$

Another important feature of those doublets is the different isospin property as discussed by Adelberger and Haxton (1985). The mixing of the doublet with the isospin $T = 0$ and 1 in ^{18}F leads to the PNC interaction with $\Delta T = 1$. The doublet with $T = 1/2$ and $T_3 = -1/2$ in ^{19}F and that with $T = 1/2$ and $T_3 = 1/2$ in ^{19}Ne are used to study the $\Delta T = 0$ and $\Delta T = 1$ PNC interactions, the relative signs of which are different in the two doublets.

7.1.3b *Parity mixing in* ^{18}F

The mixing of the $T = 0$ and $T = 1$ levels in ^{18}F has been extensively studied by measuring the circular polarization P_γ of the 1081 keV γ rays (Ahrens *et al.*, 1982; Evans *et al.*, 1985; Bini *et al.*, 1985). The doublet has been populated by the ^{16}O(^{3}He, p) reaction. The energy spacing between the 1081 keV state with $J^\pi = 0^-$ and $T = 0$ and the 1042 keV state with $J^\pi = 0^+$ and $T = 1$ is only 39 keV. The M1 decay from the 1042 keV state to the ground state is favoured, while the E1 decay from the 1081 keV state is isospin-forbidden. Thus the doublet has a large enhancement (ratio of the two γ-matrix elements of around 110, and a large nuclear sensitivity of around $6 \cdot 10^{-3}$/eV.

The experimental set-up used by the Firenze group (Bini *et al.*, 1985) is shown in Fig. 7.11. Gamma rays were detected by four high-purity Ge

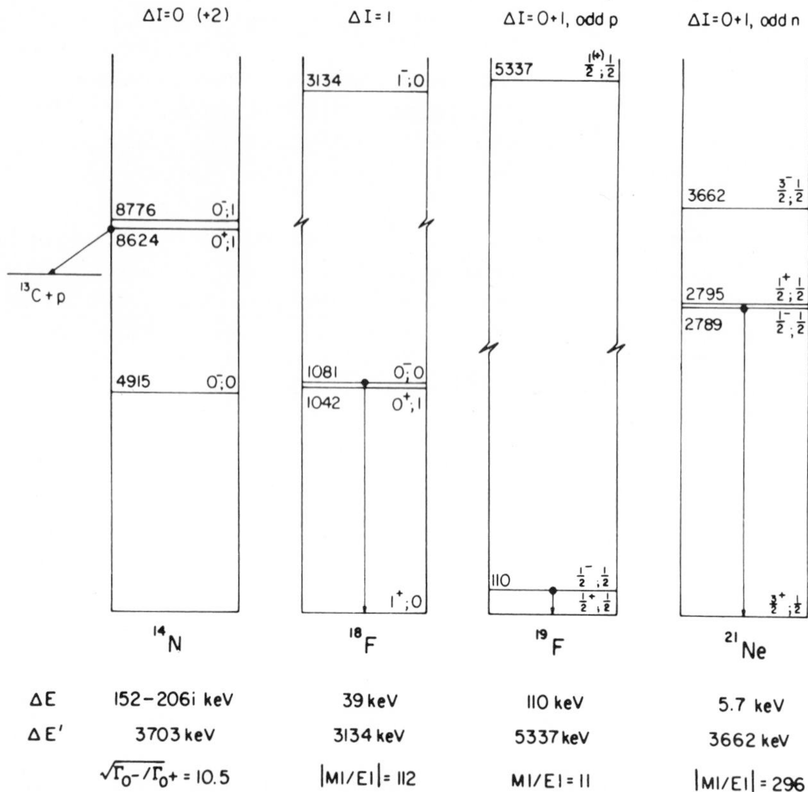

Fig. 7.10. Parity-mixed doublets in light nuclei. I is the isospin (T in the text). The transitions displaying the amplified PNC effect are indicated. The quantities ΔE and $\Delta E'$ are the smallest and next-smallest energy denominators governing the parity mixing. The excitation energies in units of keV, spins and parities J^π, and isospins for relevant states are given. The quantities shown in the bottom row are 'amplification factors'. (Adelberger and Haxton, 1985.)

detectors. The circular polarization was obtained by measuring the transmission of the γ rays through the magnetized Fe. The observed γ-ray spectrum and asymmetry with respect to the direction of the magnetization are shown in Fig. 7.12. The polarization is obtained as $P_\gamma = (2.7 \pm 5.7)\cdot 10^{-4}$. The polarization measured by the Queen's University group (Evans *et al.*, 1985) is $P_\gamma = (1.6 \pm 5.6)\cdot 10^{-4}$. These data, combined with other recent data, set an upper-limit of around 1.5 to $1.2\cdot 10^{-7}$ on the isovector weak πNN coupling constant.

7.1.3c *Parity mixing in* ^{19}F

The parity mixing of the doublet states with $J^\pi = 1/2^+$ $T = 1/2$ and $J^\pi = 1/2^-$ $T = 1/2$ in ^{19}F (see Fig. 7.10) is measured by observing the asymmetry A_γ of

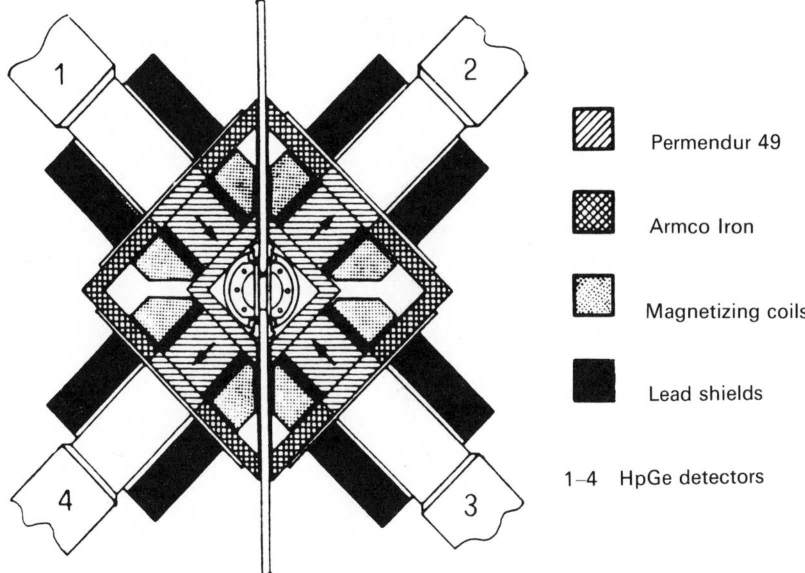

Fig. 7.11. The Four-prong polarimeter. In the centre of the polarimeter the water-jet target assembly is schematically shown. (Bini *et al.*, 1985.)

the γ-decay between the doublet states. The ^{22}Ne (p, α) reaction induced by transversely-polarized protons populates the spin-polarized 110 keV state with $J^\pi = 1/2^-$ in ^{19}F (Adelberger *et al.*, 1975). The experimental layout is shown in Fig. 7.13. The asymmetry of the γ ray between the doublet states is written

$$A_\gamma = \mathrm{Re}\left[\frac{2}{\Delta E}\langle \psi_0 | \mathbf{H}_{\mathrm{PNC}} | \psi_1 \rangle \cdot \frac{\langle \psi_0 | \mathbf{T}_\gamma(\mathrm{M1}) | \psi_0 \rangle - \langle \psi_1 | \mathbf{T}_\gamma(\mathrm{M1}) | \psi_1 \rangle}{\langle \psi_0 | \mathbf{T}_\gamma(\mathrm{E1}) | \psi_1 \rangle}\right], \tag{7.1.24}$$

where $\Delta E = 110$ keV, ψ_0 is the ground state with $J^\pi = 1/2^+$ and ψ_1 is the 110 keV state with $J^\pi = 1/2^-$. In this case $\mathbf{T}_\gamma(\mathrm{M1})$ stands for the M1 moment operator (Adelberger and Haxton, 1985). The E1 transition matrix element is deduced from the lifetime of the 110 keV state, and the M1 matrix element for the ground state from the magnetic moment. Since the E1 transition is retarded and the M1 moment is normal, the doublet has a large nuclear sensitivity of around $2 \cdot 10^{-4}$/eV. The asymmetry measured by the Seattle group is $A_\gamma = (-8.5 \pm 2.6)\, 10^{-5}$ (Adelberger *et al.*, 1983). Elsener *et al.* (1984) obtained $A_\gamma = (-6.8 \pm 1.8)\, 10^{-5}$. These data lead to $\langle \psi_0 | \mathbf{H}_{\mathrm{PNC}} | \psi_1 \rangle = 0.38 \pm 0.10$ eV (Adelberger and Haxton, 1985).

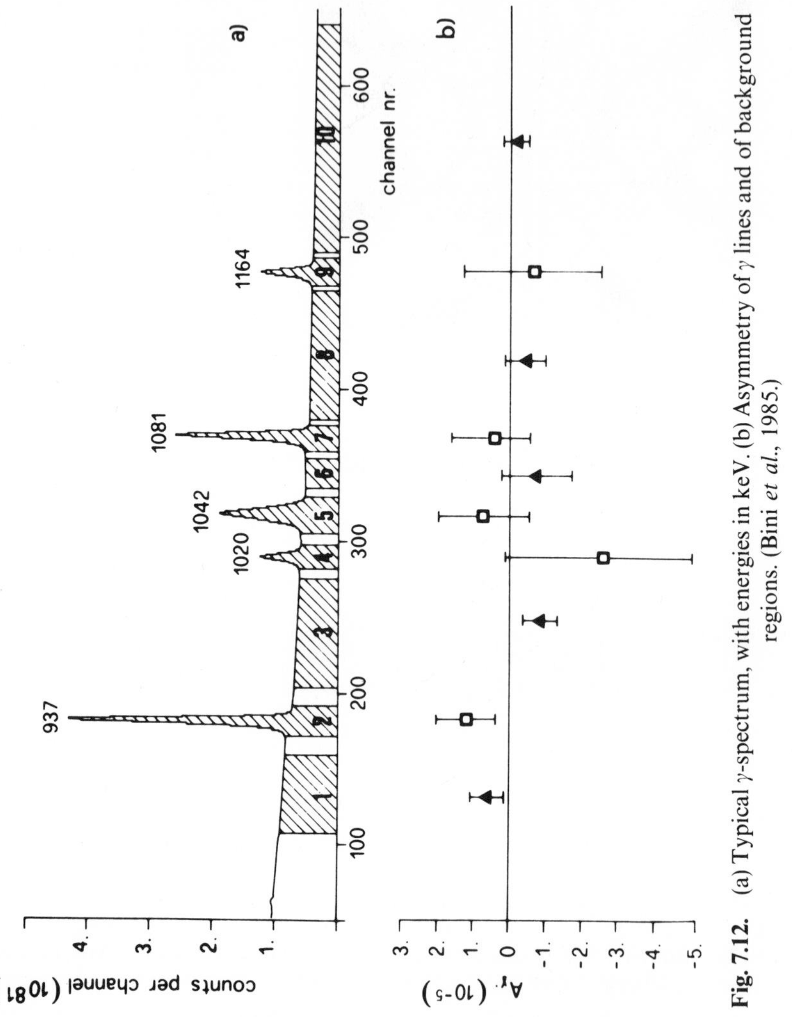

Fig. 7.12. (a) Typical γ-spectrum, with energies in keV. (b) Asymmetry of γ lines and of background regions. (Bini *et al.*, 1985.)

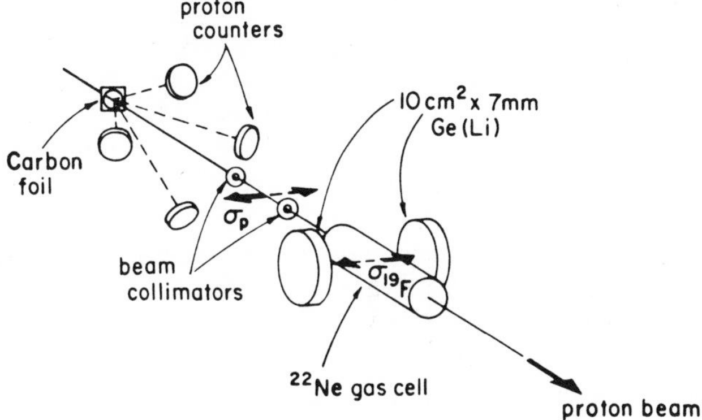

Fig. 7.13. Schematic view of the Seattle ^{19}F PNC experiment. The four proton counters view a thin carbon foil onto which a layer of Au has been evaporated. An on-line computer monitors continuously the transverse polarization by comparing the scattering yields from C and Au. (Adelberger and Haxton, 1985; Earle *et al.*, 1983.)

7.1.3d *Parity mixing in ^{19}Ne*

The doublet with $J^{\pi} = 1/2^{\pm}$ and $T = 1/2$ in ^{19}Ne (see Fig. 7.10) has a large nuclear sensitivity of $S_N = 1 \cdot 10^{-1}$/eV because the energy spacing is as small as $\Delta E = 5.74$ keV and the $1/2^- \rightarrow 3/2^+$ E1 transition is retarded by a factor 10^{-6}. Earle *et al.* (1983) measured the circular polarization of the γ-transition from the $1/2^-$ 2789 keV state to the $3/2^+$ ground state. The polarized 2789 keV state was populated by the ^{21}Ne(p, p′) reaction. The polarization was found to be quite small, as $P_{\gamma} = (8 \pm 14) \cdot 10^{-4}$, which corresponds to $|\langle \psi_0 | \mathbf{H}_{\text{PNC}} | \psi_1 \rangle| = (9 \pm 16) \cdot 10^{-3}$ eV. This is much smaller than the value in the mirror nucleus ^{19}F, and suggests destructive and constructive interferences of the $\Delta T = 0$ and $\Delta T = 1$ terms of the PNC interaction in ^{19}Ne and ^{19}F, respectively (Adelberger and Haxton, 1985).

7.2 Nuclear gamma-ray spectroscopy with K-mesons and hypernuclei

7.2.1 *Gamma-ray spectroscopy and hypernuclei*

Nuclear spectroscopy discussed so far is concerned with ordinary nuclei with nucleons and pions, which bear no strangeness (S) degree of freedom. These nucleons and pions are composed of up (u) and down (d) quarks. New types of nuclear spectroscopies involving particles with the strangeness flavour have been opened up since intense beams of K-mesons and pions in the intermediate energy region have become available for nuclear spectroscopy.

Hyperons such as Λ and Σ are baryons with a strangeness $S = -1$. They consist of three quarks with s and u and (or) d quarks. Hyperons are quite different from nucleons in their properties; their quantum numbers and quark structures are given in Table 7.2.1. Hypernuclei, in which one nucleon is replaced by a hyperon, are produced by (K^-, π^-), (π^+, K^+), and (γ, K^+) reactions. Gamma-ray spectroscopy with hypernuclei aims at studying new features of hypernuclei with the strangeness degree of freedom, fundamental hyperon–nucleon interactions, and their effective interactions in hypernuclei.

Strangeness exchange (K^-, π^-) and (π^+, K^+) reactions have been extensively used to produce hypernuclear states. So far their level structures have been studied by measuring singles spectra in the (K^-, π^-) and (π^+, K^+) reactions (Povh, 1978/1980/1981; Brückner *et al.*, 1978; Bertini *et al.*, 1980, 1981; Chrien *et al.*, 1979, 1984; May *et al.*, 1981; Piekarz *et al.*, 1982). The shell structures of hypernuclear states have been investigated theoretically (Dalitz and Gal, 1978, Auerbach *et al.*, 1983; Motoba *et al.*, 1983; Millener *et al.*, 1985). Gamma decays and weak decays have been studied by measuring decay particles (photons, mesons, nucleons) from hypernuclear states produced by (K^-, π^-) reactions (May *et al.*, 1983; Barnes, 1986; Grace *et al.*, 1985).

The unique features of the hyper nuclear γ-ray spectroscopy are summarized as follows:

1. Strangeness transfer reactions such as (K^-, π^-), (π^+, K^+) and (γ, K^+) reactions are feasible for exciting nucleon–hole (N^{-1}) hyperon–particle (Y) states in hypernuclei in a wide angular-momentum range of $J\hbar = 0$ to $12\hbar$. Some of them are bound Λ-hypernuclear states with excitation energies below the threshold energies of strong decays (nucleon or hyperon escape). They can decay by the weak process with strangeness change or by the electromagnetic (γ) process. The mesonic weak decays of $\Lambda \rightarrow p + \pi^-$ and $\Lambda \rightarrow n + \pi^0$ are reduced in hypernuclei because of the Pauli blocking for low-energy nucleons produced in the decays. On the other hand the non-mesonic weak decay of $\Lambda + N \rightarrow N + N$ becomes important. The half-lives of such weak decays are of the order of 10^{-10} sec. The γ-decay half-lives of 10^{-15} to 10^{-12} s for typical low-multipole transitions are much shorter than those of the weak decays, and so they are studied by means of γ-spectroscopy.

2. Unbound Λ hypernuclear states decay partly by emitting Λ and partly by emitting a nucleon (p or n) or a nucleon cluster (α, etc.). The latter leaves another hypernucleus in various kinds of excited states, which are studied by measuring the decay γ rays.

3. There are many states which can be studied by means of γ-ray spectroscopy but not by direct strangeness transfer reactions. The beam intensity for hypernuclear production is limited and/or the production cross-section is small. Thus the targets have to be fairly thick to get adequate yields of hypernuclei. The energy resolution for the analysis of the reaction product then suffers from the energy loss in the target. They γ-ray spectroscopy,

Table 7.2.1 Properties of nucleons and hyperons

		Mass (MeV)	J^π	Isospin	Strangeness	Quark structure	Life(s)	Decay†
Nucleon	p	938.3	$1/2^+$	1/2	0	uud	$>10^{32}$	?
	n	939.6	$1/2^+$	1/2	0	udd	917	$pe^-\bar{\nu}$
Hyperon	Λ	1115.6	$1/2^+$	0	−1	uds	$2.63\cdot10^{-10}$	$p\pi^-$, $n\pi^0$
	Σ^+	1189.4	$1/2^+$	1	−1	uus	$0.80\cdot10^{-10}$	$p\pi^0$, $n\pi^+$
	Σ^0	1192.5	$1/2^+$	1	−1	uds	$5.8\cdot10^{-20}$	$\Lambda\gamma$
	Σ^-	1197.3	$1/2^+$	1	−1	dds	$1.48\cdot10^{-10}$	$n\pi^-$
	Ξ^0	1314.9	$1/2^+$	1/2	−2	uss	$2.90\cdot10^{-10}$	$\Lambda\pi^0$
	Ξ^-	1321.3	$1/2^+$	1/2	−2	dss	$1.64\cdot10^{-10}$	$\Lambda\pi^-$

† Main decay modes.

however, can resolve most of the hypernuclear levels, as it is not affected by the target thickness.

4. There are two types of γ-decays in hypernuclei as shown in Fig. 7.14. One is a γ-transition of Λ between Λ shells and another is that of nucleon (or nucleon cluster) in the core. Since Λ is in a good quantal state of the hypernucleus, γ-ray spectroscopy provides crucial data on the effective and fundamental interactions between the hyperon and the nucleon (nucleon cluster) in the hypernucleus.

5. The accurate determination of excitation energies by γ-ray analysis gives central (spin-dependent as well as spin-independent), spin–orbit and tensor interactions between the hyperon and the nucleon.

6. The electromagentic multipole transition rates are sensitive to the hyperon–core interactions. Comparison with the corresponding nuclear γ transitions may elucidate the effect of the hypercharge (strangeness) and accordingly the role of the strange quark in the hypernucleus.

7. Λ has no isospin, and consequently the isospin-associated nuclear features, such as isospin-polarization effects on isovector transitions and isovector giant resonances, may disappear in hypernuclei. The spin of Λ is due to the s quark spin and the u–d quarks are coupled to spin O. Thus spin-associated nuclear features such as spin polarization, the Δ isobar effect and so on, may be quite different in hypernuclei. Therefore hypernuclear spectroscopy is crucial not only for studying hyperon–nucleon interactions but it may also contribute to a better understanding of the ordinary nuclear structure.

7.2.2 *Nuclear reactions for hypernuclear γ-ray spectroscopy*

7.2.2a *Hypernuclear production*

Gamma-ray spectroscopy is commonly applied to Λ-hypernuclei because Λ is the lowest-mass hyperon and its weak decay is slower than most of low-multipole γ decays. (Bando and Takai, 1984/1985; Bando *et al.*, 1985;

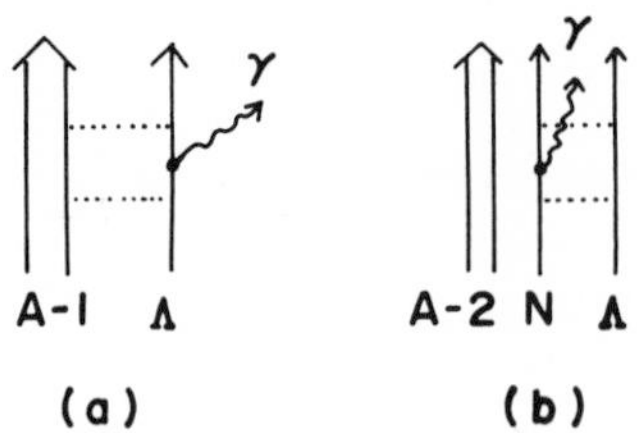

Fig. 7.14. Gamma de-excitation diagrams of a Λ hypernucleus. (a): γ-transition of Λ in a hypernucleus, and (b): γ-transition of a core nucleon (nucleon cluster).

Kisslinger, 1980). Nuclear reactions used for the production (excitation) of hypernuclei involve $K^{\pm}$ mesons with $S = \pm 1$ either in the initial channel or in the final channel. The (K^-, π^-), (π^+, K^+) and (γ, K^+) reactions, which transfer strangeness $S = -1$ to the nucleus, can be used for production of hypernuclear levels. The momentum transfers q associated with the reaction have to be in the order of the Fermi momentum for hypernuclei. The momentum transfers in the (K^-, π^-), (π^+, K^+) and (γ, K^+) reactions are shown in Fig. 7.15. Another important factor is the cross-section $(d\sigma/d\Omega)_f$ of hyperon production for a free nucleon. The momentum transfer and the hyperon production cross-section depend strongly on the type of the reaction and the incident momentum.

7.2.2b *Direct strangeness transfer reaction*

A hypernucleus is produced by a direct strangeness transfer process as shown in Fig. 7.16. The (K^-, π) reaction produces Λ and Σ hypernuclei via the strangeness transfer processes as follows:

$$K^- + p \rightarrow \pi^+ + \Sigma^-, \pi^0 + \Sigma^0, \pi^0 + \Lambda, \pi^- + \Sigma^+, \tag{7.2.1a}$$

$$K^- + n \rightarrow \pi^- + \Lambda, \pi^- + \Sigma^0, \pi^0 + \Sigma^-. \tag{7.2.1b}$$

A unique feature of the (K^-, π^-) reaction is the fairly low momentum transfer. A proper choice of the incident kaon momentum around $P_K \approx 0.4$ to 0.8 GeV/c leads to a recoil-less (zero momentum-transfer) strangeness-transfer process at $\theta_\pi = 0°$ (see Fig. 7.15). This enhances the excitation of a simple substitutional state of $[j_N^{-1} j_\Lambda]_0$, where a nucleon in a j orbit is simply substituted by Λ in the same j orbit (see Fig. 7.17). The (K^-, π^-) reaction at $P_k \approx 700$ to 800 MeV/c and the $(K^-, \pi^{\pm})$ reaction at $q_k \approx 400$ to 500 MeV/c have been used to excite Λ and Σ hypernuclei, respectively. (Povh, 1978/1980/1981; Chrien, 1984, 1986; and refs therein). Other Λ-particle N-hole states $[j_N^{-1} j'_\Lambda]_J$ with $j_N \neq j'_\Lambda$ and/or $J \neq 0$ are excited well at $\theta_\pi \neq 0°$ (Chrien, 1979; May, 1981). The (K^-, π) reaction at $P_k = 0$ (stopped K, π reaction) has the advantage of the large K^- capture cross-section to produce hypernuclei. The momentum transfer is fairly large ($q = 0.1 \sim 0.2$ GeV/c), particularly in the case of Λ-hypernucleus production (see Fig. 7.15). Thus it may be used to excite non-substitutional states with angular momenta of $qR = 2$ to $3\hbar$. Another feature of the (K^-, π^-) reaction is the short mean free path λ of about 2 fm for both incoming K^- and outgoing π^- mesons. Thus the (K^-, π^-) reaction is a peripheral process, where one of the valence neutrons and an outer-shell hyperon are involved in the strangeness-transfer process.

The (π, K) reaction produces Λ and Σ hypernuclei by the reactions

$$\pi^+ + n \rightarrow K^+ + \Lambda, K^+ + \Sigma^0, K^0 + \Sigma^+, \tag{7.2.2a}$$

$$\pi^- + p \rightarrow K^0 + \Lambda, K^0 + \Sigma^0, K^+ + \Sigma^-. \ \pi^+ + p \rightarrow K^+ + \Sigma^+ \tag{7.2.2b}$$

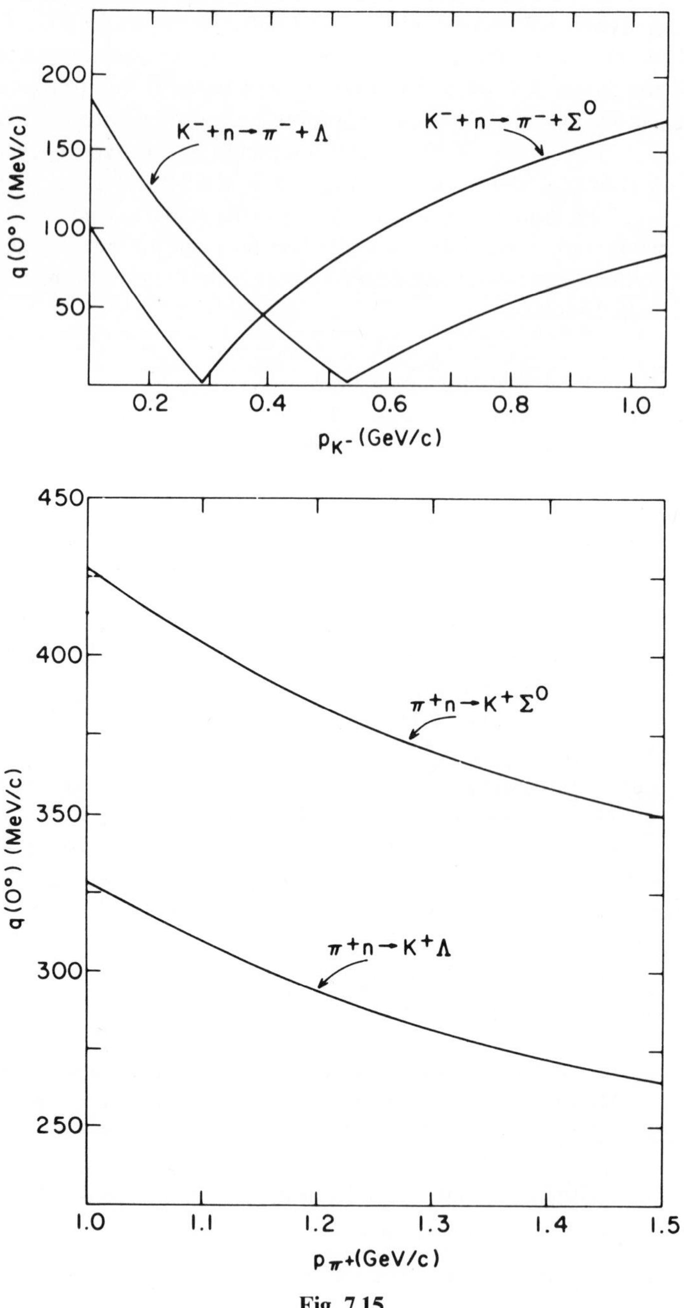
200
150
100
50
q(0°) (MeV/c)
K⁻+n→π⁻+Λ
K⁻+n→π⁻+Σ⁰
0.2
0.4
0.6
0.8
1.0
p_{K⁻} (GeV/c)
450
400
350
300
250
q(0°) (MeV/c)
π⁺n→K⁺Σ⁰
π⁺n→K⁺Λ
1.0
1.1
1.2
1.3
1.4
1.5
p_{π⁺}(GeV/c)

Fig. 7.15.

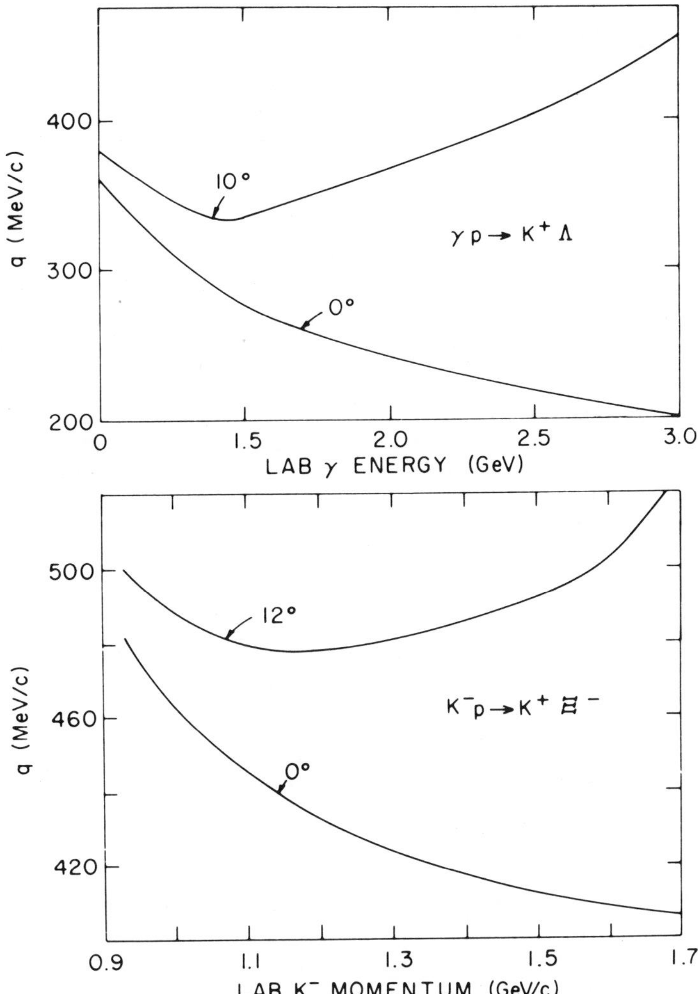

Fig. 7.15. Left-hand side: The lab. momentum transfer q at $\theta_L = 0^\circ$ as a function of incident lab. momentum for processes involving Λ or Σ production. Many-body kinematics (large A) are used, neglecting binding energy changes. Right-hand side: Momentum transfer q as a function of incident momentum for $\gamma p \to K^+\Lambda$ and $K^-p \to K^+\Xi^-$, for two different lab. angles as shown . (Dover and Walker, 1982.)

The (π^+, K^+) reaction, which is endothermic, gives rise to a large momentum transfer of about q = 300 to 350 MeV/c at $\theta_k = 0$, in contrast to the exothermic (K^-, π^-) reaction with a small momentum transfer (see Fig. 7.15). Therefore the high spin states of configurations $[j_n^{-1} j'_\Lambda]_J$ with the stretched spin of $J \approx j_N + j'_\Lambda = 6$ to 12 are preferentially excited (Dover *et al.*, 1980; Thiessen *et*

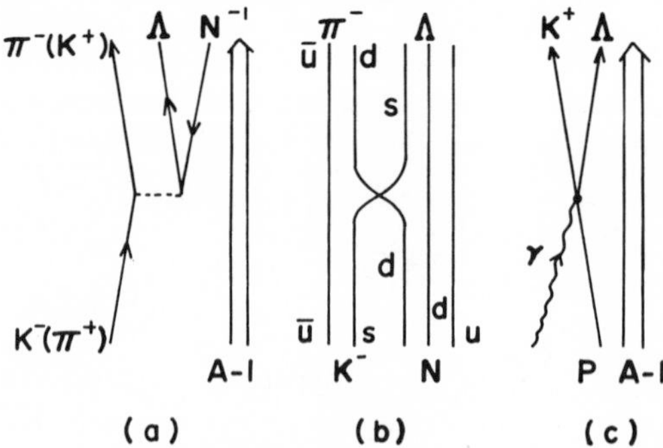

Fig. 7.16. Schematic diagrams of direct strangeness transfer reactions. (a): (K^-, π^-) and (π^+, K^+) reactions converting a neutron to a Λ to produce the nucleon–hole Λ-particle state. (b): Quark exchange in a (K^-, π^-) reaction. (c): (γ, K^+) reaction to excite a proton-hole Λ-particle state.

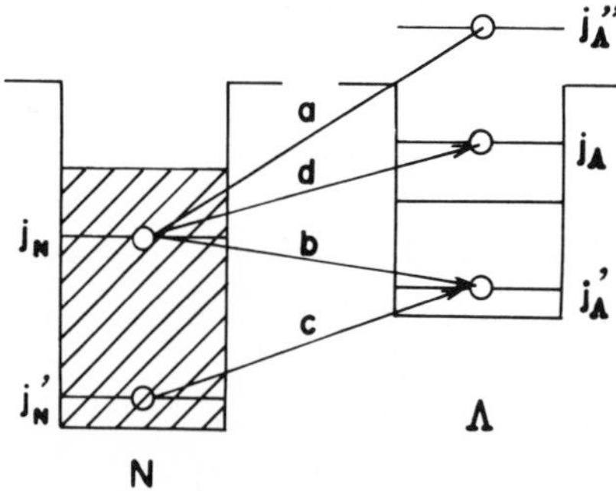

Fig. 7.17. Schematic diagrams of particle–hole excitations in Λ-hypernucleus productions. a and b are non-substitutional excitations to an unbound Λ and a deeply bound Λ states, respectively. c and d are the substitutional excitation from an inner shell j'_N to j'_Λ with $j'_N = j'_\Lambda$, and that from a surface-shell j_N to j_Λ with $j_N = j_\Lambda$, respectively.

al., 1983). Non-substitutional states with $j_n \neq j'_\Lambda$ are also excited by the (π^+, K^+) reaction (see Fig. 7.17). Since K^+ with $S = +1$ has a large mean free path of about 5 fm., in contrast to the K^- a neutron and a hyperon in inner shells are involved in the (π^+, K^+) process (see Fig. 7.17). Thus various kinds of neutron–hole hyperon–particle states in a wide range of angular momentum can be excited (Bando *et al.*, 1985).

The peak cross-section for the two-body process $\pi^+ + n \rightarrow K^+ + \Lambda$ is only 1 mb/sr at $\theta_\pi = 0°$ at $P(\pi^+) = 1.05$ GeV/c, while those of $K^- + n \rightarrow \pi^- + \Lambda$ at $P(K^-) = 0.7$ and 1.8 GeV/c are 5 and 3.5 mb/sr, respectively. Consequently the (π^+, K^+) cross-section is about one order of magnitude smaller than that for the typical (K^-, π^-) reaction. This factor, however, is compensated by the higher intensity of the π^+ beam relative to that of the K^+ beam.

The production of hypernuclei by (γ, K) or $(e, e' K)$ reactions are given by the reactions

$$\gamma + p \rightarrow \Lambda + K^+, \Sigma^0 + K^+, \Sigma^+ + K^0. \quad (7.2.3a)$$

$$\gamma + n \rightarrow \Lambda + K^0, \Sigma^0 + K^0, \Sigma^- + K^+. \quad (7.2.3b)$$

The momentum transfer of the Λ-hypernucleus production is $q = 200$ to 350 MeV/c at $\theta_K = 0°$ as shown in Fig. 7.15. Thus the (γ, K^+) process excites well the non-substitutional states and the high-spin states, as does the (π^+, K^+) process. Since the K^+ has a long mean free path, photoproduction favours the conversion of a deeply-bound nucleon to an inner-shell hyperon.

The photoproduction of hypernuclei has features quite different from those of mesonic productions (Bernstein *et al.*, 1981). The (γ, K^+) reaction gives rise to a spin–flip excitation as well as to a non-spin–flip one, while the (K^-, π^-) reaction populates natural parity states. From the experimental point of view, the (γ, K^+) cross-section is only a few hundred nb/sr, which is 3 to 4 orders of magnitude smaller than that of the (π^+, K^+) process. The small cross-section may be compensated by the high intensity of the incident photon beam even though tagging is introduced to get monoenergetic photons.

7.2.2c *Multistep (compound) hypernucleus production*

Hypernuclei can also be excited by multistep processes as shown in Fig. 7.18. Here the first doorway state is a nucleon–hole ($1h_N$) Λ-particle ($1p_\Lambda$) configuration produced by the strangeness-exchange process discussed in 7.2.2b. The interaction of either $1p_\Lambda$ or $1h_N$ with another nucleon gives rise to the excitation of another type of nucleon (hyperon) particle–hole state. Thus the internucleon–hyperon cascade process results in complex nucleon–hyperon excitons. This is spreading of the particle–hole configuration over many p–h states. If the excitation energy exceeds a nucleon threshold (or that of a nucleon cluster such as α, etc) the nucleon may eventually escape. Spreading and nucleon escape processes produce new hypernuclei. This may be called multistep hypernucleus production, being analogous to the pre-equilibrium and equilibrium (compound or statistical) reactions used extensively for in-beam e–γ spectroscopy (see Section 4.2).

The multi-step hypernucleus production provides various kinds of Λ-particle states coupled with core (A-1 nucleon system) excited states, which are hardly populated by the direct process. The genuine hypernucleus state with a new symmetry configuration of $(j_N^{2j+1} \cdot j_\Lambda)$ with $j_N = j_\Lambda = j$ (Bando *et al.*, 1985) is not excited by a direct $[j_N^{-1}, j_\Lambda]$ doorway state, but is reached from the (j_N^{-1}, j_Λ) state by the decay of a j'_N nucleon to the j_N hole.

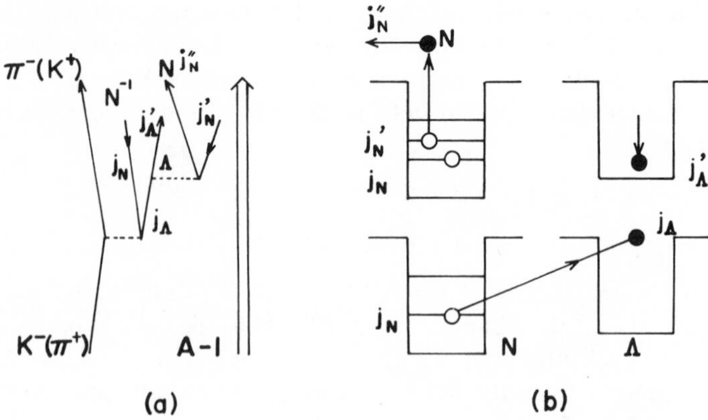

Fig. 7.18. Schematic diagram of multistep hypernucleus excitations (a) and corresponding Λ-particle N–hole picture (b). The initial doorway is $j_\Lambda j_N^{-1}$, which is spread to $j'_\Lambda j_N^{-1} j'^{-1}_N j''_N$ and so on.

7.2.2d *Spin alignment and spin polarization*

The excitation of spin-aligned and/or spin-polarized hypernuclei opens up new fields of γ-decay and weak-decay spectroscopies of hypernuclei. Spins, parities, transition multipolarities, magnetic moments, weak decay mechanisms and other important properties of hypernuclei can be studied by measuring electromagnetic or weak decays of hypernuclei with aligned (polarized) spins.

Spin alignment will be discussed by analogy with the spin alignment in the statistical reaction (see Section 4.2). The transfer of a momentum q by a strangeness transfer (K^-, π^-) or (π^+, K^+) reaction introduces angular momentum $J\hbar \sim \langle \boldsymbol{R} \times \boldsymbol{q} \rangle$ aligned in a plane perpendicular to the transferred momentum. Detection of the outgoing particle at 0° gives spin alignment in a plane perpendicular to the beam as in statistical reactions used for in-beam gamma spectroscopy. A bound state with Λ in an excited orbit j_Λ may decay to a lower orbit j'_Λ by an electric γ-transition mainly by a non-spin–flip process. Thus the spin alignment of such a state, as produced by the (π^+, K^+) or (γ, K^+) reaction with several units of angular momentum, may be kept through electric γ-decays. The nucleons emitted from multiparticle–hole and Λ-particle exciton states in the multistep hypernucleus production are mostly low energy neutrons, and do not carry off much angular momentum. Consequently, the spin alignment is likely to be preserved through γ-decays and multistep nucleon decay, provided that the spin of the initial (doorway) state is in the range of $J\hbar \approx 3$ to $10\hbar$.

The angular momentum polarization of the hypernuclear state is produced in two ways. One is due to the distortions (absorptions) of the initial (projectile) wave and the final (ejectile) one, and an other is due to the spin-dependent elementary process of the strangeness-exchange reaction. The K^-, π^- and π^+, which are involved in the (K^-, π^-) and (π^+, K^+) reactions, are strongly absorbed in nuclei with mean free paths around $\lambda(\pi^\pm) \approx \lambda(K^-) \approx 2$ fm, while the K^+ is moderately absorbed with $\lambda(K^+) \approx 5$ fm. Such absorptions give rise to the polarization of the orbital angular momenta introduced by the (K^-, π^-) and (π^+, K^+) reactions. The elementary process of the (K^-, π^-) reaction at $P(K^-) = 0.6 \sim 0.8$ GeV/c produces little spin polarization of the hyperon (Armenteros *et al.*, 1970), while that of the (π^+, K^+) one at $P(\pi^+) \approx 1$ GeV/c produces well-polarized hyperons (Rush, 1968; Baker *et al.*, 1977/1978). Therefore the polarization and alignment for hypernuclear states produced by the (K^-, π^-) reaction are due to the polarization of the transferred orbital angular momentum, while those in the case of the (π^+, K^+) reaction are composed of both the polarization of the orbital angular momentum and that of the intrinsic hyperon spin.

The spin polarization of hypernuclei has been studied in terms of the DWBA calculation and of the classical calculation (Ejiri *et al.*, 1986/1987). It is shown that the Λ hypernuclei produced by the (π^+, K^+) reaction have sizable amounts of spin-polarization as well as of spin alignment. The DWBA calculations are shown in Fig. 7.19, and the polarizations of the orbital angular momenta and the Λ spin are schematically illustrated in Fig. 7.20.

Polarized hypernuclei are produced by (K^-, π^-) reactions on polarized target nuclei (Masaike and Ejiri, 1983). The direct strangeness transfer (K^-, π^-) reaction replaces the polarized neutron by the corresponding polarized Λ, producing the polarized hypernucleus with the spin polarized Λ. This is schematically illustrated in Fig. 7.21. It is interesting to note here that the polarized target nucleus has a spin polarized nucleon in a framework of the *j–j* coupling shell model and the resulting polarized hypernucleus has a spin-polarized hyperon.

7.2.3 *Hypernuclear gamma rays in p-shell hypernuclei*

Hypernuclear γ rays have been studied so far by means of $(K^-, \pi^-\gamma)$ reactions. Gamma-ray spectra were first obtained by measuring γ rays following K-mesons stopped in light nuclei. The γ spectrum following K-mesons stopped in ^{6}Li and ^{7}Li targets exhibited a 1.09 MeV line, which was ascribed to the γ transition in one of the mass 4 hypernuclei $^4_\Lambda$H and $^4_\Lambda$He (Bamberger *et al.*, 1973). Hypernuclear γ rays in $^4_\Lambda$H, $^4_\Lambda$He and $^8_\Lambda$Li have been observed by means of (stopped K^-, π^-) reactions (Bedjidian *et al.*, 1976,

1979/1980). The reactions induced by the K^- meson stopped in the ^7Li are

$$^7Li + K^- \rightarrow {}^7_\Lambda Li^* + \pi^- \tag{7.2.4a}$$

$${}^7_\Lambda Li^* \rightarrow {}^4_\Lambda H^* + {}^3He \text{ or } {}^4_\Lambda He^* + {}^3H \tag{7.2.4b}$$

$${}^4_\Lambda H^* \rightarrow \gamma + {}^4_\Lambda H \rightarrow \gamma + \pi^- + {}^4He \tag{7.2.4c}$$

$${}^4_\Lambda He^* \rightarrow \gamma + {}^4_\Lambda He \rightarrow \gamma + \pi^0 + {}^4He. \tag{7.2.4d}$$

Hypernuclear γ rays are then identified by measuring γ rays in coincidence with π^- or π^0 mesons following weak decays of the hypernuclei. The experimental arrangement for the γ ray and pion coincidence experiment is shown in Fig. 7.22 (Bedjidian *et al.*, 1979/1980). The K-mesons from the CERN PS, identified by Čerenkov (C) and scintillation (S) counters, were

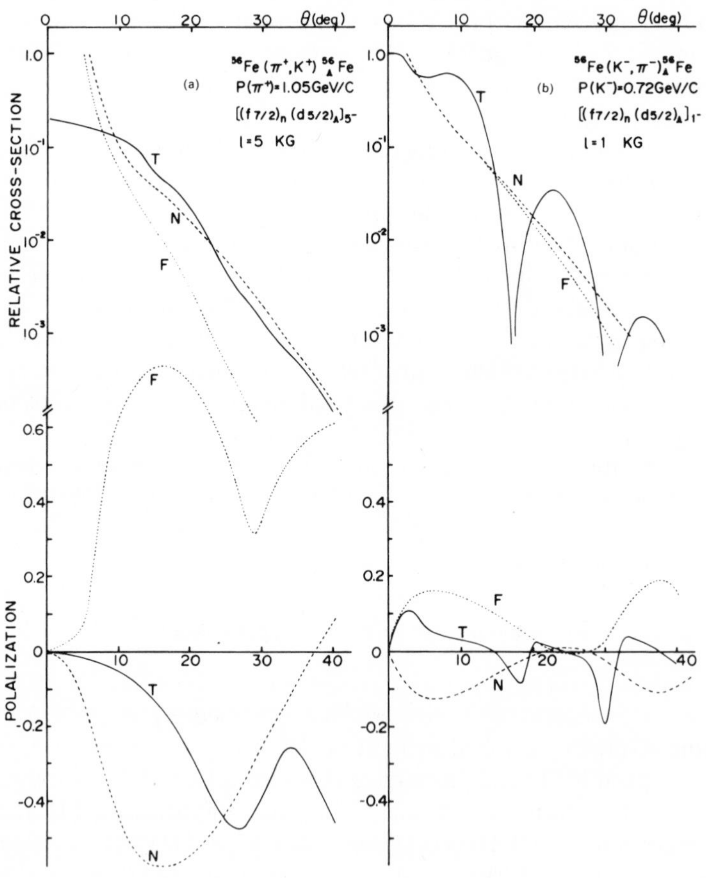

Fig. 7.19 (a and b).

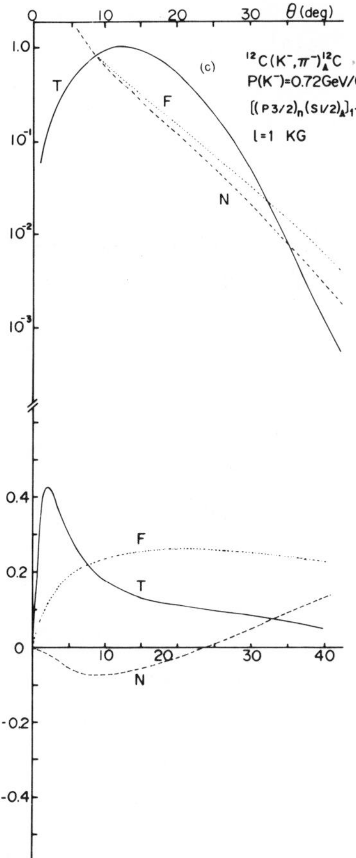

Fig. 7.19. (a) DWBA calculation for the ^{56}Fe(π^+, K$^+$)$^{56}_{\Lambda}$Fe reaction at $P(\pi^+)$ = 1.05 GeV/c, leading to the $\{(1f_{7/2})_n^{-1}\ (1d_{5/2})_\Lambda\}_{5-}$ state. Top: Differential cross-sections. Bottom: Polarization of the orbital angular momentum. The dashed lines with N and dotted lines with F are the near-side and the far-side contributions, respectively. The solid lines with T are for the coherent sum of these two contributions. (b) DWBA calculation for the ^{56}Fe (K$^-$, π^-) reaction leading to the $\{(1f_{7/2})_n^{-1}\ (1d_{5/2})_\Lambda\}_{1-}$ state. P(K$^-$) = 0.72 GeV/c. See caption in (a). (c) DWBA calculation for the ^{12}C(K$^-$, π^-) reaction leading to the $\{(1p_{3/2})_N^{-1}\ (1s_{1/2})_\Lambda\}_{1-}$ state. P(K$^-$) = 0.72 GeV/c. See caption in (a). (Ejiri *et al.*, 1986/1987.)

stopped in the thick ^{7}Li target. The gamma rays were detected by using a 4 in. × 3 in. NaI(Tl) crystal. The π^- mesons decaying from the $^4_\Lambda$H* were detected by a counter telescope consisting of a pair of proportional chambers (MWPC), hodoscope scintillation counters (H) and scintillation counters (S). The π^0 mesons decaying from the $^4_\Lambda$He* were measured by observing the

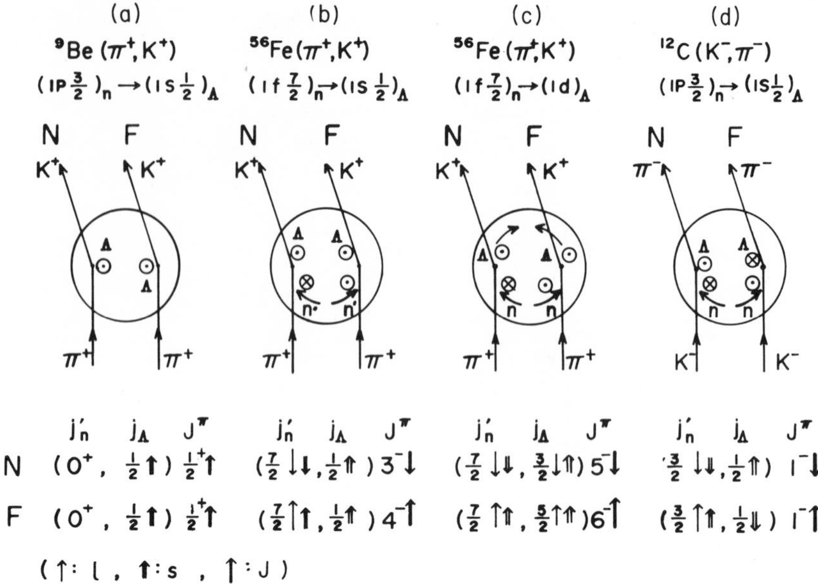

Fig. 7.20. Schematic diagram of polarization of hypernuclei produced by (π^+, K^+) and (K^-, π^-) reactions in which a neutron with j_n is transferred to a Λ hyperon with j_Λ. The configuration of the orbital angular momentum and the intrinsic nucleon (Λ) spin in the hypernucleus is given at the bottom, in which j'_n and j_Λ stand for angular momenta of the neutron and the Λ, respectively, and J^π stands for the total angular momentum and the parity. (Ejiri *et al.*, 1986/1987.)

opening angle of the two γ rays from the π^0-decay. These γ rays were detected through the conversion into electrons in the Pb-scintillator sandwich counters (N_1–N_{21}). The 1.04 ± 0.04 MeV and 1.15 ± 0.04 MeV γ rays measured in the π^- and π^0 coincidence spectra are assigned as the $1^+ \to 0^+$ spin–flip M1 transitions for the Λ and nucleon in the s-state in ${}^4_\Lambda\mathrm{H}$ and ${}^4_\Lambda\mathrm{He}$, respectively, (Bedjidian *et al.*, 1979).

The spectrum for the ^{9}Be target shows the 1.22 ± 0.04 MeV γ ray, which was tentatively assigned to the $1^- \to 1^-$ γ-transition in ${}^8_\Lambda\mathrm{Li}$ (Bedjidian *et al.*, 1980). So far the γ-spectroscopy with stopped K-mesons employed multi-step hypernucleus production, as given in eqn 7.2.4a, since the large momentum transfer tends to excite hypernuclei in highly excited unbound states.

The direct strangeness transfer (K^-, π^-) process is a straightforward method of populating a particular state for the γ-spectroscopy. A BNL group has first observed discrete γ rays from excited states in ${}^7_\Lambda\mathrm{Li}$ and ${}^9_\Lambda\mathrm{Be}$ by measuring γ rays in coincidence with pions from the (K^-, π^-) reaction. A kaon beam with an intensity of $\sim 10^{10}\,\mathrm{s}^{-1}$ from the AGS low energy (820 MeV/c) K channel (LESBI) has been used to bombard ^{7}Li and ^{9}Be

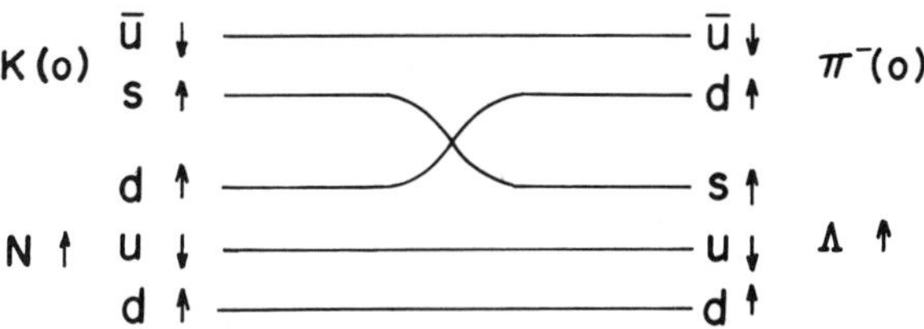

Fig. 7.21. Strangeness transfer (K^-, π^-) process on a polarized neutron in a polarized target nucleus to produce a polarized hypernucleus with a polarized Λ.

targets. Pions from the direct strangeness (K^-, π^-) reaction were momentum-analysed by means of the Moby Dick magnetic spectrometer. Gamma rays were detected by six 5 in. $\phi \times 3$ in. and two 8 in. $\phi \times 6$ in. NaI(Tl) crystals. The experimental layout is shown in Fig. 7.23. The hypernuclear γ rays in ${}^7_\Lambda$Li and ${}^9_\Lambda$Be are clearly seen in the γ-ray spectra gated by π-mesons with appropriate energy windows, and they feed the low-lying bound states in these hypernuclei. The γ-ray spectra are shown in Figs 7.24 and 7.25 (May *et al.*, 1983).

The γ-transitions in the ${}^7_\Lambda$Li and ${}^9_\Lambda$Be hypernuclei are core transitions where the Λ stays in the lowest $(s_{1/2})$ orbit and the excited states in the nucleon cores decay to the ground states. The 2.034 ± 0.023 MeV γ transition in ${}^7_\Lambda$Li has been assigned to the $5/2^+ \to 1/2^+$ transition, corresponding to the $3^+ \to 1^+$ nucleon core transition in ^{6}Li. This is based on the large population of the $5/2^+$ state of the configuration

$$[\{j_N^{-1}(p_{3/2})j_p(p_{3/2})\}_{3^+} \cdot \{j_\Lambda(s_{1/2})\}]_{5/2^+}$$

from the ^{7}Li target ground state of $\{j_p(p_{3/2})\}$ and the strong E2 $3^+ \to 1^+$ core transition. The 3.079 ± 0.040 MeV γ-transition in ${}^9_\Lambda$Be is assigned to the doublets of $5/2^+$ and $3/2^+$ core-coupled states with

$$[\{j_N(p_{3/2}) \cdot j_N(p_{3/2})\}_{2^+} \cdot \{j_\Lambda(s_{1/2})\}]_{5/2^+, 3/2^+}$$

decaying to the ground state with $j_\Lambda(s_{1/2})$ since the doublet states are almost equally populated by the direct (K^-, π^-) reaction.

It is interesting to note that these γ-transition energies are very close to the energies of the corresponding core transitions. Thus the presence of the Λ in the $s_{1/2}$ orbit does not greatly affect the excitation energy of the core nucleus. This leads to a small value of the two-body spin–orbit interactions $(\boldsymbol{\sigma}_\Lambda \pm \boldsymbol{\sigma}_N)\mathbf{l}_{n\Lambda}$. These observations suggest that the Λ spin–flip (doublet) transition energy is less than a few 100 keV. Consequently a good resolution Ge detector has to be used to resolve such low energy γ rays (May *et al.*, 1983).

The γ-transition rates for p-shell hypernuclei with the $(1s)_\Lambda$ hyperons have been derived in the framework of the Λ nuclear-core weak-coupling scheme. The M1 transition rate for the spin–flip (doublet) transition shown in Fig. 7.26

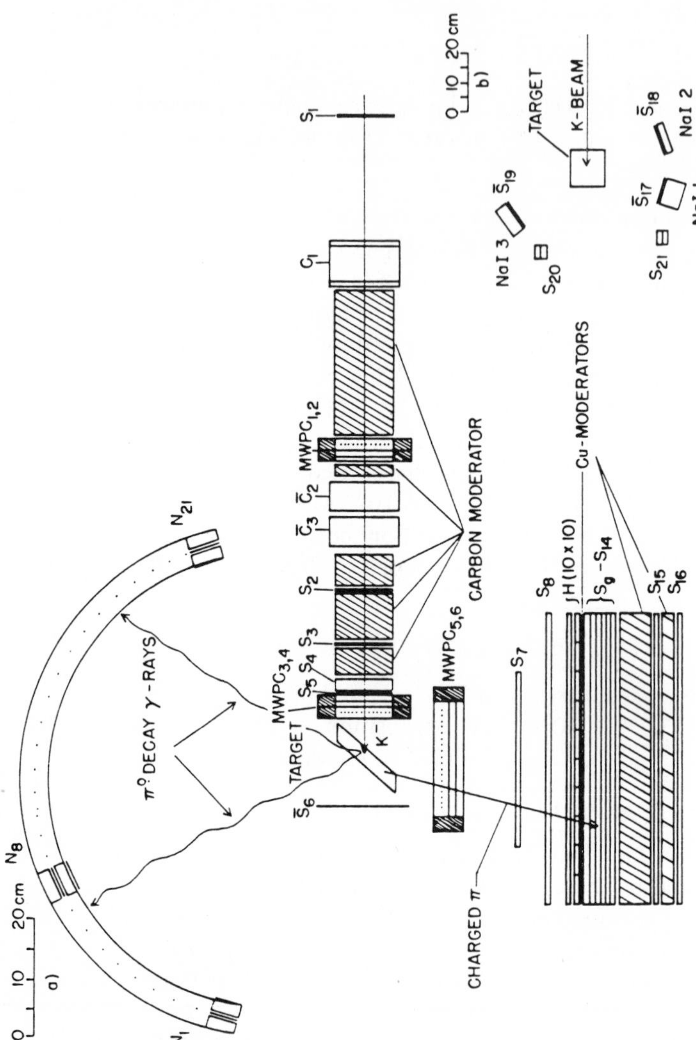

Fig. 7.22. Experimental arrangement. A top view of the kaon and pion range telescopes and the π counter with scale (a). A side view of the γ-detector and the kaon telescope is inserted in the right-hand corner with scale (b). C, S, MWPC, H and NaI stand for Čerenkov, scintillation, multiwire-proportional, hodoscope and NaI(Tl) counters, respectively. A and $\bar{A}$ denote coincidence and anti-coincidence modes, respectively. (Bedjidian *et al.*, 1979, 1980.)

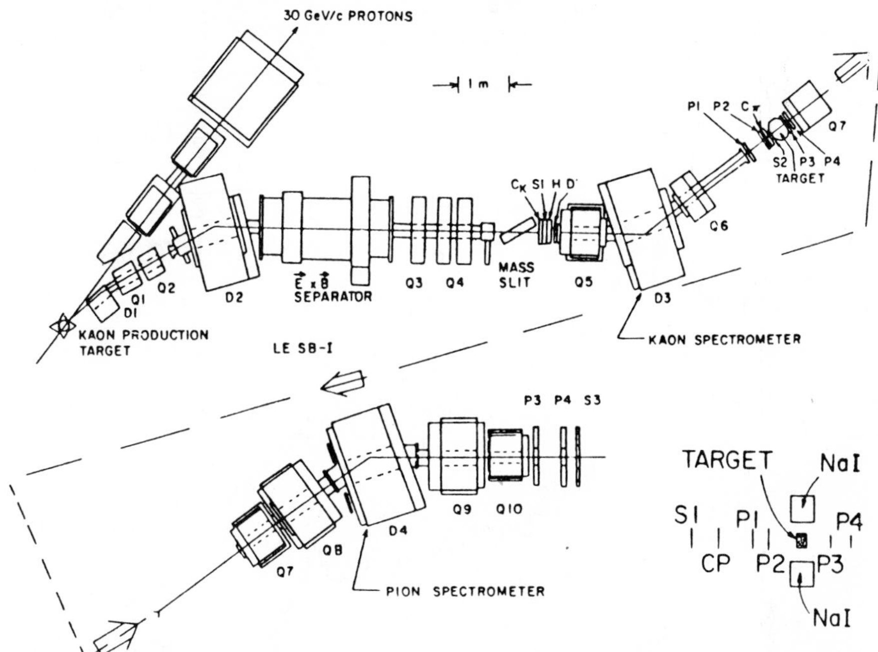

Fig. 7.23. Experimental layout of the LESBI kaon beam channel, the Moby Dick magnetic spectrometer and the γ-detectors at BNL. Q; quadrupole magnet, D; dipole magnet, P; multiwire proportional chamber,C_K and C_π; Cêrenkov counters, H; hodoscope, and S; scintillator. (Chrien, 1984.)

is given by (Dalitz and Gal, 1978)

$$\Gamma_{M1} = \tfrac{3}{2}\alpha_1(\Delta E)^3(2J_f+1)\begin{bmatrix} J_i & J_f & 1 \\ \frac{1}{2} & \frac{1}{2} & J_N \end{bmatrix}^2 (g_N - g_\Lambda)^2, \qquad (7.2.5)$$

where $\alpha_1 = 4.2\cdot 10^{12}\,\mathrm{s}^{-1}\,\mathrm{MeV}^{-3}$ and g_N and g_Λ are the g-factors for the nuclear core and Λ, respectively. The nuclear core $J_n \to J'_n$ transition has the M1, E2, or E1 character. The transition rate $\Gamma(k)$ is simply related to that of the core transition $\Gamma_N(k)$ by

$$\Gamma(k)/\Gamma_N(k) = (2J_N+1)(2J_f+1)\begin{bmatrix} J'_N & J_f & \frac{1}{2} \\ J_i & J_N & k \end{bmatrix}^2 \cdot \left(\frac{\Delta E}{\Delta E_N}\right)^{2k+1}, \qquad (7.2.6)$$

where k is the multipolarity of the transition. Therefore precise studies of the γ-transition rates may elucidate the effective g-factor for Λ in the hypernucleus and accordingly the possible Λ-core polarization effect.

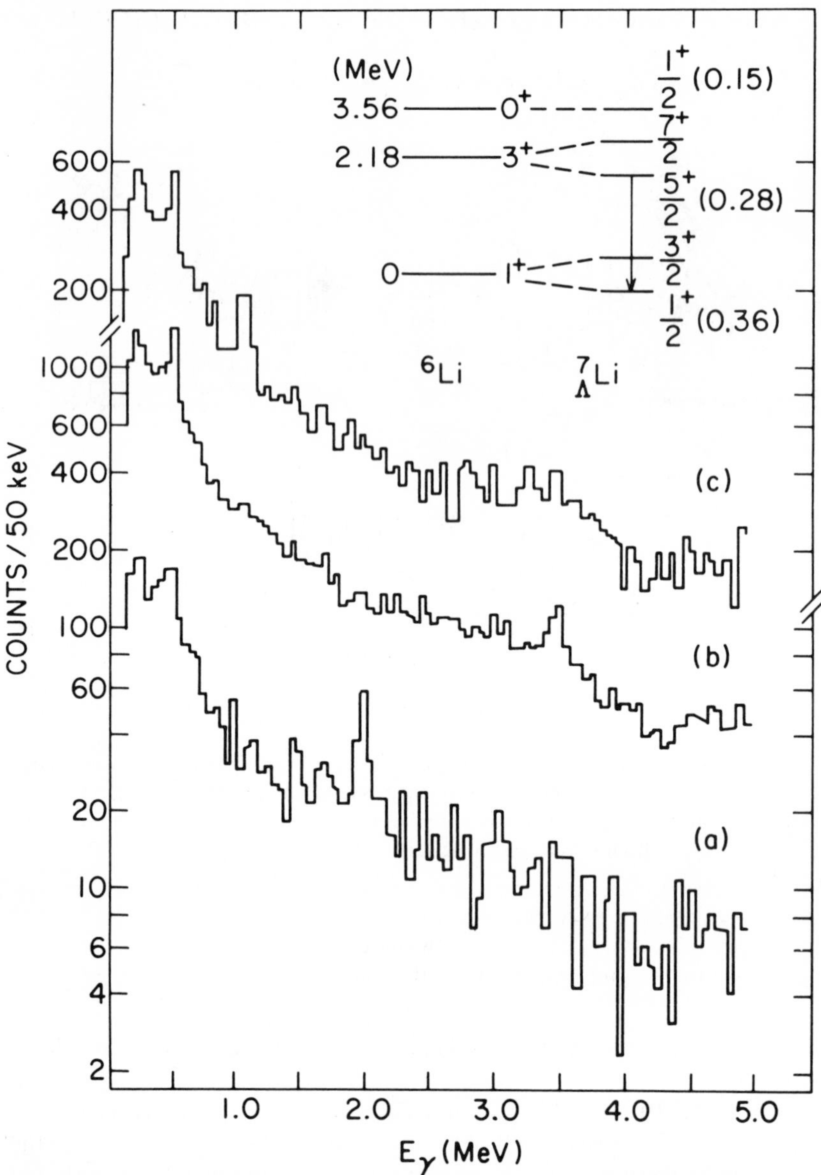

Fig. 7.24. Gamma-ray energy spectra observed in coincidence with the ^{7}Li(K$^-$, π^-) reaction at 0° for $P_K = 820$ MeV/c, for 10^{10} kaons on target. Spectra are shown for excitation-energy intervals of (a) -2 to 6 MeV, (b) 6 to 22 MeV, and (c) 22 to 39 MeV. The insert shows the energy spectrum of excited states of ^{6}Li which give rise to doublets in the hypernucleus $^7_\Lambda$Li. The spectrum (a), corresponding to the γ-rays from $^7_\Lambda$Li, shows the 2.034 MeV γ transition as indicated in the insert. The expected relative populations of the ^{7}Li hypernuclear states in the (K$^-$, π^-) reaction at 0° are shown in brackets. (May *et al.*, 1983.)

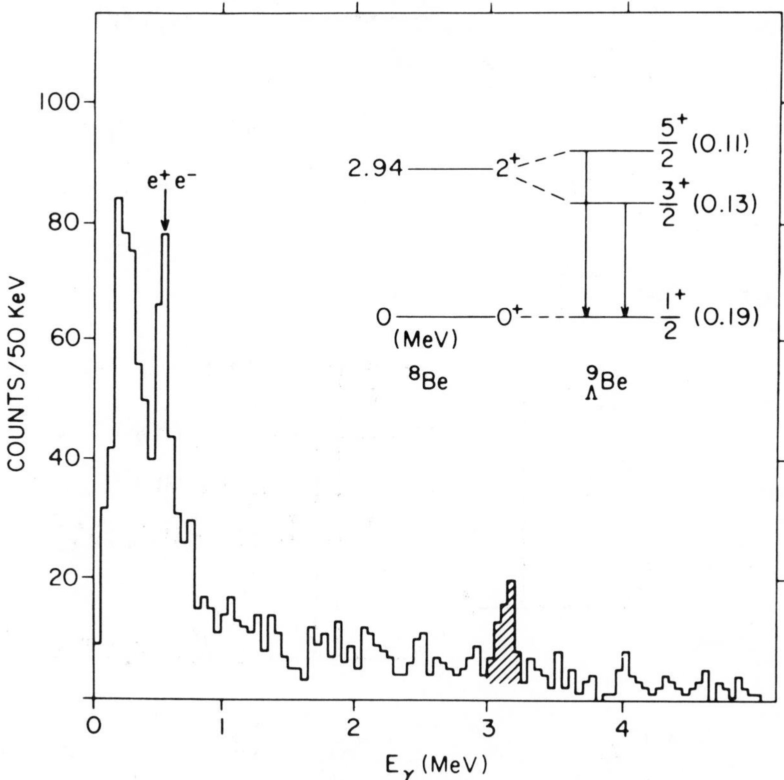

Fig. 7.25. The gamma-ray energy spectrum observed in coincidence with the bound-state region of the $^{9}_{\Lambda}$Be excitation-energy spectrum, for ^{9}Be (K^-, π^-) at 0°, with 1.9 $\times 10^{10}$ kaons on the target. The insert shows the energy levels for ^{8}Be and the expected levels in $^{9}_{\Lambda}$Be. The numbers in brackets are the relative formation rates for the reaction $K^- + {}^{9}Be \rightarrow \pi^- + {}^{9}_{\Lambda}Be^*$ at 0°. (May *et al.*, 1983.)

7.2.4 *Perspectives of hypernuclear γ-spectroscopy*

So far γ-ray spectroscopy has been limited to low-spin states in light hypernuclei produced by (K^-, π^-) reactions. Extension of γ-spectroscopy to higher spin states and to medium-heavy nuclei can be made by using the (π^+, K^+) reaction. As discussed in Section 7.2.2, the (π^+, K^+) reaction may take advantage of the population of a variety of excited states in a wide range of angular momentum.

High-resolution studies of γ-transitions by means of Ge-detectors may resolve fine level structures. The excitation energies are sensitive to the effective Λ–N interactions. A phenomenological analysis of energy levels in

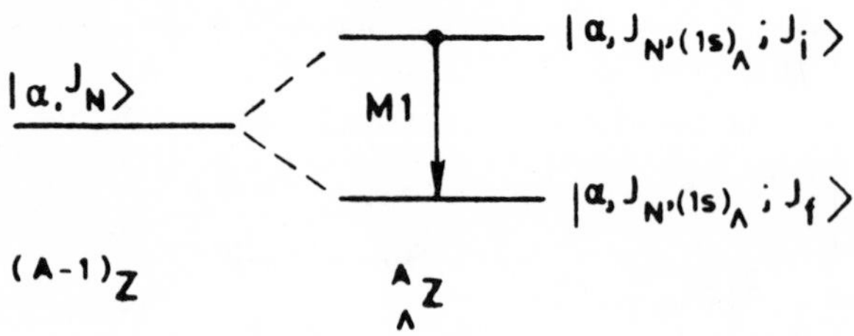

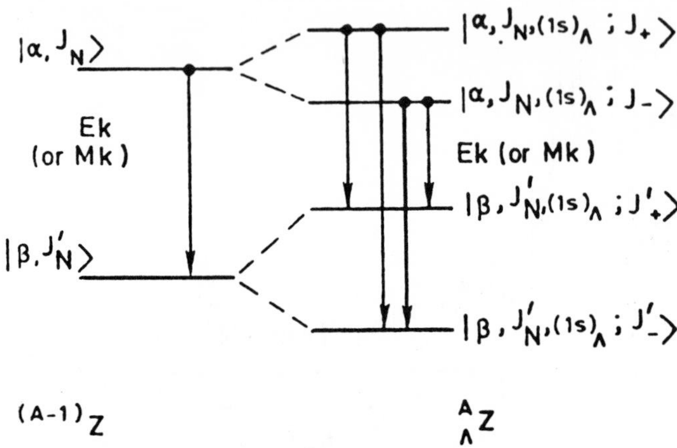

Fig. 7.26. The upper figure illustrates an M1 spin–flip transition $J_i \rightarrow J_f$ between the two states with spins $J_N \pm \frac{1}{2}$ formed by the attachment of a Λ-particle in the $1s_{1/2}$ orbital to the same core nucleus with spin J_N. The lower figure illustrates the Ek (or Mk) transitions which are generally possible between the states of two doublets (J_+, J_-) and (J'_+, J'_-) of the hypernucleus ${}^{A}_{\Lambda}Z$, which are built on two states (α, J_N) and (β, J'_N) of its core nucleus ${}^{(A-1)}Z$, as consequences of an Ek (or Mk) transition between these two states in ${}^{(A-1)}Z$. (Dalitz and Gal, 1978; Dover and Walker, 1982.)

the p-shell hypernuclei has been performed in terms of the two body N–Λ interaction as given by

$$V_{\Lambda N} = V_0(r) + V_\sigma \mathbf{s}_N \cdot \mathbf{s}_\Lambda + V_\Lambda \mathbf{l}_{N\Lambda} \cdot \mathbf{s}_\Lambda + V_N \mathbf{l}_{N\Lambda} \cdot \mathbf{s}_N + V_T \cdot s_{12} \tag{7.2.7a}$$

$$s_{12} = 3(\boldsymbol{\sigma}_N \cdot \mathbf{r})(\boldsymbol{\sigma}_\Lambda \cdot \mathbf{r}) - \boldsymbol{\sigma}_N \cdot \boldsymbol{\sigma}_\Lambda, \quad \mathbf{r} = \mathbf{r}_N - \mathbf{r}_\Lambda. \tag{7.2.7b}$$

Here the five terms in the interaction potential are the central, the spin–spin, the Λ spin–orbit, the induced nuclear spin–orbit and the tensor terms, respectively (Dalitz and Gal, 1978; Millener *et al.*, 1985).

The angular distributions of γ rays following the decay of spin-aligned states and/or spin-polarized states are very useful to determine magnetic moments, multipolarities of γ-transitions and spin assignments. The branching ratio is quite sensitive to the hyperon nuclear-core interactions. Hypernuclear γ-ray spectroscopy now constitutes a new kind of frontier in nuclear structure studies. Meson beams with high quality and high intensity, combined with high-efficiency counter arrays, will certainly open up this fascinating field of hypernuclear spectroscopy. Kaon factories which provides 10^6 to 10^7 kaons s^{-1} at $P_K = 0.4 - 1.8$ GeV/c and pion factories with 10^7 to 10^{10} pions s^{-1} at $P_\pi = 1 - 2$ GeV/c may facilitate detailed spectroscopic studies of hypernuclei with $S = -1$ and that of even double hypernuclei with $S = -2$ (Chrien, 1985, 1986). The Ξ hypernuclei with $S = -2$ can be produced by the (K^-, K^+) reaction. The $\Lambda\Lambda$ hypernuclei with $S = -2$ are also quite interesting.

7.3 References

Adelberger, E. G., Swanson, H. E., Cooper, M. D., Tape, J. W. and Trainor, T. A. (1975). *Phys. Rev. Lett.* **34,** 402.

Adelberger, E. G., Hindi, M. M., Hoyle, C. D., Swanson, H. E., Lintig, R. D. von and Haxton, W. C. (1983). *Phys. Rev.* **C27,** 2833.

Adelberger, E. G. and Haxton, W. C. (1985). *Ann. Rev. nucl. part. Sci.* **35,** 501.

Ahrens, G., Harfst, W., Kass, J. R., Mason, E. V. and Schober, M. (1982). *Nucl. Phys.* **A390,** 486.

Armenteros, R. *et al.* (1970). *Nucl. Phys.* **B21,** 15.

Auerbach, E. H., Baltz, A. J., Dover, C. B., Kahana, S. H., Ludeking, L. and Millener, D. J. (1983). *Ann. Phys.* **148,** 381.

Avignone, III, F. T., Brodzinski, R. L., Brown, D. P., Evans, Jr., J. C., Hensley, W. K., Miley, H. S., Reeves, J. H. and Wogman, N. A. (1983/1985). *Phys. Rev. Lett.* **50,** 721; *Phys. Rev. Lett.* **54,** 2309.

Baker, R. D. *et al.* (1977/1978). *Nucl. Phys.* **B126,** 365; **B141,** 29.

Bamberger, A., Faessler, M. A., Lynen, U., Piekarz, H., Piekarz, J., Pniewski, J., Povh, B. and Ritter, H. G. (1973). *Nucl. Phys.* **60B,** 1.

Bando, H., Ikeda, K. and Motoba, T. (1981). *Progr. theor. Phys.* **66,** 1344.

Bando, H. (1983). *Progr. theor. Phys.* **69,** 1731.

Bando, H. and Takaki, H. (1984/1985). *Progr. theor. Phys.* **72,** 106; *Phys. Lett.* **150B,** 409.

Bando, H., Ikeda, K. and Yamada, T. (1985). *Progr. theor. Phys.* Suppl. No. 81.

Bardin, B. K., Gollon, P. J., Ullman, J. D. and Wu, C. S. (1970). *Nucl. Phys.* **A158,** 337.

Barnes, P. D. (1986). *Nucl. Phys.* **A450,** 43C.

Bedjidian, M., Filipkowski, A., Grossiord, J. Y., Guichard, A., Gusakow, M., Majewski, S., Piekarz, H., Piekarz, J. and Pizzi, J. R. (1976). *Phys. Lett.* **62B,** 467.

Bedjidian, M., Descroix, E., Grossiord, J. Y., Guichard, A., Gusakow, M., Jacquin, M., Kudla, M. J., Piekarz, H., Piekarz, J., Pizzi, J. R. and Pniewski, J. (1979/1980). *Phys. Lett.* **83B,** 252; **94B,** 480.

Bellotti, E., Cremonesi, O., Fiorini, E., Liguori, C., Pullia, A., Sverzellati, P. and Zanotti, L. (1984, 1986). *Phys. Lett.* **146B,** 450. *Nuovo Cimento* **95A,** 1

Bernstein, A. M., Donnelly, T. W. and Epstein, G. N. (1981). *Nucl. Phys.* **A358,** 195C.

Bertini, R. *et al.* (1980). *Phys. Lett.* **90B,** 375.

Bertini, R. *et al.* (1981). *Nucl. Phys.* **A360,** 315; **A368,** 365.

Bini, M., Fazzini, T. F., Poggi, G. and Taccetti, N. (1985). *Phys. Rev. Lett.* **55,** 795.

Brückner, W. *et al.* (1978). *Phys. Lett.* **79B,** 157.

Caldwell, D. O., Eisberg, R. M., Grumm, D. M., Hale, D. L., Witherell, M. S., Goulding, F. S., Landis, D. A., Madden, N. W., Malone, D. F., Pehl, R. H. and Smith, A. R. (1985, 1986). *Phys. Rev. Lett.* **54,** 281; *Phys. Rev.* **D33,** 2737.

Caldwell, D. O., Eisberg, R. M., Grumm, D. M., Witherell, M. S., Goulding, F. S. and Smith, A. R. (1987). *Phys. Rev. Lett.* **59,** 419.

Chrien, R. E. *et al.* (1979). *Phys. Lett.* **89B,** 31.

Chrien, R. E. (1984). *Proc. Int. Symp. Nucl. Spectroscopy and Nucl. Interactions,* March 1984, Osaka (eds. H. Ejiri and T. Fukuda), p. 569. World Scientific, Singapore.

Chrien, R. E. (1985). Brookhaven National lab. Report BNL 36082.

Chrien, R. E. (1986). *Nucl. Phys.* **A450,** 1. (Proc. Int. Symp. Hypernuclear and Kaon Physics, Sept. 1985 BNL).

Cleveland, B. T., Leo, W. R., Wu, C. S., Kasday, L. R., Rushton, A. M., Gollon, P. J. and Ullman, J. D. (1975). *Phys. Rev. Lett.* **35,** 757.

Dalitz, R. H. and Gal, A. (1978). *Ann. Phys.* **116,** 167.

Doi, M., Kotani, T., Nishiura, H., Okuda, K. and Takasugi, E. (1981). *Phys. Lett.* **103B,** 219.

Doi, M., Kotani, T., Nishiura, H. and Takasugi, E. (1983). *Progr. theor. Phys.* **69,** 602; ibid. **70,** 1353.

Doi, M., Kotani, T. and Takasugi, E. (1985). *Theor. Phys.* Suppl. No. 83.

Dover, C. B. and Walker, G. E. (1979). *Phys. Rev.* **C19,** 1393.

Dover, C. B., Ludeking, L. and Walker, G. E. (1980). *Phys. Rev* **C22,** 2073.

Dover, C. B. and Walker, G. E. (1982). *Phys. Rep.* **89,** 1.

Earle, E. D., McDonald, A. B., Adelberger, E. G., Snover, K. A., Swanson, H. E., Linting, R. von, Mak, H. B. and Barnes, C. A. (1983). *Nucl. Phys.* **A396,** 221C.

Ejiri, H., Takahashi, N., Shibata, T., Nagai, Y., Okada, K., Kamikubota, N. and Watanabe, T. (1984a). *Proc. Int. Symp. Nuclear Spectroscopy and Nuclear Interactions,* Osaka, (ed. H. Ejiri and T. Fukuda), p. 284. World Scientific, Singapore.

Ejiri, H., Kamikubota, N., Nagai, Y., Okada, K., Shibata, T., Takahashi, N. and Watanabe, T. (1984b). *Proc. Int. Conf. Neutrino-mass and Low Energy Weak Interactions,* Telemark, Wisconsin, AIP Conf. Proc. (ed. V. Barger and D. Cline), p. 383. World Scientific, Singapore.

Ejiri, H., Takahashi, N., Shibata, T., Nagai, Y., Okada, K., Kamikubota, N., Watanabe, T., Irie, T., Itoh, Y. and Nakamura, T. (1986a). *Nucl. Phys.* **A448,** 271.

Ejiri, H. (1986b). *Proc. Seventh Workshop on Grand Unification/ICOBAN'86,* Toyama, Japan (ed. J. Arafune), p. 215. World Scientific, Singapore.

Ejiri, H., Kamikubota, N., Nagai, Y., Nakamura, T., Okada, K., Shibata, T., Shima, T., Takahashi, N. and Watanabe, T. (1986c). *Proc. Int. Conf. Neutrino Physics and Astrophysics,* Sendai (ed. T. Kitagaki and H. Yuta), p. 101. World Scientific, Singapore.

Ejiri, H., Fukuda, T., Shibata, T., Bando, H. and Kubo, K.-I. (1986/1987). *Proc. INS Int. Symp. on Hypernuclear Physics,* Tokyo (ed. H. Bando, O. Hahimoto and K. Ogawa), INS, Univ. Tokyo; *Phys. Rev.* **C36,** 1435.

Ejiri, H., Kamikubota, N., Nagai, Y., Nakamura, T., Okada, K., Shibata, T., Shima, T., Takahashi, N. and Watanabe, T. (1987). *J. Phys. G. nucl. Phys.* **13,** 839.

Elliott, S. R., Hahn, A. A. and Moe, M. K. (1987). *Phys. Rev. Lett.* **59,** 2020.

Elsener, K., Grüebler, W., König, V., Schmelzbach, P. A., Ulbricht, J., Singy, D., Forstner, Ch., Zhang, W. Z. and Vuaridel, B. (1984). *Phys. Rev. Lett.* **52,** 1476.

Evans, H. C., Ewan, G. T., Kwan, S. P., Leslie, J. R., MacArthur, J. D., Mak, H. B., McLatchie, W., Page, S. A., Skensved, P., Wang, S. S., McDonald, A. B., Barnes, C. A., Alexander, T. K. and Clifford, E. T. H. (1985). *Phys. Rev. Lett.* **55,** 791.

Fiorini, E., Pullia, A., Bertolini, G., Cappellani, F. and Restelli, G. (1973). *Nuovo Cim.* **13A,** 747.

Forster, A., Kwon, H., Markey, J. K., Boehm, F. and Henrikson, H. E. (1984). *Phys. Lett.* **138B,** 301.

Gelmini, G. B. and Roncadelli, M. (1981). *Phys. Lett.* **99B,** 411.

Georgi, H. M., Glashow, S. L. and Nussinov, S. (1981). *Nucl. Phys.* **B193,** 297.

Grace, R. *et al.* (1985). *Phys. Rev. Lett.* **55,** 1055.

Grotz, K. and Klapdor, H. V. (1985). *Phys. Lett.* **157B,** 242.

Haxton, W. C., Stephenson, Jr., G. J. and Strottman, D. (1981/1982). *Phys. Rev. Lett.* **47,** 153; *Phys. Rev.* **D25,** 2360.

Haxton, W. C. and Stephenson, Jr., G. J. (1984). *Progr. part. nucl. Phys.* **12,** 409.

Hennecke, E. W., Manuel, O. K. and Sabu, D. O. (1975). *Phys. Rev.* **C11,** 1378.

Hennecke, E. W. (1978). *Phys. Rev.* **C17,** 1168.

Holstein, B. R. and Treiman, S. B. (1977). *Phys. Rev.* **D16,** 2369.

Kamikubota, N., Ejiri, H., Shibata, T., Nagai, Y., Okada, K., Watanabe, T., Irie, T., Itoh, Y., Nakamura, T. and Takahashi, N. (1986). *Nucl. Instr. Meth.* **A245,** 379.

Kirsten, T., Richter, H. and Jessberger, E. (1983). *Phys. Rev. Lett.* **50,** 474.

Kisslinger, L. S. (1980). *Phys. Rev. Lett.* **44,** 968.

Klapdor, H. V. and Grotz, K. (1984). *Phys. Lett.* **142B,** 323.

Klinken, J. van, Stam, K., Venema, W. Z., Wichers, V. A. and Atkinson, D. (1983). *Phys. Rev. Lett.* **50,** 94.

Kotani, T., Ejiri, H. and Takasugi, E. (1986). *Proc. Int. Symp. Nucl. Beta Decays and Neutrino*, Osaka, World Scientific, Singapore.

Leccia, F., Hubert, Ph., Dassie, D., Mennrath, P., Villard, M. M., Morales, A., Morales, J. and Nunez-Lagos, R. (1983/1985). *Nuovo Cim.* **78A,** 50; ibid., **85A,** 19.

Marlow, D. *et al.* (1982). *Phys. Rev.* **C25,** 2619.

Masaike, A. and Ejiri, H., *Proc. Third Lamph II Workshop*, Los Alamos, July 1983, LA-9933-C Vol. II, p. 823.

May, M. *et al.* (1981). *Phys. Rev. Lett.* **47,** 1106.

May, M. *et al.* (1983). *Phys. Rev. Lett.* **51,** 2085.

Millener, D. J., Gal, A., Dover, C. B. and Dalitz, R. H. (1985). *Phys. Rev.* **C31,** 499.

Miyahara, K., Ikeda, K. and Bando, H. (1983). *Progr. theor. Phys.* **69,** 1717.

Moe, M. K. and Lowenthal, D. D. (1980). *Phys. Rev.* **C22,** 2186.

Mohapatra, R. N. and Vergados, J. D. (1981). *Phys. Rev. Lett.* **47,** 1713.

Motoba, T., Bando, H. and Ikeda, K. (1983). *Progr. theor. Phys.* **70,** 189.

Piekarz, H. *et al.* (1982). *Phys. Lett.* **110B,** 428.

Povh, B. (1978/1980/1981). *Ann. Rev. nucl. Sci.* **28,** 1; *Nucl. Phys.* **A335,** 223, *Prog. Part. Nucl. Phys.* **5,** 245.

Primakoff, H. and Rosen, S. P. (1969/1981). *Phys. Rev.* **184,** 1925; *Ann. Rev. nucl. Sci.* **31,** 145.

Rush, Jr. J. E. (1968). *Phys. Rev.* **73,** 1776.

Senjanović, G. (1979). *Nucl. Phys.* **B153,** 334.

Simpson, J. J., Jagam, P., Campbell, J. L., Malm, H. L. and Robertson, B. C. (1984). *Phys. Rev. Lett.* **53,** 141.

Strovink, M. (1980). *Weak Interactions as Probes of Unification,* (ed. G. B. Collins, L. N. Chang and J. R. Ficenec), p. 46. AIP Conference Proceedings N. 72, American Institute of Physics, New York.

Takaoka, N. and Ogata, K. (1966). *Z. Naturf.* **21A,** 84.

Thiessen, H. A. *et al.* (1983). AGS Exp. 758.

Watanabe, T., Ejiri, H., Shibata, T., Nagai, Y., Okada, K., Kamikubota, N., Shima, T., Tanaka, J. and Taniguchi, T. (1987) to be published in *Nucl. Inst. Meth.*

Wichers, V. A., Armbrust, T. F. L., Atkinson, D., Hageman, T., Kamphuis, D., Klinken, J. van and Wilschut, H. W. (1985). KVI Annual Report 1985, Kernfysisch Versneller Instituut, Groningen, p. 59.

Wichers, V. A., Hageman, T. R., Klinken, J. van, Wilschut, H. W. and Atkinson, D. (1987). *Phys. Rev. Lett.* **58,** 1821.

Wolfenstein, L. (1982). *Phys. Rev.* **D26,** 2507.

INDEX